W9-BWO-467

ANNUAL REVIEW OF
ECOLOGY AND SYSTEMATICS

EDITORIAL COMMITTEE (2001)

JAMES CLARK
DAPHNE GAIL FAUTIN
DOUGLAS J. FUTUYMA
RICHARD G. HARRISON
H. BRADLEY SHAFFER
SCOTT WING

RESPONSIBLE FOR THE ORGANIZATION OF VOLUME 32
(EDITORIAL COMMITTEE, 1999)

DAPHNE GAIL FAUTIN
DOUGLAS J. FUTUYMA
RICHARD G. HARRISON
FRANCES C. JAMES
JUDY L. MEYER
H. BRADLEY SHAFFER
JERRY COYNE (GUEST)
DAVID JABLONSKI (GUEST)
JOY ZEDLER (GUEST)

Production Editor: ANNE E. SHELDON
Bibliographic Quality Control: MARY A. GLASS
Color Graphics Coordinator: EMÉ O. AKPABIO
Subject Indexer: BRUCE TRACY

ANNUAL REVIEW OF
ECOLOGY AND SYSTEMATICS

VOLUME 32, 2001

DAPHNE GAIL FAUTIN, *Editor*
University of Kansas

DOUGLAS J. FUTUYMA, *Associate Editor*
State University of New York at Stony Brook

H. BRADLEY SHAFFER, *Associate Editor*
University of California, Davis

www.AnnualReviews.org   science@AnnualReviews.org   650-493-4400

ANNUAL REVIEWS
4139 El Camino Way • P.O. BOX 10139 • Palo Alto, California 94303-0139

℟

ANNUAL REVIEWS
Palo Alto, California, USA

COPYRIGHT © 2001 BY ANNUAL REVIEWS, PALO ALTO, CALIFORNIA, USA. ALL RIGHTS RESERVED. The appearance of the code at the bottom of the first page of an article in this serial indicates the copyright owner's consent that copies of the article may be made for personal or internal use, or for the personal or internal use of specific clients. This consent is given on the condition that the copier pay the stated per-copy fee of $14.00 per article through the Copyright Clearance Center, Inc. (222 Rosewood Drive, Danvers, MA 01923) for copying beyond that permitted by Section 107 or 108 of the US Copyright Law. The per-copy fee of $14.00 per article also applies to the copying, under the stated conditions, of articles published in any *Annual Review* serial before January 1, 1978. Individual readers, and nonprofit libraries acting for them, are permitted to make a single copy of an article without charge for use in research or teaching. This consent does not extend to other kinds of copying, such as copying for general distribution, for advertising or promotional purposes, for creating new collective works, or for resale. For such uses, written permission is required. Write to Permissions Dept., Annual Reviews, 4139 El Camino Way, P.O. Box 10139, Palo Alto, CA 94303-0139 USA.

International Standard Serial Number: 0066-4162
International Standard Book Number: 0-8243-1432-8
Library of Congress Catalog Card Number: 71-135616

Annual Review and publication titles are registered trademarks of Annual Reviews.
⊗ The paper used in this publication meets the minimum requirements of American National Standards for Information Sciences—Permanence of Paper for Printed Library Materials. ANSI Z39.48-1992.

Annual Reviews and the Editors of its publications assume no responsibility for the statements expressed by the contributors to this *Annual Review*.

TYPESET BY TECHBOOKS, FAIRFAX, VA
PRINTED AND BOUND IN THE UNITED STATES OF AMERICA

Annual Review of Ecology and Systematics
Volume 32, 2001

CONTENTS

ERRATA
An online log of corrections to *Annual Review of Ecology and Systematics* chapters may be found at http://ecolsys.AnnualReviews.org/errata.shtml

RELATED ARTICLES

From the *Annual Review of Microbiology*, Volume 55 (2001)

A Community of Ants, Fungi, and Bacteria: A Multilateral Approach to Studying Symbiosis, Cameron R. Currie

Ammonia-Oxidizing Bacteria: A Model for Molecular Microbial Ecology, George A. Kowalchuk and John R. Stephen

Recombination and the Population Structures of Bacterial Pathogens, Edward J. Feil and Brian G. Spratt

Horizontal Gene Transfer in Prokaryotes: Quantification and Classification, Eugene V. Koonin, Kira S. Makarova, and L. Aravind

From the *Annual Review of Phytopathology*, Volume 39 (2001)

Variability and Genetic Structure of Plant Virus Populations, Fernando Garcia-Arenal, Aurora Fraile, and José M. Malpica

Pathogen Fitness Penalty as a Predictor of Durability of Disease Resistance Genes, Jan E. Leach, Casiana M. Vera Cruz, Jianfa Bai, and Hei Leung

Resistance Gene Complexes: Evolution and Utilization, Scot H. Hulbert, Craig A. Webb, Shavannor M. Smith, and Qing Sun

ANNUAL REVIEWS is a nonprofit scientific publisher established to promote the advancement of the sciences. Beginning in 1932 with the *Annual Review of Biochemistry*, the Company has pursued as its principal function the publication of high-quality, reasonably priced *Annual Review* volumes. The volumes are organized by Editors and Editorial Committees who invite qualified authors to contribute critical articles reviewing significant developments within each major discipline. The Editor-in-Chief invites those interested in serving as future Editorial Committee members to communicate directly with him. Annual Reviews is administered by a Board of Directors, whose members serve without compensation.

2001 Board of Directors, Annual Reviews

Richard N. Zare, *Chairman of Annual Reviews*
   Marguerite Blake Wilbur Professor of Chemistry, Stanford University
John I. Brauman, *J. G. Jackson–C. J. Wood Professor of Chemistry, Stanford University*
Peter F. Carpenter, *Founder, Mission and Values Institute*
W. Maxwell Cowan, *Vice President and Chief Scientific Officer, Howard Hughes
   Medical Institute, Bethesda*
Sandra M. Faber, *Professor of Astronomy and Astronomer at Lick Observatory,
   University of California at Santa Cruz*
Eugene Garfield, *Publisher*, The Scientist
Samuel Gubins, *President and Editor-in-Chief, Annual Reviews*
Daniel E. Koshland, Jr., *Professor of Biochemistry, University of California at Berkeley*
Joshua Lederberg, *University Professor, The Rockefeller University*
Gardner Lindzey, *Director Emeritus, Center for Advanced Study in the Behavioral
   Sciences, Stanford University*
Sharon R. Long, *Professor of Biological Sciences, Stanford University*
J. Boyce Nute, *President and CEO, Mayfield Publishing Co.*
Michael E. Peskin, *Professor of Theoretical Physics, Stanford Linear Accelerator Ctr.*
Harriet A. Zuckerman, *Vice President, The Andrew W. Mellon Foundation*

Management of Annual Reviews

Samuel Gubins, President and Editor-in-Chief
Richard L. Burke, Director of Production
Paul J. Calvi, Jr., Director of Information Technology
Steven J. Castro, Chief Financial Officer
John W. Harpster, Director of Sales and Marketing

Annual Reviews of

Anthropology	Fluid Mechanics	Physiology
Astronomy and Astrophysics	Genetics	Phytopathology
Biochemistry	Genomics and Human Genetics	Plant Physiology and Plant
Biomedical Engineering	Immunology	Molecular Biology
Biophysics and Biomolecular	Materials Research	Political Science
Structure	Medicine	Psychology
Cell and Developmental	Microbiology	Public Health
Biology	Neuroscience	Sociology
Earth and Planetary Sciences	Nuclear and Particle Science	
Ecology and Systematics	Nutrition	SPECIAL PUBLICATIONS
Energy and the Environment	Pharmacology and Toxicology	Excitement and Fascination of
Entomology	Physical Chemistry	Science, Vols. 1, 2, 3, and 4

Annu. Rev. Ecol. Syst. 2001. 32:1–23
Copyright © 2001 by Annual Reviews. All rights reserved

CHEMICAL DETECTION OF NATURAL ENEMIES BY ARTHROPODS: An Ecological Perspective

Marcel Dicke and Paul Grostal

*Laboratory of Entomology, Wageningen University, P.O. Box 8031, NL-6700 EH,
Wageningen, The Netherlands; e-mail: marcel.dicke@users.ento.wau.nl*

Key Words   enemy avoidance, infochemicals, food webs, trade-offs, learning

■ **Abstract**   Food webs are overlaid with infochemical webs that mediate direct
and indirect interactions. The infochemicals may result in shifts in trait values, which
affect the strength of species interactions. As a consequence, population dynamics and
evolutionary changes can be affected. Chemical information can mediate the interactions between animals and their resources, competitors and enemies. Of all chemical
information gathered by animals, cues about predation risk are of special significance
because predation risk usually has important and immediate consequences on fitness.
In this paper we selectively review the role of chemical information in enemy avoidance by arthropods. Arthropods not only constitute important components of food
webs, being the largest group in numbers and species diversity; they also make excellent models for ecological studies. We discuss the evidence, the key mechanisms, and
the trade-offs involved in chemical detection of enemies by potential arthropod prey.
Further, we address the variation in prey responses and the evidence for learning in
avoiding enemies by arthropods. Finally, we identify and prioritize major questions to
be tackled by future studies.

INTRODUCTION

Information flows within food webs are important processes that influence behavior, ecology, and population dynamics of animals (e.g., Abrams et al. 1996, Krebs
& Davies 1987). Animals are well adapted to gather information about their environment, considering the number and often fine tuning of the sensory structures
they possess. The extrinsic information that plays a crucial role in the survival of
animals includes cues on the availability of food and mates, abiotic factors (e.g.,
temperature, shelter), as well as the presence of competitors and natural enemies.
Although the definition of natural enemies is broad, we limit our use to animals
that can individually kill other animals (i.e., predators and parasitoids). Parasites
and pathogens are not discussed in this review. Consequently, we use the term
prey to describe the attacked animal. Extrinsic information is typically gathered
via three pathways: visual, mechanical, and chemical.

0066-4162/01/1215-0001$14.00

Of all types of extrinsic information, cues about natural enemies are especially important for survival; the failure to detect enemies may have more serious consequences for the future fitness of animals than temporary failure to find food or mates (Lima & Dill 1990). Thus, efficiency in recognizing enemies and timely engagement in defensive responses, e.g., avoidance, activation of direct defenses, or the protection of offspring, has obvious advantages, especially for species that are under strong selective pressure from organisms at higher trophic levels.

In this review, we discuss the chemical detection of natural enemies by arthropods. Of all organisms, arthropods not only comprise the largest group in numbers and species diversity (Berenbaum 1995), they also constitute important components of food webs (Schoonhoven et al. 1998, Strong et al. 1984). Also, many arthropods, due to their ubiquity, small size, short life span, and relatively easy captive breeding, make excellent models for manipulative studies. Finally, a large number of arthropod species have important consequences for agriculture and human health. Understanding the interaction between these arthropods and their natural enemies is imperative for the successful application of biological control. However, although information flow from higher to lower trophic levels has been previously discussed for animals in general (Kats & Dill 1998, Lima & Dill 1990, Tollrian & Harvell 1999), to the best of our knowledge no reviews that focus on arthropod enemy detection have been published to date.

The reliance on chemical information is particularly strong for many arthropods. Most large-scale studies of perception of chemical cues by arthropods focus on foraging for resources or mates (e.g., Bell & Cardé 1984, Brönmark & Hansson 2000, Cardé & Bell 1995, Cardé & Minks 1997, Chadwick & Goode 1999, Hay 1996, Nordlund et al. 1981, Roitberg & Isman 1992). Because of the differential emphasis of arthropod studies on infochemical use in foraging (see all references above) versus enemy detection (no reviews to date), in spite of the relatively higher importance of the latter for arthropod survival, this review is timely.

CHEMICAL INFORMATION

Chemical information can be acquired by animals through olfaction, through taste, or through a combination of both. Typically, olfaction is the detection of chemical cues that are dissolved in media such as air and water, away from their original source (Mustaparta 1984), whereas taste is the acquisition of cues by direct contact with the source (Städler 1984) or with an undissolved product. Although some forms of chemical information can be short lived (e.g., volatiles), other forms can be detected for a relatively long period of time after they are deposited or released (e.g., nonvolatile products).

Potential arthropod prey can perceive chemical cues about the presence of their enemies from either a direct or an indirect source (Figure 1). Cues from direct sources consist of infochemicals that are produced by the enemy and then recognized by the prey (kairomones, Table 1). Thus, direct cues may be contained in the exuviae, eggs, excreta, marking pheromones, or any other product of the

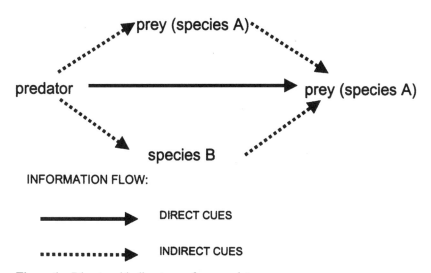

Figure 1   Direct and indirect cues from predators.

enemy (e.g., Dial & Schwenk 1996, Grostal & Dicke 2000, Hoffmeister & Roitberg 1997, Nolte et al. 1994).

Cues from indirect sources are not produced by the enemy. These cues may include pheromones (Table 1) from alarmed, injured, or dead conspecifics (Chivers & Smith 1994, Manzoli-Palma et al. 1998, Pijanowska 1997), but perhaps also allelochemicals (Table 1) from heterospecifics (Chivers & Smith 1998, Huryn & Chivers 1999). Although byproducts produced by plants or microorganisms (e.g., in contact with the enemy's body) may enhance the amount of information available to potential prey, this review is concerned with the primary sources of information. Thus, metabolic products of microorganisms that decompose predator excreta may constitute an indirect cue; but for the sake of simplicity, we encompass all predator secretions as a direct cue. Similarly, predator excreta that may contain chemicals from consumed prey are discussed here as a direct cue.

The effects of predator kairomones and alarm signaling on prey from diverse taxonomic groups have already been thoroughly summarized by Kats & Dill (1998) and Chivers & Smith (1998). Consequently, in the section below we elaborate on a few of the examples that involve arthropods.

Direct Cues

There is abundant evidence that direct chemical cues from natural enemies trigger defensive reactions in potential prey. These reactions have been documented for a wide diversity of vertebrates, including mammals, amphibians, reptiles, and fish, but also for simpler animals such as arthropods, molluscs, cnidarians, and rotifers (Chivers & Smith 1998, Kats & Dill 1998). Even very simple organisms

TABLE 1   Infochemical terminology[a]

Infochemical

A chemical that, in the natural context, conveys information in an interaction between two individuals, evoking in the receiver a behavioral or physiological response that is adaptive either to one or both of the interactants.

Pheromone

An infochemical that mediates an interaction between organisms of the same species whereby the benefit is to the origin-related organism $(+, -)$, to the receiver $(-, +)$, or to both $(+, +)$.

Allelochemical

An infochemical that mediates an interaction between two individuals that belong to different species.

Allomone

An allelochemical that is pertinent to the biology of an organism (organism 1) and that, when it contacts an individual of another species (organism 2), evokes in the receiver a behavioral or physiological response that is adaptively favorable to organism 1 but not to organism 2.

Kairomone

An allelochemical that is pertinent to the biology of an organism (organism 1) and that, when it contacts an individual of another species (organism 2), evokes in the receiver a behavioral or physiological response that is adaptively favorable to organism 2 but not to organism 1.

Synomone

An allelochemical that is pertinent to the biology of an organism (organism 1) and that, when it contacts an individual of another species (organism 2), evokes in the receiver a behavioral or physiological response that is adaptively favorable to both organism 1 and 2.

[a]From Dicke & Sabelis (1988).

such as protozoans have been observed to react to predator kairomones (Kats & Dill 1998). For example, ciliates (*Euplotes* sp.) exhibit defensive avoidance behavior in response to chemical cues from predatory amoebae, *Amoeba proteus* (Kusch 1999). Among arthropods, similar behavior has been widely observed in aquatic systems, especially among zooplankton (e.g., Dawidowicz 1999, Loose et al. 1993, Pijanowska 1997, Sakwinska 1998, Von Elert & Pohnert 2000).

Most of the evidence gathered on arthropod reactions to predator kairomones focuses on changes in the distribution and activity levels of prey, especially in aquatic environments. For example, nymphs of stream mayfly larvae (*Baetis bicaudatus*) reduce their levels of exposure (resting on stone tops) and drift in response to kairomones from brook trout, *Salvelinus fontinalis* (McIntosh & Peckarsky 1999). In addition, the level of response of *B. bicaudatus* depends on mayfly size

(McIntosh et al. 1999). Activity of mayfly larvae (*Siphlonurus* sp.) can also be reduced in response to chemical cues from other predatory mayfly larvae (*Siphlonica* sp.) (Huryn & Chivers 1999). Similarly, chemical predator cues induce diel vertical migration of *Daphnia* and other zooplankton away from the cue source (e.g., Dawidowicz 1999, Sakwinska 1998, Von Elert & Pohnert 2000). The amphipod *Gammarus pulex* adaptively adjusts drifting behavior in the presence of odors from caged trout, *Salmo trutta* (Dahl et al. 1998). The amphipods avoid drifting into cages that contain their predators and are capable of actively generating backflows of water current, which facilitates the detection of cues from predators that are located upstream (Dahl et al. 1998). Further, in response to kairomones from predatory fish, benthic isopods (*Sadunia entomon*) decrease their activity and foraging levels by remaining buried in the sediment (Ejdung 1998). Also, settlement of postlarvae of blue crabs (*Callinectes sapidus*) on seagrass is reduced in response to cues from predatory crabs, consequently reducing the foraging success of the predators (Welch et al. 1997).

Similar behavioral responses were observed for terrestrial arthropods. Fruit flies (*Rhagoletis basiola*) delay oviposition when exposed to chemical information from their egg parasitoids (Hoffmeister & Roitberg 1997). The herbivorous spider mite *Tetranychus urticae*, when given a choice between feeding arenas (leaf discs) that were previously exposed to their predators (*Phytoseiulus persimilis*) and unexposed arenas, preferred to forage and oviposit on arenas with no previous predator exposure (Grostal & Dicke 1999, Kriesch & Dicke 1997). Further, after being initially placed on leaf discs that previously contained predators (*P. persimilis*) and discs that previously contained nonpredatory mites (the fungivorous *Tyrophagus putrescentiae*), more spider mites emigrated from predator-exposed discs (Grostal & Dicke 1999). Also, large groups of spider mites avoided the colonization of whole plants when exposed to volatiles related to the presence of *P. persimilis* (Pallini et al. 1999). Finally, the distribution of spider mites was compared on discs exposed/unexposed to mites from three feeding groups: (*a*) predators of spider mites, (*b*) predators/parasites of other animals, and (*c*) fungivores/pollen-feeders (Grostal & Dicke 2000). Here, the spider mites showed a significant avoidance of discs exposed to the two predatory/parasitic groups, but no avoidance of discs exposed to fungivores or pollen feeders (Figure 2, see color insert). The spotted cucumber beetle (*Diabrotica undecimpunctata howardi*) reduced feeding when exposed to chemical information from the wolf spider *Hogna helluo*, but not when exposed to information from three species of other, less dangerous, predators (Snyder & Wise 2000).

Additionally, predator chemical cues can affect the reproductive behavior and morphology of potential prey. For example, *Chaoborus* sp. (Diptera) prefer to oviposit in water bodies that are free of kairomones from predatory fish (Berendonk 1999). In turn, *Daphnia pulex* delay their reproduction and increase their offspring size when exposed to water that previously contained their predators (*Chaoborus*), an adaptation that apparently reduces the success of the size-selective predation by the midge (Repka & Walls 1998, Tollrian 1995). Further, exudates of fish (a predator capable of attacking all life stages of *Daphnia*) elicit the production of smaller and more numerous *Daphnia* offspring (Sakwinska 1998, Stibor 1992). Among terrestrial arthropods, parasitic wasps (*Aphidius uzbekistanicus*) fly away from breeding areas that are exposed to odors from their hyperparasitoid enemy, *Alloxysta victrix* (Höller et al. 1994). Also, pea aphids (*Acyrthosiphon pisum*) develop an increased number of winged morphs in response to chemical cues from ladybird beetle larvae, *Adalia bipunctata* (Dixon & Agarwala 1999, see "Trade-offs"). The winged morphs appear to be produced at the cost of reduced fecundity (Dixon & Agarwala 1999) but are capable of avoiding predation through rapid dispersal. Dixon & Agarwala (1999) also noted that two other aphid species, *Megoura viciae* and *Aphis fabae fabae*, which constitute poor quality or toxic prey for the ladybird beetles, did not produce more winged morphs in the presence of predator cues. The most spectacular changes in response to predator kairomones are those involving prey morphology. For example, *Daphnia lumholtzi* responds to fish kairomones with increased production of head spines (Boersma et al. 1998, Tollrian 1994), a phenotypic trait that decreases predation success (Kolar & Wahl 1998). Similar morphological responses to predator odors include an increased production of neck-teeth among *Daphnia* (Parejko & Dodson 1990) and changes in shell shape by barnacles (Lively 1986); for a more extensive list see Kats & Dill (1998).

Indirect Cues

To date, evidence for the effect of indirect chemical cues on potential prey has largely been gathered in studies on conspecific alarm pheromones. The use of alarm pheromones by arthropods has been recognized for at least 30 years (Blum 1969). Similar to the effects of direct cues, most evidence for the effect of alarm pheromones centers on changes in prey behavioral responses. For instance, in aquatic systems, larvae of the mayfly *Siphlonurus* sp. significantly reduce their movement when exposed to chemicals from injured conspecifics: a response that reduces exposure to visual predators (Huryn & Chivers 1999). Aphids stop feeding and walk away from feeding sites in response to the synthetic alarm pheromone (E)-β-farnesene (Shah et al. 1999). Another synthetic alarm pheromone, citral, triggers an active dispersal of four mite species: *Carpoglyphus lactis*, *Lardoglyphus konoi*, *Tyrophagus putrescentiae*, and *Aleuroglyphus ovatus* (Matsumoto et al. 1998). Also, spider mites (*Tetranychus urticae*) avoid foraging and ovipositing in areas that contain damaged conspecifics consisting of either adults or eggs (Grostal & Dicke 1999). Further, spider mites, when given a choice, were found in higher numbers on discs with live healthy conspecifics (either adults or eggs) than on discs with artificially injured conspecifics (Grostal & Dicke 1999). Similarly, thrips (*Frankliniella occidentalis*) increased the use of refuge when exposed to odors from conspecifics that were attacked by the predatory bug *Orius laevigatus* (Venzon et al. 2000). In turn, odors from the bugs that were previously fed on a different diet had no significant effect on the thrips distribution (Venzon et al. 2000).

Alarm pheromones also trigger an aggressive response (nest defense) in vespid wasps. For example, *Polybia paulista* workers initiate stinging attacks in the presence of pheromones extracted from their venom reservoirs (Manzoli-Palma et al. 1998). Similarly, honeybees (*Apis mellifera scutellata*) prolong their stinging response when exposed to the alarm pheromone isopentyl acetate (Nuñez et al. 1997).

Little has been done to assess whether potential arthropod prey utilize information about enemy presence from heterospecifics, especially from nonrelated groups (Figure 1). Some animals, including fishes and salamanders, have been recorded displaying defensive behavior in response to chemical alarm signals from heterospecifics (Chivers et al. 1995, Lutterschmidt et al. 1994, Wisenden & Sargent 1997).

Considering the plethora of mutualistic interactions that arthropods are involved in (Boucher et al. 1982), interspecific information on enemy presence is likely to be utilized by many species. For example, taxonomically unrelated mutualists could exchange chemical information about the presence of an enemy of one or both species. This is discussed in more detail in "Pressing Questions and Future Prospects."

Source and Identity of the Infochemicals

There is strong evidence that a wide range of arthropods can recognize predator risk through chemical cues. Experimentally manipulated chemical information can trigger the potential prey to adjust behavior, morphology, and natural history, which in turn decreases the predation success of natural enemies (Chivers & Smith 1998, Kats & Dill 1998) (Table 2). Furthermore, these induced responses often come at a relatively high cost to the prey, through loss of feeding sites, decreased reproduction, and/or high energy investment. Assuming the optimization of prey response, the evidence outlined in the previous two sections and the associated costs (see "Trade-offs") suggest that recognition of predation risk through chemical information is crucial for the fitness of many arthropods.

The above line of evidence leads to two important questions. First, how accurate is the direct chemical recognition of predators by potential prey? For example, could cues from nonenemies or from nonorganic sources also trigger defensive responses? Second, what is the source and identity of the kairomones or alarm pheromones that trigger these responses? Few studies to date investigated these two questions. The accuracy of chemical enemy recognition is overlooked by many studies that manipulate direct predator information. For example, prey responses are often investigated by exclusively manipulating the presence/absence of predator cues. The result of these studies (if significant) may be confounded by a mere chemical change (coincidentally induced by the predators) in the prey environment. Would the same prey show a similar response to previously unencountered chemical compounds from nonpredator sources? Can potential prey distinguish the chemical cues from their predators from those emitted by animals that exclusively prey on other species? Answers to these questions are crucial to understand the ecology underlying predator-prey interactions in general, but to

TABLE 2 Effects of direct and indirect cues on prey responses[a]

Response	Prey	Predator	Reference
Direct cues			
Avoidance/escape	*Daphnia*	Midge	Dodson (1988)
		Fish	Loose & Dawidowicz (1994)
	Copepod	Midge	Neill (1990)
	Copepod	Fish	Short & Holomuzki (1992)
	Barnacle	Whelk	Johnson & Strathmann (1989)
	Mayfly	Stoneflies	Peckarsky (1980)
	Mosquito	Fish, Frog	Petranka & Fakhoury (1991)
	Midge	Fish	Dawidowicz (1993)
Reduction of activity	Amphipod	Fish	Holomuzki & Hoyle (1990)
	Mayfly	Fish	McIntosh & Peckarsky (1996)
	Caddisfly	Stonefly	Malmqvist (1992)
Use of refugia	Ostracod	Fish	Roca et al. (1993)
	Lobster	Fish	Wahle (1992)
	Crayfish	Fish	Blake & Hart (1993)
Change of oviposition site	Mite	Mite	Grostal & Dicke (2000)
Defensive posture	Crayfish	Fish	Shave et al. (1994)
Morphological changes	*Daphnia*	Midge	Tollrian (1994)
	Barnacle	Gastropod	Lively (1986)
Indirect cues			
Avoidance/escape	Mosquito Larvae		Sih (1986)
	Mites		Grostal & Dicke (2000)
Reduction of activity	Mayfly Larvae		Huryn & Chivers (1999)
	Damselfly Larvae		Chivers et al. (1996)
	Amphipods		Mathis & Hoback (1997)
	Crayfish		Hazlett (1994)
	Pebble Crabs		McKillup & McKillup (1992)
Use of refugia	Amphipods		Wudkevich et al. (1997)
	Hermit Crabs		Kats & Rittschof (1993)
	Thrips		Venzon et al. (2000)
Change of oviposition site	Mites		Grostal & Dicke (2000)
Watchful behavior	Hermit Crabs		Hazlett (1990a)
	Crayfish		Hazlett (1985, 1990b)
Increased aggression	Wasps		Manzoli-Palma et al. (1998)
	Bees		Nunez et al. (1997)

[a]For more information on direct and indirect cues, see Kats & Dill (1998), Chivers & Smith (1998), respectively.

provide convincing evidence they require thorough experimental controls, e.g., those that involve "benign," but foreign cues.

One of the bottlenecks constraining the present-day studies is the relative lack of knowledge about the identity of the infochemicals that are responsible for evasive reactions by prey. Often, the exact source of chemical cues can only be speculated on, as many studies (including those performed by us) use the predators themselves as a source of information for the observed potential prey. For example, in several studies, defensive responses of *Daphnia* were stimulated with liquid mixtures extracted from dead predatory fish. In these experiments, the exact source of information, i.e., predator secretions versus blood (or other potential alarm substances) is impossible to determine. Further, potential prey may react not only to the specific type of kairomone but also to its concentration (Brönmark & Hansson 2000). Because the amount of secreted kairomones may differ between predator species, the effects of kairomone type and concentration may often be difficult to separate. Finally, a single species of predator can vary the secretion of kairomones under specific environmental conditions. For example, pike can avoid defecating in its foraging area, an adaptation that appears to minimize the recognition of this predator's presence by its prey (Brown et al. 1995).

Some workers (e.g., Höller et al. 1994, Loose et al. 1993, Tollrian & Von Elert 1994, Von Elert & Pohnert 2000) have investigated the predator kairomone and used it for experimental manipulation on potential prey. Thus, the compounds identified as active components for a single prey may constitute only part of the chemical cocktail used by prey to detect their enemies (Brönmark & Hansson 2000, Von Elert & Pohnert 2000). Needless to say, this area of study is open to much more investigation.

RELIABILITY OF CUES

How important is it to pinpoint the cues from an enemy as opposed to cues from nonenemy species? Any defensive response (including fleeing or migration) by potential prey is costly (Chase 1999, "Trade-offs"), and frequent responses to cues from sources that have no consequences on prey fitness would be counteradaptive. The range of cues that a potential prey may react to is likely to be controlled by the range and predictability of enemies that the prey has. For example, accurate perception of enemies may be difficult if the potential prey is exposed to many, or changing, generalist enemies. For gregarious species, recognition of enemies through injured conspecifics (alarm pheromones) may constitute an effective strategy. Direct cues from generalist predators can be efficiently assessed through recognition of previous predator diet. For instance, potential prey may recognize protein metabolites (derived from carnivorous diet), present in the excreta of predators (Nolte et al. 1994). This generalized assessment of potential predator presence is supported by evidence of potential prey showing defensive reactions in response to cues from a range of predators that feed exclusively on other animals (e.g., Grostal & Dicke 2000). Additionally, prey can respond to information from conspecifics that is present within predator excreta. Evidence for the discrimination between the previous (carnivorous) diet of predators has been collected for mayflies. For example, Huryn & Chivers (1999) found that *Siphlonurus* sp. responded defensively to cues from predators (brook trout) that were previously fed with mayflies, but not to cues from trout fed with shrimp. Nevertheless, it is important to determine whether this interesting result is not constrained by different rates of cue secretion by the predator (i.e., resulting in varying kairomone concentration) in response to different diets.

Although a predator is more likely to be in the vicinity when its kairomone intensity is increased (Von Elert & Pohnert 2000), a higher accuracy in predator detection can be achieved in the presence of more than one source of chemical information. Here, the reaction of prey to multiple information sources (e.g., direct and indirect cues) can be more pronounced than the reaction to a single source. For example, Slusarczyk (1999) found that the production of resting eggs by *Daphnia* was affected only in the presence of both predator kairomones and conspecific alarm chemicals. Similarly, it has been noted that crayfish cryptic behavior increased more in the presence of cues from predators and injured conspecifics than in response to either of these cues alone (Hazlett 1999, Hazlett & Schoolmaster 1998).

TRADE-OFFS

Optimal foraging theory predicts that animals are under selection to make "decisions" that maximize reproductive success (Krebs & Davies 1987, Stephens & Krebs 1986). This relates to decisions during food and mate selection, as well as the avoidance of predation and of competition. These decisions are interrelated and complex and lead to trade-offs. One of the most prominent trade-offs in ecology is that between the acquisition of resources and the avoidance of predation (Bake et al. 1977, Krebs & Davies 1984, Stephens & Krebs 1986). Optimal foraging theory assumes that animals are omniscient, which allows them to make decisions that maximize fitness (Stephens & Krebs 1986). Although omniscience of arthropods has often been considered unrealistic, there is increasing evidence that arthropods can be remarkably well-informed about prevailing conditions (Krebs & Kacelnik 1991, Vet 1996). Many internal as well as external factors affect the responses of animals to their environment (Figure 3). For example, chemical information from resources (Bernays & Chapman 1994, Chadwick & Goode 1999, Dicke & Van Loon 2000, Nordlund et al. 1981, Schoonhoven et al. 1998, Vinson 1976, Visser 1986), natural enemies (Kats & Dill 1998, Lima & Dill 1990, this review), competitors (Hilker & Klein 1989, Janssen et al. 1997, Janssen et al. 1995, Prokopy et al. 1984, Schoonhoven et al. 1989), and mates (Cardé & Minks 1997) are of considerable importance to arthropods in making optimal foraging decisions. In

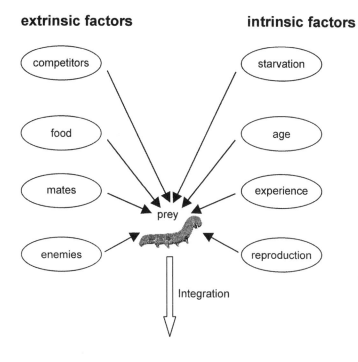

Figure 3   Extrinsic and intrinsic factors that may affect the responses of animals.

fact, recent evidence shows that food webs are superimposed and often controlled by infochemical webs (e.g., Abrams et al. 1996, Dicke & Vet 1999, Janssen et al. 1998, Schmitz et al. 1997, Turner et al. 2000, Werner 1991).

Trade-offs between resource acquisition and avoiding predation risks may be constitutively present (Bernays & Graham 1988, Bernays 1998a, Müller & Godfray 1999, Ohsaki & Sato 1994). Some animals select suboptimal resources that represent enemy-free space independently of information on enemy presence (Ohsaki & Sato 1994). However, this will not be a profitable strategy when the presence of enemies is unpredictable. The selection of an enemy-free space as a feeding site is very important because feeding can be risky. For example, Bernays (1997) showed that for caterpillars, mortality can be 3–100 times higher during feeding than between feeding bouts. Thus, crucial decisions and adaptations are necessary to optimize the conflict between feeding and predation risk. One of these adaptations may include feeding specialization, common among many herbivorous arthropods (e.g., Bernays & Chapman 1994, Rosenthal & Berenbaum 1992, Schoonhoven et al. 1998). Initially, the driving force behind this specialization was thought to be plant chemistry [see Bernays & Chapman (1994) and Schoonhoven et al. (1998) for reviews], but evidence for the influence of natural enemies on this adaptation is increasing (e.g., Bernays & Graham 1988, Bernays 1988, 1998a). For instance, among closely related herbivorous arthropods, specialists spend less time in making foraging decisions and in foraging activities than do generalists (Bernays 1998b, Bernays & Funk 1999, Janz & Nylin 1997), a difference that can have important consequences for fitness.

In addition to intrinsic food preferences, animals can exploit different types of chemical information to select the most profitable foraging sites. When arthropods show the ability to perceive the presence, quantity, and identity of food, competitors, and natural enemies, the most obvious question that arises is, What decisions are made by these animals when they are confronted with conflicting information? For instance, how does a herbivorous insect optimize its behavior when the alternatives are an inferior host plant without predator cues versus a superior host plant with abundant predator cues? Are the decisions different for specialist and generalist herbivores? Such research questions are an exciting part of behavioral ecology (e.g., Anholt & Werner 1995, Bouskila et al. 1998, Courtney 1986, Godin & Sproul 1988, Gotceitas et al. 1995, Krebs & Davies 1984) and deserve the incorporation of studies on chemical ecology.

The most direct trade-off of predator avoidance is that avoidance responses to predator-related cues can result in reduced food intake (Kats & Dill 1998). The animal may shift from acquiring resources to other behaviors, such as hiding or emigration. Moreover, the avoidance behavior may even result in new types of mortality. For instance, aphids that drop from a plant in response to a predator attack may encounter carabid beetles on the ground (Bryan & Wratten 1984, Losey & Denno 1998). Predator avoidance may also lead to competitive changes. For example, when exposed to upwind odors from patches of conspecifics that were invaded by predators, thrips larvae hide in webs produced by spider mites. Within spider-mite webs they compete with the spider mites for food and space and experience lower reproductive success (Pallini et al. 1998). Predator avoidance may also result in reduced competitive ability (Werner & Anholt 1996). Nevertheless, in the trade-off between predator avoidance and competition, the benefits of avoiding a predator likely outweigh the costs of reduced resource allocation due to altered competitive conditions.

Life history changes in response to predation risk can also impose a cost on potential prey. The increased production of winged morphs by aphids (Dixon & Agarwala 1999, Weisser 1999, discussed in the "Direct Cues" section) constitutes one example. The winged morphs are superior dispersers, and dispersion facilitates escape, but they also take longer to develop and have reduced fecundity (Dixon 1998, Dixon & Wratten 1973), and this results in a significant reproductive cost.

Optimal defense theory assumes that resources can be allocated to different types of defense and that trade-offs exist not only between defense and resource acquisition but also between different types of defenses (e.g., Fagerström et al. 1987, van Dam et al. 1996, Yamamura & Tsuji 1995). For instance, aphids may experience trade-offs among defenses such as dispersal through the production of alates, production of toxins, or recruitment of ant bodyguards (Dixon & Agarwala 1999).

Antipredator decisions of animals are expected to reflect the relative risk levels under field conditions (e.g., Rigby & Jokela 2000). Different predator species, with distinct foraging strategies, may select for different antipredator responses in potential prey (McIntosh & Peckarsky 1999, Pitt 2000). For instance, larvae of the mayfly *Baetis bicaudatus* may be preyed on by stoneflies and brook trout that represent different types of foragers. *Baetis* responded in contrasting ways when presented with cues from either of the two predator species. When exposed to chemical cues from nonfeeding stoneflies, the larvae drifted away; but in contrast, they responded to cues from nonfeeding trout with reduced exposure and drift levels (McIntosh & Peckarsky 1999). When cues from both predators were present simultaneously, responses to trout cues dominated during the day, when visually-foraging trout are active and pose the largest risk. Responses were intermediate during the night, when stoneflies forage and visually-hunting trout pose a reduced risk.

VARIATION IN RESPONSES

Behavioral responses to predator cues may vary geographically; prey living in sympatry with a given predator are often more responsive than prey that live in allopatry with this predator (Kats & Dill 1998). For instance, antipredator behavior of copepods from different pools with different predation pressures differed in antipredator behavior. These differences have a genetic basis (Neill 1992). Pea aphids, when parasitized by a parasitoid, commit suicide in response to alarm pheromones by jumping off the plant. This results in the death of the parasitoid larva as well. As a consequence, this parasitoid will not be able to attack the genetically identical aphids of the local population. However, the suicidal response is only found for pea aphids from populations that live in a region with a dry, hot climate, where jumping off the plant has a high probability of mortality, and not for pea aphids from a region with a more moderate climate, where jumping off the plant has a low probability of mortality (McAllister & Roitberg 1987, McAllister et al. 1990).

The response to an information source may depend on e.g., physiological state, previous experiences, and abiotic conditions (for reviews see e.g., Bernays 1995, Dicke et al. 1998, Jaenike 1988, Jaenike & Papaj 1992, Papaj & Lewis 1993, Papaj & Prokopy 1989, Robertson et al. 1995, Vet et al. 1995). There is abundant evidence for learning of chemical information by arthropods during foraging for resources (see Papaj & Lewis 1993, Vet et al. 1995). With increased exposure, the animal may show an increased response through sensitization. Furthermore, arthropods are capable of Pavlovian conditioning, whereby the animals learn to associate a conditioned stimulus to an unconditioned stimulus (Papaj & Lewis 1993, Vet et al. 1995). For instance, parasitic wasps can learn to associate previously unencountered odors (conditioned stimulus) with oviposition in a host or with exposure to host feces (unconditioned stimulus) (Lewis & Tumlinson 1988, Vet & Groenewold 1990). Consequences of learning may include faster foraging decisions (Papaj & Vet 1990), which contribute to a reduced risk of predation (Bernays & Funk 1999).

Initially, it may seem that learning cannot be involved in the avoidance response to predator kairomones. However, some studies have shown that learning can occur (Brown & Godin 1999, Chivers et al. 1996, Dukas 1998, Mallet et al. 1987, Rochette et al. 1998, Wisenden et al. 1997). For instance, *Drosophila* larvae that experience a novel odor in combination with the crushing of conspecific larvae avoid the novel odor in subsequent tests (Dukas 1998). In this way, arthropods may learn that certain environments currently represent predator-dense space, resulting in changes within prey distribution and affecting the interactions among species in food webs. Additionally, responses to predator kairomones might be intensified after the kairomones are associated with indirect cues, such as alarm chemicals from conspecifics (Grostal & Dicke 1999). For instance, damselfly larvae from an area where pike do not occur respond to chemical cues from injured conspecifics but do not respond to chemical stimuli from pike. However, after a single exposure to stimuli from pike and injured conspecifics they learned to recognize pike stimuli, leading to antipredator response (Wisenden et al. 1997). Learning to respond to predator kairomones is likely to result in more efficient defensive responses, which eventually result in a reduced mortality from predators. The influence of learning on avoidance of natural enemies by arthropods is an interesting aspect of predator-prey interactions and deserves further study.

Another question is whether prey habituate to chemical information related to their enemies in the absence of predation per se. After all, predator kairomones can remain active for quite some time after the predator has left a patch. Therefore, when predator kairomones are not encountered in combination with cues from (recently) killed conspecifics, the predator cues may not represent a real danger.

PRESSING QUESTIONS AND FUTURE PROSPECTS

In spite of the wealth of evidence that arthropods recognize, and react to, chemical cues related to enemy presence, there are considerable gaps in our understanding of the interaction. One of the key areas that still requires research includes the diversity of direct cues that prey react to, as well as the link between this diversity and the range of natural enemies of that prey species. For example, do prey that are under evolutionary pressure from a narrow group of specialist predators possess more accurate, predator-specific recognition? The second area of interest includes the role of trade-offs on prey response, once the enemy is recognized. The third area involves the role of learning and conditioning of prey in response to (*a*) novel cues associated with active enemies, and (*b*) potential enemy cues in the prolonged absence of predation. Yet another area that is relatively unexplored is the role of indirect cues. Here, points of special interest include the response of predators to alarm cues by prey, as well as the potential presence of interspecific warning signals among sympatric mutualists. The above research areas cannot be thoroughly investigated without knowledge of the source and identity of the infochemicals involved in these interactions.

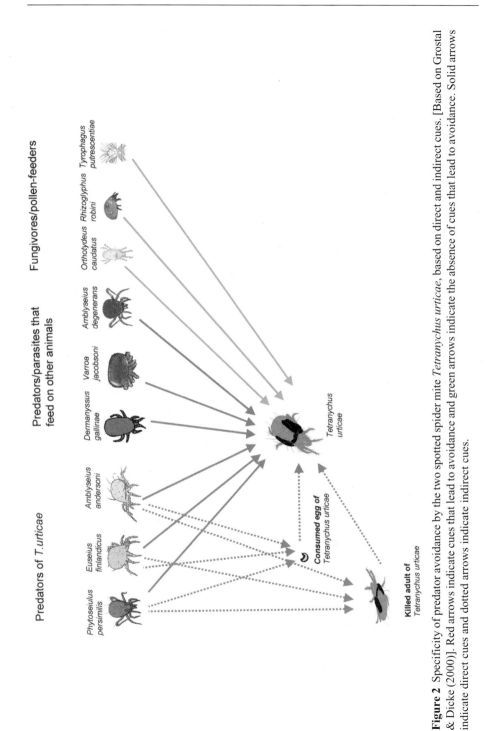

Figure 2 Specificity of predator avoidance by the two spotted spider mite *Tetranychus urticae*, based on direct and indirect cues. [Based on Grostal & Dicke (2000)]. Red arrows indicate cues that lead to avoidance and green arrows indicate the absence of cues that lead to avoidance. Solid arrows indicate direct cues and dotted arrows indicate indirect cues.

Although abundant knowledge is available on the identity of infochemicals related to arthropods' resources (Cardé & Bell 1995, Chadwick & Goode 1999, Nordlund et al. 1981), relatively little information exists on the identity of direct and indirect cues that influence arthropod prey behavior (see "Source and Identity of the Infochemicals"). More knowledge on mechanistic aspects, such as the origin, the site of production, and the biosynthetic pathways involved, as well as the conditions that lead to production and to emission will provide tools for manipulative studies to unravel the ecological aspects of the interaction. For instance, synthetic cues may serve as a tool in behavioral ecological analyses, where the combination of cues from different sources, including resources and enemies, can be used to simulate different ecological conditions. Furthermore, availability of synthetic cues can allow the application of these cues in controlled amounts. In this way, chemical ecology may be combined with a behavioral ecological approach to allow a more thorough investigation of chemical information flows on the foraging decisions of animals.

Food web and population dynamics studies have traditionally concentrated on population or species densities in relation to the abundance and quality of food and enemies. However, food availability and risk from predators affect not only birth and death rates but also behavioral, morphological, physiological, and life-history characteristics. Food webs are overlaid with infochemical webs that are even more complex than the food webs themselves because infochemicals can have many indirect effects that operate independently of trophic interactions (e.g., Dicke & Vet 1999, Janssen et al. 1998). Because the infochemicals may result in a shift in trait values, they can have an effect on the strength of species interactions. For instance, a predator kairomone may affect prey behavior and distribution, consequently influencing the intensity of the interaction between predator and prey. In turn, these changes in animal traits may have indirect yet important consequences on population dynamics (e.g., Scheffer 1997, Schmitz et al. 1997, Sih 1987, Turner et al. 2000). In general, trait-mediated effects are difficult to separate from density effects. With infochemical-mediated trait shifts, however, one may manipulate the system by applying the infochemical without affecting birth and death rates. Alternatively, one may want to manipulate the system such that the infochemical is absent while the predator is present. Although no satisfactory approach is known to us to date, the progress in molecular genetics may provide tools for this approach in the future. A limited number of studies have incorporated the effects of predator-related cues on interactions between prey and the prey's food and reported strong effects of predator presence and predator identity on herbivores and their food (e.g., Turner et al. 2000). Carrying out such studies will provide insight into the potential role of chemical predator avoidance on population dynamics. The next question to be answered is to what extent predator avoidance varies within and between populations. Information on the variation among populations is especially likely to provide insight into the strength of selection on the trait by predator pressure (De Meester et al. 1995, Neill 1992).

Infochemicals have found various applications in biological control of arthropods. For instance, sex pheromones are used to monitor pest populations or to disrupt mating (Cardé & Minks 1997), and allelochemicals may be applied in improving biological control (Dicke et al. 1990). Ideas for exploiting predator avoidance to control mammal pests have been presented in the past (Sullivan et al. 1988a, 1988b). Such ideas have also been put forward for the control of herbivorous arthropods (e.g., Snyder & Wise 2000). However, when no alternatives are presented, strategies involving repellents or deterrents are doomed to fail: Any mutant that no longer responds to the cue in the absence of predators will have greater reproductive success than genotypes that do respond. One of the approaches in the application of infochemicals in control of arthropods is the push-pull strategy or stimulo-deterrent diversion strategy (Miller & Cowles 1990), in which a combination of a deterrent and an attractant is used to direct the pest species to a selected site. Knowledge of the behavioral responses of pest species when presented with attractants and deterrents/repellents in different choice situations will be valuable in investigating the potential for such an application of predator-related cues that are avoided by prey. Thus, information from the behavioral ecological approach advertised above may provide useful directions for the application of predator-related cues.

In conclusion, predator avoidance through exploitation of chemical information is widespread among arthropods. This, in turn, affects other aspects of evolutionary, behavioral, and population ecology that involve arthropods. To achieve a better understanding of the influence of chemically mediated predator avoidance on food webs, we need more studies that address the role of the specific infochemicals. To reach this goal, we need more information on mechanistic issues, such as infochemical identity. Conversely, ecological studies will require studies on the responses to multiple chemical inputs. The integration of mechanistic and ecological studies in the field of predator avoidance will provide a fertile research area that will make significant contributions to our understanding of food webs.

ACKNOWLEDGMENTS

We thank Thomas Hoffmeister, Ralph Tollrian, Jeff Harvey, Rieta Gols, and Luc De Meester for constructive comments on an earlier version of the manuscript. MD was partly funded by the Uyttenboogaart-Eliasen Foundation (Amsterdam).

Visit the Annual Reviews home page at www.AnnualReviews.org

LITERATURE CITED

Abrams PA, Menge BA, Mittelbach GG. 1996. The role of indirect effects in food webs. In *Food Webs: Integration of Patterns and Dynamics*, ed. GA Polis, KO Wine-miller, pp. 371–95. New York: Chapman & Hall

Anholt BR, Werner EE. 1995. Interaction between food availability and predation mortality mediated by adaptive behavior. *Ecology* 76:2230–34

Bake GH, Pulliam HR, Charnov EL. 1977. Optimal foraging: a selective review of theory and tests. *Q. Rev. Biol.* 52:137–55

Bell WJ, Cardé RT, eds. 1984. *Chemical Ecology of Insects*. London: Chapman & Hall. 512 pp.

Berenbaum MR. 1995. *Bugs in the System.* Reading, MA: Addison-Wesley. 377 pp.

Berendonk TU. 1999. Influence of fish kairomones on the ovipositing behavior of *Chaoborus imagines*. *Limnol. Oceanogr.* 44:454–58

Bernays EA. 1988. Host specificity in phytophagous insects: selection pressure from generalist predators. *Entomol. Exp. Appl.* 49:131–40

Bernays EA. 1995. Effects of experience on host-plant selection. In *Chemical Ecology of Insects 2*, ed. RT Carde, WJ Bell, pp. 47–64. New York: Chapman & Hall

Bernays EA. 1997. Feeding by lepidopteran larvae is dangerous. *Ecol. Entomol.* 22:121–23

Bernays EA. 1998a. Evolution of feeding behavior in insect herbivores. Success seen as different ways to eat without being eaten. *BioScience* 48:35–44

Bernays EA. 1998b. The value of being a resource specialist: behavioral support for a neural hypothesis. *Am. Nat.* 151:451–64

Bernays EA, Chapman RL. 1994. *Host-Plant Selection by Phytophagous Insects*. New York: Chapman & Hall. 312 pp.

Bernays EA, Funk DJ. 1999. Specialists make faster decisions than generalists: experiments with aphids. *Proc. R Soc. London Ser. B* 266:151–56

Bernays E, Graham M. 1988. On the evolution of host specificity in phytophagous arthropods. *Ecology* 69:886–92

Blake MA, Hart PJB. 1993. The behavioral responses of juvenile signal crayfish *Pacifastacus leniusculus* to stimuli from perch and eels. *Freshw. Biol.* 29:89–97

Blum MS. 1969. Alarm pheromones. *Annu. Rev. Entomol.* 14:57–80

Boersma M, Spaak P, Meester Ld. 1998.

Predator-mediated plasticity in morphology, life history, and behavior of *Daphnia*: the uncoupling of responses. *Am. Nat.* 152:237–48

Boucher DH, James S, Keeler KH. 1982. The ecology of mutualism. *Annu. Rev. Ecol. Syst.* 13:315–47

Bouskila A, Robinson ME, Roitberg BD, Tenhumberg B. 1998. Life-history decisions under predation risk: importance of a game perspective. *Evol. Ecol.* 12:701–15

Brönmark C, Hansson LA. 2000. Chemical communication in aquatic systems: an introduction. *Oikos* 88:103–9

Brown GE, Chivers DP, Smith RJF. 1995. Localized defecation by pike: a response to labelling by cyprinid alarm pheromone? *Behav. Ecol. Sociobiol.* 36:105–10

Brown GE, Godin JGJ. 1999. Who dares, learns: chemical inspection behaviour and acquired predator recognition in a characin fish. *Anim. Behav.* 57:475–81

Bryan KM, Wratten SD. 1984. The responses of polyphagous predators to prey spatial heterogeneity: aggregation by carabid and staphylinid beetles to their cereal aphid prey. *Ecol. Entomol.* 9:251–59

Cardé RT, Bell WJ, eds. 1995. *Chemical Ecology of Insects 2*. New York: Chapman & Hall

Cardé RT, Minks AK, eds. 1997. *Insect Pheromone Research: New Directions*. New York: Chapman & Hall. 684 pp.

Chadwick DJ, Goode JA, eds. 1999. *Insect-Plant Interactions and Induced Plant Defence (Novartis Foundation Symposium 223)*. Chichester, UK: Wiley. 281 pp.

Chase JM. 1999. To grow or to reproduce? The role of life-history plasticity in food web dynamics. *Am. Nat.* 154:571–86

Chivers DP, Smith RJF. 1994. The role of experience and chemical alarm signalling in predator recognition by fathead minnow, *Pimephales promelas*. *J. Fish Biol.* 44:272–85

Chivers DP, Smith RJF. 1998. Chemical alarm signalling in aquatic predator-prey systems: a review and prospectus. *Ecoscience* 5:338–52

Chivers DP, Wisenden BD, Smith RJF. 1995.

The role of experience in the response of fathead minnows (*Pimephales promelas*) to skin extracts of Iowa darters (*Etheostoma exile*). *Behaviour* 132:665–74

Chivers DP, Wisenden BD, Smith RJF. 1996. Damselfly larvae learn to recognize predators from chemical cues in the predator's diet. *Anim. Behav.* 52:315–20

Courtney SP. 1986. Why insects move between host patches: some comments on 'risk spreading'. *Oikos* 47:112–14

Dahl J, Nillson PA, Pettersson LB. 1998. Against the flow: chemical detection of downstream predators in running waters. *Proc. R. Soc. London Ser. B* 265:1339–44

Dawidowicz P. 1993. Diel vertical migration in *Chaoborus flavicans*: population patterns versus individual tracks. *Arch. Hydrobiol. Beih. Erg. Limnol.* 39:19–28

Dawidowicz P. 1999. Costs of behavioural defense against predation: a model of diel vertical migration in zooplankton. *Wiad. Ekol.* 45:3–16

De Meester L, Weider LJ, Tollrian R. 1995. Alternative antipredator defences and genetic polymorphism in a pelagic predator-prey system. *Nature* 378:483–85

Dial BE, Schwenk K. 1996. Olfaction and predator detection in *Coleonyx brevis* (Squamata, Eublepharidae), with comments on the functional significance of buccal pulsing in geckos. *J. Exp. Zool.* 276:415–24

Dicke M, Sabelis MW. 1988. Infochemical terminology: based on cost-benefit analysis rather than origin of compounds? *Funct. Ecol.* 2:131–39

Dicke M, Takabayashi J, Posthumus MA, Schütte C, Krips OE. 1998. Plant-phytoseiid interactions mediated by prey-induced plant volatiles: variation in production of cues and variation in responses of predatory mites. *Exp. Appl. Acarol.* 22:311–33

Dicke M, Van Lenteren JC, Minks AK, Schoonhoven LM. 1990. Semiochemicals and pest control, prospects for new applications. *J. Chem. Ecol.* 16:3015–212

Dicke M, Van Loon JJA. 2000. Multitrophic effects of herbivore-induced plant volatiles in an evolutionary context. *Entomol. Exp. Appl.* 97:237–49

Dicke M, Vet LEM. 1999. Plant-carnivore interactions: evolutionary and ecological consequences for plant, herbivore and carnivore. In *Herbivores: Between Plants and Predators*, ed. H Olff, VK Brown, RH Drent, pp. 483–520. Oxford, UK: Blackwell Sci.

Dixon AFG. 1998. *Aphid Ecology*. London: Chapman & Hall

Dixon AFG, Agarwala BK. 1999. Ladybird-induced life-history changes in aphids. *Proc. R. Soc. London Ser. B* 266:1549–53

Dixon AFG, Wratten SD. 1973. Laboratory studies on aggregation, size and fecundity in the black bean aphid, *Aphis fabae* Scop. *Bull. Entomol. Res.* 61:97–111

Dodson SI. 1988. The ecological role of chemical stimuli for the zooplankton: predator-avoidance behavior in *Daphnia*. *Limnol. Oceanogr.* 33:1431–39

Dukas R. 1998. Ecological relevance of associative learning in fruit fly larvae. *Behav. Ecol. Sociobiol.* 19:195–200

Ejdung G. 1998. Behavioural responses to chemical cues of predation risk in a three-trophic-level Baltic sea food chain. *Mar. Ecol. Prog. Ser.* 165:137–44

Fagerström T, Larsson S, Tenow O. 1987. On optimal defence in plants. *Funct. Ecol.* 1:73–81

Godin JJ, Sproul CD. 1988. Risk taking in parasitized sticklebacks under threat of predation: effects of energetic need and food availability. *Can. J. Zool.* 66:2360–67

Gotceitas V, Fraser S, Brown JA. 1995. Habitat use by juvenile Atlantic cod (*Gadus morhua*) in the presence of an actively foraging and non-foraging predator. *Mar. Biol.* 123:421–30

Grostal P, Dicke M. 1999. Direct and indirect cues of predation risk influence behavior and reproduction of prey: a case for acarine interactions. *Behav. Ecol.* 10:422–27

Grostal P, Dicke M. 2000. Recognising one's enemies: a functional approach to risk assessment by prey. *Behav. Ecol. Sociobiol.* 47:258–64

Hay ME. 1996. Marine chemical ecology: What's known and what's next? *J. Exp. Mar. Biol. Ecol.* 200:103–34

Hazlett BA. 1985. Disturbance pheromones in the crayfish *Orconectes virilis. J. Chem. Ecol.* 11:1695–711

Hazlett BA. 1990a. Disturbance pheromone in the hermit crab *Calcinyus laevimanus* (Randall, 1840). *Crustaceana* 58:316–18

Hazlett BA. 1990b. Source and nature of disturbance-chemical system in crayfish. *J. Chem. Ecol.* 16:2263–75

Hazlett BA. 1994. Alarm response in the crayfish *Orconectes virilis* and *Orconectes propinquus. J. Chem. Ecol.* 20:1525–35

Hazlett BA. 1999. Responses to multiple chemical cues by the crayfish *Orconectes virilis. Behaviour* 136:161–77

Hazlett BA, Schoolmaster DR. 1998. Responses of cambarid crayfish to predator odor. *J. Chem. Ecol.* 24:1757–70

Hilker M, Klein B. 1989. Investigation of oviposition deterrent in larval frass of *Spodoptera littoralis* (Boisd.). *J. Chem. Ecol.* 15:929–37

Hoffmeister TS, Roitberg BD. 1997. Counterespionage in an insect herbivore-parasitoid system. *Naturwissenschaften* 84:117–19

Höller C, Micha SG, Schulz S, Francke W, Pickett JA. 1994. Enemy-induced dispersal in a parasitic wasp. *Experientia* 50:182–85

Holomuzki JR, Hoyle JD. 1990. Effect of predatory fish presence on habitat use and diel movement of the stream amphipod, *Gammarus minus. Freshw. Biol.* 24:509–17

Huryn AD, Chivers DP. 1999. Contrasting behavioral responses by detritivorous and predatory mayflies to chemicals released by injured conspecifics and their predators. *J. Chem. Ecol.* 25:2729–40

Jaenike J. 1988. Effects of early adult experience on host selection in insects: some experimental and theoretical results. *J. Insect Behav.* 1:3–15

Jaenike J, Papaj DR. 1992. Behavioral plasticity and patterns of host use by insects. In *Chemical Ecology of Insects: An Evolutionary Approach,* ed. BD Roitberg, MB Isman, pp. 245–64. New York: Chapman & Hall

Janssen A, Bruin J, Jacobs G, Schraag R, Sabelis MW. 1997. Predators use volatiles to avoid prey patches with conspecifics. *J. Anim. Ecol.* 66:223–32

Janssen A, Pallini A, Venzon M, Sabelis MW. 1998. Behaviour and indirect food web interactions among plant inhabiting arthropods. *Exp. Appl. Acarol.* 22:497–521

Janssen A, Van Alphen JJM, Sabelis MW, Bakker K. 1995. Specificity of odour mediated avoidance of competition in *Drosophila* parasitoids. *Behav. Ecol. Sociobiol.* 36:229–35

Janz N, Nylin S. 1997. The role of female search behaviour in determining host plant range in plant feeding insects: a test of the information processing hypothesis. *Proc. R. Soc. London Ser. B* 264:701–7

Johnson LE, Strathmann RR. 1989. Settling barnacle larvae avoid substrata previously occupied by a mobile predator. *J. Exp. Mar. Biol. Ecol.* 128:87–103

Kats JN, Rittschof D. 1993. Alarm/investigation responses of hermit crabs as related to shell fit and crab size. *Mar. Behav. Physiol.* 22:171–82

Kats LB, Dill LM. 1998. The scent of death: chemosensory assessment of predation risk by prey animals. *Ecoscience* 5:361–94

Kolar CS, Wahl DH. 1998. Daphnid morphology deters fish predators. *Oecologia* 116:546–54

Krebs JR, Davies NB. 1984. *Behavioural Ecology, An Evolutionary Approach.* Oxford, UK: Blackwell. 493 pp. 2nd ed.

Krebs JR, Davies NB, eds. 1987. *An Introduction to Behavioural Ecology.* Oxford, UK: Blackwell. 389 pp.

Krebs JR, Kacelnik A. 1991. Decision making. In *Behavioural Ecology: An Evolutionary Approach,* ed. JR Krebs, NB Davies, pp. 105–36. Oxford, UK: Blackwell

Kriesch S, Dicke M. 1997. Avoidance of predatory mites by the two-spotted spider mite *Tetranychus urticae*: the role of infochemicals. *Proc. Exp. Appl. Entomol.* 8:121–26

Kusch J. 1999. Self-recognition as the original function of an amoeban defense-inducing kairomone. *Ecology* 80:715–20

Lewis WJ, Tumlinson JH. 1988. Host detection by chemically mediated associative learning in a parasitic wasp. *Nature* 331:257–59

Lima SL, Dill LM. 1990. Behavioral decisions made under the risk of predation: a review and prospectus. *Can. J. Zool.* 68:619–40

Lively CM. 1986. Predator-induced shell dimorphism in the acorn barnacle, *Chthamalus anisopoma. Evolution* 40:232–42

Loose CJ, Dawidowicz P. 1994. Trade-offs in vertical migration by zooplankton: the costs of predator avoidance. *Ecology* 75:2255–63

Loose CJ, Von Elert E, Dawidowicz P. 1993. Chemically-induced diel vertical migration in *Daphnia*: a new bioassay for kairomones exuded by fish. *Arch. Hydrobiol.* 126:329–37

Losey JE, Denno RF. 1998. Positive predator-predator interactions: enhanced predation rates and synergistic suppression of aphid populations. *Ecology* 79:2143–52

Lutterschmidt WI, Marvin GA, Hutchison VH. 1994. Alarm response by a plethodontid salamander (*Desmognathus ochrophaeus*): conspecific and heterospecific "Schreckstoff." *J. Chem. Ecol.* 20:2751–59

Mallet J, Longino JT, Murawski D, Murawski A, De Gamboa AS. 1987. Handling effects in *Heliconius.* Where do all the butterflies go? *J. Anim. Ecol.* 56:377–86

Malmqvist B. 1992. Stream grazer responses to predator odour: an experimental study. *Nordic J. Freshw. Res.* 67:27–34

Manzoli-Palma MF, Gobbi N, Palma MS. 1998. Alarm pheromones and the influence of pupal odor on the aggressiveness of *Polybia paulista* (Ihering) (Hymenoptera: Vespidae). *J. Venom. Anim. Toxins* 4:61–69

Mathis A, Hoback WW. 1997. The influence of chemical stimuli from predators on precopulatory pairing by the amphipod, *Gammarus pseudolimnaeus. Ethology* 103:33–40

Matsumoto K, Okamoto M, Horikawa H, Nakagawa K, Yamamaura H, Kuwahara Y. 1998. The effects of the different environmental conditions on the dispersion of grain and house dust mites (Acari: Astigmata). *Med. Entom. Zool.* 49:291–300

McAllister MK, Roitberg BD. 1987. Adaptive suicidal behaviour in pea aphids. *Nature* 328:797–99

McAllister MK, Roitberg BD, Weldon KL. 1990. Adaptive suicide in pea aphids: decisions are cost sensitive. *Anim. Behav.* 40:167–75

McIntosh AN, Peckarsky BL, Taylor BW. 1999. Rapid size-specific changes in the drift of *Baetis bicaudatus* (Ephemeroptera) caused by alterations in fish odour concentration. *Oecologia* 118:256–64

McIntosh AR, Peckarsky BL. 1996. Differential behavioral responses of mayflies from streams with and without fish to trout odour. *Freshw. Biol.* 35:141–48

McIntosh AR, Peckarsky BL. 1999. Criteria determining behavioural responses to multiple predators by a stream mayfly. *Oikos* 85:554–64

McKillup SC, McKillup RV. 1992. Inhibition of feeding in response to crushed conspecifics by the pebble crab *Philyra laevis* (Bell). *J. Exp. Mar. Biol. Ecol.* 161:33–43

Miller JR, Cowles RS. 1990. Stimulo-deterrent diversion: a concept and its possible application to onion maggot control. *J. Chem. Ecol.* 16:3197–212

Müller CB, Godfray HCJ. 1999. Predators and mutualists influence the exclusion of aphid species from natural communities. *Oecologia* 119:120–25

Mustaparta H. 1984. Olfaction. In *Chemical Ecology of Insects*, ed. WJ Bell, RT Cardé, pp. 37–72. London: Chapman & Hall

Neill WE. 1990. Induced vertical migration in copepods as a defence against invertebrate predation. *Nature* 345:524–26

Neill WE. 1992. Population variation in the ontogeny of predator-induced vertical migration of copepods. *Nature* 356:54–57

Nolte DL, Mason JR, Epple G, Aronov E, Campbell DL. 1994. Why are predator urines aversive to prey? *J. Chem. Ecol.* 20:1505–16

Nordlund DA, Jones RL, Lewis WJ, eds. 1981.

Semiochemicals, Their Role in Pest Control. New York: Wiley. 306 pp.

Nuñez J, Almeida L, Balderrrama N, Giurfa M. 1997. Alarm pheromone induces stress analgesia via an apioid system in the honeybee. *Physiol. Behav.* 63:75–80

Ohsaki N, Sato Y. 1994. Food plant choice of *Pieris* butterflies as a trade-off between parasitoid avoidance and quality of plants. *Ecology* 75:59–68

Pallini A, Janssen A, Sabelis MW. 1998. Predators induced interspecific competition for food in refuge space. *Ecol. Lett.* 1:171–77

Pallini A, Janssen A, Sabelis MW. 1999. Spider mites avoid plants with predators. *Exp. Appl. Acarol.* 23:803–15

Papaj DR, Lewis AC, eds. 1993. *Insect Learning: Ecological and Evolutionary Perspectives.* New York: Chapman & Hall. 398 pp.

Papaj DR, Prokopy RJ. 1989. Ecological and evolutionary aspects of learning in phytophagous insects. *Annu. Rev. Entomol.* 34:315–50

Papaj RD, Vet LEM. 1990. Odor learning and foraging success in the parasitoid. *Leptopilina heterotoma. J. Chem. Ecol.* 16:3137–50

Parejko K, Dodson S. 1990. Progress towards characterization of a predator/prey kairomone: *Daphnia pulex* and *Chaoborus americanus. Hydrobiology* 198:51–59

Peckarsky BL. 1980. Predator-prey interactions between stoneflies and mayflies: behavioral observations. *Ecology* 61:932–43

Petranka JW, Fakhoury K. 1991. Evidence of a chemically-mediated avoidance response of ovipositing insects to bluegills and green frog tadpoles. *Copeia* 1991:234–39

Pijanowska J. 1997. Alarm signals in *Daphnia*? *Oecologia* 112:12–16

Pitt WC. 2000. Effects of multiple vertebrate predators on grasshopper habitat selection: trade-offs due to predation risk, foraging, and thermoregulation. *Evol. Ecol.* 13:499–515

Prokopy RJ, Roitberg BD, Averill AL. 1984.

Resource partitioning. In *Chemical Ecology of Insects*, ed. WJ Bell, RT Cardé, pp. 301–30. New York: Chapman & Hall

Repka S, Walls M. 1998. Variation in the neonate size of *Daphnia pulex*: the effects of predator exposure and clonal origin. *Aquatic Ecol.* 32:203–9

Rigby MC, Jokela J. 2000. Predator avoidance and immune defence: costs and trade-offs in snails. *Proc. R. Soc. London Ser. B* 267:171–76

Robertson IC, Roitberg BD, Williamson I, Senger SE. 1995. Contextual chemical ecology: an evolutionary approach to the chemical ecology of insects. *Am. Entomol.* 41:237–39

Roca JR, Baltanas A, Uiblein F. 1993. Adaptive responses in *Cypridopsis vidua* (Crustacea: Ostracoda) to food and shelter offered by a macrophyte *(Chara fragilis). Hydrobiology* 262:127–31

Rochette R, Arsenault DJ, Justome B, Himmelman JH. 1998. Chemically-mediated predator recognition learning in a marine gastropod. *Ecoscience* 5:353–60

Roitberg BD, Isman MB, eds. 1992. *Insect Chemical Ecology. An Evolutionary Approach.* New York: Chapman & Hall 359 pp.

Rosenthal GA, Berenbaum MR, eds. 1992. *Herbivores: Their Interaction with Secondary Plant Metabolites.* 2nd ed. I:468 pp.; II:493 pp. New York: Academic

Sakwinska O. 1998. Plasticity of *Daphnia magna* life history traits in response to temperature and information about a predator. *Freshw. Biol.* 39:681–87

Scheffer M. 1997. On the implications of predator avoidance. *Aquatic Ecol.* 31:99–107

Schmitz OJ, Beckerman AP, O'Brien KM. 1997. Behaviorally mediated trophic cascades: effects of predation risk on food web interactions. *Ecology* 78:1388–99

Schoonhoven LM, Beerling EAM, Klijnstra JW, Vugt Yv. 1989. Two related butterfly species avoid oviposition near each other's eggs. *Experientia* 46:527–28

Schoonhoven LM, Jermy T, Van Loon JJA. 1998. *Insect-Plant Biology. From*

Physiology to Evolution. London: Chapman & Hall. 409 pp.

Shah PA, Pickett JA, Vandenberg JD. 1999. Responses of Russian wheat aphid (Homoptera: Aphididae) to aphid alarm pheromone. *Environ. Entomol.* 28:983–85

Shave CR, Townsend CR, Crowl TA. 1994. Anti-predator behaviours of a freshwater crayfish (*Paranephrops zealandicus* White) to a native and an introduced predator. *N. Z. J. Ecol.* 18:1–10

Short TM, Holomuzki JR. 1992. Indirect effects of fish on foraging behaviour and leaf processing by the isopod *Lirceus fontinalis*. *Freshw. Biol.* 27:91–97

Sih A. 1986. Antipredator responses and the perception of danger by mosquito larvae. *Ecology* 67:434–41

Sih A. 1987. Prey refuges and predator-prey stability. *Theor. Popul. Biol.* 31:1–12

Slusarczyk M. 1999. Predator-induced diapause in *Daphnia magna* may require two chemical cues. *Oecologia* 119:159–65

Snyder WE, Wise DH. 2000. Antipredator behavior of spotted cucumber beetles (Coleoptera: Chrysomelidae) in response to predators that pose varying risks. *Environ. Entomol.* 29:35–42

Städler E. 1984. Contact chemoreception. In *Chemical Ecology of Insects*, ed. WJ Bell, RT Cardé, pp. 3–36. London: Chapman & Hall

Stephens DW, Krebs JR. 1986. *Foraging Theory*. Princeton, NJ: Princeton Univ. Press. 247 pp.

Stibor H. 1992. Predator induced life-history shifts in a freshwater cladoceran. *Oecologia* 92:162–65

Strong DR, Lawton JH, Southwood TRE. 1984. *Insects on Plants: Community Patterns and Mechanisms*. Cambridge: Harvard Univ. Press. 313 pp.

Sullivan TP, Crump DR, Sullivan DS. 1988a. Use of predator odors as repellents to reduce feeding damage by herbivores. III. Montane and meadow voles (*Microtus montanus* and *Microtus pennsylvanicus*). *J. Chem. Ecol.* 14:363–77

Sullivan TP, Crump DR, Sullivan DS. 1988b. Use of predator odors as repellents to reduce feeding damage by herbivores. IV. Northern pocket gophers (*Thomomys talpoides*). *J. Chem. Ecol.* 14:379–89

Tollrian R. 1994. Fish-kairomone induced morphological changes in *Daphnia lumholtzi* (Sars). *Arch. Hydrobiol.* 130:69–75

Tollrian R. 1995. Predator-induced morphological defenses: costs, life history shifts, and maternal effects in *Daphnia pulex*. *Ecology* 76:1691–705

Tollrian R, Harvell CD, eds. 1999. *The Ecology and Evolution of Inducible Defenses*. Princeton, NJ: Princeton Univ. Press. 383 pp.

Tollrian R, Von Elert E. 1994. Enrichment and purification of *Chaoborus* kairomone from water: further steps toward its chemical characterization. *Limnol. Oceanogr.* 39:788–92

Turner AM, Bernot RJ, Boes CM. 2000. Chemical cues modify species interactions: the ecological consequences of predator avoidance by freshwater snails. *Oikos* 88:148–58

van Dam NM, de Jong TJ, Iwasa I, Kubo T. 1996. Optimal distribution of defences: Are plants smart investors? *Funct. Ecol.* 10:128–36

Venzon M, Janssen A, Pallini A, Sabelis MW. 2000. Diet of a polyphagous arthropod predator affects refuge seeking of its thrips prey. *Anim. Behav.* 60:369–75

Vet LEM. 1996. Parasitoid foraging: the importance of variation in individual behaviour for population dynamics. In *Frontiers of Population Ecology*, ed. RB Floyd, AW Sheppard, pp. 245–56. Melbourne: CSIRO

Vet LEM, Groenewold AW. 1990. Semiochemicals and learning in parasitoids. *J. Chem. Ecol.* 16:3119–35

Vet LEM, Lewis WJ, Cardé RT. 1995. Parasitoid foraging and learning. In *Chemical Ecology of Insects*, ed. RT Carde, WJ Bell, pp. 65–101. New York: Chapman & Hall

Vinson SB. 1976. Host selection by insect parasitoids. *Annu. Rev. Entomol.* 21:109–34

Visser JH. 1986. Host odor perception in phytophagous insects. *Annu. Rev. Entomol.* 31:121–44

Von Elert E, Pohnert G. 2000. Predator specificity of kairomones in diel vertical migration of *Daphnia*: a chemical approach. *Oikos* 88:119–28

Wahle RA. 1992. Body-size dependent antipredator mechanisms of the American lobster. *Oikos* 65:52–60

Weisser WW. 1999. Predator-induced morphological shift in the pea aphid. *Proc. R. Soc. London Ser. B* 266:1175–81

Welch JM, Rittschof D, Bullock TM, Forward RB. 1997. Effects of chemical cues on settlement behavior of blue crab *Callinectes sapidus* postlarvae. *Mar. Ecol. Progr. Ser.* 154:143–53

Werner EE. 1991. Nonlethal effects of a predator on competitive interactions between two anuran larvae. *Ecology* 72:1709–20

Werner EE, Anholt BR. 1996. Predator-induced behavioral indirect effects: consequences to competitive interactions in anuran larvae. *Ecology* 77:157–69

Wisenden BD, Chivers DP, Smith RJF. 1997. Learned recognition of predatory risk by *Enallagma* damselfly larvae (Odonata, Zygoptera) on the basis of chemical cues. *J. Chem. Ecol.* 23:137–51

Wisenden BD, Sargent RC. 1997. Antipredator behaviour and suppressed aggression by convict cichlids in response to injury-related chemical cues of conspecifics but not to those of an allopatric heterospecific. *Ethology* 103:283–91

Wudkevich K, Wisenden BD, Chivers DP, Smith RJF. 1997. Reactions of *Gammarus lacustris* (Amphipoda) to chemical stimuli from natural predators and injured conspecifics. *J. Chem. Ecol.* 23:1164–73

Yamamura N, Tsuji N. 1995. Optimal strategy of plant antiherbivore defense: implications for apparency and resource-availability theories. *Ecol. Res.* 10:19–30

Annu. Rev. Ecol. Syst. 2001. 32:25–49
Copyright © 2001 by Annual Reviews. All rights reserved

SEX CHROMOSOME MEIOTIC DRIVE

John Jaenike

Department of Biology, University of Rochester, Rochester, New York 14627;
e-mail: joja@mail.rochester.edu

Key Words intragenomic conflict, levels of selection, evolutionary genetics, segregation distortion, sex ratio

■ **Abstract** Sex chromosome drive refers to the unequal transmission of X and Y chromosomes from individuals of the heterogametic sex, resulting in biased sex ratios among progeny and within populations. The presence of driving sex chromosomes can reduce mean fitness within a population, bring about intragenomic conflict between the X chromosome, the Y, and the autosomes, and alter the intensity or mode of sexual selection within species. Sex chromosome drive, or its genetic equivalent, is known in plants, mammals, and flies. Many species harboring driving X chromosomes have evolved Y-linked and autosomal suppressors of drive. If a drive polymorphism is not stable, then driving chromosomes may spread to fixation and cause the extinction of a species. Certain characteristics of species, such as population density and female mating rate, may affect the probability of fixation of driving chromosomes. Thus, sex chromosome drive could be an agent of species-level selection.

INTRODUCTION

In most species, individuals of the heterogametic sex produce equal numbers of X- and Y-bearing gametes, thus resulting in approximately equal numbers of male and female offspring. The general tendency to produce 1:1 sex ratio, as well as many deviations from it, can often be explained in terms of adaptive sex ratio theory, with individuals producing a sex ratio that maximizes the transmission of autosomal alleles to future generations (Fisher 1930, Hamilton 1967, Trivers & Willard 1973, Charnov 1982).

In some species, however, a fraction of the individuals produce highly unequal and, for autosomal genes, nonadaptive offspring sex ratios. For instance, some maternally inherited microorganisms enhance their own transmission by killing male offspring, inducing parthenogenesis, or causing feminization of males (Stouthamer 1997, Rigaud 1997, Hurst & Jiggins 2000). This review focuses on a different non-adaptive mechanism of sex-ratio distortion: unequal transmission of functional X- and Y-bearing gametes by individuals of the heterogametic sex. The genes responsible are typically linked to one of the sex chromosomes and act by preventing or interfering with the production of functional gametes bearing the other sex chromosome. In most cases, the sex-ratio distortion results in the production of an excess of female offspring and depends only on the genotype of the male parent. Because *sex-ratio* genes usually act during spermatogenesis, the unequal production of X- and Y-bearing gametes is due to meiotic drive, or segregation distortion (Lyttle 1991).

The presence of driving sex chromosomes can be of considerable ecological and evolutionary importance in those groups where it occurs. For simplicity, the following discussion focuses on X drive, which is much more common than Y drive, but the arguments presented apply similarly. The most important ecological consequences of sex chromosome drive stem from the biased sex ratio that arises at the population level. The presence of a driving X chromosome within a population is expected to lead to a female-biased population (Hamilton 1967), and this has been found to be the case in natural populations (Bryant et al. 1982, James & Jaenike 1990). At moderate frequencies of X^D (the driving X chromosome), there may be enough males to inseminate all the females in a population. In such cases, a population harboring X^D chromosomes may have a greater intrinsic rate of increase and thus be capable of rebounding more quickly from population declines. However, if an X^D chromosome reaches a very high frequency, which is expected to occur in the absence of countervailing selection, the female bias will become so great that many females will go unmated and the population may go extinct (Hamilton 1967; see also Hatcher et al. 1999). Thus, the effect of X^D on population persistence probably depends on the frequency of these chromosomes.

Driving X chromosomes can also affect mating behavior and sexual selection in several ways. Because populations harboring X^D chromosomes are more female biased, this will affect the operational sex ratio among sexually active individuals, and one might expect the intensity of sexual selection on males to be less in such species, perhaps even selecting for male mate choice, as Randerson et al. (2000) have suggested for sex-ratio distortion brought about by male-killing bacteria. The female bias at the population level imposes selection on females to mate in a way that maximizes the proportion of sons among their offspring (Fisher 1930), and there are several ways this might be achieved. Lande & Wilkinson (1999) have shown that the presence of X drive could lead to the evolution of female preference for male traits encoded by genes linked to wild-type allele at the *sex-ratio* locus. Thus, linkage between genes for the male trait and the drive locus provides the basis for a "good genes" type of sexual selection.

Multiple mating by females may also result in the production of less female-biased offspring sex ratios. As discussed below, sperm competition can adversely affect offspring production by Sex-ratio males relative to Standard males. Consequently, females that mate randomly with several males will produce, on average, more male offspring than females that mate infrequently and for which there is no sperm competition. Thus, the presence of a meiotic drive gene can favor the evolution of multiple mating by females (Haig & Bergstrom 1995). Similarly, selection might favor females that are not susceptible to an "insemination reaction," a phenomenon known in many species of *Drosophila* (Patterson 1946). The insemination reaction, which is caused by some substance transmitted with a male's ejaculate, renders females less capable of remating for some period of time. Thus, susceptibility to the insemination reaction lowers the level of sperm competition within females and could result in their producing a lower percentage of male offspring. These possibilities remain to be tested within a comparative phylogenetic context.

The broader evolutionary consequences of sex chromosome drive are equally profound, as many general principles of population genetic theory are violated by non-Mendelian segregation (Charlesworth & Hartl 1978). For example, meiotic drive at any genetic locus reduces the mean fitness of the population, because variation is maintained by a balance between meiotic drive and countervailing natural selection (Hiraizumi et al. 1960). X drive also sets the stage for intense intragenomic conflicts. This topic has recently been reviewed by Hurst et al. (1996) and Werren & Beukeboom (1998), and I highlight only some salient points here. Because enhanced transmission of driving X chromosomes depends on preventing the production of functional Y-bearing sperm, there is strong selection on the Y to resist this effect. The presence of driving X chromosomes is a plausible explanation for the maintenance of a Y chromosome polymorphism in natural populations (Clark 1987, Carvalho et al. 1997, Jaenike 1999). The autosomes are also selected to suppress the X-linked drive, for two reasons. First, because the presence of X^D chromosomes results in a female-biased population, males have a higher mean fitness than females (Fisher 1930). Autosomal genes that suppress X drive will be passed more frequently to male offspring and thus will experience enhanced transmission to subsequent generations. Furthermore, because polymorphism at a drive locus requires countervailing selection against carriers of the drive allele, autosomal suppressors are less likely than nonsuppressors to be associated with low-fitness individuals carrying the drive allele in future generations.

Meiotic drive has also been hypothesized to have major effects on the structure of the genome. In all cases that have been well characterized, drive involves interactions between a drive locus on one chromosome and a sensitive responder locus on the homologous chromosome (Lyttle 1991). For a newly arisen drive allele to spread, it is necessary that there be little or no recombination between the drive and responder loci. Haig & Grafen (1991) therefore suggested that genome-wide recombination between homologous chromosomes is an evolutionary response to the threat of meiotic drive genes.

Frank (1991) and Hurst & Pomiankowski (1991) postulated that sex chromosome drive could be responsible for Haldane's rule, the greater inviability or sterility of heterogametic-sex hybrids. There is, however, little empirical support for this idea (Coyne et al. 1991, Coyne & Orr 1993). It has also been suggested that the presence of X chromosome drive could lead to evolutionary changes in sex determination mechanisms to restore a more equal sex ratio (Lyttle 1981, Haig 1993a,b, Werren & Beukeboom 1998). As seen below, there are several species groups in which a change in the genetics of sex determination is at least correlated with the presence of sex chromosome drive. These considerations highlight the potential ecological and evolutionary ramifications of sex chromosome drive, from intragenomic conflict to species-level extinction.

EMPIRICAL PATTERNS

Table 1 lists species in which sex chromosome drive has been documented, or for which there is strongly suggestive evidence. Although several species exhibit Y drive, X drive is far more common. Sex chromosome drive has a spotty taxonomic distribution, having been found so far in insects, mammals, and angiosperms. The vast majority of known cases are from insects, within which sex chromosome drive is phylogenetically concentrated, being known with certainty only from Diptera, although there is one possible example in butterflies. Within flies, most examples are from the acalyptrate families, and within these families, certain clades (e.g., the *Drosophila obscura* group, stalk-eyed flies) seem particularly prone to *sex-ratio* polymorphisms. Among flies, sex chromosome drive has been discovered serendipitously in a number of species that are studied because of their medical (mosquitoes, tsetse flies), agricultural (medflies), or genetic importance (*Drosophila*). This suggests that sex chromosome drive is probably very common, at least within certain taxonomic groups. A more detailed discussion of the individual species can be obtained elsewhere (http://www.annualreviews.org/supmat/supmat.asp).

There is considerable variability among species in the specifics of sex chromosome drive. The following discussion is aimed at determining whether any patterns apply broadly across these systems, as such patterns would facilitate an understanding of the evolution and dynamics of sex chromosome drive in general. A consistent system of nomenclature has not been applied to sex chromosome drive, resulting in confusion and misunderstanding. For instance, the term sex-ratio has been variously applied to the phenomenon of X chromosome drive, the driving X chromosome itself, males that carry such chromosomes, and the loci responsible for the drive. In Table 2, I suggest terms that could be used to more clearly distinguish among the different aspects of X chromosome drive.

MECHANISMS OF DRIVE

Cytological studies of several species of flies with strong X drive reveal that Sex-ratio males are characterized by abnormal Y chromosome behavior in meiosis II. These males typically produce about half as many functional sperm as do Standard males, with the remaining sperm failing to individualize properly. Such patterns have been found in *Drosophila pseudoobscura* (see Novitsky et al. 1965, Policansky & Ellison 1970, Cobbs et al. 1991), *Drosophila subobscura* (see Hauschteck-Jungen & Maurer 1976), *Drosophila simulans* (see Montchamp-Moreau & Joly 1997, Cazemajor et al. 2000), *Drosophila neotestacea* (K. Dyer, personal communication), and *Cyrtodiopsis whitei* (see Wilkinson & Sanchez 2001, unpublished manuscript). Using fluorescence in situ hybridization with a Y-specific marker, Cazemajor et al. (2000) demonstrated that Y-bearing spermatids do not elongate properly.

Males of the *D. obscura* group produce both long and short sperm. Bircher et al. (1995) found that Sex-ratio males actually produce more short sperm than do Standard males and suggested that the excess short sperm compensates for the reduction in numbers of long sperm produced by these males. This is unlikely, because only the long sperm in this species are involved in fertilization (Snook & Karr 1998). Bircher et al. (1995) also hypothesized that sperm dimorphism is a necessary prerequisite for the evolution of a *sex-ratio* polymorphism, but this cannot be a general prerequisite, as sperm are not dimorphic in the *sex-ratio* polymorphic *D. simulans* (see Joly et al. 1989).

Although the mechanism outlined above appears to be the most common, several other mechanisms of sex chromosome drive have been uncovered. In *Drosophila melanogaster*, males carrying large deletions of the X heterochromatic region (denoted Xh^-) sire a significant excess of female offspring (McKee 1991). X drive in this species is due to mispairing of the X and Y chromosomes in meiosis I (McKee et al. 1998). Because spermatid viability is inversely related to the amount of chromatin a spermatid contains and because Xh^- males produce spermatids carrying smaller X chromosomes, they sire relatively more female offspring than do Xh males (McKee 1991).

In *Aedes aegypti*, the target of the driving D allele is thought to be the m allele of the sex-determining locus, with drive brought about by preferential breakage of the m-bearing chromosome during male meiosis. This effectively Y drive in *Aedes* is also associated with abnormal spermiogenesis and premature senescence of spermatozoa (Newton et al. 1976, Owusu-Daaka et al. 1997).

Cytological studies of wood lemmings reveal that although the somatic cells of X*Y females carry both the X* and Y chromosomes, their oocytes are X*X*, presumably as a result of selective nondisjunction and elimination of the Y chromosome from the germ line (Fregda et al. 1976, 1977). Thus, these females produce only X*-bearing ova. Because all males are XY, X*Y females produce X*X and X*Y progeny, all of which will develop as females. Thus, the sex-ratio distortion in this species, which leads to production of all-female progeny, results from germline selection (Hastings 1991).

In the plant *Silene alba*, biased offspring sex ratios are produced even at low pollen densities, with distorter and nondistorter males siring equal numbers of progeny (Taylor et al. 1999), indicating that distorter males produce a greater proportion of X-bearing pollen, and that differential fertilization success of X- and Y-bearing pollen cannot alone be responsible for the sex-ratio bias. The fraction of female progeny sired by distorter males increases with pollen density, indicating that Y-bearing pollen do poorly in pollen competition (Taylor et al. 1999). Furthermore, the genotype of a female crossed to a distorter male affects offspring sex ratio (Taylor 1994, 1999, Taylor et al. 1999). These observations indicate that sex-ratio distortion in this species is due to both meiotic and postmeiotic events.

IS SEX CHROMOSOME MEIOTIC DRIVE COMMON?

In addition to the examples shown in Table 1, female-biased offspring sex ratios have been found in several genera of dioecious angiosperms in which males are the heterogametic sex (Lloyd 1974). Other than *Silene*, it is not known whether these

TABLE 1 Examples of sex chromosome drive. Detailed discussion of individual case studies available at http://www.annualreviews.org/supmat/supmat.asp

Species	Drive Type[a]	Known Suppressors	Notes	Selected References
Drosophila obscura (Drosophilidae)	X m		First example of X drive.	Gershenson 1928
D. pseudoobscura	X m	None found, despite extensive search	Polymorphism depends on effect on females, as well as males. Sperm competition in multiply-mated females reduces transmission of X^D. Loci associated with SR inversions are genetically depauperate compared to ST-associated alleles. Cryptic sex-ratio system uncovered in crosses between mainland and Bogotá subspecies.	Wallace 1948; Curtsinger & Feldman 1980; Policansky 1974, 1979; Beckenbach 1978, 1981, 1983, 1996; Policansky & Ellison 1970; Wu 1983a,b; Prakash & Merritt 1972; Babcock & Anderson 1996; Beckenbach et al. 1982
D. persimilis	X m		Drive requires presence of several X-linked loci	Wu & Beckenbach 1983
D. azteca	X m			Sturtevant & Dobzhansky 1936
D. subobscura	X m	Y and/or autosomal	X^D present in North African, but not European populations. Drive is stronger in normal genetic background. Interpopulation crosses result in hybrid male sterility if male carries X^D.	Hauschteck-Jungen & Maurer 1976; Hauschteck-Jungen 1990; Bircher et al. 1995
D. affinis	X m	Y	X^DO males produce only sons. Y chromosome types differ in susceptibility to the different X^D's	Voelker 1972
D. athabasca	X m		X drive occurs in two eastern semispecies, but not in western-northern semispecies. Its presence in the closely related *D. affinis* and *D. azteca* suggests that X^D has been lost from the western-northern semispecies.	Miller & Voelker 1969
D. melanogaster	X m		*Stellate* hypothesized to be relic X-linked drive locus. Males carrying large deletions from heterochromatic region sire excess of females → abnormal behavior of Y during meiosis I.	Hurst 1992, 1996; McKee 1991

D. simulans	X m	Y, autosomes	Populations carrying X^D also carry suppressors, so drive is rarely expressed. A different and completely suppressed X drive system uncovered in interspecific crosses with *D. sechellia*.	Faulhaber 1967; Mercot et al. 1995; Atlan et al. 1997; Cazemajor et al. 1997, 2000; Montchamp-Moreau & Joly 1997; Capillon & Atlan 1999; Montchamp-Moreau et al. 2001; Dermitzakis et al. 2000
D. quinaria	X m	Y, autosomes	Local polymorphism for susceptible and partially resistant Y chromosome types.	Jaenike 1996, 1999
D. recens	X m	Y, autosomes	For both *D. recens* and *D. quinaria*, multiple mating by males reduces fertility of SR males more than that of ST males.	Jaenike 1996
D. paramelanica	X m	Y, autosomes	Geographic variation in frequencies of two types of X^D and two types of Y. X^D's drive best the Y with which they are usually associated locally.	Stalker 1961
D. mediopunctata	X m	Y, autosomes	X drive due to at least two loci, one of which may be primary drive locus, and the 2nd is an enhancer. Little geographic variation in frequencies of suppressor and non-suppressor Y chromosome types.	Carvalho et al. 1989, 1997, 1998; Carvalho & Klaczko 1993, 1994; Varandas et al. 1997
D. neotestacea	X m	none found	Unlike most other *Drosophila* species, SR not associated with inversions. X^D occurs at 20% to 30% frequency in all sampled populations in eastern United States.	James & Jaenike 1990
Cyrtodiopsis dalmanni (Diopsidae: stalk-eyed fly)	X m	Y and/or autosomes	Sex-ratio males carrying X^D and suppressors sire significant excess of sons. Sperm competition in multiply-mated females reduces transmission of X^D.	Presgraves et al. 1997; Wilkinson et al. 1998a,b, Wilkinson & Sanchez 2000
C. whitei	X m		In both *C. whitei* and *C. dalmanni*, males may mate many times per day. Females in these species remate much more rapidly than congeneric species that lacks X drive. For both species, interpopulation crosses yield hybrid males with defects in spermatogenesis similar to that seen in sex-ratio males.	Presgraves et al. 1997

(Continued)

TABLE 1 (*Continued*)

Species	Drive Type[a]	Known Suppressors	Notes	Selected References
Diasemopsis sylvatica (Diopsidae)	X m			Lande & Wilkinson 1999
Sphyracephala beccarii (Diopsidae)	X m			Lande & Wilkinson 1999
Ceratitis capitata (Tephritidae: medfly)	Y m		Driving Y arose in laboratory population after X irradiation treatment.	Wood 1995
Musca domestica (Muscidae: housefly)	Y m		Drive associated with neo-Y chromosome, which arose following translocation of part of Y to the 3rd chromosome.	M. Clark, personal communication
Glossina morsitans (Muscidae: tsetse fly)	X m		Sex-ratio males known from several areas in Africa; associated with female-biased emergence sex ratios.	Rawlings & Maudlin 1984; Gooding 1986; Gooding et al. 1989
Aedes aegypti (Culicidae)	Y m	X	Drive locus (D) closely linked to sex determining locus. Responder thought to be *m* allele at the sex-determining locus. Drive associated with breakage of *m*-bearing chromosome. Drive allele (D) tends to be geographically associated with resistant X chromosomes resistant to drive.	Hickey & Craig 1966; Hickey 1970; Newton et al. 1976; Hastings & Wood 1978; Wood & Newton 1991; Owusu-Daaku et al. 1997
Culex pipiens (Culicidae)	Y m		Sex ratio distortion occurs only in individuals homozygous for drive allele *d*. Drive allele therefore gains no obvious transmission advantage.	Sweeny & Barr 1978
Eucheira socialis (Lepidoptera: Pieridae)	X f		Strongly male-biased primary sex ratios in some egg masses. Bias toward males indicates that distortion is not caused by a cytoplasmic factor such as *Wolbachia*.	Underwood & Shapiro 1999

Species	Driving chromosome	Notes	References
Dicrostonyx torquatus (Cricetidae: varying lemming)	Y m	Includes both XX and X*Y females. Weak Y drive in males.	Bull & Bulmer 1981; Gileva et al. 1982; Gileva 1987
Myopus schisticolor (Cricetidae: wood lemming)	X f	Includes XX, XX*, and X*Y females. Drive results from germline selection in X*Y females, yielding only X*-bearing ova.	Fregda et al. 1976, 1977
Akodon azarae (Muridae: field mouse)	Y m and Y* f / Y m and Y* f	Includes both XX and XY* females. Weak Y drive occurs in males, and weak Y* drive occurs in XY* females.	Hoekstra & Edwards 2000, Hoekstra & Hoekstra 2000
Silene alba (Caryophyllaceae: white campion)	X m Y	Dioecious plant. Female-biased sex ratios appear due to both meiotic drive and pollen competition. Maternal genotype can affect success of X- and Y-pollen. Considerable variation among local populations in sex-ratio frequencies. Although X chromosome enjoys transmission advantage, it is the Y chromosome that determines this. Y may be polymorphic for resistance to fixed X drive in this species. Pollen competition reduces fertilization success of pollen from distorter males.	Taylor 1993, 1994, 1996, 1999; Taylor et al. 1999
Datisca cannabina (Datiscaceae)	X m	Dioecious plant. X drive may have been a factor in the evolution to androdioecy in the sister species *D. glomerata*.	Wolf & Rieseberg 2000

[a]X and Y indicate driving chromosome; m and f indicate whether drive occurs in males or females.

TABLE 2   Proposed terminology for X chromosome drive

Variable	Associated with Drive or Suppression	Wild Type
General phenomenon	X drive	
Chromosome	X^D (D = drive): If >1 type, X^{D1}, X^{D2} ...	X^S (S = Standard)
Gene arrangement	SR: If >1 type, SR(1), SR(2) ...	ST
Locus	*sex-ratio*: If >1 locus, *sex-ratio*1, *sex-ratio*2 ...	
Alleles	*sr*: If >1 locus, *sr1*, *sr2* ... , where numbers refer to loci. Superscripts would refer to different *sr* alleles at a given locus.	sr^+: $sr1^+$, $sr2^+$ ...
Male type	Sex-ratio (carries an X^D chromosome)	Standard
Y chromosome	Y^S (S = suppressor)	Y (susceptible to X^D drive)
	Y^{S1} (specific suppressor of X^{D1})	

species harbor driving sex chromosomes. Among insects, sex chromosome drive has been unequivocally demonstrated only in Diptera, although within this group, drive has evolved repeatedly. In other insects, sex-ratio distortion is often due to maternally inherited endosymbionts, which can act by killing male offspring, feminizing them, or inducing parthenogenesis. Jiggins et al. (1999) have argued that, statistically at least, such endosymbionts are a more important cause of sex-ratio bias than meiotic drive outside the Diptera. However, it remains to be seen how taxonomically widespread sex chromosome drive is; only a tiny, and taxonomically nonrandom, sample of species has been examined in a way that could reveal such drive. In species that have them, driving sex chromosomes are often present at low frequency and their expression can be masked by suppressors. Therefore, ruling out their existence within a species may require large sample sizes of wild-caught males bred individually to virgin females, perhaps from different populations that may lack suppressors.

IS X CHROMOSOME DRIVE MORE COMMON THAN AUTOSOMAL DRIVE?

Whereas numerous examples of sex chromosome drive have been documented, far fewer examples of autosomal drive are known, with the best-studied examples being Segregation Distorter in *D. melanogaster*, the *t* locus in mice, and spore killer in *Neurospora* (reviewed in Lyttle 1991). One obvious reason for the discrepancy is that X (or Y) drive leads to distorted offspring sex ratios, which are easily detected. In contrast, autosomal drive is evident only when one can follow the unequal segregation of genetic markers that happen to be linked to the drive locus. Thus, autosomal drive had previously been discovered only in species that served as model organisms for genetic research. Ongoing Quantitative Trait Loci (QTL) studies should provide a basis for detection of autosomal drive, although if the drive alleles occur at low frequency, then numerous crosses may have to be carried out. More recently, QTL analyses using molecular markers have revealed that what appears to be autosomal segregation distortion may be more common than previously thought (reviewed in Doebley & Stec 1991). However, because many of these cases involve plant hybrids, the apparent segregation distortion might actually be due to differential pollen inviability or fertilization success. The increasing number of QTL studies in animals should provide data with which to assess the relative frequency of autosomal and sex-linked meiotic drive.

There are two reasons why sex chromosome drive may be more common than autosomal drive. First, in all cases that have been examined in detail, drive entails linkage between the driving allele (e.g., *SD* of *D. melanogaster*) and an insensitive allele at a responder locus (e.g., *Rsp*^{i} of *D. melanogaster*). Such linkage disequilibrium can occur if the drive and responder loci are located near the centromere and/or are associated with inversions, as is the case for *SD*, the *t* complex of *Mus*, and spore killer of *Neurospora* (Temin et al. 1991, Lyon et al. 1988, Turner & Perkins 1991). As Lyttle (1991) and Hurst & Pomiankowski (1991) have pointed out, the lack of recombination between the X and Y chromosomes means that, in terms of linkage relationships, any gene on the X (or Y) can drive against any responder site on the Y (or X). Thus, there may be fewer genetic constraints for the evolution of a driving sex chromosome. If low levels of recombination are the limiting factor for the evolution of drive, then one might expect inversion-rich species, such as *D. pseudoobscura* and many other drosophilids, to exhibit autosomal drive. Yet, aside from *SD* drive of *D. melanogaster*, no cases of autosomal drive have been discovered in *Drosophila*. It is possible that autosomal inversions do harbor loci that drive, but the lack of genetic markers has precluded their discovery.

Another reason X chromosome drive may be more common than autosomal drive concerns the nature of the responder sites. *Responder*, the target of drive by the *SD* locus in *D. melanogaster*, is a repetitive sequence located within the centromeric heterochromatin (Wu et al. 1988). Similarly, the X-linked *m* allele of *Aedes aegypti* is thought to be the target of the drive (*D*) locus and it is completely linked to the centromere (Wood & Newton 1991). In both cases, therefore, the responder locus occurs within a heterochromatic region. Because the Y chromosome of many species appears to be composed entirely of heterochromatin (White 1973), it may offer a much larger array of targets for various X-linked drive loci. If target sites are the limiting factor, then X drive may be particularly common in species with heterochromatic Y chromosomes.

Y CHROMOSOME DRIVE

Several species exhibit what is effectively Y drive, leading to male-biased offspring sex ratios. The Y drive appears to be of two types, which may be termed primary and redirected drive. The mosquitoes *A. aegypti* and *Culex pipiens* exhibit primary drive. In these species, drive of the Y-linked *D* allele requires that the X chromosome have a sensitive allele at the responder locus (Hickey 1970). If the X chromosome has an insensitive responder, then males sire unbiased offspring sex ratios, whether they carry the D allele or not. Genetically, the situation is somewhat similar to the autosomal SD drive of $D. melanogaster$, in which drives occurs when the SD allele on one chromosome drives against a sensitive responder allele on the homologous chromosome.

The only known examples of primary Y chromosome drive among insects occur in mosquitoes. Because sex is determined by a single locus or chromosomal region, the X and Y chromosomes are largely homomorphic. In this sense, the X and Y chromosomes are similar to autosomes and would seem equally likely to evolve drive. Why then has Y, but not X, chromosome drive arisen in these species? One reason could be that there is stronger selection for Y drive, as such a chromosome is passed only to sons and can therefore express drive every generation, whereas a driving X is carried by both males and females but is expressed only in males (Hamilton 1967).

Male-biased offspring sex ratios can also result from what I would call redirected drive. In the stalk-eyed fly *Cyrtodiopsis dalmanni*, males that carry particular Y chromosomes exhibit what is effectively Y drive, but this occurs only if that male also carries a driving X^D chromosome. Males that carry the same Y chromosomes in conjunction with standard X^S chromosomes sire normal offspring sex ratios (Presgraves et al. 1997). *Drosophila affinis* exhibits a similar phenomenon: Whereas X^DY males sire only daughters, X^DO males sire only sons (Voelker 1972). In males that carry the nondriving X^S chromosome, both X^SY and X^SO males sire normal offspring sex ratios. In both *C. dalmanni* and *D. affinis*, therefore, Y (or O) drive requires the presence of an X^D chromosome. One possible explanation for the type of male bias seen in X^D males of *C. dalmanni* and *D. affinis* is the following. Suppose the driving X^D encodes a product that can bind to certain target sites. If a typical Y chromosome has many such target sites and the X chromosome has a few, then the Y chromosome will suffer disproportionately and thus be driven against. However, if there are Y chromosome variants that have fewer target sites than the X—and XO males would have no Y chromosome targets at all—then the X chromosome would be the target of its own drive, resulting in an underproduction of X-bearing sperm and few female offspring. The fact that X^DO males of *D. affinis* sire all male offspring indicates that a Y chromosome is not required to bring about reduced transmission of X^D.

MECHANISMS TO EXPLAIN THE POLYMORPHISM

Any X- or Y-linked factor that exhibits complete drive will cause the extinction of a population or species if it spreads to fixation. The persistence of a species with a driving sex chromosome could thus depend on the action of countervailing selection against these elements. Thus, there has been considerable effort devoted to understanding the selective factors that lead to stable *sex-ratio* polymorphisms.

Wallace (1968) proposed a group selection mechanism that requires a metapopulation structure. He argued that if X^D chromosomes spread rapidly to fixation within individual populations, these populations will exhibit higher extinction rates and send out fewer colonists than populations founded by individuals lacking X^D. Wallace did not present a formal analysis of his hypothesis, and it is not clear that such a group selection process could result in a stable metapopulation-wide polymorphism. Empirically, patterns of X^D frequencies in nature provide little support for this hypothesis. In *D. pseudoobscura, D. subobscura*, and *D. neotestacea*, long-term studies reveal (*a*) no directional changes in the frequency of X^D and (*b*) either uniform or clinal changes in gene frequency across the range of a species, rather than a checkerboard pattern of presence and absence (Sturtevant & Dobzhansky 1936, Dobzhansky & Epling 1944, Hauschteck-Jungen 1990, James & Jaenike 1990, Beckenbach 1996). Thus, for *Drosophila* at least, there is no evidence that population-level selection facilitates the maintenance of sex-ratio polymorphisms.

Based on the cytological finding that Sex-ratio males produce only half as many sperm as do Standard males in *D. pseudoobscura* (see Policansky & Ellison 1970), Policansky (1974) argued that the 50% reduction in the number of sperm produced by Sex-ratio males exactly balanced the twofold meiotic drive advantage of the X^D chromosome in these males. Consequently, he suggested that X^D chromosomes would not experience greater transmission than nondriving X^S chromosomes and should therefore show no tendency to increase. This hypothesis assumes that offspring production is limited by sperm rather than egg production, an assumption at variance with most life history and sexual selection theory. Furthermore, the polymorphism thus brought about would be neutrally stable (Thomson & Feldman 1975), and the frequency of X^D would drift to fixation or loss.

Edwards (1961) and Curtsinger & Feldman (1980) derived the conditions for maintenance of a stable polymorphism of an X-linked drive locus. In their formulations, the fitnesses of all genotypes were constant, i.e., independent of the frequency of X^D. Their models assumed that there was no Y-linked or autosomal variation for suppression of drive. An important result of their models was that a stable polymorphism at the drive locus could be maintained only if the fitness of females were affected by the *sex-ratio* loci or genes linked to these loci. For instance, a polymorphism could be maintained via heterozygote superiority in females or by a balance between drive in males and adverse fitness effects in females.

One difficulty with maintenance of an X^D polymorphism based on effects in females is that the primary effect of *sex-ratio* genes is on male spermatogenesis. In *D. melanogaster*, the vast majority of genes that affect male fertility have no discernible effect on female fertility or viability (Lindsley & Zimm 1992). It is therefore unlikely that effects of *sex-ratio* genes in females can serve as a general mechanism for the maintenance of X^D polymorphisms. It is more likely that any fitness effects in females, as documented in *D. pseudoobscura*, are due to other loci tied up within inversions associated with the SR gene arrangement (i.e., that characteristic of the X^D chromosome).

Before such inversions evolve, the dynamics of driving X^D chromosomes is probably governed primarily through their effects on male fitness. Because males are hemizygous for the X, maintenance of a drive polymorphism via heterosis in males is not possible. An alternative possibility is negative frequency-dependent selection. One such mechanism involves the increased proportion of females in a population, and consequently an increased rate of male mating and greater sperm depletion, that results from the spread of X^D chromosomes (Jaenike 1996). It has been shown in several species that Sex-ratio males produce about half as many sperm as do Standard males (*D. pseudoobscura*, Policansky & Ellison 1970; *D. simulans*, Montchamp-Moreau & Joly 1997; *D. subobscura*, Hauschteck-Jungen & Maurer 1976; *Cyrtodiopsis dalmanni*, Presgraves et al. 1997); thus, a high rate of male mating is likely to cause greater sperm depletion in Sex-ratio males than in Standard males. For several species, including *D. pseudoobscura*, *D. neotestacea*, *D. recens*, *D. quinaria*, and *Aedes aegypti*, laboratory studies have shown that virgin Sex-ratio and Standard males are about equally fertile but that multiple mating by these males results in a proportionately greater reduction in the fertility of Sex-ratio males (Wu 1983a,b, James 1992, Jaenike 1996, Hickey & Craig 1966).

Wu (1983a) has shown that Sex-ratio males suffer the greatest reduction in relative fertility when both they and their mates have mated several times. How will female multiple mating and the resulting sperm competition among males affect the dynamics of X^D? Suppose that the female mating rate is simply proportional to the male-female encounter rate. When the frequency of X^D is low, there are relatively many males per female, resulting in more frequent female remating. Thus, as a result of sperm competition, Sex-ratio males will suffer the greatest reduction in relative fertility at low frequencies of X^D, which will impede the initial spread of X^D within a population. At high frequencies, however, females will remate less often, thus reducing the intensity of selection against X^D. Consequently, if sperm competition plays a role in preventing the spread of X^D at low frequencies, there may be an unstable equilibrium frequency above which X^D spreads to fixation.

Alternatively, a female's propensity to remate may depend on the fertility of the last male with which she mated. Suppose a female remates sooner if the last male she mated with was sperm depleted, as shown experimentally in *D. pseudoobscura* (see Beckenbach 1981). Under these conditions, when the frequency of X^D and the rate of male mating are high, most females will have mated with a sperm-depleted male. As a result, these females will remate more often than when X^D is at low frequency, thus intensifying selection against Sex-ratio males when X^D is common. In this manner, multiple mating by females and the resulting sperm competition could serve to stabilize X^D frequencies. However, there is still likely to be a frequency of X^D above which (and/or a population density below which) the ratio of females to males in a population is so high that multiple mating by females becomes improbable. This possibility remains to be explored both theoretically and empirically.

With respect to stabilization of a driving Y, Maffi & Jayakar (1981) examined the conditions under which a polymorphism at a Y-linked drive locus can be maintained in species where sex is determined by a single locus, rather than by separate X and Y chromosomes. They considered a situation, as in *Aedes aegypti*, in which the drive locus is linked to, but separate from, the sex determining locus (Newton et al. 1978). They assumed that neither locus affects male or female fertility and viability. The analysis showed that some recombination between the two loci is necessary to bring about a protected polymorphism. This analysis revealed that for species in which there is no recombination between the X and Y chromosomes, a driving Y polymorphism cannot be maintained with frequency-independent fitnesses.

EVOLUTION OF SUPPRESSORS

An X^D chromosome spreads at the expense of the nondriving X^S chromosomes in a population. However, because such drive is typically expressed only in males, X drive necessarily entails a reduction in Y chromosome transmission by Sex-ratio males. If X^D exhibits complete drive, the selective difference between Y chromosomes that are completely susceptible versus completely resistant is proportional to the frequency of X^D. The female-biased population-level sex ratio resulting from X drive favors the evolution of autosomal suppressors. In the absence of other fitness effects, the mean fitness of females relative to males is $W_f = (1 - P)/(1 + P)$, where P is the frequency of X^D. In Sex-ratio males, nonsuppressor alleles at autosomal loci are passed exclusively to females, whereas suppressor alleles can be transmitted to males at a rate dependent on the degree of suppression. The selective difference between autosomal suppressor and nonsuppressor alleles can be shown to be proportional to P^2. Thus, when X^D is at high frequency, selection on Y-linked and autosomal suppressors of drive will be of comparable magnitude, but at low to moderate frequencies of X^D, selection will be considerably stronger on the Y.

In an attempt to understand the conditions under which an autosomal suppressor of drive may or may not evolve, Wu (1983c) considered the population genetics of an autosomal locus that suppresses X^D drive but otherwise has no effect on the fitness of its carriers. Wu assumed that a population is at a stable equilibrium frequency of X^D and that all males are completely sensitive to drive. His model focuses on the conditions under which an autosomal suppressor can spread within a population, not the conditions for a stable polymorphism at a suppressor locus or fixation of a suppressor allele. Although one might intuitively expect that a cost-free suppressor would always be favored, Wu showed that under some conditions such a suppressor does not invade. Specifically, if Sex-ratio males have very low fitness and if there is heterozygote superiority at the drive locus in females, the suppressor will not spread. The reason for this appears to be twofold. First, if there is strong selection against Sex-ratio males, then the equilibrium frequency of X^D will be low (Edwards 1961). As a result, the population will be only slightly female biased, and selection in favor of the suppressor will be correspondingly weak. Second, when P is low, most of the offspring of Sex-ratio males with unsuppressed drive will be heterozygous females, which exhibit above-average fitness. Suppression of drive in these males would result in their siring fewer daughters, which would be heterozygous at the drive locus, resulting in reduced mean fitness of a Sex-ratio male's offspring.

With respect to Y-linked suppressors of drive, Thomson & Feldman (1975) showed that if a population is polymorphic for driving and nondriving X chromosomes, then a Y chromosome that is less susceptible to drive will replace a more susceptible Y, as long as males carrying the two Y chromosome types are equally fertile. Carvalho et al. (1997) relaxed the assumption that the Y chromosome type does not affect male fitness and showed by computer simulation that the presence of a *sex-ratio* polymorphism can often lead to the maintenance of a stable Y chromosome polymorphism. Their models assume that X^D is equally deleterious in males and homozygous females. Jaenike (1999) obtained similar results analytically and found that if there is a stable Y chromosome polymorphism, then the equilibrium frequency of the X^D chromosome depends primarily on Y chromosome parameters.

Y-linked and/or autosomal suppressors of X drive are common, having been documented in *Drosophila subobscura, D. affinis, D. simulans, D. paramelanica, D. mediopunctata, D. quinaria, Cyrtodiopsis dalmanni, Aedes aegypti,* and *Silene alba.* Species lacking suppressors, such as *D. pseudoobscura* and *D. neotestacea,* appear to be the exception. Population cage experiments with *D. simulans* and *D. mediopunctata* clearly show the selective advantage of suppressors when a population harbors driving X^D chromosomes (Capillon & Atlan 1999, Carvalho et al. 1998). Nothing is yet known about the molecular mechanism of suppression, although Cazemajor et al. (1997) speculate that variation among Y chromosomes in sensitivity to X drive in *D. simulans* could be related to the size of a heterochromatic responder region. It is known that in *D. melanogaster,* the size of the *responder* locus, a tandemly repeated array, determines sensitivity to drive at the *SD* locus (Wu et al. 1988).

Because a driving X favors the evolution of Y-linked and autosomal suppressors, one might expect there to be a positive association among populations between the frequency of X^D and these suppressors. In *D. mediopunctata,* two widely separated natural populations that have been examined are both polymorphic for X^D and suppressors of drive (Carvalho et al. 1997). For *D. simulans* and *A. aegypti,* populations where driving sex chromosomes are present also have high levels of drive suppression (Atlan et al. 1997, Wood & Newton 1991). In *Drosophila paramelanica,* each of the two X^D chromosome types (Northern and Southern) exhibits strong drive against the Y with which it is normally associated in natural populations (Stalker 1961). However, the Southern-type Y chromosome is capable of suppressing drive by the Northern-type X^D, with which it rarely co-occurs. This pattern could have arisen as a result of the Southern-type Y excluding the Northern-type X^D in areas where the Southern Y occurs. This raises the question of why the Southern Y does not spread into areas where the Northern X^D and the drive-susceptible Northern Y occur. In *D. subobscura,* drive by the North African X^D chromosome is stronger when the Y and autosomes are also from North Africa than when they are from European populations (Hauschteck-Jungen 1990). Thus, it is not yet clear whether drive and suppressors of drive tend to be positively or negatively associated across populations. Such patterns may yield insights into the dynamics of these genetic arms races. The various species harboring resistant Y chromosomes provide excellent models for the study of Y chromosome polymorphism, which is generally expected to be rare (Clark 1987).

EVOLUTION OF SEX-RATIO INVERSIONS

In most species examined, X^D chromosomes differ from their standard, nondriving counterparts by one or more inversions. Drive is not due to a position effect, since the X^D chromosome of *D. persimilis* and the nondriving X^S of *D. pseudoobscura* have the same gene arrangement (Sturtevant & Dobzhansky 1936). Furthermore, in both *D. neotestacea* and *D. simulans*, the X^D and X^S chromosomes appear to be homosequential (James & Jaenike 1990, Cazemajor et al. 1997).

Unlike the situation for autosomal drive, association of an X-linked *sex-ratio* locus with an inversion is not necessary for the maintenance of linkage disequilibrium between the drive and responder loci. With the exception of *Aedes* and *Culex*, the X and Y chromosomes of the insect species considered here do not undergo recombination. Thus, inversions probably tie up several loci that together yield the drive characteristic of X^D chromosomes. Wu & Beckenbach (1983) clearly demonstrated that several loci are required for the expression of drive in *Drosophila persimilis*.

The evolution of SR inversions may have occurred via two selective paths. Thomson & Feldman (1975) showed that if fertility and viability are unaffected, an X chromosome with stronger drive will replace one with weaker drive. Suppose there exist several freely recombining X-linked *sex-ratio* loci, each with an allele exhibiting weak drive. If the drive effects are more or less additive across loci, then a chromosome in which two or more of these loci are tied up within a newly arisen inversion will be favored. Thus, a strongly driving X^D chromosome can evolve through the gradual recruitment of *sr* alleles that exhibit only weak drive individually. Similar ideas have been put forward by Sturtevant & Dobzhansky (1936) and Babcock & Anderson (1996). Cazemajor et al. (1997) have shown that X chromosome drive in *D. simulans* is probably due to the additive effects of two (or more) closely linked loci, which individually exhibit only partial drive.

Although this scenario may apply in some cases, several observations indicate that strong X drive is not always due to the additive effects of several weakly driving loci. First, the existence of strong drive in *D. neotestacea* and *D. simulans*, which lack SR inversions, indicates that one or a few closely linked loci are sufficient to achieve drive equivalent to that seen in other species with as many as five SR inversions. Second, in their genetic dissection of the sex-ratio trait in *D. persimilis*, Wu & Beckenbach (1983) found that all of the genetically marked chromosome regions were required for expression of drive: Individual chromosome regions did not express even partial drive. This finding suggests that the different *sex-ratio* loci interact epistatically, rather than additively. In *D. pseudoobscura*, the SR gene arrangement of X^D chromosomes differs from the ST gene arrangement of X^S chromosomes by three inversions. Beckenbach (1996) discovered a very rare X chromosome type that carries only the terminal sex-ratio inversion, lacking the medial and subbasal inversions, and males carrying this X chromosome type sired normal offspring sex ratios. Another rare X chromosome type in this species lacks the terminal inversion but carries the other two. Males carrying this X chromosome type sire strongly female-biased progeny (Wallace 1948, Beckenbach 1996), indicating that a full complement of the SR inversions is not required for drive. Thus, in both *D. persimilis* and *D. pseudoobscura*, X chromosome drive results from something other than the additive effects of weak drive at individual loci.

Inversions might also be favored as a result of interactions between autosomal or Y-linked suppressors of drive and X-linked loci that counteract the effect of those suppressors (Jaenike 1996). According to this hypothesis, SR inversions would include one locus that drives against the Y plus other loci that prevent suppression of drive. Thus, drive would be expressed only if the primary drive locus as well as the suppressors-of-suppressors were present on an X^D chromosome. This would account for the all-or-none drive of partial X^D chromosomes of *D. persimilis* and *D. pseudoobscura* (Wu & Beckenbach 1983, Beckenbach 1996). This hypothesis is also consistent with the observation that the strength of drive is not correlated across species with the number of SR inversions. For instance, within the obscura group of *Drosophila*, effectively complete drive is associated with as few as one (*D. persimilis*, *D. affinis*) to as many as five (*D. athabasca*) inversions.

Regardless of how inversions evolved, once large blocks of the X^D chromosome became tied up within inversions, the ST and SR gene arrangements would differ at numerous loci besides those affecting X chromosome drive. Some of these are likely to affect the fertility and viability of females, thus allowing maintenance of a *sex-ratio* polymorphism via the frequency-independent mechanism modeled by Edwards (1961) and Curtsinger & Feldman (1980). Thus, the maintenance of the polymorphism through this frequency-independent mechanism may be a derived condition, possibly requiring coevolution between the driving X^D and the suppressors in the rest of the genome (Jaenike 1996).

MACROEVOLUTION AND EXTINCTION

The sporadic taxonomic distribution of X chromosome drive indicates that it has evolved on numerous occasions, although the phylogenetic limits of its occurrence remain to be determined. Within the genus *Drosophila*, the loci responsible for X chromosome drive probably differ among species. Chromosomal evolution in *Drosophila* has proceeded largely through fusions and inversions among six chromosomal elements (Patterson & Stone 1952). In *D. simulans* and many species of the subgenus *Drosophila*, including *D. neotestacea*, *D. quinaria*, and *D. medio-punctata*, any X-linked drive loci would be part of element A. In contrast, the SR inversions of *D. pseudoobscura* and *D. persimilis* are located on the right arm of the X, which in these species is equivalent to element D (Patterson & Stone 1952). In *D. affinis* and *D. paramelanica*, the X^D chromosomes differ from X^S by inversions in both the right and left arms of the X, i.e., elements A and D (Stalker 1961). Thus, at least two distinct loci, or sets of loci, appear capable of causing X chromosome drive in *Drosophila*. Although a molecular analysis of X chromosome drive has not yet been conducted in any species, such studies could shed light on the number of genetically independent pathways by which such drive can evolve.

Once a driving X has evolved, it can persist for extended evolutionary periods. For instance, the sequence difference between ST- and SR-associated alleles of *Esterase-5* in *D. pseudoobscura*, which is located within one of the SR inversions, indicate a divergence time of about 1 million years (Babcock & Anderson 1996, Kovacevic & Schaeffer 2000). Based on mtDNA sequence data, Spicer & Jaenike (1996) estimate that *Drosophila quinaria* and *D. recens* split about 1.5 million years ago. The presence of X^D chromosomes in both of these species suggests that the *sex-ratio* polymorphism may be at least this old in the *quinaria* species group. Similarly, X^D chromosomes are present in two species of stalk-eyed flies, *Cyrtodiopsis dalmanni* and *C. whitei* (Presgraves et al. 1997), which molecular evidence indicates are sister species (Wilkinson et al. 1998a). Finally, the Eastern A and Eastern B semispecies of *Drosophila athabasca* share only one X chromosome gene arrangement, the multiple-inversion SR chromosome, showing that this driving X^D arose before these semispecies diverged (Yoon & Aquadro 1994). All of these findings indicate that once an X^D chromosome has invaded a species, it has the potential to persist for extended evolutionary periods. Because X drive can bring about intense intragenomic conflict, X^D chromosomes may impose major long-term selective pressures of the sort discussed in the introduction.

Although X^D chromosomes can persist for long periods, they may be much more ephemeral in many cases. An X^D chromosome could fail evolutionarily either because it did not spread within a population, or because it spread to fixation and caused population- or species-level extinction. Suppose the dynamics of X^D is initially governed by sperm depletion in multiply mated males and sperm competition within multiply mated females. If so, then some species may be more susceptible to invasion and spread of an X^D chromosome than others. For example, in a high-density species with frequent male and female mating, there could be strong selection against X^D from the outset, whereas a low-density species would be more susceptible to invasion by X^D (Jaenike 1996). Alternatively, consider the insemination reaction of *Drosophila*, whose effect is to prevent a female from remating for various lengths of time (Patterson 1946). If the insemination reaction reduces the average intensity of sperm competition within females, this could facilitate the spread of a driving X^D chromosome. These considerations suggest that species-level characteristics could influence the fate of a newly arisen *sr* allele.

At the other extreme, an X^D chromosome could fail to persist by spreading to high frequency, bringing about a critical deficiency of males and causing extinction. Carvalho & Vaz (1999) have argued that newly arisen X^D chromosomes often will not have selection coefficients that lead to a stable polymorphism and, as a result, may spread to such a high frequency that population extinction is likely. Consequently, extant *sex-ratio* polymorphisms might represent only a select subset of all those that have arisen. In particular, only those X^D chromosomes having substantial deleterious effects on their carriers will fail to spread to fixation. A *sex-ratio* polymorphism could also be stabilized by the evolution of suppressors. Thus, whether X^D spreads to fixation or exists at a stable equilibrium could depend on how rapidly suppressors evolve. Carvalho & Vaz (1999) conclude that if X^D chromosomes have fitness effects comparable to random X chromosomes in *Drosophila*, then it is likely that hundreds of *Drosophila* species have been driven extinct by the spread of X^D chromosomes whose deleterious effects were too weak to stabilize the polymorphism.

Such extinction could bring about species-level selection for traits related to susceptibility to *sr* invasion. Alternatively, an *sr* allele may be stabilized at an intermediate frequency, resulting in an unequal population-level sex ratio and instigating a prolonged genetics arms race between the driving X^D chromosome, the Y, and the autosomes. Because of these considerable effects—species-level selection and extended intragenomic conflicts—it is important to determine the range of taxa susceptible to sex chromosome drive and the factors that determine the dynamics of these chromosomes in natural populations.

ACKNOWLEDGMENTS

I would like to thank Anne Atlan, Bernardo Carvalho, Michael Clark, Jerry Coyne, Kelly Dyer, Hopi Hoekstra, Catherine Montchamp-Moreau, Allen Orr, Daven Presgraves, Doug Taylor, Jack Werren, Jerry Wilkinson, Diana Wolf, and Roger Wood for helpful comments. This work was supported by the National Science Foundation (grant DEB 9615065), the Underwood Fund (UK), and the NERC Centre for Population Biology, Imperial College at Silwood Park.

Visit the Annual Reviews home page at www.AnnualReviews.org

LITERATURE CITED

Atlan A, Mercot H, Landre C, Montchamp-Moreau C. 1997. The *sex-ratio* trait in *Drosophila simulans*: geographical distribution of distortion and resistance. *Evolution* 51:1886–95

Babcock CS, Anderson WW. 1996. Molecular evolution of the sex-ratio inversion complex in *Drosophila pseudoobscura*: analysis of the *Esterase-5* gene region. *Mol. Biol. Evol.* 13:297–308

Beckenbach AT. 1978. The "sex-ratio" trait

in *Drosophila pseudoobscura*: fertility relations of males and meiotic drive. *Am. Nat.* 112:97–117

Beckenbach AT. 1981. Multiple mating and the "sex-ratio" trait in *Drosophila pseudoobscura. Evolution* 35:275–81

Beckenbach AT. 1983. Fitness analysis of the "sex ratio" polymorphism in experimental populations of *Drosophila pseudoobscura. Am. Nat.* 121:630–48

Beckenbach AT. 1996. Selection and the

"sex-ratio" polymorphism in natural populations of *Drosophila pseudoobscura*. *Evolution* 50:787–94

Beckenbach A, Curtsinger JW, Policansky D. 1982. Fruitless experiments with fruit flies: the "sex ratio" chromosomes of *D. pseudoobscura*. *Drosophila Inf. Serv.* 58:22

Bircher U, Jungen H, Burch R, Hauschteck-Jungen E. 1995. Multiple morphs of sperm were required for the evolution of the Sex Ratio trait in *Drosophila*. *J. Evol. Biol.* 8:575–88

Bryant SH, Beckenbach AT, Cobbs GA. 1982. "Sex-ratio" trait, sex composition, and relative abundance in *Drosophila pseudoobscura*. *Evolution* 36:27–34

Bull JJ, Bulmer MG. 1981. The evolution of XY females in mammals. *Heredity* 47:347–65

Capillon C, Atlan A. 1999. Evolution of driving X chromosomes and resistance factors in experimental populations of *Drosophila simulans*. *Evolution* 53:506–17

Carvalho AB, Klaczko LB. 1993. Autosomal suppressors of sex-ratio in *Drosophila mediopunctata*. *Heredity* 71:546–51

Carvalho AB, Klaczko LB. 1994. Y-linked suppressors of sex-ratio in *Drosophila mediopunctata*. *Heredity* 73:573–79

Carvalho AB, Peixoto AA, Klaczko LB. 1989. Sex-ratio in *Drosophila mediopunctata*. *Heredity* 62:425–28

Carvalho AB, Sampaio MC, Varandas FR, Klaczko LB. 1998. An experimental demonstration of Fisher's principle: evolution of sexual proportion by natural selection in *Drosophila mediopunctata*. *Genetics* 148:719–31

Carvalho AB, Vaz SC. 1999. Are *Drosophila* SR chromosomes always balanced? *Heredity* 83:221–28

Carvalho AB, Vaz SC, Klaczko LB. 1997. Polymorphism for Y-linked suppressors of *sex-ratio* in two natural populations of *Drosophila mediopunctata*. *Genetics* 146:891–902

Cazemajor M, Joly D, Montchamp-Moreau C. 2000. *Sex-ratio* meiotic drive in *Drosophila simulans* is related to equational nondis-

junction of the Y chromosome. *Genetics* 154:229–36

Cazemajor M, Landre C, Montchamp-Moreau C. 1997. The *sex-ratio* trait in *Drosophila simulans*: genetic analysis of distortion and suppression. *Genetics* 147:635–42

Charlesworth B, Hartl DL. 1978. Population dynamics of the Segregation Distorter polymorphism of *Drosophila melanogaster*. *Genetics* 89:171–92

Charnov EL. 1982. *The Theory of Sex Allocation*. Princeton, NJ: Princeton Univ. Press

Clark AG. 1987. Natural selection and *Y*-linked polymorphism. *Genetics* 115:569–77

Coyne JA, Charlesworth B, Orr HA. 1991. Haldane's rule revisited. *Evolution* 45:1710–14

Coyne JA, Orr HA. 1993. Further evidence against meiotic-drive models of hybrid sterility. *Evolution* 47:685–87

Curtsinger JW, Feldman MW. 1980. Experimental and theoretical analyses of the "sex ratio" polymorphism in *Drosophila pseudoobscura*. *Genetics* 94:445–66

Dermitzakis ET, Maly JP, Waltrip HM, Clark AG. 2000. NonMendelian segregation of sex chromosomes in heterospecific *Drosophila* males. *Genetics* 154:687–94

Dobzhansky T, Epling C. 1944. Contributions to the genetics, taxonomy, and ecology of *Drosophila pseudoobscura* and its relatives. *Carnegie Inst. Wash. Publ.* 554

Doebley J, Stec A. 1991. Genetic analysis of the morphological differences between maize and teosinte. *Genetics* 129:285–95

Edwards AWF. 1961. The population genetics of "sex-ratio" in *Drosophila pseudoobscura*. *Heredity* 16:291–304

Faulhaber SH. 1967. An abnormal sex ratio in *Drosophila simulans*. *Genetics* 56:189–213

Fisher RA. 1930. *The Genetical Theory of Natural Selection*. Oxford, UK: Oxford Univ. Press

Frank SA. 1991. Divergence of meiotic drive-suppression systems as an explanation for sex-biased hybrid sterility and inviability. *Evolution* 45:262–67

Fredga K, Gropp A, Winking H, Frank F. 1977. A hypothesis explaining the exceptional sex ratio in the wood lemming (*Myopus schisticolor*). *Hereditas* 85:101–4

Fregda K, Gropp A, Winking H, Frank F. 1976. Fertile XX- and XY-type females in the wood lemming *Myopus schisticolor*. *Nature* 261:225–27

Gershenson S. 1928. A new sex ratio abnormality in *Drosophila obscura*. *Genetics* 13:488–507

Gileva EA. 1987. Meiotic drive in sex chromosome system of the varying lemming. *Dicrostonyx torquatus*. (Rodentia: Microtinae). *Heredity* 59:383–89

Gileva EA, Benenson IE, Konopistseva LA, Puchkov VF, Makaranets IA. 1982. XO females in the varying lemming, *Dicrostonyx torquatus*: reproductive performance and its evolutionary significance. *Evolution* 36:601–9

Gooding RH. 1986. Evidence for genetic control of sex ratio distortion in two colonies of *Glossina morsitans submorsitans* Newstead (Diptera: Glossinidae). *Quaest. Entomol.* 22:19–28

Gooding RH, Rolseth BM, Nesbitt SAT. 1989. Mapping four loci in *Glossina morsitans submorsitans* Newstead (Diptera: Glossinidae). *Can. Entomol.* 121:823–24

Haig D. 1993a. The evolution of unusual chromosomal systems in sciarid flies: intragenomic conflict and the sex ratio. *J. Evol. Biol.* 6:249–61

Haig D. 1993b. The evolution of unusual chromosomal systems in coccoids: extraordinary sex ratios revisited. *J. Evol. Biol.* 6:69–77

Haig D, Bergstrom CT. 1995. Multiple mating, sperm competition and meiotic drive. *J. Evol. Biol.* 8:265–82

Haig D, Grafen A. 1991. Genetic scrambling as a defence against meiotic drive. *J. Theor. Biol.* 153:531–58

Hamilton WD. 1967. Extraordinary sex ratios. *Science* 156:477–88

Hastings I. 1991. Germline selection: population genetic aspects of the sexual/asexual life cycle. *Genetics* 129:1167–76

Hastings RJ, Wood RJ. 1978. Meiotic drive at the $D(M^D)$ locus and fertility in the mosquito. *Aedes aegypti* (L.). *Genetica* 49:159–63

Hatcher MJ, Taneyhill DE, Dunn AM. 1999. Population dynamics under parasitic sex ratio distortion. *Theor. Popul. Biol.* 56:11–18

Hauschteck-Jungen E. 1990. Postmating reproductive isolation and modification of the 'sex ratio' trait in *Drosophila subobscura* induced by the sex chromosome gene arrangement $A_{2+3+5+7}$. *Genetica* 83:31–44

Hauschteck-Jungen E, Maurer B. 1976. Sperm dysfunction in sex-ratio males of *Drosophila subobscura*. *Genetica* 46:459–77

Hickey WA. 1970. Factors influencing the distortion of sex ratio in *Aedes aegypti*. *J. Med. Entomol.* 7:727–35

Hickey WA, Craig GB. 1966. Genetic distortion of sex ratio in a mosquito. *Aedes aegypti*. *Genetics* 53:1177–96

Hiraizumi Y, Sandler L, Crow JF. 1960. Meiotic drive in natural populations of *Drosophila melanogaster*. III. Populational implications of the segregation-distorter locus. *Evolution* 14:433–44

Hoekstra HE, Edwards SV. 2000. Multiple origins of XY female mice (genus *Akodon*): phylogenetic and chromosomal evidence. *Proc. R. Soc. London Ser. B* 276:1825–31

Hoekstra HE, Hoekstra JM. 2001. An unusual sex-determination system in South American field mice (genus *Akodon*): the role of mutation, selection and meiotic drive in maintaining XY females. *Evolution* 55:190–97

Hurst GDD, Jiggins FM. 2000. Male-killing bacteria in insects: mechanisms, incidence, and implications. *Emerg. Infect. Dis.* 6:329–36

Hurst LD. 1992. Is *Stellate* a relict meiotic driver? *Genetics* 130:229–30

Hurst LD. 1996. Further evidence consistent with *Stellate's* involvement in meiotic drive. *Genetics* 142:641–43

Hurst LD, Atlan A, Bengtsson BO. 1996. Genetic conflicts. *Q. Rev. Biol.* 71:317–64

Hurst LD, Pomiankowski A. 1991. Causes of sex-ratio bias may account for unisexual sterility in hybrids: a new explanation of

Haldane's rule and related phenomena. *Genetics* 128:841–58

Jaenike J. 1996. Sex-ratio meiotic drive in the *Drosophila quinaria* group. *Am. Nat.* 148:237–54

Jaenike J. 1999. Suppression of sex-ratio meiotic drive and the maintenance of Y chromosome polymorphism in *Drosophila*. *Evolution* 53:164–74

James AC. 1992. "Sex ratio" meiotic drive in *Drosophila neotestacea*. PhD thesis. Univ. Rochester, Rochester, NY

James AC, Jaenike J. 1990. "Sex ratio" meiotic drive in *Drosophila testacea*. *Genetics* 125:651–56

Jiggins FM, Hurst GDD, Majerus MEN. 1999. How common are meiotically driving sex chromosomes in insects? *Am. Nat.* 154:481–83

Kovacevic M, Schaeffer SW. 2000. Molecular population genetics of X-linked genes in *Drosophila pseudoobscura*. *Genetics* 156:155–72

Lande R, Wilkinson GS. 1999. Models of sex-ratio meiotic drive and sexual selection in stalk-eyed flies. *Genet. Res.* 74:245–53

Lindsley DL, Zimm GG. 1992. *The genome of Drosophila melanogaster*. London: Academic

Lloyd DG. 1974. Female-predominant sex ratios in angiosperms. *Heredity* 32:35–44

Lyon MF, Zenthon J, Evans EP, Burtenshaw MD, Willison KR. 1988. Extent of the mouse *t* complex and its inversions shown by in situ hybridization. *Immunogenetics* 27:375–82

Lyttle TW. 1981. Experimental population genetics of meiotic drive systems. III. Neutralization of sex-ratio distortion in *Drosophila* through sex chromosome aneuploidy. *Genetics* 98:317–34

Lyttle TW. 1991. Segregation distorters. *Annu. Rev. Genet.* 25:511–57

Maffi G, Jayakar SD. 1981. A two-locus model for polymorphism for sex-linked meiotic drive modifiers with possible applications to *Aedes aegypti*. *Theor. Popul. Biol.* 19:19–36

McKee BD. 1991. X-Y pairing, meiotic drive, and ribosomal DNA in *Drosophila melanogaster* males. *Am. Nat.* 137:332–39

Merçot H, Atlan A, Jacques M, Montchamp-Moreau C. 1995. Sex-ratio distortion in *Drosophila simulans*: co-occurrence of a meiotic drive and suppressor of drive. *J. Evol. Biol.* 8:283–300

Miller DD, Voelker RA. 1969. Salivary gland chromosome variation in the *Drosophila affinis* subgroup. IV. The short arm of the X chromosome in "western" and "eastern" *Drosophila athabasca*. *J. Hered.* 60:307–11

Montchamp-Moreau C, Ginhoux V, Atlan A. 2001. The Y chromosomes of *Drosophila simulans* are highly polymorphic for their ability to suppress *sex-ratio* drive. *Evolution* 55:728–37

Montchamp-Moreau C, Joly D. 1997. Abnormal spermiogenesis is associated with the X-linked sex-ratio trait in *Drosophila simulans*. *Heredity* 79:24–30

Newton ME, Wood RJ, Southern DI. 1976. A cytological analysis of meiotic drive in the mosquito *Aedes aegypti* (L.). *Genetica* 46:297–318

Newton ME, Wood RJ, Southern DI. 1978. Cytological mapping of the *M* and *D* loci in the mosquito. *Aedes aegypti* (L.). *Genetica* 48:137–43

O'Neill SL, Hoffmann AA, Werren JW, eds. 1997. *Influential Passengers*. Oxford, UK: Oxford Univ. Press

Owusu-Daaku K, Wood RJ, Butler RD. 1997. Selected lines of *Aedes aegypti* with persistently distorted sex ratios. *Heredity* 79:388–93

Patterson JT. 1946. A new type of isolating mechanism in *Drosophila*. *Proc. Natl. Acad. Sci. USA* 32:202–8

Patterson JT, Stone WS. 1952. *Evolution in the Genus Drosophila*. New York: Macmillan

Policansky D. 1974. "Sex ratio," meiotic drive, and group selection in *Drosophila pseudoobscura*. *Am. Nat.* 108:75–90

Policansky D. 1979. Fertility differences as a factor in the maintenance of the "sex ratio" polymorphism in *Drosophila pseudoobscura*. *Am. Nat.* 114:672–80

Policansky D, Ellison J. 1970. "Sex ratio" in *Drosophila pseudoobscura*: spermiogenic failure. *Science* 169:888–89

Prakash S, Merritt RB. 1972. Direct evidence of genic differentiation between sex-ratio and standard gene arrangements of the X chromosome in *Drosophila pseudoobscura*. *Genetics* 72:169–75

Presgraves DC, Severance E, Wilkinson GS. 1997. Sex chromosome meiotic drive in stalk-eyed flies. *Genetics* 147:1169–80

Randerson JP, Jiggins FM, Hurst LD. 2000. Male killing can select for male mate choice: a novel solution to the paradox of the lek. *Proc. R. Soc. London Ser. B* 267:867–74

Rawlings P, Maudlin I. 1984. Sex ratio distortion in *Glossina morsitans submorsitans* Newstead (Diptera: Glossinidae). *Bull. Entomol. Res.* 74:311–15

Rigaud T. 1997. Inherited microorganisms and sex determination of arthropod hosts. See O'Neill et al. 1997, pp. 81–101

Spicer G, Jaenike J. 1996. Phylogenetic analysis of breeding site use and α-amanitin tolerance within the *Drosophila quinaria* group. *Evolution* 50:2328–37

Stalker HD. 1961. The genetic systems modifying meiotic drive in *Drosophila paramelanica*. *Genetics* 46:177–202

Stouthamer R. 1997. *Wolbachia*-induced parthenogenesis. See O'Neill et al. 1997, pp. 102–24

Sturtevant AH, Dobzhansky T. 1936. Geographical distribution and cytology of "sex ratio" in *Drosophila pseudoobscura* and related species. *Genetics* 21:473–90

Sweeny TL, Barr AR. 1978. Sex ratio distortion caused by meiotic drive in a mosquito. *Culex pipiens* L. *Genetics* 88:427–46

Taylor DR. 1993. Sex ratio hybrids between *Silene alba* and *Silene dioica*: evidence for Y-linked restorers. *Heredity* 74:518–26

Taylor DR. 1994. The genetic basis of sex ratio in *Silene alba* (=*S. latifolia*). *Genetics* 136:641–51

Taylor DR. 1996. Parental expenditure and offspring sex ratios in the dioecious plant *Silene alba* (=*Silene latifolia*). *Am. Nat.* 147:870–79

Taylor DR. 1999. Genetics of sex ratio variation among natural populations of a dioecious plant. *Evolution* 53:55–62

Taylor DR, Saur MJ, Adams E. 1999. Variation in pollen performance and its consequences for sex ratio evolution in a dioecious plant. *Evolution* 53:1028–36

Temin RG, Ganetzky B, Powers PA, Lyttle TW, Pimpinelli S, et al. 1991. Segregation distortion in *Drosophila melanogaster*: genetic and molecular analyses. *Am. Nat.* 137:287–331

Thomson GJ, Feldman MW. 1975. Population genetics of modifiers of meiotic drive. IV. On the evolution of sex-ratio distortion. *Theor. Popul. Biol.* 8:202–11

Trivers RL, Willard DE. 1973. Natural-selection of parental ability to vary sex-ratio of offspring. *Science* 179:90–92

Turner BC, Perkins DD. 1991. Meiotic drive in *Neurospora* and other fungi. *Am. Nat.* 137:416–29

Underwood DA, Shapiro AM. 1999. A male-biased primary sex ratio and larval mortality in *Eucheira socialis* (Lepidoptera: Pieridae). *Evol. Ecol. Res.* 1:703–17

Varandas FR, Sampaio MC, Carvalho AB. 1997. Heritability of sexual proportion in experimental sex-ratio populations of *Drosophila mediopunctata*. *Heredity* 79:104–12

Voelker RA. 1972. Preliminary characterization of "sex ratio" and rediscovery and reinterpretation of "male sex ratio" in *Drosophila affinis*. *Genetics* 71:597–606

Wallace B. 1948. Studies on "sex ratio" in *Drosophila pseudoobscura*. I. Selection and "sex ratio." *Evolution* 2:189–217

Wallace B. 1968. *Topics in Population Genetics*. New York: Norton

Werren JH, Beukeboom LW. 1998. Sex determination, sex ratios, and genetic conflict. *Annu. Rev. Ecol. Syst.* 29:233–61

White MJD. 1973. *Animal Cytology and Evolution*. Cambridge, UK: Cambridge Univ. Press. 3rd ed.

Wilkinson GS, Kahler H, Baker RH. 1998a. Evolution of female mating preferences in stalk-eyed flies. *Behav. Ecol.* 9:525–33

Wilkinson GS, Presgraves DC, Crymes L. 1998b. Male eye span in stalk-eyed flies indicates genetic quality by meiotic drive suppression. *Nature* 391:276–79

Wilkinson GS, Sanchez MI. 2001. Sperm development, age, and sex chromosome drive in the stalk-eyed fly, *Cyrtodiopsis whitei. Heredity.* In press

Wolf DE, Rieseberg LH. 2001. The genetics of sex determination in the androdioecious plant, *Datisca glomerata*, and its dioecious sister species, *D. Cannabina. Genetics.* In press

Wood RJ. 1995. Potential for distorting sex ratio by meiotic drive genes. In *Zoologia—Sviluppi a Pavia nel XX secolo*, ed. M Fasolo, pp. 67–80. Cisalpino, Pavia, Italy

Wood RJ, Newton ME. 1991. Sex-ratio distortion caused by meiotic drive in mosquitoes. *Am. Nat.* 137:79–91

Wu C-I. 1983a. Virility deficiency and the sex-ratio trait in *Drosophila pseudoobscura.* I. Sperm displacement and sexual selection. *Genetics* 105:651–62

Wu C-I. 1983b. Virility deficiency and the sex-ratio trait in *Drosophila pseudoobscura.* II. Multiple mating and overall virility selection. *Genetics* 105:663–79

Wu C-I. 1983c. The fate of autosomal modifiers of the sex-ratio trait in *Drosophila* and other sex-linked meiotic drive systems. *Theor. Popul. Biol.* 24:107–20

Wu C-I, Beckenbach AT. 1983. Evidence for extensive genetic differentiation between the sex-ratio and the standard arrangement of *Drosophila pseudoobscura* and *D. persimilis* and identification of hybrid sterility factors. *Genetics* 105:71–86

Wu C-I, Lyttle TW, Wu M-L, Lin G-F. 1988. Association between a satellite DNA sequence and the Responder of Segregation Distorter in *Drosophila melanogaster. Cell* 54:179–89

Yoon CK, Aquadro CF. 1994. Mitochondrial DNA variation among the *Drosophila athabasca* semispecies and *Drosophila affinis. J. Hered.* 85:421–26

Annu. Rev. Ecol. Syst. 2001. 32:51–93

ENVIRONMENTAL INFLUENCES ON REGIONAL DEEP-SEA SPECIES DIVERSITY*

Lisa A. Levin,[1] Ron J. Etter,[2] Michael A. Rex,[2] Andrew J. Gooday,[3] Craig R. Smith,[4] Jesús Pineda,[5] Carol T. Stuart,[2] Robert R. Hessler,[1] and David Pawson[6]

[1]*Scripps Institution of Oceanography, University of California, San Diego, La Jolla, California 92093-0218; e-mail: llevin@ucsd.edu; rhessler@ucsd.edu*
[2]*Department of Biology, University of Massachusetts, Boston, Massachusetts 02125; e-mail: ron.etter@umb.edu; michael.rex@umb.edu; carol.stuart@umb.edu*
[3]*Southampton Oceanography Centre, European Way, Southampton SO14 3ZH United Kingdom; e-mail: ang@soc.soton.ac.uk*
[4]*Department of Oceanography, University of Hawaii, Honolulu, Hawaii 96822; e-mail: csmith@soest.hawaii.edu*
[5]*Department of Biology, MS 34, Woods Hole Oceanographic Institution, Woods Hole, Massachusetts 02543; e-mail: jpineda@whoi.edu*
[6]*National Museum of Natural History, Smithsonian Institution, MRC 106, Washington, DC 20560; e-mail: pawson.david@nmnh.si.edu*

Key Words Biodiversity, benthos, environmental gradients, depth gradients, diversity measures, bathyal, abyssal, sediments

■ **Abstract** Most of our knowledge of biodiversity and its causes in the deep-sea benthos derives from regional-scale sampling studies of the macrofauna. Improved sampling methods and the expansion of investigations into a wide variety of habitats have revolutionized our understanding of the deep sea. Local species diversity shows clear geographic variation on spatial scales of 100–1000 km. Recent sampling programs have revealed unexpected complexity in community structure at the landscape level that is associated with large-scale oceanographic processes and their environmental consequences. We review the relationships between variation in local species diversity and the regional-scale phenomena of boundary constraints, gradients of productivity, sediment heterogeneity, oxygen availability, hydrodynamic regimes, and catastrophic physical disturbance. We present a conceptual model of how these interdependent environmental factors shape regional-scale variation in local diversity. Local communities in the deep sea may be composed of species that exist as metapopulations whose regional distribution depends on a balance among global-scale, landscape-scale, and small-scale dynamics. Environmental gradients may form geographic patterns of diversity by influencing local processes such as predation, resource partitioning, competitive exclusion,

*The US Government has the right to retain a nonexclusive, royalty-free license in and to any copyright covering this paper.

and facilitation that determine species coexistence. The measurement of deep-sea species diversity remains a vital issue in comparing geographic patterns and evaluating their potential causes. Recent assessments of diversity using species accumulation curves with randomly pooled samples confirm the often-disputed claim that the deep sea supports higher diversity than the continental shelf. However, more intensive quantitative sampling is required to fully characterize the diversity of deep-sea sediments, the most extensive habitat on Earth. Once considered to be constant, spatially uniform, and isolated, deep-sea sediments are now recognized as a dynamic, richly textured environment that is inextricably linked to the global biosphere. Regional studies of the last two decades provide the empirical background necessary to formulate and test specific hypotheses of causality by controlled sampling designs and experimental approaches.

INTRODUCTION

Regional-scale sampling studies have provided the primary evidence for spatial patterns of deep-sea community structure and their causes. There has been no synoptic review of regional diversity patterns published since the early 1980s (Rex 1981, 1983). During the last two decades, our knowledge has expanded dramatically from descriptions of geographic gradients in diversity and indirect inferences about the biotic and abiotic factors that may shape them, to more precise understanding of associations between diversity and specific environmental conditions. These findings have transformed our understanding of the deep-sea ecosystem. The deep-sea, soft-sediment environment is highly complex at the landscape level of biodiversity as well as at the local community level, and is dynamically linked to oceanographic processes at the surface.

Here we summarize regional-scale geographic trends of deep-sea species diversity and propose a conceptual model to explain their causes. We are primarily concerned with ecological structuring agents that function on generational rather than evolutionary time scales. The stochasticity or patchiness observed at very small scales in deep-sea communities (Jumars & Eckman 1983, Grassle & Morse-Porteous 1987, Grassle & Maciolek 1992) becomes resolved at larger scales into measurable geographic patterns. The key to understanding deep-sea species diversity lies in documenting these patterns and in discerning the scales at which various causes operate to generate them. Patterns of diversity in the deep sea are much more complicated than previously thought. Early attention centered on bathymetric gradients in the western North Atlantic, the most intensively sampled region of the deep sea. Qualitative (Rex 1981) and quantitative (Etter & Grassle 1992) sampling studies indicated that diversity-depth patterns in the deep sea are unimodal with a peak at intermediate depths and depressed diversity at upper bathyal and abyssal depths. However, unimodal patterns do not appear to be universal (Rex et al. 1997, Stuart et al. 2001), and where they do occur in other basins have been attributed to varied environmental gradients (Paterson & Lambshead 1995, Cosson-Sarradin et al. 1998). Also, a variety of unanticipated oceanographic conditions at specific depths interrupt and modify bathymetric horizontal diversity trends (Gage 1997, Levin & Gage 1998, Vetter & Dayton 1998).

We relate diversity patterns to the regional-scale phenomena of geographic boundary effects, variation in sediment grain size, productivity gradients, oxygen minimum zones, current regimes, and catastrophic disturbances. Regional gradients of diversity represent the combined effects of these ecological factors in governing the rates of local processes. Regional-scale processes are, in turn, a consequence of global oceanography and climate. Compared to other major ecosystems, deep-sea communities appear to be controlled by the same basic mechanisms of energy availability, biological interactions, disturbance, and heterogeneity, but exerted through a very distinctive set of environmental circumstances that is new to ecology. We then consider the important issue of measuring diversity in the deep sea—a basic and persistent problem in comparisons among deep-sea studies and between deep-sea and surface environments (Gage & May 1993, Rex et al. 2000). Finally, a conceptual model is offered to explain how environmental factors interact with biotic processes to generate observed regional patterns.

It is difficult to define, in general terms, a physically or biologically meaningful regional scale in the deep sea. A region is often envisioned topographically as a major deep basin, such as the North American Basin of the North Atlantic that is bounded by the North American Continent and the mid-Atlantic Ridge system (Stuart & Rex 1994). However, deep basins are confluent at considerable depths and connected by the deep thermohaline circulation. Their faunas are not entirely or even largely endemic, suggesting that they neither have distinctive ecologies nor act as isolated theaters of evolution. For our purposes, regions represent areas of roughly 100s to 1000s km^2 that have been well sampled and encompass potentially significant and measurable environmental gradients or ecotones. Most regional studies have taken place at bathyal depths (200–4000 m) on continental margins; much less is known about large-scale spatial variation in abyssal (>4000 m) communities.

We focus on the deep-sea macrofauna of soft sediments because it is the most diverse and well-studied component of the benthos. The macrofauna is composed of animals retained on a 300 μm sieve. It includes most familiar invertebrate phyla, and is dominated particularly by polychaete worms, peracarid crustaceans, and mollusks (Gage & Tyler 1991). Where there are sufficient data, we also include the smaller meiofaunal elements. We do not consider reducing environments (hydrothermal vents and seeps); Van Dover (2000) provides an excellent and extensive treatment of these remarkable deep-sea habitats and their ecology.

This review developed from discussion by the Working Group on Deep-Sea Biodiversity supported by the National Center of Ecological Analysis and Synthesis (NCEAS). The Working Group's main objectives are to integrate pattern and scale of biodiversity in deep-sea benthic communities and to apply this to the design of future research programs. Mechanisms of species coexistence on local scales (<1 m^2) are addressed for the Working Group by P. V. Snelgrove & C. R. Smith (unpublished manuscript). Later reviews will cover the historical development and global spread of deep-sea faunas and propose strategies for conservation based on our current understanding of pattern and scale.

PATTERNS AND CAUSES OF DIVERSITY

Boundary Constraints and Species Ranges

Deep-sea ecology has focused primarily on the role of local processes in regulating community structure. Explaining species coexistence at small scales in the deep sea remains a major challenge to ecological theory (Gage 1996; P. V. Snelgrove & C. R. Smith, unpublished manuscript). It has proven difficult to extend mechanisms of local community structure to regional patterns in any ecological system (Ricklefs 1987). An alternative approach is to view diversity within regions as a composite of overlapping species distributional ranges (MacArthur 1972, Stevens 1989). Recently, there has been considerable interest in how physiographic and physiological boundaries affect species ranges and consequently patterns of diversity within geographic domains (Colwell & Hurtt 1994, Pineda & Caswell 1998, Willig & Lyons 1998, Lees et al. 1999, Colwell & Lees 2000). Interestingly, a stochastic placement of geographic ranges between boundaries will produce within the bounded region a unimodal pattern of diversity (Colwell & Hurtt 1994) similar to bathymetric gradients found in the deep sea (Figure 1). It is therefore possible to construct null models to test whether observed species diversity gradients depart significantly from those generated randomly by boundary constraints alone. If not, there is little justification for invoking an environmental gradient as the cause.

Pineda (1993) and Pineda & Caswell (1998) applied the geometric constraints model to patterns of deep-sea species diversity along depth gradients in the western North Atlantic. As noted earlier, the macrofauna of this region tend to show unimodal patterns of diversity with depth. The frequency distribution of bathymetric ranges of marine species resembles latitudinal ranges found in terrestrial taxa (Pineda 1993, Brown et al. 1996). Are unimodal diversity depth trends caused solely by boundary constraints imposed by the sea-air interface and the abyssal plain? Pineda & Caswell (1998) tested this possibility for gastropods and polychaetes by using a null model that randomly placed species ranges between the presumed upper and lower boundaries. Actual and randomly simulated diversities were compared for three components of the unimodal pattern: position of peak diversity along the depth gradients, peakedness of the diversity curve, and height of the peak. The null model explained only one aspect of the diversity pattern for each taxon, suggesting that whereas boundaries may affect the general shape of diversity-depth patterns, many important features of the patterns were decidedly nonrandom and attributable to other processes associated with depth.

Sediment Heterogeneity

The remarkable number of species that coexist within deep-sea assemblages is surprising when one realizes that the overwhelming majority of these species are deposit feeders that rely on organic detritus for food (Jumars & Eckman 1983). How can so many species coexist while exploiting the same limited resource? Because the primary food resource is detritus, the nature of the sediments ought

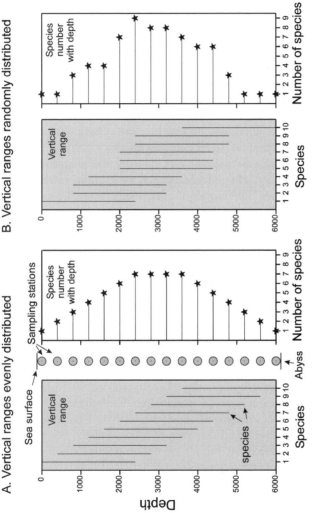

Figure 1   Schematic representation of the potential effects of boundaries on bathymetric patterns in species diversity. The gray panels show ten species evenly and randomly distributed with depth, from 0 to 6000 m (to the left and to the right, respectively). In this representation, all species have a vertical range of 2400 m. Sampling these distributions every 400 m (gray circles) yields bathymetric patterns in species diversity peaking at intermediate depths.

to play an important role in structuring deep-sea communities. Numerous studies have shown that the structure and composition of soft-sediment communities are related to sediment characteristics (e.g., Petersen 1913, Sanders 1968, Rhoads 1974, Gray 1981), but the explanations for these relationships are varied and remain controversial (Snelgrove & Butman 1994).

On the northwest Atlantic slope, spatiotemporal variation in species diversity is correlated with the heterogeneity of sediment grain size across a wide variety of spatial scales (Etter & Grassle 1992). Where sediment grain size is more varied, more species coexist. This is consistent with the hypothesis that species partition the sediments with respect to size. There is abundant evidence that deposit feeders selectively ingest sediments of particular size classes (Taghon 1982, Whitlatch 1980, Wheatcroft & Jumars 1987, Self & Jumars 1988, Wheatcroft 1992). In addition, several species exhibit interspecific differences in particle size preference (Fenchel et al. 1975, Fenchel & Kofoed 1976, Whitlatch 1980), suggesting that the sediments may be partitioned by size in some shallow-water communities. However, this has not been tested for deep-sea species.

The strong correlation between species diversity and sediment heterogeneity does not imply causality. The relationship may be spurious or reflect more important proximal factors. For example, the activities of more diverse communities might actually increase sediment heterogeneity. Also, the size ranges used by Etter & Grassle (1992) in the correlations were of disaggregated grains, and thus may bear little resemblance to food diversity or to the aggregated sediments organisms experience in situ. Definitive answers will require manipulative experiments that tease apart the various potential mediating processes.

Productivity and Food Supply

Spatial gradients in productivity are widely believed to influence species diversity (Waide et al. 1999). In terrestrial systems, the number of animal and plant species often appear to vary unimodally with productivity or nutrient availability (e.g., Tilman 1982, Rosenzweig & Abramsky 1993, Rosenzweig 1995). Diversity within a functional group or taxon increases from regions of low to moderate productivity, and then declines toward regions of higher productivity. Similarly, the unimodal species diversity-depth gradient that attends the exponential decrease in benthic standing stock with depth has been attributed to productivity and its potential mediation of biological interactions (Rex 1973, 1976, 1981).

The deep sea lacks in situ primary production, apart from chemoautotrophic production in reducing environments such as hydrothermal vent and seep habitats (Van Dover 2000). Most food material sinks from the euphotic zone to the benthos in the form of small particles. Particulate organic-carbon (POC) flux can be measured directly with sediment traps, and records integrating annual time scales now exist for at least 37 sites in the open ocean (Lampitt & Antia 1997). Energy availability in deep-sea benthic habitats is also positively correlated with, in order of decreasing strength: 1. sediment-community respiration (Jahnke 1996, Berelson et al. 1997), 2. rate of organic carbon burial within the sediment (Jahnke 1996), 3. benthic biomass and abundance (Rowe et al. 1991, Smith et al. 1997, Cosson et al. 1997), and 4. overlying primary productivity (Deuser et al. 1990, Watts et al. 1992, Lampitt & Antia 1998). It is negatively correlated with depth of water through which phytodetrital food sinks (Suess 1980, Martin et al. 1987). In some cases, the concentrations of organic carbon and chlorophyll a in surface sediments may be directly related to the flux of particulate organic carbon to the seafloor (Emerson 1985, Stephens et al. 1997). But these positive relationships are not necessarily maintained, and may even reverse, over large spatial scales (Jahnke 1996).

A number of productivity gradients have been characterized in the deep sea. There is a general decrease in POC flux (as well as sediment-community respiration and benthic standing crop) from shelf depths to the abyssal plain (Smith & Hinga 1983, Rowe et al. 1991). This depth-related decrease in productivity is well substantiated for gradually sloping and well oxygenated margins such as in the northwest Atlantic (Rowe et al. 1991), but becomes more complicated on irregular slopes or where oxygen minimum zones intersect continental margin (Reimers et al. 1992) and seamounts (Levin et al. 1991). A second productivity gradient is the decrease in seafloor POC flux from the productive coastal zone to the open ocean (Smith & Hinga 1983, Cosson et al. 1997). A third is the reduction in POC flux, sediment-community respiration, and standing crop observed at the abyssal seafloor from the Pacific equatorial zone and its associated nutrient upwelling northward or southward into the oligotrophic central gyres (Smith et al. 1997). Of these three productivity gradients, only that in the abyssal equatorial Pacific varies independently of many other variables that may influence diversity such as the hydrodynamic regime, bottom-water oxygen concentration, and physical environmental stability.

As in other environments (Waide et al. 1999), the relationships between POC flux or proxy variables for productivity and diversity are complicated and scale-dependent. POC flux and sediment-community respiration drop roughly fivefold at the abyssal seafloor from 0°N to 23°N latitude in the central Pacific Ocean (Smith et al. 1997). Mean local nematode diversity, expressed as the number of species rarefied to 51 individuals, exhibits a significant monotonic decline with decreasing POC flux along this gradient [P. J. D. Lambshead, C. J. Brown et al., unpublished manuscript) (Figure 2)]. Within this same general equatorial region, polychaete species richness, normalized to 163 individuals, was weakly positively correlated with total polychaete density, and by inference, POC flux (A. Glover, C. Smith, et al., unpublished manuscript, Figure 2). Neither taxon exhibits a unimodal diversity pattern comparable to those documented for terrestrial productivity gradients (Rosenzweig 1995). However, it is unclear from a comparative standpoint whether we are examining diversity along the ascending or descending part of the diversity-productivity curve, and diversity has been assessed for only one meiofaunal taxon and one macrofaunal group at the family level. Whereas productivity levels beneath central-gyre waters (e.g., 23°N) are extremely low, the productivity levels attained beneath equatorial upwelling are still moderate by deep-sea standards, so

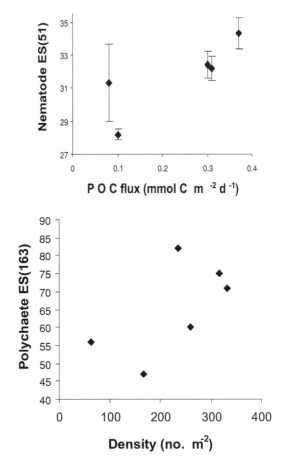

Figure 2 *Top*: Mean local rarefaction species richness (normalized to 51 individuals) as a function of POC flux for nematode samples collected along the equatorial Pacific POC flux gradient. Stations were located at water depths ranging from 4300 to 5000 m, 0°N 140°W; 2°N 140°W; 5°N 140°W; 9°N 140°W; and 23°N 158°W. The correlation between POC flux and rarefaction diversity is high ($r = 0.83$) but not statistically significant ($P = 0.09$) most likely due to small sample size ($n = 5$). Data from Lambshead et al. (2001a, in press). *Bottom*: Rarefaction diversity for macrofaunal polychaetes (normalized to 163 individuals) as a function of polychaete numerical density in the equatorial Pacific. Polychaete density is used as a proxy for POC flux and benthic productivity. Data are for pooled box-core samples (3 to 47 per station) collected from depths of 4300–5000 m at the following locations: 0°N 140°W; 2°N 140°W; 5°N 140°W; 8°27′N 150°47′W; 12°57′N 128°19′W; and 14°40′N 126°25′W. For these data, $r = 0.616$, and $P = 0.193$. Data from A. Glover, C. Smith, et al. (unpublished manuscript).

it is quite conceivable that we are examining diversity along the ascending portion of a more general unimodal diversity-productivity curve.

Cosson-Sarradin et al. (1998) studied polychaete species diversity along a transect of three stations from the continental shelf (<100 m depth) to the abyss (4600 m) off the west coast of central Africa. POC flux varied more than 16-fold from the shelf site to the oligotrophic abyssal station. Local polychaete diversity, based on the Shannon-Wiener index, exhibited a unimodal pattern as a function of POC flux, with a peak in diversity occurring at the mid-slope site (1700 m). However, this transect also represents a strong gradient in physical disturbance resulting from current scour, and possibly physiological stress from organic loading on the continental shelf. Paterson et al. (1998) studied local polychaete diversity (using rarefaction) at six abyssal stations in the Atlantic and Pacific with presumed (but unmeasured) differences in seafloor POC flux. Whereas polychaete abundance covaried with putative POC flux, species diversity showed no obvious relationship. Tietjen (1984, 1989) evaluated local nematode species diversity at six sites in the deep northwest Atlantic. The highest nematode diversity occurred at the sites presumed to have the highest POC flux and productivity.

On the North Carolina slope, the burial rate of organic carbon varies roughly 80-fold along the 850-m contour (Schaff et al. 1992, Blair et al. 1994, Levin et al. 1999). Schaff et al. (1992) and Levin et al. (1994b) examined local polychaete diversity at three stations spaced equidistantly along this productivity gradient and found reduced macrofaunal rarefaction diversity at the station with the highest carbon burial rate. Similar patterns were observed for macrofaunal foraminifera (>300 μm) (Gooday et al. 2001). Levin & Gage (1998) used sedimentary organic-matter content as a proxy of food availability to examine diversity-productivity relationships within existing data sets from the deep Indo-Pacific. Their study included a broad range in sedimentary organic carbon (<0.5% to >6%), and, presumably, habitat productivity. They found negative correlations between sedimentary organic-carbon concentrations and the local diversity of total macrofauna and polychaetes. Dominance in particular, was positively correlated with sediment POC concentrations, suggesting that competitive interactions may shift along POC gradients.

Sampling studies in the western North Atlantic provide, by far, the largest and most geographically extensive database available to examine diversity-productivity relationships. The unimodal diversity-depth patterns revealed by 1. rarefying large qualitative samples (Rex 1981) and 2. tabulations of species number from intensive quantitative sampling (Etter & Grassle 1992) parallel an exponential decline in benthic standing stock with increased depth. As mentioned above, benthic biomass and abundance are assumed to reflect the rate of nutrient input to the seafloor, but this has not been directly measured (Smith et al. 1997). Rex (1973) suggested that depressed diversity in the abyss was imposed by extremely low population densities, essentially as a chronic Allee Effect. There seems to be general agreement among ecologists that this kind of mechanism is responsible for low diversity at very low productivity (Rosenzweig & Abramsky 1993). The drop in diversity in the upper bathyal zone, close to high surface production and terrestrial runoff, could be driven by pulsed nutrient inputs and its potential to accelerate competitive exclusion (Rex 1976, Huston 1979).

Diversity trends in deep-sea microfossils over geological time scales mirror some of the spatial patterns described for modern productivity gradients and corroborate the inference that energy supply can regulate species diversity. Fluctuations in Shannon Wiener (H′) diversity of benthic ostracods correspond to orbitally driven glaciation cycles over a 550,000 year period in the late Pliocene in the North Atlantic (Cronin & Raymo 1997). Diversity was highest during interglacial periods and declined during glacial advances, when surface production may have been lower in the North Atlantic. Thomas & Gooday (1996) proposed that the establishment of a Southern Hemisphere latitudinal gradient in foraminiferan diversity in the late Eocene around 38 my ago, which persists today (Culver & Buzas 1998, 2000), was created by an increase in the seasonally pulsed organic flux that accompanied high latitude, Southern Hemisphere cooling, and the buildup of ice on the Antarctic continent. This temporal diversity pattern was driven by fluctuating dominance of opportunistic foraminifera that consume sinking phytodetritus.

The recent studies discussed above that use direct measures of productivity, such as POC flux, sediment organic content, and carbon burial rates, appear to give conflicting results; diversity varies positively, negatively, or unimodally with productivity. However, these findings are not necessarily contradictory. The productivity/input gradients examined may represent ranges of food availability that occupy different segments of a unimodal diversity-productivity relationship. The increase in diversity from areas of low to high POC flux in the abyssal equatorial Pacific may correspond to the ascending part of the diversity-productivity curve, much as proposed for the elevation of diversity from abyssal to the bathyal regions in the western North Atlantic. The pattern reported for the eastern tropical Atlantic spans a large vertical and horizontal geographic range that may experience productivity inputs broad enough to realize the full unimodal trend. Regions that include exceptionally high nutrient input, such as the North Carolina slope, may fall on the descending part of the curve where diversity and productivity are negatively correlated. Other observations suggest that depressed diversity in the deep sea is associated with periodic high organic loading such as might be associated with areas subject to intense upwelling (Sanders 1969), deposition from lateral transport of nutrients (Blake & Hilbig 1994), benthic storms that expose reactive sediments (Aller 1997), or bottom topography that concentrates food (Jumars & Hessler 1976, Vetter & Dayton 1998). However, all involve multiple factors that could affect diversity, and because the level and variation of production may covary, it is difficult to separate their influence. In general, a unimodal relationship between diversity and productivity at large scales in the deep sea is plausible, but is not well substantiated. Just as in other environments (Waide et al. 1999), it has proven difficult to accurately place available studies on a continuous productivity gradient and to identify clearly the underlying mechanisms through which productivity influences species diversity.

At low levels of productivity, food limitation is thought to constrain the number of species that can survive. Declines in diversity at higher productivity levels may result from four possible causes. 1. Differential numerical responses among species to nutrient loading (varying population growth rates) such that a small number of opportunistic species take over. This will elevate dominance and lower diversity measures that incorporate evenness. In many cases local species richness may not change much (see Levin et al. 2000). 2. Faster rates of competitive exclusion as envisioned by Huston (1979). 3. Increased variability in productivity, which is often correlated with amount of productivity. This variability may bring about declines due to demographic stochasticity. 4. Excess oxygen demand creating hypoxia and leading to declines in both richness and evenness due to physiological stress.

Bottom-Water Oxygen

Bottom-water oxygen concentrations in the deep oceans vary from near 0 to over 7 ml/l (Tyler 1995). Although much of the ocean has oxygen values near saturation, there are extensive midwater regions where oxygen is depleted; these typically occur between 100 and 1200 m depth. They are usually formed beneath highly productive, upwelled waters by degradation of organic matter. The resulting hypoxic zones, referred to as oxygen minimum zones (OMZs) and operationally defined as areas where $O_2 < 0.5$ ml/l, persist over geologic time. OMZs occur in much of the eastern Pacific Ocean, in the Arabian Sea, and off West Africa (Kamykowski & Zentara 1990). Certain deep basins (e.g., off southern California) and fjords also contain permanently hypoxic or anoxic waters. Where these low oxygen regions intercept the continental seabed, the benthos experiences either permanent hypoxia or an oxygen gradient, which may fluctuate daily with internal tides (e.g., Levin et al. 1991), interannually (e.g., with ENSO events, Arntz et al. 1991, Gallardo 1985), or over geologic time (den Dulk et al. 1998, Rogers 2000).

Sediments having oxygen-depleted overlying bottom water typically exhibit substantially reduced macrofaunal diversity. Within OMZs the macrofauna exhibit low species richness and very high dominance (Table 1). This pattern was first reported on the West African margin off Walvis Bay by Sanders (1969), and has since been observed in the eastern Pacific on a seamount off Mexico (Levin et al. 1991), on the Peru and Chile margins (Levin et al. unpublished data), and in the NW Arabian Sea off Oman (Levin et al. 1997b, 2000). Among the macrofauna, many molluscs, crustaceans, echinoderms, and cnidarians appear less tolerant of hypoxia than other taxa (Diaz & Rosenberg 1995), although there are exceptions (Levin & Gage 1998). No single taxon dominates the macrofauna of low oxygen settings, although annelid species are often prevalent. At upper slope depths, within the least oxygenated portions of OMZs, dominance of the most abundant species typically ranges from 40% to 85% (Levin & Gage 1998) (Table 1). Less information is available concerning the diversity responses to reduced oxygen concentrations of bacteria, small protists (nanofauna), meiofauna, or megafauna. Smaller organisms

TABLE 1 Community structure of macrofauna and foraminifera at hypoxic sites within oxygen minimum zones of the eastern Pacific and Indian Oceans

Location	Depth (m)	Bottom-water oxygen (ml/l)	Dominant taxon	R1D (%)	$E_{s\,(100)}$	H'	Density	Reference
Oman margin	400	0.13	Spionidae (Polychaeta)	63	5.6	1.45 ($\log_2$)	12,362 m^{-2}	Levin et al. 2000
Oman margin	412	0.13	Foraminifera >125 μm, 0–1 cm	27	23.9	1.23 ($\log_{10}$)	2533 10 cm^{-2}	Gooday et al. 2000
Santa Barbara Basin	550	0.06	Tubificidae (Oligochaeta)	44	N/A	1.77 ($\log_2$)	1,691 m^{-2}	Levin et al. unpublished
Santa Barbara Basin	590 / 610	0.05 / 0.15	Foraminifera >63 μm, 0–1 cm	49 / 25	8.9 / 21.4	0.72 ($\log_{10}$) / 1.10 ($\log_{10}$)	N/A	Gooday et al. 2000
Volcano 7, Mexico	750	0.08	Aplacophora (Mollusca)	47	10.8	2.59 ($\log_2$)	1,854 m^{-2}	Levin et al. 1991
Peru margin	300	0.02	Tubificidae (Oligochaeta)	83	4.7	0.84 ($\log_2$)	13,539 m^{-2}	Levin et al. unpublished
N. Chile margin (Iquique)	300	0.26	Dorvilleidae (Polychaeta)	73	N/A	N/A	1,834 m^{-2}	Levin et al. unpublished
Central Chile margin (Concepcion)	364	0.52	Amphinomidae (Polychaeta)	31	17.6	2.46 ($\log_2$)	14,206 m^{-2}	Gallardo et al. unpublished

living entirely within the sediments and with no access to the surface may be confined to hypoxic or even anoxic pore-waters, even when the overlying bottom water is oxic (Corliss & Emerson 1990, Jorissen et al. 1998). Yet foraminifera and a variety of larger metazoans (polychaetes, crustaceans, molluscs, echinoderms) all display abundance peaks close to OMZ boundaries (Mullins et al. 1985, Levin et al. 1991).

Where a range of faunal size groups have been compared, larger taxa (megafauna and macrofauna) exhibit density reductions within the most hypoxic portions of OMZs (Wishner et al. 1990, Levin et al. 1991, 2000) that are not evident in bacteria and metazoan meiofauna (Levin et al. 1991, Cook et al. 2000, Neira et al. in press). Metazoan meiofauna of the eastern Pacific OMZ exhibit loss of harpacticoid cope-pods, and dominance by nematodes (Levin et al. 1991, Neira et al. 2001), but diversities have not been recorded. Foraminiferal assemblages characterized by high dominance and few species are reported from modern, low-oxygen basins in the eastern Pacific and elsewhere (Phleger & Soutar 1973, Douglas et al. 1980, Hermelin & Schimmield 1990, SenGupta & Machain-Castillo 1993, Bernhard et al. 1997). The fossil record yields evidence for similar responses among ancient deep-sea faunas (e.g., den Dulk et al. 1998, Jorissen 1999). However, precise diversity data are rarely reported. At a site within the NW Arabian Sea OMZ, foraminifera exhibited reduced species richness and elevated dominance compared with assem-blages from a deep site below the OMZ (Gooday et al. 2000). These low-oxygen faunas consisted largely of small calcareous forms, a trend observed in the OMZ off Peru as well (A. Rathburn, unpublished manuscript). In contrast to fully oxic deep-sea sites, monothalamous and other delicate agglutinated and allogromiid taxa are rare compared to calcareous forms within OMZs (Gooday et al. 2000).

These limited deep-sea observations are consistent with field (Josefson & Widbom 1988, Murrell & Fleeger 1989, Radziejewska & Maslowski 1997, Luth & Luth 1997) and experimental (Moodley et al. 1997) studies conducted in shallow water that suggest that meiofauna, particularly nematodes and foraminifera, are less affected by hypoxia than the macrofauna, at least at higher taxonomic lev-els. There are, however, shallow-water examples in which nematode diversity is reduced by severe hypoxia (Keller 1986, Austin &Widbom 1991).

Historical records support a role for oxygenation in control of deep-sea diversity. In the northern Arabian Sea, foraminiferal diversity appears to have been strongly influenced during the Quaternary by changes in the flux and quality of organic matter to the seafloor, which led to variations in the thickness and intensity of the oxygen minimum zone (Hermelin & Shimmield 1995, den Dulk 2000). den Dulk et al. (1998) studied a 120,000 years-long Quaternary record from the Pakistan margin in the Northern Arabian Sea. Two foraminiferal assemblages were recognized. A low diversity assemblage with high dominance recurred every 23,000 years, possibly as a result of enhanced summer surface productivity and therefore intensified OMZ development, linked to the precessional component of orbital forcing. A more sustained period of low diversity occurred under glacial conditions, perhaps reflecting a strengthening of the NE monsoon, leading to higher winter productivity. In a detailed multiproxy study of shorter cores (spanning the last 30,000 y) from the same margin, von Rad et al. (1999) detected a switch from low to high foraminiferal diversity on the Pakistan margin during brief, late Quaternary to early Holocene climatic oscillations (Younger Dryas, Heinrich events 1 and 2) when surface productivity was unusually low and presumably the OMZ intensity diminished. Evidence for a close coupling between foraminiferal benthic community structure (including dominance) and bottom-water oxygenation is seen in the Santa Barbara Basin on even shorter time scales. There, fluctuations on decadal to millennial time scales are associated with major climate oscillations that change thermolhine circulation and ventilation (Cannariato et al. 1999).

During the Late Cretaceous, anoxia sometimes occurred on a much larger scale than in modern oceans, and caused widespread benthic mortality (Rogers 2000). Most of these events are far too ancient to have any residual effect on modern diversity patterns. A possible exception occurs in the eastern Mediterranean, which was subject to repeated episodes of basin-wide, deep-water anoxia during the Late Quaternary and most recently around 6000 years ago. Anoxic episodes, represented in the sedimentary record by clearly defined, dark horizons (sapropels), had a profound effect on the benthic foraminiferal faunas (Jorissen 1999). Distinct regional-scale differences in foraminiferal species richness and composition between the eastern and western Mediterranean basins may reflect these major disturbances in the recent geological past (de Rijk et al. 1999).

There is a difficulty with interpreting diversity responses to bottom-water oxygen conditions that derives from the tight linkage between bottom-water oxygen concentration and organic matter inputs in the deep sea (Levin & Gage 1998). Multiple regression analyses of a large macrofaunal data set from the Indian and eastern Pacific Oceans suggest that oxygen exerts a strong effect on species richness, although organic matter availability (evaluated from sediment POC) has a greater influence on dominance (Levin & Gage 1998). Together, these factors lower diversity within OMZs. Although food availability may affect diversity over a broad range of oxygen values, significant reduction of macrofaunal species richness by low oxygen may not occur until concentrations fall below 0.4 or 0.3 ml l^{-1} (Figure 3); this value may be even lower for annelids, the taxon most tolerant to hypoxia. Sulfides, which are toxic at high concentrations to most marine organisms (Bagarinao 1992), are associated with high inputs of labile organic matter, and also vary inversely with oxygen concentration. Their role in reducing diversity within OMZs has yet to be explored.

A major challenge is unraveling the relative importance of physiological stress and biotic interactions in creating deep-sea responses to low oxygen. Although stress may cause loss of species and dominate diversity responses at the lowest oxygen levels, facilitation may also be important. Animal structures such as tubes or burrows that conduct seawater into the sediments, or irrigation activities that mix solutes may enhance oxygen availability and dilute sulfides, facilitating colonization by other taxa. At higher oxygen levels, competitive abilities and predation may regulate diversity.

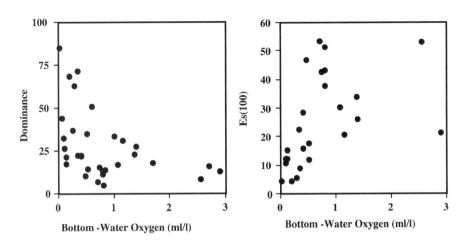

Figure 3   Macrofaunal Rank 1 dominance (% of the total accounted for by the top ranked species) and species richness ($E[S_{100}]$) plotted as a function of bottom-water oxygen concentration for bathyal stations within and beneath oxygen minimum zones in the eastern Pacific and northern Indian Oceans.

Deep-Sea Currents

Near-bottom flow rates in the deep ocean are typically a few cm/sec, too weak to erode the seabed (Munk 1970, Tyler 1995). The ocean floor is not uniformly quiescent, however, and numerous areas are subject to currents strong enough to erode and transport sediments and disturb soft-bottom communities on scales of 10s or 100s of km, (Heezen & Hollister 1971, Hollister et al. 1984). Episodic benthic storms characterize areas beneath western boundary currents where surface kinetic energy is transmitted through the water column to the sea floor; near-bottom currents reach speeds of 15–40 cm/sec and persist for several days (Gross & Williams 1991, Hollister & Nowell 1991, Aller 1997, Weatherley & Kelley 1985). First described in the HEBBLE area on the Nova Scotian continental rise, benthic storms occur along the western margins of the North and South Atlantic, around South Africa, and in regions around the Antarctic continent (Rowe & Menzies 1968, Flood & Shor 1988, Hollister & Nowell 1991, Richardson et al. 1993). Similar transient, high-energy episodes have been reported in abyssal areas distant from continental margins, for example in the NE Atlantic (Klein 1988) and NE Tropical Pacific (Koutar & Sokov 1994).

There are many other sources of strong current activity in the deep sea. Thermo-haline-driven bottom currents transport vast amounts of sediment around the deep ocean, depositing them as huge sediment drifts (contourites), for example, west of Scotland (Hollister et al. 1984, Stow & Holbrook 1984, Kidd & Hill 1986, Stow & Faugères 1993, Viana et al. 1998). In the Rockall Trough, current speeds >15 cm/sec are frequently recorded by current meters moored on the upper slope above 1000 m (Paterson & Lambshead 1995) with peak flows reaching 48 cm/sec (Viana et al. 1998). Along continental margins, more localized hydrographic phenomena such as internal tides, water column instability, and storm-driven eddies may create strong, erosional currents on the upper slope (Dickson & McCave 1986, Pingree & New 1989, Rice et al. 1990, Gage 1997). Complex interactions between steeply sloping topographic features and local hydrography create regions of intensified near-bottom flow on seamounts (Noble & Mullineaux 1989), as well as in canyons (Shepard et al. 1979, Gage et al. 1995, Gage 1997) where turbidity currents also may be active (Jorissen et al. 1994).

Near-bottom currents are among the agents of disturbance that can modify the structure and composition of benthic faunas (Hall 1994). As Levin et al. (1994a) emphasize, however, the critical factor may be sediment mobility, which depends on sedimentary characteristics as well as the current flow itself. Gage et al. (1995) and Gage (1997) demonstrate with rarefaction curves, rank abundance plots, and univariate diversity indices that polychaete diversity is highest at tranquil sites on the Tagus Abyssal Plain and in the central North Pacific, lower in the Rockall Trough and in the hydrodynamically active Sebutal Canyon, and much reduced at the HEBBLE site. Bivalve diversity exhibits a similar trend, although the differences are much less pronounced. Foraminifera at the main (Mid) HEBBLE site were significantly less diverse (H') than at the relatively tranquil shallow site (Kaminski 1985). Allen & Sanders (1996) attribute anomalously low protobranch bivalve mollusc diversity in the North American Basin to possible benthic storms. Paterson & Lambshead (1995) observed a strong linear relationship between the frequency of current velocities > 15 m/sec in the Rockall Trough and the equitability statistic V for polychaetes. Negative values of V, indicating high dominance, were associated with high current speeds on the upper slope in the Rockall Trough. As at HEBBLE, polychaete taxa generally considered to be opportunists predominated in parts of the Rockall Trough where physical disturbance was high. On Fieberling Guyot, Shannon-Wiener (H') values for macrofauna were very similar at contrasting sites with daily and infrequent, episodic sediment transport, but species richness measured by rarefaction was somewhat higher at the more stable SPR site (Levin et al. 1994a, see also Levin & DiBacco 1995).

In the case of meiofaunal taxa, strong erosive currents do not necessarily depress diversity. Thistle (1983) found no difference in harpacticoid copepod diversity in samples from the energetic HEBBLE site and the tranquil San Diego Trough. Contrary to his initial expectation, Thistle (1998) observed that harpacticoid copepod diversity was actually higher at the Fieberling Guyot WSS site, where the *Globigerina* sands are mobile on a daily basis, than at the more stable SPR site. Nematode diversity was very similar at three stations (545 m, 835 m, 1474 m depth) in the Rockall Trough and at 1050 m in the San Diego Trough, despite differences in current velocities both between these two basins and with depth within the Rockall Trough (Lambshead et al. 1994). Severe benthic storms do appear, however, to depress nematode species richness [measured as $E(S_{51})$] at the HEBBLE site compared with values from relatively tranquil localities on the Porcupine and Hatteras

Abyssal Plains (Lambshead et al. 2001). This may reflect a greater intensity of current disturbance at HEBBLE than in the Rockall Trough.

The observations reviewed above suggest that currents can modify benthic diversity both locally and regionally, although the mechanisms involved are not well understood. The effects can be either positive or negative, and there are indications that macrofauna are more strongly impacted than meiofauna. Possible mechanisms can be divided into those having a direct impact and those having an indirect impact on diversity. Strong currents may depress diversity directly by eroding surficial sediments and carrying away the animals living in them (Aller 1997). Observations at the HEBBLE site (Thistle et al. 1985, 1991, Lambshead et al. 2001) and the WSS site subject to daily sediment transport on Fieberling Guyot (Levin & DiBacco 1995) suggest that episodic disturbance by erosive flow creates repeated opportunities for recolonization. Constant reworking of the sediment ensures that the benthic fauna remains in an early successional state, favoring opportunists, and keeping diversity low.

Near-bottom currents have the potential to impact benthic faunal diversity in a variety of indirect ways. Moderate currents can enhance the food supply by delivering organic matter and stimulating bacterial production (Thistle et al. 1985, Aller 1989). As long as these inputs are not excessive, they should lead to an increase in both abundance and diversity of macro- and meiofauna. Currents may also entrain larval and subadult organisms, allowing animals to colonize disturbed patches of sediment and enhance local diversity. Moderate flow conditions may potentially increase sediment heterogeneity by creating sedimentary structures and by concentrating organic matter in localized patches. At the White Sand Swale (WSS) site on Fieberling Guyot, active ripples generate heterogeneity by successively burying and exposing organic matter accumulated in the troughs. This process may explain why harpacticoid copepod diversity was higher at WSS than at the more stable Sea Pen Rim site (Thistle 1998). When flow velocities were enhanced experimentally at WSS for a 6-week period, however, there was no detectable change in harpacticoid or total macrofaunal diversity or equitability, possibly because the assemblages were already adapted to strong erosive flow (Levin et al. 1994a, Thistle & Levin 1998).

As indicated above, near-bottom flows in excess of 20–25 cm/s can potentially depress diversity by eroding epifaunal species, but they may also impact diversity indirectly by smoothing out and reducing physical heterogeneity. On a regional scale, erosive bottom currents will tend to homogenize the fauna by dispersing juveniles and subadults. Given the large areas of seafloor swept by erosive currents (Hollister et al. 1994, Hollister & Nowell 1991), the impact of hydrodynamics on regional deep-sea species diversity and biogeography may be considerable.

Hydrodynamic processes are probably involved in the creation of larger habitat patches that increase seafloor heterogeneity. On the upper slope around the NW European continental margin, interactions between the seafloor and internal tides and waves are apparently linked to conspicuous concentrations of hexactinellid sponges (Rice et al. 1990) and corals (Fredrickson et al. 1992). In the Porcupine

Seabight (1000–1300 m water depth), sponges and spicule mats derived from them enhance the abundance and modify the taxonomic composition of the macrofauna (Bett & Rice 1992). Both corals and sponges have numerous associated organisms (Klitgaard 1995, Jensen & Frederiksen 1992, Bartel & Gutt 1992). However, at least in the NE Atlantic, most of the associated organisms are also present in the background community (Klitgaard 1995, Jensen & Frederiksen 1992), suggesting that regional diversity may not be increased substantially by the presence of these large, habitat-creating organisms.

Catastrophic Disturbance

Over geological time scales, continental margin sediments have been disrupted by gravity-driven mass movements, including slumps, slides, debris flows, and turbidity currents (Masson et al. 1994, 1996). At their distal extremities, disturbance by turbidity currents is probably similar to that caused by severe erosive currents such as benthic storms. In the Atlantic Ocean, mass movements are well documented off NW Europe, NW Africa, southern Africa, the United States between New York and Cape Hatteras, and Brazil (Emery & Uchupi 1984). Off NW Africa debris flows, particularly the unusually large Canary and Saharan flows (Jacobi & Hayes 1982, Simms et al. 1991, Embley 1976, Masson et al. 1994, Masson 1996), have displaced 600 km^3 of sediment on the upper continental rise. Elsewhere in the NE Atlantic, the mid-Norwegian margin (the "Storegga" area; 62°N) is notable for a series of enormous slides, the most recent of which occurred about 7000 years ago (Bugge et al. 1987). These slides, and associated debris flows and turbidity currents, have transported 6000 km^3 of sediment from the shelf to depths of 3500 m over horizontal distances of 800 km. The flanks of volcanic islands such as the Hawaiian Islands in the Pacific and the Canary Islands in the Atlantic are prone to catastrophic collapses that give rise to massive avalanche deposits on the adjacent deep-sea floor (Lipman et al. 1988, Moore et al. 1989, Cochonat et al. 1990, Masson 1996). These mass movements of sediments are often associated with turbidite deposition, an important mechanism in the formation of abyssal plains adjacent to continents. The best known modern turbidity flow occurred in 1929 when an earthquake on Grand Banks, Newfoundland, triggered a sediment slump that developed into a turbidity current extending 800 km from its source across the abyssal plain. More recently, Thunnell et al. (1999) provided the first direct, real-time documentation of earthquake-generated suspended sediment flows.

Although major mass movements of sediment must devastate the benthic fauna, most of these events occurred thousands of years ago and are unlikely to affect modern faunas directly. Indirect effects, however, may persist for much longer. Debris avalanches and similar chaotic deposits will introduce long-lasting physical heterogeneity on spatial scales up to kilometers (Masson 1996). Turbidites have granulometric characteristics and total organic carbon (TOC) values that differ from those of pelagic sediments (Huggett 1987) and can potentially influence diversity. There is evidence for such an effect at a location on the Madeira Abyssal Plain (MAP; 4950 m water depth) that was swept by a turbidite about 1000 years ago. Here, polychaetes exhibit lower abundance, much lower species richness [expressed as $E(S_n)$ and per unit area], and higher dominance compared to other abyssal NE Atlantic sites, including the equally oligotrophic EUMELI site, where sedimentation is entirely pelagic (Glover et al. 2001). The sedimentary characteristics of the turbidite deposit, which is overlain by only a thin veneer of pelagic sediment, may have confined recolonization to a relatively small suite of opportunistic polychaete species. Nematodes, which lack a dispersive larval phase, also exhibit lower diversity [$E(S_{51})$] at the MAP than at nonturbidite sites on the Porcupine and Hatteras Abyssal Plains (Lambshead et al. 2001b in press). Foraminiferal diversity, on the other hand, is not noticeably depressed at the MAP (Gooday 1996). There is no evidence for low nematode diversity at a site in the Venezuela Basin (5054 m depth) subject to periodic turbidite impacts. This is probably because the sedimentation rate ($7.2\,\text{cm} \cdot \text{ky}^{-1}$) is much higher and the turbidites are older in the Venezuela Basin than in the MAP (Lambshead et al. 2001b in press).

Submarine and subaerial volcanic eruptions may impact deep-sea benthic faunas directly, through deposition of lava or ash layers (Cita & Podenzani 1980), and indirectly through changes in climate and water column stratification (Genin et al. 1995). Fossil foraminiferal evidence from the eastern Mediterranean suggests that the benthic fauna was obliterated by an ash deposit 35,000 years ago but that a very similar assemblage reestablished rapidly (Cita & Podenzani 1980). We are aware of only one species-level study of the effect of a modern ashfall in the deep sea on a benthic taxon (Hess & Kuhnt 1996). During 1991, an ash layer >2 cm thick was deposited over 36,000 km^2 in the South China Sea following the eruption of Mt. Pinatubo, Philippines. Rarefaction curves based on data from Hess & Kuhnt (1996) indicate that foraminiferal assemblages were severely affected by a 6 cm ash deposit at 2503–2506 m; most foraminifera died, although epifaunal specimens found below the ash layer may have survived for some time in a starved, quiescent state. After almost three years, the ash layer had been recolonized by a low-diversity assemblage of infaunal opportunists. In a 2-cm deep ash deposit at 4226 m, however, many mobile infaunal taxa survived and species richness decreased only slightly.

MEASURING THE LEVEL OF DEEP-SEA SPECIES DIVERSITY

Very basic questions remain about how diversity should be measured and interpreted. A wide variety of diversity measures exist that incorporate both richness and evenness but differ in how these components of diversity influence the magnitude of the index (Magurran 1988). Because most diversity indices are sensitive to both evenness and richness, differences can reflect changes in either or both; changes in evenness should not be interpreted as changes in richness. The choice of an index depends on the nature of the question and the type of data available. In most cases, it will be useful to use a variety of indices that together provide greater insight into how and why diversity varies than does a single index.

The most widely used measure of diversity in deep-sea ecology is $E_{(Sn)}$, the Expected Number of Species (Sanders 1968, Hurlbert 1971). This index estimates the number of species in samples normalized (rarefied) to successively smaller sample sizes, providing an interpolated curve of the relationship between the number of species (S) and the number of individuals (N) based on resampling the relative abundance distribution of an actual sample. It is sensitive to both evenness and richness (Gage & May 1993), and was originally designed to allow comparisons of nonquantitative samples taken by epibenthic sleds. The sleds are towed from a surface vessel for a variable distance of about a kilometer and are apt to bounce along the bottom. The only way to compare diversity among such samples is by normalizing them to the same number of individuals, typically the lowest common number of individuals. Quantitative coring devices (Hessler & Jumars 1974) sample a precise area of bottom and accurately measure faunal density. Should quantitative samples be rarefied? This is problematic if the habitats sampled differ greatly in density, which they often do. For example, a boxcore may yield 10,000 individuals from a continental shelf habitat, but fewer than 100 at greater bathyal depths. Normalizing the two samples to 100 individuals effectively compares the number of species at 9 cm^2 on the shelf to 900 cm^2 in the deep sea. This procedure will mask the differences in density between sites, which may be critical for understanding why diversity differs. Because no shallow-water samples may yield as few as 100 individuals, the number of species is predicted from the lower portion of the species-individuals curve where the slope is steepest and the error for estimating S is largest. Finally, because rarefaction assumes a random distribution between S and N, replicate samples from a site are often pooled (e.g., Grassle & Maciolek 1992, Etter & Grassle 1992). This obscures the actual relationship between S and N at the sampling scale, and how differences in heterogeneity among sites may contribute to variation in diversity. Rarefaction has been an important and useful tool for quantifying and comparing community structure, but for a comprehensive understanding of regional patterns of diversity it should be used in combination with other measures.

A better approach for comparing regional species richness with quantitative samples is species accumulation curves with randomly pooled samples (Etter & Mullineaux 2000). This requires replicate samples at each location, but maintains the relationship between S and N, which preserves differences in heterogeneity and allows one to compare diversity based on the number of individuals or area sampled. Most importantly, it can be used to estimate the asymptotic S (the number of species in the community if it were completely sampled) and how well the community has been sampled (the position on the species accumulation curve), both of which are crucial for comparing and interpreting differences in S. Randomized species accumulation curves will typically be lower than rarefaction curves (Figure 4) because they maintain the nonrandom distribution of S and N at the scale of the sampling device. The difference between the two curves reflects the amount of heterogeneity at the scale of the sampler. Two programs that allow one easily to

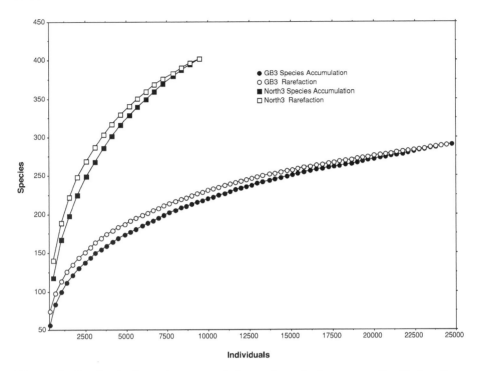

Figure 4   Randomized species accumulation and rarefaction curves for Station 3 on Georges Bank (GB3, 100 m, 40°53.7'N, 66°46.5'W) and Station 3 from the North ACSAR (Atlantic Continental Slope and Rise) (North 3, 1350 m, 41°1.40'N, 66°20.20'W). The randomized species accumulation curves are the average cumulative number of species from 50 randomizations of pooling successively larger numbers of replicates at each station. Data are from Maciolek et al. (1985) and Maciolek et al. (1987b).

compute randomized species accumulation curves are Estimate S (Colwell 1997) and Rosenzweig's (1995).

Gray (1994) and Gray et al. (1997) provide a recent example of the problems encountered when comparing diversity among habitats. Benthic samples from a variety of shallow-water habitats (<200 m) were compared to those collected from the deep northwest Atlantic by Grassle & Maciolek (1992) to argue that shallow-water and deep-sea communities may be quite similar in diversity. However, these comparisons were confounded by the shallow-water samples being collected from broader geographic areas (Gray 1994) at different latitudes and by using different sieve sizes (1 mm rather than 300 μm), both of which can influence diversity (Bachelet 1990, Warwick & Clarke 1996, Rex et al. 1993). More importantly, because estimates of species richness are highly sensitive to sampling effort (May 1975, Colwell & Coddington 1994, Rozenzweig 1995), comparisons should be made only at the asymptote of a species accumulation curve. Although the shallow-water samples of Gray et al. (1997) approach an asymptote, the deep-sea samples show no sign of leveling off (Grassle & Maciolek 1992). The problem of assessing diversity when no asymptote has been reached is often overlooked in spatial and temporal comparative studies of richness. Rarefying the samples to a common number of individuals does not eliminate the sensitivity of richness estimates to sampling effort. For example, samples of 100 individuals in one community may contain 90% of the species, although another may contain only 10%. The only way to overcome this potential artifact is to restrict comparisons of S to the asymptotic S.

To reexamine the question of whether deep-sea habitats support more species than shallow-water habitats, we can compare diversity between a series of deep-water samples collected off the coast of Massachusetts [the North data collected as part of the Atlantic Continental Slope and Rise Study (ACSAR)] (Maciolek et al. 1987b) to a very similar set of samples collected from the nearby shallow waters of Georges Bank (Maciolek et al. 1985). This comparison was selected because the shallow and deep samples are geographically adjacent (both just off the coast of Massachusetts), were collected in a similar way, were sorted by sieves of the same size, have consistent taxonomy (done by the same individuals), and represent the entire macrofaunal community. The Georges Bank samples were collected with 0.04 m^2 Ekman grabs from 38 to 167 m. The ACSAR samples were collected with 0.25 m^2 boxcores from 250 to 2500 m, but only the center 0.09 m^2 was used for community analysis. No other shallow- and deep-water data bases are as comparable.

For the deep-water samples (those >200 m), on average 278 species coexist in an area of 1 m^2 and the expected number of species was 156 when normalized to 1000 individuals (ES_{1000}). In contrast, the shallow-water samples produced an average of only 165 species m^{-2} and 68.8 species per 1000 individuals (Table 2).

TABLE 2   Average measures of diversity and density for the Georges Bank and the ACSAR samples. $E_{(S1000)}$ are Hurlbert's (1971) expected number of species normalized to 1000 individuals. ACSAR North (130 m) are averages based only on samples collected between 1220–1350 m from the North. Data are from Maciolek et al. (1987b) and Maciolek-Blake et al. (1985)

	Georges Bank	**ASCAR North**	**ASCAR North (130 m)**
Samples	1149	191	63
Depth Range (m)	38–167	250–2180	1220–1350
Individuals	680,600	95,140	27,906
Species m^{-2}	165	278	319
$E_{(S1000)}$	68.8	156	188
Shannon Wiener Index ($\log_2$)	4.09	5.59	6.70
Species Richness	680	952	599

This comparison of averages is conservative because the deep-sea samples span a 2000-m depth gradient and include relatively shallow depths that are lower in diversity. If the deep-sea samples are restricted to a 130-m depth interval (similar to the range of Georges Bank samples) centered at 1300 m, the differences are even more pronounced (Table 2). Randomized species accumulation curves based on either individuals or area also indicate that diversity is greater in the deep-sea samples (Figure 5). Asymptotic S values estimated from the species accumulation curves at each site using Chao1 (Colwell & Coddington 1994) indicate that richness varies unimodally with depth (Figure 6), as suggested previously based on rarefied qualitative (Rex 1981, 1983) and quantitative samples (Etter & Grassle 1992).

Pooling the ACSAR samples and Georges Bank samples by depth produces estimates of species richness that are quite similar (Table 2), as Gray [(1994), Gray et al. (1997)] suggested for his studies. In fact, when pooled over 130 m, richness appears to be greater on Georges Bank, but the species accumulation curves (Figure 7) clearly show that the shallow-water estimates are near an asymptote

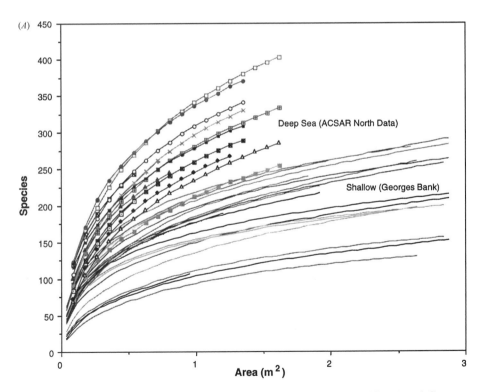

Figure 5   Randomized species accumulation curves for all North (ACSAR) and Georges Bank stations as a function of (A) area sampled and (B) numbers of individuals. The Georges Bank curves lack symbols while the North stations have symbols. Data are from Maciolek et al. (1985) and Maciolek et al. (1987b).

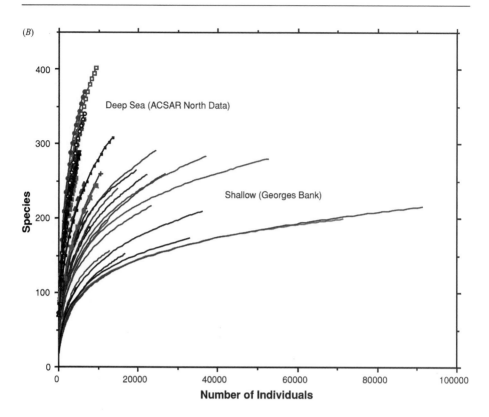

Figure 5   (*Continued*)

although the deep-sea samples are not. This demonstrates the fallacy of comparing richness among locations while ignoring how well each community has been sampled (location on species accumulation curve). When comparisons are restricted to contiguous geographic regions, sampled in similar ways, as are the Georges Bank and North (ACSAR) data, diversity (richness, species/area, species/individual) is much lower on the shelf than at bathyal depths. Although the diversity of shallow-water communities in some parts of the World Ocean may equal or exceed that of the deep sea, if comparisons are not controlled geographically, taxonomically, by habitat, and by sampling methods, they can tell us little about what forces are shaping species diversity in marine ecosystems.

INTERRELATIONSHIPS OF LOCAL AND REGIONAL DIVERSITY

Spatiotemporal variation in deep-sea species diversity represents an integration of ecological and evolutionary processes that operate at different spatial and temporal scales (Figure 8). Smaller-scale processes are embedded hierarchically within

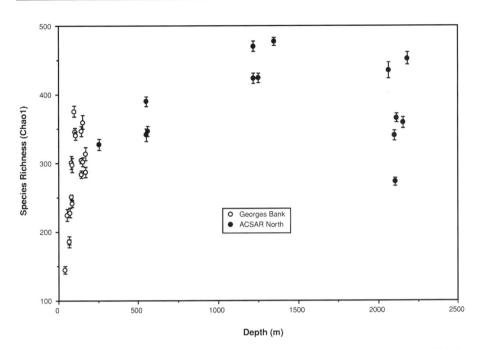

Figure 6   Asymptotic species richness estimated by Chao1 (Estimate S [Colwell 1997]) as a function of depth for Georges Bank and ACSAR North stations. Error bars are 95% confidence limits. Data are from Maciolek et al. (1985) and Maciolek et al. (1987b).

larger-scale processes, and tend to occur at faster rates. Species within a local assemblage (1–10 m²) are controlled by small-scale processes involving resource partitioning, competitive exclusion, predation, facilitation, physical disturbance, recruitment, and physiological tolerances, all of which are mediated by the nature and degree of heterogeneity. How these processes might regulate local diversity has been reviewed elsewhere (Etter & Mullineaux 2000, P. V. Snelgrove & C. R. Smith, unpublished manuscript), but they will tend to occur on much shorter times scales than landscape or regional-level changes examined here. At regional scales of 100s to 1000s of m, several environmental gradients, dispersal, metapopulation dynamics, and gradients in habitat heterogeneity are likely to be important (Figure 8). We suggest that the environmental gradients discussed in this paper essentially control variation in local diversity by accelerating or decelerating local processes (e.g., patch dynamics). We describe in more detail below how each gradient may influence these local processes. Dispersal among patches will also be important at landscape or regional scales because it determines the potential for membership in local assemblages and plays a critical role in metapopulation dynamics and species persistence at local and larger scales (Caswell & Cohen 1993, Caswell & Etter 1999). For instance, low dispersal rates can decrease the intensity of biological interactions (Menge & Sutherland 1987) and potentially foster coexistence through recruitment limitation (Tilman 1994, Hubbell et al. 1999). At regional

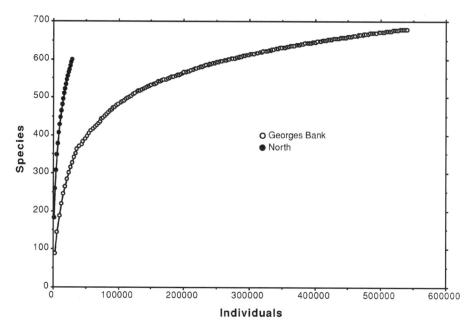

Figure 7 Species accumulation curves for pooled samples over a depth range of 130 m for Georges Bank and ACSAR North studies. The North samples were pooled from stations between 1220–1350 m to compare to a similar depth span in the Georges Bank study. All Georges Bank samples were used (38–167 m). See Table 2 for other diversity metrics comparing these pooled samples.

scales, many species will exist as metapopulations; their presence locally or regionally will be determined by local dynamics, dispersal among patches, and the degree of reproductive asynchrony among populations (e.g., Hanski 1998). Variation in local diversity within regions is also likely to be influenced by the number and different types of habitats at intermediate scales (e.g., 1–10 km) through a mass effect (Shmida & Wilson 1985). We know little about how mass effects contribute to local diversity, but the large number of singletons (species represented by a single individual) within natural deep-sea assemblages suggests that it may be extensive.

It is clear from the foregoing discussion that local diversity varies regionally along environmental gradients in the deep sea. How the numerous environmental factors act to control local diversity is not well understood, but we have developed a conceptual model of how they may directly and indirectly regulate the number of species within communities (Figure 9). These environmental factors do not work in isolation; they are often very interdependent such that changes in one precipitate changes in others, producing cascade effects, all of which modify the rates of local processes.

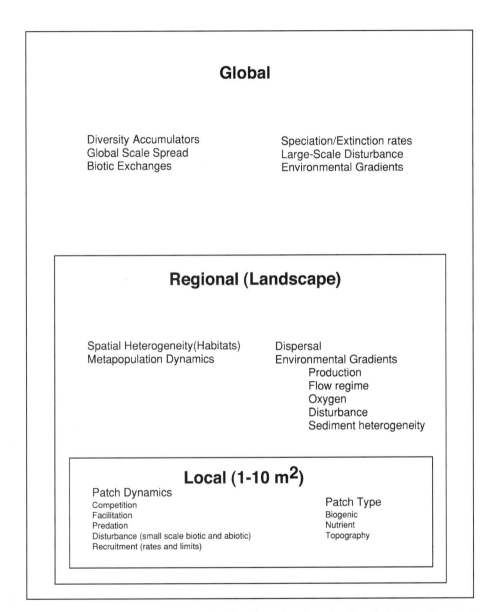

Figure 8 Processes regulating species diversity at local, regional, and global scales. Each box represents one scale and is embedded within the larger scales.

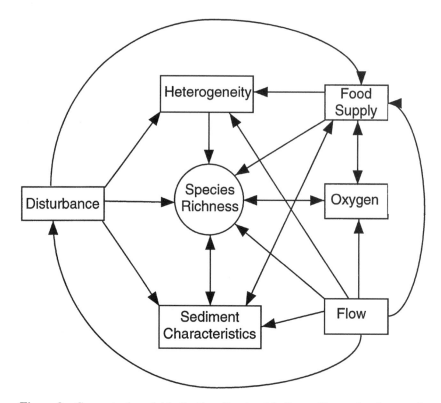

Figure 9 Conceptual model indicating direct and indirect effects of various environmental factors on species richness of local communities.

Near-bottom flow has pervasive effects; it can influence the number of species directly as well as indirectly through modification of other environmental factors (Gage 1997). Near-bottom current speed can directly control species number through larval supply, and at sufficiently high speeds will erode the sediments and many of the organisms on or within them. As current speeds vary, they can also indirectly influence diversity by changing food supply (POM flux and bacterial production), sediment characteristics, oxygen levels above and within the seabed, spatial heterogeneity, and levels of biotic and abiotic disturbance (Figure 9). Each of these factors can also affect the number of species within communities directly as well as indirectly.

Hydrodynamic influences on diversity can be extensive and involve numerous paths of different length and intensity. A unimodal relationship between diversity and flow strength is predicted (Figure 10), but has not been well documented. Low flows may be associated with depositional regimes and reduced sediment heterogeneity. Very strong flows are expected to reduce effectiveness of predators, and eliminate heterogeneity in particle size and surface features such as ripples.

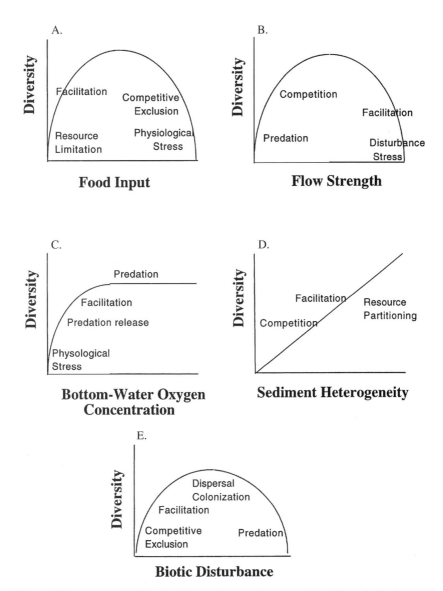

Figure 10 Patterns of diversity change along environmental gradients in the deep sea and biotic interactions hypothesized to be responsible for generating diversity patterns. (A) Productivity, (B) Flow, (C) Bottom-water oxygen concentration, (D) Sediment heterogeneity, (E) Biotic disturbance.

Disturbance resulting from substrate instability or the stress of maintaining feeding activities and dwelling position are likely to reduce species richness and evenness when currents are sufficiently strong to resuspend sediments, although these effects may be ameliorated by large or stabilizing biogenic structures. Intermediate or variable flows are expected to promote maximal diversity through resource partitioning that is maximized by interactions of food and larval supply (particle flux) and sediment heterogeneity.

POM flux or food supply may ultimately play the most significant role in regulating the number of species. As with flow, effects may be direct or be mediated through currents, sediments, disturbance, or hydrologic properties (Figure 9). At low food levels, diversity will be low because there are insufficient resources to support viable populations of many species. As food supply increases, diversity will increase because more species can maintain viable populations. As food availability continues to increase, diversity may decline, producing a unimodal relationship between diversity and food inputs (Figure 10). Why diversity declines at high levels of food supply is not well understood but may reflect a decrease in habitat heterogeneity, differential numerical responses leading to increased dominance by a few species (Rosenzweig & Abramsky 1993, Waide et al. 1999), and/or increased physiological stress due to oxygen limitation or sulfide toxicity. A strong negative correlation of evenness with sediment POC (Levin & Gage 1998) may provide some clues about the importance of species interactions in reducing diversity. Because increases in food supply are often coupled with variability (seasonality) in supply, decline may also reflect increased extinction rates of more specialized forms, due to either demographic instabilities or accelerating rates of competitive exclusion (Rex 1976).

If deposit feeders select sediment particles according to characteristics such as their size, shape, composition, and quality, as has been argued by Wheatcroft et al. (1990), Etter & Grassle (1992), and Smith et al. (1993), diversity will respond positively to increased sediment heterogeneity (Figure 10). Any environmental or biotic factor that alters sediment structure (hydrodynamics, bioturbation, disturbance) has the potential to regulate diversity indirectly. The distribution of particle sizes within the sediments also influences organic content (Milliman 1994); thus changes in diversity due to alterations of sediment heterogeneity may be mediated through food resources.

Oxygen is a requirement for most metazoan life; reduced oxygen availability can exert both direct and indirect negative effects on diversity (Figures 9, 10). At low oxygen concentrations, species richness may be reduced directly through differential tolerance to hypoxia or to the sulfides that build up in organic-rich hypoxic sediments (Diaz & Rosenberg 1995, Bagarinao 1992). Hypoxic conditions may impose habitat homogeneity, contributing to spatially uniform assemblages of low species richness (Levin et al. 2000). Such effects probably occur mainly at bottom-water oxygen concentrations $\ll 0.5$ ml/l (Levin & Gage 1998). Differential numerical responses and competitive exclusion may explain increased dominance and reduced species richness where oxygen is low but food availability remains

high. Tolerance to hypoxia varies with body size; meiofauna are much more tolerant to low oxygen than most macrofauna or megafauna. Organisms that oxygenate
sediments or reduce sulfide concentrations through feeding, dwelling structures,
and burrowing may indirectly facilitate other organisms, ameliorating physiological stress and potentially elevating diversity (e.g., Reise 1982, Aller & Aller 1986).
Predation pressure and its potential diversifying effects (Hall et al. 1994, Dayton
& Hessler 1972) are expected to increase with increasing oxygen saturation.

The processes of disturbance and positive interactions (facilitation) pervade the
environmental factors considered above and merit further consideration. Disturbance, a well-known mechanism regulating community structure in other ecosystems (Connell 1978, Petraitis et al. 1989, Pickett & White 1985), undoubtedly plays
a similar role in the deep sea. Disturbance to the seafloor is manifested through a
variety of biotic and abiotic factors that can affect diversity by regulating levels
of competition, predation, and physiological stress. Indirectly, disturbance can increase heterogeneity at low levels and reduce it at high levels by smoothing out
biogenic or topographic structures or by inducing stress. Disturbance can also alter
the structure of the sediments and thus can influence diversity indirectly through
the generation or removal of habitat heterogeneity (Figure 9).

Facilitative interactions, like disturbance, can be important over a range of conditions and are predicted to have strong diversifying effects where conditions are
most extreme (Bertness & Callaway 1994). We hypothesize that these effects are
especially important in marine sediments, where animal feeding, dwelling, and
respiratory activities alter the surrounding milieu. When food is scarce, animal
activities that enhance or redistribute food may elevate species richness (Bianchi
et al. 1989, Levin et al. 1997a). Stress or disturbance associated with low redox
conditions may be ameliorated by infaunal activities (solute pumping, burrowing)
that oxygenate sediments (Aller 1982), thereby allowing more species to persist
(Reise 1982). Structures formed by animals can provide shelter and a stable substrate, and may serve as hotspots of diversity under strong flow conditions (e.g.,
tests of xenophyophores on seamounts, Levin et al. 1988).

CONCLUSIONS

Two important conclusions result from our survey of diversity patterns at regional
scales. First, the deep-sea ecosystem, long known to support high local species
diversity, also shows high biodiversity at the landscape level. High diversity is
related to environmental gradients and habitat shifts. Second, there appear to be
multiple forces that shape patterns of diversity on regional scales. Variation in
species diversity is associated with large-scale variation in sediment grain size diversity, nutrient input, and productivity as well as oxygen availability, hydrologic
conditions, and catastrophic events. Disturbance and facilitation are predicted to
be agents of particular importance in mediating environmental effects on diversity in the deep sea. These environmental factors are highly interdependent and

ultimately reflect global-scale oceanographic processes and climate. It seems likely that regional-scale phenomena interact in complex ways to form biogeographic patterns of local diversity by controlling gradients of ecological opportunity that affect population dynamics at small scales. Exactly how this occurs is uncertain, but, as shown here, the basic features of the regional ecological system are becoming clearer. Importantly, the recognition of conspicuous variation in biodiversity at the landscape level and the discovery of how relevant processes are geographically distributed now make it possible to design controlled sampling studies and manipulative, experimental approaches to better understand the causality underlying regional diversity trends.

Although several large data sets on the composition of deep-sea communities are available, taxonomic and geographic coverage is still very limited. Most large-scale studies of deep-sea species diversity have taken place in the Atlantic. It is reasonable to expect that new biogeographic patterns and ecological forces of fundamental importance will continue to be discovered. The vast majority of diversity studies center on the macrofauna and are often restricted to one or a few taxa. Much less is known about the meiofauna and megafauna. Meiofauna and annelids often demonstrate parallel trends along gradients of oxygen and organic matter (e.g., Cook et al. 2000, Gooday et al. 2000, Gooday et al. 2001), although other groups may respond to environmental pressures in different ways (Lambshead et al. 2001a in press). Thus, whereas recent regional studies provide important direction for mechanistic studies of diversity in the deep sea, documenting the structure and function of whole deep-sea communities on a planet-wide basis remains an important priority.

Sampling methodology and the measurement of species diversity have important implications for conceptualizing patterns of deep-sea diversity and how they compare to those of other environments. Rarefaction, when critically applied, can be a useful way to estimate relative diversity from samples collected by large qualitative trawls and epibenthic sleds. Because the deep-sea fauna is sparsely distributed, large samples are vital for taxonomic discrimination of species and for assessing regional species pools. However, quantitative sampling with sufficient replication to adequately characterize local communities is necessary to measure the actual number of species, standing stock, and the environmental parameters that appear to influence diversity directly. The first controlled comparison of this kind, between the continental shelf and upper bathyal zone, confirms that the deep sea in the Northwest Atlantic region supports considerably higher species richness than in shelf communities. Although the shelf fauna is fairly well sampled, the level of replication at bathyal depths is still inadequate to reveal an asymptote in the species accumulation curve. An important objective for future sampling designs is to determine the level and spatial scale of replication necessary to represent local diversity. Also, because rarefaction of large qualitative samples suggests a marked decline in diversity at abyssal depths (e.g., in the western North Atlantic), quantitative sampling should be extended to the lower bathyal zone and abyssal plain to firmly establish a diversity-depth gradient.

Contemporary ecological theory is based exclusively on the history of observations made in terrestrial, aquatic, and surface marine systems. Most of Earth's largest environment, the deep sea, remains unexplored. It is a vast world of extreme physical circumstances, novel adaptations, and very basic permutations of the carbon cycle that support all life (Gage & Tyler 1991, Rowe & Pariente 1992, Van Dover 2000). Explaining the extraordinarily high level of species coexistence encountered at very small scales in the deep sea remains a fundamental problem (P. V. Snelgrove & C. R. Smith, unpublished manuscript; Etter & Mullineaux 2000). Although the regional studies reviewed here represent scattered areas of the deep sea, they significantly enrich our knowledge of global biodiversity at the landscape level, and should ultimately contribute to a comprehensive theoretical understanding of community structure and function.

ACKNOWLEDGMENTS

The Deep-Sea Biodiversity Working Group is supported by the National Center for Ecological Analysis and Synthesis, a Center funded by the National Science Foundation (DEB-94–21535), the University of California, Santa Barbara, and the State of California. The authors' research is supported by the National Science Foundation grants OCE-9811925 (RJE and MAR), OCE 98-03861 (LAL), OCE 99–86627 (JP), OCE-9811925 (RJE), OPP 98–15823 (CRS), and the NOAA National Undersea Research Program grant UAF 00–0050 (LAL). We thank D. Morrow for assistance with manuscript preparation.

Visit the Annual Reviews home page at www.AnnualReviews.org

LITERATURE CITED

Allen JA, Sanders HL. 1996. The zoogeography, diversity and origin of the deep-sea protobranch bivalves of the Atlantic: the epilogue. *Prog. Oceanog.* 38:95–153

Aller J, Aller R. 1986. Evidence for localized enhancement of biological activity associated with tube and burrow structures in deep-sea sediments at the Hebble site; western North Atlantic. *Deep-Sea Res.* 33:755–90

Aller JY. 1989. Quantifying sediment disturbances by bottom currents and its effect on benthic communities in a deep-sea western boundary zone. *Deep-Sea Res.* 36:901–34

Aller JY. 1997. Benthic community response to temporal and spatial gradients in physical disturbance within a deep-sea western boundary region. *Deep-Sea Res. Part 1 - Oceanog. Res.* 44:39

Aller RC. 1982. The effects of macrobenthos on chemical properties of marine sediment and overlying water. In *Animal-Sediment Relations*, ed. PL McCall, MJ Tevesz, pp. 53–102. New York: Plenum

Arntz WE, Tarazona J, Gallardo VA, Flores LA, Salzwedel H. 1991. Benthos communities in oxygen deficient shelf and upper slope areas of the Peruvian and Chilean Pacific coast, and changes caused by El Niño. In *Modern and Ancient Continental Shelf Anoxia*, ed. RV Tyson, TH Pearson, pp. 131–54. Tulsa, OK: Geol. Soc. Spec. Publ. No. 58

Austin MC, Widbom B. 1991. Changes in and

slow recovery of a meiobenthic nematode assemblage following a hypoxic period in the Gullmar Fjord basin, Sweden. *Mar. Biol.* 111:139–45

Bachelet G. 1990. The choice of sieving mesh size in quantitative assessment of marine macrobenthos: a necessary compromise between aims and constraints. *Mar. Environ. Res.* 30:21–35

Bagarinao T. 1992. Sulfide as an environmental factor and toxicant: tolerance and adaptations in aquatic organisms. *Aquat. Toxicol.* 24:21–62

Bartel D, Gutt J. 1992. Sponge associations in the eastern Weddell Sea. *Antarct. Sci.* 4:137–50

Berelson WM, Anderson RF, Dymond J, Demaster D, Hammond DE, et al. 1997. Biogenic budgets of particle rain, benthic remineralization and sediment accumulation in the equatorial Pacific. *Deep-Sea Res. II* 44:2251–82

Bernhard JM, Sen Gupta BK, Bourne PF. 1997. Benthic foraminiferal proxy to estimate dysoxic bottom-water oxygen concentrations: Santa Barbara Basin, U.S. Pacific continental margin. *J. Foraminifer. Res.* 27:301–10

Bertness MD, Callaway R. 1994. Positive interactions in communities. *TREE* 9:191–93

Bett BJ, Rice AL. 1992. The influence of hexactinellid sponge (*Pheronema carpenteri*) spicules on the patchy distribution of macrobenthos in the Porcupine Seabight (bathyal NE Atlantic) *Ophelia* 36:217–26

Bianchi TS, Jones CG, Shachak M. 1989. Positive feedback of consumer populations on resource supply *Trends Ecol. Evol.* 4:234–38

Blair NE, Plaia GR, Boehme SE, DeMaster DJ, Levin LA. 1994. The remineralization of organic carbon on the North Carolina continental slope. *Deep-Sea Res.* 41:755–66

Blake JA, Hilbig B. 1994. Dense infaunal assemblages on the continental slope off Cape Hatteras, North Carolina. *Deep-Sea Res.* 41:875–900

Brown JH, Stevens GC, Kaufman DM. 1996. The geographic range: Size, shape, boundaries, and internal structure. *Annu. Rev. Ecol. Syst.* 27:597–623

Brown TN, Kulasiri D. 1996. Validating models of complex, stochastic, biological systems. *Ecol. Model* 86:129–34

Bugge T, Befring S, Belderson RH, Eidvin T, Jansen E, et al. 1987. A giant three-stage submarine slide off Norway. *Geo-Marine Lett.* 7:191–98

Cannariato KG, Kennett JJP, Behl RJ. 1999. Biotic response to late Quaternary rapid climate switches in Santa Barbara Basin: Ecological and evolutionary implications. *Geology* 27:63–66

Caswell H, Cohen JE. 1993. Local and regional regulation of species-area relations: a patch-occupancy model. In *Species Diversity in Ecological Communities.* ed. RE Ricklefs, ED Schluter, pp. 99–107. Chicago: Univ. Chicago Press

Caswell H, Etter R. 1999. Cellular automaton models for competition in patchy environments: Facilitation, inhibition, and tolerance. *Bull. Math. Biol.* 61:625–49

Charles C. 1998. The ends of an era. *Nature* 394:422–23

Cita MB, Podenzani M. 1980. Destructive effects of oxygen starvation and ash falls on benthic life: a pilot study. *Quat. Res.* 13:230–41

Clarke KR, Warwick RM. 1994. Change in marine communities, an approach to statistical analysis and interpretation. *Natl Environ. Res. Counc. UK.* 144 pp.

Cochonat P, Lenat JF, Bachelery P, Boiwin P, Cornaglia B. 1990. Gravity events as a primary process in the construction of a submarine volcano-sedimentary system (Fournaise Volcano, Reunion Island). *C. R. Acad. Sci. Ser. II-* 311:679–86

Colwell RK. 1997. Estimate S: Statistical estimation of species richness and shared species from samples. Version 5. Users Guide and Appl. http://viceroy.eeb.uconn.edu/estimates

Colwell RK, Coddington JA. 1994. Estimating

terrestrial biodiversity through extrapolation. *Philos. Trans. R. Soc. London Ser. B* 345:101–18

Colwell RK, Hurtt GC. 1994. Nonbiological gradients in species richness and a spurious Rappoport effect. *Am. Nat.* 144:570–95

Colwell RK, Lees DC. 2000. The mid-domain effect: geometric constraints on the geography of species richness. *Trends Ecol. Evol.* 15:70–76

Connell JH. 1978. Diversity in tropical rain forests and coral reefs. *Science* 199:1302–10

Cook AA, Lambshead PJD, Hawkins LE, Mitchell N, Levin LA. 2000. Nematode abundance at the oxygen minimum zone in the Arabian Sea. *Deep-Sea Res.* 47:75–85

Corliss BH, Emerson S. 1990. Distribution of rose bengal stained deep-sea benthic foraminifera from the Nova Scotian continental margin and Gulf of Maine. *Deep-Sea Res. Part A- Oceanog. Res. Pap.* 37:381–400

Cosson N, Sibuet M, Galeron J. 1997. Community structure and spatial heterogeneity of the deep-sea macrofauna at three contrasting stations in the tropical northeast Atlantic. *Deep-Sea Res.* 44:247–69

Cosson-Sarradin N, Sibuet M, Paterson GLJ, Vangriesheim A. 1998. Polychaete diversity at tropical Atlantic deep-sea sites: environmental effects. *Mar. Ecol. Prog. Ser.* 165:173–85

Cronin TM, Raymo ME. 1997. Orbital forcing of deep-sea benthic species diversity. *Nature* 385:624–27

Culver SJ, Buzas MA. 1998. Patterns of occurrence of benthic foraminifera in time and space. In *The Adequacy of the Fossil Record*, ed. SK Donovan, CRC Paul, pp. 207–26. New York: Wiley

Culver SJ, Buzas MA. 2000. Global latitudinal species diversity gradient in deep-sea benthic foraminifera. *Deep-Sea Res. I* 47:259–75

Dayton PK, Hessler RR. 1972. Role of biological disturbance in maintaining diversity in the deep sea. *Deep-Sea Res.* 19:199–208

den Dulk M. 2000. Benthic foraminiferal response to Late Quaternary variations in surface water productivity and oxygenation in the northern Arabian Sea. *Geol. Ultraiectina*, No. 188. 205 pp.

den Dulk M, Reichart GJ, Memon GM, Roelofs EMB, Zachariasse WJ, van der Zwaan GJ. 1998. Benthic foraminiferal response to variations in surface water productivity and oxygenation in the northern Arabian Sea. *Mar. Micropaleontol.* 35:43–66

de Rijk S, Troelstra SR, Rohling EJ. 1999. Benthic foraminiferal distribution in the Mediterranean Sea. *J. Foraminifer. Res.* 29:93–103

Deuser W, Muller-Karger F, Evans R, Brown O, Esaias W, Feldman G. 1990. Surface-ocean color and deep-ocean carbon flux: how close a connection? *Deep-Sea Res.* 37:1331–43

Diaz RJ, Rosenberg R. 1995. Marine benthic hypoxia: a review of its ecological effects and the behavioural responses of benthic macrofauna. *Oceanogr. Mar. Biol. Annu. Rev.* 33:245–303

Dickson RR, McCave IN. 1986. Nepheloid layers on the continental slope west of Porcupine Bank. *Deep-Sea Res.* 33:791–818

Douglas RG, Liestman J, Walch C, Blake G, Cotton ML. 1980. The transition from live to sediment assemblage in benthic foraminifera from the Southern California Borderland. In *Quaternary Depositional Environments of the Pacific Coast*, ed. ME Field, AH Bouma, IP Colburn, RG Douglas, JC Dingle, pp. 257–80. Los Angeles: Pac. Coast Paleogeogr. Symp. 4, Pac. Sect., Soc. Econ. Paleontol. Mineral.

Embley RW. 1976. New evidence for occurrence of debris flow deposits in the deep sea. *Geology* 4:371–74

Emerson S. 1985. Organic carbon preservation in marine sediments. In *The Carbon Cycle and Atmospheric CO_2: Natural Variations Archean to Present*, ed. I Sundquist, W Broecker, 32:78–87. Washington, DC: Geophys. Monogr., AGU

Emery KO, Uchupi E. 1984. *The Geology of the Atlantic Ocean.* New York/Berlin: Springer-Verlag. Charts I–XI. 2 Vols. 1050 pp.

Etter RJ, Grassle JF. 1992. Patterns of species

diversity in the deep sea as a function of sediment particle size diversity. *Nature* 360:576–78

Etter RJ, Mullineaux L. 2000. Deep-Sea communities. In *Marine Community Ecology* ed. MD Bertness, S Gaines, M Hay, pp. 367–93, Chapter 14. Sunderland, MA: Sinauer

Fenchel T, Kofoed LH. 1976. Evidence for exploitative interspecific competition in mud snails (Hydrobiidae). *Oikos* 27:367–76

Fenchel T, Kofoed H, Lappalainen A. 1975. Particle size selection of two deposit feeders: the amphipod *Corophium volutator* and the prosobranch *Hydrobia ulvae*. *Mar. Biol.* 30:119–28

Flood RD, Shor AN. 1988. Mud waves in the Argentine Basin and their relationship to regional bottom circulation patterns. *Deep-Sea Res.* 35:943–71

Frederiksen RA, Jensen A, Westerberg H. 1992. The distribution of the scleractinian coral *Lophelia pertusa* around the Faroe Islands and the relation to internal mixing. *Sarsia* 77:157–71

Gage JD. 1996. Why are there so many species in deep-sea sediments? *J. Exp. Mar. Biol. Ecol.* 200:257–86

Gage JD. 1997. High benthic species diversity in deep-sea sediments: The importance of hydrodynamics. In *Marine Biodiversity*, ed. RFG Ormond, JD Gage, MV Angel, pp. 148–77. Cambridge, UK:Cambridge Univ. Press

Gage JD, Lamont PA, Tyler PA. 1995. Deep-sea macrobenthic communities at contrasting sites off Portugal, preliminary results: I introduction and diversity comparisons. *Int. Rev. Hydrobiol.* 80:235–50

Gage JD, May RM. 1993. Biodiversity—a dip into the deep seas. *Nature* 365:609–10

Gage JD, Tyler PA. 1991. *Deep-Sea Biology: A Natural History of Organisms at the Deep-Sea Floor.* Cambridge, UK: Cambridge Univ. Press. 504 pp.

Gallardo VA. 1985. Efectos del fenómeno de El Niño sobre el bentos sublitoral frente a Concepción, Chile. In *El Niño y su impacto en la fauna marina*, ed. W Arntz, A Landa, J Tarazona, pp. 79–85. Inst. del Mar del Perú: Lima, Peru: Bol. Extraordin.

Genin A, Lazar B, Brenner S. 1995. Vertical mixing and coral death in the Red Sea following the eruption of Mount Pinatubo. *Nature* 377:507–10

Glover A, Paterson G, Bett B, Gage J, Sibuet M, et al. 2001. Patterns in polychaete abundance and diversity from the Madeira Abyssal Plain, northeast Atlantic. *Deep-Sea Res. I* 48:217–36

Gooday AJ. 1996. Epifaunal and shallow infaunal foraminiferal communities at three abyssal NE Atlantic sites subject to differing phytodetritus regimes. *Deep-Sea Res.* 43:1395–431

Gooday AJ, Bernhard JM, Levin LA, Suhr S. 2000. Foraminifera in the Arabian Sea OMZ and other oxygen deficient settings: taxonomic composition, diversity and relation to metazoan faunas. *Deep-Sea Res.* 47:54–73

Gooday AJ, Hughes JA, Levin LA. 2001. The foraminiferal macrofauna from three North Carolina (U.S.A.) slope sites with contrasting carbon flux: a comparison with the metazoan macrofauna. *Deep-Sea Res. I.* 48:1709–39

Grassle JF, Maciolek NJ. 1992. Deep-sea species richness: regional and local diversity estimates from quantitative bottom samples. *Am. Nat.* 139:313–41

Grassle JF, Morse-Porteous LS. 1987. Macrofaunal colonization of disturbed deep-sea environments and the structure of deep-sea benthic communities. *Deep-Sea Res.* 34:1911–50

Gray JS. 1981. *The Ecology of Marine Sediments.* Cambridge, UK: Cambridge Univ. Press

Gray JS. 1994. Is deep-sea species diversity really so high? Species diversity of the Norwegian continental shelf. *Mar. Ecol. Prog. Ser.* 112:205–9

Gray JS, Poore GCB, Ugland KI, Wilson RS, Olsgard F, Johannessen O. 1997. Coastal and deep-sea benthic diversities compared. *Mar. Ecol. Prog. Ser.* 159:97–103

Gross TF, Williams AJ. 1991. Characterization of deep-sea storms. *Mar. Geol.* 99:281–301

Hall SJ. 1994. Physical disturbance and marine benthic communities: life in unconsolidated sediments. *Oceanog. Mar. Biol.: An Annu. Rev.* 32:179–239

Hall SJ, Raffaelli D, Thrush SF. 1994. Patchiness and disturbance in shallow water benthic assemblages. In *Aquatic Ecology: Scale, Pattern and Process*, ed. PS Giller, AG Hildrew, DG Raffaelli, pp. 333–75. London: Blackwell Sci.

Hanski I. 1998. Metapopulation dynamics. *Nature* 396:41–49

Heezen BC, Hollister CD. 1971. *The Face of the Deep.* New York: Oxford Univ. Press. 659 pp.

Hermelin JOR, Schimmield GB. 1990. The importance of the oxygen minimum zone and sediment geochemistry in the distribution of benthic foraminifera in the Northwestern Indian Ocean. *Mar. Geol.* 91:1–29

Hermelin JOR, Schimmield GB. 1995. Impact of productivity events on benthic foraminiferal faunas in the Arabian Sea over the last 150,000 years. *Paleoceanography* 10:85–116

Hess S, Kuhnt W. 1996. Deep-sea benthic foraminiferal recolonization of the 1991 Mt. Pinatubo ash layer in the South China Sea. *Mar. Micropaleontol.* 28:171–97

Hessler RR, Jumars PA. 1974. Abyssal community analysis from replicate box cores in the central North Pacific. *Deep-Sea Res.* 21:185–209

Hollister CD, Nowell ARM. 1991. HEBBLE epilogue. *Mar. Geol.* 99:445–60

Hollister CD, McCave IN. 1984. Sedimentation under deep-sea storms. *Nature* 309:220–25

Hollister CD, Nowell ARM, Jumars PA. 1984. The dynamic abyss. *Sci. Am.* 250:42–53

Hubbell SP, Foster RB, O'Brien ST, Harms KE, Condit R, et al. 1999. Light-cap disturbances, recruitment limitation, and tree diversity in a neotropical forest. *Science* 283:554–57

Huggett QJ. 1987. Mapping of hemipelagic versus turbidite muds by feeding traces observed in deep-sea photographs. In *Geology and Geochemistry of Abyssal Plains*, ed. PPE Weaver, J Thomson, pp. 105–12. Geol. Soc. Spec. Publ. No. 31

Hurlbert SM. 1971. The non-concept of species diversity, A critique and alternative parameters. *Ecology* 52:577–86

Huston M. 1979. A general hypothesis of species diversity. *Am. Nat.* 113:81–101

Jacobi RD, Hayes DE. 1982. Bathymetry, microphysiography and reflectivity characteristics of the West African margin between Sierra Leone and Mauritania. In *Geology of the Northwest African Margin*, ed. U von Rad, K Hinz, M Sarnthein, E Siebold, pp. 182–212. Heidelberg: Springer-Verlag

Jahnke R. 1996. The global ocean flux of particulate organic carbon, areal distribution and magnitude. *Global Biogeochem. Cycles* 10:71–88

Jensen A, Frederickson R. 1992. The fauna associated with the bank-forming deepwater coral *Lophelia pertusa* (Scleractiniaria) on the Faroe shelf. *Sarsia* 77:53–69

Jorissen FJ. 1999. Benthic foraminiferal successions across Late Quaternary Mediterranean sapropels. *Mar. Geol.* 153:91–101

Jorissen FJ, Buzas M, Culver S, Kuehl S. 1994. Vertical distribution of living benthic foraminifera in submarine canyons off New Jersey. *J. foraminiferal Res.* 24:28–36

Jorissen FJ, Wittling I, Peypouquet JP, Rabouille C, Relexans JC. 1998. Live foraminiferal faunas off Cap Blanc, NW Africa: Community structure and microhabitats. *Deep-Sea Res. I* 45:2157–88

Josefson AB, Widbom B. 1988. Differential response of benthic macrofauna and meiofauna to hypoxia in the Gullmar Fjord basin. *Mar. Biol.* 100:31–40

Jumars PA, Eckman JE. 1983. Spatial structure within deep-sea benthic communities. In *The Sea*, ed. GT Rowe, pp. 399–452. New York: Wiley

Jumars PA, Hessler RR. 1976. Hadal community structure: implications from the Aleutian Trench. *J. Mar. Res.* 34:547–60

Kaminski M. 1985. Evidence for control of

abyssal agglutinated foraminiferal community structure by substrate disturbance: results from the HEBBLE area. *Mar. Geol.* 66:113–31

Kamykowski D, Zentara SJ. 1990. Hypoxia in the world ocean as recorded in the historical data set. *Deep-Sea Res.* 37:1861–74

Keller M. 1986. Structure des peuplements méiobenthiques dans la secteur pollué par la rejet en mer de l'égout de Marseille. *Annales Inst. océanog.* 62:13–36

Kidd RB, Hill PR. 1986. Sedimentation on mid-ocean sediment drifts. In *North Atlantic Palaeoceanography*, ed. CP Summerhayes, NJ Shackleton, pp. 87–102. Geol. Soc. Spec. Pub., No. 21

Klein H. 1988. Benthic storms, vortices, and particle dispersion in the deep western European Basin. *Deutsches hydrographishes Zeitung* 40:87–102

Klitgaard AB. 1995. The fauna associated with outer shelf and upper slope sponges (Porifera, Demospongiae) at the Faroe Islands, northeastern Atlantic. *Sarsia* 80:1–22

Kontar EA, Sokov AV. 1994. A benthic storm in the northeastern tropical Pacific over the fields of manganese nodules. *Deep-Sea Res. I* 41:1069–89

Lambshead PJD, Brown CJ, Ferrero TJ, Jensen J, Smith CR, Hawkins LE, Tietjen J. 2001a. Latitudinal diversity patterns for deep sea marine nematodes and organic fluxes—a test from the central equatorial Pacific. *Mar. Ecol. Prog. Ser.* In press

Lambshead PJD, Elce BJ, Thistle D, Eckman JE, Barnett PRO. 1994. A comparison of biodiversity of deep-sea marine nematodes from three stations in the Rockall Trough, northeast Pacific. *Biodiv. Lett.* 2:95–107

Lambshead PJD, Tietjen J, Ferrero T, Jensen J. 2000. Latitudinal diversity gradients in the deep sea with special reference to North Atlantic nematodes. *Mar. Ecol. Prog. Ser.* 194:159–67

Lambshead PJD, Tietjen J, Glover A, Ferrero T, Thistle D, Gooday AJ. 2001. The impact of large-scale natural physical disturbance on the diversity of deep-sea North Atlantic nematodes. *Mar. Ecol. Prog. Ser.* 214:121–26

Lambshead PJ, Tietjen J, Moncrieff C, Ferrero T. 2001b. North Atlantic latitudinal diversity patterns in deep-sea marine nematode data. *Mar. Ecol. Prog. Ser.* In press

Lampitt R, Antia R, 1997. Particle flux in the deep seas: regional characteristics and temporal variability. *Deep-Sea Res. I* 44:1377–73

Lees DC, Kremen C, Andriamampianina L. 1999. A null model for species richness gradients: bounded range overlap of butterflies and other rainforest endemics in Madagascar. *Biol. J. Linnean Soc.* 67:529–84

Levin LA. Blair N, DeMaster DJ, Plaia G, Fornes W, et al. 1997a. Rapid subduction of organic matter by maldanid polychaetes on the North Carolina slope. *J. Mar. Res.* 55:595–611

Levin L, Blair N, Martin C, DeMaster D, Plaia G, Thomas C. 1999. Macrofaunal processing of phytodetritus at two sites on the Carolina margin: *In situ* experiments using ^{13}C-labeled diatoms. *Mar. Ecol. Progr. Ser.* 182:37–54

Levin LA, DiBacco C. 1995. The influence of sediment transport on short-term recolonization by seamount infauna. *Mar. Ecol. Progr. Ser.* 123:163–75

Levin LA, Gage JD. 1998. Relationships between oxygen, organic matter and the diversity of bathyal macrofauna. *Deep-Sea Res.* 45:129–63

Levin LA, Gage J, Lamont P, Cammidge L, Martin C, et al. 1997b. Infaunal community structure in a low-oxygen, organic rich habitat on the Oman continental slope, NW Arabian Sea. In *Responses of marine organisms to their environments*, ed. L Hawkins, S Hutchinson, pp. 223–30. Proc. 30th Eur. Mar. Biol. Symp., Univ. Southampton

Levin LA, Gage JD, Martin C, Lamont PA. 2000. Macrobenthic community structure within and beneath the oxygen minimum zone, NW Arabian Sea. *Deep-Sea Res.* 47:189–226

Levin LA, Huggett CL, Wishner KF. 1991.

Control of deep-sea benthic community structure by oxygen and organic-matter gradients in the eastern Pacific Ocean. *J. Mar. Res.* 49:763–800

Levin LA, Leithold EL, Gross TF, Huggett CL, DiBacco C. 1994a. Contrasting effects of substrate mobility on infaunal assemblages inhabiting two high-energy settings on Fieberling Guyot. *J. Mar. Res.* 52:489–522

Levin LA, Plaia GR, Huggett CL. 1994b. The influence of natural organic enhancement on life histories and community structure of bathyal polychaetes. In *Reproduction, larval biology, and recruitment of the deep-sea benthos*, ed. CM Young, KJ Eckelbarger, pp. 261–83. Columbia, SC: Columbia Univ. Press

Levin LA, Thomas CL, 1988. The ecology of xenophyophores (Protista) on eastern Pacific seamounts. *Deep-Sea Res.* 35:2003–27

Lipman PW, Normark WR, Moore JG, Wilson JB, Gutmacher CE. 1988. The giant Alika debris slide, Mauna Loa, Hawaii. *J. Geophys. Res.* 93:4279–99

Luth U, Luth CM. 1997. A benthic approach to determine long-term changes in the water column of the Black Sea. In: *The Responses of Marine Organisms to their Environment.* Southampton, UK: Proc. 30th Eur. Mar. Biol. Symp., Sept. 1995. Southampton Oceanogr. Centre. pp. 223–30

MacArthur RH. 1972. *Geographical Ecology. Patterns in the Distribution of Species*, Princeton, NJ: Princeton Univ. Press. 269 pp.

Maciolek-Blake NJ, Grassle JF, Blake JA, Neff JM. 1985. *Georges Bank Infauna Monitoring Program:Final report for the third year of sampling.* Washington, DC: US Dept. Int., Minerals Mgmt. Ser.

Maciolek NJ, Grassle JF, Hecker B, Boehm PD, Brown B et al. 1987a. *Study of biological processes on the U.S. mid-Atlantic slope and rise.* Phase 2. Washington, DC: US Dept. Int., Minerals Mgmt. Ser.

Maciolek NJ, Grassle JF, Hecker B, Brown B, Blake JA et al. 1987b. *Study of biological processes on the U.S. North Atlantic slope*

and rise. Washington, DC: US Dept. Int., Minerals Mgmt. Ser.

Magurran A. 1988. *Ecological Diversity and Its Measurement.* Princeton, NJ: Princeton Univ. Press, 179 pp.

Martin JH, Knauer GA, Karl DM, Broenkow WW. 1987. VERTEX: Carbon cycling in the northeast Pacific. *Deep-Sea Res.* 34:267–85

Masson DG. 1996. Catastrophic collapse of the volcanic island of Hierro 15 ka ago and the history of landslides in the Canary Islands. *Geology* 24:231–34

Masson DG, Kidd RB, Gardner JV, Huggett QJ, Weaver PPE. 1994. Saharan continental rise: facies distribution and sediment slides. In *Geological Evolution of Atlantic Continental Rises*, ed. VW Poag, PC Degrocianski, pp. 3–10. New York: Van Nostrand Reinhold

Masson DG, Kenyon NH, Weaver PPE. 1996. Slides, debris flows, and turbidity currents, In *Oceanography: An Illustrated Guide*, ed. CP Summerhayes, SA Thorpe, pp. 136–51. London: Manson

May RM. 1975. Patterns of species abundance and diversity. In *Ecology and Evolution of Communities.* ed. ML Cody, M Diamond. Cambridge, MA: Harvard Univ. Press. pp. 81–120

Menge BA, Sutherland JP. 1987. Community regulation: variation in distribution, competition, and predation in relation to environmental stress and recruitment. *Am Nat.* 130:730–57

Milliman JD. 1994. Organic matter content in U.S. Atlantic continental slope sediments: decoupling the grain size factor. *Deep-Sea Res.* 41:797–808

Moodley L, van der Zwaan GJ, Herman PMJ, Kempers L, van Breugel P. 1997. Differential response of benthic meiofauna to anoxia with special reference to the Foraminifera (Protista: Sarcodina). *Mar. Ecol. Progr. Ser.* 158:151–63

Mullins HT, Thompson JB, McDougall K, Vercoutere TL. 1985. Oxygen-minimum zone edge effects, evidence from the central California coastal upwelling system. *Geology* 13:491–94

Munk W. 1970. The circulation of the oceans. In *Adventures in Earth History*, ed. P Cloud, p. 235–40. San Francisco: Freeman

Murrell MC, Fleeger JW. 1989. Meiofaunal abundances on the Gulf of Mexico continental shelf affected by hypoxia. *Cont. Shelf Res.* 9:1049–62

Neira C, Sellanes J, Levin LA, Arntz WA. 2001. Meiofaunal distributions on the Peru margin: relationship to oxygen and organic matter availability. *Deep-Sea Res.* 48:2453–72

Noble M, Mullineaux LS. 1989. Internal tidal currents over the summit of Cross Seamount. *Deep-Sea Res.* 36:1791–1802

Paterson GLJ, Wilson GDF, Cosson N, Lamont PA. 1998. Hessler and Jumars (1974) revisited: abyssal polychaete assemblages from the Atlantic and Pacific. *Deep-Sea Res. II* 45:225–51

Paterson GLJ, Lambshead PJD. 1995. Bathymetric patterns of polychaete diversity in the Rockall Trough, northeast Atlantic. *Deep-Sea Res. I* 42:1199–1214

Petersen C. 1913. Valuation of the sea. II. The animal communities of the sea bottom and their importance for marine zoogeography. *Rep. Danish Biol. Stn. to Board Ag.* 21:1–44

Petraitis PS, Latham RE, Niesenbaum RA. 1989. The maintenance of species diversity by disturbance. *Quar. Rev. Biol.* 64:393–418

Phleger PB, Soutar A. 1973. Production of benthic foraminifera in three east Pacific oxygen minima. *Micropaleontology* 19:110–15

Pickett STA, White PS. 1984. *Natural Disturbance: An Evolutionary Perspective.* New York: Academic

Pineda J. 1993. Boundary effects on the vertical ranges of deep-sea benthic species. *Deep-Sea Res. I* 40:2179–92

Pineda J, Caswell H. 1998. Bathymetric species-diversity patterns and boundary constraints on vertical range distribution. *Deep-Sea Res. II,* 45:83–101

Pingree RD, New AL. 1989. Downward propagation of internal tidal energy in the Bay of Biscay. *Deep-Sea Res.* 36:735–58

Radziejewska T, Maslowski J. 1997. Macro- and meiobenthos of the Arkona Basin (western Baltic Sea): differential recovery following hypoxic events. In *Responses of Marine Organisms to Their Environments*, ed. L.E. Hawkins, S. Hutchinson, pp. 251–262. Proc. 30th Eur. Mar. Biol. Symp., Univ. Southampton

Reimers CE, Jahnke RA, McCorkle DC. 1992. Carbon fluxes and burial rates over the continental slope and rise off central California with implications for the global carbon cycle. *Global Biogeochemical Cycles* 6:199–224

Reise K. 1985. *Tidal Flat Ecology.* Berlin: Springer-Verlag 191 pp.

Rex MA. 1973. Deep-sea species diversity: decreased gastropod diversity at abyssal depths. *Science* 181:1051–53

Rex MA. 1976. Biological accommodation in the deep-sea benthos: comparative evidence on the importance of predation and productivity. *Deep-Sea Res.* 23:975–87

Rex MA. 1981. Community structure in the deep-sea benthos. *Annu. Rev. Ecol. Syst.* 12:331–53

Rex MA. 1983. Geographic patterns of species diversity in deep-sea benthos. In *The Sea*, ed. GT Rowe, Vol. 8, pp. 453–72. New York: Wiley

Rex MA, Stuart CT, Coyne G. 2000. Latitudinal gradients of species richness in the deep-sea benthos of the North Atlantic. *Proc. Natl. Acad. Sci. USA* 97:4082–85

Rex MA, Stuart CT, Etter R. 1997. Large-scale patterns of species diversity in the deep-sea benthos. In *Marine Biodiversity: Patterns and Processes*, ed. R Ormond, J D Gage, MV Angel, pp. 94–121. Cambridge, UK: Cambridge Univ. Press

Rex MA, Stuart CT, Hessler RR, Allen JA, Sanders HL, Wilson GDF. 1993. Global-scale latitudinal patterns of species diversity in the deep-sea benthos. *Nature* 365:636–39

Rhoads DC. 1974. Organism sediment relations on the muddy sea floor. *Oceanogr. Mar. Biol. Ann. Rev.* 12:263–300

Rice A, Thurston M, New A. 1990. Dense aggregations of a hexactinellid sponge, *Pheronema carpenteri*, in the Porcupine

Seabight (northeast Atlantic Ocean), and possible causes. *Prog. Oceanog.* 24:179–96

Richardson MJ, Weatherly GL, Gardner WD. 1993. Benthic storms in the Argentine Basin. *Deep-Sea Res. II,* 40:975–87

Ricklefs RE. 1987. Community diversity: relative roles of local and regional processes. *Science* 235:167–71

Rogers A. 2000. The role of the oxygen minima in generating biodiversity in the deep sea. *Deep-Sea Res. II* 47:119–48

Rosenzweig ML. 1995. *Species Diversity in Space and Time.* Cambridge: Cambridge Univ. Press. 436 pp.

Rosenzweig ML, Abramsky H. 1993. How are diversity and productivity related? In *Species Diversity in Ecological Communities: Historical and Geographical Perspectives,* ed. RE Ricklefs, D Schluter, pp. 53–65. Chicago: Univ. Chicago Press

Rowe GT, Menzies RJ. 1968. Deep bottom currents off the coast of North Carolina. *Deep-Sea Res.* 15:711–19

Rowe G, Pariente V. 1992. *Deep-Sea Food Chains and the Global Carbon Cycle,* pp. 560. Dordrecht, Netherlands: Kluwer

Rowe GT, Sibuet M, Deming J, Khripounoff A, Tietjen J, et al. 1991. 'Total' sediment biomass and preliminary estimates of organic carbon residence time in deep-sea benthos. *Mar. Ecol. Prog. Ser.* 79:99–114

Sanders HL. 1968. Benthic studies in Buzzards Bay. I: Animal-sediment relationships. *Limnol. Oceanog.* 3:245–58

Sanders HL. 1969. Benthic marine diversity and the time-stability hypothesis. *Brookhaven Symp. Biol.* 22:71–81

Schaff T, Levin L, Blair N, DeMaster D, Pope R, Boehme S. 1992. Spatial heterogeneity of benthos on the Carolina continental slope: large (100-km)-scale variation. *Mar. Ecol. Prog. Ser.* 88:143–60

Self RFL, Jumars PA. 1988. Cross-phyletic patterns of particle selection by deposit feeders. *J. Mar. Res.* 46:119–143

Sen Gupta BK, Machain-Castillo ML. 1993. Benthic foraminifera in oxygen-poor habitats. *Mar. Micropaleontology* 20:183–201

Shepard FP, Marshall NF, McLoughlin PA, Sullivan GG. 1979. Currents in submarine canyons and other seavalleys. Tulsa, OK: AAPG Stud. Geol., Vol. 8

Shmida A, Wilson MV. 1985. Biological determinants of species diversity. *J. Biogeogr.* 12:1–20

Simms RW, Weaver PPE, Kidd RB, Jones EJW. 1991. Late Quaternary mass movement on the lower continental rise and abyssal plain off western Sahara. *Sedimentology* 38:27–40

Smith CR, Berelson W, Demaster DJ, Dobbs FC, Hammond D, et al. 1997. Latitudinal variations in benthic processes in the abyssal equatorial Pacific: control by biogenic particle flux. *Deep-Sea Res.* II:2295–317

Smith CR, Pope RH, DeMaster DJ, Magaard L. 1993. Age-dependent mixing of deep-sea sediments. *Geochim. Cosmochim. Acta* 57:1473–88

Smith KL, Carlucci AF, Jahnke RA, Craven DB. 1987. Organic carbon mineralization in the Santa Catalina Basin: benthic boundary layer metabolism. *Deep-Sea Res.* 34:185–211

Smith KL, Hinga KR. 1983. Sediment community respiration in the deep sea. In *Deep Sea Biology,* ed. GT Rowe, pp. 331–70. New York: Wiley

Snelgrove PV, Butman CA. 1994. Animal-sediment relationships revisited: cause vs. effect. *Oceanogr. Mar. Biol. Ann. Rev.* 32:111–177

Stephens MP, Kadko DC, Smith CR, Latasa M. 1997. Chlorophyll-*a* and pheopigments as tracers of labile organic carbon at the central equatorial Pacific seafloor. *Geochim. et Cosmochim. Acta* 61:4605–19

Stevens GC. 1989. The latitudinal gradient in geographical range: how so many species coexist in the tropics. *Am. Natur.* 133:240–56

Stow DAV, Faugères JCF, eds. 1993. Contourites and Bottom Currents. *Sed. Geol.* 82:310 pp.

Stow DAV, Holbrook JA. 1984. North Atlantic contourites: an overview. In *Fine-Grained*

Sediments: Deep-Water Processes and Facies, ed. DAV Stow, DJW Piper, pp. 245–56. Oxford: Blackwell Sci.

Stuart CT, Rex MA. 1994. The relationship between developmental pattern and species diversity in deep-sea prosobranch snails. In *Reproduction, Larval Biology, and Recruitment of the Deep-sea Benthos*, ed. CM Young, KJ Eckelbarger, pp. 118–36. New York: Columbia Univ. Press

Stuart CT, Rex MA, Etter RJ. 2001. Large scale spatial and temporal patterns of deep-sea benthic species diversity. In *Ecosystems of the World: Ecosystems of Deep Oceans*, ed. PA Tyler. Amsterdam: Elsevier. In press

Suess E. 1980. Particulate organic carbon flux in the oceans—surface productivity and oxygen utilization. *Nature* 288:260–263

Taghon GL. 1982. Optimal foraging by deposit-feeding invertebrates: roles of particle size and organic coating. *Oecologia* 52:295–304

Thistle D. 1998. Harpacticoid copepod diversity at two physically reworked sites in the deep sea. *Deep-Sea Res. II* 45:13–24

Thistle D. 1983. The role of biologically produced habitat heterogeneity in deep-sea diversity maintenance. *Deep-Sea Res.* 30:1235–45

Thistle D, Ertman SC, Fauchald K. 1991. The fauna of the HEBBLE site: patterns in standing stock and sediment-dynamic effects. *Mar. Geol.* 99:413–22

Thistle D, Levin L. 1998. The effect of experimentally increased near-bottom flow on metazoan meiofauna at a deep-sea site, with comparison data on macrofauna. *Deep-Sea Res.* 45:625–38

Thistle D, Yingst UY, Fauchald K. 1985. A deep-sea benthic community exposed to strong near-bottom currents on the Scotian Rise (western Atlantic). *Mar. Geol.* 66:91–112

Thomas E, Gooday AJ. 1996. Cenozoic deep-sea benthic foraminifers: tracers for changes in oceanic productivity? *Geology* 24:355–58

Thunnell R, Tappa E, Varela R, Llano M, Astor Y. 1999. Increased marine sediment suspension and fluxes following an earthquake. *Nature* 398:233–36

Tietjen JH. 1984. Distribution and species diversity of deep-sea nematodes in the Venezuela Basin. *Deep-Sea Res.* 31:119–32

Tietjen JH. 1989. Ecology of deep-sea nematodes from the Puerto Rico Trench area and the Hatteras Abyssal Plain. *Deep-Sea Res.* 36:1579–94

Tilman D. 1982. *Resource Competition and Community Structure*. Princeton, NJ: Princeton Univ. Press

Tillman D. 1994. Competition and biodiversity in spatially structured habitats. *Ecology* 75:2–16

Tyler PA. 1995. Conditions for the existence of life at the deep-sea floor: an update. *Oceanogr. Mar. Biol. Annu. Rev.* 33:221–44

Van Dover C. 2000. *The Ecology of Deep-sea Hydrothermal Vents*, Princeton, NJ: Princeton Univ. Press. 424 pp.

Vetter EW, Dayton PK. 1998. Macrofaunal communities within and adjacent to a detritus-rich submarine canyon system. *Deep-Sea Res.* 45:25–54

Viana AR, Faugères J-C, Stow DAV. 1998. Bottom-current-controlled sand deposits—a review of modern shallow- to deep-water environments. *Sed. Geol.* 115:53–80

Von Rad U, Schulz H, Riech V, den Dulk M, Berner U, Sirocko F. 1999. Multiple monsoon-controlled breakdown of oxygen-minimum conditions during the past 30,000 years documented in laminated sediments off Pakistan. *Palaeogeog., Palaeoclim., Palaeoecol.*, 152:129–61

Waide RB, Willig MR, Steiner CF, Mittelbach G, Gough L. 1999. The relationship between productivity and species richness. *Annu. Rev. Ecol. Syst.* 30:257–300

Warwick RM, Clarke KR. 1996. Relationships between body-size, species abundance and diversity in marine benthic assemblages: facts or artifacts? *J. Exp. Mar. Biol. Ecol.* 202:63–71

Watts MC, Etter RJ, Rex MA. 1992. Effects of spatial and temporal scale on the relationship of surface pigment biomass to community structure in the deep-sea benthos. In *Deep-sea Food Chains and the Global Carbon Cycle*, ed. GT Rowe, V Pariente, pp. 245–54. Dordrecht, The Netherlands: Kluwer

Weatherley GL, Kelley EA. 1985. Storms and flow reversals at the HEBBLE site. *Mar. Geol.* 66:205–18

Wheatcroft RA. 1992. Experimental tests for particle size-dependent bioturbation in the deep ocean. *Limn. Oceanogr.* 37:90–104

Wheatcroft RA, Jumars PA, Smith CR, Nowell ARM. 1990. A mechanistic view of the particulate biodiffusion coefficient-step lengths, rest periods and transport directions *J. Mar. Res.* 48:177–208

Wheatcroft RA, Jumars PA. 1987. Statistical reanalysis for size dependent bioturbation in the deep ocean. *Limnol. and Oceanogr.* 37:90–104

Whitlatch RB. 1980. Patterns of resource utilization and coexistence in marine intertidal deposit-feeding communities. *J. Mar. Res.* 38:743–65

Willig MR, Lyons SK. 1998. An analytical model of latitudinal gradients of species richness with an empirical test for marsupials and bats in the New World. *Oikos* 81:93–98

Wishner K, Levin L, Gowing M, Mullineaux L. 1990. Involvement of the oxygen minimum in benthic zonation on a deep sea mount. *Nature* 346:57–59

Annu. Rev. Ecol. Syst. 2001. 32:95–126
Copyright © 2001 by Annual Reviews. All rights reserved

THE PHYSIOLOGY OF LIFE HISTORY TRADE-OFFS IN ANIMALS

Anthony J. Zera and Lawrence G. Harshman

School of Biological Sciences, University of Nebraska, Lincoln, Nebraska 68588;
e-mail: azera@unlserve.unl.edu and lharsh@unlserve.unl.edu

Key Words   allocation, energetics, hormones, wing polymorphism, *Drosophila*

■ **Abstract**   The functional causes of life history trade-offs have been a topic of interest to evolutionary biologists for over six decades. Our review of life history trade-offs discusses conceptual issues associated with physiological aspects of trade-offs, and it describes recent advances on this topic. We focus on studies of four model systems: wing polymorphic insects, *Drosophila*, lizards, and birds. The most significant recent advances have been: (*a*) incorporation of genetics in physiological studies of trade-offs, (*b*) integration of investigations of nutrient input with nutrient allocation, (*c*) development of more sophisticated models of resource acquisition and allocation, (*d*) a shift to more integrated, multidisciplinary studies of intraspecific trade-offs, and (*e*) the first detailed investigations of the endocrine regulation of life history trade-offs.

INTRODUCTION

Life history traits are often negatively associated with each other (Clutton-Brock et al. 1982, Reznick 1985, Stearns 1989, 1992, Roff 1992, Rose et al. 1996). Classic examples include decreased early fecundity in lines of *Drosophila melanogaster* selected for increased longevity, and reduced overwintering survivorship in lactating (reproductive) red deer, *Cervus elaphus*. These negative associations, referred to as life history trade-offs, have played a prominent role in theory and interpretation of life history studies. For example, trade-offs are a key assumption of optimality models of life history evolution, and they provide an explanation for the widespread occurrence of variable life history traits in natural populations (Reznick 1985, Roff 1992, Stearns 1992).

The physiological causes of life history trade-offs have been a central topic in life history studies for more than six decades (Fisher 1930, Tinkle & Hadley 1975, Townsend & Calow 1981, Dunham et al. 1989, Adolph & Porter 1993, Stearns 1992, Roff 1992, Zera et al. 1998). The ultimate goal of physiological studies has been to illuminate the mechanisms of life history evolution by identifying functional interactions among the various components of life history traits. In many cases, life history trade-offs have been thought to result from competition among

0066-4162/01/1215-0095$14.00

95

different organismal functions for limited internal resources. Hence, traditional physiological studies of life history trade-offs have focused almost exclusively on the differential allocation of limiting internal nutrients to reproduction, maintenance metabolism, growth, and storage within single species or variation in these processes among populations or species (Calow 1979, Townsend & Calow 1981, Congdon et al. 1982).

During the past decade, significant advances have been made in understanding the physiological mechanisms that underlie life history trade-offs. For example, while retaining focus on individual species or variation between populations or species (Bernardo 1994, Niewiarowski 2001), studies of trade-off physiology have recently expanded to include comparisons between phenotypes or genotypes within populations (Rose et al. 1996, Zera & Huang 1999, Salmon et al. 2001, Zera & Cisper 2001). This expansion has provided the first data on genetic variation and covariation for physiological traits and their relationship to trade-offs at the demographic level. In addition, trade-off studies have recently expanded to include non-energetic aspects of resource allocation, such as the hormonal control of antagonistic traits that comprise trade-offs (Ketterson & Nolan 1992, 1999, Sinervo & Basolo 1996, Sinervo 1999, Zera & Cisper 2001).

The past decade also has seen the development of more complex models of trade-offs (Houle 1991, de Jong 1993, Reznick et al. 2000). These models underscore the importance of functional studies as opposed to the sole use of phenotypic or genetic correlation to identify the existence of trade-offs, and the necessity of taking into account nutrient input in studies of nutrient allocation. Finally, more sophisticated genetic, environmental, endocrine, and molecular approaches have been used to investigate classic trade-offs such as the cost of reproduction (Ketterson & Nolan 1992, 1999, Sinervo & Basolo 1996, Zera et al. 1998, Sinervo 1999, Salmon et al. 2001, Stearns & Kaiser 1993).

Physiological aspects of life history trade-offs have been reviewed previously (e.g., Townsend & Calow 1981, Bell & Koufopanou 1986, Sibley & Calow 1986, Ricklefs 1991, Sibley & Antonovics 1992). However, many of the recent developments in trade-off physiology described above have never been reviewed or have been reviewed in a limited manner. The literature on functional aspects of life history trade-offs is considerable, and no single review can adequately cover all aspects of this topic in all major groups of organisms. In this review, we focus on three topics that have been especially prominent in recent physiological studies of intraspecific trade-offs: (*a*) the influence of nutrient acquisition on the trade-off of internal resources, (*b*) hormonal control of trade-offs, and (*c*) genetic and experimental analyses of trade-offs. We also focus on four model systems that have been especially prominent in physiological studies of life history trade-offs: wing-polymorphic crickets, *Drosophila melanogaster*, lizards, and birds. These models were chosen because they illustrate a diversity of approaches that have been used to investigate a range of central issues in trade-off physiology in taxonomically diverse organisms studied in the laboratory or in the field.

CONCEPTUAL BACKGROUND

Definitions and Basic Principles

Definitions of key trade-off terms such as costs, constraints, and trade-off often vary among studies, which could lead to confusion (Antonovics & van Tienderen 1991). For example, in some cases trade-off is defined as the result of physiological or fitness costs (Leroi et al. 1994b, see below), while in other cases the term cost is used to define a trade-off (Reznick 1985, 1992). The classic example of the latter usage is the cost of reproduction, which is the term used to define the trade-off between current and expected future reproduction (Reznick 1985, 1992). The term cost of reproduction requires special attention because it has been used in many different ways. For example, this term can refer to either a price (e.g., amount of calories required to produce an egg) or penalty of reproduction, that is measured in physiological (i.e., calories), demographic (survivorship), or fitness units. Moreover, the cost of reproduction can refer to a direct penalty of current reproduction or to a penalty exacted in the future.

If internal resources are limited and are insufficient to pay all construction and maintenance costs for two life history traits that share a common resource pool, then a trade-off results: an increment of resources allocated to one trait necessitates a decrement of resources to another trait (the traditional "Y" model of resource allocation; van Noordwijk & de Jong 1986) (Figure 1). In the absence of variation in resource input, two traits linked in a functional trade-off are negatively correlated (more complex situations involving more than two traits and variation in resource input are described below). For example, if internal reserves allocated to current reproduction limit resources available for future reproduction, a trade-off between current and future reproduction exists for physiological reasons (Calow 1979, Bell & Koufopanou 1986). Trade-offs can occur between physiological traits expressed during the same or different times in the life cycle (Chippindale et al. 1996, Zera et al. 1998, Stevens et al. 1999), and they can result from variation in genetic factors (e.g., pleiotropy), environmental factors, or combinations of these two types of factors that give rise to negative interactions between traits. If the trade-off results from a negative genetic correlation, then short-term evolutionary change in one phenotype constrains (i.e., limits) evolutionary change in the other phenotype. Ecological factors, such as predation (Reznick et al. 1990), or behavioral factors, such as time-based conflicts between activities (Marler & Moore 1988, Marler et al. 1995) can be a primary cause of life history trade-offs. Thus, assessing the importance of variation in internal physiological factors on a life history trade-off is most appropriately done by taking into account the relative importance of variation in external ecological or behavioral factors.

Appropriate Use of the Term "Trade-Off"

Implicit in the term trade-off, used in a physiological context, is the notion that a negative functional interaction is the cause of the negative association between

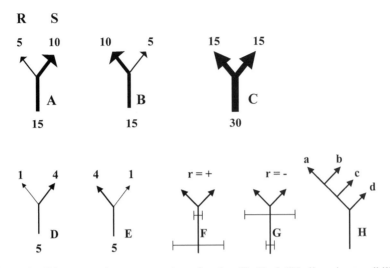

Figure 1 Diagrammatic representation of trade-offs. Each "Y allocation tree" illustrates the amount of resource input (acquisition; number at the base) and the pattern of resource allocation (numbers at the tips of the branches) for a particular phenotype or genotype. R denotes allocation to reproduction, while S denotes allocation to soma. In trees *A–E*, maximal allocation to (maximal physiological cost of) R or S is 15 resource units. Trees *A* and *B* illustrate a standard trade-off (differential allocation of a limiting internal resource). Relative to trees *A* and *B*, tree *C* illustrates the obviating effect of increased resource input on a trade-off (resource input matches physiological costs of both traits), while trees *D* and *E* illustrate the exacerbating effect of decreased nutrient input on a trade-off. Trees *A* and *B*, relative to *D* and *E*, illustrate plasticity of a resource-based trade-off. Trees *F* and *G* illustrate the influence of relative variation in resource input (length of the bar at the base of the tree) versus variation in resource allocation (length of bar at the branch of the tree) on the sign of the correlation between two traits involved in a trade-off. Tree *H* illustrates a more complex allocation tree with multiple dichotomous branches. See text for additional explanation and references. Trees *F* and *G* were redrawn from van Noordwijk & de Jong (1986), while tree *H* was redrawn from de Jong (1993).

two traits under consideration. However, the term trade-off has often been applied to trait associations for which only minimal or no information is available as to whether the traits interact functionally (e.g., see Mole & Zera 1993). Traits might be negatively associated for a variety of reasons other than functional interaction, such as genetic linkage (Mole & Zera 1993, Zera et al. 1998, Zera & Cisper 2001). In our view, the term trade-off, used in physiological studies of life history variation, should represent a hypothesis concerning the cause of a negative trait association, just as the term adaptation is a hypothesis concerning the role of natural selection in shaping the form or function of a trait. It is inappropriate to apply the term evolutionary adaptation to a trait for which there is no strong evidence that its form or function was shaped by natural selection (Lauder 1996). In the future, we anticipate that the term trade-off will be restricted to cases where data indicate a negative functional interaction between traits. When little or no functional information is available, the terms negative association, or potential trade-off, are preferable to trade-off. Research on trade-offs should be increasingly directed toward understanding the underlying mechanism of the negative association between traits.

TRADE-OFF ARCHITECTURE AND EFFECT OF NUTRIENT ACQUSITION

Y Model and Variable Nutrient Input

During the past decade, the properties of trade-offs have been studied using quantitative-genetic and optimization models. Most of these models are more complex versions of the standard "Y" model of allocation discussed previously. An important result of these studies is that, for a variety of reasons, a positive correlation can exist between traits that are linked in a functional trade-off. For example, once more than two traits are involved in a trade-off (see Figure 1H), positive correlations can occur between subsets of those traits (Charlesworth 1990, Houle 1991, Roff 1992). Furthermore, if variability in nutrient input among individuals, due to either genetic variation in loci that control nutrient acquisition or to environmental variation in available resources, is greater than variability in nutrient allocation, positive correlations between traits that comprise a functional trade-off can occur (Figure 1F–G; van Noordwijk & de Jong 1986, de Laguerie et al. 1991, Houle 1991, de Jong & van Noordwijk 1992, de Jong 1993). de Laguerie et al. (1991) and de Jong (1993) developed allocation tree models of trade-offs, involving successive dichotomous branches (trade offs) in a tree with nutrient input as the trunk (See Figure 1H). Both the position of a branch point on an allocation tree and the extent of resource transformations strongly affect whether positive or negative correlations exist between traits that trade off.

Theoretical studies of trade-offs clearly show that the sign of a correlation between two traits cannot be used as an unambiguous indicator of whether those traits interact functionally in a negative manner (i.e., whether they trade-off). This important result underscores that functional trade-offs can be validated only by direct physiological studies rather than by purely phenotypic or genetic approaches. Theoretical studies also illustrate the importance of controlling or quantifying nutrient input in functional studies of nutrient-based internal trade-offs (Figure 1). Quantifying nutrient input has long been regarded as an essential aspect of energy budget studies (Congdon et al. 1982, Nagy 1983, Withers 1992), but has been neglected in many recent energetics studies of trade-offs. This omission is a likely cause of unexpected results obtained in some trade-off studies, such as the failure to observe increased metabolic costs when reproduction is increased (Rose &

Bradley 1998; see also Zera et al. 1998, Reznick et al. 2000). Trade-off models discussed above also provide an explanation for the common observation that traits that are expected to trade-off, and hence exhibit negative correlations, are often positively correlated (Mole & Zera 1993, Reznick et al. 2000). Although these trade-off models are more complex versions of the original "Y" model of allocation (van Noordwijk & de Jong 1986), they are still rudimentary descriptors of the physiology of internal resource allocation.

Trade-Off Plasticity and Priority Rules

Trade-offs are influenced not only by variation among individuals in nutrient input, but also by the absolute amount of nutrient input. Reduced nutrient availability can substantially magnify, while increased nutrient availability can diminish or obviate an apparent trade-off (Figure 1) (Kaitala 1987, Chippindale et al. 1993, Simmons & Bradley 1997, Nijhout & Emlen 1998, Zera et al. 1998, Zera & Brink 2000; see clutch size studies in "Birds" section). These plastic responses of a trade-off are determined by priority rules, which govern relative allocation to organismal processes as a function of nutrient input. Recent experiments have only begun to identify the broad outlines of these rules at the whole-organism level. For example, in laboratory and field experiments on bivalves, cladocerans, insects, and mammals, allocation to maintenance or storage was found to take precedence over allocation to reproduction under nutrient-poor or stressful conditions (Rogowitz 1996, Perrin et al. 1990, Boggs & Ross 1993, Jokela & Mutikainen 1995, Zera et al. 1998). Priority rules are shaped by ecological factors. A more thorough understanding of the evolution of priority rules requires a deeper synthesis among physiological studies of allocation, ecological studies of nutrient acquisition (e.g., foraging), and life history studies in the field (Boggs 1992, 1997).

Timing of Nutrient Input and Trade-Off Dynamics

Another important aspect of nutrient acquisition on a trade-off is the relative timing of these two processes. The most extensively studied aspect of this topic concerns capital versus income breeding. In income breeding, resources used for reproduction are acquired during the reproductive period, while in capital breeding, resources are derived from stores acquired during an earlier period (Drent & Daan 1980, Jonsson 1997, Bonnet et al. 1998). The relative advantage of capitol versus income breeding, and the relative demographic and energetic costs and benefits of energy storage are a matter of debate (e.g., Jonsson 1997, Bonnet et al. 1998). Capital breeding can give rise to trade-offs between different stages of the life cycle, nearly all physiological aspects of which are not well understood. Recent experiments have documented the use of nutrients in reproduction that were acquired during earlier stages; in some cases, unexpectedly long time lags (e.g., >12 months in lizards) between resource acquisition and expenditure on reproduction were noted (e.g., Reznick & Yang 1993, Boggs 1997, Doughty & Shine 1998).

Especially noteworthy are the feeding and radiotracer studies by Boggs and colleagues (Boggs 1997). These studies investigated the relative use of larval and adult-acquired nutrients in reproduction in butterflies as a function of life history and foraging. The relationship between the timing of nutrient acquisition and use of energy stores for reproduction also has important implications for the identification of the ecological causes of trade-offs. For example, a long time lag between resource input and use for reproduction will tend to obscure the influence of specific aspects of resource availability in the field on reproductive output (Doughty & Shine 1997, 1998).

Trade-offs are not static; they can change during development and can evolve (Leroi et al. 1994b). Many life history models have explored dynamical aspects of resource allocation (Perrin & Sibly 1993, Noonburg et al. 1998, Heino & Kaitala 1999). For example, such models predict that the ratio (P/m), where P = productivity (growth and reproduction), and where m = mortality becomes more sensitive to mortality as a function of age. Assuming a trade-off between mortality and productivity, organisms should initially invest in productivity, then later in survival. This might explain why growth rate decreases with age (Perrin & Sibly 1993). One problem with these and a broad class of related models is that their empirical validity is not well established owing to the paucity of direct mechanistic information on putative resource allocation trade-offs.

MEASURING PHYSIOLOGICAL TRADE-OFFS

Three main empirical approaches have been used to investigate life history trade-offs and their physiological causes: (*a*) measurement of phenotypic correlations on unmanipulated individuals, (*b*) genetic analyses, and (*c*) experimental manipulation of phenotypes. The relative merits of these approaches have been extensively debated (e. g., Reznick 1985, 1992, Bell & Koufopanou 1986, Partridge & Sibly 1991, Partridge 1992, Sinervo & Basolo 1996, Rose & Bradley 1998, Harshman & Schmid 1998, Zera et al. 1998). Phenotypic correlations measured on unmanipulated individuals in the field or lab are important in that they can suggest physiological causes of trade-offs. However, there is broad agreement that phenotypic correlations, by themselves, can provide only limited information on the physiological mechanisms that underlie trade-offs. The main problem is that uncontrolled variables can reduce the magnitude of a trade-off or can lead to spurious correlations. For example, as mentioned above, variable nutrient intake can cause two traits involved in a trade-off to be positively correlated (van Noordwijk & de Jong 1986, de Jong 1993), a problem that is expected to be particularly acute in field studies (Haukioja & Hakala 1986, Tuomi et al. 1983). Simple changes in experimental design can sometimes obviate the masking effect of uncontrolled variables on trade-offs. For example, a negative correlation between somatic lipid reserves and reproductive effort in a lizard species was identified by quantifying lipid levels before and after reproduction in the same individuals using a non-invasive method (Doughty & Shine 1997). Individual variation in lipid levels had previously obscured this negative association.

One school of thought is that genetically based covariation should be a focus of attention in trade-off studies because only genetically based traits can evolve (Reznick 1985, Bell & Koufopanou 1986, Partridge & Sibly 1991, Rose et al. 1996). Physiological-genetic studies of life history trade-offs represent one of the most important advances in functional studies of trade-offs during the past decade. Nevertheless, genetic approaches also have important drawbacks (Bell & Koufopanou 1986, Partridge & Sibly 1991). Most notably, in the process of measuring the genetic basis of a trade-off (e.g., via artificial selection), the trade-off itself and its relationship to its physiological causes can be altered (Moller et al. 1989, Partridge & Sibly 1991, Rose & Bradley 1998). Furthermore, genetic studies are limited by standing genetic variation and only indicate how traits may evolve in the short-term; they can only be performed on a subset of organisms, and require much more time and resources than do phenotypic studies (Partridge & Sibly 1991).

The most controversial empirical approach in life history studies is environmental manipulation (sometimes called phenotypic manipulation) (Reznick 1985, 1992, Partridge & Sibly 1991, Sinervo & Basolo 1996, Chippindale et al. 1997). This approach essentially involves altering an environmental variable (e.g., nutrient level) to generate a phenotypic trade-off whose properties can then be studied. The main advantages of this approach are that a great range of character values can be produced, and trade-offs between genetically invariant traits can be investigated. The most serious disadvantage is the typical lack of specificity of the manipulation. Traits can be modified independently of the mechanisms that link them in a trade-off, thus giving rise to altered trade-off functions (Moller et al. 1989, Partridge & Sibly 1991, Sinervo & Basolo 1996). Some of the problems with environmental manipulation can be circumvented by using more targeted surgical or endocrine manipulations (i.e., physiological manipulations or phenotypic engineering (Sinervo & McEdward 1988, Marler & Moore 1988, 1991, Ketterson & Nolan 1992, 1999, Landwer 1994, Sinervo & Basolo 1996, Zera et al. 1998, Zera & Cisper 2001). These manipulations represent a powerful approach to the study of functional aspects of life history trade-offs. They not only shed light on the specific physiological mechanisms that underlie trade-offs, but they can also be used to produce phenotypes to test various trade-off hypotheses (phenotypic engineering) (Ketterson & Nolan 1992, 1999, Zera et al. 1998, Sinervo 1999).

Like other approaches, physiological manipulation has its limitations. For example, surgical removal of an organ does not necessarily abolish the energetic cost of producing or maintaining the function performed by that organ. Removal of ovaries does not abolish physiological costs of reproduction since yolk proteins are still synthesized in other organs, sometimes in enormous quantities (Chinzei & Wyatt 1985). Applied hormones can have pharmacological effects or can alter trade-offs in unsuspected ways by inducing the production of unknown regulators

(Zera et al. 1998). Thus, in the absence of direct measures of endocrine traits (e.g., in vivo hormone titers), only limited conclusions can be drawn from hormone manipulation experiments concerning specific endocrine mechanisms that underlie trade-offs (e.g., Zera et al. 1998, Zera & Cisper 2001). A developing consensus is that each empirical approach has its strengths and weaknesses, and that the use of multiple complementary approaches is essential for a thorough investigation of the physiological causes of life history trade-offs (Moller et al. 1989, Partridge & Sibly 1991, Sinervo & Basolo 1996, Rose & Bradley 1998; Zera et al. 1998; Zera and Cisper 2001).

MODEL SYSTEMS

Wing-Polymorphic Crickets

For over four decades, wing polymorphism has been viewed as a classic example of a life history trade-off that results from the differential allocation of internal reserves to ovarian growth versus somatic growth, maintenance, or storage (i.e., growth and maintenance of fight muscles, accumulation of flight fuels, Johnson 1969, Bell & Koufopanou 1986, Zera & Denno 1997). During the past decade, detailed energetic, endocrine, and genetic studies in wing polymorphic crickets have provided strong support for this notion (Mole & Zera 1993, Tanaka 1993, Zera et al. 1994, 1997, 1998, Zera & Denno 1997, Roff et al. 1997, Zera & Brink 2000, Zera & Cisper 2001). Wing polymorphism is now one of the life history trade-off models most thoroughly studied from a physiological perspective.

Wing polymorphic species consist of a flight-capable morph that has large functional flight muscles and large quantities of lipid flight-fuel, and of flightless morphs that have small, nonfunctional flight muscles and much reduced lipid stores. Early fecundity is negatively associated with flight capability (Roff 1986, Zera & Denno 1997, Zera et al. 1998). Differences between morphs in these dispersal and reproductive traits are often large, making wing polymorphism an exceptional experimental model for investigating trade-off physiology. For example, at the end of the first week of adulthood, flight muscles are typically reduced by 40%, ovaries are enlarged by 200%–400%, and whole-body triglyceride reserves are reduced by 30% in flightless versus flight-capable morphs of *Gryllus* (crickets) (Mole & Zera 1993, Zera et al. 1994, 1998, Zera & Larsen 2001).

Feeding studies of *Gryllus* species have provided some of the best documented examples of the relative importance of nutrient acquisition versus nutrient allocation in a life history trade-off. Naturally occurring or hormonally engineered flightless morphs of two *Gryllus* species consumed and assimilated the same amount of nutrients as the flight-capable morph (Mole & Zera 1993, Zera et al. 1998). Thus, the increased ovarian growth of the flightless morph in these species must have resulted, at least in part, from the greater allocation of absorbed nutrients, rather than from increased food consumption or assimilation by the flightless morph. In a third species, *G. firmus*, increased ovarian growth in the flightless morph was due to both increased nutrient allocation and greater food intake (Zera et al. 1998, Zera & Brink 2000, A.J. Zera & T. Rooneem, unpublished data).

Only by quantifying nutrient consumption and assimilation by the morphs could increased allocation of nutrients (internal trade-off) as opposed to increased acquisition of nutrients (no internal trade-off) have been identified as a potential cause of the difference in ovarian growth between morphs (Mole & Zera 1993, Zera & Denno 1997, Zera et al. 1998). The study of *G. firmus* is the only case in which genetically based differences in total nutrient consumption, assimilation, and allocation have been documented between phenotypes that differ in life histories (Zera & Brink 2000, Zera & Rooneem, unpublished data). One limitation of these investigations in *Gryllus* is that indices of nutrient allocation and acquisition were measured in units of mass rather than in units of energy (Zera et al. 1998, Zera & Brink 2000). However, more recent studies indicate that morphs differ to similar degree in these indices measured in units of energy (A.J. Zera & T. Rooneem, unpublished data). Although nutrient input is an integral component of energy budgets (Congdon et al. 1982, Withers 1992), many recent energetic studies of intraspecific life history trade-offs have not measured nutrient consumption or assimilation.

Some of the most detailed information on specific physiological causes of life history trade-offs has come from studies of *Gryllus*. Biosynthesis of triglyceride flight fuel may be an especially important physiological cost of flight capability that results in reduced ovarian growth (Zera & Larsen 2001). As mentioned above, whole-body triglyceride content was 30%–40% higher in flight-capable versus flightless morphs of two *Gryllus* species on a variety of diets (Zera & Denno 1997, Zera et al. 1994, 1998, Zera & Brink 2000, Zera & Larsen 2001). In *G. firmus*, a genetically based elevation in triglyceride stores in the flight-capable morph (*a*) was produced during a period of reduced ovarian growth in the flight-capable morph, (*b*) did not result from increased lipid intake from the diet, and (*c*) paralleled increased activities of enzymes of lipid biosynthesis and rate of lipid biosynthesis in the flight-capable morph (Zera & Brink 2000, Zera & Larsen 2001, Zhao & Zera 2001). The extent to which the negative association between lipid biosynthesis and ovarian growth results from an energetic trade-off (differential allocation of limited internal reserves to these two organismal traits) is unknown. Other possibilities include a regulatory trade-off in which antagonistic aspects of the control of lipid versus protein biosynthesis preclude the accumulation of elevated amounts of both of these compounds (Zhao & Zera 2001). Finally, the trade-off between lipid accumulation and ovarian growth could result from limited space within the abdomen.

Increased maintenance of flight muscles, but not flight muscle growth, has also been implicated as a potentially important energetic cost of flight capability. The large flight muscles of the flight-capable morph exhibited a 300%–350% greater respiration rate in vitro compared with the reduced flight muscles of the flightless morphs, and may contribute to the higher whole-body respiration rate of the flight-capable morph (Zera et al. 1997, 1998, Zera & Brink 2000). By contrast, on high nutrient diets, flight muscle growth, as indicated by increased mass or rate of muscle protein biosynthesis, is restricted to the first 2 to 3 days of adulthood in *Gryllus* species, before significant ovarian growth commences (Gomi et al. 1995, Zera et al. 1998, A.J. Zera & T. Rooneem, unpublished data). Thus, flight muscle growth does not appear to directly trade-off with ovarian growth. Finally, some potentially important contributors to the energetic cost of flight capability, such as increased activity of the flight-capable morph in the absence of flight, have yet to be assessed for any case of wing polymorphism (Zera et al. 1998).

Feeding studies in *Gryllus* also have provided information on the physiological mechanisms by which trade-offs may be exacerbated when nutrient input is decreased (Figure 1). In both *G. assimilis* and *G. firmus*, ovarian growth in the flight-capable versus the flightless morph was reduced to a greater degree on a low-nutrient diet compared with a high-nutrient diet. Whole-organism respiration was proportionally higher in the flight-capable morph compared with the flightless morph when on a low-nutrient diet compared with a high-nutrient diet (Zera et al. 1998, Zera & Brink 2000, A.J. Zera & T. Rooneem, unpublished data). These data suggest that some aspect of maintenance metabolism of the flight-capable morph, possibly flight-muscle maintenance, consumes a greater proportion of the reduced energy budget on the low-nutrient diet. This would reduce nutrients available for ovarian growth in the flight-capable versus the flightless morph to a greater degree on the low-nutrient diet, compared with the high-nutrient diet. The Colorado potato beetle histolyzes its flight muscles prior to overwintering, suggesting that maintenance metabolism of flight muscles may be a significant energy drain under prolonged low-nutrient conditions (El-Ibrashy 1965, Zera & Denno 1997). Alternatively, other factors such as increased activity of the flight-capable morph of *Gryllus* species may account for its increased respiratory metabolism, relative to the flight-capable morph, under low-nutrient conditions.

Because hormones regulate many key components of life histories (e.g., egg production, growth, metabolism), variation in endocrine regulation has long been suspected to be a primary physiological cause of life history trade-offs (Stearns 1989, Ketterson & Nolan 1992, 1999). Yet detailed studies of the mechanisms by which hormones regulate life history trade-offs have only begun during the past decade. Wing polymorphism in *Gryllus* is currently the most intensively studied life history trade-off with respect to endocrine regulation and is one of the only cases where genetic information is available on this topic (Zera & Tiebel 1989, Zera & Denno 1997, Roff et al. 1997, Zera & Huang 1999, Zera & Cisper 2001, Zera & Bottsford 2001). Thus far, hormonal studies have focused almost exclusively on the regulation of growth and degeneration of the key organs involved in the trade-off between flight capability and fecundity: ovaries and flight muscles. An elevated hemolymph titer of juvenile hormone (JH), a major gonadotropin in insects, has long been suspected to cause the elevated ovarian growth during early adulthood in the flightless morph of wing polymorphic species (Nijhout 1994, 1999, Zera & Denno 1997). This notion is supported by the strong positive effect of a JH mimic on ovarian growth in several *Gryllus* species (Zera et al. 1998, Zera &

Cisper 2001). Recent direct measures of the in vivo JH titer in *G. firmus* indicate that the hormonal regulation of morph-specific ovarian growth is more complex than previously suspected. The hemolymph JH titer is, in fact, lower in the flight-capable morph relative to the flightless morph in the morning. However, the level of this hormone rises dramatically (10- to 100-fold) during the day in the flight-capable morph, but not in the flightless morph, which leads to a higher JH titer in the flight-capable versus flightless morphs in the evening (Zera & Cisper 2001). The short-term elevation in the JH titer possibly regulates nocturnal flight in the long-winged morph of *G. firmus*. Although JH is likely involved in the increased ovarian growth of the flightless morph, it clearly cannot be the only factor involved and may not be the primary factor (Zera & Cisper 2001). This level of complexity in the endocrine regulation of a life history trade-off could only have been identified by direct measurement of in vivo hormone levels.

Discontinuous variation in the size of flight muscles, the other key organ participating in the trade-off between flight capability and ovarian growth, is produced by two different mechanisms. Differences in muscle growth between morphs can occur during the juvenile stage. Variation in the titers of JH and 20-hydroxy ecdysone, which are important developmental as well as reproductive hormones in insects, may regulate differences in muscle growth. Extensive endocrine-genetic studies suggest that the catabolic enzyme, juvenile hormone esterase, which is thought to regulate the JH titer in many insects, may modulate the JH titer in nascent *Gryllus* morphs, leading to variation in the development of flight muscles and wings (Zera & Tiebel, 1989, Zera & Denno 1997, Roff et al. 1997, Zera & Huang 1999, Zera 1999). Studies of Zera & Tiebel (1989) and Roff et al. (1997) are among the few investigations in which genetic covariance has been documented between an important component of a life history trait (i.e., wing and flight muscle phenotype) and a putative endocrine regulator of that trait.

Discontinuous variation in flight muscle mass in *Gryllus* and many other insects also results from histolysis (degeneration) of fully developed muscles during adulthood. Flight muscle degeneration occurs coincident with ovarian growth, which leads to a strong negative correlation between the size of these two organs (Pener 1985, Zera & Denno 1997). Hormonal manipulation in *G. firmus* and other insects strongly points to JH as the regulator of this trade-off (Pener 1985, Zera & Denno 1997, Zera & Cisper 2001). However, as mentioned above, an assessment of the exact mechanisms by which JH regulates the trade-off between ovaries and flight muscles is complicated by the large diurnal change in the JH titer in the flight-capable morph (Zera & Cisper 2001). Endocrine factors other than JH may also be involved in regulating this trade-off. To summarize, energetic and endocrine aspects of life history trade-offs have been extensively studied in species of *Gryllus*. However, the relative contribution of differential allocation of limited internal reserves, antagonistic aspects of hormonal regulation, or other factors, such as limited space within the organism, to the negative association between components of flight capability and ovarian growth is unclear and remains a major challenge to future research.

Drosophila

The distinguishing feature of studies using Drosophila is the range and power of genetic approaches that have been used to investigate life history trade-offs. As one example, artificial or natural selection in the laboratory is especially useful for studying life history evolution and mechanisms underlying apparent trade-offs (Rose et al. 1990, Huey & Kingsolver 1993, Rose et al. 1996, Gibbs 1999, Harshman & Hoffmann 2000a). Selected and control lines are often markedly divergent and can be useful for physiological studies of life history trade-offs, while indirect responses to selection can identify genetically based trade-offs (Rose et al. 1990).

A range of life history trade-offs have been implicated using *Drosophila* selection experiments. For example, selection for a relatively short larval period is correlated with reduced larval viability, diminished adult size, and reduced fecundity (Nunney 1996, Chippindale et al. 1997). Relative to selected lines (i.e., in control lines), an extended period of larval development was correlated with an accumulation of lipid and increased dry weight, as well as increased conservation of larval reserves in young adults (Chippindale et al. 1996, 1997).

The trade-off between extended longevity and early age reproduction has been a focus of numerous physiological-genetic studies. Selection for longevity and late life reproduction consistently produces lines characterized by longer-lived flies with decreased early age fecundity (Rose 1984, Luckinbill et al. 1984, Partridge et al. 1999). Sgro & Partridge (1999) sterilized females using irradiation or by crossing to a dominant sterile mutation, and in both cases they circumvented the age-specific mortality cost of reproduction. They concluded that aging was a function of the damaging effects of earlier reproduction. Underlying delayed early reproduction in one set of selected lines was a decreased rate of early age vitellogenic oocyte maturation (Carlson et al. 1998, Carlson & Harshman 1999). Moreover, there is preliminary evidence for an early age diminution in whole-body ecdysteroid titers in one set of lines selected for longevity and late life reproduction (Harshman 1999).

Differential respiration does not appear to play a role in the trade-off between longevity and early age reproduction. Variable results on respiration rates were obtained using one set of lines selected for longevity (summarized in Rose & Bradley 1998). When respiration was measured in small chambers, Service (1987) found relatively higher respiration rates at young ages in the control (unselected) lines, but not at later ages. However, Djawdan et al. (1996) found no difference between selected and control lines when respiration was measured in cages. When adults from both lines were provided with supplementary yeast, the control line females exhibited a slightly higher metabolic rate than selected line females (Simmons & Bradley 1997). Djawdan et al. (1997) found no difference in respiration rate when the mass of selected and control flies was adjusted by removing the weight of water, lipid, and carbohydrate.

The accumulation of energy storage compounds in flies selected for greater longevity suggests that there could be an energetic basis underlying the trade-off between fecundity and longevity (Service 1987, Djawdan et al. 1996). Djawdan et al. (1996) and Simmons & Bradley (1997) have determined that the selected (long-lived) females store more energy in the body and that the control line females, which are more fecund early in life, allocate much more energy to reproduction. However, energy might not be limiting in the laboratory environment because flies may consume as much food as needed for a given level of reproduction (Rose & Bradley 1998).

Stress resistance has been documented as a correlated response to selection for extended longevity and late-life reproduction. One set of lines selected for extended longevity exhibited desiccation resistance and increased glycogen content, as well as starvation resistance and increased lipid content (Rose 1984, Service et al. 1985). Another set of lines selected for extended longevity did not have substantially increased starvation resistance or desiccation resistance but was clearly resistant to oxidative stress (Luckinbill et al. 1984, Force et al. 1995). Increased oxidative stress resistance might be a consistent response to selection for extended longevity (Harshman & Haberer 2000). An association between longevity and starvation resistance is supported by extensively replicated selection experiments, on each of the traits (Service et al. 1985, Rose et al. 1992). Moreover, selection for reduced longevity was correlated with decreased lipid levels (Zwann et al. 1995). However, selection for extended longevity did not result in significant increased starvation resistance in two selection experiments (Zwann et al. 1995, Force et al. 1995). Selection on the basis of differential adult mortality did not affect starvation resistance as an indirect response to selection, but the response to selection did indicate trade-offs between early fecundity, late fecundity, and starvation resistance that were mediated by lipid allocation (Gasser et al. 2000). Selection for female starvation resistance was conducted using a wild-type stock maintained in the laboratory for approximately 35 generations prior to selection and using relatively large numbers of replicate-selected and control lines (Harshman & Schmid 1998). Selection for female starvation resistance resulted in multiple stress resistance as an indirect response to selection, but increased longevity was not a correlated response (Harshman & Schmid 1998, Harshman et al. 1999).

Relevant selection experiments identify a prospective trade-off between fecundity and adult stress resistance. Selection for increased desiccation resistance using *D. melanogaster* resulted in reduced early age fecundity in one study (Hoffmann & Parsons 1989), but not in another study (Chippindale et al. 1993). Selection for cold resistance in *D. melanogaster* and *D. simulans* was correlated with decreased early age fecundity (Watson & Hoffmann 1996). Selection for female starvation resistance resulted in reduced early age fecundity (Wayne et al. 2001). Lines selected for extended longevity and reverse selection also provide evidence for a negative genetic relationship between fecundity and starvation resistance (Service et al. 1985, Service et al. 1988, Leroi et al. 1994a), although this relationship changed in the course of a long-term selection experiment (Leroi et al. 1994b). A negative relationship between fecundity and ethanol or desiccation resistance was not observed in the reverse selection experiment (Service et al. 1988). Sib analysis revealed a negative genetic correlation between fecundity and starvation resistance (Service & Rose 1985). Phenotypic manipulation experiments also provide evidence for a trade-off between fecundity and stress resistance (Chippindale et al. 1993, Salmon et al. 2001, Wang et al. 2001). Perhaps stress susceptibility associated with reproduction contributes to decreased survival as a cost of reproduction.

There are problems associated with selection in the laboratory (Rose et al. 1990, Harshman & Hoffmann 2000a). Supernumerous correlations among selected traits could be the consequence of strong unrelenting directional selection in the laboratory (Harshman & Hoffmann 2000a). Other artifacts may arise from constraints on normal behavioral options in culture containers (Huey et al. 1991) and from copious food availability in the laboratory (Harshman & Hoffmann 2000a). Moreover, there is substantial variation in the quality of different selection experiments, and there are problems associated with experiments conducted on fly stocks recently derived from the field (Service & Rose 1985, Harshman & Hoffmann 2000a,b, Matos et al. 2000). For a variety of reasons, the indirect responses observed in similar selection experiments are often heterogeneous (Tower 1996, Harshman & Hoffmann 2000a). Consistent indirect responses (robust responses) are most informative with respect to implicating mechanisms underlying life history trade-offs (Harshman & Hoffmann 2000a, Harshman & Haberer 2000).

Another genetic approach has been to use lines of *D. melanogaster* that have been made homozygous for chromosomes isolated from natural populations for life history and physiological measurements (Clark 1989). In this study, there was a positive correlation between viability and fecundity, as well as associations between these traits and metabolic enzyme activities. A caveat is that inbreeding (inbreeding depression) can generate positive correlations among traits because of the pleiotropic effects of recessive deleterious alleles. As another genetic approach, P element mutagenesis has been used to induce mutational variation in longevity, age-specific reproduction, and metabolic enzyme activities (Clark & Guadalupe 1995, Clark et al. 1995). Finally, mutation analysis of life history characters is potentially informative for the study of the physiology of trade-offs (Lin et al. 1998).

Transgenic Drosophila can play an important role in investigating the physiology of life history trade-offs (Tatar 2000). For example, extra copies of a heat shock protein (Hsp70) gene in transgenic *D. melanogaster* confer heat stress resistance and reduced age specific mortality, but also result in decreased larval growth and survival (Tatar et al. 1997, Krebs & Feder 1997, Feder & Hoffman 1999). Stearns & Kaiser (1993) increased the expression of elongation factor EF-1 alpha by introducing an extra copy of the gene. The effects on a trade-off between reproduction and longevity were marginal and varied depending upon the experimental design. However, when enhanced EF-1 alpha expression increased lifespan, it also decreased fecundity.

Lizards

Studies of lizards have been at the forefront of research on life history physiology for over two decades (e.g., Tinkle & Hadley 1975, Congdon et al. 1982,

Schwarzkopf 1994, Marler & Moore 1991, Sinervo 1999). Many lizard species are amenable to experimental manipulation and field studies, thus allowing integrated physiological and ecological investigations of trade-offs (Marler & Moore 1991, Marler et al. 1995, Sinervo 1999). Lizards also grow significantly after sexual maturity and thus are a useful vertebrate model to study trade-offs between adult growth and reproduction. Many lizard species also store fat reserves for long periods of time and thus are useful for investigating trade-offs between reproduction and storage. Lizards were the focus of pioneering studies in which detailed energy budgets were constructed in interspecific investigations of life history evolution (reviewed in Congdon et al. 1982). More recently, studies on lizards were among the first to investigate, and currently provide some of the most important information on endocrine aspects of trade-offs (Marler & Moore 1988, Marler et al. 1995, Sinervo 1999, Sinervo et al. 2000).

The negative correlation between egg number and egg size is a classic life history trade-off (Stearns 1992, Roff 1992, Schwarzkopf et al. 1999) that has been documented in a variety of lizard species (Sinervo 1990, Schwarzkopf 1994, Doughty & Shine 1997). However, correlations were derived mainly from uncontrolled and unmanipulated field studies and thus provide little information on underlying physiological mechanisms. In a breakthrough series of studies, Sinervo and colleagues (reviewed in Sinervo 1999) used experimental manipulations to study the proximate mechanisms controlling the trade-off between egg size and egg number in the side-blotched lizard, *Uta stansburiana*. Application of the gonadotrophin, FSH (follicle-stimulating hormone), increased egg number but decreased egg size, whereas ablation of a proportion of the egg follicles resulted in a smaller number of larger eggs. Experimental results were interpreted in terms of competition by the developing eggs for limited yolk or space in the maternal abdomen. These studies were among the first to experimentally investigate the endocrine causes of a classic life history trade-off and to suggest that variation in the titer of, or tissue response to, FSH in field populations may be a physiological cause of the trade-off between egg size and egg number. However, these hormone manipulation experiments need to be interpreted with caution since FSH titers or receptors have yet to be directly measured in individuals of this species.

Phenotypic manipulation allowed Sinervo and colleagues to investigate several aspects of the evolution of the trade-off between egg size and egg number (reviewed in Sinervo 1994, 1999). Lizards of different size produced by endocrine or surgical manipulation were returned to field sites, and their survivorship to adult maturity was monitored. Results showed that the trade-off between egg size and egg number was influenced by natural selection. A morphological constraint on egg size in *Uta* was also identified. Unusually large eggs became lodged in the oviduct or burst upon oviposition in the laboratory (Sinervo & Licht 1991).

Behaviors that enhance current reproduction but that decrease future reproduction are important components of the cost of reproduction. Marler and colleagues (Marler & Moore 1988, 1991, Marler et al. 1995) investigated the endocrine regulation of this type of behavior-based trade-off in the field using the lizard *Sceloporus jarrovi*. In this species, testosterone is elevated during the nonbreeding season and functions to elicit behaviors involved in territorial defense (Moore 1986). A comprehensive set of field experiments involving measurement of energy consumption using doubly labeled water, supplemental feeding, and measurement of time-energy budgets showed that lizards with experimentally elevated testosterone (but within physiological levels) had increased activity and reduced survivorship. This increased activity resulted from an increased number of territorial displays and movements, with consequent decreased time spent feeding (Marler et al. 1995). Increased testosterone level was also associated with a higher parasite load (C.A. Marler, personal communication). These and other studies (e.g., Sinervo et al. 2000, Veiga et al. 1998) are important because they identify potential direct negative effects of the hormonal regulation of reproduction and behavior.

Lizard studies were at the forefront of interspecific investigations of life history energetics during the 1970s and early 1980s (reviewed in Congdon et al. 1982). More recent studies have focused on the effect of variation in the thermal environment on seasonal or interpopulational variation in life histories (Dunham et al. 1989, Adoph & Porter 1993, Niewiarowski 2001). However, there have been surprisingly few detailed studies of the energetic correlates of life history variation within populations of lizards. A few studies have focused on variation in energy reserves, growth, and metabolism within populations, especially in the context of the cost of reproduction. For example, metabolic rates measured in the field were higher in reproductive versus nonreproductive females of several species (Nagy 1983, Anderson & Karasov 1988). A number of correlational and experimental studies indicate that a common effect of current reproduction may be the reduction of internal reserves, which impacts future reproduction either by reducing survivorship or by reducing nutrients required for future reproduction (Landwer 1994, Schwarzkopf 1994, Doughty & Shine 1997, 1998, Wilson & Booth 1998). The trade-off between reproductive effort and energy storage in many purely correlational studies may have been underestimated or missed entirely because of the masking effect of variation in storage reserves (Doughty & Shine 1997).

The physiological cost of current reproduction on future reproduction could be mitigated if nutrient reserves are replenished before or during future reproduction. However, field studies indicate that up to 90% of the energy used for egg production comes from fat stores rather than recently consumed food (Karasov & Anderson 1998). Doughty & Shine (1998) also showed that variation in nutrient input affected variation in lipid stores and reproductive output during the next year. Thus, long time lags can occur between variation in nutrient acquisition and variation in reproduction, further complicating physiological studies of trade-offs in the field.

On average, about 30% of growth in lizards occurs after sexual maturity, and fecundity in lizards is correlated with body size (Schwarzkopf 1994). Hence, current reproduction can reduce future reproduction by diverting nutrients away from current growth. Such a trade-off between current reproduction and growth is suggested by increased growth during years in which reproduction does not occur in some lizard species (Schwarzkopf 1993). Landwer (1994) found that *Urosaurus ornatus*, in which egg production was experimentally reduced, showed greater growth in the field compared with unmanipulated controls.

Ecological factors such as age-dependent survivorship can have a strong influence on physiological aspects of trade-offs, such as the prioritization of nutrient allocation. For example, short-lived mainland anoles allocated a greater amount of supplemental food to current reproduction, compared with long-lived island anoles which converted supplemental food mainly into fat reserves (Guyer 1988, Schwarzkopf 1994). Tail breakage is an important adaptation for escape from predation in lizards (Dial & Fitzpatrick 1981, Wilson & Booth 1998). Allocation of energy to tail regeneration versus current reproduction in lizards depends upon whether the species is short- or long-lived (Dial & Fitzpatrick 1981, Wilson & Booth 1998). The relative importance of ecological factors such as predation, and physiological factors such as allocation, in the evolution of life history traits such as age-specific fecundity is the subject of debate in lizard studies. Results of various mathematical models have led workers to support either physiological or ecological factors as being of primary importance (Shine & Schwarzkopf 1992, Niewiarowski & Dunham 1994, Shine et al. 1996).

Finally, studies of skinks indicate how lineage-specific attributes can magnify an apparent life history trade-off. In lizards, tail loss allows escape from predators but reduces future reproduction because the tail is an important lipid storage organ. Certain lineages of skinks lack abdominal fat bodies and store nearly 100% of body lipid reserves in the tail. The impact of tail loss is especially pronounced in skinks that lack abdominal fat bodies, reducing clutch sizes 50% to 100% (Wilson & Booth 1998).

Birds

Three ornithologists were the intellectual predecessors of modern life history thinking: Lack, Skutch, and Moreau (Ricklefs 2000a). Lack's insight (Lack 1947, 1954), that clutch size evolves to maximize individual reproductive success, became a cornerstone of life history theory and motivated physiological studies on life histories. Continuing to the present, extensive knowledge about bird demography and intensive field studies provide an important context for physiological studies of life history trade-offs. However, it has been argued that our knowledge of avian life histories is still rudimentary, and thus basic studies of biochemistry and physiology of life history traits and the controlling endocrine mechanisms are badly needed (Ricklefs 2000b). Because of space constraints, this section focuses on two topics: energetics, especially in relation to the cost of reproduction, and immunocompetence, a topic of relatively intense recent study on birds.

Birds have the highest core body temperature of any endotherm (Nagy 1987) and flying birds have size constraints that presumably limit the amount of energy storage (Calder 1984). Thus, it would appear that energetics must play an important role in shaping bird life histories (Walsberg 1983, Paladino 1989, McNab 1997). Birds expend relatively high levels of energy during reproduction and thermoregulation

(Weiner 1992). Energy constraints, which can arise from the competing costs of reproduction, thermoregulation, growth, energy assimilation, and energy assimilation capacity (Karasov 1986, Weiner 1992, Konarzewski 1995, Jackson & Diamond 1996), could be integral to life history evolution. As an alternative, the cost of foraging and the energetic demands of other external activities could be limiting, and various aspects of physiology might simply adjust to meet energetic requirements at any particular point in the life history (Ricklefs 1991, 1996). From another perspective, predation appears to be a major factor in shaping bird life histories (Conway & Martin 2000, Ghalambor & Martin 2000). In general, it is not clear to what degree life histories are defined by endogenous trade-offs and physiological constraints.

Either from observational data or by experimental clutch manipulation, there is evidence for a cost of reproduction in birds (Linden & Moller 1989, Dijkstra et al. 1990, Styrsky et al. 1999). This cost can be manifest in terms of parental survival, future reproduction and/or offspring condition-fitness. The cost of reproduction is often observed as an effect on offspring (Linden & Moller 1989). Egg mass representing the energetic investment of reproduction has an effect on initial offspring growth and development of passerines, but the effect of egg mass often does not extend to older offspring (Styrsky et al. 2000). Other factors that are positively correlated with egg mass could have an extended effect on offspring (Styrsky et al. 2000). In lesser black-backed gulls (*Larus fuscus*), the order of egg laying was correlated with abundance of antioxidants (carotenoid and vitamin E), arachidonic acid, and cholesterol ester fractions in the eggs (Royle et al. 1999). Fat storage could partially underlie the trade-off between reproduction and parental survival, especially in passerines where energy storage reserves tend to be quite limited. Increased fat storage could support reproduction, but there can be a survival cost (Lima 1986, Witter & Cuthill 1993). Specifically, stored fat might incur acquisition and maintenance costs (Houston et al. 1997) as well as flight energetics or agility costs (Rogers & Smith 1993, Gosler et al. 1995).

Overall, metabolism and energy budgets have been the focus of physiological work on the cost of reproduction. Historically, evidence accumulated for an energetic cost of bird reproduction (King 1973, Ricklefs 1974, Drent & Daan 1980). For many species, the preponderance of this cost is due to parental care rather than egg production (Trillmich 1986, Ward 1996). The ecology of some birds can make this cost acute. For example, the low foraging efficiency of two diving petrels (*Pelecanoides urinatrix* and *P. georgicus*) and an auklet (*Aethia pusilla*) results in relatively high daily energy expenditure during the breeding season (Roby & Ricklefs 1986). Parental weight loss is common at the time of brood provisioning, and this loss can reflect energy demands on the parents (Martin 1987). When clutch sizes were experimentally increased during incubation, there was an increase in parental energy expenditure in eight of nine studies, but diminished adult body condition was observed in only two of five studies (Thompson et al. 1998). In the glaucous-winged gull (*Larus glaucescens*), experimental brood enlargement resulted in increased parental reproductive effort, increased loss of weight during the breeding season, and reduced winter survival (Reid 1987). Is the cost of reproduction mediated by energetics? In general, it is conceptually and empirically difficult to demonstrate that an energy budget deficit is directly responsible for the costs of reproduction.

A decline in immunocompetence and potential damage from microbes and parasites could be a physiological cost of reproduction. In general, there is considerable current interest in avian immune function in relation to physiological trade-offs and life histories (Sheldon & Verhulst 1996, Horak et al. 1999, Owens & Wilson 1999). In the collared flycatcher (*Ficedula albicollins*) and other bird species, there tends to be a negative relationship between brood size and immune system indicators (Gustafsson et al. 1994). The allocation of endogenous energy might play an important role in a trade-off between reproduction and immune system function (Apanius 1998). However, immunocompetence can be measured in numerous ways, and it is important to be aware of the assumptions, limitations, and consequences of various methods (Norris & Evans 2000). Moreover, studies relating immunocompetence to fitness and disease resistance or resource limitation in the field are rare (Owens & Wilson 1999, Norris & Evans 2000). One prospect for the future is to directly manipulate immunocompetence (Sheldon & Verhulst 1996) in relation to the cost of reproduction or other life history trade-offs.

Is immune function costly? When a passerine (*Parus caeruleus*) was subjected to cold stress, the increase in daily energy expenditure was associated with decreased antibody responses (Svensson et al. 1998). The energetic cost of the antibody response was estimated to be 8% to 13% of the basal metabolic rate. Although there is direct and indirect support for the hypothesis of a costly immune response (Sheldon & Verhulst 1996), a number of studies have failed to find a significant relationship between energetics and the immune response (Owens & Wilson 1999). Moreover, it has been argued that the energetic cost of immune system function might be relatively low (Hillgarth & Wingfield 1997, Owens & Wilson 1999).

The relative cost of life history traits is relevant to a prospective trade-off with the immune system. Hoglund & Sheldon (1998) argue that allocation to sexual display, which is apparently costly, should be considered in the same manner as allocation to a classic life history trait. A relatively high level of cost for sexual ornamentation could trade-off with immune function (Sheldon & Verhulst 1996, Norris & Evans 2000). Hamilton & Zuk (1982) formulated the idea that the intensity of secondary sexual ornamentation would signal resistance to parasites. There is evidence for positive correlations between sexual ornamentation and resistance to parasites and pathogens (reviewed in Moller 1990, Hamilton & Poulin 1997, Lindstrom & Lundstrom 2000). However, there also is countervailing evidence for a negative relationship, perhaps a trade-off, between sexual ornamentation and infection (Moller 1997, Norris et al. 1994, Richner et al. 1995, Dale 1996, Sheldon & Verhulst 1996). Hormones could play a role in such a trade-off by intensifying sexual display with a concomitant decrease in immune function (Folstad & Karter 1992). Zuk et al. (1995) have shown that testosterone level is negatively associated with leukocyte counts in the red jungle fowl (*Gallus gallus*), but no such relationship was detected in the barn swallow (*Hirundo rustica*) by Saino et al. (1995). However, Saino et al. (1995) found reduced antibody production when testosterone was administered to the barn swallows in the field. No suppression of the avian immune system was observed in a laboratory study of the effects of testosterone on red-winged blackbirds (*Agelaius phoeniceus*) (Hasselquist et al. 1999), nor was there a correlation between elevated testosterone titers and parasite abundance in the same species (Weatherhead et al. 1993). Similarly, Ros et al. (1997) did not find a negative correlation between testosterone titer and antibody production in black-headed gulls (*Larus ridibundus*). In a study on the house sparrow (*Passer domesticus*), testosterone administration appeared to decrease antibody production (Evans et al. 2000). In this study, the suppressive effect of decreased corticosterone was taken into account, and consequently the net effect of testosterone was to stimulate antibody production hypothetically by "dominance influencing access to resources" (Evans et al. 2000). Testosterone administration to captive and field dark-eyed juncos (*Junco hyemalis*) resulted in decreased antibody production and increased corticosterone abundance (Castro et al. 2001). Testosterone could affect immune function indirectly via corticosteroids (Evans et al. 2000, Castro et al. 2001). Corticosteroids have been associated with increased parasite loads (Hillgarth & Wingfield 1997). Gonadal hormones, including testosterone and metabolites, can play a role in controlling adult plumage (Hillgarth & Wingfield 1997). The components of an endocrine mediated trade-off between the immune system and sexual ornamentation have been documented in some studies, and differential resource allocation could play a role in this prospective trade-off.

Avian endocrinology studies have identified mechanisms whereby life history traits are linked. Bird corticosteroids can increase gluconeogenesis and survival under stress conditions, but suppress reproduction and the immune system (Wingfield 1988, Wingfield et al. 1998). Testosterone administered to dark-eyed junco (*Junco hyemalis*) males resulted in physiological changes including accelerated loss of mass, reduced subcutaneous fat, and a delay in the molt (Ketterson et al. 1991, Nolan et al. 1992). Studies on the dark-eyed junco and a diversity of other bird species indicate that testosterone can stimulate mating effort, suppress parental effort, and could be associated with a diminished life span (Ketterson & Nolan 1999). As mentioned previously in this review, hormones often regulate multiple antagonistic processes, and thus are notable candidate mediators of life history trade-offs (Ketterson & Nolan 1992, 1999, Finch & Rose 1995).

SUMMARY AND FUTURE DIRECTIONS

The past decade has seen significant advances in physiological studies of life history trade-offs. With respect to the topic of energetics, the most important advances have been (*a*) the first genetic analyses of energetic components of trade-offs, (*b*) the integration of investigations of nutrient input and allocation, (*c*) development of more sophisticated models of resource acquisition and allocation, and (*d*) more detailed investigations of specific physiological costs associated with life history traits. Advances have largely come about owing to a shift to more focused, comprehensive, and integrated studies.

Although hormones have long been suspected to be key regulators of life history trade-offs (Stearns 1989, Ketterson & Nolan 1992, 1999), only during the past decade have the first detailed endocrine studies of trade-offs been undertaken. These investigations also constitute one of the most significant recent advances in functional studies of life history trade-offs. Thus far endocrine investigations have involved manipulation or quantification of the titers of hormones or the activities of enzymes that regulate hormones, that potentially control life history trade-offs. This approach will almost certainly become more common in the future. Endocrine studies will also likely expand to other important topics, such as investigations of the endocrine regulation of trade-offs at the level of hormone receptors, and ultimately the hormonal control of gene expression.

A central unresolved issue in trade-off physiology is the relative importance of energetic, regulatory, and other (e.g., space) constraints as causes of life history trade-offs. Although energetics has been the primary focus of functional studies of trade-offs, in no case is it certain that the differential allocation of limiting internal resources is the primary cause of a life history trade-off. There is an increasing appreciation that the physiological mechanisms that underlie life history trade-offs are complex and that their elucidation will require a much deeper understanding of basic energetic, regulatory, and structural aspects of organismal function and how these aspects interact.

Molecular genetic and genomic studies of organismal trade-offs, typically in model species, have just begun. It is now possible to study gene regulation in terms of transcription factors and regulatory DNA elements to some degree in any organism. Genomic and cDNA library sequences, microarrays, general transformation procedures, and bioinformatics are "democratizing the genome." Thus, extension of molecular genetic studies beyond a few model species to many species with life histories of interest will likely constitute an important future development relevant to functional studies of trade-offs.

A potentially important avenue of research includes traits and processes that have not traditionally been investigated in a life history context. For example, oxidative damage may be one of the universal challenges of life on earth and oxidative stress resistance may trade-off with life history characters (Salmon et al. 2001). The cost of acclimation (Hoffmann 1995, Huey & Berrigan 1996) and development (Ricklefs 1979, Lindstrom 1999) are additional examples of topics that could be important in life history studies, but which have received scant attention.

This is an exciting time in evolutionary biology, in which the confluence of methodological and conceptual advances in molecular genetics, development, and physiology will result in an increasingly mechanistic understanding of evolution in model systems and in the diversity of living organisms. In particular, delineation of mechanisms underlying life history trade-offs will allow us to understand how organisms do and can evolve.

ACKNOWLEDGMENTS

A.J. Zera acknowledges support from the National Science Foundation (DEB-9107429, IBN-9507388, and IBN-9808249) and the Research Council of the University of Nebraska. L.G. Harshman acknowledges support from the National Institute of Aging (NIA AG08761) and the Research Council of the University of Nebraska. We thank the following for their comments on this manuscript or for providing the first sentence of the final paragraph of the review: A. Chippindale, R. Huey, E. Ketterson, R. Gibson, J. Kingsolver, C. Marler, R. Ricklefs, and L. Schwarzkopf.

Visit the Annual Reviews home page at www.AnnualReviews.org

LITERATURE CITED

Adoph SC, Porter WP. 1993. Temperature, activity, and lizard life histories. *Am. Nat.* 142:273–95

Anderson RA, Karasov WH. 1988. Energetics of the lizard *Cnemidophorus tigris* and life history consequences of food-acquisition mode. *Ecol. Monogr.* 58:79–110

Antonovics J, van Tienderen PH. 1991. Ontoecogeniphyloconstraints? The chaos of constraint terminology. *Trends Ecol. Evol.* 6: 166–68

Apanius V. 1998. Stress and immune defense. *Adv. Stud. Behav.* 27:133–53

Bell G, Koufopanou V. 1986. The cost of reproduction. In *Oxford Surveys in Evolutionary Biology*, ed. R Dawkins, M Ridley, 3:83–131 Oxford, UK: Oxford Univ. Press

Bernardo J. 1994. Experimental analysis of allocation in two divergent, natural salamander populations. *Am. Nat.* 143:14–38

Boggs C. 1992. Resource allocation: exploring connections between foraging and life history. *Funct. Ecol.* 6:508–18

Boggs C. 1997. Dynamics of reproductive allocation from juvenile and adult feeding: radiotracer studies. *Ecology* 78:192–202

Boggs C, Ross CL. 1993. The effect of adult food limitation on life history traits in *Speye-*

ria mormonia (Lepidoptera: Nymphalidae). *Ecology* 74:433–41

Bonnet X, Bradshaw D, Shine R. 1998. Capital vs. income breeding: an ectothermic perspective. *Oikos* 83:333–42

Calder WA. 1984. *Size, Function and Life History*. Cambridge, MA: Harvard Univ. Press

Calow P. 1979. The cost of reproduction–a physiological approach. *Biol. Rev.* 54:23–40

Carlson KA, Harshman LG. 1999. Extended longevity lines of *Drosophila melanogaster*: characterization of oocyte stages and ovariole numbers as a function of age and diet. *J. Gerontol.* 54:B432–40

Carlson KA, Nusbaum TJ, Rose MR, Harshman LG. 1998. Oocyte maturation and ovariole numbers in lines of *Drosophila melanogaster* selected for postponed senescence. *Funct. Ecol.* 52:514–20

Castro JM, Nolan V, Ketterson ED. 2001. Steroid hormones and immune function: experimental studies in wild and captive dark-eyed juncos (*Junco hyemalis*). *Am. Nat.* 157:408–20

Charlesworth B. 1990. Optimization models, quantitative genetics, and mutation. *Evolution* 44:520–38

Chinzei Y, Wyatt GR. 1985. Vitellogenin

titre in haemolymph of *Locusta migratoria* in normal adults, after ovariectomy, and in response to methoprene. *J. Insect Physiol.* 31:441–45

Chippindale AK, Alipaz JA, Chen HW, Rose MR. 1997. Experimental evolution of accelerated development in *Drosophila*. 1I. Developmental speed and larval survival. *Evolution* 51:1536–51

Chippindale AK, Chu TJ, Rose MR. 1996. Complex trade-offs and the evolution of starvation resistance in *Drosophila melanogaster*. *Evolution* 50:753–66

Chippindale AK, Leroi AM, Kim SB, Rose MR. 1993. Phenotypic plasticity and selection in *Drosophila* life history evolution. I. Nutrition and the cost of reproduction. *J. Evol. Biol.* 6:171–93

Clark AG. 1989. Causes and consequences of variation in energy storage in *Drosophila melanogaster*. *Genetics* 123:131–44

Clark AG, Guadalupe RN. 1995. Probing the evolution of senescence in *Drosophila melanogaster* with P-element tagging. *Genetica* 96:225–34

Clark AG, Wang L, Hullenberg T. 1995. P-element-induced variation in metabolic regulation in *Drosophila*. *Genetics* 139:337–48

Clutton-Brock TH, Guinness FE, Albon S. 1982. *Red Deer: Behavior and Ecology of Two Sexes*. Chicago, IL: Univ. Chicago Press

Congdon JD, Dunham AE, Tinkle DW. 1982. Energy budgets and life histories of reptiles. In *Biology of the Reptilia*, ed. C Gans, pp. 233–71. New York: Academic

Conway CJ, Martin TE. 2000. Evolution of passerine incubation behavior: influence of food, temperature, and nest predation. *Int. J. Evol.* 54:670–85

Dale S, Kruszenski A, Slagsvold T. 1996. Effects of brood parasites on sexual and natural selection in the pied flycatcher. *J. Zool.* 238:373–93

de Jong G. 1993. Covariances between traits deriving from successive allocations of a resource. *Funct. Ecol.* 7:75–83

de Jong G, van Noordwijk AJ. 1992. Acquisition and allocation of resources: genetic (co)variances, selection, and life histories. *Am. Nat.* 139:749–70

de Laguerie P, Olivieri I, Atlan A, Gouyon PH. 1991. Analytic and simulation models predicting positive genetic correlations between traits linked by trade-offs. *Evol. Ecol.* 5:361–69

Dial BE, Fitzpatrick LC. 1981. The energetic costs of tail autotomy to reproduction in the lizard *Coleonyx brevis* (Saura: Gekkonidae). *Oecologia* 51:310–17

Dijkstra C, Bult A, Bulsma S, Daan S, Meier T, et al. 1990. Brood size manipulations in the kestral (*Falco tinnunculus*): effects on offspring and parent survival. *J. Anim. Ecol.* 59:269–85

Djawdan M, Rose MR, Bradley TJ. 1997. Does selection for stress resistance lower metabolic rate? *Ecology* 78:828–37

Djawdan M, Sugiyama TT, Schlaeger LK, Bradley TJ, Rose MR. 1996. Metabolic aspects of the trade-off between fecundity and longevity in *Drosophila melanogaster*. *Physiol. Zool.* 69:1176–95

Doughty P, Shine R. 1997. Detecting life history trade-offs: measuring energy stores in "capital" breeders reveals costs of reproduction. *Oecologia* 110:508–13

Doughty P, Shine R. 1998. Reproductive energy allocation and long-term energy stores in a viviparous lizard (*Eulamprus tympanum*). *Ecology* 79:1073–83

Drent RH, Daan S. 1980. The prudent parent: energetic adjustments in avian breeding. *Ardea* 68:225–52

Dunham AE, Grant BW, Overall KL. 1989. Interfaces between biophysical and physiological ecology and the population ecology of terrestrial vertebrate ectotherms. *Physiol. Zool.* 62:335–55

El-Ibrashy MT. 1965. A comparative study of metabolic effects of the corpus allatum in two adult Coleoptera, in relation to diapause. *Meded. Landbouwhogesch. Wageningen* 65:275–97

Evans MR, Goldsmith AR, Norris SRA. 2000. The effects of testosterone on antibody production and plumage coloration in male house sparrows (*Passer domesticus*). *Behav. Ecol. Sociobiol.* 47:156–63

Feder ME, Hofmann GH. 1999. Heat-shock proteins, molecular chaperones, and the stress response: evolutionary and ecological physiology. *Annu. Rev. Physiol.* 61:243–82

Finch CE, Rose MR. 1995. Hormones and the physiological architecture of life history evolution. *Q. Rev. Biol.* 70:1–51

Fisher RA. 1930. *The Genetical Theory of Natural Selection.* New York: Dover

Folstad I, Karter X. 1992. Parasites, bright males and the immuno-competence handicap. *Am. Nat.* 139:603–22

Force AG, Staples T, Soliman S, Arking R. 1995. Comparative biochemical and stress analysis of genetically selected *Drosophila* strains with different longevities. *Dev. Genet.* 17:340–51

Gasser M, Kaiser M, Berrigan D, Stearns SC. 2000. Life-history correlates of evolution under high and low adult mortality. *Evolution* 54:1260–72

Ghalambor CK, Martin TE. 2000. Parental investment strategies in two species of nuthatch vary with stage-specific predation risk and reproductive effort. *Anim. Behav.* 60:263–67

Gibbs AG. 1999. Laboratory selection for the comparative physiologist. *J. Exp. Biol.* 202:2709–18

Gomi T, Okuda T, Tanaka S. 1995. Protein synthesis and degradation in the flight muscles of adult crickets (*Gryllus bimaculatus*). *J. Exp. Biol.* 198:1071–77

Gosler AG, Greenwood JD, Perrins C. 1995. Predation risk and the cost of being fat. *Nature* 377:621–23

Gustafsson L, Nordling D, Andersson MS, Sheldon BC, Ovarstrom A. 1994. Infectious diseases, reproductive effort and the cost of reproduction in birds. *Philos. Trans. R. Soc. London Ser. B* 346:323–31

Guyer C. 1988. Food supplementation in a tropical mainland anole, *Norops humilis*: Effects on individuals. *Ecology* 69:362–69

Hamilton WD, Zuk M. 1982. Heritable true fitness and bright birds: a role for parasites? *Science* 218:384–87

Hamilton WJ, Poulin R. 1997. The Hamilton and Zuk hypothesis revisited: a meta-analytical approach. *Behavior* 134:299–320

Harshman LG. 1999. Investigation of the endocrine system in extended longevity lines of *Drosophila melanogaster*. *Exp. Gerontol.* 34:997–1106

Harshman LG, Haberer BA. 2000. Oxidative stress resistance: a robust correlated response to selection in extended longevity lines of *Drosophila melanogaster*? *J. Gerontol. A. Biol. Sci. Med. Sci.* 55:B415–17

Harshman LG, Hoffmann AA. 2000a. Laboratory selection experiments using *Drosophila*: what do they really tell us? *Trends Ecol. Evol.* 15:32–36

Harshman LG, Hoffmann AA. 2000b. Reply from LG Harshman and AA Hoffmann. *Trends Ecol. Evol.* 15:207

Harshman LG, Moore KM, Sty MA, Magwire MM. 1999. Stress resistance and longevity in selected lines of *Drosophila melanogaster*. *Neurobiol. Aging* 20:521–29

Harshman LG, Schmid JL. 1998. Evolution of starvation resistance in *Drosophila melanogaster*: aspects of metabolism and counter-impact selection. *Evolution* 52:1679–85

Hasselquist DJ, Sherman PW, Wingfield JC. 1999. Is avian humoral immunocompetence suppressed by testosterone? *Behav. Ecol. Sociobiol.* 45:167–75

Haukioja E, Hakala T. 1986. Life history evolution in *Anodonta piscinalis*. *Oecologia* 35:253–66

Heino M, Kaitala V. 1999. Evolution of resource allocation between growth and reproduction in animals with indeterminate growth. *J. Evol. Biol.* 12:423–29

Hillgarth N, Wingfield JC. 1997. Parasite-mediated sexual selection: endocrine aspects. In *Parasite-Mediated Sexual Selection: Endocrine Aspects*, ed. DH Clayton, J Moore, pp. 78–104. Oxford, UK: Oxford Univ. Press

Hoffmann AA. 1995. Acclimation: increasing survival at a cost. *Trends Ecol. Evol.* 10:1–2

Hoffmann AA, Parsons PA. 1989. An integrated approach to environmental stress tolerance and life history variation. Desiccation tolerance in *Drosophila*. *Biol. J. Linn. Soc.* 37:117–36

Hoglund J, Sheldon BC. 1998. The cost of reproduction and sexual selection. *Oikos* 83:478–83

Horak P, Tegelmann L, Ots I, Moller AP. 1999. Immune function and survival of great tit nestlings in relationship to growth conditions. *Oecologica* 121:316–22

Houle D. 1991. Genetic covariance of fitness correlates: what genetic correlations are made of and why it matters. *Evolution* 45:630–48

Houston AI, Welton NJ, McNamara JM. 1997. Acquisition and maintenance costs in the long-term regulation of avian fat reserves. *Oikos* 78:331–40

Huey RB, Berrigan D. 1996. Testing evolutionary hypotheses of acclimation. In *Animals and Temperature: Phenotypic and Evolutionary Adaptation*, ed. IA Johnston, AF Bennett, pp. 205–37. Cambridge, UK: Cambridge Univ. Press

Huey RB, Kingsolver JG. 1993. Evolution of resistance to high temperature in ectotherms. *Am. Nat.* 142:521–46

Huey RB, Partridge L, Fowler K. 1991. Thermal sensitivity of *Drosophila melanogaster* responds rapidly to laboratory natural selection. *Evolution* 43:751–56

Jackson S, Diamond J. 1996. Metabolic and digestive responses to artificial selection in chickens. *Evolution* 50:1638–50

Johnson CG. 1969. *Migration and Dispersal of Insects by Flight*. London: Methuen

Jokela J, Mutikainen P. 1995. Phenotypic plasticity and priority rules for energy allocation in a freshwater clam: a field experiment. *Oecologia* 104:122–32

Jonsson KI. 1997. Capital and income breeding as alternative tactics of resource use in reproduction. *Oikos* 78:57–66

Kaitala A. 1987. Dynamic life-history strategy of the waterstrider *Gerris thoracicus* as an adaptation to food and habitat variation. *Oikos* 48:125–31

Karasov WH. 1986. Energetics, physiology, and vertebrate ecology. *Trends Ecol. Evol.* 1:101–4

Karasov WH, Anderson RA. 1998. Correlates of average daily metabolism of field-active zebra-tailed lizards (*Callisaurus draconoides*). *Physiol. Zool.* 71:93–105

Ketterson ED, Nolan V. 1992. Hormones and life histories: an integrative approach. *Am. Nat.* 140:S33–62

Ketterson ED, Nolan V. 1999. Adaptation, exaptation, and constraint. *Am. Nat.* 154:S4–25

Ketterson ED, Nolan V, Wolf L, Ziegenfus A, Dufty M, et al. 1991. Testosterone and avian life histories: the effect of experimentally elevated testosterone on corticosterone and body mass in dark-eyed juncos (*Junco hyemalis*). *Horm. Behav.* 25:489–503

King JR. 1973. Energetics of reproduction in birds. In *Breeding Biology in Birds*, ed. DS Farner, pp. 78–107. Washington, DC: Natl. Acad. Sci.

Konarzewski M. 1995. Allocation of energy to growth and respiration in avian postembryonic development. *Ecology* 76:8–19

Krebs RA, Feder ME. 1997. Natural variation in the expression of the heat-shock protein HSP70 in a population of *Drosophila melanogaster* and its correlation with tolerance of ecologically relevant thermal stress. *Evolution* 51:173–79

Lack D. 1947. The significance of clutch size. *Ibis* 89:302–52

Lack D. 1954. *The Natural Regulation of Animal Numbers*. Oxford, UK: Clarendon

Landwer AJ. 1994. Manipulation of egg production reveals costs of reproduction in the tree lizard (*Urosaurus ornatus*). *Oecologia* 100:243–49

Lauder GV. 1996. The argument from design. In *Adaptation*, ed. MR Rose, GV Lauder, pp. 55–91. San Diego, CA: Academic

Leroi AM, Chen WR, Rose MR. 1994a. Long-term laboratory evolution of a genetic life history trade-off in *Drosophila melanogaster*. 2. Stability of genetic correlations. *Evolution* 48:1258–68

Leroi AM, Chippindale AK, Rose MR. 1994b.

Long-term laboratory evolution of a genetic life-history trade-off in *Drosophila melanogaster*. 1. The role of genotype-by-environment interaction. *Evolution* 48:1244–57

Lima SL. 1986. Predation risk and unpredictable feeding conditions: determinants of body mass in birds. *Ecology* 67:366–76

Lin YJ, Seroude L, Benzer S. 1998. Extended life-span and stress resistance in the *Drosophila* mutant *methuselah*. *Science* 282:943–46

Linden M, Moller AP. 1989. Cost of reproduction and covariation of life history traits in birds. *Trends Ecol. Evol.* 6:183–85

Lindstrom J. 1999. Early development and fitness in birds and mammals. *Trends Ecol. Evol.* 14:343–48

Lindstrom K, Lundstrom J. 2000. Male greenfinches (*Carduelis chloris*) with brighter ornaments have higher virus infection clearance rate. *Behav. Ecol. Sociobiol.* 48:44–51

Luckinbill LS, Arking R, Clare MJ, Cirocco WC, Buck SA. 1984. Selection for delayed senescence in *Drosophila melanogaster*. *Evolution* 38:996–1004

Marler CA, Moore MC. 1988. Evolutionary costs of aggression revealed by testosterone manipulations in free-living male lizards. *Behav. Ecol. Sociobiol.* 23:21–26

Marler CA, Moore MC. 1991. Supplementary feeding compensates for testosterone-induced costs of aggression in male mountain spiny lizards. *Anim. Behav.* 42:209–19

Marler CA, Walsberg G, White ML, Moore M. 1995. Increased energy expenditure due to increased territorial defense in male lizards after phenotypic manipulation. *Behav. Ecol. Sociobiol.* 37:225–31

Martin TE. 1987. Food as a limit on breeding birds: a life-history perspective. *Annu. Rev. Ecol. Syst.* 18:453–87

Matos M, Rego C, Levy A, Teotonio H, Rose MR. 2000. An evolutionary no man's land. *Trends Ecol. Evol.* 15:206

McNab BK. 1997. On the utility of uniformity in the definition of basal rate of metabolism. *Physiol. Zool.* 70:718–20

Mole S, Zera AJ. 1993. Differential allocation of resources underlies the dispersal-reproduction trade-off in the wing-dimorphic cricket, *Gryllus rubens*. *Oecologia* 93:121–27

Moller AP. 1990. Parasites and sexual selection: current status of the Hamilton and Zuk hypothesis. *J. Evol. Biol.* 3:319–28

Moller AP. 1997. Parasitism and evolution of host life history. In *Host-Parasite Evolution: General Principles and Avian Models*, ed. DH Clayton, J Moore, pp. 105–27. Oxford, UK: Oxford Univ. Press

Moller H, Smith RH, Sibly RM. 1989. Evolutionary demography of a bruchid beetle. I. Quantitative genetical analysis of the female life history. *Funct. Ecol.* 3:673–81

Moore MC. 1986. Elevated testosterone levels during non-breeding season territoriality in a fall-breeding lizard, *Sceloporus jarrovi*. *J. Comp. Physiol.* 158:159–63

Nagy K. 1983. Ecological energetics. In *Lizard Ecology: Studies of a Model Organism*, ed. RB Huey, ER Pianka, TW Schoener, pp. 24–54. Cambridge, MA: Harvard Univ. Press. 501 pp.

Nagy K. 1987. Field metabolic rate and food requirement scaling in mammals and birds. *Ecol. Monogr.* 57:111–28

Niewiarowski PH. 2001. Energy budgets, growth rates, and thermal constraints: toward an integrative approach to the study of life-history variation. *Am. Nat.* 157:421–33

Niewiarowski PH, Dunham AE. 1994. The evolution of reproductive effort in squamate reptiles: costs, trade-offs, and assumptions reconsidered. *Evolution* 48:137–45

Nijhout HF. 1994. *Insect Hormones*. Princeton, NJ: Princeton Univ. Press 267 pp.

Nijhout HF. 1999. Control mechanisms of polyphenic development in insects. *BioScience* 49:181–92

Nijhout HF, Emlen DJ. 1998. Competition among body parts in the development and evolution of insect morphology. *Proc. Natl. Acad. Sci. USA* 95:3685–89

Nolan V, Ketterson ED, Ziegenfus C, Cullen DP, Chandler CR. 1992. Testosterone and avian life histories: effects of experimentally elevated testosterone on prebasic molt and survival in male dark-eyed juncos. *Condor* 94:364–70

Noonburg EG, Nisbet RM, McCauley E, Gurney WSC, Murdoch WW, et al. 1998. Experimental testing of dynamic energy budget models. *Funct. Ecol.* 12:211–22

Norris K, Anwar N, Read AF. 1994. Reproductive effort influences the prevalence of haematozoan parasites in great tits. *J. Anim. Ecol.* 63:601–10

Norris K, Evans MR. 2000. Ecological immunology: life history trade-offs and immune defense in birds. *Behav. Ecol.* 11:19–26

Nunney L. 1996. The response to selection for fast development in *Drosophila melanogaster*. *Evolution* 50:1193–204

Owens IPF, Wilson K. 1999. Immunocompetence: a neglected life history trait or conspicuous red herring? *Trends Ecol. Evol.* 14:170–72

Paladino FV. 1989. Constraints of bioenergetics on avian population dynamics. *Physiol. Zool.* 62:410–72

Partridge L. 1992. Measuring reproductive costs. *Trends Ecol. Evol.* 7:99–100

Partridge L, Prowse N, Pignatelli P. 1999. Another set of responses and correlated responses to selection on age at reproduction in *Drosophila melanogaster*. *Proc. R. Soc. London Ser. B* 266:255–61

Partridge L, Sibly R. 1991. Constraints in the evolution of life histories. *Philos. Trans. R. Soc. London Ser. B* 332:3–13

Pener MP. 1985. Hormonal effects on flight and migration. In *Comprehensive Insect Physiology, Biochemistry and Pharmacology*, ed. G Kerkut, LI Gilbert, 8:491–550. Oxford, UK: Pergamon

Perrin N, Bradley MC, Calow P. 1990. Plasticity of storage allocation in *Daphnia magna*. *Oikos* 59:70–74

Perrin N, Sibly RM. 1993. Dynamic models of energy allocation and investment. *Annu. Rev. Ecol. Syst.* 24:379–410

Reid WV. 1987. The cost of reproduction in the glaucous-winged gull. *Oecologia* 74:458–67

Reznick D, Nunny L, Tessier A. 2000. Big houses, big cars, superfleas, and the costs of reproduction. *Trends Ecol. Evol.* 15:421–25

Reznick D, Yang AP. 1993. The influence of fluctuating resources on life history: patterns of allocation and plasticity in female guppies. *Ecology* 74:2011–19

Reznick DN. 1985. Costs of reproduction: an evaluation of the empirical evidence. *Oikos* 44:257–67

Reznick DN. 1992. Measuring the costs of reproduction. *Trends Ecol. Evol.* 7:42–45

Reznick DN, Bryga H, Endler JA. 1990. Experimentally-induced life history evolution in a natural population. *Nature* 346:357–59

Richner H, Christie P, Oppliger A. 1995. Paternal investment affects prevalence of malaria. *Proc. Natl. Acad. Sci. USA* 92:1192–94

Ricklefs RE. 1974. Energetics of reproduction in birds. In *Avian Energetics*, ed. RA Paynter, pp. 152–297. Cambridge, UK: Nutall Ornithol. Club

Ricklefs RE. 1979. Adaptation, constraint, and compromise in avian postnatal development. *Biol. Rev.* 54:269–90

Ricklefs RE. 1991. Structures and transformations of life histories. *Funct. Ecol.* 5:174–83

Ricklefs RE. 1996. Avian energetics, ecology, and evolution. In *Avian Energetics*, ed. C Carey, pp. 1–30. New York: Chapman & Hall

Ricklefs RE. 2000a. Lack, Skutch and Moreau: the early development of life-history thinking. *Condor* 102:3–8

Ricklefs RE. 2000b. Density dependence, evolutionary optimization and the diversification of avian life histories. *Condor* 102:9–22

Roby DD, Ricklefs RE. 1986. Energy expenditure in adult least auklet and diving petrels during the chick-rearing period. *Physiol. Zool.* 59:661–78

Roff DA. 1986. The evolution of wing dimorphism in insects. *Evolution* 40:1009–20

Roff DA. 1992. *The Evolution of Life Histories*. New York: Chapman & Hall. 535 pp.

Roff DA, Stirling G, Fairbairn DJ. 1997. The

evolution of threshold traits: a quantitative genetic analysis of the physiological and life-history correlates of wing dimorphism in the sand cricket. *Evolution* 51:1910–19

Rogers CM, Smith JNM. 1993. Life-history theory in the nonbreeding period: trade-offs in avian fat reserves. *Ecology* 74:419–26

Rogowitz GL. 1996. Trade-offs in energy allocation during lactation. *Am. Zool.* 36:197–204

Ros AF, Groothius TTG, Apanius V. 1997. The relationship among gonadal steroids, immunocompetence, body mass, and behavior in young black-headed gulls (*Larus ridibundus*). *Am. Nat.* 150:201–19

Rose MR. 1984. Laboratory evolution of postponed senescence in *Drosophila melanogaster*. *Evolution* 38:1004–10

Rose MR, Bradley TJ. 1998. Evolutionary physiology of the cost of reproduction. *Oikos* 83:443–51

Rose MR, Graves JL, Hutchinson EW. 1990. The use of selection to probe patterns of pleiotropy in fitness characters. In *Insect Life Cycles*, ed. F Gilbert, pp. 29–42. New York: Springer-Verlag

Rose MR, Nusbaum TJ, Chippendale AK. 1996. Laboratory evolution: the experimental wonderland and the Cheshire Cat Syndrome. In *Adaptation*, ed. MR Rose, GV Lauder, pp. 221–44. San Diego, CA: Academic

Rose MR, Vu LN, Park SU, Graves JL. 1992. Selection on stress resistence increases longevity in *Drosophila melanogaster*. *Exp. Gerontol.* 27:241–50

Royle NJ, Surai PF, McCartney RJ, Speake BK. 1999. Parental investment and egg yolk lipid composition. *Funct. Ecol.* 13:298–306

Saino N, Moller AP, Bolzern. 1995. Testosterone effects on the immune system and parasite infestations in the barn swallow (*Hirundo rustica*): an experimental test of the immunocompetence hypothesis. *Behav. Ecol.* 6:397–404

Salmon AB, Marx DB, Harshman LG. 2001. A cost of reproduction: stress susceptibility. *Evolution.* In press

Schwarzkopf L. 1994. Measuring trade-offs: a review of studies of costs of reproduction in lizards. In *Lizard Ecology. Historical and Experimental Perspectives*, ed. LJ Vitt, ER Pianka, pp. 7–29 Princeton: Princeton Univ. Press. 403 pp.

Schwarzkopf L, Blows MW, Caley MJ. 1999. Life-history consequences of divergent selection on egg size in *Drosophila melanogaster*. *Am. Nat.* 29:333–40

Service PM. 1987. Physiological mechanisms of increased stress resistance in *Drosophila melanogaster* selected for postponed senescence. *Physiol. Zool.* 60:321–26

Service PM, Hutchinson EW, MacKinley MD, Rose MR. 1985. Resistance to environmental stress in *Drosophila melanogaster* selected for postponed senescence. *Evolution* 42:708–16

Service PM, Hutchinson EW, Rose MR. 1988. Multiple genetic mechanisms for the evolution of senescence in *Drosophila melanogaster*. *Evolution* 42:708–16

Service PM, Rose MR. 1985. Genetic covariation among life history components: the effects of novel environments. *Evolution* 39:943–45

Sgro CM, Partridge L. 1999. A delayed wave of death from reproduction in *Drosophila*. *Science* 286:2521–24

Sheldon BC, Verhulst S. 1996. Ecological immunity: costly parasite defenses and trade-offs in evolutionary ecology. *Trends Ecol. Evol.* 11:317–21

Shine R, Schwarzkopf L. 1992. The evolution of reproductive effort in lizards and snakes. *Evolution* 46:62–75

Shine R, Schwarzkopf L, Caley MJ. 1996. Energy, risk and reptilian reproductive effort: a reply to Niewiarowski and Dunham. *Evolution* 50:2111–14

Sibly R, Antonovics J. 1992. Life-history evolution. In *Genes in Ecology*, ed. RJ Berry, TJ Crawford, GM Hewitt, pp. 87–121. Oxford, UK: Blackwell Sci. 535 pp.

Simmons FH, Bradley TJ. 1997. An analysis of resource allocation in response to dietary yeast in *Drosophila melanogaster*. *J. Insect Physiol.* 43:779–88

Sinervo B. 1990. The evolution of maternal investment in lizards: an experimental and comparative analysis of egg size and its effects on offspring performance. *Evolution* 44:279–94

Sinervo B. 1994. Experimental tests of allocation paradigms In *Lizard Ecology*, ed. L Vitt, ER Pianka, pp. 73–90. Princeton, NJ: Princeton Univ. Press. 403 pp.

Sinervo B. 1999. Mechanistic analysis of natural selection and a refinement of Lack's and Williams's principles. *Am. Nat.* 154:S26–42

Sinervo B, Basolo AL. 1996. Testing adaptation using phenotypic manipulations. In *Adaptation*, ed. ML Rose, GV Lauder, pp. 149–85. San Diego, CA: Academic. 511 pp.

Sinervo B, Licht P. 1991. Proximate constraints on the evolution of egg size, egg number and total clutch mass in lizards. *Science* 252:1300–2

Sinervo B, McEdward LR. 1988. Developmental consequences of an evolutionary change in egg size: an experimental test. *Evolution* 42:885–99

Sinervo B, Miles DB, DeNardo DF, Frankino WA, Klukrowski M. 2000. Testosterone, endurance, and Darwinian fitness: natural and sexual selection on the physiological basis of alternative male behaviors in side-blotched lizards. *Horm. Behav.* 38:222–33

Stearns SC. 1989. Trade-offs in life history evolution. *Funct. Ecol.* 3:259–68

Stearns SC. 1992. *The Evolution of Life Histories*. Oxford, UK: Oxford Univ. Press. 249 pp.

Stearns SC, Kaiser M. 1993. The effects of enhanced expression of elongation factor EF-1 alpha on lifespan in *Drosophila melanogaster*. IV. A summary of three experiments. *Genetica* 91:167–82

Stevens DJ, Hansell MH, Freel JA, Monaghan P. 1999. Developmental trade-offs in caddis flies: increased investment in larval defense alters adult resource allocation. *Proc. R. Soc. London Ser. B* 266:1049–54

Styrsky JD, Dobbs RC, Thompson CF. 2000. Food supplementation does not override the effect of egg mass on fitness-related traits of nestling house wrens. *J. Anim. Ecol.* 69:690–702

Styrsky JD, Eckerle KP, Thompson CF. 1999. Fitness-related consequences of egg mass in nestling house wrens. *Proc. R. Soc. Biol. Sci. Ser. B* 266:1253–58

Svensson E, Raberg L, Koch C, Hasselquist D. 1998. Energetic stress, immunosuppression and the costs of an antibody response. *Funct. Ecol.* 12:912–19

Tanaka S. 1993. Allocation of resources to egg production and flight muscle development in a wing dimorphic cricket, *Modicogryllus confirmatus*. *J. Insect Physiol.* 39:493–98

Tatar M. 2000. Transgenic organisms in evolutionary ecology. *Trends Ecol. Evol.* 15:207–11

Tatar M, Khazeli AA, Curtsinger JW. 1997. Chaperoning extended life. *Nature* 390:30

Thompson DL, Monaghan P, Furness RW. 1998. The demands of incubation and avian clutch size. *Biol. Rev.* 73:293–304

Tinkle DW, Hadley NF. 1975. Lizard reproductive effort: caloric estimates and comments on its evolution. *Ecology* 56:427–34

Tower J. 1996. Aging mechanisms in fruit flies. *BioEssays* 18:799–807

Townsend CR, Calow P. 1981. *Physiological Ecology. An Evolutionary Approach to Resource Use*. Oxford, NY: Blackwell Sci. 393 pp.

Trillmich F. 1986. Are endotherms emancipated? Some considerations on the cost of reproduction. *Oceologia* 69:631–33

Tuomi J, Haukioja E, Hakala T. 1983. Alternative concepts of reproductive efforts, costs of reproduction and selection on life history evolution. *Ecology* 56:427–34

van Noordwijk AJ, de Jong G. 1986. Acquisition and allocation of resources: their influence on variation in life history tactics. *Am. Nat.* 128:137–42

Veiga JP, Salvador A, Merino S, Puerta M. 1998. Reproductive effort affects immune and parasite infection in a lizard: a phenotypic manipulation using testosterone. *Oikos* 82:313–18

Walsberg GE. 1983. Avian ecological genetics.

In *Avian Biology*, ed. DS Farner, pp. 161–219. New York: Cold Springs Harbor Lab. Press

Wang Y, Salmon AB, Harshman LG. 2001. Loss of stress resistance is associated with increased egg production in *Drosophila melanogaster*. *Exp. Gerontol.* In press

Ward S. 1996. Energy expenditure of female barn swallows, *Hirundo rustica*, during egg formation. *Physiol. Zool.* 69:930–51

Watson MJO, Hoffmann AA. 1996. Acclimation, cross-generation effects, and the response to selection for increased cold resistance in *Drosophila*. *Evolution* 50:1182–92

Wayne ML, Soundararajan U, Harshman LG. 2001. Correlated responses to selection for female starvation resistance in *Drosophila melanogaster*: ovariole number and age-specific egg production. *Evolution*. In review

Weatherhead PJ, Metz KJ, Bennett GF, Irwin RE. 1993. Parasite faunas, testosterone and secondary sexual traits in male red-winged blackbirds. *Behav. Ecol. Sociobiol.* 33:12–23

Weiner J. 1992. Physiological limits to sustainable energy budgets in birds and mammals: ecological implications. *Trends Ecol. Evol.* 7:384–88

Wilson RS, Booth DT. 1998. Effect of tail loss on reproductive output and its ecological significance in the skink *Eulamprus quoyii*. *J. Herpetol.* 32:128–31

Wingfield JC. 1988. Changes in reproductive function of free-living birds in direct response to environmental perturbations. In *Processing of Environmental Information in Vertebrates*, ed. MH Stetson, pp. 520–28. Berlin: Springer-Verlag

Wingfield JC, Maney DL, Breuner CW, Jacobs JD, Lynn S, et al. 1998. Ecological bases of hormone-behavior interactions: the "emergency life history stage." *Am. Zool.* 38:191–206

Withers PC. 1992. *Comparative Animal Physiology*. Fort Worth, TX: Saunders College. 949 pp.

Witter MS, Cuthill IC. 1993. The ecological costs of avian fat storage. *Philos. Trans. R. Soc. London Ser. B* 340:73–92

Zera AJ. 1999. The endocrine genetics of wing polymorphism in *Gryllus*: critique of recent studies and state of the art. *Evolution* 53:972–76

Zera AJ, Bottsford J. 2001. The endocrine-genetic basis of life-history variation: relationship between the ecdysteroid titer and morph-specific reproduction in the wing-polymorphic cricket, *Gryllus firmus*. *Evolution* 55:538–49

Zera AJ, Brink T. 2000. Nutrient absorption and utilization by wing and flight muscle morphs of the cricket *Gryllus firmus*: implications for the trade-off between flight capability and early reproduction. *J. Insect Physiol.* 46:1207–18

Zera AJ, Cisper GL. 2001. Genetic and diurnal variation in the juvenile hormone titer in a wing polymorphic cricket: implications for the evolution of life histories and dispersal. *Physiol. Biochem. Zool.* 74:293–306

Zera AJ, Denno RF. 1997. Physiology and ecology of dispersal polymorphism in insects. *Annu. Rev. Entomol.* 42:207–31

Zera AJ, Huang Y. 1999. Evolutionary endocrinology of juvenile hormone esterase: functional relationship with wing polymorphism in the cricket, *Gryllus firmus*. *Evolution* 53:837–47

Zera AJ, Larsen A. 2001. The metabolic basis of life history variation: genetic and phenotypic differences in lipid reserves among life history morphs of the wing-polymorphic cricket, *Gryllus firmus*. *J. Insect Physiol.* In press

Zera AJ, Mole S, Rokke K. 1994. Lipid, carbohydrate and nitrogen content of long- and short-winged *Gryllus firmus*: implications for the physiological cost of flight capability. *J. Insect Physiol.* 40:1037–44

Zera AJ, Potts J, Kobus K. 1998. The physiology of life history trade-offs: experimental analysis of a hormonally-induced life history trade-off in *Gryllus assimilis*. *Am. Nat.* 152:7–23

Zera AJ, Sall J, Grudzinski K. 1997. Flight-muscle polymorphism in the cricket *Gryllus firmus*: muscle characteristics and their influence on the evolution of flightlessness. *Physiol. Zool.* 70:519–29

Zera AJ, Tiebel KC. 1989. Differences in juvenile hormone esterase activity between presumptive macropterous and brachypterous *Gryllus rubens*: implications for the hormonal control of wing polymorphism. *J. Insect Physiol.* 35:7–17

Zhao Z, Zera AJ. 2001. Enzymological and radiotracer studies of lipid metabolism in the flight-capable and flightless morphs of the wing-polymorphic cricket, *Gryllus firmus. J. Insect Physiol.* In press

Zuk M, Johnsen TS, Maclarty T. 1995. Endocrine-immune interactions, ornaments and mate choice in red jungle fowl. *Proc. R. Soc. London Ser. B* 260:205–10

Zwann B, Bijlsma R, Hoekstra RF. 1995. Direct selection on life span in *Drosophila melanogaster. Evolution* 49:649–59

Annu. Rev. Ecol. Syst. 2001. 32:127–57

URBAN ECOLOGICAL SYSTEMS: Linking Terrestrial Ecological, Physical, and Socioeconomic Components of Metropolitan Areas[*]

S. T. A. Pickett,[1] M. L. Cadenasso,[1] J. M. Grove,[2] C. H. Nilon,[3] R. V. Pouyat,[4] W. C. Zipperer,[4] and R. Costanza[5]

[1]Institute of Ecosystem Studies, Millbrook, New York; e-mail: picketts@ecostudies.org, cadenassom@ecostudies.org
[2]USDA Forest Service, Northeastern Research Station, Burlington, Vermont; e-mail: jmgrove@att.net
[3]Fisheries and Wildlife, University of Missouri, Columbia, Missouri; e-mail: nilonc@missouri.edu
[4]USDA Forest Service, Northeastern Research Station, Syracuse, New York; e-mail: rpouyat@aol.com, wzipperer@fs.fed.us
[5]Center for Environmental and Estuarine Studies, University of Maryland, Solomans, Maryland; e-mail: costza@cbl.cees.edu

Key Words city, hierarchy theory, integration, patch dynamics, urban ecology

■ **Abstract** Ecological studies of terrestrial urban systems have been approached along several kinds of contrasts: ecology in as opposed to ecology of cities; biogeochemical compared to organismal perspectives, land use planning versus biological, and disciplinary versus interdisciplinary. In order to point out how urban ecological studies are poised for significant integration, we review key aspects of these disparate literatures. We emphasize an open definition of urban systems that accounts for the exchanges of material and influence between cities and surrounding landscapes. Research on ecology in urban systems highlights the nature of the physical environment, including urban climate, hydrology, and soils. Biotic research has studied flora, fauna, and vegetation, including trophic effects of wildlife and pets. Unexpected interactions among soil chemistry, leaf litter quality, and exotic invertebrates exemplify the novel kinds of interactions that can occur in urban systems. Vegetation and faunal responses suggest that the configuration of spatial heterogeneity is especially important in urban systems. This insight parallels the concern in the literature on the ecological dimensions of land use planning. The contrasting approach of ecology of cities has used a strategy of biogeochemical budgets, ecological footprints, and summaries of citywide species richness. Contemporary ecosystem approaches have begun to integrate organismal,

---

[*]The U.S. Government has the right to retain a nonexclusive, royalty-free license in and to any copyright covering this paper.

nutrient, and energetic approaches, and to show the need for understanding the social dimensions of urban ecology. Social structure and the social allocation of natural and institutional resources are subjects that are well understood within social sciences, and that can be readily accommodated in ecosystem models of metropolitan areas. Likewise, the sophisticated understanding of spatial dimensions of social differentiation has parallels with concepts and data on patch dynamics in ecology and sets the stage for comprehensive understanding of urban ecosystems. The linkages are captured in the human ecosystem framework.

INTRODUCTION: JUSTIFICATION FOR URBAN ECOLOGICAL STUDIES

Urbanization is a dominant demographic trend and an important component of global land transformation. Slightly less than half of the world's population now resides in cities, but this is projected to rise to nearly 60% in the next 30 years (United Nations 1993). The developed nations have more urbanized populations; for example, close to 80% of the US population is urban. Urbanization has also resulted in a dramatic rise in the size of cities: over 300 cities have more than 10^6 inhabitants and 14 megacities exceed 10^7. The increasing population and spatial prominence of urban areas is reason enough to study them, but ecologists must also inform decision makers involved in regional planning and conservation. Proper management of cities will ensure that they are reasonable places to live in the future.

In addition to its global reach, urbanization has important effects in regional landscapes. For example, in industrialized nations, the conversion of land from wild and agricultural uses to urban and suburban occupancy is growing at a faster rate than the population in urban areas. Cities are no longer compact, isodiametric aggregations; rather, they sprawl in fractal or spider-like configurations (Makse et al. 1995). Consequently, urban areas increasingly abut and interdigitate with wild lands. Indeed, even for many rapidly growing metropolitan areas, the suburban zones are growing faster than other zones (Katz & Bradley 1999). The resulting new forms of urban development, including edge cities (Garreau 1991) and housing interspersed in forest, shrubland, and desert habitats, bring people possessing equity generated in urban systems, expressing urban habits, and drawing upon urban experiences, into daily contact with habitats formerly controlled by agriculturalists, foresters, and conservationists (Bradley 1995).

Urban habitats constitute an open frontier for ecological research. Ecologists have come to recognize that few ecosystems are totally devoid of direct or subtle human influence (McDonnell & Pickett 1993). Yet urban systems are relatively neglected as an end member with which to compare the role of humans in ecosystems. Notably, many classic geographic studies of cities, which offer valuable insights to ecologists, are based on outmoded ecological theory such as deterministic models of succession and assumptions of equilibrium dynamics of ecosystems. Hence classical ecological approaches and the geographic studies that have relied on them have not been as useful as they would be otherwise (Zimmerer 1994).

Although the ecology of urban areas has long elicited the academic attention of ecologists, physical and social scientists, and regional planners, there is much opportunity to extend and integrate knowledge of the metropolis using an ecological lens. The purpose of this paper is to review the status of ecological knowledge of the terrestrial components of urban areas and to present a framework for continued ecological research and integration with social and economic understanding. This paper complements the review of aquatic components of urban systems by Paul & Meyer (2001).

Definition and Roots of Urban Ecology

Urban ecosystems are those in which people live at high densities, or where the built infrastructure covers a large proportion of the land surface. The US Bureau of the Census defines urban areas as those in which the human population reaches or exceeds densities of 186 people per km^2. However, an ecological understanding of urban systems also must include less densely populated areas because of reciprocal flows and influences between densely and sparsely settled areas. Comparisons along gradients of urbanization can capture the full range of urban effects as well as the existence of thresholds. Therefore, in the broadest sense, urban ecosystems comprise suburban areas, exurbs, sparsely settled villages connected by commuting corridors or by utilities, and hinterlands directly managed or affected by the energy and material from the urban core and suburban lands.

The boundaries of urban ecosystems are often set by watersheds, airsheds, commuting radii, or convenience. In other words, boundaries of urban ecosystems are set in the same ways and for the same reasons as are the boundaries in any other ecosystem study. In the case of urban ecosystems, it is clear that many fluxes and interactions extend well beyond the urban boundaries defined by political, research, or biophysical reasons. Urban ecology, as an integrative subdiscipline of the science of ecology, focuses on urban systems as broadly conceived above. There is little to be gained from seeking distinctions between "urban" and abutting "wild" lands, as a comprehensive, spatially extensive, systems approach is most valuable for science (Pickett et al. 1997) and management (Rowntree 1995).

There are two distinct meanings of urban ecology in the literature (Sukopp 1998). One is a scientific definition, and the other emerges from urban planning. In ecology, the term urban ecology refers to studies of the distribution and abundance of organisms in and around cities, and on the biogeochemical budgets of urban areas. In planning, urban ecology has focused on designing the environmental amenities of cities for people, and on reducing environmental impacts of urban regions (Deelstra 1998). The planning perspective is normative and claims ecological justification for specific planning approaches and goals. We review key aspects of these complementary approaches and then frame a social-ecological approach to integrate these two approaches.

BIOGEOPHYSICAL APPROACHES

There are two aspects to the biogeophysical approach to urban ecological studies. One, the pioneering and most common approach, examines ecological structure and function of habitats or organisms within cities. This approach is called ecology in cities. The second, more recent and still emerging approach, examines entire cities or metropolitan areas from an ecological perspective. The second approach is labeled ecology of cities (Grimm et al. 2000). Although the differences in the prepositions in the phrases identifying the two contrasting approaches may appear subtle, the understanding achieved by identifying them as poles between which urban ecological studies sort out, is crucial to understanding the history of urban ecology, and the integration it is now poised to make. We review literature that has taken these two contrasting approaches in turn.

Ecology in the City

The study of ecology in the city has focused on the physical environment, soils, plants and vegetation, and animals and wildlife. These studies are the foundation for understanding urban ecosystems. The literature in this area has taken a case study approach, and unifying themes are still to emerge. We highlight key examples from among the many cases.

URBAN PHYSICAL ENVIRONMENT   The urban heat island constitutes climate modification directly related to urban land cover and human energy use (Oke 1995). The urban heat island describes the difference between urban and rural temperatures. Such differences often are negligible in the daytime but develop rapidly after sunset, peaking 2–3 h later. Ambient air temperatures may reach maxima of 5–10°C warmer than hinterlands (Zipperer et al. 1997). For example, New York City is, on average, 2–3°C warmer than any other location along a 130-km transect into surrounding rural areas (McDonnell et al. 1993). The duration and magnitude of the temperature differential depend on the spatial heterogeneity of the urban landscape. As the percentage of artificial or human-made surfaces increases, the temperature differential increases. Hence, the urban core is warmer than neighboring residential areas, which are warmer than neighboring farmlands or forests. The differences also change seasonally. For example, cities in mid-lattitudes of the United States are typically 1–2°C warmer than the surroundings in winter, and 0.5–1.0°C warmer in summer (Botkin & Beveridge 1997).

The heat island effect varies by region, as seen in a comparison between Baltimore, Maryland, and Phoenix, Arizona (Brazel et al. 2000). During the summer, mean maximum temperatures in Baltimore were greater than in the rural landscape. Phoenix, on the other hand, became an oasis, with cooler temperatures than the surrounding desert. The cooling of Phoenix is due to the watering of mesic plantings in the city. In contrast, the mean minimum temperatures were warmer in both cities than in the respective neighboring rural landscape, although the differential in Phoenix was greater.

The heat island intensity also is related to city size and population density (Oke 1973, Brazel et al. 2000). For example, Baltimore's mean minimum temperature differential increased until the 1970s when the city experienced a decline in population. Since 1970, the mean minimal temperature differential has leveled off. Phoenix also showed an increase in mean minimum temperature differential with an increase in population. However, Phoenix is the second fastest growing metropolitan area in the United States, so the differential has continued to increase with population growth. In general, a nonlinear relationship exists between mean minimum temperature differential and population density (Brazel et al. 2000).

The differences in climate between city and countryside have biological implications. For example, as a result of climatic modification in temperate zone cities, leaf emergence and flowering times are earlier, and leaf drop is later than in the surrounding countryside (Sukopp 1998). Increased temperatures in and around cities enhance ozone formation, and increase the number of officially recognized pollution days and trace gas emissions (Sukopp 1998). Ozone concentrations tend to be highest in and around urban areas. As urban areas have expanded through processes of suburban sprawl, the spatial influence of urbanization has increased. Within regions of ozone pollution, agricultural crops may be adversely affected and yields decrease 5–10% (Chameides et al. 1994). Crop type and stage of development, and the degree, spatial extent, and duration of ozone exposure may all influence the decrease in production (Chameides et al. 1994).

Precipitation is enhanced in and downwind of cities as a result of the higher concentrations of particulate condensation nuclei in urban atmospheres. Precipitation can be up to 5–10% higher in cities, which can experience greater cloudiness and fog (Botkin & Beveridge 1997). The probability of precipitation increases toward the end of the work week and on weekends due to a buildup of particulates resulting from manufacturing and transportation (Collins et al. 2000).

Urban hydrology is drastically modified compared to agricultural and wild lands. This topic is covered more fully in a companion review (Paul & Meyer 2001). Relativizing a water budget to 100 units of precipitation, and comparing urban to nonurban areas, evapotranspiration decreases from a value of 40% to 25%, surface runoff increases from 10% to 30%, and groundwater decreases from 50% to 32% (Hough 1995). Forty-three percent of precipitation exits the urban area via storm sewers, with 13% of that having first fallen on buildings. The role of impervious surfaces is crucial to the functioning of urban watersheds (Dow & DeWalle 2000). The hydrology in urban areas can be further modified by ecological structures. For example, reduced tree cover in urban areas increases the rate of runoff and decreases the time lag between initiation of storms and initiation of runoff (Hough 1995). The increased runoff in urban areas changes the morphology of urban streams, which become deeply incised in their floodplains. Remnant riparian vegetation may suffer as a result of isolation from the water table.

URBAN SOILS   Soils in urban landscapes retain and supply nutrients, serve as a growth medium and substrate for soil fauna and flora, and absorb and store water.

Soils also intercept contaminants such as pesticides and other toxics generated through human activities (Pouyat & McDonnell 1991). However, in urban settings soils are modified by human activity and, consequently, are functionally altered (Effland & Pouyat 1997). In addition, completely new substrates are created by deposition of debris, soil, and rock in urban sites. Such new substrates are called made land.

As land is converted to urban use, both direct and indirect factors can affect the functioning of soils. Direct effects include physical disturbances, burial of soil by fill material, coverage by impervious surfaces, and additions of chemicals and water (e.g., fertilization and irrigation). Direct effects often lead to highly modified substrates in which soil development then proceeds (Effland & Pouyat 1997). Indirect effects change the abiotic and biotic environment, which in turn can influence soil development and ecological processes in intact soils. Indirect effects include the urban heat island (Oke 1995), soil hydrophobicity (White & McDonnell 1988), introductions of nonnative plant and animal species (Airola & Buchholz 1984, Steinberg et al. 1997), and atmospheric deposition of pollutants (Lovett et al. 2000). Moreover, toxic, sublethal, or stress effects of the urban environment on soil decomposers and primary producers can significantly affect the quality of organic matter and subsequent soil processes (Pouyat et al. 1997).

The results from a transect along an urban-rural land-use gradient in the New York City metropolitan area show the influence of urban environments on intact forest soils (McDonnell et al. 1997). Along this transect, human population density, percent impervious surface, and automobile traffic volume were significantly higher at the urban than the rural end of the gradient (Medley et al. 1995). Soil chemical and physical properties, soil organism abundances, and C and N processes were investigated along this gradient to assess the sometimes complex and sometimes contradictory interactions between urban soil chemistry, local leaf litter quality, and exotic species.

Soil chemistry significantly correlated with measures of urbanization. Higher concentrations of heavy metals (Pb, Cu, Ni), organic matter, salts, and soil acidity were found in the surface 10 cm of forest soils at the urban end of the transect (Pouyat et al. 1995). The most probable factor is metropolitan-wide atmospheric deposition.

Litter decomposition studies in both the field and the laboratory determined that urban-derived oak litter decomposed more slowly than rural-derived oak litter under constant conditions (Carreiro et al. 1999), with lignin concentration explaining 50% of the variation. Moreover, a site effect was measured for reciprocally transplanted litter, as decomposition was faster in the urban than in the rural sites, regardless of litter origin. This result is surprising because the lower litter fungal biomass and microinvertebrate abundances found in urban stands compared to rural (Pouyat et al. 1994) would lead to an expectation of slower litter turnover rates in urban forests. However, abundant earthworms (Steinberg et al. 1997) and higher soil temperatures in the urban stands may compensate for lower litter quality, lower fungal biomass, and lower microinvertebrate abundances in the urban stands

(R.V. Pouyat & M.M. Carreiro, submitted for publication). Such compensation likely explains the faster decomposition rates in the urban stands.

Litter quality, site environment, and soil organism differences between the urban and rural sites also affected C and N pools and processes in the soil. Urban litter was found to be of lower quality (higher C:N ratio) than rural litter. Typically, poor quality litter either decreases the rate at which labile C mineralizes N or increases the amount of organic matter transferred to recalcitrant pools, or both. Hence, urban litter is expected to be more recalcitrant than rural litter. Indeed, measurements of soil C pools along the urban-rural transect suggest that recalcitrant C pools are higher and passive pools lower in urban forest soils relative to rural forest soils (Groffman et al. 1995). Consequently, N mineralization rates were expected to be lower in urban stands. However, the opposite was found. Net potential N-mineralization rates in the A-horizon were higher in the urban stands than in the rural stands (Pouyat et al. 1997). Soil cores taken from urban areas accumulated more NH_4^+ and NO_3^- than soil from rural areas. In addition, from a reciprocal transplant experiment, soil cores incubated at urban sites accumulated more inorganic N than cores incubated at the rural sites, regardless of where the cores originated. These net N mineralization rates contradict both the litter decomposition results from the transplant experiment discussed above and expectations derived from measurements of soil C pools.

In contrast, when net N-mineralization rates were measured for mixed O, A, and B horizon material (15 cm depth), the total inorganic N pool accumulated was higher in rural than in urban sites, although less NO_3^- was accumulated in rural samples (Goldman et al. 1995). The mechanisms behind these results have yet to be elucidated, though it has been hypothesized that methane consumption rates and the biomass of methanotrophs are important regulators in overall soil C and N dynamics (Goldman et al. 1995).

Intense soil modifications resulting from urbanization may potentially alter soil C and N dynamics. To assess the potential effect on soil organic C, data from "made" soils (1 m depth) from five different cities, and surface (0–15 cm) soils from several land use types in Baltimore were analyzed (R.V. Pouyat, P.M. Groffman, I. Yesilonis & L. Hernandez, submitted for publication). Soil pedons from the five cities showed the highest soil organic C densities in loamy fill (28.5 kg m^{-2}) with the lowest in clean fill and old dredge materials (1.4 and 6.9 kg m^{-2}, respectively). Soil organic C for residential areas (15.5 $\pm$ 1.2 kg m^{-2}) was consistent across cities. A comparison of land-use types showed that low density residential and institutional land had 44% and 38% higher organic C densities than commercial land, respectively. Therefore, made soils, with their physical disturbances and inputs of various materials by humans, can greatly alter the amount of C stored in urban systems.

The complex patterns of C and N dynamics that have emerged from the studies reviewed above indicate interactions between key soil and organism processes in urban environments. Simple predictions based on trends in pollution, stress, or exotic species alone are inadequate to understand the complex feedbacks between these three governing factors of urban soil dynamics. Studies of soil C and N dynamics in unmanaged urban forests, highly disturbed soils, and surface soils of various urban land-use types all show that urbanization can directly and indirectly affect soil C pools and N-transformation rates. Our review also suggests that soil C storage in urban ecosystems is highly variable. How generalizable these results are across cities located in similar and dissimilar life zones needs to be investigated. In addition, more data are needed on highly disturbed soils, such as landfill, managed lawns, and covered soils to make regional and global estimates of soil C storage and N-transformation rates in urban ecosystems. Specific uncertainties include the quality of the C inputs governed by the input of litter from exotic plant species and by stress effects on native species litter, the fate of soil C in covered soils, measurements of soil C densities at depths greater than 1 m, particularly in made soils, and the effects of specific management inputs on N-transformation rates.

VEGETATION AND FLORA IN CITIES   The assessment of vegetation in urban landscapes has a long history. For example, in Europe, studies by De Rudder & Linke (1940) documented the flora and fauna of cities during the early decades of the twentieth century. After World War II, Salisbury (1943) examined vegetation dynamics of bombsites in cities. At the same time, ecologists in the United States focused on describing flora in areas of cities minimally altered by humans, such as parks and cemeteries. One of the first comprehensive studies of urban vegetation and environments was conducted by Schmid (1975) in Chicago.

The structure and composition of vegetation has been one of the foci of ecological studies in cities, and these studies have documented large effects of urbanization on forest structure. Urban stands tend to have lower stem densities, unless those stands are old-growth remnants in large parks or former estates (Lawrence 1995). In the Chicago region, street trees and residential trees tend to be larger than those in forest preserves, natural areas, and wild lands (McPherson et al. 1997), although individual trees in urban sites are often stressed, especially on or near streets (Ballach et al. 1998). Street trees in Chicago account for 24% of the total Leaf Area Index (LAI) and 43.7% of the LAI in residential areas (Nowak 1994). In mesic forest regions of the United States, tree cover of cities is approximately 31%, compared to the nearly continuous forest cover the areas would have supported before settlement. Brooks & Rowntree (1984) quantified the forest cover in counties classified as nonmetropolitan, peripheral to a central city, and encompassing a central city. They found that the proportion of total land area in forest decreased from nonmetropolitan to central city counties, with the steepest reductions between nonmetropolitan and peripheral counties. In contrast, for prairie-savanna and desert regions, tree cover in cities is greater than in pre-urban conditions (Nowak et al. 1996).

There are also local effects on urban vegetation. For example, forest patches adjacent to residential areas have increased edge openness (Moran 1984), and their margins have retreated because of recreational use, especially by children, and damage to regeneration (Bagnall 1979). Such regeneration failure is frequent in urban and suburban stands, owing to reduced natural disturbance or gap formation, substitution for natural disturbances of unfavorable anthropogenic disturbances such as frequent ground fires, trampling, and competition with exotics (Guilden et al. 1990).

The composition of urban and suburban forests differs from that of wild and rural stands in several ways. Species richness has increased in urban forests as a whole, but this is because of the increased presence of exotics (Zipperer et al. 1997). Urban areas show a preponderance of trees of wetland or floodplain provenance, owing to the lower oxygen tensions shared by wetlands and impervious urban soils (Spirn 1984). Even when the tree composition remains similar between urban and rural forests, the herbaceous flora of urban forests is likely to differ between the two types of forest (Wittig 1998). Such compositional trends reflect the context and configuration of the forest stands in urban areas. For example, vascular plant diversity increases with the area of the stand (Iida & Nakashizuka 1995, Hobbs 1988). Furthermore, in some urban stands, the adjacent land use affects species composition. For example, the interior of forests in residential areas often has more exotics than forests abutting either roads or agricultural zones (Moran 1984).

The role of exotic species has received particular attention in urban studies. The percentage of the flora represented by native species decreased from urban fringes to center city (Kowarik 1990). Along the New York City urban-to-rural gradient, the number of exotics in the seedling and sapling size classes of woody species was greater in urban and suburban oak-dominated stands (Rudnicky & McDonnell 1989). Rapoport (1993) found the number of noncultivated species decreased from fringe toward urban centers in several Latin American cities for various reasons exemplified in different cities. In Mexico City, there was a linear decrease in the number of species from 30–80 ha^{-1} encountered in suburbs to 3–10 ha^{-1} encountered in the city center. The social context also influenced species richness. At a given housing density, more affluent neighborhoods had more exotic species than less affluent ones in Bariloche, Argentina. In Villa Alicura, Argentina, exotic species increased with increasing local site alteration by humans. Near homes there were 74% exotics, while there were 48% along river banks. Although exotics were present along all roads, the number decreased on roads less frequently used. No exotic species were found outside the town. Pathways in rural recreation areas (Rapoport 1993) and in urban parks (Drayton & Primack 1996) have enhanced the presence of exotics. In an urban park in Boston, of the plant species present in 1894, 155 were absent by 1993, amounting to a decrease from 84% to 74% native flora. Sixty-four species were new. In addition to trails, Drayton & Primack (1996) blamed fire and trampling for the change in exotics.

Structural and compositional changes are not the only dynamics in urban biota. Plants and animals in cities have evolved in response to the local conditions in cities (Sukopp 1998, Bradshaw & McNeilly 1981). The famous population genetic differentiation in copper and zinc tolerance in creeping bent grass (*Agrostis stolonifera*) and roadside lead tolerance of ribwort plantain (*Plantago lanceolata*) in urban sites is an example (Wu & Antonovics 1975a, b). Industrial melanism and other forms of melanism are also urban evolutionary responses (Bishop & Cook 1980).

A major feature of urban vegetation is the spatial heterogeneity in urban landscapes created by the array of building densities and types, different land uses, and different social contexts. To characterize this heterogeneity, plant habitats have been subjected to various classification systems. Forest Stearns (Stearns 1971), one of the first American ecologists to call for research in urban landscapes, identified three major vegetation types—ruderal, managed, and residual. In addition to assessing spatial heterogeneity, classification systems recognize the importance of characterizing natural habitats in urban landscapes. Rogers & Rowntree (1988) developed a system to classify vegetation using life forms. This process was used to assess natural resources in New York City (Sisinni & Emmerich 1995). Based on site histories, Zipperer et al. (1997) classified tree-covered habitats as planted, reforested, or remnant. These approaches allow for spatially explicit comparisons between vegetation and other variables of interest for a single point in time.

The vegetation and floristic studies of nature in cities share key characteristics. They are largely descriptive, but illustrate spatial heterogeneity as a source of diversity, and suggest a functional role for landscape structure (Rebele 1994). Although it is legitimate to view the city as an open and dynamic ecosystem, few of the plant ecological studies document successional processes (Matlack 1997) or expose the functional relationships of the vegetation (Mucina 1990).

ANIMALS AND WILDLIFE   Douglas (1983), Gilbert (1989), VanDruff et al. (1994), and Nilon & Pais (1997) have summarized the literature on the animal life of cities. They reveal a long history of research on the fauna of European and North American cities. Although much of the research has been descriptive, several studies focused on processes that are also major foci of research in ecology as a whole. We review animal studies starting with coarse scale or gradient comparisons, followed by patch-oriented studies, including some that incorporate socioeconomic aspects, then show how wildlife biologists have contributed to urban ecology, and end with several examples of process studies involving animals.

Birds, mammals, and terrestrial invertebrates are the best studied taxonomic groups, with aquatic fauna, reptiles, and amphibians less studied (Luniak & Pisarski 1994). In addition, the fauna of green spaces are relatively well studied, with built-up and derelict areas and water bodies less well known. Andrzejewski et al. (1978) and Klausnitzer & Richter (1983) described how urban-to-rural gradients defined by human population density and building density impact the occurrence and abundance of various animal species. Among mammals, there is a shift to medium sized, generalist predators such as racoons and skunks in urban areas (Nilon & Pais 1997). The predator fauna contrasts between cities and nonurban environments. For example, ants were the most important predators in some urban areas although vertebrates were more significant in nonurban habitats (Wetterer 1997). Changes in fauna may influence vegetation. For example, invertebrate pest densities increased in urban trees compared to wild forest stands, reflecting the additional stress and damage to which urban trees are subjected (Nowak & McBride 1992).

Animal ecologists have recognized spatial heterogeneity in a variety of ways. Heterogeneity is typically expressed as patches or contrasting biotopes. The approach to spatial heterogeneity of urban fauna in Germany has been shaped by a national program of biotope mapping (Sukopp 1990, Werner 1999). The German biotope approach is based on phytosociologic-floristic and faunal characteristics of sites in cities (Sukopp & Weiler 1988). Biotope mapping can be used to document dynamics of urban fauna such as the changes in the bird species composition, and abundance of different types of biotopes (Witt 1996). Brady et al.'s (1979) general typology of urban ecosystems and Matthews et al.'s (1988) habitat classification scheme for metropolitan areas in New York State are examples of schemes that recognize how patterns of land-use history and resulting changes in land cover create ecologically distinct habitat patches.

Use of patch-oriented approaches is particularly well developed in studies of mammals. For example, land use and cover in 50-ha areas surrounding patches in Syracuse, New York, were the best predictors of small mammal species composition (VanDruff & Rowse 1986). Similarly, patch configuration in Warsaw, Poland, affected mammal populations. Urbanization blocked the dispersal of field mice and altered population structure and survivorship in isolated patches (Andrzejewski et al. 1978). Larger mouse populations were associated with increased percentages of built and paved areas, barriers to emigration, and decreases in patch size (Adamczewska-Andrzejewski et al. 1988). In addition, mammalian predators of seeds may be increased in urban patches compared to rural areas (Nilon & Pais 1997). The change in patch configuration resulting from suburban sprawl has led to the widespread increase in deer densities on urban fringes. Suburbanization has caused increased juxtaposition of forest habitats in which deer shelter with field or horticultural patches in which they feed (Alverson et al. 1988). The positive feedback of changing patch configuration on deer population density is magnified by reduced hunting and predator pressure associated with suburbanization. Greater deer densities have the potential to change both the animal community and vegetation (Bowers 1997).

Some biologists have looked to social causes of animal and plant distribution and abundance in cities. Studies of flora and fauna of metropolitan Liverpool, England, showed that changes in land use and technology have influenced habitat change since the industrial revolution (Greenwood 1999), and studies of fauna in suburban Warsaw included land-use information dating from the 1700s (Mackin-Rogalska et al. 1988). Key to the Warsaw study was the recognition that suburban areas have different socioeconomic processes than do either urban or rural areas, and that these processes influence land use, land-cover characteristics, and habitats for biota. Examples of the relationships of animal populations to land-cover types include those for birds. For a city as a whole, exotic generalists such as pigeons, starlings, and sparrows can constitute 80% of the bird community in the summer, and 95% in winter (Wetterer 1997). At finer scales, contrasting land-cover patches of residential areas, commercial sites, and parks supported different avian assemblages (Nilon & Pais 1997). Even differences in plant cover among residential neighborhoods affected bird community composition. Likewise, the land uses adjacent to urban green spaces strongly influence bird communities (Nilon & Pais 1997).

In Europe and North America, much of the research on the fauna of cities has been conducted by applied ecologists to support conservation and natural resource management. This work recognized that cities are areas worthy of study, and more importantly, that people and their activities create a unique context for natural resource management in and around cities (Waggoner & Ovington 1962, Noyes & Progulske 1974). In particular, the patch dynamics approach was foreshadowed by wildlife ecologists in cities. The activity of urban wildlife biologists is illustrated by the conferences of the National Institute for Urban Wildlife (Adams & Leedy 1987, 1991, Adams & VanDruff 1998). These conferences focused on (a) ecology of urban wildlife, (b) planning and design, (c) management issues and successes, and (d) public participation and education. The participatory approach to studying animals in cities involves urban residents in the process (Adams & Leedy 1987) and serves as a model for other ecological research in cities. Furthermore, the conferences focused on planning and management as activities that can change habitats at both citywide and local scales.

An additional stress on animal populations in cities is the direct or indirect effect of domestic pets. Building density is associated with an increase in the numbers of pets in an area (Nilon & Pais 1997). The amount of food energy available to free-ranging domestic cats depends on the affluence of the immediate neighborhood (Haspel & Calhoon 1991). Cats have a significant impact on bird populations in suburbs (Churcher & Lawton 1987).

An example of an ecological process familiar in wild and production landscapes that also appears in cities is animal succession. After the establishment of the new town of Columbia, Maryland, birds such as bobwhite and mourning dove, which are associated with agriculture, gave way to starlings and house sparrows, which had been absent before urbanization (Hough 1995). In an urban park near Dortmund, Germany, a 35-year study found an increase in generalist species richness and density at the expense of specialists. The species turnover rate of birds in the park between 1954 and 1997—42.1%—was higher than in a forest distant from the city over the same period (Bergen et al. 1998).

This overview of animals in urban areas has confirmed the diversity of exotic and native species in cities. These species, along with the planted and volunteer vegetation in cities, are in some cases an important amenity, and in others a significant health or economic load. They can serve to connect people with natural processes through educational activities. Although there is a great deal of descriptive knowledge about the biota of cities, there is the need to compare food web models for green and built parts of the city, to link these data with ecosystem function, and to quantify the relationships between infrastructural and human behavioral features of the metropolis (Flores et al. 1997). In addition, long-term studies are required.

The case studies we report here are a foundation for generalizations concerning the structure and function of biota within cities.

Ecology of the City

The knowledge of nature in cities is a firm foundation for understanding ecological processes in metropolitan areas. Yet it is not sufficient (Flores et al. 1997). If scientists, planners, and decision makers are to understand how the social, economic, and ecological aspects of cities interact, the feedbacks and dynamics of the ecological linkages must be assessed. We therefore turn to a review of systems-oriented approaches to urban ecology. These represent a shift to the perspective of ecology of cities, as contrasted with the literature we have reviewed so far, which focused on ecology in cities.

The diverse spatial mosaics of metropolitan areas present a variety of ecological situations in which to examine ecological structure and dynamics. For example, several of the conditions in cities are analogous to major predictions of global climate change. Increased temperatures, altered rainfall patterns, and drying of soils anticipate trends projected for some wild lands. Examination of existing urban assemblages or experimentation with novel assemblages of native and exotic species may be useful for assessing the effects of climate change on biodiversity. The stranded riparian zones of urban sites, resulting from the downcutting of streams associated with impervious surfaces, can be used to examine altered environmental drivers of system function. Plant community regeneration and the response of ecosystems to soil nitrogen cycling under altered moisture and temperature conditions may be investigated in such areas. Examining succession in vacant lots may inform practical vegetation management and suggest strategies for changing land use as the density of humans and buildings decline in some city centers. Finally, patterns of adjacency of managed and wild patches in and around cities can be used to examine landscape function.

The ecology of the entire city as a system is represented by research relating species richness to the characteristics of cities. For instance, the number of plant species in urban areas correlates with the human population size. Species number increases with log number of human inhabitants, and that relationship is stronger than the correlation with city area (Klotz 1990). Small towns have from 530 to 560 species, while cities having 100,000 to 200,000 inhabitants have upwards of 1000 species (Sukopp 1998). The age of the city also affects the species richness; large, older cities have more plant species than large, younger cities (Sukopp 1998, Kowarik 1990). These plant assemblages are characteristic throughout Europe, with 15% of species shared among cities (Sukopp 1998).

BIOGEOPHYSICAL BUDGETS   One of the earliest modern ecological approaches to urban systems was the assessment of biogeochemical budgets of whole cities (Odum & Odum 1980). It is clear that urban areas are heterotrophic ecosystems that depend on the productivity from elsewhere (Collins et al. 2000). Cities in industrial countries may use between 100,000 and 300,000 Kcal m^{-2} yr^{-1}, whereas natural ecosystems typically expend between 1,000 and 10,000 Kcal m^{-2} yr^{-1} (Odum 1997). The energy budgets of cities are driven by fossil fuel subsidies, which contribute to the urban heat island effect.

How the green component of cities affects biogeochemical processing for the city as a whole has been examined. For example, in Chicago, trees have been estimated to sequester 5575 metric tons of air pollution, and 315,800 metric tons of C per year at an average rate of 17 metric tons ha^{-1} y^{-1} (McPherson et al. 1997). In the Mediterranean climate of Oakland, California, trees sequester 11 metric tons of C ha^{-1} y^{-1} (Nowak 1993). In contrast to urban forests, natural forests on average sequester 55 metric tons ha^{-1} y^{-1} (Zipperer et al. 1997). The capacity of trees to filter particulates from urban air is based on leaf size and surface roughness (Agrawal 1998).

Urban areas also concentrate materials from elsewhere. For example, the elevation of the surface in old cities is generally higher than the surrounding areas as a result of importing construction materials (Sukopp 1998). A carbon dioxide dome accumulates over cities in association with combustion of fossil fuels (Brazel et al. 2000). Anthropogenically produced forms of N are concentrated in (Lovett et al. 2000) and downwind of cities (Chamiedes et al. 1994). Nitrogen accumulated from human metabolism and fertilizers concentrates downstream of cities. Hence, population density of the watersheds is statistically correlated with N loading in the major rivers of the world (Caraco & Cole 1999).

A useful way to quantify the dependence of urban systems on ecosystems beyond their borders is the concept of the ecological footprint. The ecological footprint of an urban area indexes the amount of land required to produce the material and energetic resources required by, and to process the wastes generated by, a metropolis (Rees 1996). The city of Vancouver, Canada, requires 180 times more land to generate and process materials than the city actually occupies. The concept is highly metaphorical, because the actual networks from which any particular city draws resources, and the areas affected by its waste, may extend around the globe. An analysis of the growth of Chicago (Cronon 1991) showed that a network of resource acquisition extended throughout the western regions of the United States in the late nineteenth century. The metropolis in the postindustrial, information age in nations enjoying a high fossil fuel subsidy has different connections with the hinterland than did the industrially and agriculturally anchored Chicago of a century ago (Bradley 1995). Telecommuting, materially and energetically subsidized recreation, and the alteration of land values for urban uses in the countryside represent a footprint based on urban capital.

ECOSYSTEM PATTERN AND PROCESS   There are three ways in which contemporary studies of ecology of cities, or more properly entire metropolitan areas, are prepared to move beyond the classical ecological approaches and to support integration with social and physical sciences. 1. The contemporary assumptions of ecosystem function are more inclusive than the classical assumptions. 2. The net effects approach to ecosystem budgets has evolved to consider multiple processes and spatial heterogeneity. Finally, 3. the narrow theories that were used to bridge disciplines in the past have been broadened or replaced.

The first tool for integration is the contrast between contemporary and classical assumptions about ecosystem function. Classically, ecologists based studies of urban areas on the assumptions that ecosystems were materially closed and homeostatic systems. Such assumptions were a part of the ecosystem theory used by many early geographers (Zimmerer 1994). These assumptions have been replaced (Zimmerer 1994, Pickett et al. 1992). Consequently, there is a new theory of ecosystems that was not available to those who pioneered the budgetary approach to urban systems (e.g., Boyden et al. 1981). What remains is the basic concept of the ecosystem as a dynamic, connected, and open system (Likens 1992), which can serve the various disciplines (Rebele 1994) that need to be integrated to form a more comprehensive theory to support joint ecological, social, and physical study of urban systems.

Contemporary ecology propounds a systems view that builds on the rigorous budgetary approach to ecosystems (e.g., Jones & Lawton 1995). Of course ecologists necessarily continue to exploit the laws of conservation of matter and energy to generate budgets for ecosystem processes. However, many contemporary studies of ecosystem budgets do not treat systems as though they were black boxes. Rather, the structural details and richness of processes that take place within the boundaries of the system are a major concern of contemporary ecosystem analysis. Contemporary ecosystem ecology exposes the roles of specific species and interactions within communities, flows between patches, and the basis of contemporary processes in historical contingencies. These insights have not been fully exploited in urban ecological studies.

The third feature of contemporary ecology is the breadth of key theories that can be used in integration. Classical ecological theory provided social scientists with only a narrow structure for integration with ecology. The social scientists of the Chicago School, which was active in the early decades of the twentieth century, used ecologically motivated theories of succession and competition, for example. At the time, only the relatively general, deterministic, and equilibrium versions of those theories were available. Contemporary theories of interaction account for both positive and competitive effects, and predictions are based on the actual mechanisms for interaction rather than net effects. In contemporary succession theory, mechanisms other than facilitation are included, and the sequence of communities may not be linear or fixed. In addition, ecologists have come to recognize that those theories have hierarchical structure, which allows them to address different levels of generality and mechanistic detail depending on the scale of the study or the scope of the research questions (Pickett et al. 1994). Hence, social scientists and ecologists now can select the most appropriate levels of generality in a theoretical area for integration. No longer must integration rely on general theories of net effects of such processes as competition and succession.

URBAN ECOLOGY AS A PLANNING APPROACH

Although the first volume of the journal *Ecology* contained a scientific paper devoted to the effect of weather on the spread of pneumonia in New York and Boston (Huntington 1920), the interactions of humans with the urban environment have been primarily the province of planners and landscape architects. For example, Central Park in New York City and other urban parks designed by Frederick Law Olmsted seem intuitively to link environmental properties to human well being in cities. In particular, Olmsted's design for the Boston Fens and Riverway shows ecological prescience in its sophisticated combination of wastewater management and recreational amenity (Spirn 1998). Ian McHarg's (1969) *Design with Nature* alerted planners and architects to the value of incorporating knowledge of ecological and natural features among the usual engineering, economic, and social criteria when developing a regional plan. In McHarg's approach, environmental risks and amenities of different types are mapped on separate layers. The composite map suggests where certain types of development should or should not occur. This approach presaged the technology of Geographic Information Systems (GIS), which has become an important tool to incorporate multiple criteria in planning (Schlutnik 1992, Grove 1997) such as those proposed by McHarg (1969). A more explicit ecological approach is that of Spirn (1984), who examined how natural processes are embedded in cities, and how the interaction between the built environment and natural processes affected economy, health, and human community. For instance, she showed how the forgotten environmental template of drainage networks continued to affect infrastructure and the social structure of a Philadelphia neighborhood.

The planning perspective of human ecology is especially strong in Europe (Sukopp 1998). Planning in Germany has been heavily influenced by a national program of biotope mapping that includes cities (Sukopp 1990, Werner 1999). This program includes descriptions of the flora and fauna of biotopes as a key to identifying types of habitats that are significant for 1. protecting natural resources, 2. quality of life, and 3. a sense of place and identity in the city (Werner 1999). In addition to identifying specific biotopes, researchers in Mainz have mapped the distribution of flora and fauna, natural phenomena, and recreational activities within the biotopes (Frey 1998). Similar research by the Polish Academy of Sciences has focused on urban and suburban areas. The research included studies of soils and abiotic ecosystem components, and research by social scientists in a mosaic of habitats with different degrees of development (Zimny 1990). Building upon the foundation of vegetation classification in cities, Brady et al. (1979) proposed a continuum of habitats from the natural to the highly artificial. Dorney (1977), using a similar approach, proposed an urban-rural continuum from a planning perspective and identified six representative land zones—central business district, old subdivisions, new subdivisions, urban construction zones, urban fringe, and rural. Each zone was characterized by three components or subsystems: cultural history,

Figure 1 Patchiness in the Rognel Heights neighborhood of Baltimore, Maryland. The spatial heterogeneity of urban systems presents a rich substrate for integrating ecological, socioeconomic, and physical patterns using Geographic Information Systems. The patch mosaic discriminates patches based on the structures of vegetation and the built environment. The base image is a false color infrared orthorectified photo from October 1999. (M.L. Cadenasso, S.T.A Pickett & W.C. Zipperer, unpublished data.)

abiotic characteristics, and biotic features. We review the status and implication of such integrated classifications later.

Urban ecology manifest as city planning is contrasted with spatial planning (de Boer & Dijst 1998) in which primary motivations are the degree of segregation or aggregation of different economic and social functions, efficiency of transportation and delivery of utilities, and efficient filling of undeveloped space. Additional components of urban planning said to have ecological foundations include life cycle analysis of products, utility planning based on use rather than medium, efficiency of resource use, exploitation of green infrastructure, and requirements for monitoring of the results (Breuste et al. 1998).

Although the planning described above is ecologically motivated, and it relies on mapping to describe environmental amenities, it is rarely based on data concerning ecological function. It therefore relies on general ecological principles and assumptions, and on the success of prior case histories (Flores et al. 1997). The insights of urban ecology as planning are summarized in manuals and codified in zoning and planning practice. However, like other environmental practices, these insights may not be applicable in novel ecological circumstances. Given the changing forms of cities in both Europe and the United States, novel ecological circumstances may be in the offing.

AN INTEGRATED FRAMEWORK FOR URBAN ECOLOGICAL STUDIES

We see three opportunities for improving the theory to understand urban systems. First, rather than modeling human systems and biogeophysical systems separately, understanding will be improved by using integrated frameworks that deal with social and biogeophysical processes on an equal footing (Groffman & Likens 1994). Second, knowing that the spatial structure of biogeochemical systems can be significant for their function (Pickett et al. 2000), we hypothesize that the spatial heterogeneity so obvious in urban systems also has ecological significance (Figure 1, see color insert). Third, insights from hierarchy theory can organize both the spatial models of urban systems and the structure of the integrated theory developed to comprehend them. We explore and combine these themes below.

Social Ecology and Social Differentiation

The study of social structures, and how those structures come to exist, are the key social phenomena to support integrated study of the ecology of urban systems. It is increasingly difficult to determine where biological ecology ends and social ecology begins (Golley 1993). Indeed, the distinction between the two has diminished through the convergence of related concepts, theories, and methods in the biological, behavioral, and social sciences. Social ecology is a life science focusing on the ecology of various social species such as ants, wolves, or orangutans. We may also study *Homo sapiens* as an individual social species or comparatively with the ecology of other social species. The subject matter of social ecology, like that of biological ecology, is stochastic, historic, and hierarchical (Grove & Burch 1997). In other words, living systems are not deterministic; they exhibit historical contingencies that cannot be predicted from physical laws alone (Botkin 1990, Pickett et al. 1994).

The underlying basis for this life science approach to the study of human ecological systems depends upon three points (Grove & Burch 1997):

1. *Homo sapiens*, like all other species, are not exempt from physical, chemical, or biological processes. Biophysical and social characteristics of humans are shaped by evolution and, at the same time, shape the environment in which *Homo sapiens* live;

2. *Homo sapiens*, like some other species, exhibit social behavior and culture; and

3. Social and cultural traits are involved fundamentally in the adaptation of social species to environmental conditions.

Human ecology must reconcile social and biological facts to understand the behavior of *Homo sapiens* over time (Machlis et al. 1997). Such a biosocial approach to human ecological systems (Burch 1988, Field & Burch 1988, Machlis et al. 1997) stands in contrast to a more traditional geographic or social approach (see Hawley 1950, Catton 1994). This is not to say that social sciences such as psychology, geography, anthropology, sociology, economics, and political science are not important to social ecology. They are, because the most fundamental trait that distinguishes humans and their evolutionary history from other species—both social and nonsocial—is that human social development has enabled the species to escape local ecosystem limitations so that local ecosystems no longer regulate human population size, structure, or genetic diversity (Diamond 1997). Nowhere is this more apparent than in urban ecosystems.

One of the major tools for integration between social and biogeophysical sciences is in the phenomenon of social differentiation. All social species are characterized by patterns and processes of social differentiation (van den Berghe 1975). In the case of humans, social differentiation or social morphology has been a central focus of sociology since its inception (Grusky 1994). In particular, social scientists have used concepts of social identity (i.e., age, gender, class, caste, and clan) and social hierarchies to study how and why human societies become differentiated (Burch & DeLuca 1984, Machlis et al. 1997).

Social differentiation is important for human ecological systems because it affects the allocation of critical resources, including natural, socioeconomic, and cultural resources. In essence, social differentiation determines "who gets what, when, how and why" (Lenski 1966, Parker & Burch 1992). Being rarely equitable, this allocation of critical resources results in rank hierarchies. Unequal access to and control over critical resources is a consistent fact within and between households, communities, regions, nations, and societies (Machlis et al. 1997). Five types of sociocultural hierarchies are critical to patterns and processes of human ecological systems: wealth, power, status, knowledge, and territory (Burch & DeLuca 1984). Wealth is access to and control over material resources in the form of natural resources, capital, or credit. Power is the ability to alter others' behavior through explicit or implicit coercion (Wrong 1988). The powerful have access to resources that are denied the powerless. One example is politicians who make land-use decisions or provide services for specific constituents at the expense of others. Status is access to honor and prestige and the relative position of an individual (or group) in an informal hierarchy of social worth (Lenski 1966). Status is distributed unequally, even within small communities, but high-status individuals may not necessarily have access to either wealth or power. For instance, a minister or an imam may be respected and influential in a community even though he or she is neither wealthy nor has the ability to coerce other people's behavior. Knowledge is access to or control over specialized types of information, such as technical, scientific, and religious. Not everyone within a social system has equal access to all types of information. Knowledge often provides advantages in terms of access to and control over the critical resources and services of social institutions. Finally, territory is access to and control over critical resources through formal and informal property rights (Burch et al. 1972, Bromley 1991).

Social differentiation of human ecological systems has a spatial dimension characterized by patterns of territoriality and heterogeneity (Morrill 1974, Burch 1988). As Burch (1988) noted, "Intimate and distant social relations, high and low social classes, favored and despised ethnic, occupational, and caste groupings all have assigned and clearly regulated measures as to when and where those relations should and should not occur." When ecosystem and landscape approaches are combined, the research changes from a question of "who gets what, when, how and why?" to a question of "who gets what, when, how, why and where?" and, subsequently, what are the reciprocal relationships between spatial patterns and sociocultural and biophysical patterns and processes of a given area (Grove 1997)?

Various processes of social differentiation occur at different scales and have corresponding spatial patterns and biophysical effects (Grove & Hohmann 1992). Based on existing social and ecological theory, examples include global and regional urban-rural hierarchies (Morrill 1974), the distribution of land uses within urban areas (Guest 1977), the stratification of communities within residential land uses (Logan & Molotch 1987), and the social differentiation of ownerships and households within communities (Burch & Grove 1993, Grove 1995).

Spatial Heterogeneity

A human landscape approach may be understood as the study of the reciprocal relationships between patterns of spatial heterogeneity and sociocultural and biophysical processes. Further, when human ecosystem and landscape approaches are combined, human ecosystem types are defined as homogeneous areas for a specified set of sociocultural and biophysical variables within a landscape. Analyses then focus on two primary issues: 1. the development and dynamics of spatial heterogeneity, and 2. the influences of spatial patterns on cycles and fluxes of critical ecosystem resources (e.g., energy, materials, nutrients, genetic and non-genetic information, population, labor, capital, organizations, beliefs, or myths). For instance, the development and dynamics of heterogeneity in a watershed spanning urban to rural conditions may influence and be influenced by sociocultural and biophysical processes. Patches within the watershed may function as either sources or sinks as well as to regulate flows and cycles of critical resources between other patches. The delineation and classification of these relatively homogeneous patches is based on a limited number of representative sociocultural and biophysical indicators (Burch & DeLuca 1984, Parker & Burch 1992), and the patches are studied as black boxes with fluxes and cycles of critical resources between areas (Zonneveld 1989). The spatial linkages between the social and ecological differentiation of the watershed and the relationship of the linkages to different types of allocation mechanisms at different scales are important for understanding the flows and cycles of critical resources within the watershed.

Hydrologists have recognized mosaics of spatial heterogeneity in the variable source area (VSA) approach. They examine how the abiotic attributes of different patches within a watershed–such as temperature and physical characteristics including topography, soil properties, water table depth, and antecedent soil moisture–contribute variable amounts of water and nutrients to streamflow, depending upon their spatial location in the watershed (Black 1991). This VSA approach can be integrated with a delineation of patches based upon the biotic attributes of the watershed, such as vegetation structure and species composition (Bormann & Likens 1979), and the social attributes of the watershed, such as indirect effects from land-use change and forest/vegetation management, and direct effects from inputs of fertilizers, pesticides, and toxins, to examine how the abiotic, biotic, and social attributes of different patches within a watershed contribute variable amounts of water and nutrients to streamflow, depending upon their spatial location in the watershed (Grove 1996). This integrated VSA approach combines nested hierarchies of land use and land cover, sociopolitical structures, and watershed heterogeneity (Figure 2). GIS is a useful tool for analyzing nested hierarchies of spatial heterogeneity.

VSA approaches can be linked to additional social processes. For example, catchments can be examined via hydrological, land-use, and economic models (Costanza et al. 1990). The three components can be combined into an ecosystem model composed of grid cells within a catchment. The integrated model is built up from a basic model that has hydrologic and ecologic components, but no economic components. Therefore, in the basic model, human behavior causing land-use change must be considered as a factor external to the focal catchment. To construct truly integrated ecological-economic models, major nutrient, water, productivity, and successional components of the basic model must be combined with land-use and economic valuations (Bockstael et al. 1994). The integrated model calculates

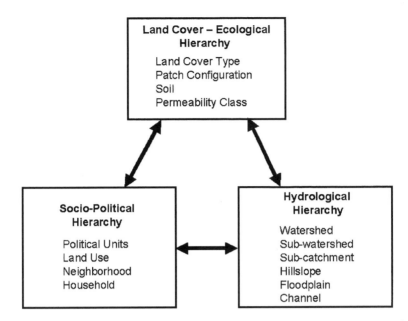

Figure 2   Three interacting nested hierarchies of spatial heterogeneity representing key disciplinary perspectives used in modeling watershed function in urban areas. Land cover represents nested ecological structures that control interception and runoff. The sociopolitical hierarchy contains nested units of environmental decision making and resource use. The hydrological hierarchy indicates the nesting of spatially differentiated units connected by runoff and runon dynamics.

land-use designation through a habitat-switching module that determines when, through natural succession or weather-driven ecological catastrophe (e.g., floods), the habitat shifts from one type to another. Hypothetical human-caused land-use changes can be imposed exogenously using the integrated model. Recognizing that the ecological effects of human activity are driven by the choices people make concerning stocks of natural capital, the economic modeling uses an understanding of how land-use decisions are made by individuals and how they are based on both the ecological and economic features of the landscape. Again, GIS serves as a tool for manipulating the spatial data on which the model depends, and for assessing the model output over space.

The Human Ecosystem Framework and Urban Ecological Systems

An integrated framework for analyzing urban systems as social, biological, and physical complexes now emerges. Social scientists have focused on interactions between humans and their environments since the self-conscious origins of their disciplines. However, the explicit incorporation of the ecosystem concept within the social sciences dates to Duncan's (1961, 1964) articles "From Social System to Ecosystem" and "Social Organization and the Ecosystem." Recently, the social sciences have focused increasingly on the ecosystem concept because it has been proposed and used as an organizing approach for natural resource policy and management (Rebele 1994).

The ecosystem concept and its application to humans is particularly important because of its utility as a framework for integrating the physical, biological, and social sciences. The ecosystem concept owes its origin to Tansley (1935), who noted that ecosystems can be of any size, as long as the concern is with the interaction of organisms and their environment in a specified area. Further, the boundaries of an ecosystem are drawn to answer a particular question. Thus, there is no set scale or way to bound an ecosystem. Rather, the choice of scale and boundary for defining any ecosystem depends upon the question asked and is the choice of the investigator. In addition, each investigator may place more or less emphasis on the chemical transformations and pools of materials drawn on or created by organisms; or on the flow, assimilation, and dissipation of biologically metabolizable energy; or on the role of individual species or groups of species on flows and stocks of energy and matter. The fact that there is so much choice in the scales and boundaries of ecosystems, and how to study and relate the processes within them, indicates the profound degree to which the ecosystem represents a research approach rather than a fixed scale or type of analysis.

Although the ecosystem concept is flexible enough to account for humans and their institutions (Tansley 1935, Rebele 1994), the application of an ecosystem approach to the study of human ecosystems requires additional analytical components. The analytical framework (Figure 3) we use here (see Burch & DeLuca 1984, Machlis et al. 1997, Pickett et al. 1997) is not itself a theory. As Machlis et al. (1997: p. 23) noted,

"This human ecosystem model is neither an oversimplification nor caricature of the complexity underlying all types of human ecosystems in the world. Parts of the model are orthodox to specific disciplines and not new. Other portions of the model are less commonplace—myths as a cultural resource, justice as a critical institution. Yet we believe that this model is a reasonably coherent whole and a useful organizing concept for the study of human ecosystems as a life science."

Several elements are critical to the successful application of this framework. First, it is important to recognize that the primary drivers of human ecosystem dynamics are both biophysical and social. Second, there is no single determining driver of anthropogenic ecosystems. Third, the relative significance of drivers may vary over time. Fourth, components of this framework need to be examined simultaneously in relationship to each other (Machlis et al. 1997). Finally, researchers need to examine how dynamic biological and social allocation mechanisms such as ecological constraints, economic exchange, authority, tradition, and knowledge affect the distribution of critical resources including energy, materials, nutrients,

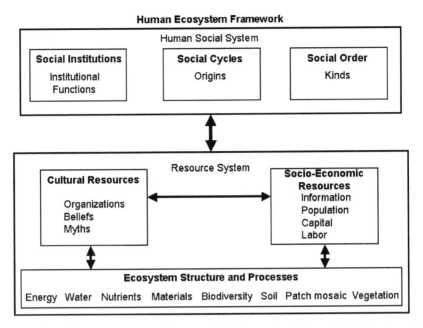

Figure 3   A human ecosystem framework for integrating biogeophysical and social structures and processes. This conceptual structure shows the most general components of any ecosystem that includes or is affected by humans. The nesting of the boxes indicates the inclusion of specific structures or processes within more general phenomena. The social system contains social institutions which function for provision of government, administration of justice, delivery of health services, provision of sustenance, etc. Social order is determined by factors of individual and group identity, formal and informal norms of behavior, and hierarchies that determine allocation of resources. The resource system is founded on bioecological structures and processes that include the standard subjects of ecology texts. The resource system also includes cultural and socioeconomic resources, which interact with bioecological resources in determining the dynamics of the social system. Based loosely on the work of Machlis et al. (1997). Further ecological details added by Pickett et al. (1997).

population, genetic and nongenetic information, labor, capital, organizations, beliefs, and myths within any human ecosystem (Parker & Burch 1992).

CONCLUSIONS

Although there is a wealth of information on the terrestrial components of urban ecological sytstems, much of it is organized from the perspective of ecology in cities. This perspective stands in contrast to a more comprehensive perspective identified as the ecology of cities (Grimm et al. 2000). Studies of ecology in cities have exposed the environmental stresses, subsidies, and constraints that affect urban biota and have documented that the biotic components of metropolitan areas have considerable predictability. In addition, the capacity of certain organisms to adapt to urban environments results in characteristic assemblages.

An alternative approach to urban ecology exists in landscape architecture and planning. This professional practice is motivated by a desire to incorporate ecological principles, to make environmental amenities available to metropolitan residents, and to decrease the negative impacts of urban resource demand and waste on environments elsewhere. Although floristic and faunistic descriptions from urban sites are frequently used in design and planning, there are few data available on ecological functions in cities that can inform such practice. Furthermore, the rapidly changing spatial forms of urban growth and change, and the complex of environmental factors that interact in and around cities, make simple environmental extrapolations risky. Although most of the urban ecological research that has been motivated by planning is of the sort that can be labeled ecology in cities, the field often takes a more comprehensive approach that expresses the ideal of ecology of cities.

In basic ecological research the ecology of the city was first addressed by budgetary studies. Classically this approach has been informed by a biogeochemical perspective based on closed, homeostatic systems. Material and energy budgets of urban systems have been estimated under this rubric. Of course the budgetary approach works whether systems are closed or not, or externally regulated or not. However, assumptions about spatial uniformity and that social agents are external to the ecological processes are questionable.

The distinct ecological approaches classically applied to urban research, and the parallel planning approach, point to the need for integration of different disciplinary perspectives. We have presented a framework that uses middle level theories from ecology and social sciences to identify key factors that should govern the structure and function of biotic, abiotic, and socioeconomic processes in and around cities. This framework is identified as a human ecosystem model, in which both social and ecological processes are integral. Furthermore, the spatial heterogeneity in the biogeophysical and social components of urban systems can be portrayed as patch dynamics. Because patch dynamics can be addressed over many nested scales, structure-function relationships in the human ecosystem can be examined from household to region. The integrative tools and existing data prepare urban ecological studies to continue to benefit from the nature in cities approach, as well as to expoit the contemporary concerns of ecology with spatially heterogeneous, adaptive, self-organizing networks of entire metropolitan systems.

ACKNOWLEDGMENTS

We are grateful to the National Science Foundation, DEB 97–48135 for support of the Baltimore Ecosystem Study (BES) Long-Term Ecological Research program. We also thank the EPA STAR Grant Program for support through the Water and Watersheds program, GAD R825792. This paper has benefited from insights gained through interaction with generous collaborators, students, and community partners in New York City and in Baltimore over the last decade.

Visit the Annual Reviews home page at www.AnnualReviews.org

LITERATURE CITED

Adamczewska-Andrzejewska K, Mackin-Rogalska R, Nabaglo L. 1988. The effect of urbanization on density and population structure of *Apodemus agrarius* (Pallas, 1771). *Pol. Ecol. Stud.* 14:171–95

Adams LW, Leedy DL, eds. 1987. *Integrating Man and Nature in the Metropolitan Environment.* Columbia, MD: Natl. Inst. Urban Wildl.

Adams LW, Leedy DL, eds. 1991. *Wildlife Conservation in Metropolitan Environments.* Columbia, MD: Natl. Inst. Urban Wildl.

Adams LW, VanDruff LW. 1998. Introduction. *Urban Ecosyst.* 2:3

Agrawal M. 1998. Relative susceptibility of plants in a dry tropical urban environment. See Breuste et al. 1998, pp. 603–7

Airola TM, Buchholz K. 1984. Species structure and soil characteristics of five urban sites along the New Jersey Palisades. *Urban Ecol.* 8:149–64

Alverson WS, Waller DM, Solheim SL. 1988. Forests too deer: edge effects in northern Wisconsin. *Conserv. Biol.* 2:348–58

Andrzejewski R, Babinska-Werka J, Gliwicz J, Goszczynski J. 1978. Synurbization processes in population of *Apodemus agrarius.* I. Characteristics of populations in an urbanization gradient. *Acta Theriol.* 23:341–58

Bagnall RG. 1979. A study of human impact on an urban forest remnant: Redwood Bush, Tawa, near Wellington, New Zealand. *N. Z. J. Bot.* 17:117–26

Ballach H-J, Goevert J, Kohlmann S, Wittig R. 1998. Comparative studies on the size of annual rings, leaf growth and structure of treetops of urban trees in Frankfurt/Main. See Breuste et al. 1998, pp. 699–701

Bergen F, Schwerk A, Abs M. 1998. Long-term observation of the fauna of two man-made nature habitats in cities of the Ruhr-Valley area. See Breuste et al. 1998, pp. 618–22

Bishop JA, Cook LM. 1980. Industrial melanism and the urban environment. *Adv. Ecol. Res.* 11:373–404

Black PE. 1991. *Watershed Hydrology.* Englewood Cliffs, NJ: Prentice Hall

Bockstael N, Costanza R, Strand I, Boynton W, Bell K, Wainger L. 1995. Ecological economic modeling and valuation of ecosystems. *Ecol. Econ.* 14:143–59

Bormann FH, Likens GE. 1979. *Pattern and Process in a Forested Ecosystem.* New York: Springer-Verlag

Botkin DB. 1990. *Discordant Harmonies: a New Ecology for the Twenty-first Century.* New York: Oxford Univ. Press

Botkin DB, Beveridge CE. 1997. Cities as environments. *Urban Ecosyst.* 1:3–19

Bowers MA. 1997. Influence of deer and other factors on an old-field plant community. In *The Science of Overabundance: Deer Ecology and Population Management,* ed. WJ McShea, BH Underwood, JH Rappole, pp. 310–26. Washington, DC: Smithson. Inst. Press

Boyden S, Millar S, Newcombe K, O'Neill B. 1981. *The Ecology of a City and its People: the Case of Hong Kong.* Canberra: Aust. Natl. Univ. Press

Bradley GA, ed. 1995. *Urban Forest Landscapes: Integrating Multidisciplinary Perspectives.* Seattle: Univ. Wash. Press

Bradshaw AD, McNeilly T. 1981. *Evolution and pollution.* London: Edward Arnold

Brady RF, Tobias T, Eagles PFJ, Ohrner R,

Micak J et al. 1979. A typology for the urban ecosystem and its relationship to larger biogeographical landuse units. *Urban Ecol.* 4:11–28

Brazel A, Selover N, Vose R, Heisler G. 2000. The tale of two cities—Baltimore and Phoenix urban LTER sites. *Climate Res.* 15:123–35

Breuste J, Feldman H, Uhlmann O, eds. 1998. *Urban Ecology.* Berlin: Springer-Verlag

Bromley DW. 1991. *Environment and Economy: Property Rights and Public Policy.* Cornwall, UK: TJ

Brooks RT, Rowntree RA. 1984. Forest area characteristics for metropolitan and non-metropolitan counties of three northeastern states of the United States. *Urban Ecol.* 8:341–46

Bullock P, Gregory PJ. 1991. Soils: a neglected resource in urban areas. In *Soils in the Urban Environment*, ed. P Bullock, PJ Gregory, pp. 1–5. Oxford, UK: Blackwell Sci.

Burch WR Jr. 1988. Human ecology and environmental management. In *Ecosystem Management for Parks and Wilderness*, ed. JK Agee, J Darryll, pp. 145–59. Seattle: Univ. Wash. Press

Burch WR Jr, Cheek NH Jr, Taylor L, eds. 1972. *Social Behavior, Natural Resources, and the Environment.* New York: Harper & Row

Burch WR Jr, DeLuca D. 1984. *Measuring the Social Impact of Natural Resource Policies.* Albuquerque: Univ. New Mexico Press

Burch WR Jr, Grove JM. 1993. People, trees, and participation on the urban frontier. *Unasylva* 44:19–27

Caraco NF, Cole JJ. 1999. Human impact on nitrate export: an analysis using major world rivers. *Ambio* 28:167–70

Carreiro MM, Howe K, Parkhurst DF, Pouyat RV. 1999. Variations in quality and decomposability of red oak litter along an urban-rural land use gradient. *Biol. Fert. Soils* 30: 258–68

Catton WR Jr. 1994. Foundation of human ecology. *Sociol. Perspect.* 37:75–95

Chameides WL, Kasibhatla PS, Yienger J,

Levy H. 1994. Growth of continental-scale metro-agro-plexes, regional ozone pollution, and world food production. *Science* 264:74–77

Churcher PB, Lawton JH. 1987. Predation by domestic cats in an English village. *J. Zool.* 212:439–55

Collins JP, Kinzig A, Grimm NB, Fagan WF, Hope D et al. 2000. A new urban ecology. *Am. Sci.* 88:416–25

Costanza R, Sklar FH, White ML. 1990. Modeling coastal landscape dynamics. *BioScience* 40:91–107

Cronon W. 1991. *Nature's Metropolis: Chicago and the Great West.* New York: Norton

de Boer J, Dijst M. 1998. Urban development and environmental policy objectives—an outline of a multi-disciplinary research program. See Breuste et al. 1998, pp. 38–42

Deelstra T. 1998. Towards ecological sustainable cities: strategies, models and tools. See Breuste et al. 1998, pp. 17–22

De Rudder B, Linke F. 1940. *Biologie der Großstadt.* Dresden: Leipzig

Diamond JM. 1997. *Guns, Germs, and Steel: the Fate of Human Societies.* New York: Norton

Dorney RS. 1977. Biophysical and cultural-historic land classification and mapping for Canadian urban and urbanizing landscapes. In *Ecological/biophysical Land Classification in Urban Areas*, ed. Environ. Can., pp. 57–71. Ottawa: Environ. Can.

Douglas I. 1983. *The Urban Environment.* Baltimore, MD: Edward Arnold

Dow CL, DeWalle DR. 2000. Trends in evapotranspiration and Bowen ration on urbanizing watersheds in eastern United States. *Water Resour. Res.* 7:1835–43.

Drayton B, Primack RB. 1996. Plant species lost in an isolated conservation area in metropolitan Boston from 1894 to 1993. *Conserv. Biol.* 10:30–39

Duncan OD. 1961. From social system to ecosystem. *Sociol. Inq.* 31:140–49

Duncan OD. 1964. Social organization and the ecosystem. In *Handbook of Modern*

Sociology, ed. REL Faris, pp. 37–82. Chicago: Rand McNally

Effland WR, Pouyat RV. 1997. The genesis, classification, and mapping of soils in urban areas. *Urban Ecosyst.* 1:217–28

Field DR, Burch WR Jr. 1988. *Rural Sociology and the Environment.* Middleton, WI: Soc. Ecol.

Flores A, Pickett STA, Zipperer WC, Pouyat RV, Pirani R. 1997. Adopting a modern ecological view of the metropolitan landscape: the case of a greenspace system for the New York City region. *Landsc. Urban Plan.* 39:295–308

Frey J. 1998. Comprehensive biotope mapping in the city of Mainz—a tool for integrated nature conservation and sustainable urban planning. See Breuste et al. 1998, pp. 641–47

Garreau J. 1991. *Edge City: Life on the New Frontier.* New York: Doubleday

Gilbert OL. 1989. *The Ecology of Urban Habitats.* New York/London: Chapman & Hall

Goldman MB, Groffman PM, Pouyat RV, McDonnell MJ, Pickett STA. 1995. CH_4 uptake and N availability in forest soils along an urban to rural gradient. *Soil Biol. Biochem.* 27:281–86

Golley FB. 1993. *A History of the Ecosystem Concept in Ecology: More than the Sum of the Parts.* New Haven: Yale Univ. Press

Greenwood EF, ed. 1999. *Ecology and Landscape Development: a History of the Mersey Basin.* Liverpool: Liverpool Univ. Press

Grimm NB, Grove JM, Pickett STA, Redman CL. 2000. Integrated approaches to long-term studies of urban ecological systems. *BioScience* 50:571–84

Groffman PM, Likens GE, eds. 1994. *Integrated Regional Models: Interactions Between Humans and Their Environment.* New York: Chapman & Hall

Groffman PM, Pouyat RV, McDonnell MJ, Pickett STA, Zipperer WC. 1995. Carbon pools and trace gas fluxes in urban forest soils. In *Soil Management and Greenhouse Effect,* ed. R Lal, J Kimble, E Levine, BA Stewart, pp. 147–58. Boca Raton: CRC Lewis

Grove JM. 1995. Excuse me, could I speak to the property owner please? *Comm. Prop. Resour. Dig.* 35:7–8

Grove JM. 1996. *The relationship between patterns and processes of social stratification and vegetation of an urban-rural watershed.* PhD dissertation. Yale Univ., New Haven. pp. 109

Grove JM. 1997. New tools for exploring theory and methods in human ecosystem and landscape analyses: computer modeling, remote sensing and geographic information systems. In *Integrating Social Sciences and Ecosystem Management,* ed. HK Cordell, JC Bergstrom. Champaign: Sagamore

Grove JM, Burch WR Jr. 1997. A social ecology approach and application of urban ecosystem and landscape analyses: a case study of Baltimore, Maryland. *Urban Ecosyst.* 1:259–75

Grove JM, Hohmann M. 1992. GIS and social forestry. *J. For.* 90:10–15

Grusky DB, ed. 1994. *Social Stratification: Class, Race, and Gender in Sociological Perspective.* Boulder: Westview

Guest AM. 1977. Residential segregation in urban areas. In *Contemporary Topics in Urban Sociology,* ed. KP Schwirian, pp. 269–336. Morristown: Gen. Learn.

Guilden JM, Smith JR, Thompson L. 1990. Stand structure of an old-growth upland hardwood forest in Overton Park, Memphis, Tennessee. In *Ecosystem Management: Rare Species and Significant Habitats,* ed. RS Mitchell, CJ Sheviak, DJ Leopold, pp. 61–66. Albany: New York State Mus.

Haspel C, Calhoon RE. 1991. Ecology and behavior of free-ranging cats in Brooklyn, New York. See Adams & Leedy 1991, pp. 27–30

Hawley D. 1950. *Human Ecology: A Theory of Community Structure.* New York: Ronald

Hobbs ER. 1988. Species richness in urban forest patches and implications for urban landscape diversity. *Landsc. Ecol.* 1:141–52

Hough M. 1995. *Cities and Natural Processes.* London: Routledge

Huntington E. 1920. The control of pneumonia an influenza by weather. *Ecology* 1:1–23

Iida S, Nakashizuka T. 1995. Forest fragmentation and its effect on species diversity in suburban coppice forests in Japan. *For. Ecol. Manage.* 73:197–210

Jones CG, Lawton JH, eds. 1995. *Linking Species and Ecosystems.* New York: Chapman & Hall

Katz B, Bradley J. 1999. Divided we sprawl. *Atl. Mon.* 284:26–42

Klausnitzer B, Richter K. 1983. Presence of an urban gradient demonstrated for carabid associations. *Oecologia* 59:79–82

Klotz S. 1990. Species/area and species/inhabitants relations in European cities. See Sukopp 1990a, pp. 100–12

Kowarik I. 1990. Some responses of flora and vegetation to urbanization in Central Europe. See Sukopp 1990a, pp. 45–74

Lawrence HW. 1995. Changing forms and persistent values: historical perspectives on the urban forest. See Bradley 1995, pp. 17–40

Lenski GE. 1966. *Power and Privilege: a Theory of Social Stratification.* New York: McGraw-Hill

Likens GE. 1992. *Excellence in Ecology, Vol. 3. The Ecosystem Approach: Its Use and Abuse.* Oldendorf/Luhe: Ecol. Inst.

Logan JR, Molotch HL. 1987. *Urban Fortunes: the Political Economy of Place.* Berkeley: Univ. Calif. Press

Lovett GM, Traynor MM, Pouyat RV, Carreiro MM, Zhu W, Baxter J. 2000. Nitrogen deposition along an urban-rural gradient in the New York City metropolitan area. *Environ. Sci. Technol.* 34:4294–300

Luniak M, Pisarski B. 1994. State of research into the fauna of Warsaw (up to 1990). *Mem. Zool.* 49:155–65

Machlis GE, Force JE, Burch WR Jr. 1997. The human ecosystem part I: the human ecosystem as an organizing concept in ecosystem management. *Soc. Nat. Res.* 10:347–67

Mackin-Rogalska R, Pinowski J, Solon J, Wojick Z. 1988. Changes in vegetation, avifauna, and small mammals in a suburban habitat. *Pol. Ecol. Stud.* 14:239–330

Makse HA, Havlin S, Stanley HE. 1995. Modelling urban growth patterns. *Nature* 377:608–12

Matlack GR. 1997. Land use and forest habitat distribution in the hinterland of a large city. *J. Biogeogr.* 24:297–307

Matthews MJ, O'Connor S, Cole RS. 1988. Database for the New York State urban wildlife habitat inventory. *Landsc. Urban Plan.* 15:23–37

McDonnell MJ, Pickett STA, eds. 1993. *Humans as Components of Ecosystems: the Ecology of Subtle Human Effects and Populated Areas.* New York: Springer-Verlag

McDonnell MJ, Pickett STA, Groffman P, Bohlen P, Pouyat RV et al. 1997. Ecosystem processes along an urban-to-rural gradient. *Urban Ecosyst.* 1:21–36

McDonnell MJ, Pickett STA, Pouyat RV. 1993. The application of the ecological gradient paradigm to the study of urban effects. See McDonnell & Pickett 1993, pp. 175–89

McHarg I. 1969. *Design with Nature.* Garden City, NJ: Doubleday/Nat. Hist.

McPherson EG, Nowak D, Heisler G, Grimmond S, Souch C et al. 1997. Quantifying urban forest sturcture, function, and value: the Chicago urban forest climate project. *Urban Ecosyst.* 1:49–61

Medley KE, McDonnell MJ, Pickett STA. 1995. Human influences on forest- landscape structure along an urban-to-rural gradient. *Prof. Geogr.* 47:159–68

Moran MA. 1984. Influence of adjacent land use on understory vegetation of New York forests. *Urban Ecol.* 8:329–40

Morrill RL. 1974. *The Spatial Organization of Society.* Duxbury, MA: Duxbury

Mucina L. 1990. Urban vegetation research in European COMECON countries and Yougoslavia: a review. See Sukopp 1990a, pp. 23–43

Nilon CH, Pais RC. 1997. Terrestrial vertebrates in urban ecosystems: developing hypotheses for the Gwynns Falls Watershed in Baltimore, Maryland. *Urban Ecosyst.* 1:247–57

Nowak DJ. 1993. Atmospheric carbon reduction by urban trees. *J. Environ. Manage.* 37:207–17

Nowak DJ. 1994. Urban forest structure: the state of Chicago's urban forest. In *Chicago's Urban Forest Ecosystem: Results of the Chicago Urban Forest Climate Project*, ed. EG McPherson, D Nowak, RA Rowntree, pp. 140–64. Radnor, PA: USDA Forest Serv.

Nowak DJ, McBride JR. 1992. Differences in Monterey pine pest populations in urban and natural forests. *For. Ecol. Manage.* 50:133–44

Nowak DJ, Rowntree R, McPherson EG, Sisinni SM, Kerkmann E, Stevens JC. 1996. Measuring and analyzing urban tree cover. *Landsc. Urban Plan.* 36:49–57

Noyes JH, Progulske DR. 1974. *Wildlife in an Urbanizing Environment.* Amherst: Mass. Coop. Ext. Serv., Univ. Mass.

Odum EP. 1997. *Ecology: a Bridge Between Science and Society.* Sunderland, MA: Sinauer

Odum HT, Odum EC. 1980. *Energy Basis for Man and Nature.* New York: McGraw-Hill

Oke TR. 1973. City size and urban heat island. *Atmos. Environ.* 7:769–79

Oke TR. 1995. The heat island of the urban boundary layer: charcteristics, causes and effects. In *Wind Climate in Cities*, ed. JE Cermak, pp. 81–107. Netherlands: Kluwer Acad.

Parker JK, Burch WR Jr. 1992. Toward a social ecology for agroforestry in Asia. In *Social Science Applications in Asian Agroforestry*, ed. WR Burch Jr, JK Parker, pp. 60–84. New Delhi: IBH

Paul MJ, Meyer JL. 2001. Riverine ecosystems in an urban landscape. *Annu. Rev. Ecol. Syst.* 32: In press

Pickett STA, Burch WR Jr, Dalton SD, Foresman TW. 1997. Integrated urban ecosystem research. *Urban Ecosyst.* 1:183–84

Pickett STA, Cadenasso ML, Jones CG. 2000. Generation of heterogeneity by organisms: creation, maintenance, and transformation. In *Ecological Consequences of Habitat Heterogeneity*, ed. M Hutchings, L John, A Stewart, pp. 33–52. New York: Blackwell

Pickett STA, Kolasa J, Jones CG. 1994. *Ecological Understanding: The Nature of Theory and the Theory of Nature.* San Diego: Academic

Pickett STA, Parker VT, Fiedler PL. 1992. The new paradigm in ecology: implications for conservation biology above the species level. In *Conservation Biology: The Theory and Practice of Nature Conservation, Preservation, and Management*, ed. PL Fiedler, SK Jain, pp. 65–88. New York: Chapman & Hall

Pouyat RV, McDonnell MJ. 1991. Heavy metal accumulation in forest soils along an urban to rural gradient in southern NY, USA. *Water Soil Air Pollut.* 57/58:797–807

Pouyat RV, McDonnell MJ, Pickett STA. 1997. Litter decomposition and nitrogen mineralization along an urban-rural land use gradient. *Urban Ecosyst.* 1:117–31

Pouyat RV, McDonnell MJ, Pickett STA, Groffman PM, Carreiro MM et al. 1995. Carbon and nitrogen dynamics in oak stands along an urban-rural gradient. In *Carbon Forms and Functions in Forest Soils*, ed. JM Kelly, WW McFee, pp. 569–87. Madison, WI: Soil Sci. Soc.Am.

Pouyat RV, Parmelee RW, Carreiro MM. 1994. Environmental effects of forest soil-invertebrate and fungal densities in oak stands along an urban-rural land use gradient. *Pedobiologica* 38:385–99

Rapoport EH. 1993. The process of plant colonization in small settlements and large cities. See McDonnell & Pickett 1993, pp. 190–207

Rebele F. 1994. Urban ecology and special features of urban ecosystems. *Global Ecol. Biogeogr. Lett.* 4:173–87

Rees WE. 1996. Revisiting carrying capacity: area-based indicators of sustainability. *Popul. Environ.* 17:195–215

Rogers GF, Rowntree RA. 1988. Intensive surveys of structure and change in urban natural areas. *Landsc. Urban Plan.* 15:59–78

Rowntree RA. 1995. Toward ecosystem management: shifts in the core and the context of urban forest ecology. See Bradley 1995, pp. 43–59

Rudnicky JL, McDonnell MJ. 1989. Forty-eight years of canopy change in a hardwood-hemlock forest in New York City. *Bull. Torrey Bot. Club* 116:52–64

Salisbury EJ. 1943. The flora of bombed areas *Proc. R. Inst. G. B.* 32:435–55

Schlutink G. 1992. Integrated remote sensing, spatial information systems, and applied models in resource assessment, economic development, and policy analysis. *Photogramm. Eng. Remote Sens.* 58:1229–37.

Schmid JA. 1975. *Urban Vegetation: a Review and Chicago Case Study.* Chicago: Univ.Chicago Dep. Geogr.

Sissini SM, Emmerich A. 1995. Methodologies, results and applications of natural resource assessment in New York City. *Nat. Areas J.* 15:175–88

Spirn AW. 1984. *The Granite Garden: Urban Nature and Human Design.* New York: Basic Books

Sprin AW. 1998. *The Language of Landscape.* New Haven: Yale Univ. Press

Stearns FW. 1971. Urban botany—an essay on survival. *Univ. Wis. Field Stn. Bull.* 4:1–6

Steinberg DA, Pouyat RV, Parmelee RW, Groffman PM. 1997. Earthworm abundance and nitrogen mineralization rates along an urban-rural land use gradient. *Soil Biol. Biochem.* 29:427–30

Sukopp H. 1998. Urban ecology—scientific and practical aspects. See Breuste et al. 1998, pp. 3–16

Sukopp H, ed. 1990a. *Urban Ecology: Plants and Plant Communities in Urban Environments.* The Hague: SPB Acad.

Sukopp H. 1990b. Urban ecology and its application in Europe. See Sukopp 1990a, pp. 1–22

Sukopp H, Weiler S. 1988. Biotope mapping and nature conservation strategies in urban areas of the Federal Republic of Germany. *Landsc. Urban Plan.* 15:39–58

Tansley AG. 1935. The use and abuse of vegetational concepts and terms. *Ecology* 16:284–307

United Nations. 1993. *World Population Prospects.* New York: United Nations

van den Berghe PL. 1975. *Man in Society: a Biosocial View.* New York: Elsevier

VanDruff LW, Bolen EG, San Julian GJ. 1994. Management of urban wildlife. In *Research and Management Techniques for Wildlife and Habitats,* ed. TA Bookhout, pp. 507–30. Bethesda, MD: Wildl. Soc.

VanDruff LW, Rowse RN. 1986. Habitat association of mammals in Syracuse, New York. *Urban Ecol.* 9:413–34

Velguth PH, White DB. 1998. Documentation of genetic differences in a volunteer grass, *Poa annua* (annual meadow grass), under different conditions of golf course turf, and implications for urban landscape plant selection and management. See Breuste et al. 1998, pp. 613–17

Waggoner PE, Ovington JD, eds. 1962. *Proceedings of the Lockwood Conference on the Suburban Forest and Ecology.* New Haven: Conn. Agric. Exp. Stn.

Werner P. 1999. Why biotope mapping in populated areas? *Deinsea* 5:9–26

Wetterer JK. 1997. *Urban Ecology. Encyclopedia of Environmental Sciences.* New York: Chapman & Hall

White CS, McDonnell MJ. 1988. Nitrogen cycling processes and soil characteristics in an urban versus rural forest. *Biogeochemistry* 5:243–62

Witt K. 1996. Species/plot turnovers from reapeated atlas mapping of breeding birds in southern Berlin 1980 and 1990. *Acta Orn.* 31:81–84

Wittig R. 1998. Urban development and the integration of nature: reality or fiction? See Breuste et al. 1998, pp. 593–99

Wrong DH. 1988. *Power: its Forms, Bases, and Uses.* Chicago: Univ. Chicago Press

Wu L, Antonovics J. 1975a. Zinc and copper uptake by *Agrostis stolonifera* tolerant to zinc and copper. *New Phytol.* 75:231–37

Wu L, Antonovics J. 1975b. Experimental ecological genetics in *Plantago.* II. Lead tolerance in *Plantago lanceolata* and *Cynodon dactylon* from a roadside. *Ecology* 57:205–8

Zimmerer KS. 1994. Human geography and the "new ecology:" the prospect and promise of integration. *Ann. Assoc. Am. Geogr.* 84:108–25

Zimny H. 1990. Ecology of urbanized systems-problems and research in Poland. In *Urban Ecological Studies in Central and Eastern Europe*, ed. M Luniak, pp. 8–18. Warsaw: Pol. Acad. Sci. Inst. Zool.

Zipperer WC, Foresman TW, Sisinni SM, Pouyat RV. 1997. Urban tree cover: an ecological perspective. *Urban Ecosyst.* 1:229–47

Zonneveld IS. 1989. The land unit—a fundamental concept in landscape ecology and its applications. *Landsc. Ecol.* 3:67–89

Annu. Rev. Ecol. Syst. 2001. 32:159–81
Copyright © 2001 by Annual Reviews. All rights reserved

DISPERSAL IN FRESHWATER INVERTEBRATES

David T. Bilton,[1] Joanna R. Freeland,[2] and Beth Okamura[3]

[1]Benthic Ecology Research Group, Department of Biological Sciences,
University of Plymouth, Drake Circus, Plymouth PL4 8AA, United Kingdom;
e-mail: dbilton@plymouth.ac.uk
[2]Department of Biological Sciences, The Open University, Milton Keynes,
United Kingdom MK7 6AA; e-mail: J.R.Freeland@open.ac.uk
[3]School of Animal and Microbial Sciences, University of Reading, Whiteknights,
P.O. Box 228, Reading RG6 6AJ, United Kingdom; e-mail: b.okamura@reading.ac.uk

Key Words gene flow, colonization, passive dispersal, active dispersal, propagule banks

■ **Abstract** Movement between discrete habitat patches can present significant challenges to organisms. Freshwater invertebrates achieve dispersal using a variety of mechanisms that can be broadly categorized as active or passive, and which have important consequences for processes of colonization, gene flow, and evolutionary divergence. Apart from flight in adult freshwater insects, active dispersal appears relatively uncommon. Passive dispersal may occur through transport by animal vectors or wind, often involving a specific desiccation-resistant stage in the life cycle. Dispersal in freshwater taxa is difficult to study directly, and rare but biologically significant dispersal events may remain undetected. Increased use of molecular markers has provided considerable insight into the frequency of dispersal in freshwater invertebrates, particularly for groups such as crustaceans and bryozoans that disperse passively through the transport of desiccation-resistant propagules. The establishment of propagule banks in sediment promotes dispersal in time and may be particularly important for passive dispersers by allowing temporal escape from unfavorable conditions. Patterns that apply to dispersal in freshwater invertebrates can be readily extended to other freshwater taxa, since common challenges arise from the colonization of isolated aquatic systems.

INTRODUCTION

Freshwater invertebrates occur in habitats that represent discrete sites surrounded by an inhospitable terrestrial landscape. Despite this lack of obvious connectivity among sites, many freshwater taxa have broad geographical ranges, as was noted by Darwin (1859). Some organisms achieve wide distributions through active

0066-4162/01/1215-0159$14.00

means such as aerial flight across the intervening landscape. Many, however, are incapable of dispersing themselves and rely on agents such as animal vectors, wind, or water flow to provide passive transport between sites. It is notable that, despite our longstanding appreciation that freshwater organisms achieve dispersal, the extent and modes of dispersal remain poorly understood. Direct study of dispersal is notoriously difficult, since it involves detection of movements by capturing, marking, and recapturing individuals (Southwood & Henderson 2000), an approach that can easily overlook infrequent but biologically significant levels of interpopulation exchange. In aquatic environments, assessment of dispersal can also pose practical problems. Methods of marking individuals may require modification for use in water, and many dispersive stages do not lend themselves to such methods.

Despite the formidable obstacles and problems associated with studying dispersal, the combination of ecological study with new molecular approaches is providing a better understanding of patterns of dispersal in freshwater invertebrates. This review synthesizes what is known of these patterns and emphasizes the ecological and evolutionary consequences of dispersal. We consider empirical studies and theoretical approaches, practical means of investigation, and the comparative biology of dispersing invertebrates that inhabit both lotic and lentic habitats. Because it is beyond the scope of this review to provide a comprehensive survey of the dispersal of invertebrates across the spectrum of freshwater habitats, we focus on systems that have received the widest investigation. In addition, we review evidence that some taxa achieve two-dimensional dispersal by exploiting both temporal and spatial scales, and we note the recent influence of humans on dispersal. Given the general nature of the issues discussed, the diversity of invertebrate life histories, and the ubiquity of invertebrates in freshwater habitats, the various patterns, processes, and predictions that arise should allow appropriate extension to other groups of organisms and to freshwater habitats not explicitly considered.

WHAT CONSTITUTES DISPERSAL?

The term dispersal has been used broadly, with the definition varying between areas of research (Dingle 1996). *The Dictionary of Ecology, Evolution and Systematics* defines dispersal as "[the] outward spreading of organisms or propagules from their point of origin or release; one-way movement of organisms from one home site to another" (Lincoln et al. 1998). As a working definition, we equate dispersal to the movement of individuals or propagules between spatially (or temporally) discrete localities or populations, and we focus mainly on evidence for the overland movement of freshwater invertebrates. Defined in this way, dispersal may or may not entail migration, colonization, or gene flow. Stenseth & Lidicker (1992) provide further discussion of the various uses and definitions of this term.

THE EVOLUTION, HERITABILITY, AND MAINTENANCE OF DISPERSAL

Why Disperse?

From an individual's point of view, there are both advantages and disadvantages to dispersing from one site to another (Stenseth & Lidicker 1992). Advantages include inbreeding avoidance, the possibility of locating a new site with low-density occupation and few resource competitors, and a potential escape from unfavorable conditions such as limited resources, predators, pathogens, and parasites. Disadvantages include an inability to locate a suitable new site, predation en route, failure to locate a mate, and outbreeding depression. The most widely perceived risk when sexually reproducing organisms fail to disperse is inbreeding, whereas overcrowding, predation, and an inability to contend with pathogens and parasites (Bell 1982) are the most obvious threats to clonally reproducing organisms.

Theoretical investigation into the evolution of dispersal has resulted in numerous mathematical models (reviewed in Johnson & Gaines 1990). Recent theoretical developments are dealt with by Clobert et al. (2001), Dieckmann et al. (1999), and Ferrière et al. (2000). Most models identify evolutionarily stable strategies based on game theory (Comins et al. 1980) and focus on variables such as risk of habitat extinction, competition among kin, temporal and spatial variability in habitat quality, costs of dispersal, and avoidance of inbreeding. Adaptive dynamics theory provides an alternative approach through assessing the potential for evolutionary invasion based on the population dynamics of mutant and resident individuals (Dieckmann et al. 1999). Although theoretical developments continue to outpace practical support (Dieckmann et al. 1999, Johnson & Gaines 1990), empirical studies are on the increase. This imbalance no doubt reflects the fact that many of the mechanisms and model parameters are exceedingly difficult to test.

Heritability and Maintenance of Dispersal

For dispersal to be selected, traits related to dispersal must be variable and heritable. Wing length is such a trait in at least two groups of freshwater insects. In the water beetle *Helophorus granularis*, the occurrence of long-winged and flightless morphs is controlled by a single locus system (Angus 1970). In a number of water skaters, the inheritance of wing-length polymorphisms appears to be polygenic (Vepsäläinen 1974, Zera et al. 1983), and in some *Gerris* species, seasonal polyphenism acts in conjunction with genetic polymorphism to determine wing length and therefore dispersal ability (Vepsäläinen 1978). In other species, environmental switches alone appear to determine wing length (Andersen 1982). Heritability of dispersal in other freshwater groups remains poorly understood.

Dispersal is likely associated with the long-term persistence of freshwater taxa, since most freshwater sites are ephemeral relative to species' lifetimes. The relationship between dispersal and persistence of freshwater species has received increasing attention (Avise 1992, Hogg et al. 1998), since dispersal may be a critical predictor of a species' ability to escape threats posed by global atmospheric change (Hogg et al. 1998, Hogg & Williams 1996).

CONSEQUENCES OF DISPERSAL

Dispersal can affect many aspects of the evolution and population genetics of a species, but only if successful colonization is followed by breeding of subsequent generations. Such colonization will result in gene flow, the transfer of genes from one population to another. The effects of dispersal and gene flow are varied and profound. For instance, dispersal can promote range expansion following colonization of new sites. Intercontinental dispersal of freshwater cladocerans (Berg & Garton 1994, Havel et al. 2000), copepods (Cordell & Morrison 1996), and snails (Zaranko et al. 1997) provides dramatic examples of such range expansions. Dispersal may also alter the probability of extinction within local populations by introducing new colonists and increasing genetic diversity, as has been demonstrated in populations of a freshwater bryozoan (Freeland et al. 2000a). In addition, dispersal can strongly reduce the amount of genetic differentiation among populations (Bohonak 1999a). Finally, in the absence of appreciable levels of gene flow, evolutionary independence of populations may result, and this may lead to reproductive isolation and speciation (see Howard & Berlocher 1998 for a recent review).

METHODS OF ASSESSING DISPERSAL

Feasibility studies

The difficulties in quantifying dispersal mean that much of the evidence remains anecdotal, including collections of aquatic insects in traps (Zalom et al. 1980); observations of aquatic insects in flight (Freeman 1945); observations of adults and propagules attached externally to vertebrates and invertebrates (see Figure 1) and in vertebrate digestive tracts (see Table 1); demonstration that propagules remain viable after passing through vertebrate digestive tracts (see Table 1); and colonization of new sites (Jenkins 1995, Maguire 1963). With the exception of colonization events, this body of evidence confirms the feasibility of dispersal but provides little information on its extent or frequency.

Mark-Recapture Techniques

Mark-recapture techniques were developed in ecology principally for estimates of population size (Southwood & Henderson 2000) but can also be applied to assessments of dispersal (Service 1993). Physical marking methods for invertebrates include the use of dyes, stains, and fluorescent or colored powders and pigments (Conrad et al. 1999, Nürnberger & Harrison 1995, Service 1993) applied either to the entire organism or as discrete spots, letters, and numbers (see Figure 2, color

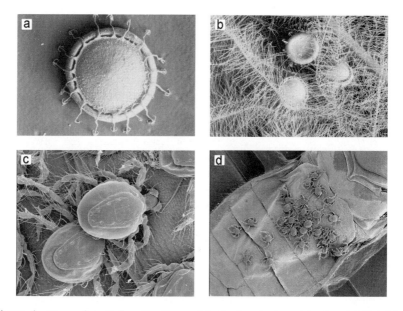

Figure 1 Examples of phoretic dispersal in freshwater invertebrates. (*a*) Statoblast of the bryozoan *Cristatella mucedo*. Gas-filled cells confer buoyancy, making floating statoblasts available for entanglement in, e.g., fur and feathers. Note marginal hooks and spines that are suitable for such attachment. (*b*) Statoblasts of *Cristatella mucedo* entangled in a moulted feather. (*c*, *d*) Larvae of the water mite, *Eylais* sp., attached to the water boatman *Sigara falleni*. Larvae are parasitic, but this relationship also allows phoretic transport of mites between waterbodies.

insert). Such methods are particularly suited to terrestrial adults of aquatic insects. Methods suitable for fully aquatic organisms with hardened external surfaces include the use of a modified dentist's drill or a fine abrasive needle to produce a permanent mark (Svensson 1998). Such marks will not be lost if organisms reenter the water, are unlikely to disappear due to abrasion, and do not introduce toxic organic solvents (present in many waterproof pigments). The incorporation of colored beads in caddisfly larval cases provides a special means of following movements of marked individuals (Erman 1986, Jackson et al. 1999). One potential problem with direct marking is that it may change the behavior of marked individuals, a possibility that is seldom explored. More fundamental problems include the practicalities of marking sufficient numbers of individuals to be able to detect relatively infrequent dispersal events.

Radioactive and stable isotopes have been used to investigate dispersal distances of freshwater insects (Service 1993). Discharge of ^{65}Zn from atomic reactor cooling water demonstrated upstream dispersal in caddisflies (Coutant 1982), and stable isotopes (^{15}N) revealed upstream flight in *Baetis* mayflies (Hershey et al. 1993). In recent years, miniaturized transponder tags have been exploited to track

TABLE 1   Some observations and experiments providing evidence that animal vectors conceivably disperse freshwater invertebrates among sites

Invertebrate	Animal Association	References
Larval and juvenile stages of zebra mussel (*Dreissena polymorpha*)	Zebra mussel stages found on mallard ducks (*Anas platyrhynchos*)	Johnson & Carlton 1996
Adults and juveniles of sphaeriid bivalves (*Pisidium* & *Sphaerium* spp.)	Mussels found attached to the limbs of a range of freshwater insects and amphibians	Kew 1893, Rees 1952, Lansbury 1955
Juvenile mussel (*Anodonta cygnea*)	Individual attached to plover's foot	Adams 1905
Eggs and adults of the river limpet (*Ancylus fluviatilis*)	Limpet eggs and adults found attached to the wing cases of a water beetle (*Acilius sulcatus*)	Kew 1893
Juvenile pond snails (*Lymnaea stagnalis, Stagnicola elodes, Helisoma triviolis*)	Adherence to whistling swan (*Olor colombianus*) feathers and survival in air	Boag 1986
Adult Ostracoda	Viable adults recovered from the gut of the least sandpiper (*Erolia minutilla*) and found attached to the abdominal hair fringe of water boatmen (*Notonecta glauca* & *N. obliqua*)	Proctor et al. 1967, Lansbury 1955
Adult Amphipoda (*Hyallela azteca* & *Gammarus lacustris*)	Living amphipods collected from the fur of beavers and muskrats	Peck 1975
Adult and cocoon stages of ectoparasitic leeches, *Theromyzon rude* & *Placobdella papillifera*	Attachment of nonfeeding adult leeches to domestic ducks (*Anas platyrhynchos*); viability of some cocoons fed to ducks	Davies et al. 1982
Larval water mites (Hydracarina)	Attachment of parasitic larval stages to winged adult freshwater insects	Bohonak 1999b
Various crustacean eggs	Viability of eggs recovered from digestive tract and feces of domesticated and wild ducks	Proctor & Malone 1965, Proctor et al. 1967
Eggs of brine shrimp (*Artemia salina*)	Exposure of eggs to digestive enzymes has no effect on hatching	Horne 1966
Statoblasts of freshwater bryozoans	Some statoblasts remain viable after passing through digestive tracts of ducks, amphibians, and reptiles	Brown 1933

insects moving close to the ground. These tags capture and reradiate some of the energy emitted from hand-held radar, producing a characteristic signal (Riley et al. 1996). Such methods have obvious potential for studying short-distance movement and dispersal in adult aquatic insects and will no doubt see increased use in the future.

Population Genetic Approaches

Population genetic studies provide indirect methods for studying dispersal. Levels of gene flow among populations can be inferred from the genetic characterization of individuals using molecular markers such as allozymes (Crease et al. 1997, Hughes et al. 1999), mitochondrial DNA (Crease et al. 1997, Taylor et al. 1998), randomly amplified polymorphic DNA (RAPD) (Thomas et al. 1998), and microsatellite loci (Freeland et al. 2000a,b). Indirect assessment of gene flow from genetic data is based on the spatial distributions of alleles or chromosomal segments (reviewed in Slatkin 1985). High levels of dispersal will result in shared alleles and little genetic subdivision between populations, whereas low levels of dispersal will result in genetic divergence among populations as a result of drift and/or selection (but see Nürnberger & Harrison 1995). An important caveat is that the apparent genetic similarity of populations will depend to some extent on the relative mutation rate and the mode of inheritance (e.g., nuclear versus mitochondrial) of the molecular markers that are employed. Different regions of the genome evolve at different rates, which can influence, sometimes profoundly, our interpretation of dispersal patterns (Avise 1994).

The most common approach for inferring gene flow (Nm) from genetic data is based on the variance in allele frequencies among populations, e.g., Wright's F_{ST} (Wright 1951), Nei's G_{ST} (Nei 1972), and Weir's θ (Weir & Cockerham 1984). Although widely used, such derivations of Nm should be interpreted with caution, since numerous assumptions such as constant population size, an infinite number of populations, and a constant rate of migration are frequently violated, particularly in nonequilibrium populations (Whitlock & McCauley 1999). Furthermore, F-statistics and their analogs can yield varying estimates of population differentiation and gene flow (Freeland et al. 2000c, Raybould et al. 1998), and there is no consensus on which method is most appropriate. Despite these drawbacks, estimates of Nm can often provide a useful indication of the relative amount of gene flow between pairs of populations (Slatkin 1993). They should ideally be supplemented by alternate methods for calculating gene flow, however. These include assignment tests, in which multilocus genotypes are assigned to populations based on frequency (Paetkau et al. 1995), Bayesian (Rannala & Mountain 1997) or genetic distance methods (Cornuet et al. 1999), and discriminant function analyses (Freeland et al. 2000a,b). None of these methods is infallible. For example, assignment tests assume Hardy-Weinberg proportions and a complete lack of linkage disequilibrium and can yield different results depending on the molecular marker(s) that is employed and the mutation pattern that is assumed

(Cornuet et al. 1999). Another drawback common to these methods is that all individuals are assigned to a population regardless of whether their population of origin was sampled.

Alternative approaches to Nm for inferring gene flow from mitochondrial data include nested clade analysis, which can offer greater power than traditional F_{ST} approaches in detecting geographical associations and hence estimating gene flow (Templeton 1998). This technique may also differentiate between ongoing and historical gene flow and can provide a means of detecting range expansions. However, as with any mitochondrial-based inference of gene flow, the influences of stochastic change and/or selection on a single genetic locus may provide misleading results. Furthermore, since mitochondrial DNA is uniparentally inherited, the dispersal patterns of one gender in gonochoristic taxa will be largely ignored. It is therefore advisable to use data from multiple genetic loci and to subject these data to several analytical methods in order to maximize the accuracy of gene flow estimates.

MODES AND MECHANISMS OF DISPERSAL

The many different mechanisms of dispersal can be broadly classified into two modes: active and passive. Active dispersal entails self-generated movements of individual organisms, while passive dispersal entails movements achieved by use of an external agent. Consideration of dispersal as a result of human activities is deferred to a later section.

Passive Dispersal

In lotic habitats, passive dispersal of invertebrates by water currents or downstream drift can displace from 1% to 2% of benthic stream organisms (Waters 1972) and can result in movement of individuals between spatially discrete populations. Drift is the most common means of transport for many stream invertebrates, such as baetid mayflies and amphipod crustaceans. Passive transport of freshwater invertebrates to new water bodies may be achieved using both animal vectors and wind (Maguire 1963). Passive transport via hitchhiking or phoresy is achieved by the movement of resistant resting stages or of individuals that become attached to mobile animal vectors such as waterfowl or other aquatic vertebrates (see Figure 1). Such vector-mediated dispersal may also occur through transport of adults or resting stages in the guts of animal vectors followed by defecation of viable stages. Examples of such dispersal are reviewed in Table 1.

Wind dispersal may result in short-distance transport of anostrocan eggs (Brendonck & Riddoch 1999) and, by extension, may disperse other small desiccation-resistant stages. Longer-distance dispersal such as aerial plankton is likely in small weakly flying insects, such as stoneflies, mayflies, caddisflies, and some members of Diptera (Kelly et al. 2001). However, for many organisms we question the reliability of dispersing by wind to suitable freshwater habitats owing to the high likelihood of terrestrial deposition.

Active Dispersal

In freshwater invertebrates, active dispersal results predominantly from flight in adult insects that show varying degrees of dispersal according to taxonomic group, situation, and prevailing environmental conditions. An additional means of active dispersal is seen in some molluscs (Kerney 1999) and flightless beetles (Balfour-Browne 1958) that can move between sites via intervening areas of wet habitat.

APPARENT ADAPTATIONS FOR DISPERSAL

Active dispersal may be triggered by environmental conditions that are changing or may be predicted to change in the near future. For example, both increasing temperature and decreasing water depth play a role in triggering the dispersal of aquatic members of Heteroptera and Coleoptera (Velasco et al. 1998). In general, however, the cues that trigger aquatic insects to disperse are poorly understood. Achieving passive transport is likely to be a relatively rare event, and many taxa display features that appear to increase its likelihood. These include hooks on bryozoan statoblasts (Wood 1991) (see Figure 1); sticky envelopes, knobs, and spines on branchiopod crustacean eggs (Fryer 1996) and cladoceran ephippia (Dole-Olivier et al. 2000); and release of large numbers of small dispersing stages to coincide with the timing of peak waterfowl migrations (Okamura & Hatton-Ellis 1995).

Active and passive dispersal stages must be equipped for at least brief exposure to terrestrial/aerial conditions, apart from the special case of internal transport in digestive tracts. For passive dispersers this is accomplished with small dormant stages with reduced metabolic rates and resistance to desiccation and extremes of temperature (Williams 1987). Such stages occur in a wide range of taxa, including sponges (Simpson & Fell 1974), monogonont rotifers (Gilbert 1974), bryozoans (Bushnell & Rao 1974), tardigrades (Nelson & Marley 2000), cladocerans (Dodson & Frey 1991), copepods (Dahms 1995), and branchiopods (Korovchinsky & Boikova 1996). Although suited for overland dispersal, some of the characteristics of these propagules may have evolved under a variety of selection regimes, for example following the drying of temporary ponds or a seasonal reduction in food and temperature.

The exploitation of aquatic habitats is a derived condition in insects and most commonly occurs in pterygote larvae, which often represent the main feeding stage in the life cycle. The adult insect is typically winged, and selection for dispersal between habitats is often used to explain the retention of the terrestrial adult phase. With the exception of the Coleoptera and Heteroptera, an alternative explanation is that wings would be unable to function after periods of inundation. Fully aquatic adult beetles and bugs are able to protect their functional hindwings by sclerotized forewings, the elytra and hemielytra, respectively.

VARIATION IN DISPERSAL ABILITY

Intraspecific differences in dispersal abilities are found in some passively dispersing species. For example, *Branchipodopsis wolfi*, the southern African fairy shrimp, produces both sticky and smooth eggs, with the latter showing a higher likelihood of dispersal among sites (Brendonck & Riddoch 1999). The freshwater bryozoan *Plumatella repens* produces two types of statoblasts: stationary sessoblasts and dispersing floatoblasts. The higher proportion of floatoblasts produced by larger colonies suggests that dispersal provides an escape from local resource depletion zones (Karlson 1992).

Variation in dispersal ability is also found within some actively dispersing insect species that contain both winged and wingless forms (Harrison 1980, Roff & Fairbairn 1991). Presumably such wing dimorphism evolved as a result of the differing relative fitness of dispersing offspring under different environmental scenarios. A relatively high frequency of flightless insect morphs is associated with stable habitats (Roff 1990, Vepsäläinen 1974). Short-wingedness is the optimal within-site strategy in a range of water striders of the genus *Gerris*, owing to the greater local reproductive efficiency of this morph (Vepsäläinen 1974, 1978). Long-wingedness is the optimal between-sites strategy, since dispersing individuals can establish new populations when sites temporarily dry out. The persistence of wing dimorphism within populations may occur as a result of dispersal within the metapopulation even though dimorphism is not a locally optimal strategy.

The evolutionary maintenance of flight ability in insects, and the trade-offs between dispersal and other life-history parameters, have been reviewed by Roff (1990, 1994). Since flight in insects is energetically expensive (Chapman 1999) and reduces egg production in some taxa (Roff 1977), it can be expected that aquatic insects will show variation in apparent trade-offs between flight ability and other life-history parameters related to fitness. Some aquatic taxa appear to conform to the oogenesis-flight syndrome (Johnson 1969), in which dispersal occurs early in adult life, and energy for reproduction is then obtained through the autolysis of flight musculature (Hocking 1952). Examples of trade-offs between dispersal and fitness parameters in aquatic insects are reviewed in Table 2.

THE COMPARATIVE BIOLOGY OF DISPERSAL

In order to attain a broad overview of dispersal in freshwater invertebrates, we now focus on several relatively well-studied systems. In particular, we compare passive dispersers that have contrasting life histories in standing waters with actively dispersing taxa that inhabit continuous and discontinuous riverine habitats.

TABLE 2   Trade-offs between dispersal and fitness parameters in aquatic insects and apparent cases of oogenesis-flight syndrome

Invertebrate	Observation	References
Pondskaters (*Limnoporus canaliculatus*, *Gerris* spp.)	Brachypterous females have higher egg output than their macropterous conspecifics	Vespäläinen 1978, Zera 1984
Water boatmen (*Sigara* spp.)	Individuals lacking functional flight musculature have higher fecundity than those capable of flight	Young 1965
Whirligig beetles (*Gyrinus opacus*)	Lifetime reproductive success maximized by remaining resident in first year, then dispersing to other pools	Svensson 1998
Aquatic Heteroptera (*Mesovelia*, *Gerris*), water beetle (*Helophorus strigifrons*)	Autolysis of flight musculature on commencement of reproductive activity (oogenesis-flight syndrome)	Galbreath 1975, Vespäläinen 1978, Landin 1980
Diving beetles (Dytiscidae)	Variable flight muscle development; developed musculature present only in tenerals; suggests oogenesis-flight syndrome	Jackson 1956, Bilton 1994

It will become apparent that even in fully winged invertebrates, dispersal between sites may be rather limited. Darwin's observation of the widespread distribution of freshwater species is tempered by the recognition of cryptic allopatric taxa in some groups.

Dispersal of Zooplankton and Bryozoans from Ponds and Lakes

Studies of zooplankton and bryozoans do not reveal any consistent patterns of either historical or ongoing dispersal. Cladoceran populations, for example, commonly show high levels of genetic differentiation across short geographical distances (Boileau et al. 1992, Crease et al. 1990), and there have been several discoveries of species complexes and cryptic endemics (Hebert & Finston 1997, Weider et al. 1999). Rotifers also show strong phylogeographical structuring over several hundred kilometers (Gómez et al. 2000). However, genetic lineages in zooplankton may also show little divergence over several thousand kilometers (Hann 1995, Hebert & Finston 1996), a discrepancy that suggests very different levels of connection across varying spatial scales, or varying mutation rates. Similarly contradictory patterns were found in the freshwater bryozoan *Cristatella mucedo*. A study of *C. mucedo* in Europe revealed evidence for gene flow among populations over broad spatial scales (Freeland et al. 2000a), whereas conspecific populations in North America remained genetically isolated from one another (Freeland et al. 2000b).

Connection among populations of freshwater species that rely on passive dispersal is influenced by the vectors that transport their propagules. For example, the *C. mucedo* populations studied in Europe were located along a transect that roughly corresponds to a commonly used waterfowl migration route, and therefore waterfowl may act as vectors that link subpopulations within a large metapopulation in northwest Europe (Freeland et al. 2000a). Conversely, the North American *C. mucedo* populations were located along multiple and divergent waterfowl migratory routes, and connection among these sites should be less predictable (Freeland et al. 2000b). Waterfowl have also been implicated in the dispersal of zooplankton propagules. The distribution of genetic lineages within North American populations of the cladoceran *Daphnia laevis* is roughly concordant with three major waterfowl flyways (Taylor et al. 1998). Similar agreement between mitochondrial haplotype distribution and waterfowl migratory routes suggests that waterfowl have played an important role in the postglacial expansion of *Daphnia pulex* in Greenland and Iceland (Weider et al. 1996), and Beringia (Weider & Hobæk 1997).

Although waterfowl migrations apparently play at least an occasional role in the dispersal of freshwater zooplankton and bryozoans, there are several reasons why it will be difficult to fully determine the extent that waterfowl link sites through genetic studies. First, there are likely to be additional vectors of dispersal, e.g., animals, boats, and lotic channels. Second, it is logistically impossible to sample an appreciable proportion of zooplankton or bryozoan populations from most sites, and therefore the genetic identity assigned to a population may depend on which individuals are sampled. In a similar vein, many species can be found in a large number of sites, and those sites that share relatively high levels of gene flow may not have been sampled. Third, it is worth reiterating here that data from different molecular markers will influence conclusions. It was not until both microsatellite and mitochondrial data were obtained from North American *C. mucedo* that a pattern suggesting two cryptic species emerged (Freeland et al. 2000c). Estimates of gene flow will be artificially reduced when data from two species are combined, and this may at least partially explain the apparent lack of gene flow among North American *Cristatella* populations. Similar situations may at times influence gene flow estimates among zooplankton, as, for example, molecular data have suggested that some morphologically similar cladoceran populations are actually distinct species or subspecies (Crease et al. 1997, Taylor et al. 1998). Studies of dispersal have only recently been enhanced by molecular data, and these have targeted a small proportion of taxa. Therefore, some element of caution should be retained when generalizations based on inferences of gene flow are proposed.

Active Dispersal in Riverine Taxa

Running waters are comprised of drainage networks (Banarescu 1990), with the extent and position in the landscape dependent on local topography. Individual watersheds may be viewed as discrete freshwater systems that are only occasionally interconnected by processes such as river capture (Bishop 1995). Given the often highly disturbed nature of stream environments, the populations and assemblages present within local reaches may be strongly determined by dispersal movements of individuals, both in stream and overland (Palmer et al. 1996). In stream-dwelling insects with flying adults, the possibility of lateral dispersal between streams and catchments may lead to gene flow between populations in different watercourses. To date, however, direct studies of dispersal in adult aquatic insects have largely focused on testing Müller's (1954) colonization cycle, whereby upstream adult dispersal compensates for the downstream losses of individuals due to drift (Hershey et al. 1993, Williams & Williams 1993). Significantly higher rates of upstream dispersal have been reported in a range of Plecoptera, Ephemeroptera, and Trichoptera, although the generality of these results has recently been questioned (Petersen et al. 1999). Upstream distances traveled by individual insects have rarely been estimated, but mark-recapture using stable isotopes has revealed them to be on the order of 1 km in *Baetis* mayflies (Hershey et al. 1993). For insects in short headwater streams, such dispersal may allow individuals to move from one headwater to another without lateral dispersal. This can happen if larvae drift downstream to a position below the confluence of two first-order streams, and then adults move upstream into an adjacent headwater (Griffith et al. 1998).

Direct estimates of lateral dispersal away from streams have relied on malaise traps (Griffith et al. 1998), sticky traps (Jackson & Resh 1989), light traps (Kovats et al. 1996), or a combination of approaches (Collier & Smith 1998). These studies suggest limited lateral dispersal in the majority of stream-dwelling insects, particularly weakly flying species of Trichoptera and Plecoptera. Population structure at the drainage level would therefore be expected to follow a stepping-stone model with isolation by distance (Kimura & Weiss 1964). However, the potentially homogenizing effects of upstream dispersal and downstream drift will likely result in less isolation by distance within streams than between streams. Unfortunately, genetic studies of stream insects have rarely involved a sampling design that would allow testing of the above hypothesis through hierarchical analysis of the spatial distribution of genetic diversity. To date, most genetic studies of stream insect populations have relied on allozyme data, and most have found that populations in different stream systems show moderate to high levels of genetic differentiation, including evidence of cryptic speciation (Hughes et al. 1999, Jackson & Resh 1992, Preziosi & Fairbairn 1992).

The only direct attempts to compare population differentiation within and among individual streams and drainages have been undertaken by Bunn and colleagues (Bunn & Hughes 1997, Schmidt et al. 1995), mostly in small rainforest streams in northern Australia. Population genetic data for a mayfly (*Baetis* sp.), a water strider (*Rheumatometra* sp.), and a caddisfly (*Tasiagma ciliata*) all reveal a striking and rather unexpected relationship between levels of genetic differentiation and spatial scale. In all three taxa, genetic differentiation (F_{ST}) decreases with level in the stream hierarchy, being highest between populations in different reaches of the same stream and lowest between separate catchments. Such a pattern is counterintuitive, as relatively low genetic differentiation between catchments points to regular adult dispersal, a scenario that must be reconciled with high differentiation within streams. Schmidt et al. (1995) suggest that unexpectedly high F_{ST} values at smaller spatial scales may be explained if instream movement is restricted, and if individuals within a reach are the offspring of a limited number of females. In these Australian systems, unpredictable climatic conditions lead to asynchronous adult emergence, and therefore relatively few breeding adult insects are present at any given time. These patterns appear to be mirrored in the subtropical Canary Islands (Kelly et al. 2001) but may not hold for stream insects with synchronized emergences. This hypothesis is supported by a study of the synchronously emerging stonefly *Yoraperla brevis* (see Hughes et al. 1999), in which among-stream differentiation (F_{ST}) was an order of magnitude higher than that within streams. Such findings are in keeping with our understanding of adult movement in most stream insect taxa in the northern hemisphere, where, in general, dispersal rates within a drainage network appear to be higher than those between separate catchments, even in organisms capable of active flight.

DISPERSAL IN TIME

Recent evidence indicates that some invertebrate taxa may achieve temporal dispersal through the accumulation and subsequent release of dormant propagules that create a reservoir of genetic material analogous to seed banks in plants. Such dispersal in time will be a function of the dormant period of viable propagules and therefore could vary considerably between taxa. Here we must distinguish between the routine year-to-year continuity of populations maintained via dormant overwintering stages produced in the preceding year, and the recruitment of stages that have remained dormant over longer periods. The latter can result in temporal dispersal that may promote the long-term persistence of both genotypes and local populations during extended adverse conditions (Gómez & Carvalho 2000, Freeland et al. 2001).

The dormancy period of resting stages has been studied in a number of species. In copepods, there is evidence that recruitment of genotypes from sediments occurs when favorable conditions return (Hairston & Caceres 1996). Recent studies provide evidence that egg banks of rotifers and cladocerans contribute to population genetic structure following the hatching of dormant eggs throughout the growing season, a process that contributes to the maintenance of genetic diversity in these populations (Caceres 1998, Gómez & Carvalho 2000). Temporal dispersal has also been inferred from the differential hatching regime of stonefly eggs (Zwick 1996) and from genetic analyses of bryozoan populations sampled over several different timescales (Freeland et al. 2001).

The importance of temporal dispersal via propagule banks is suggested by the fact that some diapausing zooplankton eggs can remain viable in sediment for 200 years or more (Hairston et al. 1995, Caceres 1998). Such an extension of generation time may profoundly influence genetic structure, for example by influencing the rate and direction of microevolution following directional and temporally fluctuating selection (Hairston et al. 1996). Temporal dispersal may particularly benefit predominantly clonal or inbred organisms that have a limited ability to produce novel genotypes and therefore would otherwise remain at a disadvantage in the Red Queen race (Tooby 1982). Furthermore, temporal gene flow may be important to taxa that are incapable of actively dispersing among sites. Notably, many zooplankton and benthic taxa are both clonal and incapable of active dispersal, and these taxa present most of the evidence for temporal gene flow via propagule banks.

HUMAN-MEDIATED DISPERSAL

Dispersal of freshwater organisms through human activities has recently been the focus of considerable investigation (reviewed by Claudi & Leach 2000). Introductions of nonindigenous invertebrates have been achieved through intentional stocking programs for fisheries and for biocontrol, release of organisms from the aquarium or horticultural trade, release of organisms along with bait fish by fishermen, release of ballast water, deliberate establishment of exotic food sources for human consumption, and the creation of humanmade water channels. Clearly, some of these mechanisms of dispersal will have resulted in the dispersal of indigenous species as well. In many cases, patterns of occurrence strongly implicate human-mediated dispersal but, because many invasions occurred before biological surveys, the number of species involved is not known.

Introduction of a number of nonindigenous species has resulted in dramatic postinvasion spread and severe economic and ecological impacts. For instance, release of larvae in ballast water into Lake St. Clair in 1985 or 1986 was the original mechanism by which the zebra mussel, *Dreissena polymorpha*, was introduced to North America (Hebert et al. 1989). Since then *D. polymorpha* has invaded most of the major North American river systems (Mackie 2000) through a combination of further release of larvae in ballast, bilge, and engine cooling water, transport of adults and juveniles on boats and macrophytes entangled on boat trailers, and downstream dispersal of planktonic larval stages. Severe ecological and socioeconomic impacts have resulted (Kinzelbach 1992, Mackie 2000).

Human activities may also impede dispersal among freshwaters. The disruption and fragmentation of rivers by a series of impoundments (Englund & Malmqvist 1996) may diminish dispersal, as is suggested by the relative development of riparian floras and patterns of movement of fish and plant diaspores (Jansson et al. 2000).

KEY DIRECTIONS FOR FUTURE RESEARCH

Integrative Approaches

Few studies combine direct and indirect approaches to estimate levels of dispersal in freshwater invertebrates (but see Nürnberger & Harrison 1995). Consequently, our understanding of local dispersal rates within metapopulations, and the influence of landscape features, is severely limited. Such integrated studies should be conducted across a range of phylogenetically independent taxa, with the simultaneous application of several molecular markers targeting regions of nuclear and mitochondrial DNA that have inherently variable mutation rates. Key areas for such study include lateral dispersal of stream insects, particularly those with aquatic adult stages, and short-range dispersal of pond insects with differing mobilities.

Temporal Dispersal and Propagule Survival in Passive Dispersers

Recent developments indicate that temporal gene flow via propagule banks plays an important role in the metapopulation biology of many freshwater invertebrates. Understanding the extent to which dispersal in both space and time allows the long-term persistence of metapopulations will provide important insights into the maintenance of biodiversity in freshwater systems and may be of particular relevance to conservation issues. Newly created sites, or sites with histories of disturbance, may not be the best habitats for protection if temporal gene flow is important. An awareness of dispersal through time and space highlights the fallacy of viewing populations as discrete units and should serve as a warning that the effects of restricting or otherwise altering patterns of dispersal may have unforeseen and potentially far-reaching consequences.

While passive dispersal via animal vectors has been inferred from both direct and indirect observations, the survival of passive dispersers under different regimes is poorly understood. More systematic study is required before we can determine whether particular vector species and habitat types favor passive transport.

Historical Versus Ongoing Gene Flow

Molecular markers have contributed enormously to our understanding of the movements of many freshwater taxa. However, such approaches can lead to new sets of problems. For example, range extensions of freshwater taxa at the end of the last ice age may have introduced multiple genetic lineages into individual lakes and ponds (Stemberger 1995), and such historical events may not be readily distinguishable from ongoing gene flow without the use of rapidly evolving molecular markers (Freeland et al. 2000c). Notably few studies have used mitochondrial DNA data to infer historical patterns of range expansion and dispersal in freshwater invertebrates (Avise 2000; but see Bilton 1994, Nürnberger & Harrison 1995, Meyran et al. 1997, Weider & Hobæk 1997, Gómez et al. 2000). Disentangling the extent to which population structure results from recurrent forces, such as gene flow, versus historical events, such as fragmentation and range expansion, will be crucial to understanding the frequency of dispersal in freshwater taxa. With the further development of molecular and analytical approaches and the streamlining of laboratory practices, it will become easier to genetically characterize individuals at multiple loci and to conduct more suitable analyses of the data through adoption of increasingly refined models.

Figure 2 Mark-recapture in action. Adult *Libellula quadrimaculata* showing individual paint-marking.

Understanding Anthropogenic Influences

While a large proportion of original wetlands have been lost (Carp 1980, Rackham 1986), humans have created new water bodies through damming of rivers (Moss 1998), creation of small farm ponds (Gerking 1966) and ornamental lakes (Bennion et al. 1997), and digging of gravel pits. The distribution of freshwater invertebrates in many, if not most, places on earth is now a result of contraction from natural habitats and expansion into humanmade environments. It is unclear how this changing array of habitats will influence dispersal patterns of freshwater invertebrates. Global climate change and habitat degradation may mean that the persistence of populations and species will rely on increasingly frequent dispersal events. At the same time, human-mediated dispersal often provides a new degree of connectivity between populations. Many predictions regarding the fate of freshwater taxa under rapidly changing environmental conditions remain speculative. In order to refine these predictions, we must improve our understanding of dispersal patterns and processes among freshwater populations.

ACKNOWLEDGMENTS

We are grateful to Gary Carvalho, Dagmar Frisch, Peter Hammond, Lucy Kelly, Björn Malmqvist, Simon Rundle, and Steve Threlkeld for editing and commenting on the manuscript. Roy Moate and Paul Taylor kindly produced the SEM photographs used in Figure 1, and Antonio Di Sabatino identified the mite larvae. Kelvin Conrad provided the photo of the marked dragonfly for Figure 2 (see color insert). Thanks also to Lucy Kelly, Ron Carr, and David Sims for providing references. BO and JRF wish to thank the Natural Environment Research Council Work for funding to study dispersal in *Cristatella mucedo* (grants GR3/11068, GR3/8961).

Visit the Annual Reviews home page at www.AnnualReviews.org

LITERATURE CITED

Adams LE. 1905. A plover with *Anodonta cygnea* attached to its foot. *J. Conchol.* 11: 175

Andersen NM. 1982. *The Semiaquatic Bugs.* Klampenborg, Denmark: Scandinavian Sci. 455 pp.

Angus RB. 1970. Genetic experiments on *Helophorus* F. (Coleoptera: Hydrophilidae). *Trans. R. Entomol. Soc. London* 122:257–76

Avise JC. 1992. Molecular population structure and the biogeographic history of a regional fauna—a case history with lessons for conservation biology. *Oikos* 63:62–76

Avise JC. 1994. *Molecular Markers, Natural History and Evolution.* New York: Chapman & Hall. 511 pp.

Avise JC. 2000. *Phylogeography. The History and Formation of Species.* Cambridge, MA: Harvard Univ. Press. 384 pp.

Balfour-Browne WAF. 1958. *British Water Beetles.* London: Ray Soc. 210 pp.

Banarescu P. 1990. *Zoogeography of Fresh Waters. General Distribution and Dispersal of Freshwater Animals.* Wiesbaden: Aula–Verlag

Bell G. 1982. *The Masterpiece of Nature: The Evolution and Genetics of Sexuality.* Berkeley: Univ. Calif. Press

Bennion HR, Harriman R, Battarbee R. 1997. A chemical survey of standing waters in south-east England, with reference to acidification and eutrophication. *Freshwater Forum* 8:28–44

Berg DJ, Garton DW. 1994. Genetic differentiation in North American and European populations of the cladoceran *Bythotrephes. Limnol. Oceanogr.* 39:1503–16

Bilton DT. 1994. Phylogeography and recent historical biogeography of *Hydroporus glabriusculus* Aubé (Coleoptera: Dytiscidae) in the British Isles and Scandinavia. *Biol. J. Linn. Soc.* 51:293–307

Bishop P. 1995. Drainage rearrangement by river capture, beheading and diversion. *Prog. Phys. Geogr.* 19:449–73

Boag DA. 1986. Dispersal in pond snails: potential role of waterfowl. *Can. J. Zool.* 64:904–9

Bohonak AJ. 1999a. Dispersal, gene flow, and population structure. *Q. Rev. Biol.* 74:21–45

Bohonak AJ. 1999b. Effect of insect-mediated dispersal on the genetic structure of postglacial water mite populations. *Heredity* 82:451–61

Boileau MG, Hebert PDN, Schwartz SS. 1992. Nonequilibrium gene frequency divergence: persistent founder effects in natural populations. *J. Evol. Biol.* 5:25–39

Brendonck L, Riddoch BJ. 1999. Wind-borne short-range egg dispersal in anostracans (Crustacea: Branchiopoda). *Biol. J. Linn. Soc.* 67:87–95

Brown CJD. 1933. A limnological study of certain fresh-water Polyzoa with special reference to their statoblasts. *Trans. Am. Microsc. Soc.* 52:271–314

Bunn SE, Hughes JM. 1997. Dispersal and recruitment in streams: evidence from genetic studies. *J. N. Am. Benth. Soc.* 16:338–46

Bushnell JH, Rao KS. 1974. Dormancy or quiescent stages and structures among the Ecto-

procta: physical and chemical factors affecting viability and germination of statoblasts. *Trans. Am. Microsc. Soc.* 93:524–43

Caceres CE. 1998. Interspecific variation in the abundance, production, and emergence of *Daphnia* diapausing eggs. *Ecology* 79:1699–710

Carp E. 1980. *A Directory of Western Palaearctic Wetlands.* Gland: IUCN

Chapman RF. 1999. *The Insects: Structure and Function.* London: Chapman & Hall

Claudi R, Leach JR, eds. 2000. *Nonindigenous Freshwater Organisms. Vectors, Biology, and Impacts.* Boca Raton, FL.: Lewis. 464 pp.

Clobert J, Danchin E, Dhont AA, Nichols JD. 2001. *Dispersal.* Oxford: Oxford Univ. Press

Collier KJ, Smith BJ. 1998. Dispersal of adult caddisflies (Trichoptera) into forests alongside three New Zealand streams. *Hydrobiologia* 361:53–65

Comins HM, Hamilton WD, May RM. 1980. Evolutionary stable dispersal strategies. *J. Theor. Biol.* 82:205–30

Conrad KF, Wilson KH, Harvey IF, Thomas CJ, Sherratt TN. 1999. Dispersal characteristics of seven odonate species in an agricultural landscape. *Ecography* 22:524–31

Cordell JR, Morrison SM. 1996. The invasive Asian copepod *Pseudodiaptomus inopinus* in Oregon, Washington, and British Columbia estuaries. *Estuaries* 19:629–38

Cornuet JM, Piry S, Luikart G, Estoup A, Solignac M. 1999. New methods employing multilocus genotypes to select or exclude populations as origins of individuals. *Genetics* 153:1989–2000

Coutant CC. 1982. Evidence for upstream dispersal of adult caddisflies (Trichoptera: Hydropsychidae) in the Colombia River. *Aquat. Insects* 4:61–66

Crease TJ, Lee SK, Yu SI, Spitze K, Lehman N, Lynch M. 1997. Allozyme and mtDNA variation in populations of the *Daphnia pulex* complex from both sides of the Rocky mountains. *Heredity* 79:242–51

Crease TJ, Lynch M, Spitze K. 1990. Hierarchical analysis of population genetic variation in mitochondrial and nuclear genes of *Daphnia pulex*. *Mol. Biol. Evol.* 7:444–58

Dahms H-U. 1995. Dormancy in the Copepoda—an overview. *Hydrobiologia* 306:199–211

Darwin C. 1859. *On the Origin of Species by Means of Natural Selection*. London: John Murray

Davies RW, Linton LR, Wrona FJ. 1982. Passive dispersal of four species of freshwater leeches (Hirudinoidea) by ducks. *Fresh. Invert. Biol.* 1:40–44

Dieckmann U, O'Hara B, Weisser W. 1999. The evolutionary ecology of dispersal. *TREE* 14:88–90

Dingle H. 1996. *Migration. The Biology of Life on the Move*. Oxford, UK: Oxford Univ. Press. 480 pp.

Dodson SI, Frey DG. 1991. Cladocera and other Branchiopoda. See Thorp & Covich 1991, pp. 723–86

Dole-Olivier M-J, Galassi DMP, Marmonier P, Creuzé des Châtelliers M. 2000. The biology and ecology of lotic microcrustaceans. *Freshwater Biol.* 44:63–91

Englund G, Malmqvist B. 1996. Effects of flow regulation, habitat area and isolation on the macroinvertebrate fauna of rapids in north Swedish rivers. *Regul. Rivers Res. Manage.* 12:433–45

Erman NA. 1986. Movements of self-marked caddisfly larvae, *Chyranda centralis* (Trichoptera: Limnephilidae), in a Sierran spring stream, California, U.S.A. *Freshwater Biol.* 16:455–64

Ferriere R, Belthoff JR, Olivieri I, Krackow S. 2000. Evolving dispersal: where to go next? *TREE* 15:5–7

Freeland JR, Noble LR, Okamura B. 2000a. Genetic consequences of the metapopulation biology of a facultatively sexual freshwater invertebrate. *J. Evol. Biol.* 13:383–95

Freeland JR, Noble LR, Okamura B. 2000b. Genetic diversity of North American populations of *Cristatella mucedo*, inferred from microsatellite and mitochondrial DNA. *Mol. Ecol.* 9:1375–89

Freeland JR, Romualdi C, Okamura B. 2000c.

Gene flow and genetic diversity: a comparison of freshwater bryozoan populations in Europe and North America. *Heredity* 85:498–508

Freeland JR, Rimmer VK, Okamura B. 2001. Genetic changes within freshwater bryozoan populations suggest temporal gene flow from statoblast banks. *Limnol. Oceanogr.* 46:1121–29

Freeman JA. 1945. Studies in the distribution of insects by aerial currents. The insect population of the air from ground level to 300 feet. *J. Anim. Ecol.* 14:128–54

Fryer G. 1996. Diapause, a potent force in the evolution of freshwater crustaceans. *Hydrobiologia* 320:1–14

Galbreath JE. 1975. Thoracic polymorphism in *Mesovelia mulsanti* (Hemiptera, Mesoveliidae). *Kansas Univ. Sci. Bull.* 50:457–82

Gerking SD. 1966. Central states. In *Limnology in North America*, ed. DG Frey, pp. 239–68. Madison: Univ. Wisc. Press. 752 pp.

Gilbert JJ. 1974. Dormancy in rotifers. *Trans. Am. Microsc. Soc.* 93:490–513

Gómez A, Carvalho GR. 2000. Sex, parthenogenesis and genetic structure of rotifers: microsatellite analysis of contemporary and resting egg bank populations. *Mol. Ecol.* 9:203–14

Gómez A, Carvalho GR, Lunt DH. 2000. Phylogeography and regional endemism of a passively dispersing zooplankter: mitochondrial DNA variation in rotifer resting egg banks. *Proc. R. Soc. London Ser. B* 267:2189–97

Griffith MB, Barrows EM, Perry SA. 1998. Lateral dispersal of adult insects (Plecoptera, Trichoptera) following emergence from headwater streams in forested Appalachian catchments. *Ann. Entomol. Soc. Am.* 91:195–201

Hairston NGJ, Van Brunt RA, Kearns CM, Engstrom DR. 1995. Age and survivorship of diapausing eggs in a sediment egg bank. *Ecology* 76:1706–11

Hairston NGJ, Caceres CE. 1996. Distribution of crustacean diapause: micro- and macroevolutionary pattern and process. *Hydrobiologia* 320:27–44

Hairston NGJ, Kearns CM, Ellner SP. 1996. Phenotypic variation in a zooplankton egg bank. *Ecology* 77:2382–92

Hann BJ. 1995. Genetic variation in *Simocephalus* (Anomopoda, Daphniidae) in North America: patterns and consequences. *Hydrobiologia* 307:9–14

Harrison RG. 1980. Dispersal polymorphisms in insects. *Annu. Rev. Ecol. Syst.* 11:95–118

Havel JE, Colbourne JK, Hebert PDN. 2000. Reconstructing the history of intercontinental dispersal of *Daphnia lumholtzi* by use of genetic markers. *Limnol. Oceangr.* 45:1414–19

Hebert PDN, Finston TL. 1996. Genetic differentiation in *Daphnia obtusa*: a continental perspective. *Freshwater Biol.* 35:311–21

Hebert PDN, Finston TL. 1997. A taxonomic reevaluation of North American Daphnia (Crustacea: Cladocera). 3. The *D. catawba* complex. *Can. J. Zool.* 75:1254–61

Hebert PDN, Wilson CC, Murdoch MH, Lazar R. 1989. Ecological and genetic studies on *Dreissena polymorpha* (Pallas): a new mollusc in the Great Lakes. *Can. J. Zool.* 59:405–9

Hershey AE, Pastor J, Peterson BJ, Kling GJ. 1993. Stable isotopes resolve the drift paradox for *Baetis* mayflies in an arctic river. *Ecology* 74:2415–25

Hocking B. 1952. Autolysis of flight muscle in a mosquito. *Nature* 169:1101

Hogg ID, Eadie JM, De Lafontaine Y. 1998. Atmospheric change and the diversity of aquatic invertebrates: Are we missing the boat? *Environ. Mon. Assess.* 49:291–301

Hogg ID, Williams DD. 1996. Response of stream invertebrates to a global-warming thermal regime: an ecosystem-level manipulation. *Ecology* 77:395–407

Horne FR. 1966. The effect of digestive enzymes on the hatchability of *Artemia salina* eggs. *Trans. Am. Microsc. Soc.* 85:271–74

Howard DJ, Berlocher S. 1998. *Endless Forms: Species and Speciation.* Oxford, UK: Oxford Univ. Press

Hughes JM, Mather PB, Sheldon AL, Allendorf FW. 1999. Genetic structure of the stonefly, *Yoraperla brevis*, populations: the extent of gene flow among adjacent montane streams. *Freshwater Biol.* 41:63–72

Jackson DJ. 1956. Observations on flying and flightless water beetles. *J. Linn. Soc. London* 43:18–43

Jackson JK, McElravy EP, Resh VH. 1999. Long-term movements of self-marked caddisfly larvae (Trichoptera: Sericostomatidae) in a California coastal mountain stream. *Freshwater Biol.* 42:525–36

Jackson JK, Resh VH. 1989. Distribution and abundance of adult aquatic insects in the forest adjacent to a northern Californian stream. *Environ. Entomol.* 18:278–83

Jackson JK, Resh VH. 1992. Variation in genetic structure among populations of the caddisfly *Helicopsyche borealis* from three streams in northern California, U.S.A. *Freshwater Biol.* 27:29–42

Jansson R, Nilson C, Renofalt B. 2000. Fragmentation of riparian floras in rivers with multiple dams. *Ecology* 81:899–903

Jenkins DG. 1995. Dispersal-limited zooplankton distribution and community composition in new ponds. *Hydrobiologia* 313:15–20

Johnson CG. 1969. *Migration and Dispersal of Insects by Flight.* London: Methuen. 763 pp.

Johnson LE, Carlton JT. 1996. Post-establishment spread in large-scale invasions: dispersal mechanisms of the zebra mussel *Dreissena polymorpha. Ecology* 77:1686–90

Johnson ML, Gaines MS. 1990. Evolution of dispersal: theoretical models and empirical tests using birds and mammals. *Annu. Rev. Ecol. Syst.* 21:449–80

Karlson RH. 1992. Divergent dispersal strategies in the freshwater bryozoan *Plumatella repens*—ramet size effects on statoblast numbers. *Oecologia* 89:407–11

Kelly LC, Bilton DT, Rundle SD. 2001. Genetic differentiation and dispersal in the Canary Island caddisfly *Mesophylax aspersus* (Trichoptera: Limnephilidae). *Heredity* 86:370–77

Kerney MP. 1999. *Atlas of the Land and Freshwater Molluscs of Britain and Ireland.* Colchester: Harley. 264 pp.

Kew HW. 1893. *The Dispersal of Shells.* London: Kegan Paul, Trench, Trübner

Kimura M, Weiss GH. 1964. The stepping stone model of population structure and the decrease of genetic correlation with distance. *Genetics* 49:561–76

Kinzelbach R. 1992. The main features of the phylogeny and dispersal of the zebra mussel *Dreissena polymorpha.* In *The Zebra Mussel* Dreissena polymorpha. *Ecology, Biological Monitoring and First Applications in the Water Quality Management,* ed. D Neumann, HA Jenner, pp. 5–17. Stuttgart: Gustav Fischer. 280 pp.

Korovchinsky NM, Boikova OS. 1996. The resting eggs of Ctenopoda (Crustacea: Branchiopoda): a review. *Hydrobiologia* 320: 131–40

Kovats ZE, Ciborowski JJH, Corkum LD. 1996. Inland dispersal of adult aquatic insects. *Freshwater Biol.* 36:265–76

Landin J. 1980. Habitats, life histories, migration and dispersal by flight of two water beetles *Helophorus brevipalpis* and *H. strigifrons* (Hydrophilidae). *Holarct. Ecol.* 3:190–201

Lansbury I. 1955. Some notes on invertebrates other than Insecta found attached to water bugs (Hemipt.-Heteroptera). *Entomologist* 88:139–40

Lincoln R, Boxshall G, Clark P. 1998. *A Dictionary of Ecology, Evolution and Systematics.* Cambridge, UK: Cambridge Univ. Press. 350 pp.

Mackie GL. 2000. Ballast water introductions of Mollusca. See Claudi & Leach 2000, pp. 219–54

Maguire BJ. 1963. The passive dispersal of small aquatic organisms and their colonization of isolated bodies of water. *Ecol. Monogr.* 33:161–85

Meyran JC, Monnerot M, Taberlet P. 1997. Taxonomic status and phylogenetic relationships of some species of the genus *Gammarus* (Crustacea, Amphipoda) deduced from mitochondrial DNA sequences. *Mol. Phyl. Evol.* 8:1–10

Moss B. 1998. *Ecology of Fresh Waters. Man and Medium.* Oxford, UK: Blackwell Sci.

Müller K. 1954. Investigations on the organic drift in North Swedish streams. *Rep. Inst. Freshwater Res. Drottingholm* 34:133–48

Nei M. 1972. Genetic distance between populations. *Am. Nat.* 106:283–92

Nelson DR, Marley NJ. 2000. The biology and ecology of lotic Tardigrada. *Freshwater Biol.* 44:93–108

Nürnberger B, Harrison RG. 1995. Spatial population structure in the whirligig beetle *Dineutus assimilis*: evolutionary influences based on mitochondrial DNA and field data. *Evolution* 49:266–75

Okamura B, Hatton-Ellis T. 1995. Population biology of bryozoans—correlates of sessile, colonial life-histories in fresh-water habitats. *Experientia* 51:510–25

Paetkau D, Calvert W, Stirling I, Strobeck C. 1995. Microsatellite analysis of population structure in Canadian polar bears. *Mol. Ecol.* 4:347–54

Palmer MA, Allan JD, Butman CA. 1996. Dispersal as a regional process affecting the local dynamics of marine and stream benthic invertebrates. *TREE* 11:322–26

Peck SB. 1975. Amphipod dispersal in the fur of aquatic mammals. *Can. Field Nat.* 89:181–82

Petersen I, Winterbottom JH, Orton S, Hildrew AG. 1999. Does the colonization cycle exist? In *Biodiversity in Benthic Ecology. Proc. Nordic Benthol. Meet.,* ed. N Friberg, JD Carl, pp. 59–62. Silkeborg, Denmark: NERI Tech. Rep. No. 226

Preziosi RF, Fairbairn DJ. 1992. Genetic population structure and levels of gene flow in the stream dwelling waterstrider. *Aquarius* (= *Gerris*) *remigis* (Hemiptera, Gerridae). *Evolution* 46:430–44

Proctor VW, Malone C. 1965. Further evidence of the passive dispersal of small aquatic organisms via the intestinal tract of birds. *Ecology* 46:728–29

Proctor VW, Malone CR, DeVlaming VL.

1967. Dispersal of aquatic organisms: viability of disseminules recovered from the intestinal tract of captive killdeer. *Ecology* 48:672–76

Rackham O. 1986. *The History of the Countryside*. London: Dent. 464 pp.

Rannala B, Mountain JL. 1997. Detecting immigration by using multilocus genotypes. *Proc. Natl. Acad. Sci. USA* 94:9197–201

Raybould AF, Mogg RJ, Aldam C, Gliddon CJ, Thorpe RS, Clarke RT. 1998. The genetic structure of sea beet (*Beta vulgaris* ssp. *maritima*) populations. III. Detection of isolation by distance at microsatellite loci. *Heredity* 80:127–32

Rees WJ. 1952. The role of amphibia in the dispersal of bivalve molluscs. *Br. J. Herpetol.* 1:125–29

Riley JR, Smith AD, Reynolds DR, Edwards AS, Osborne JL, et al. 1996. Tracking bees with harmonic radar. *Nature* 379:29–30

Roff DA. 1977. Dispersal in dipterans: its costs and consequences. *J. Anim. Ecol.* 46:443–56

Roff DA. 1990. The evolution of flightlessness in insects. *Ecol. Monogr.* 60:389–421

Roff DA. 1994. The evolution of flightlessness: Is history important? *Evol. Ecol.* 8:639–57

Roff DA, Fairbairn DJ. 1991. Wing dimorphisms and the evolution of migratory polymorphisms among the insects. *Am. Zool.* 31:251

Schmidt SF, Hughes JM, Bunn SE. 1995. Gene flow among conspecific populations of *Baetis* (Ephemeroptera): adult flight and larval drift. *J. N. Am. Benth. Soc.* 14:47–57

Service MW. 1993. *Mosquito Ecology Field Sampling Methods*. London: Elsevier. 988 pp.

Simpson TL, Fell PE. 1974. Dormancy among the Porifera: gemmule formation and germination in fresh-water and marine sponges. *Trans. Am. Microsc. Soc.* 93:544–77

Slatkin M. 1985. Gene flow in natural populations. *Annu. Rev. Ecol. Syst.* 16:393–430

Slatkin M. 1993. Isolation by distance in equilibrium and non-equilibrium populations. *Evolution* 47:264–79

Southwood TRE, Henderson PA. 2000. *Ecological Methods*. Oxford, UK: Blackwell Sci.

Stemberger RS. 1995. Pleistocene refuge areas and postglacial dispersal of copepods of the northeastern United States. *Can. J. Fish. Aquat. Sci.* 52:2197–210

Stenseth NC, Lidicker WC Jr. 1992. The study of dispersal: a conceptual guide. In *Animal Dispersal: Small Mammals as a Model*, ed. NC Stenseth, WC Lidicker Jr, pp. 5–20. New York/London: Chapman & Hall

Svensson BW. 1998. Local dispersal and its life-history consequences in a rock-pool population of a gyrinid beetle. *Oikos* 82:111–22

Taylor DJ, Finston TL, Hebert PDN. 1998. Biogeography of a widespread freshwater crustacean: pseudocongruence and cryptic endemism in the North American *Daphnia laevis* complex. *Evolution* 52:1648–70

Templeton A. 1998. Nested clade analyses of phylogeographic data: testing hypotheses about gene flow and population history. *Mol. Ecol.* 7:381–97

Thomas EP, Blinn DW, Keim P. 1998. Do xeric landscapes increase genetic divergence in aquatic landscapes? *Freshwater Biol.* 40:587–93

Thorp JH, Covich AP, eds. 1991. *Ecology and Classification of North American Freshwater Invertebrates*. New York: Academic

Tooby J. 1982. Pathogens, polymorphism, and the evolution of sex. *J. Theor. Biol.* 97:557–76

Velasco J, Suarez ML, Vidal-Abarca MR. 1998. Factores que determinan la colonizacion de insectos acuaticos en pequenos estanques. *Oecol. Aquat.* 11:87–99

Vepsäläinen K. 1974. Determination of wing length and diapause in water striders (*Gerris* Fabr., Heteroptera). *Hereditas* 77:163–76

Vepsäläinen K. 1978. Wing dimorphism and diapause in *Gerris*: determination and adaptive significance. In *Evolution of Insect Migration and Diapause*, ed. H Dingle, pp. 218–53. New York: Springer-Verlag. 284 pp.

Waters TF. 1972. The drift of stream insects. *Annu. Rev. Entomol.* 17:253–72

Weider LJ, Hobæk A. 1997. Postglacial dispersal, glacial refugia, and clonal structure in Russian/Siberian populations of the arctic *Daphnia pulex* complex. *Heredity* 78:363–72

Weider LJ, Hobæk A, Crease TJ, Stibor H. 1996. Molecular characterization of clonal population structure and biogeography of arctic apomictic *Daphnia* from Greenland and Iceland. *Mol. Ecol.* 5:107–18

Weider LJ, Hobæk A, Colbourne JK, Crease TJ, Dufresne F, Hebert PDN. 1999. Holarctic phylogeography of an asexual species complex. I. Mitochondrial DNA variation in Arctic *Daphnia. Evolution* 53:777–92

Weir BS, Cockerham CC. 1984. Estimating F-statistics for the analysis of population structure. *Evolution* 38:1358–70

Whitlock MC, McCauley DE. 1999. Indirect measures of gene flow and migration: Fst ≠ 1/(4Nm+1). *Heredity* 82:117–25

Williams DD. 1987. *The Ecology of Temporary Waters*. London: Croom Helm. 205 pp.

Williams DD, Williams NE. 1993. The upstream/downstream movement paradox of lotic invertebrates: quantitative evidence from a Welsh mountain stream. *Freshwater Biol.* 30:199–218

Wood TS. 1991. Bryozoans. See Thorp & Covich 1991, pp. 481–99

Wright S. 1951. The genetical structure of populations. *Eugenics* 15:323–54

Young EC. 1965. Flight muscle polymorphism in British Corixidae: ecological observations. *J. Anim. Ecol.* 34:353–90

Zalom FG, Grigarick AA, Way MO, Grigarick AA, Way MO. 1980. Diel flight periodicities of some Dytiscidae (Coleoptera) associated with California rice paddies. *Ecol. Entomol.* 5:183–87

Zaranko DT, Farara DG, Thompson FG. 1997. Another exotic mollusc in the Laurentian Great Lakes: the New Zealand native *Potamopyrgus antipodarum* (Gray 1843) (Gastropoda, Hydrobiidae). *Can. J. Fish. Aquat. Sci.* 54:809–14

Zera AJ. 1984. Differences in survivorship, developmental rate and fertility between the long-winged and wingless morphs of the water strider *Limnoporus canaliculatus. Evolution* 38:1023–32

Zera AJ, Innes DT, Saks ME. 1983. Genetic and environmental determinants of wing polymorphism in the water strider *Limnoporus canaliculatus. Evolution* 37:513–22

Zwick P. 1996. Variable egg development of *Dinocras* spp. (*Plecoptera, Perlidae*) and the stonefly seed bank theory. *Freshwater Biol.* 35:81–100

Annu. Rev. Ecol. Syst. 2001. 32:183–217
Copyright © 2001 by Annual Reviews. All rights reserved

APPLIED EVOLUTION

J. J. Bull[1] and H. A. Wichman[2]

[1]Section of Integrative Biology, Institute of Cellular and Molecular Biology, University of Texas, Austin, Texas 78712-1023; e-mail: bull@bull.biosci.utexas.edu
[2]Department of Biological Sciences, University of Idaho, Moscow, Idaho 83844-3051; e-mail: hwichman@uidaho.edu

Key Words artificial selection, directed evolution, phylogenetics, resistance, evolutionary computation

■ **Abstract** Evolutionary biology is widely perceived as a discipline with relevance that lies purely in academia. Until recently, that perception was largely true, except for the often neglected role of evolutionary biology in the improvement of agricultural crops and animals. In the past two decades, however, evolutionary biology has assumed a broad relevance extending far outside its original bounds. Phylogenetics, the study of Darwin's theory of "descent with modification," is now the foundation of disease tracking and of the identification of species in medical, pharmacological, or conservation settings. It further underlies bioinformatics approaches to the analysis of genomes. Darwin's "evolution by natural selection" is being used in many contexts, from the design of biotechnology protocols to create new drugs and industrial enzymes, to the avoidance of resistant pests and microbes, to the development of new computer technologies. These examples present opportunities for education of the public and for nontraditional career paths in evolutionary biology. They also provide new research material for people trained in classical approaches.

OVERVIEW

Evolutionary biology has undergone an expansion and transformation in the past few decades. Despite occasional claims to the contrary, the big changes in evolutionary biology have come from improvements in understanding mechanisms that are fully compatible with Darwinism; descent with modification and natural selection are still the conceptual foundations of the discipline. For example, a veritable explosion of studies estimating the relationships among different species has refined our understanding of evolutionary history, but the modern version of the tree of life has many similarities to old ones and certainly supports a Darwinian model. The sequencing of genes and genomes has yielded insights into genetic mechanisms underlying evolution, and we are even progressing toward a genetic understanding of the major developmental and morphological transitions—advances that augment earlier ideas about these transitions. Theories based on natural selection have led to revolutions in understanding behavior, parasitism, and a wealth of

0066-4162/01/1215-0183$14.00

genetic and physiological mechanisms that benefit neither the individual nor the population.

Some of the revolution in evolution has occurred outside the traditional academic boundaries of evolutionary biology. Those who started as evolutionary biologists two or more decades ago are beholding a transformation of the field into one of broad social relevance. Much of the biotechnology industry is concerned with creating biological molecules that have specific functions. This goal-oriented enterprise has quickly embraced evolutionary principles to direct the evolution of molecules in test tubes and, in so doing, has profoundly expanded the horizon and relevance of evolutionary biology. Evolutionary principles are suddenly the material of multimillion dollar patents, leading industrial biochemists to new drugs and other commercial molecules. On a different front, the medical establishment, after long ignoring evolution, is faced with an onslaught of drug-resistant microbes, has seen monkey viruses jump into humans and accelerate into epidemics, and must now use evolutionary principles to understand the worldwide dynamics of pathogens.

The theme that unites the examples in this paper is that evolution and evolutionary biology are socially relevant. There are two main reasons for writing such a paper. First, public perception of evolutionary biology is not up to date with the discipline. "Evolution" is still a bad word to many people, not only because it is perceived as conflicting with some religious views, but also because it is widely viewed as an irrelevant science with no social value. Acceptance of evolutionary biology is far more likely when the public realizes that it holds the key to many social improvements [a view that motivated G.C. Williams in his work on applications of evolutionary biology to medicine (Nesse & Williams, 1994; G.C. Williams, personal communication)]. As evolutionary biologists, we need to use these examples of relevance when explaining evolution to our students and the public. A second reason for writing this paper is that historical inertia in the training of evolutionary biologists has resulted in a lack of exposure to socially relevant applications. Individuals trained as evolutionary biologists will have much to offer in solving these problems, but they need to be aware of these applications. Career opportunities for evolutionary biologists may already be more plentiful outside academia than inside it.

This paper is an introduction to some examples of socially relevant evolutionary biology. We have chosen topics with which we are familiar and in which evolutionary biology has already been used to produce an outcome or to affect a policy: phylogenetics, artificial selection (in biotechnology), resistance management, and computation. Other applications of evolutionary principles to socially relevant problems include Darwinian medicine (Lappé 1994, Nesse & Williams 1994, Trevathan et al. 1999), infectious diseases (Ewald 1994, Morse 1994), and human impact on evolution (Palumbi 2001). An excellent overview of the social relevance of evolution has been assembled and endorsed by eight scientific societies (Futuyma 1999).

The topics in this review fit logically into the conceptual framework of evolutionary biology. The first two sections are based on Darwin's theory of descent with modification: The first is about estimating evolutionary history for biomedical and other applications, and the second discusses the role of evolutionary models in interpreting the molecular variation seen in the burgeoning genome databases. The last three sections are based on the neo-Darwinian framework for natural selection. The first describes examples in which humans have modified the elements of natural selection to produce various biological commodities. The next is somewhat the reverse of that, in which an understanding of natural selection is recruited to try to block evolution. The last section is a brief introduction to uses of models of natural selection in designing computer programs.

PHYLOGENETICS: USING THE TREE OF LIFE

The premise that all life shares common ancestry has been a central tenet of evolutionary biology since Darwin. If we go back far enough in time, the genealogy of every organism alive today can be traced to a point that unites it with the genealogy of any other organism alive today. By definition, closely related species have recent common ancestors, whereas the common ancestors of distantly related species go far back in time. Furthermore, the process is continual. Today's species are themselves comprised of lineages that continue to diverge.

Phylogenetics is the study of these evolutionary genealogies. Despite the antiquity of the common-ancestry principle, the field of phylogenetics has matured immensely in the past 10 to 15 years, owing to advances in computer technology and DNA sequencing as well as to the development of explicit theories and methodologies for phylogenetic reconstruction. Most phylogenetic methods use DNA sequences, protein sequences, or RNA sequences, but some methods can use morphological data, which are vital to the analysis of fossils. Not only do methods differ in the types of data that can be analyzed, methods using the same types of data may also differ in the assumptions used to convert the data into evolutionary history.

The output of a phylogenetic analysis is a branching tree that represents the evolutionary history of the lineages being studied. The tree provides not only a nested hierarchy of common ancestors going back in time, but also quantitative information on the amount of change between the different points in the tree. The tree may incorporate taxa whose common ancestors reach back over a billion years, or the tree may be limited to a group of viruses whose common ancestor existed only weeks or months ago. The applications of phylogenetic methods to socially relevant problems likewise occur at several timescales.

Disease Tracking: Molecular Epidemiology

Phylogenetics has become indispensable in identifying disease reservoirs and in tracking the step-by-step transmission of some viruses. The conceptual basis of this work is as follows. We want to know the source of a virus infecting person X. Suppose that four different possible sources have been identified: A, B, C, and D.

These four sources could be different infected individuals who had contact with X, they could be different geographic locations that person X visited, or they could be four species of mammals living in the village of person X that are sometimes infected with the type of virus in X. The problem in figuring out the source is that none of the viruses in A, B, C, and D will necessarily have the same sequence as the virus in X. Phylogenetic analysis gets around that problem by establishing the evolutionary relatedness of viruses from each of the possible sources (Figure 1). The analysis not only indicates which of the viruses in A–D are most similar to that in X (source D in Figure 1), it also indicates how closely related the viruses are and, hence, whether the source might be other than A–D.

Phylogenetic analysis is now a standard part of any disease epidemiology. Below we offer a few of the many applications.

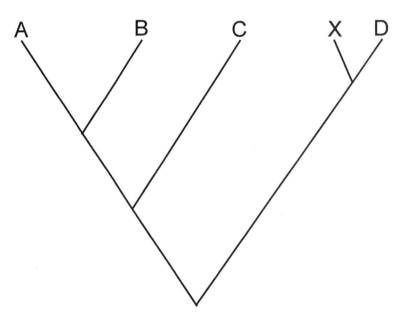

Figure 1 The virus acquired by individual X is compared by phylogenetic methods with viruses from four possible sources (A–D). The context for this could be any of the following. (*a*) A–D are different individuals infected with HIV who had sex or shared needles with individual X, and we are trying to find out who transmitted the virus to X. (*b*) A–D are different mammal species with rabies viruses circulating in their populations, and X is a human who died of rabies. Which mammal species transmitted the infection? (*c*) A–D are mice with hantavirus from New Mexico, Texas, California, and Nevada. X is a Texas resident who recently traveled across the United States and became infected with hantavirus. Where did X contract the virus? In this hypothetical example, the virus in X is most closely related to the virus from source D, which gives information about the likely source (and place) of transmission.

ERADICATION OF WILD-TYPE POLIOVIRUS FROM THE AMERICAS   Poliomyelitis is a paralyzing, occasionally fatal, disease caused by an RNA virus. On recovery from infection, a person carries lifetime immunity, but there are three forms of the virus, and immunity to one form does not confer immunity to the others. Less than 1% of those infected by the virus actually develop the disease, but the incidence of disease per infection is thought to increase with age. It has thus been speculated that the widespread outbreaks of polio disease that occurred in the first part of the twentieth century were a consequence of social hygiene, because improvements in social hygiene delayed the average age of infection (Nathanson et al. 1995, Garrett 2000).

Over 38,000 cases of poliomyelitis were reported in the United States in 1954. The first vaccine was approved a year later, and within 2 decades, the native polio virus was thought to have been eradicated from the Americas. The only known host for poliovirus was human beings, making eradication seem a feasible goal. However, isolated cases of polio continue to occur in the Americas. With the exception of an outbreak in a religious community that avoided vaccines and a recent outbreak in the Dominican Republic (Greensfelder 2000), these isolated cases failed to materialize into epidemics because the population maintained high levels of vaccination (which continues today), but they raised the possibility that native polio strains might still be present. The alternative explanation for these sporadic cases was either nonnative poliovirus introduced from parts of the world where it was still endemic or a vaccine strain. The vaccine in use after 1961 was a live, attenuated virus that was capable not only of transmission, but also of evolving into a more virulent form. Phylogenetic analysis showed that the viruses that gave rise to isolated cases and epidemics were invariably either vaccine derived or wild strains originating from outside the Americas; no native, American virus has been found, and it is presumed to have been eradicated (Rico-Hesse et al. 1987). An assault to eradicate poliovirus worldwide continues to this day, aided by the knowledge that the virus does not lay hidden in nonhuman reservoirs.

ORIGIN OF HIV   The most notorious human infectious disease to have arisen in the 1900s is AIDS (acquired immune deficiency syndrome). Unknown until about 1981, 20 years later it was estimated that more than 30 million people were infected worldwide and more than 16 million had already died. This disease is caused by the retrovirus HIV (human immunodeficiency virus), which exists in two basic forms, HIV-1 and HIV-2. Although the HIV epidemic is now driven by human-human transmission, phylogenetic evidence indicates that HIV-1 originally came from chimpanzees and HIV-2 (which is less abundant and less often fatal than HIV-1) came from a type of monkey known as the sooty mangabey (Gao et al. 1999, Hahn et al. 2000). Phylogenetic evidence supports multiple introductions of HIV-1 and HIV-2 into humans this century (Korber et al. 2000).

HIV TRANSMISSION BETWEEN PEOPLE   One of HIV's unusual properties is that it evolves rapidly, even for a virus. That property is an unfortunate one from the perspective of curing an infection or creating a vaccine, because the virus evolves quickly in response to drug treatment and immune attack (Coffin 1996, Colgrove & Japour 1999, Mosier 2000). However, its rapid evolution enables fine-scale analysis with molecular epidemiology that is not possible with most other viruses. Starting from a single virus in a person, the infection will blossom into a miniature tree of life, the viral lines ever expanding as the infection continues. Consequently, two viruses from one person usually have different genome sequences, but sequences of virus from one person will be more similar to each other than they are to sequences of viruses from other people. This property potentially allows determination of the individual who transmitted the virus.

One of the early analyses of HIV transmission revolved around a Florida dentist with AIDS whose patients exhibited an unusually high incidence of HIV infection (10 patients were eventually discovered to be HIV$^+$). By itself, the clustering of HIV+ patients associated with this dentist might have been regarded as evidence of dentist-to-patient transmission, but this cluster was discovered in the earliest days of understanding HIV, and the possibility of unknown routes of infection had to be considered. If the dentist was not the source of the infection, it was important to discover the source to halt further transmissions. On the other hand, if the dentist was the source, the implications for health care practices were enormous. The stakes were very high either way. Phylogenetic analysis suggested that the patient viruses were close relatives of the dentist's virus in all but two cases, and those two patients had other known risk factors (Ou et al. 1992, Hillis & Huelsenbeck 1994, Hillis et al. 1994). Thus the dentist was likely the source of infection for the eight patients with no known risk factors.

Perhaps the most sensational case of HIV molecular epidemiology was a criminal case in Lafayette, LA. A physician was accused of injecting his former mistress with blood containing HIV. He had been giving her vitamin B injections, and it was supposedly the final injection in August 1994 that contained the blood with HIV. When the woman was diagnosed with HIV and hepatitis C in December 1994, she suspected the physician's injection as the source; at the time of a blood donation in April 1994, she had been negative for both viruses. This case was unusual in that the person infecting the woman (the physician) was not himself infected, so it was necessary both to locate the patient whose HIV infected the woman and to demonstrate that the physician had access to this patient's blood. Records were discovered in the physician's office indicating that blood had been drawn from two patients during the week in question; one patient was previously known to be HIV$^+$ and the other positive for hepatitis-C. Phylogenetic analyses showed that the HIV sequences of the ex-mistress clustered within those of the patient with HIV, supporting her story (State of Louisiana Criminal Dockett #96CR73313; D. Hillis, personal communication). The physician was convicted of attempted second-degree murder and sentenced to a 50-year term in prison in this first use of phylogenetics in a U.S. criminal court.

Predicting Evolution

A novel step in phylogenetic analysis—predicting evolution—was introduced in work on influenza A, the respiratory virus that causes flu epidemics every year. At any one time, several influenza types are circulating in the human population. A viral type is defined by how the virus reacts with a set of antibodies (Webster 1993). Two viral proteins, hemagglutinin and neuraminidase, determine the viral type.

For some decades, influenza isolates have been stored by the Centers for Disease Control; these isolates include lineages that are now extinct in human populations. When techniques were developed to obtain viral gene sequences, it was possible not only to estimate the evolutionary history from currently circulating lineages, but also to use the stored isolates to sequence extinct types to obtain detailed insight into ancestral states. An unusual property of influenza discovered from this work is that the level of sequence diversity has not progressively expanded over time (Fitch et al. 1997, Bush et al. 1999). At any point over the past 15 years, variation existed around a recent common ancestor that was never more than a few years old. Thus, viral extinctions occurred as fast as new types evolved, with only one ultimate winner. Each winner would continue to generate new variants that competed in the population, and only one of them would win, and so on. The phylogeny of this process was a tree with a long trunk and lots of short (dead-end) side branches.

The study that predicted influenza used the hemagglutinin gene for a single viral type, HA3. The fact that only one lineage of HA3 survives means that it may be possible to predict which of the variants circulating in a population will give rise to the viruses of the future. Fitch et al. (1997) observed rapid evolution at a few residues in the hemagglutinin gene. This finding was facilitated by the analysis of historical samples, because the repeated evolution at common sites had otherwise erased earlier evolution at those sites. Those rapidly evolving residues provided an obvious yardstick by which to compare different viruses: Of the viruses present at any one time, the ones with the most evolution at those sites were the candidates as progenitors of future lineages. Bush et al. (1999) developed this predictive statistic and applied it retrospectively to the existing data. The prediction worked exceedingly well, although it awaits a truly a priori test (which will certainly be carried out in the coming years).

Will this method lead to better predictions for flu vaccines? Not necessarily. The difficulty in deciding each year which flu vaccine to produce lies in predicting which of the circulating viral types will infect the most people. The phylogenetic prediction is, instead, of which viral type will prevail in the long term and, thus, covers a longer timescale and a smaller magnitude of viral variation than is necessary for vaccine design. Furthermore, the worldwide pandemics of influenza (that infect and kill more people than average) are caused by a fundamentally different type of sequence change than was studied in the HA3 predictions (Webster 1993). However, the phylogenetic prediction method points toward a new level of utility for this kind of evolutionary biology.

Species Identification on the Tree of Life:
Ribotyping and Other Methods

The cells of all organisms from bacteria to humans (but not viruses) carry genes that build ribosomes, the scaffolding to assemble proteins. Parts of the ribosome are proteins, and other parts (rRNA) are RNA molecules whose ancestry may be traced all the way back to the hypothetical "RNA world," before DNA was the genetic material. Portions of these rRNA genes evolve slowly, so their sequences are very similar even between distantly related organisms; other portions are more variable. The essential nature of rRNA genes combined with the wide range of variation in rates of evolution across different stretches of the molecules makes them ideally suited for estimating the tree of life. This approach to assessing the diversity of and relationships among all life on the same scale was pioneered by Woese (2000). This tree of life shows three main domains: Bacteria, Archaea, and Eukarya. The impressive diversity of plants and animals is confined to a modest branch in one of these three domains, the Eukarya (Figure 2).

Over the years, the rRNA tree of life has been filled in densely, covering thousands of taxa. Of course, the relationships of many of the taxa in this tree of life are consistent with earlier theories, but many taxa have been added whose relationships were previously obscure. Equally important is the fact that this tree provides a universal standard for comparing any life form to any other (which was not possible when using morphology, for example). The utility of the rRNA tree of life has been facilitated by the development of different sets of polymerase chain reaction (PCR) primers that can be used to amplify parts of rRNA genes from any organism or amplify them from only specific taxa. It is now routine to acquire DNA or RNA from an unknown taxon and to use its rRNA sequences to identify it, or at least identify its closest known relatives. This technique (known as ribotyping) can be extremely useful because it allows identification of the different species of organisms present in a community, even though it may be impossible to culture the organisms or recognize them by any other method (Pace 1997).

Ribotyping is not the only molecular method used for species identification. Perhaps the main advantage of ribotyping is that it can be used across the full spectrum of life (excepting viruses) without knowledge of the organisms being typed. But other methods may be preferred when working within narrow taxonomic groups (e.g., mammals, or at a finer level, whales). Thus, noncoding regions of mitochondrial DNA are often more sensitive than rRNA sequences in distinguishing closely related species and subspecies, because these mitochondrial sequences evolve faster. These sensitive methods are not only useful in identifying a species, they may also provide information about the geographic location of the taxon as well (the DNA "zip codes" of Baker 1994).

USING THE TREE OF LIFE: PATHOGEN IDENTIFICATION AND DISEASE ETIOLOGY   Several diseases have no known infectious causes. Ribotyping and other genotyping

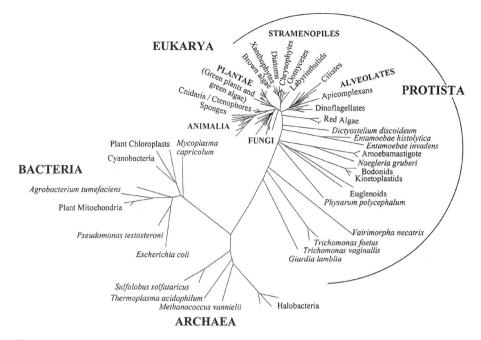

Figure 2   A tree of life, based on 18S rRNA sequences. The tree shows the three domains (Bacteria, Archaea, and Eukarya) as well as many representatives in each domain. Note what little divergence is represented by the animals and plants relative to the entire tree. The 18S sequences from hundreds of life forms (but not viruses) have been obtained and can be resolved on this tree, with closest relatives sharing the most recent common ancestors (the figure is simply too crowded to portray the many species that have been analyzed in this fashion). The 18S sequences from samples of unknown identity can thus be mapped onto this tree to discover what types of life forms were present in the sample. (Figure courtesy of Mitch Sogin.)

methods offer a way of identifying whether diseased tissues are associated with any microbe (Relman & Falkow 1992, Relman 1998). If the causative agent is a previously unknown species, evolutionary biology plays a role in identifying what type of organism it is, i.e., in identifying its closest known relatives. Once a suspect microbe has been identified, steps can be taken to treat with appropriate drugs. Had this approach been available at the time, it might have greatly facilitated the discovery that ulcers were associated with a bacterium. Relman (1999) listed five infectious diseases with causative agents recently discovered by genotyping methods. An extension of this approach is to use ribotyping to assess the microbial composition of a community of organisms, as in animal guts and other passages. Changes in the composition of a community may foreshadow a predisposition to disease or the onset of disease (Kroes et al. 1999).

USING THE TREE OF LIFE: CONSERVATION AND FORENSICS   Concern for endangered species has led to widespread adoption of laws limiting the harvesting, marketing, and even possession of tissues and other products of those organisms. Although some derivatives of a species are unmistakable (e.g., a sea turtle shell, an elephant tusk), others are not so obvious. In many cases, an endangered species has close relatives that are not endangered and are legitimately marketed. If the harvesting occurs at a remote site, far from the eyes of concerned observers, by the time the meat of an endangered species reaches market, it may be indistinguishable from the meat of a legal species. Using genotypic methods, Baker & Palumbi (1996) and Baker et al. (2000) reported the sale of meat from endangered species of whales in Asian markets. Ribotyping and similar molecular methods allow simple and portable means of identifying endangered species in the marketplace.

GENOMICS AND BIOINFORMATICS

From the perspective of human health, a major goal of genomics work is to understand not only the function of human genes, but also the impact of mutations in those genes, and how drugs can be designed to modify or repair those functions. Yet we humans are neither sufficiently genetically variable nor amenable to experimentation that the function of most genes could be ascertained from just our species. An immeasurable benefit to understanding human genetics comes from work on other species—model organisms. As we now know, work on the genetics of other eukaryotes can often be extrapolated to humans. One recent example is the identification of a human gene responsible for a sleep disorder based on the Drosophila circadian clock gene *per* (Toh et al. 2001).

The extrapolation of information between species will accelerate now that complete genome sequences are available for many species. Rapid advances in biotechnology have taken us from the first complete sequence of a small DNA viral genome (Sanger et al. 1977) to the complete sequence of the human genome in less than 25 years (Venter et al. 2001). During the same period, the number of transistors per computer processor has increased 6000-fold. The marriage of increased computing power and large biological data sets has spawned the field of bioinformatics, which is dominated by the analysis of nucleic acid and protein data. Bioinformatics is firmly rooted in evolutionary biology. From the initial step of carrying out a BLAST search (Altschul et al. 1990) of a database to identify related sequences, to multiple sequence alignment (Higgins & Sharp 1988), to the identification of orthologous genes in other species (DeBry & Seldin 1996, Tatusov et al. 1997, Chambers et al. 2000), bioinformatics is a comparative exercise, and descent with modification is one of its inherent and most essential assumptions.

Molecular evolution, specifically sequence divergence, is both a plus and a minus in bioinformatics. A gene in humans and its counterpart in yeast will not have the same DNA sequence even if the function of both has remained the same since their common ancestor. Divergence can be so great that it is difficult to align or even recognize homologous genes. However, once alignments of homologous genes are achieved, evolutionary divergence can be very informative. The divergence of genes with the same function is the evolutionary equivalent of an experiment in which gene positions are mutated to identify the functionally important ones—neutral or nearly neutral changes accumulate over evolutionary time, but changes that disrupt gene function are weeded out by the filter of natural selection. The concept that the most highly conserved amino acid residues are important for gene function is firmly entrenched in molecular biology (Benner 1995, Golding & Dean 1998). But not all residues that are important for function are highly conserved—adaptive evolution is achieved by genetic change. Statistical signatures of past adaptive evolution can be recognized as high rates of nonsynonymous to synonymous substitutions among homologous sequences (McDonald & Kreitman 1991, Boyd & Hartl 1998, Crill et al. 2000) or as correlated changes between different residues (Gutell et al. 2000).

ARTIFICIAL SELECTION

Artificial Selection in the Past

Nearly all the common animals and plants we use today were domesticated thousands of years ago, some (sheep, goats, dogs, wheat, and rice) at least 9000 years ago. Domestication probably started as a process of taming, then of captive breeding, and finally of selecting for specific traits. These early domestications may well have been the first experiments in applied evolution. Their impact was so profound as to make civilization possible by enabling societies to switch from hunting and gathering to agriculture (Ucko & Dimbleby 1969, Clutton-Brock 1999). Although few new species have been domesticated in the past millennium, we have continued to refine the old ones. The success of artificial selection is evident in its ultimate creation of selected phenotypes well outside the extremes of the original species. For example, Chihuahuas, Saint Bernards, pit bulls, and golden retrievers differ in both appearance and behavior, and nothing like any of these breeds would have been found in a population of wild dogs. Fruits of modern strains of domestic plants are much larger than those of their ancestors; corn (maize) is profoundly different from its ancestor, teosinte.

A model of evolution by natural or artificial selection has three components: (*a*) variation, (*b*) inheritance, and (*c*) differential reproductive success (Lewontin 1970). Darwin's theory was formulated largely on an understanding of variation and differential reproductive success but a relatively poor understanding of inheritance. The inheritance void began to be filled early in the twentieth century with the rediscovery of Mendel's work, enhanced (in the West) by a close association of evolutionary biology with agriculture to improve methods of artificial selection. This productive marriage continues today with the use quantitative trait locus (QTL) mapping in both fields.

Another cultural relationship between evolutionary theory, genetics and agriculture, was not so productive, however. In the 1930s, Trofim Lysenko rejected Darwin's theory of natural selection in favor Lamarck's theory of the inheritance of acquired traits. Lysenko's subsequent rise in power as the head of the Soviet Ministry of Agriculture directly destroyed genetic research and decimated agriculture in the Soviet Union (Soyfer 1994, Garrett 2000). The history of Lysenkoism highlights the danger of letting political and other nonscientific ideology dictate scientific practice.

Historically, artificial selection was a manipulation of differential reproductive success. Parents closest to the desired phenotype were chosen to produce the next generation, and individuals that fell short of the ideal were omitted from the breeding population. The components of variation and inheritance were present but were often not manipulated from their natural states, except for the occasional introduction of novel strains or wild relatives into the breeding stock. Mutation rates were seldom, if ever, purposely manipulated, but it is possible that some of the "wide crosses" used to introduce variation into the breeding population could have increased mutation rate by inducing transposition or other mutagenic mechanisms. Although the mechanisms of inheritance in artificial selection were not fundamentally different from inheritance during natural selection, the manipulation of crosses frequently increased the level of inbreeding and thus the ability to select for the expression of recessive traits.

Artificial selection moved into a new realm with biotechnology (Kauffman 1993). In contrast to agriculture, biotechnology specializes in evolving small things, i.e., molecules and microbes. Evolutionary methods in biotechnology have much in common with classical artificial selection. The same three factors—variation, inheritance, and differential reproductive success—are manipulated. But it is no longer the artificial selection of old. The methods used in biotechnology, including DNA sequencing of the entire evolved genome, determination of molecular structures, and monitoring gene expression levels, provide unparalleled levels of analysis of evolution. The combination of experiments, replication, product-oriented research, and analysis of results has allowed rapid attainment of a new level of scientific accomplishment in evolutionary biology.

Directed Evolution

Various protocols based on evolution are currently used to fashion nucleic acids (ribozymes and aptamers) or proteins with specific functions. The directed evolution of novel biological pathways and evolutionary engineering of whole genomes may not be far off. These accomplishments were made possible by changing the components of variation, inheritance, and differential reproductive success. Nature comes close to some of these methodologies, and indeed, several are borrowed from nature, using microbes and various types of parastic genetic elements. Biotechnology has, nonetheless, created unnatural means of evolving molecules.

Perhaps the most extreme form of directed evolution is the in vitro replication of nucleic acids outside of life-forms. The earliest system of this sort was Spiegelman's replication of the Qβ genome with the Qβ RNA-dependent RNA polymerase (Spiegelman et al. 1968). Qβ is a phage whose genome encodes four genes, one being the polymerase enzyme that makes RNA copies of the phage's RNA genome. By isolating the phage's replication enzyme and then placing the RNA genome in a cocktail of nucleotides plus the enzyme, the genome was replicated on its own. In that environment none of its genes were being expressed—the genome was just a molecule that was evolving to copy itself in an environment where its genes were irrelevant. It rapidly evolved to a small size.

The Spiegelman in vitro system did not allow evolution to be directed toward any goal other than fast self-replication. Perhaps the first purely in vitro system that did allow directed evolution was the Systematic Evolution of Ligands by EXponential enrichment (SELEX) system that was simultaneously developed by two different labs (Ellington & Szostak 1990, Tuerk & Gold 1990). This elegant system allowed one to start with a synthesized pool of nucleic acids (Figure 3). The ends of those nucleic acids were "constant" regions, the sequences of which matched sequences of primers for PCR amplification. The middle regions were randomized. This pool of molecules was then washed across many copies of a particular target molecule (e.g., a protein). The molecules that bound the target stayed behind, and the rest were washed off. The bound molecules were then eluted into a separate tube, amplified by PCR, and passed through the cycle again. In this way, a nucleic acid-binding species ("aptamer") could be obtained for any particular target molecule (Tuerk & Gold 1990, Gold et al. 1995, Osborne & Ellington 1997, Famulok & Mayer 1999).

A variety of interesting variations on this theme has since been developed. Beaudry & Joyce (1992) used an in vitro selection and amplification scheme to modify the function of a ribozyme, an RNA molecule with enzymatic activity. The starting ribozyme cleaved RNA molecules, but directed evolution produced a ribozyme that cleaved DNA molecules. Breaker & Joyce (1994) then evolved DNA molecules that could cleave RNA. These methods were similar to the SELEX method in using PCR to amplify molecules that survived the selection, but the methods differed from SELEX in the selection itself. Table 1 lists the impressive variety of unnatural ribozymes (not known from nature) that have been evolved by these methods. Although these few examples highlight nucleic acids, directed evolution also manipulates proteins and entire microbes. In many ways, these new technologies are so different from natural evolution as to warrant a special term (e.g., techno-evolution or technovolution?). However, the term " directed evolution," which is now in wide use, does have the advantage of conveying the message that these socially beneficial applications are based on the foundations of standard evolutionary biology.

DIFFERENTIAL REPRODUCTIVE SUCCESS: REPRODUCTION   Reproduction of the things being selected is essential to evolutionary progress. In the past, one had

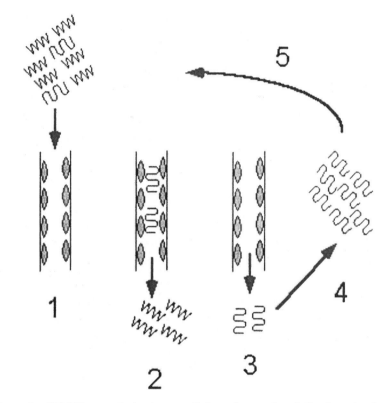

Figure 3 SELEX, a purely in vitro, nonliving scheme of evolution by natural selection. (*1*) A heterogeneous pool of oligonucleotides (the squiggles at the *top*) is passed through a column of anchored target molecules (*dark ovals*). All oligos have identical sequences on their ends but vary in the sequences of their middle region. (*2*) Oligo variants whose sequences facilitate binding remain behind in the column, whereas variants that do not bind pass through and are washed away. (*3*) Binding oligos are eluted, by changing the salt concentration or pH of the solution. (*4*) The oligos from step *3* are amplified by PCR. (*5*) The cycle is repeated by passing the amplified pool of oligos over the column. After a few cycles, the oligo sequences remaining are limited to those capable of binding the target molecules.

to work with an organism capable of reproducing, but now we can literally create and then evolve molecules that reproduce themselves in the right cocktail of enzymes and other nutrients. The most common method of reproducing nucleic acid molecules is PCR. A similar but less popular method is self-sustained sequence replication (3SR). In PCR, DNA molecules are the parents, and their progeny are complementary DNA molecules. In 3SR, RNA molecules are the parents, cDNA molecules are the progeny, and transcription of these DNA molecules creates RNA copies as the grandchildren. PCR uses synchronized cycles of reproduction, in which all molecules in the tube reproduce once and then stop until the next cycle

TABLE 1   Unnatural ribozymes evolved by directed evolution[a]

Ribosyl 2'-O–mediated cleavage
$\quad$ Mg^{2+}-dependent cleavage
$\quad$ Pb^{2+}-dependent cleavage
Ribosyl 2',3'-cyclic P hydrolysis
$\quad$ Pb^{2+} dependent
RNA ligation
$\quad$ 3',5' Ligation (class 1)
$\quad$ 2',5' Ligation
$\quad$ 5',5' Ligation
RNA phosphorylation
$\quad$ Class 1 (ATP-S)
$\quad$ Class 1 (ATP)
Self-aminoacylation
Acyl transfer reaction
$\quad$ Acyl transfer
$\quad$ Aminoacyl transfer
Self-nitrogen alkylation
Suflur alkylation
Biphenyl isomerization
Prophyrin metalation

[a]RNA molecules with these activities are unknown from nature. (From Jaeger 1997.)

is initiated. 3SR is a method in which everything happens continuously (Fahy et al. 1991).

For many purposes, the directed evolution of proteins and many other types of molecules still requires an organism. A gene is expressed in an organism (typically a bacterium or yeast), but the gene may be removed from that organism and subjected to various manipulations before being returned to the same or a different organism. Biotechnology thus allows a gene to hitchhike as part of an organism but then be divorced from that organism at will. However, two new methods allow in vitro translation to couple a protein to its mRNA, enabling proteins to be evolved in a SELEX-like fashion (Wilson et al. 2001). Limited forms of peptide self-replication have also been produced, although the chemistry of the peptide copying mechanisms is fundamentally different from that of nucleic acids (Lee et al. 1997).

DIFFERENTIAL REPRODUCTIVE SUCCESS: DIFFERENTIAL SUCCESS   A challenge in carrying out directed evolution is to come up with a powerful method to manipulate the differential reproductive success of the molecules under selection. The goal is to choose genes whose phenotypes offer the greatest improvement as parents for the next round of selection. This can be done directly through selection or indirectly through screening. True selection occurs when molecules with the desired phenotypes reproduce faster than molecules with less desired phenotypes; in the extreme case, molecules with the desired phenotypes are the only ones to survive. In screening, the good and bad survive; the phenotype of each molecule is individually assessed—for example, through a colorimetric enzyme assay—and genes for the most desired molecules are chosen as parents for the next generation. The distinction between selection and screening in directed evolution thus appears to be functionally the same as between natural selection and artificial selection (D. Futuyma, personal communication). True selection is generally much more powerful than screening because the number of variant molecules that can be subjected to selection is generally much larger than for screening (Arnold & Volkov 1999). However, schemes for true selection can be difficult to design.

The form of selection used in much of biotechnology is what quantitative geneticists refer to as truncation selection—reproducing only those molecules that meet a certain standard, in the same way that a livestock breeder mates only animals that achieve a certain weight or fat content. In biotechnology, the selection might be determined by the ability of a bacterial cell to grow on a toxic or antibiotic substrate. In other cases, as in aptamer selection, reproductive success is based on adherence to a target molecule. In some protocols for DNA enzyme evolution, reproductive success is determined in the reverse fashion—escaping from a bound state is the gateway to future reproduction.

One of the advantages of evolution in biotechnology is that truncation selection can achieve astronomical levels. The limiting factors in any truncation selection are (*a*) the range of variation, which is limited by the number of different genotypes that exist, and (*b*) the fecundity of those organisms between bouts of selection. If an asexual organism produces only two offspring between bouts of selection, for example, then a 50% culling is the long-term upper limit, whereas if the fecundity is 1000, then a 99.9% culling is possible. The latter allows much more rapid evolution, genetic variation permitting. Methods such as cloning genes into selectable bacterial plasmids and PCR allow amplifications (reproduction) of as high as 10^9 between bouts of selection. Such levels could be achieved only in special cases of natural selection (e.g., with microbes evolving in a new environment or invading a new host species).

VARIATION AND INHERITANCE: ELEVATING LEVELS OF GENETIC VARIATION   Perhaps the greatest deviations from natural evolutionary processes used in biotechnology have come from manipulating genetic variation. These advances include elevating genetic variation and recombining molecules.

MUTATION   The rate of evolution in many experiments is limited by genetic variation, largely because the high levels of truncation selection that may be applied can take advantage of levels of variation that lie orders of magnitude beyond anything natural. To raise mutation rates, a variety of approaches have been developed.

The most extreme is simply to incorporate high levels of variation in synthesized molecules. Pools of oligonucleotides can be synthesized with arbitrary base compositions at specific sites. These molecules can be used as the direct target of selection (as in aptamer evolution studies) or as material cloned into vectors that then becomes translated into protein. At the extreme, the sequence of the initial selective population is completely randomized and all bases have equal frequencies. Current technology limits the pool size to about 10^{16}, so an exhaustive search of "deep random" is limited to a sequence of 27 bases (Landweber 1999). When larger stretches of randomized sequences are used, this limitation means that independent experiments using the same protocol may arrive at different solutions because only a small fraction of the possible sequences are present in the initial pool of molecules. Evolution in these cases is dominated by the "tyranny of small motifs" pattern (Ellington 1994), such that any solutions to the selective challenge that are specified by a small number of residues will certainly be present in the pool of randomized molecules, whereas solutions specified by a large number of residues may not. So the evolved solutions tend to be the simpler ones, not necessarily the best ones.

Oligo synthesis is not the only means of elevating mutation rates. If molecules are not synthesized, then they are copied from preexisting templates. Natural enzymes and processes are invariably used for this replication, and their inherent error rates are usually much lower than wanted. Other methods of generating variation rely on inflating the error rate during replication or introducing errors into the template itself. The standard method of elevating whole-organism mutation rates has been to expose the organism to a chemical or physical mutagen. That method does not allow precise control over the genomic locations of the mutations, although it allows some control over the types of genetic changes, because different agents tend to cause different types of mutations. More recent methods involve inflating the error rates during replication: (*a*) PCR with unnatural bases, with asymmetric base compositions, and/or with manganese; (*b*) amplification that involves alternately copying between DNA and RNA because the transcription step (DNA into RNA) is highly error prone (e.g., Fromant et al. 1995).

RECOMBINATION   One of the most powerful techniques used in directed evolution is recombination among variants of the same or similar molecules. Several methods for performing in vitro recombination are now routine (known in the field as gene shuffling or molecular breeding). Pim Stemmer of Maxygen developed the first purely in vitro shuffling method (Stemmer 1994), which involves pooling DNA templates, nicking them to cut strands, denaturing them, and letting them come back together before patching them up. Molecules require only short regions of identity to realign, so recombination (strand exchange) can occur between molecules that differ substantially. The variants to be shuffled can be those already improved during earlier rounds of selection, or they can be naturally occurring homologous genes from a diversity of species—family shuffling (Crameri et al. 1998). Family shuffling has the same advantage as increasing genetic variation by introducing novel strains or wild relatives into the genetic stock: Natural selection has already weeded out the deleterious mutations from these variants.

In at least some cases family shuffling radically accelerates the process of directed evolution. For example, when cephalosporinase genes from four microbial species were independently evolved through one round of DNA shuffling, they showed up to an eightfold improvement in their resistance to the antibiotic moxalactam. A single cycle of family shuffling that combined all four genes yielded a 540-fold improvement in resistance. The evolved gene showing the greatest increase in resistance was made of eight fragments from three of the four genes and had an additional 33 amino acid substitutions (Crameri et al. 1998).

VARIATION AND INHERITANCE: VARIANTS WITH REDUCED CONSTRAINTS   Reproductive success is an ultimate necessity in any form of evolution by natural or artificial selection. Yet in a narrow sense, reproduction may have nothing to do with the phenotype sought by artificial selection. The reproductive requirement often limits the progress that can be achieved in artificial selection. For example, most people prefer to eat seedless watermelons, but seeds are needed for the next generation. (Seedless watermelons are created as triploid hybrids from crosses between different strains.) One of the big successes in biotechnology has been the reduction of such constraints, so that molecules with desired phenotypes can be propagated without concern for their correlated negative effects on reproduction. Several tricks facilitate the divorce between reproductive constraints and selection on the desired phenotype.

REPRODUCTION BY PCR   When a nucleic acid (RNA or DNA) molecule itself is selected for the phenotype, as in aptamer or ribozyme evolution, amplification by PCR is not only the easiest method of reproducing the molecule, it also entails minimal constraints. The ends of the molecule must match primer sequences, but intervening sequences are largely irrelevant. Earlier methods of amplification (cloning) incorporated the DNA molecule in a plasmid, and a plasmid's reproduction is tied to that of its host. Not only was cloning a cumbersome method of amplification, if the cloned sequences were incompatible with cell growth, the amplification would not work. PCR is a nonselective amplification method, such that the entire pool of nucleic acids (with suitable ends) experiences minimal evolution during the amplification (Bull & Pease 1995). Thus with PCR, differential reproductive success can be virtually removed from the amplification/reproduction stage.

LIMITING THE DELETERIOUS CONSEQUENCES OF PHENOTYPE EXPRESSION   In many cases, the desired phenotype requires protein expression. The standard way to create proteins is to put the gene inside cells and let the cellular machinery build the protein. This approach can be a problem if the protein is incompatible with cellular function. A compatibility problem can be overcome by placing the gene in vectors that offer control over gene expression. Gene expression can be suppressed until the host has reproduced, so that even if the cell is killed by expression, the DNA vector from selected colonies can be recovered and put into new cells. The new methods for in vitro translation and coupling of the peptide to its mRNA avoid the problem entirely, however (Wilson et al. 2001). Another mechanism for limiting deleterious consequences applies at the protein level instead of the genome level. When evolving peptide sequences, it is often desirable to couple the evolvable peptide with a reporter peptide (for screening or perhaps as part of the selection). Activity of a reporter peptide will be incompatible with sequence insertions at many positions, but there are often regions that can tolerate small or even large peptide additions without destroying activity, such as protein ends or loops. Thus a fusion protein is created that has both the reporter activity and a variable region that can be selected for a different function. This rationale underlies the highly successful methods of the yeast two-hybrid assay (Chien et al. 1991), phage display (Smith 1985), and some other systems.

VARIATION AND INHERITANCE: UNNATURAL MOLECULES   The synthesis of unnatural molecules, which is now routine chemistry, has affected artificial selection in two ways: New types of molecules can be evolved, and old types of molecules can be evolved toward new targets. Replication of new kinds of nucleic acids is possible with modified bases, sometimes known as base analogs. The ribonucleotides in natural RNA contain the bases adenine, uracil, guanine, and cytosine. Several types of modified bases have been incorporated into ribonucleotides, and a few types have been incorporated into deoxyribonucleotides. These modified bases have novel groups attached that do not interfere with hydrogen bonding but that can alter the characteristics of the nucleic acid in other ways. Provided the polymerase will accept the base analog, it is straightforward to create a mix of bases that will result in a nucleic acid with one or more of the natural bases replaced by the analog. These nucleic acids may then be subjected to the usual selections, in the hope that the product will be superior to the natural nucleic acids (Tarasow et al. 1997, Wiegand et al. 1997, Sakthivel & Barbas 1998, Battersby et al. 1999).

Novel molecules may also serve as selective agents. This phenomenon has been visited many times in human history, as pesticides were applied to control pests or synthetic drugs were used to treat microbes, and the offended organisms responded with their own evolution (see section below). One of the more interesting uses of novel molecules as a selective agent is based on the symmetry of mirror-image molecules. Life uses the L-forms of amino acids and D-forms of nucleic acids, but the mirror image forms of both can be synthesized chemically. The novel method uses a synthesized D-peptide as a target against a pool of D-nucleic acids as potential aptamers; the D-peptide is the unnatural form, whereas the D-nucleic acids are of the natural form so that they can be evolved through directed evolution. When a successful D-aptamer is evolved, its L-form is synthesized (which is unnatural).

Because of the mirror-image symmetry, the synthetic L-aptamer recognizes the L-peptide that is found in nature. This two-step technique was used to create an L-form DNA aptamer that inhibits vasopressin and, unlike the D-aptamer, was resistant to nuclease degradation (Klussmann et al. 1996, Williams et al. 1997).

Examples of Directed Evolution

A DNA ENZYME TO LIMIT ARTERIAL DAMAGE FROM ANGIOPLASTY   Santoro & Joyce (1997) evolved a short DNA molecule that cleaved a specific site in a target RNA molecule. From a sequence comparison between this DNA enzyme and its target, the DNA enzyme appeared to contain a catalytic core flanked on both sides by regions that formed Watson-Crick base pairs with the target RNA. This discovery suggested that a DNA enzyme could be synthesized against almost any RNA merely by changing its flanking sequences to complement the RNA sequence at the desired target. In one application, the DNA enzyme was designed to cleave the mRNA of the *Egr*-1 gene, whose expression causes unwanted proliferation of the arterial wall in response to damage from angioplasty (balloon inflation). Arterial proliferation is counterproductive because the goal of angioplasty is to expand the arterial canal, and the proliferation narrows it. Tests in a rat model showed that the DNA enzyme had the desired effect of reducing arterial proliferation (Santiago et al. 1999).

ENZYMES FOR HOUSEHOLD AND AGRICULTURAL USE   The year 2000 marked the 5th Annual World Congress on Enzyme Technologies, organized by the IBC (International Business Communications). The sponsors were the biotech companies Maxygen, Genencor, Diversa, and Thermogen. This meeting was dominated by talks on using directed evolution to improve enzyme performance in specific settings. One industrial goal is improved cellulase activity. Agricultural waste in the form of corn stalks and other plant material offers a potential windfall of ethanol production, if cellulose can be digested with cheap enzymes. Another goal is the improvement of enzymes (proteases, lipases) to use in laundry detergents to remove stains and other dirt. Cellulases, proteases, and lipases have been isolated from numerous organisms, but their activity levels are too low under applied conditions to justify their use. For example, naturally occurring proteases do not function in warm, soapy water. Directed evolution is being used to improve that performance [for an introduction to this large area of research, see Marrs et al. (1999), Schmidt-Dannert & Arnold (1999), Voigt et al. (2000)].

EVOLVING PEPTIDES THAT LIMIT CRYSTAL SIZE   Phage display is the insertion of peptide sequences into coat proteins of a bacteriophage (bacterial virus) so that the peptide insert is able to contact surfaces in the phage's environment. The size and location of the peptide insert is chosen so that it permits phage reproduction. With randomized inserts, a phage display library may contain billions of different insert sequences that can be selected for binding to many types of substrates. Using a commercial phage display library, Whaley et al. (2000) recovered peptide epitopes that bound gallium arsenate crystals; some epitopes discriminated different forms of the crystal. The peptide insert was a mere 12 amino acids long, yet it seemed that crystal recognition could be achieved with even fewer than 12 amino acids. This discovery may be an important step in miniaturizing the manufacture of semiconductors (nanotechnology) because by binding the crystal lattice, these peptides may offer a simple way of reproducibly controlling crystal growth.

BLOCKING HIV EXPRESSION   Many drugs have been developed to suppress HIV, but evolution of HIV resistance to these drugs fuels a continual demand for new drugs. Some recent approaches to inhibit HIV use technologies involving RNA molecules. These new methods include antisense RNA (RNAs that are complementary to the single-stranded viral mRNAs), ribozymes (RNA enzymes that cleave the viral mRNAs), and aptamers that bind HIV proteins. The antisense RNAs and ribozymes can be developed merely from an understanding of HIV genome sequences, whereas the aptamers need to be evolved in a SELEX-like manner. When different agents created by these three methods were compared for their efficacy against HIV in cell culture, successful inhibition was obtained only with aptamers (Good et al. 1997). Assuming that an anti-HIV aptamer could become an effective drug in vivo, the encouraging aspect of this result is that new aptamers could possibly be evolved each time the virus evolved resistance to the old aptamers.

Other Approaches

Evolution is just one of several technology-driven methods for modifying molecules for specific uses. Methods for protein improvement are generally divided into approaches known as rational design and irrational design (Arnold 1997, Arnold & Volkov 1999). Rational design consists of protein engineering by modification of specific amino acid residues based on knowledge of protein structure and mechanistic details. Although this approach is potentially powerful, we are currently far from having enough information about most proteins to apply rational design, and in general we lack the deep understanding of protein structure and function that would allow us to predict the outcome of specific amino acid substitutions (Tobin et al. 2000). The "irrational" approaches of directed evolution do not require such detailed knowledge because they rely on selection to reach the desired outcome.

These applications in biotechnology not only offer a new relevance for evolutionary biology, they also provide new opportunities for biologists with classical training in evolution. The industrial approach of large-scale experiments with a focus on products and detailed molecular analysis is often done in ignorance of the underlying classical framework. A truly golden opportunity lies ahead for the marriage of classical theory with industrial interests. What level of recombination and mutagenesis is optimal? How rugged is the fitness landscape? It should be possible to build a new level of evolutionary theory and help industry reach its goals in this new area of directed evolution, allowing for appropriate accommodation of the sometimes conflicting goals between industry and academia.

RESISTANCE MANAGEMENT

Many organisms (including humans) have been creative in developing chemistry to deter or kill unwanted competitors. For humans, the pests may be cancer cells, insects, mites, worms, weeds, or microbes. As is known all too well, virtually all our attempts to control pests with chemical agents have led to resistance to those agents (Garrett 1994). The evolution of resistance is so routine that it seems inevitable, notwithstanding a few noteworthy examples, such as the American chestnut, in which an entire species with billions of individuals apparently failed to evolve resistance to an invading pathogen (Newhouse 1990). The question is how an understanding of evolution can help us retard or even prevent the evolution of resistance.

The evolution of resistance is not a new problem. It accompanied the introduction of antibiotics and pesticides 50 years ago. Attention to this problem from evolutionary biologists has lagged, however. Part of the reason for this slow response may have been the seemingly limitless supply of new chemicals. When the first antibiotics were isolated from nature, a virtual windfall of different drugs was found. The evolution of resistance was inconsequential when new drugs became available faster than old ones failed. Likewise, new pesticides may have seemed easy to engineer in the days before government regulation. Now, however, many of the old technologies have reached limits, and the cost of obtaining government approval is enormous (e.g., on the order of half a billion dollars for a new drug). There are thus ample incentives to use chemicals wisely and prolong the lives of what works now.

One role of evolutionary biologists is to educate the public. A widespread misconception concerning antibiotics is that an individual who abuses them will develop a tolerance, so the drugs will no longer work for them personally. That is, some people mistakenly assume that the person's body changes in response to antibiotic misuse. The true problem is that misuse of antibiotics encourages evolution in the bacterium so that it is no longer affected by the drug. If the bacterium spreads, then people contracting it are at risk of an untreatable infection, no matter how conscientious they have been in their past use of antibiotics. Resistance quickly becomes a global problem caused by evolution.

The evolution of resistance is affected by the manner in which we apply the toxins. These factors are under our control and allow us at least to impede the evolution of resistance.

The Right Dose

No doubt the most widely acknowledged factor influencing the evolution of resistance is the dose. Many Americans, at least, are aware of the admonition to take the full course of prescribed antibiotics, lest the infection return in a more resistant, less easily treated form (this is especially a problem with tuberculosis infections, in which eradication of the bacterium from a person requires months of treatment). Evolutionary biologists since Darwin have been aware that weak selection can ultimately yield phenotypes that lie well outside the range of phenotypes currently in the population, whereas immediate selection for those extremes would fail and thereby extinguish the population. A low dose of a drug—low enough to allow survival of some sensitive individuals—is a form of weak selection. Its main detriment is that it favors individuals with partial resistance; as partial resistance evolves, full resistance is more easily attained. A second complication with a low dose is that it leaves sensitive survivors that may mutate to resistance before further doses are applied. The evolution of bacterial resistance is now a factor when recommending levels of some antibiotics (Blondeau et al. 2001), and variations in dose, both temporally and within the patient, are used to help understand the evolution of resistance (Baquero & Negri 1997, Baquero et al. 1997, Blondeau et al. 2001). "Dose" is likewise a consideration when planting insect-toxic, transgenic plants to avoid insect resistance (see below).

Selective Application

Larger populations are more likely than small ones to contain resistant genotypes, so limiting the size of the population treated reduces the chance that resistance will evolve. A simple way to limit the treated population yet still be effective is to treat only those pests causing damage. With antimicrobial agents, selective application would consist of treating only those patients manifesting an infection; with pesticides, selective application would consist of spraying only those crops experiencing economic injury levels of pests. Although this model is simple in principle, for social and technical reasons it is often difficult to institute. With antibiotic treatment of bacterial infections, a strict adherence to selective application would mean that patients are not given a drug until their infecting microbe has been diagnosed as a strain sensitive to the drug. This practice is neither patient-friendly nor safe in all cases, as an infection can worsen during the time required for diagnosis. Even when it is clear that antibiotic treatment is unwarranted, a patient's demands for a drug may override any physician concern for the eventual evolution of resistance. The problem is a classic example of a phenomenon described by Hardin (1968), "Tragedy of the Commons," in that the cost/benefit ratio from the individual's perspective favors antibiotic overuse, in opposition to the common good (Palumbi 2001). Industrial concerns further contribute to antibiotic overuse in our environment because antibiotic food supplements yield faster livestock growth. The agricultural interest of greater meat production has so far been the victor over proposals to restrict antibiotics in livestock food, despite clear evidence that antibiotic food supplements encourage the evolution of antibiotic resistance in human pathogens (Garrett 1994).

Does limited use (selective application) reduce resistance levels? It is of course clear that antibiotic use led directly to the widespread evolution of antibiotic resistance in many bacteria, as indicated by geographic and temporal correlations between drug resistance and antibiotic use (Garrett 1994, 2000; Granizo et al. 2000). Limiting antibiotic use during the past four decades would at least have slowed the rate at which resistance evolved. But selective application also causes current resistance levels to drop (Cristino 1999).

Combination Therapy

Simultaneous application of multiple agents may extinguish a small population, even when high doses of single agents would not work, for the same reason that high doses of a single agent are better than low doses: The chance that resistant individuals occur in the initial population decreases with the magnitude of the imposed mortality. If resistance to a single agent can be conferred by a single mutation, then multiple agents may offer the advantage that no single mutation confers complete resistance. Perhaps the original use of combination therapy was in the treatment of tuberculosis with the first antibiotics (Ryan 1993). Treatment with a single antibiotic resulted in the evolution of drug resistance within the patient. Simultaneous treatment with a combination of three drugs allowed the infection to be cured and the within-patient evolution of resistance prevented. More recently, the simultaneous use of multiple drugs to treat HIV infections has resulted in prolonged avoidance of AIDS because the virus is so effectively curtailed with the harsh treatment (Coffin 1995, Matsushita 2000). The long-term success of many antiviral vaccines may likewise stem from the immunity generated against multiple targets through both the humoral and cell-mediated components. With other goals, however, simultaneous, multidrug treatment may not be the best practice (Bonhoeffer et al. 1997).

Charting the Course of Resistance

When more drugs are available than can be given to a single patient, the choice of which drugs to take should be based on the evolution of resistance. Screening for preexisting resistance to the drugs can ensure that the combination of drugs administered is maximally effective. For example, anti-HIV drugs (of which there are now many) are so harsh on the patient that doses and numbers of drugs are often based on patient tolerance levels. There are three major classes of anti-HIV drugs (protease inhibitors and two kinds of reverse-transcriptase inhibitors), but many drug options exist within each class. A patient on highly active antiretroviral treatment will take one drug from each class to maximize the number of viral targets, but within each class, the drug that is taken is somewhat optional. Protocols are now being tested in which a patient's viral population is assessed for resistance mutations, with the drug choice based on that genetic information (O'Meara et al. 2001, MacArthur 2000). It is obvious that this kind of approach requires understanding the molecular basis of resistance. It also requires a technical means of assessing low levels of resistance in the target population (e.g., on the order of 1% or even 0.1%), as such levels are not easily detectable yet can change rapidly in response to selection.

Refugia

One of the more exciting new developments in agriculture is the engineering of transgenic crops expressing a gene from the bacterium *Bacillus thuringiensis* (Bt), the protein of which is highly toxic to butterflies and moths. This toxin (Bt) has long been a favorite of organic farmers, used as a spray prior to transgenic technology, because it selectively kills only a small subset of insects and typically does not harm the parasitic and predatory control agents. Corn and cotton are two major Bt transgenic crops in wide use now, and the potential for reducing pesticide applications to the environment from them alone is impressive. Despite a public backlash against genetically modified foods and the consequent withdrawal of one Bt corn strain from the market, approximately one quarter of the U. S. corn crop is expected to be Bt transgenic in 2001 (F. Gould, personal communication). There is thus a huge market for Bt strains.

Alleles for Bt resistance are already found at moderate frequencies in several pest populations, in some cases at frequencies greater than 10% (Gould et al. 1997, Tabashnik et al. 2000). These pests are thus poised for a rapid response to Bt transgenic plants. Fortunately, levels of resistant insects are much lower than levels of resistant alleles because the resistance alleles are recessive, hence resistant insects occur at the square of the allele frequency. These surprisingly high frequencies of resistance alleles are unexplained in at least some species, because there is no known history of exposure to agricultural Bt toxin (F. Gould, personal communication).

The commercial implications of Bt resistance are obvious and have inspired the U. S. Department of Agriculture and seed companies to anticipate and slow its evolution (Gould 1998). An antiresistance evolution strategy is being employed that depends on recessive resistance. Farmers growing Bt crops are mandated to plant a certain amount of non–Bt crops as well—until recently 4%; now it is higher. These non–Bt crops are planted in separate fields (refugia) close to the Bt crops, to provide a safe haven for pests that are not resistant to the toxin.

Refugia impact the evolution of resistance to Bt as follows. Pests in the refugia survive regardless of their resistance to the toxin. Pests in the Bt fields survive only if they are resistant, i.e., homozygous for the resistance allele. As long as resistance alleles are uncommon, most insects will be sensitive. Thus, refugia will produce many more insects than the Bt fields. With random mating between pests born in the refugia and pests born in the Bt fields, most offspring will be sensitive to the toxin (because they are either homozygous sensitive or heterozygous). Refugia will maintain resistance alleles at low frequencies for long periods of time because they continually dilute the resistance genes and keep them from being selected. Resistance will eventually ascend, but the ascent is greatly delayed with refugia (Figure 4). This result is a special case of the population genetics principle that

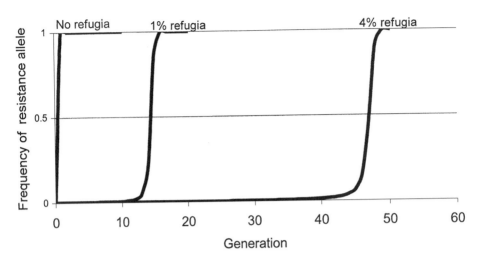

Figure 4   Effect of refugia on the delay of *Bacillus thuringiensis* (Bt) resistance evolution. Three curves of gene frequency evolution are shown: (*a*) no refugia, (*b*) 1% refugia, and (*c*) 4% refugia. The evolutionary response to no refugia is immediate, because the only survivors are recessive homozygotes. The ascent of the resistance allele is progressively delayed with larger refugia because refugia are not selective for genotype and thus weaken selection for the recessive, resistance allele. These simulations assumed complete mixing of individuals between refugia and Bt crops when mating and when ovipositing. They also assumed that the only source of selection was mortality of all heterozygotes and of sensitive homozygotes in Bt fields, that generations were discrete, and that the initial frequency of the resistance allele was 0.001. The effect of refugia would be modified to the extent that these assumptions do not hold, especially the recessivity of the resistance phenotype and the mixing. Also, the curves give only allele frequencies; the actual numbers of surviving pests would be an important factor to consider in the economic injury to a crop, but that factor is not included here. These curves were calculated by iteration of the equation $p' = (p^2 W_{aa} + pq W_{Aa})/(p^2 W_{aa} + 2pq W_{Aa} + q^2 W_{AA})$, where p is the frequency of the recessive resistance allele in generation t, and p' is its frequency in generation $t + 1$. The fitness values were determined as $W_{aa} = 1, W_{Aa} = W_{AA} = R$, where R is the proportion of crops in the refugia.

weak selection on a recessive trait is ineffective when the recessive allele is rare (Crow & Kimura 1970); the strength of selection decreases with refugium size. One reason non–Bt crops are not mixed in fields with Bt crops is to prevent insects from moving between the two types of plants and thus experiencing intermediate doses of the toxin (which might reduce the recessivity of the resistance alleles).

Evolvable Drugs

Biotechnology offers new drugs from new classes of molecules but it also offers new drugs from some old molecules—proteins and nucleic acids. (Some types of RNA-based drugs were discussed above in a paragraph about blocking HIV.) Proteins and nucleic acids in essence have their own genome, so theoretically it should be possible to evolve those drugs. Evolving a drug might provide a way of producing new forms of the drug that overcome pests resistant to the original form of that drug. At the moment, this approach is futuristic. It is analogous to the use of combinatorial chemistry to produce new generations of antibiotics by modifying subgroups of old drugs. Drug evolution was attempted in a model system of antisense RNA used to inhibit a virus (Bull et al. 1998). Antisense nucleic acids work on the simple principle of being complementary to a target sequence (in this case a regulatory sequence of the virus) and thus blocking gene expression. If the target sequence in the virus evolves to a resistant form, antisense molecules complementary to the new target are then easily created. This "arms race" can thus potentially be continued indefinitely because a new antisense RNA can be created for every step the virus takes in evolving resistance. The empirical test of this theory supported some of the assumptions but not all: New antisense RNAs could be created that inhibited individual viruses resistant to the original antisense RNA. However, the viral population exhibited a variety of escape mutants, and no single new antisense RNA could control the resistant viral population. Resistance polymorphism in the viral population thus thwarted control by the second generation of drugs.

Other variations of this approach have also been developed, again without success. Djordjevic & Klaenhammer (1997) and Bull et al. (2001) each developed a suicide plasmid cassette against a bacterial virus. Each plasmid contained a toxic gene downstream of a promoter expressed by the virus. Infection expressed the toxic gene, which, in turn, killed the infection. A priori, it seemed likely that an arms race could be waged against the virus by inserting a new toxic gene into the cassette each time the virus evolved resistance to the old toxic gene. However, in both systems the virus evolved resistance by a change in its transcriptional activity, selectively reducing expression from the plasmid regardless of which toxic gene was carried by the plasmid. In a further attempt to keep pace with the resistant virus, Djordjevic & Klaenhammer (1997) created a new plasmid that contained multiple copies of the viral promoter; this plasmid partially restored inhibition against the virus, although at some cost to the uninfected cell (perhaps through low levels of constitutive expression).

It is not clear how widely the principle of drug "evolution" can be applied to block resistance. Drugs that share a common molecular ancestor will likely use the same molecular target. If evolution in the pest can protect the target and render it inaccessible, then drug evolution will fail. The case of nonnucleoside reverse-transcriptase inhibitors (NNRTI) in HIV provides an analogy to drug evolution. Various NNRTI drugs have been developed, but they all function by similar mechanisms, binding to the same pocket in the viral reverse transcriptase (Emini 1996). Resistance to one of these drugs often confers resistance to several of the others. If one regards the different NNRTI drugs as the equivalents of evolved drugs, then an evolved drug in this case affords much less protection against a resistant virus than does a drug using a different mechanism of inhibition. All the different evolved forms of a drug may use the same basic mechanism of inhibition, just as different NNRTI drugs use similar mechanisms, in which case viral resistance to one drug will at least partly overlap with resistance to the others.

EVOLUTIONARY COMPUTATION

Introduction

Evolution is not restricted to biological systems, nor has the benefit of using evolution to solve problems been limited to biology. Although application of evolutionary principles to computer programming preceded the work of Fogel et al. (1966), their book brought the idea to the forefront. The goal of this work was lofty—"that through a replication of specific aspects of evolution, means will be found for the generation of an artificially intelligent automata ... capable of solving problems in previously undiscovered ways." The field gained sophistication almost as quickly as its biological counterpart, and by the mid-1970s, Holland (1975) was discussing coadapted gene sets and epistatic interactions in the context of computation. Many concepts and much of the language of evolutionary computation (EC) are borrowed from the biological world. Bits are "loci," potential solutions to a problem, are termed individuals, chromosomes, or genomes, and they are changed by "mutation" and "recombination." A collection of variant solutions, termed a population, is subjected to "selection" by imposing some measure of "fitness." The best solutions are selected as "parents" for the next generation, they "reproduce" with mutation and recombination to produce a new population that is subjected to selection, and so on.

EC, a subfield of artificial intelligence, exists in several forms that differ in the representation of the problem that is evolved and the mechanisms of evolution employed (Foster 2001). For example, in genetic programming, executable programs themselves are evolved, whereas in genetic algorithms, the parameters that describe a potential solution to a problem are evolved (Koza 1992). Although recombination is commonly employed in genetic algorithms and genetic programming, it is generally not employed in the other two EC varieties, evolutionary programming and evolutionary strategies, which instead rely on large-scale mutations.

Like studies of the evolution of biological systems, research in the area of EC makes use of model systems. In this case the models are problems to be solved. These include the "traveling salesman problem," an NP-complete problem (i.e., a class of computationally difficult problems) in which one tries to determine the shortest route for a salesman traveling between a large number of cities. Another interesting model problem is the "iterated Prisoner's Dilemma," in which one decides when it is best to cooperate and when it is best to defect in a series of confrontations. This problem has been explored from the perspective of the evolution of cooperation (Axelrod & Hamilton 1981) and is now a classroom standard in EC (Vogel 1995). Another common model problem in genetic programming is symbolic regression, where one must evolve a program to reproduce the input-output behavior of an unknown function given only randomly selected inputs and function values.

Evolved programs have interesting properties that in some cases are shared with biological systems. For example, evolution of a genetic program via recombination is accompanied by "code bloat" (Langdon et al. 1999). Like biological genomes, such computer genomes almost always accumulate "junk," and it is not easy to separate the functional components of the genome from the nonfunctional ones. In fact, a defense that removes junk from the genome will lead to the evolution of code that evades the editing algorithm (Soule & Foster 1998). However, genome size can be constrained directly by charging a fitness penalty for larger genomes. This is similar to biological systems in which microbial genomes maintain more stream-lined genomes than those of most multicellular organisms, presumably because of selection for rapid replication.

What Kinds of Problems are Best Explored with EC?

Although in principle evolutionary algorithms can be applied to any computational problem, they are best suited to very difficult problems—problems to which there are no other known efficient solutions. As with biological evolution, EC may not find the optimal solution because it is a stochastic approximation algorithm. For problems with a very large potential solution space, such as the traveling salesman problem in which the number of possible routes through N cities is $N!$, EC will not explore all possible solutions. But if the topology of the fitness landscape is rough, evolution is an efficient way to find and explore fitness peaks.

Biological problems are naturals for EC. On one hand, this approach can be used to build tools to solve difficult problems, such as predicting protein or RNA folding, inferring phylogenies (Lewis 1998), or aligning DNA or protein sequences (Notredame & Higgins 1996). On the other hand, it can be used to model complex biological systems, such as the immune system, ecosystems, or cells (Adami 1998).

Genetic algorithms have found applications for some very practical problems, such as scheduling and constructing complex timetables. In such cases, there may be other methods that find equally good or even better solutions, but perhaps not with the same flexibility and speed. Genetic algorithms have also been applied to nonlinear filtering problems, such as processing signals from radar, sonar, and GPS satellites (Whitely et al. 2000).

Evolved programs tend to be more robust than user-created (written) programs. They can withstand more damage without total failure, a property that is analogous to biological homeostasis. The basis for robustness is not always clear, but at least in some cases, it is not merely redundancy in evolved programs (Masner et al. 2000). Different user-created programs may tend to use the same approach to a particular problem, but evolved solutions will tend to be unique. An important potential application of the robustness and the unique nature of evolved programs is fault tolerance. Where computer failure can have catastrophic consequences (i.e., on a space shuttle or the navigation system of an airplane), agreement between multiple independent programs is used to reduce the probability of failure of the system. But if all the programs have the same underlying "fault" because they were written according to the same programming principles, agreement may not assure against failure. Evolved programs may provide greater fault tolerance than do written programs. And it is not out of the question to imagine programs that evolve to repair themselves when they detect a fault (J.A. Foster, personal communication).

CONCLUSION

The examples presented here are but a small fraction of the applications of evolutionary biology to socially relevant issues. We have limited this review to examples in which evolutionary principles and methods have actually been used to solve problems. An even richer variety of problems is being approached with evolutionary perspectives, and some of these efforts are likely to yield fruit in the near future.

A research focus on "applied evolution" will perhaps prove to be an ephemeral one. It does not offer a conceptual organization for a discipline as broad as evolutionary biology. New applications will, of course, continue to arise and prosper, and applications may be useful in garnering funding and influencing students' career choices. But at the moment, the defining basis for applied evolution is ignorance. Many people, both scientists and the public, are unaware that evolutionary biology has become very relevant, yet attacks to suppress the teaching of evolution receive widespread support at the local level. We hope that ignorance of applications will be short-lived, as texts and the news media begin to disseminate the examples, and evolutionary biology should emerge in the future with widespread public acceptance. This review has focused on positive applications of evolution, but it is also important to understand these potential applications to guard against other uses of this technology, such as production of biological weapons of mass destruction or the evolution of damaging computer viruses. The time has arrived for wide public understanding of the importance and relevance of evolutionary biology in everyday lives.

ACKNOWLEDGMENTS

Many of the examples in this paper were identified through discussions with colleagues. We thank D. Hillis, F. Gould, A. Ellington, M. Robertson, B. Levin, L. Ancel, L. Vawter, W. Maddison, I. Matsumura, S. Palumbi, D. Futuyma, R. Bush, M. Courtney, I. Eckstranol, and J. Foster for discussions, references, or, in some cases, personal accounts. Mitch Sogin provided Figure 2. Editorial comments of D. Futuyma and D. Fautin helped improve the prose. The topics in this paper also overlap with a symposium we organized on applied evolution at the 2000 meetings of the Society for the Study of Evolution (Bloomington). We also are very grateful for the support of the NIH (GM38737 and GM 57756) and the NSF (DEB 9726902) during the time we prepared this review.

Visit the Annual Reviews home page at www.AnnualReviews.org

LITERATURE CITED

Adami C. 1998. *Artificial Life VI. Proc. 6th Int. Conf. Artif. Life.* Cambridge, MA: MIT Press. 498 pp.

Altschul SF, Gish W, Miller W, Myers EW, Lipman DJ. 1990. Basic local alignment search tool. *J. Mol. Biol.* 215:403–10

Arnold FH. 1997. Design by directed evolution. *Acc. Chem. Res* 31:125–31

Arnold FH, Volkov AA. 1999. Directed evolution of biocatalysts. *Curr. Opin. Chem. Biol.* 3:54–59

Axelrod R, Hamilton WD. 1981. The evolution of cooperation. *Science* 211:1390–96

Baquero F, Negri MC. 1997. Strategies to minimize the development of antibiotic resistance. *J. Chemother.* 9(Suppl. 3):29–37

Baquero F, Negri MC, Morosini MI, Blazquez J. 1997. The antibiotic selective process: concentration-specific amplification of low-level resistant populations. *Ciba Found. Symp.* 207:93–105

Baker CS, Palumbi SR. 1996. Population structure, molecular systematics, and forensic idenfication of whales and dolphins. In *Conservation Genetics: Case Histories from Nature*, ed. JC Avise, JL Hamrick, pp. 10–49. New York: Chapman & Hall

Baker CS, Lento GM, Cipriano F, Palumbi SR. 2000. Predicted decline of protected whales based on molecular genetic monitoring of Japanese and Korean markets. *Proc. R. Soc. London Ser. B* 267:1191–99

Baker RJ. 1994. Some thoughts on conservation, biodiversity, museums, molecular characters, systematics, and basic research. *J. Mammal.* 75:277–87

Battersby TR, Ang DN, Burgstaller P, Jurczyk SC, Bowser MT, et al. 1999. Quantitative analysis of receptors for adenosine nucleotides obtained via in vitro selection from a library incorporating a cationic nucleotide analog. *J. Am. Chem. Soc.* 121:9781–89

Beaudry AA, Joyce GF. 1992. Directed evolution of an RNA enzyme. *Science* 257:635–41

Benner SA. 1995. Predicting the conformation of proteins from sequences. Progress and future progress. *J. Mol. Recognit.* 8:9–28

Blondeau JM, Zhao X, Hansen G, Krlica K. 2001. Mutant prevention concentrations of fluoroquinolones for clinical isolates of *Streptococcus pneumoniae. Antimicrob. Agents Chemother.* 45:433–38

Bonhoeffer S, Lipsitch M, Levin BR. 1997. Evaluating treatment protocols to prevent antibiotic resistance. *Proc. Natl. Acad. Sci. USA* 94:12106–11

Boyd EF, Hartl DL. 1998. Diversifying selection governs sequence polymorphism in the major adhesin proteins fimA, papA, and sfaA of *Escherichia coli. J. Mol. Evol.* 47:258–67

Breaker RR, Joyce GF. 1994. A DNA enzyme that cleaves RNA. *Chem. Biol.* 1:223–29

Bull JJ, Badgett MR, Molineux IJ. 2001. A general mechanism for viral resistance to suicide gene expression. *J. Mol. Evol.* 53:47–54

Bull JJ, Jacobson A, Badgett MR, Molineux IJ. 1998. Viral escape from antisense RNA. *Mol. Microbiol.* 28:835–46

Bull JJ, Pease CM. 1995. Why is the polymerase chain reaction resistant to in vitro evolution? *J. Mol. Evol.* 41:1160–64

Bush RM, Bender CA, Subbarao K, Cox NJ, Fitch WM. 1999. Predicting the evolution of human influenza A. *Science* 286:1921–25

Chambers JK, Macdonald LE, Sarau HM, Ames RS, Freeman K, et al. 2000. A G protein-coupled receptor for UDP-glucose. *J. Biol. Chem.* 275:10767–71

Chien CT, Bartel PL, Sternglanz R, Fields S. 1991. The two-hybrid system: a method to identify and clone genes for proteins that interact with a protein of interest. *Proc. Natl. Acad. Sci. USA* 88:9578–82

Clutton-Brock J. 1999. *A Natural History of Domesticated Mammals*. Cambridge, UK: Cambridge Univ. Press. 238 pp.

Coffin JM. 1995. HIV population dynamics in vivo: implications for genetic variation, pathogenesis, and therapy. *Science* 267:483–89

Coffin JM. 1996. HIV viral dynamics. *AIDS* 10(Suppl. 3):S75–84

Colgrove R, Japour A. 1999. A combinatorial ledge: reverse transcriptase fidelity, total body viral burden, and the implications of multiple-drug HIV therapy for the evolution of antiviral resistance. *Antiviral Res.* 41:45–56

Crameri A, Raillard SA, Bermudez E, Stemmer WP. 1998. DNA shuffling of a family of genes from diverse species accelerates directed evolution. *Nature* 391:288–91

Crill WD, Wichman HA, Bull JJ. 2000. Evolutionary reversals during viral adaptation to alternating hosts. *Genetics* 154:27–37

Cristino JM. 1999. Correlation between consumption of antimicrobials in humans and development of resistance in bacteria. *Int. J. Antimicrob. Agents* 12:199–202

Crow JF, Kimura M. 1970. *An Introduction to Population Genetics Theory*. New York: Harper & Row. 591 pp.

DeBry RW, Seldin MF. 1996. Human/mouse homology relationships. *Genomics* 33:337–51

Djordjevic GM, Klaenhammer TR. 1997. Bacteriophage-triggered defense systems: phage adaptation and design improvements. *Appl. Environ. Microbiol.* 63:4370–76

Ellington AD. 1994. Empirical explorations of sequence space: host-guest chemistry in the RNA world. *Ber. Bunsenges. Phys. Chem.* 98:1115–21

Ellington AD, Szostak JW. 1990. In vitro selection of RNA molecules that bind specific ligands. *Nature* 346:818–22

Emini E. 1996. Non-nucleoside reverse transcriptase inhibitors—mechanisms. In *Antiviral Drug Resistance*, ed. DD Richman, pp. 225–40. New York: Wiley

Ewald PW. 1994. *Evolution of Infectious Disease*. New York: Oxford Univ. Press. 298 pp.

Fahy E, Kwoh DY, Gingeras TR. 1991. Self-sustained sequence replication (3SR): an isothermal transcription-based amplification system alternative to PCR. *PCR Methods Appl.* 1:25–33

Famulok M, Mayer G. 1999. Aptamers as tools in molecular biology and immunology. *Curr. Top. Microbiol. Immunol.* 243:123–36

Fitch WM, Bush RM, Bender CA, Cox NJ. 1997. Long term trends in the evolution of H(3) HA1 human influenza type A. *Proc. Natl. Acad. Sci. USA* 94:7712–18

Fogel LJ, Owens AJ, Walsh MJ. 1966. *Artificial Intelligence through Simulated Evolution*. New York: Wiley. 170 pp.

Foster JA. 2001. Evolutionary computation. *Nat. Rev. Genet.* 2:428–36

Fromant M, Blanquet S, Plateau P. 1995. Direct random mutagenesis of gene-sized DNA fragments using polymerase chain reaction. *Anal. Biochem.* 224:347–53

Futuyma D, ed. 1999. *Evolution, Science, and Society*. Rutgers: Off. Univ. Publ., State Univ. NJ. 46 pp.

Gao F, Bailes E, Robertson DL, Chen Y, Rodenburg CM, et al. 1999. Origin of HIV-1 in the chimpanzee *Pan troglodytes troglodytes*. *Nature* 397:436–41

Garrett L. 1994. *The Coming Plague: Newly Emerging Diseases in a World Out of Balance*. New York: Farrar Straus & Giroux. 750 pp.

Garrett L. 2000. *Betrayal of Trust. The Collapse of Global Public Health*. New York: Hyperion. 754 pp.

Gold L, Polisky B, Uhlenbeck O, Yarus M. 1995. Diversity of oligonucleotide functions. *Annu. Rev. Biochem.* 64:763–97

Golding GB, Dean AM. 1998. The structural basis of molecular adaptation. *Mol. Biol. Evol.* 15:355–69

Good PD, Krikos AJ, Li SX, Bertrand E, Lee

NS, et al. 1997. Expression of small, therapeutic RNAs in human cell nuclei. *Gene Ther.* 4:45–54

Gould F. 1998. Sustainability of transgenic insecticidal cultivars: integrating pest genetics and ecology. *Annu. Rev. Entomol.* 43:701–26

Gould F, Anderson A, Jones A, Sumerford D, Heckel DG, et al. 1997. Initial frequency of alleles for resistance to *Bacillus thuringiensis* toxins in field populations of *Heliothis virescens. Proc. Natl. Acad. Sci. USA* 94:3519–23

Granizo JJ, Aguilar L, Casal J, Dal-Re R, Baquero F. 2000. *Streptococcus pyogenes* resistance to erythromycin in relation to macrolide consumption in Spain (1986–1997). *J. Antimicrob. Chemother.* 46:959–64

Greensfelder L. 2000. Infectious diseases. Polio outbreak raises questions about vaccine. *Science* 290:1867–69

Gutell RR, Cannone JJ, Konings D, Gautheret D. 2000. Predicting U-turns in ribosomal RNA with comparative sequence analysis. *J. Mol. Biol.* 300:791–803

Hahn BH, Shaw GM, De Cock KM, Sharp PM. 2000. AIDS as a zoonosis: scientific and public health implications. *Science* 287:607–14

Hardin G. 1968. The tragedy of the commons. *Science* 162:1243–48

Higgins DG, Sharp PM. 1988. CLUSTAL: a package for performing multiple sequence alignment on a microcomputer. *Gene* 73:237–44

Hillis DM, Huelsenbeck JP. 1994. Support for dental HIV transmission. *Nature* 369:24–25

Hillis DM, Huelsenbeck JP, Cunningham CW. 1994. Application and accuracy of molecular phylogenies. *Science* 264:671–77

Holland JH. 1975. *Adaptation in Natural and Artificial Systems.* Ann Arbor: Univ. Mich. Press. 183 pp.

Jaeger L. 1997. The new world of ribozymes. *Curr. Opin. Struct. Biol.* 7:324–35

Kauffman SA. 1993. *The Origins of Order: Self Organization and Selection in Evolution.* New York: Oxford Univ. Press. 709 pp.

Klussmann S, Nolte A, Bald R, Erdmann VA, Furste JP. 1996. Mirror-image RNA that binds D-adenosine. *Nat. Biotechnol.* 14:1112–15

Korber B, Muldoon M, Theiler J, Gao F, Gupta R, et al. 2000. Timing the ancestor of the HIV-1 pandemic strains. *Science* 288:1789–96

Koza JR. 1992. *Genetic Programming: On the Programming of Computers by Natural Selection.* Cambridge, MA: MIT Press

Kroes I, Lepp PW, Relman DA. 1999. Bacterial diversity within the human subgingival crevice. *Proc. Natl. Acad. Sci. USA* 96:14547–52

Landweber LF. 1999. Experimental RNA evolution. *TREE* 14:353–58

Langdon WB, Soule T, Poli R, Foster JA. 1999. The evolution of size and shape. In *Advances in Genetic Programming*, ed. L Spector, WB Langdon, U-M O'Reilly, PJ Angeline, pp. 162–91. Cambridge, MA: MIT Press

Lappé M. 1994. *Evolutionary Medicine: Rethinking the Origins of Disease.* San Francisco: Sierra Club Books. 255 pp.

Lee DH, Severin K, Yokobayashi Y, Ghadiri MR. 1997. Emergence of symbiosis in peptide self-replication through a hypercyclic network. *Nature* 390:591–94

Lewis PO. 1998. A genetic algorithm for maximum-likelihood phylogeny inference using nucleotide sequence data. *Mol. Biol. Evol.* 15:277–83

Lewontin RC. 1970. The units of selection. *Annu. Rev. Ecol. Syst.* 1:10–18

MacArthur RD. 2000. Sequencing antiretrovirals. *AIDS Read.* 10:359–64

Marrs B, Delagrave S, Murphy D. 1999. Novel approaches for discovering industrial enzymes. *Curr. Opin. Microbiol.* 2:241–45

Masner J, Cavalieri J, Frenzel J, Foster JA. 2000. Size versus robustness in evolved sorting networks: Is bigger better? In *Proc. NASA/DoD Workshop Evolvable Hardware*, pp. 81–90. Piscataway, NJ: IEEE

Matsushita S. 2000. Current status and future issues in the treatment of HIV-1 infection. *Int. J. Hematol.* 72:20–27

McDonald JH, Kreitman M. 1991. Adaptive

protein evolution at the Adh locus in Drosophila. *Nature* 351:652–54

Morse SS. 1994. *The Evolutionary Biology of Viruses.* New York: Raven. 353 pp.

Mosier DE. 2000. Virus and target cell evolution in human immunodeficiency virus type 1 infection. *Immunol. Res.* 21:253–58

Nathanson N, McGann KA, Wilesmith J. 1995. The evolution of virus diseases: their emergence, epidemicity, and control. In *Molecular Basis for Viral Evolution*, ed. A Gibbs, CH Calisher, F Garcia-Arenal, pp. 31–46. Cambridge, UK: Cambridge Univ. Press

Nesse RM, Williams GC. 1994. *Why We Get Sick: The New Science of Darwinian Medicine.* New York: Times Books. 291 pp.

Newhouse JR. 1990. Chestnut blight. *Sci. Am.* 262:106–11

Notredame C, Higgins DG. 1996. SAGA: sequence alignment by genetic algorithm. *Nucleic Acids Res.* 24:1515–24

O'Meara D, Wilbe K, Leitner T, Heideman B, Albert J, Lundeberg J. 2001. Monitoring resistance to human immunodeficiency virus type 1 protease inhibitors by pyrosequencing. *J. Clin. Microbiol.* 39:464–73

Osborne SE, Ellington AD. 1997. Nucleic acid selection and the challenge of combinatorial chemistry. *Chem. Rev.* 97:349–70

Ou CY, Ciesielski CA, Myers G, Bandea CI, Luo CC, et al. 1992. Molecular epidemiology of HIV transmission in a dental practice. *Science* 256:1165–71

Pace NR. 1997. A molecular view of microbial diversity and the biosphere. *Science* 276:734–40

Palumbi SR. 2001. *The Evolution Explosion.* New York: Norton. 288 pp.

Relman DA. 1998. Detection and identification of previously unrecognized microbial pathogens. *Emerg. Infect. Dis.* 4:382–89

Relman DA. 1999. The search for unrecognized pathogens. *Science* 284:1308–10

Relman DA, Falkow S. 1992. Identification of uncultured microorganisms: expanding the spectrum of characterized microbial pathogens. *Infect. Agents Dis.* 1:245–53

Rico-Hesse R, Pallansch MA, Nottay BK,

Kew OM. 1987. Geographic distribution of wild poliovirus type 1 genotypes. *Virology* 160:311–22

Ryan F. 1993. *The Forgotten Plague: How the Battle Against Tuberculosis was Won—and Lost.* Boston: Little, Brown

Sakthivel K, Barbas CF. 1998. Expanding the potential of DNA for binding and catalysis: highly functionalized dUTP derivatives that are substrates for thermostable polymerases. *Angew. Chem. Int. Ed.* 37:2872–75

Sanger F, Air GM, Barrell BG, Brown NL, Coulson AR, et al. 1977. Nucleotide sequence of bacteriophage ϕX174 DNA. *Nature* 265:687–95

Santiago FS, Lowe HC, Kavurma MM, Chesterman CN, Baker A, et al. 1999. New DNA enzyme targeting Egr-1 mRNA inhibits vascular smooth muscle proliferation and regrowth after injury. *Nat. Med.* 5:1264–69

Santoro SW, Joyce GF. 1997. A general purpose RNA-cleaving DNA enzyme. *Proc. Natl. Acad. Sci. USA* 94:4262–66

Schmidt-Dannert C, Arnold FH. 1999. Directed evolution of industrial enzymes. *Trends Biotechnol.* 17:135–36

Smith GP. 1985. Filamentous fusion phage: novel expression vectors that display cloned antigens on the virion surface. *Science* 228:1315–17

Soule T, Foster JA. 1998. Limiting code growth in genetic programming. *Evol. Comput.* 6:293–310

Soyfer V. 1994. *Lysenko and the Tragedy of Soviet Science.* New Brunswick, NJ: Rutgers Univ. Press. 379 pp.

Spiegelman S, Pace NR, Mills DR, Levisohn R, Eikhorn TS, et al. 1968. The mechanism of RNA replication. *Cold Spring Harbor Symp. Quant. Biol.* 33:101–24

Stemmer WP. 1994. Rapid evolution of a protein in vitro by DNA shuffling. *Nature* 370:389–91

Tabashnik BE, Patin AL, Dennehy TJ, Liu YB, Carriere Y, et al. 2000. Frequency of resistance to *Bacillus thuringiensis* in field populations of pink bollworm. *Proc. Natl. Acad. Sci. USA* 97:12980–84

Tarasow TM, Tarasow SL, Eaton BE. 1997. RNA-catalysed carbon-carbon bond formation. *Nature* 389:54–57

Tatusov RL, Koonin EV, Lipman DJ. 1997. A genomic perspective on protein families. *Science* 278:631–37

Tobin MB, Gustafsson C, Huisman GW. 2000. Directed evolution: the "rational" basis for "irrational" design. *Curr. Opin. Struct. Biol.* 10:421–27

Toh KL, Jones CR, He Y, Eide EJ, Hinz WA, et al. 2001. An hPer2 phosphorylation site mutation in familial advanced sleep phase syndrome. *Science* 291:1040–43

Trevathan W, Smith EO, McKenna JJ. 1999. *Evolutionary Medicine.* New York: Oxford Univ. Press. 480 pp.

Tuerk C, Gold L. 1990. Systematic evolution of ligands by exponential enrichment: RNA ligands to bacteriophage T4 DNA polymerase. *Science* 249:505–10

Ucko PJ, Dimbleby GW. 1969. *The Domestication and Exploitation of Plants and Animals.* Chicago: Aldine. 581 pp.

Venter JC, Adams MD, Myers EW, Li P, Mural R, et al. 2001. The sequence of the human genome. *Science* 291:1304–51

Vogel D. 1995. *Evolutionary Computation: Toward a New Philosophy of Machine Intelligence.* Piscataway, NJ: IEEE

Voigt CA, Kauffman S, Wang ZG. 2000. Rational evolutionary design: the theory of in vitro protein evolution. *Adv. Protein Chem.* 55:79–160

Webster RG. 1993. Influenza. In *Emerging Viruses*, ed. SS Morse, pp. 37–45. Oxford, UK: Oxford Univ. Press

Whaley SR, English DS, Hu EL, Barbara PF, Belcher AM. 2000. Selection of peptides with semiconductor binding specificity for directed nanocrystal assembly. *Nature* 405:665–68

Whitely D, Goldberg D, Cantú-Paz E, Spector L, Parmee I, Beyer H-G, eds. 2000. *Proceedings of the Genetic and Evolutionary Computation Conference*, San Francisco: Morgan Kauffman

Wiegand TW, Janssen RC, Eaton BE. 1997. Selection of RNA amide synthases. *Chem. Biol.* 4:675–83

Williams KP, Liu XH, Schumacher TN, Lin HY, Ausiello DA, et al. 1997. Bioactive and nuclease-resistant L-DNA ligand of vasopressin. *Proc. Natl. Acad. Sci. USA* 94:11285–90

Wilson DS, Keefe AD, Szostak JW. 2001. The use of mRNA display to select high-affinity protein-binding peptides. *Proc. Natl. Acad. Sci. USA* 98:3750–55

Woese CR. 2000. Interpreting the universal phylogenetic tree. *Proc. Natl. Acad. Sci. USA* 97:8392–96

Annu. Rev. Ecol. Syst. 2001. 32:219–49
Copyright © 2001 by Annual Reviews. All rights reserved

MISTLETOE—A KEYSTONE RESOURCE IN FORESTS AND WOODLANDS WORLDWIDE

David M. Watson

The Johnstone Centre and Environmental Studies Unit, Charles Sturt University, Bathurst New South Wales 2795, Australia; e-mail: dwatson@csu.edu.au

Key Words   Loranthaceae, Viscaceae, frugivore, plant-animal interactions

■ **Abstract**   Mistletoes are a diverse group of parasitic plants with a worldwide distribution. The hemiparasitic growth form is critical to understanding their biology, buffering variation in resource availability that constrains the distribution and growth of most plants. This is manifested in many aspects of mistletoe life history, including extended phenologies, abundant and high-quality fruits and nectar, and few chemical or structural defenses. Most mistletoe species rely on animals for both pollination and fruit dispersal, and this leads to a broad range of mistletoe-animal interactions. In this review, I summarize research on mistletoe biology and synthesize results from studies of mistletoe-animal interactions. I consolidate records of mistletoe-vertebrate interactions, incorporating species from 97 vertebrate families recorded as consuming mistletoe and from 50 using mistletoe as nesting sites. There is widespread support for regarding mistletoe as a keystone resource, and all quantitative data are consistent with mistletoe functioning as a determinant of alpha diversity. Manipulative experiments are highlighted as a key priority, and six explicit predictions are provided to guide future experimental research.

> *The facts which kept me longest scientifically orthodox are those of adaptation—the pollen-masses in Asclepias—the misseltoe, with its pollen carried by insects and seed by Birds—the woodpecker, with its feet and tail, beak and tongue, to climb the tree and secure insects. To talk of climate or Lamarckian habit producing such adaptation to other organic beings is futile. This difficulty, I believe I have surmounted.*
>
> From a letter to Asa Gray by Charles Darwin, 1857.

INTRODUCTION AND SCOPE

Interactions between mistletoes and animals have long been noted and were used by Darwin as early exemplars of evolutionary adaptation (Burkhardt & Smith 1990, p. 445). Linné is credited as the first to describe mistletoe life history (Landell 1998), noting that thrushes ate the berries and expelled the sticky seeds upon subsequent perches. Pliny recorded similar observations some 1600 years earlier (Pliny & Rackham 1960). Indeed, most species of mistletoe are dispersed by animals, chiefly birds (Calder 1983, Hawksworth 1983, Kuijt 1969, Reid 1986, Reid 1991, Snow & Snow 1988); and this close relationship has been treated as a model system in the study of fruit dispersal generally (Howe & Estabrook 1977, Howe & Smallwood 1982, Herrera 1985a,b, McKey 1975, Snow 1971, Wheelwright 1988). The various reviews of avian use of mistletoe are limited either to particular regions—India (Davidar 1978, 1985), North America (Hawksworth & Geils 1996, Stoner 1932), Australia (Reid 1986, Turner 1991), Latin America (Restrepo 1987, Sargent 1994, Skutch 1980)—or concerned solely with frugivory (Reid 1991, Snow & Snow 1988). As Darwin noted (see Burkhardt & Smith 1990), mistletoes are also pollinated by animals (mainly birds and insects; Davidar 1985, Ford et al. 1979, Penfield et al. 1976, Reid 1986, Whittaker 1984), but this aspect of their life history has received markedly less attention than fruit dispersal. Indeed, the majority of records of mistletoe-animal interactions have been incidental; a wealth of anecdotal information is contained in species-specific accounts, autecological studies, and works of natural history. Drawing on these highly dispersed data is challenging, and although similar compendia exist for other plant groups (e.g., lichens, Sharnoff & Rosentreter 1998; palms, Zona & Henderson 1989), the breadth of mistletoe-animal interactions worldwide has not been documented nor fully appreciated.

I begin with an introduction to mistletoes, summarize diversity and distribution patterns, and combine these accounts with information about life history, phenology, pollination, and dispersal syndromes. Drawing on the extensive autecological and natural history literature, I then consolidate known interactions between mistletoe and vertebrates worldwide. I propose that mistletoes function as keystone resources in forests and woodlands of many regions, providing important resources for a broad range of taxa and determining local diversities in these habitats. After evaluating this hypothesis with available data, I identify gaps in our knowledge and make explicit predictions to guide future research.

Incorporating all mistletoe-animal interactions, however, would greatly exceed the scope of this review. Therefore, I restrict my focus to vertebrates, although I recognize that this does not correspond with the breadth of known interactions. Many insects pollinate mistletoes (primarily in the Coleoptera, Diptera, Hymenoptera, and Lepidoptera; Hawksworth & Wiens 1996, Whittaker 1984); more than 200 insect species are documented pollinating a single mistletoe species (Penfield et al. 1976). Many species within the Coleoptera, Diptera, Hemiptera, Homoptera, Hymenoptera, Lepidoptera, Orthoptera, and Thysanoptera feed on mistletoe, and all orders contain mistletoe-obligate species (de Baar 1985, Hawksworth & Wiens 1996, Mushtaque & Baloch 1979, Whittaker 1984). Other than lists of species known to associate with mistletoes, there has been relatively little research on mistletoe-insect interactions (Whittaker 1984) and no comparative or synthetic studies. Indeed, compared with that on vertebrates, the literature is far from complete and any review of the subject would be premature. Mistletoe-vertebrate interactions clearly are only a partial indication of the role mistletoes play in forested ecosystems, and understanding the extent and nature of mistletoe-insect interactions represents a major challenge for future research.

MISTLETOE—AN OVERVIEW

Composition and Distribution

Mistletoes are a polyphyletic group of flowering plants comprising over 1300 species from a broad range of habitats across all continents except Antarctica (Calder 1983, Calder & Bernhardt 1983, Kuijt 1969, Watson & Dallwitz 1992, Nickrent 2001). They share a common growth form—obligate hemiparasitism—such that all water and minerals are obtained from their host via a specialized vascular attachment (Ehleringer & Marshall 1995, Lamont 1983b, 1985, Pate 1995). The group contains members of five families within the Santales (Kuijt 1968, 1969, Nickrent & Soltis 1995), and the aerial parasitic life-form is thought to have evolved independently four or five times (Nickrent & Franchina 1990, Nickrent et al. 1998, Nickrent 2001, Figure 1). Thus, the term mistletoe does not refer to a lineage of plants, but a functional group (like mangroves), and will be used hereafter to denote all hemiparasitic species within the Santales. Two of the constituent families—the poorly known Misodendronaceae and Eremolepidaceae— are restricted to neotropical forests, while the aerial parasitic genera within the paraphyletic Santalaceae are known from tropical forests in Latin America and Southeast Asia (Kuijt 1968, 1969, Watson & Dallwitz 1992). In contrast, the Loranthaceae and Viscaceae are well studied and distributed worldwide, comprising the majority (>98%) of mistletoe species: approximately 940 and 350 species, respectively (Watson & Dallwitz 1992, Nickrent 2001). These families are not sister taxa, and their aerial parasitic growth-form is thought to have evolved independently (Calder 1983, Kuijt 1969, Nickrent et al. 1998, Nickrent 2001, Figure 1). Based on fossils from the Cretaceous period, and the occurrence of relictual genera in Australia, New Zealand, and South America (Barlow 1983, Kuijt 1969), Loranthaceae is considered a Gondwanan lineage that subsequently dispersed to Africa, Europe, and North America (Barlow 1983, Polhill & Wiens 1998, Raven & Axelrod 1974). Conversely, Viscaceae is thought to have originated in eastern Asia, radiating through Laurasia in the early Tertiary period, secondarily dispersing to the southern continents (Barlow 1983, Raven & Axelrod 1974).

The Loranthaceae and Viscaceae are presently distributed widely throughout Europe, the Americas, Africa, Asia, and Australasia (except Tasmania), ranging from boreal climates to temperate, tropical, and arid zones, and absent only from extremely dry or cold regions (Barlow 1983, Kuijt 1969, Raven & Axelrod 1974). They are also well represented on oceanic islands, with the Azores, Madagascar, Aldabra, Comoros Islands, Mascarene Islands, Galapagos Islands, Hawaiian Islands, New Caledonia, Lord Howe Island, Norfolk Island, New Zealand, Fijian Islands, Henderson Island, Greater and Lesser Antilles, and Hispaniola all having

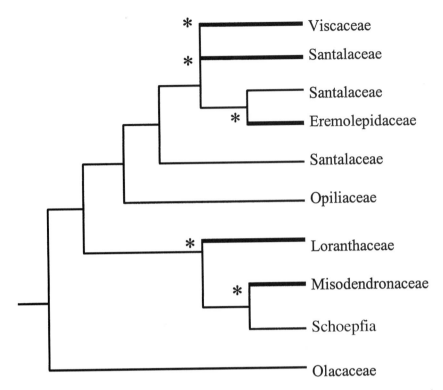

Figure 1   Consensus tree for the Santales based on data from nuclear 18S rDNA and chloroplast *rbcL* sequences (modified from Nickrent; 2001). Branches with thick lines indicate mistletoe taxa, and asterisks indicate hypothesized independent evolutionary origins of the aerial hemiparasitic habit.

representative (and principally endemic) species (Barlow 1983, Hawksworth & Wiens 1996, Kuijt 1969, Polhill & Wiens 1998). Although mistletoes are typically considered forest plants, they parasitize a wide range of hosts—coniferous trees in boreal forests (Hawksworth 1983, Hawksworth & Wiens 1996), succulent euphorbs and cacti in the deserts of Africa and Latin America (Martinez et al. 1996, Polhill & Wiens 1998), grasses and annual herbs in Australian coastal heathlands (Fineran & Hocking 1983), even orchids and ferns in Mesoamerican cloud forests (Kuijt & Mulder 1985). Most species, however, selectively parasitize trees and shrubs, and the greatest diversities of mistletoes are associated with forests and woodlands (Calder 1983, Hawksworth 1983, Kuijt 1969).

Life History and Phenology

Mistletoes are characterized by their growth habit and, excepting several root-parasitic species (Kuijt 1969, Fineran & Hocking 1983), they typically form dense clumps in the crowns of their hosts (Figures 2*a* and *c*, see color insert). These clumps are generally composed of semisucculent mistletoe stems and leaves (Figures 2*a* and *b*, see color insert), but for dwarf mistletoes (Viscaceae, *Arceuthobium* spp.), the resultant clump (termed a witch's broom) is actually the host tree's response to infection composed of thickened and twisted branches (Hawksworth & Wiens 1996, Schaffer et al. 1983; Figures 2*c* and *d*, see color insert). Although mistletoe plants are frequently regarded as detrimental to tree health (Coleman 1949, Hawksworth 1983, Heather & Griffin 1978, Reid et al. 1994), this is not necessarily the case (Reid & Lange 1988, Reid et al. 1992, Sterba et al. 1993). Mistletoe plants have low annual survivorship (Musselman & Press 1995), with estimates of 19% and 31% for two Australian species (Reid & Lange 1988), and are considered "low-grade K-selected forest parasites" (Andrews & Rouse 1982). Many species are sensitive to fire (Rowe 1983, Hawksworth & Wiens 1996, Kipfmueller & Baker 1998) and frost (Hudler & French 1976, Smith & Wass 1979); these factors are cited as limiting the distribution of mistletoe in some areas (Hawksworth 1969, Hawksworth & Wiens 1996, Spooner 1983). Mistletoe seeds require high light levels for germination, establishment, and subsequent maturation (Knutson 1983, Lamont 1982, 1983a, Polhill & Wiens 1998), and they are frequently shaded out as the host canopy develops (Lamont 1982, Richards 1952). Thus, despite high rates of dispersal and successful germination, establishment is rare, and all mistletoe species studied have narrow microsite tolerances (Hawksworth & Wiens 1996, Knutson 1983, Lamont 1982, 1983a, Overton 1994, Sargent 1994, 1995, Yan & Reid 1995).

Mistletoes obtain all of their water and minerals from the host through a vascular connection termed a haustorium (Calder 1983, Press & Graves 1995; Figure 2*b*, see color insert). This swollen holdfast serves both to attach the mistletoe plant to the host and to divert water and minerals to the parasite. The term hemiparasitic is used because most mistletoes photosynthesize, although they may obtain up to 60% of their carbohydrates from the host (Hull & Leonard 1964, Lamont 1983b). The consequence of this growth-form is that mistletoes are less affected by the edaphic, hydrological, and nutritional factors that limit the distribution, growth, and phenology of most plants; the host plant buffers the parasite against large-scale fluctuations in resource availability (Ehleringer & Marshall 1995).

Restrepo's (1987) research in Colombian cloud forests revealed that five mistletoe species had continuous fruiting seasons with a generalized peak during the wet season. Flowering seasons were similarly continuous, with nectar and fruit resources available year-round. Data from elsewhere in the Neotropics fit the same pattern, flowering and fruiting seasons ranging from 10 to 12 months (Feinsinger 1978, Leck 1972, Sargent 1994, Skutch 1980, Stiles 1985) with a pronounced wet-season peak. Mistletoes in southern Africa also have prolonged fruiting seasons, ranging from three months to year-round (Godschalk 1983a, Polhill & Wiens 1998), while those in Burkhina Faso exhibit staggered fruiting patterns throughout the year (Boussim 1991). Dwarf mistletoes in the Palearctic have more limited flowering and fruiting seasons, lasting from 4 to 6 months; flowering typically peaks in early to mid autumn, with peak fruiting in late autumn to early winter (Hawksworth & Wiens 1996). Davidar (1983) recorded phenologies for six mistletoe species in southern India, with flowering ranging from 3 to 9 months and fruiting seasons of from 4 to 7 months. These species displayed complementary peaks in fruit and nectar availability such that nectar and fruit were available from at least two species at any time. Reid (1986) summarized phenologies of 13 Australian loranthaceous species and demonstrated that the flowering season for several species extended throughout the year, with different patterns associated with regional climates. A pronounced summer peak in wetter areas graded into a winter peak in drier areas, corresponding to seasons when little nectar is available from other sources. Fruiting was similarly widespread throughout the year, and Reid (1986) noted that there was always nectar and fruit available from at least one mistletoe species within a region.

Fruit Dispersal and Pollination

Although several lineages of mistletoe (notably within the Viscaceae) use hydrostatic explosion to disperse seeds, birds play a subsequent role in long-distance dispersal (Hawksworth & Geils 1996) and mistletoes generally are regarded as bird-dispersed (Calder 1983, Hawksworth 1983, Kuijt 1969, Reid 1986). A recent evaluation of New World mistletoes (Restrepo et al, in press) suggests that vertebrate dispersers may have played a key role in the diversification of mistletoes—those lineages dispersed by vertebrates having higher diversities. Most studies of mistletoe dispersal have focused on a small number of highly specialized birds (Reid 1989, 1990, Sargent 1994, Snow & Snow 1984, Walsberg 1977). This research has involved documenting the pattern (Godschalk 1983, Liddy 1983, Snow & Snow 1988), assessing the physiological, anatomical, and behavioral traits associated with obligate frugivory (Reid 1989, Richardson & Wooller 1988, Snow & Snow 1984, Walsberg 1975), and evaluating the coevolutionary potential of such interactions (Reid 1987, 1991, Restrepo et al, in press; Snow 1971).

The fruits of many mistletoe species show a range of attributes consistent with ornithochory: large, sweet, conspicuous in color when ripe, with small peduncles or borne directly on the branch. Fruit pulp compositions vary, but most have high fractions of soluble carbohydrates, typically from 40% to 60% dry weight (Godschalk 1983b, Lamont 1983b, Restrepo 1987) but as high as 74% (Snow & Snow 1988). Fruits of some loranthaceous species are high in lipids (up to 35% dry mass; Godschalk 1983b, Restrepo 1987), with fruits of *Loranthus europaeus* containing droplets of pure fat (Chiarlo & Cajelli 1965). Fruits of viscaceous species tend to have much higher fractions of protein than other fruits (Wheelright et al. 1984)—an average of 22% in three species of *Phoradendron* (Restrepo 1987)—comprising up to 18 free amino acids (Chiarlo & Cajelli 1965, Godschalk 1983b). Indeed, detailed biochemical analyses of mistletoe fruits have identified all 10 essential amino acids as originally defined by Rose and coworkers (Rose et al. 1948, Womack & Rose 1947), especially arginine, lysine, and phenylalanine (Bushueva et al. 1990, Chiarlo & Cajelli 1965, Godshalk 1983b). As with other mistletoe tissues, fruits also contain high concentrations of minerals (notably P and K) and micronutrients (e.g., Mn and Fe) (Lamont 1983b, Pate 1995). Despite the poisonous reputation of mistletoe, toxins have been isolated from only a small number of species within the Viscaceae, and toxicity is rare within mistletoes generally (Barlow & Wiens 1977, Bushueva et al. 1990). Many mistletoe species display discontinuous ripening such that individual plants have ripe fruits available throughout the fruiting season, with the timing of peak-fruiting varying between individual plants and separate populations (Reid 1986, Restrepo 1987). Combining this within-species variation with the between-species complementarity noted earlier, fruit (and floral) resources are often available year-round within a given region (Boussim 1991, Davidar 1983, Polhill & Wiens 1998, Reid 1986)

Whereas most species of mistletoe are pollinated by birds, members of the Viscaceae are pollinated primarily by wind and insects (Hawksworth & Wiens 1996, Kuijt 1969, Polhill & Wiens 1998, Whittaker 1984). Whether anemophily or entomophily is the prime means of pollen transfer remains controversial, with many species displaying a mosaic of features that variously favors the two agents, and most species-specific studies have implicated both (Hawksworth & Wiens 1996, Penfield et al. 1976). The more speciose Loranthaceae are pollinated primarily by birds (Calder 1983, Davidar 1985, Kuijt 1969, Ladley et al. 1997, Reid 1986, Robertson et al. 1999) and display the typical suite of characteristics associated with ornithophilous pollination—large odorless flowers that are typically brightly colored (yellow, orange, red) with robust corollas, short pedicels, and often in massed inflorescences (Reid 1986, Watson & Dallwitz 1992; see Figure 2b, see color insert). Nectar secretions are typically abundant and rich in sugars—glucose, sucrose, and especially fructose (up to 60% total sugar content; Baker et al. 1998, Reid 1986, Stiles & Freeman 1993). Unlike fruit dispersers, no species (of bird, insect, or other group) is known to be wholly dependent on mistletoe nectar as a primary food source (Davidar 1983, 1985, Reid 1986). Rather, a wide range of species pollinates mistletoes, some of which may depend on the nectar during particular seasons (Davidar 1985, Feinsinger 1978, Ford et al. 1979, Ladley et al. 1997, Stiles 1985, Robertson et al. 1999, Watson 1997).

MISTLETOE-VERTEBRATE INTERACTIONS

Mistletoe as a Food Source

The importance of mistletoe as a food source was assessed for all birds and mammals, with information on other groups included incidentally. I systematically surveyed dietary information in handbooks, species accounts, and autecological studies for records of species consuming mistletoe nectar, flowers, fruit, seeds, or foliage. I restricted this review to the family level, indicating the recorded frequency of mistletoe consumption within the family and providing a species exemplar (Table 1). Whereas few families are likely to have been neglected, this summary is

TABLE 1   List of families recorded feeding on mistletoe

Family	Feeding	Frequency	Species exemplar	Citation
Characinidae	F	R	*Triportheus angulatus*	Goulding 1980
Casuariidae	F	R	*Casuarius casuarius*	Bentrupperbaumer 1997
Dromaiidae	F	—	*Dromaius novaehollandiae*	Berney 1907
Tinamidae	L	R	*Eudromia elegans*	Bohl 1970
Columbidae	F	S	*Columba flavirostris*	Skutch 1983
Cacatuidae	F	S	*Callocephalon fimbriatum*	Coleman 1949
Psittacidae	F, Fl	S	*Pionopsitta haematotis*	Eitnier et al. 1994
Loriidae	N	S	*Glosopsitta porphyrocephala*	Paton & Ford 1977
Musophagidae	F	V	*Tauraco corythaix*	Godschalk 1986
Cuculidae	F	S	*Scythrops novaehollandiae*	Lord 1956
Cracidae	F	R	*Ortalis vetula*	Lopez de Buen & Ornelas 1999
Tetraonidae	F, L	S	*Bonasa umbellus*	Skinner 1928
Odontophoridae	F	S	*Callipepla gambelii*	Rosenberg et al. 1991
Phasianidae	F, L	S	*Lophura leucomelana*	Ali & Ripley 1978
Trochilidae	N	C	*Heliodaxa jacula*	Stiles 1985
Coliidae	F	V	*Colius indicus*	Godschalk 1986
Trogonidae	F	R	*Trogon violaceus*	Dickey & van Rossem 1978
Bucerotidae	F	R	*Anthracoceros malayanus*	McConkey 1999
Ramphastidae	F	S	*Semnornis frantzii*	Sargent 1994
Lybiidae	F	C	*Pogoniulus bilineatus*	Van Someren 1956
Indicatoridae	F, I	S	*Prodotiscus zambesiae*	Fry et al. 1988
Picidae	F	C	*Colaptes auratus*	Rosenberg et al. 1991
Eurylaimidae	F, N	R	*Philepitta castanea*	Prum & Razafindrasita 1997
Tyrannidae	F, I	C	*Tyranniscus vilissimus*	Leck 1972
Cotingidae	F	C	*Phibalura flavirostris*	Snow 1982
Oxyruncidae	F	—	*Oxyruncus cristatus*	Nadkarni & Matelson 1989
Pipridae	F	S	*Manacus vitellinus*	Leck 1972
Pardalotidae	N, I	S	*Acanthiza uropygialis*	Reid 1986
Meliphagidae	F, N, I	C	*Grantiella picta*	Reid 1986
Pachycephalidae	F, N, I	S	*Mohoua ochrocephala*	O'Donnell & Dilks 1989
Vireonidae	F	S	*Vireo olivaceus*	Leck 1972
Oriolidae	F	R	*Oriolus sagittatus*	Liddy 1982
Artamidae	N	S	*Artamus superciliosus*	Barker & Vestjens 1990
Cracticidae	F	S	*Strepera graculina*	Reid 1986
Ptilonorhynchidae	F	C	*Chlamydera nuchalis*	Barker & Vestjens 1990
Corvidae	F	S	*Perisoreus canadensis*	Punter & Gilbert 1989
Paridae	F, S, I	C	*Parus caeruleus*	Heine de Balsac & Mayaud 1930
Remizidae	F	R	*Auriparus flaviceps*	Restrepo et al., unpublished data
Aegithalidae	F	R	*Psaltriparus minimus*	Sutton 1951
Sittidae	F	R	*Sitta canadensis*	Punter & Gilbert 1989
Certhiidae	F	R	*Certhia americana*	Punter & Gilbert 1989

TABLE 1 *(Continued)*

Family	Feeding	Frequency	Species exemplar	Citation
Troglodytidae	F	R	*Campylorhynchus brunneicapillus*	Austin 1970
Campephagidae	F	R	*Lalage leucomela*	Crome 1978
Pycnonotidae	F	S	*Pycnonotus capensis*	Godschalk 1986
Chloropseidae	N	S	*Chloropsis aurifrons*	Ali & Ripley 1996
Sylviidae	F, I	R	*Sylvia atricapilla*	Hein de Balsac 1930
Turdidae	F	C	*Sialia mexicanus*	Rosenberg et al. 1991
Timaliidae	N, F	S	*Turdoides squamiceps*	Cramp & Perrins 1993
Dicaeidae	F, N, I	V	*Dicaeum cruentatum*	Ali & Ripley 1999
Melanocharitidae	N	S	*Toxorhamphus poliopterus*	Mack & Wright 1996
Nectariniidae	N, I	C	*Nectarinia lotenia*	Ali & Ripley 1999
Zosteropidae	F	S	*Zosterops lateralis*	Barker & Vestjens 1990
Mimidae	F	C	*Mimus polyglottos*	Rosenberg et al. 1991
Sturnidae	F	S	*Cinnyricinclus leucogaster*	Godschalk 1986
Prunellidae	F	R	*Prunella collaris*	Cramp 1988
Bombycillidae	F	C	*Bombycilla cedrorum*	Rosenberg et al. 1991
Ptilogonatidae	F, N	V	*Ptilogonys cinereus*	Sutton 1951
Parulidae	F, I	S	*Dendroica petechia*	Ostry & Nichols 1979
Thraupidae	F, N	V	*Euphonia lauta*	Sutton 1951
Coerebidae	F, N	—	*Coereba flaveola*	Snow & Snow 1971
Emberizidae	F	S	*Junco hymenalis*	Punter & Gilbert 1989
Cardinalidae	F	S	*Pheucticus melanocephalus*	Marshall 1957
Icteridae	N	S	*Icterus bullockii*	Howell 1972
Fringillidae	F	S	*Carpodacus mexicanus*	Weathers 1983
Passeridae	F	R	*Passer montanus*	Cramp & Perrins 1994
Ploceidae	F	R	*Malimbus rubriceps*	Godschalk 1986
Estreldidae	F	R	*Lonchura punctulata*	Alam & Rahman 1988
Microbiotheriidae	F	—	*Dromiciops australis*	Amico & Aizen 2000
Phascolarctidae	L	—	*Phascolarctos cinereus*	H.E. Young, unpublished data
Phalangeridae	L, Fl	S	*Ailurops ursinus*	Dwiyarheni et al. 1999
Pseudocheiridae	L, Fl	S	*Pseudocheirus peregrinus*	Choate et al. 1987
Petauridae	Fl	S	*Petaurus australis*	Reid 1986
Pteropodidae	F	S	*Rousettus aegyptiacus*	Herzig-Streschil & Robinson 1978
Lemuridae	F, Fl, N	C	*Eulemur fulvus*	Overdorff 1993
Indriidae	L, F	C	*Propithecus diadema*	Hemmingway 1998
Cebidae	L, F	S	*Alouatta palliata*	Stoner 1996
Callitrichidae	F	C	*Saguinus fuscicollis*	Soini 1987
Cercopithecidae	L, Fl	C	*Cercopithecus aethiops*	Kavanagh 1978
Hylobatidae	L, Fl	R	*Hylobates mulleri* x *agilis*	McConkey 1999
Pongidae	L	C	*Gorilla gorilla*	Goodall 1977

(Continued)

TABLE 1   *(Continued)*

Family	Feeding	Frequency	Species exemplar	Citation
Hominidae	F, L	—	*Homo sapiens*	Morgan 1981
Procyonidae	F	S	*Bassariscus astutus*	Taylor 1954
Mustelidae	F	S	*Martes americana*	Nichols et al. 1984
Elephantidae	W	R	*Loxodonta africana*	Bax 1963
Equidae	L	S	*Equus asinus*	Fowler de Neira & Johnson 1985
Rhinocerotidae	W	R	*Diceros bicornis*	Goddard 1968
Suidae	L	R	*Sus scrofa*	Knott 1908
Camellidae	L	R	*Camellus camellus*	Morgan 1981
Giraffidae	L	R	*Giraffa camelopardalis*	Wiens 1978
Cervidae	W	C	*Oedocoileus hemionus*	Riney 1951
Antilocapridae	L	—	*Antilocapra americana*	Russel 1964
Bovidae	L	S	*Ovis canadensis*	Halloran & Crandell 1953
Sciuridae	F, S	S	*Tamiasciurus hudsonicus*	Nichols et al. 1984
Heteromyidae	F, S	S	*Perognathus parvus*	Burt 1934
Muridae	F, S, L	S	*Neotoma stephensi*	Vaughan 1982
Anomaluridae	F	R	*Anomalurus derbianus*	Kingdon 1974
Erethizontidae	W	R	*Erethizon epixanthum*	Taylor 1935

In the feeding mode column, F = fruit, Fl = flower, N = nectar, S = seed, W = whole plant, and I = insects on mistletoe. In terms of frequency, R denotes rare within the family, S denotes several records, C denotes common in family, V is very common and — is for monotypic families.

only as complete as the available dietary information. Such data are patchy, with the diets of species from some areas (e.g., Australia, North America, Europe) very well described compared with other regions (notably southeast Asia, Africa, and Latin America). In addition to regional variation, there is a taxonomic bias, with the diets of some groups (e.g., primates, galliformes) well studied compared with others. Thus, rather than considering this summary an exhaustive list, it is more appropriately viewed as an indication of the breadth of taxa that include mistletoe in their diets.

Species from 66 families of birds and 30 families of mammals have been recorded consuming mistletoe, spanning 12 and 10 orders, respectively. I also encountered a record of the Amazonian characinoid fish *Triportheus angulatus* feeding on mistletoe fruits during high seasonal floods (Goulding 1980). Frugivory and nectarivory were the most common modes of consumption among birds, whereas folivory and frugivory accounted for most of the mammalian records. Although it is generally accepted that all mistletoe-fruit specialists are birds, recently published information (Amico & Aizen 2000) suggests that the monito del monte *Dromiciops australis*—sole living member of the ancient marsupial order Microbiotheria—acts as exclusive disperser of mistletoe fruits in southern Andean cloud forests. Amico & Aizen (2000) speculated that this interaction may be indicative of an early association between mistletoes and marsupials, with both taxa originating in Gondwanaland. Although many reptiles are frugivorous or folivorous (Rand 1978), dietary composition data are incomplete for most species, and I did not find reliable records of mistletoe in the diet of any wild reptilian.

Given the year-round availability of mistletoe fruits in many regions, there is a diverse range of species that depend on mistletoe when little else is available. Many of these consumers may act as occasional dispersers, but most are more appropriately viewed as fruit predators. The diversity of these opportunistic frugivores was unexpected, given the prevailing view that mistletoes have highly specific dispersal systems, refined over evolutionary time to deter generalist frugivores (Herrera 1985a, McKey 1975, Reid 1991, Wheelwright 1988, but see Moermond & Denslow 1985). Reid (1991) noted that fruits of many mistletoe species are large, sticky, and relatively cryptic—attributes that would exclude most potential dispersers in favor of a small group of specialists (McKey 1975, Wheelwright 1988). Data summarized here suggest that many organisms regularly consume mistletoe fruit, indicating that it may not be as difficult to locate and process as previously considered. These two ideas are not mutually incompatible, the widespread consumption of mistletoe fruit is independent of the reciprocal specificity between mistletoes and dispersers as long as these coevolved specialists remain the primary dispersers. Available data suggest this is the case, with mistletoe plants in many regions reliant on a small number of specialized frugivorous birds for successful dispersal (Davidar 1987, Godschalk 1983a, 1986, Heine de Blasac & Mayaud 1930, Liddy 1983, Reid 1986, 1989, 1990, Restrepo 1987, Sargent 1994, Snow & Snow 1984, 1988, Walsberg 1975; but see Punter & Gilbert 1989).

Mistletoes also provide abundant nectar, and in addition to many nectarivorous species (Davidar 1983, Ladley et al. 1997, Reid 1986, Snow & Snow 1971, Stiles 1985), a broad range of insectivorous and generalist species have been recorded feeding from mistletoe flowers (Barker & Vestjens 1989, 1990, Crome 1978, O'Donnell & Dilks 1989, Reid 1986, Rosenberg et al. 1991). Several mammals are also known to feed on mistletoe nectar (more commonly consuming the entire flower; Dwiyarheni et al. 1999, Hemmingway 1998, Kavanagh 1978, McConkey 1999, Overdorff 1993). There are no confirmed records of mammals functioning as pollinators.

The leaves of mistletoe plants are considered highly favored browse in dietary studies of many folivorous mammals ranging from deer and rhinoceroses to gorillas and possums (Table 1). Brushtail possums, *Trichosaurus vulpeca*, introduced from Australia are thought to have decimated populations of endemic mistletoe in New Zealand (Norton 1991, Ogle & Wilson 1985), although declines in pollinating bird species may also be involved (Robertson et al. 1999). Indeed, mammalian folivory has been proposed as the primary selective force driving host-mimicry patterns displayed by many mistletoes (Barlow & Wiens 1977, Choate et al. 1987, Kavanagh & Lambert 1990, Pasteur 1982, Wiens 1978), but there are several other competing hypotheses (Atsatt 1977, 1983, Canyon & Hill 1997). No mammal is known to specialize on mistletoe foliage, but several species appear to be seasonally dependent on it (Quinton & Horejsi 1977, Riney 1951). There are several records of birds feeding on mistletoe leaves (Ali & Ripley 1978, Bohl 1970, Skinner 1928, Takatsukasa 1967), but folivory generally is rare in the group (Morton 1978).

Mistletoe clumps are often used as a foraging substrate by insectivorous species. As indicated earlier, a diverse range of insects is associated with mistletoes as both pollinators and herbivores, and mistletoe clumps often have abundant insect assemblages (Bennetts 1991, de Baar 1985, Mushtaque & Baloch 1979, Whittaker 1984). Insectivorous birds including several highly specialized species (e.g., *Prodotiscus regulus, Arachnothera longirostris*) (Ali & Ripley 1999, Fry et al. 1988) use mistletoe clumps for foraging activities (Bennetts 1991, Heine de Balsac & Mayaud 1930, O'Donnell & Dilks 1989, Turner 1991). Foraging substrate is rarely recorded in dietary studies of insectivores, however, and the extent of this foraging mode is presumably much greater than the 10 recorded families. There is also a record of a predatory lizard, *Varanus albigularis*, using mistletoe clumps as sites for hunting birds (Rose 1962). Probably a rare instance, this further highlights the importance and complexity of the habitat provided by mistletoes.

Mistletoes as Nest Sites

Mistletoes are used extensively as sites for nesting and roosting, and whereas many researchers have reported this for individual species (Bull et al. 1989, Farentinos 1972, Ralph 1975, Reynolds et al. 1982, Skutch 1969, Thompson & Owen 1964, Weathers 1983) or regions (Fry et al. 1988, Mamone 1996, North 1906, Parks et al. 1999, van Someren 1956), use of mistletoe for nest sites has never been assessed generally. I reviewed the literature on life histories of birds and mammals, collating records of species recorded using mistletoe as nesting/roosting sites. Again, I restricted this review to the family-level, with representative species listed in Table 2. Nesting records in witches' brooms as well as regular mistletoe clumps

TABLE 2   Families recorded nesting and roosting in mistletoe clumps

Family	Nesting frequency	Species exemplar	Citation
Ardeidae	R	*Egretta novaehollandiae*	Marchant & Higgins 1990
Threskiornithidae	R	*Threskiornis molucca*	Marchant & Higgins 1990
Accipitridae	C	*Accipiter cooperi*	Reynolds et al. 1982
Tetraonidae	S*	*Dendrapagus obscurus*	Pekins et al. 1991
Phasianidae	S*	*Syrmaticus soemmerringi*	Takatsukasa 1967
Alcidae	R	*Brachyramphus marmoratus*	Nelson 1997
Columbidae	S	*Zenaida macroura*	Bennetts & Hawksworth 1992
Musophagidae	V	*Corythaeola cristata*	Fry et al. 1988
Cuculidae	S	*Coccyzus pumilus*	Ralph 1975
Neomorphidae	S	*Geococcyx californianus*	Weathers 1983
Strigidae	C	*Asio otus*	Bull et al. 1989
Trochilidae	R	*Stellula calliope*	Bent 1940

TABLE 2   *(Continued)*

Family	Nesting frequency	Species exemplar	Citation
Coliidae	C	*Colius striatus*	Van Someren 1956
Tyrannidae	S	*Camptostoma imberbe*	Bleitz 1955
Cotingidae	R	*Pachyramphus polychopterus*	Skutch 1969
Maluridae	R	*Malurus lamberti*	McDonald 1973
Acanthizidae	R	*Acanthiza chrysorrhoa*	Beruldson 1980
Eupetidae	V	*Psophodes cristatus*	McDonald 1973
Meliphagidae	C	*Prosthemadera novaeseelandiae*	Soper 1976
Pachycephalidae	S	*Pachycephala rufiventris*	McDonald 1973
Laniidae	R	*Lanius ludovicianus*	Weathers 1983
Malaconotide	S	*Chlorophoneus sulfureopectus*	Harris & Franklin 2000
Vireonidae	R	*Vireo huttoni*	Davis 1995
Artamidae	S	*Artamus tenebrosus*	North 1906
Cracticidae	S	*Cracticus torquatus*	North 1906
Ptilonorhynchidae	C	*Ptilonorhynchus violaceus*	McDonald 1973
Corvidae	S	*Cyanocitta stelleri*	Brandt 1951
Remizidae	R	*Auriparus flaviceps*	Jaeger 1947
Aegithalidae	R	*Psaltriparus minimus*	Bendire 1887
Troglodytidae	S	*Campylorhynchus brunneicapillus*	Anderson & Anderson 1973
Campephagidae	R	*Lalage sueurii*	Beruldson 1980
Sylviidae	R	*Polioptila melanura*	Rosenberg et al. 1991
Muscicapidae	R	*Muscicapa adusta*	Van Someren 1956
Turdidae	S	*Catharus guttatus*	Bennetts et al. 1992
Mimidae	S	*Mimus polyglottos*	Sutton 1967
Ptilogonatidae	C	*Phainopepla nitens*	Rea 1983
Parulidae	R	*Dendroica nigrescens*	Guzy & Lowther 1997
Thraupidae	S	*Piranga rubra*	Brandt 1951
Emberizidae	S	*Pipilo aberti*	Finch 1985
Cardinalidae	S	*Cardinalis sinuatus*	Bleitz 1955
Icteridae	R	*Icterus bullockii*	Rosenberg et al. 1991
Fringillidae	S	*Carpodacus cassinii*	Bennetts et al. 1992
Estreldidae	S	*Emblema guttata*	North 1906
Erethizodontida	R[*]	*Erethizon epixanthum*	Smith 1982
Pseudocheiridae	R	*Pseudocheirus peregrinus*	Thompson & Owen 1964
Sigmodontidae	S	*Neotoma floridanum*	Mamone 1996
Arvicolidae	S	*Arborimus longicaudus*	Mamone 1996
Mustelidae	R[*]	*Martes americana*	Parks & Bull 1997
Phascolarctidae	—[*]	*Phascolarctos cinereus*	S.J. Cox, unpublished observation
Sciuridae	C	*Sciurus aberti*	Farentinos 1972

In the nesting frequency column, R denotes rare within the family, S denotes several records, C denotes common in family, V is very common with the family and — is for monotypic families.

[*]Denotes record is only of roosting—all other records pertain to nesting (rearing young in a nest).

are included. Information on nesting is highly variable across regions. Thus, although there have been several separate treatments of the nests and eggs of Australian birds (e.g., Beruldson 1980, Campbell 1900, North 1906), there is not a single such publication for Asia, Africa, or Latin America. As more research is conducted in these regions, species from many other families will probably be added and the list provided here is clearly provisional.

Species from 43 families of birds and 7 families of mammals have been recorded using mistletoe as a nesting or roosting site (Table 2). No groups are obligate mistletoe nesters, but species from several avian lineages (notably Coccyzinae, Musophagidae, Striginae, Accipitrinae, Ptilogonatidae, and Ptilonorhynchidae) favor nesting in mistletoe (Beruldson 1980, Fry et al. 1988, North 1906, Parks et al. 1999). It is unclear which attributes of mistletoe clumps are most important in nest-site selection, but many authors comment on their dense, evergreen habit and the enlarged host branches associated with the haustorium (Bleitz 1955, Jaeger 1947, McDonald 1973, Rosenberg et al. 1991, Sutton 1967, van Someren 1956). For example, long-eared owls (*Asio otus*) use witches' brooms as the primary structural support for their stick nests, with one study finding 19 of 20 nests associated with mistletoe (Bull et al. 1989). Alternatively, smaller nests can be concealed within foliose mistletoe clumps, a strategy used by many passerines (Anderson & Anderson 1973, Bennetts et al. 1996, Bleitz 1955, Brandt 1961, McDonald 1973, North 1906). In addition to such structural factors, microclimate may also be involved. Porcupines (*Erethizon epixanthum*), pine martens (*Martes americana*), and several species of squirrel have been recorded using witches' brooms as hibernaculae (Mamone 1996, Parks et al. 1999, Parks & Bull 1997, Smith 1982, Taylor 1935), and several species of birds and mammals shelter in mistletoe clumps during extremely hot weather (Brandt 1951, Jaeger 1947, Stoner 1932, SJ Cox, unpublished observations). Other than this anecdotal information, there has been only one study on the nesting of organisms in mistletoe clumps (Parks et al. 1999), so the relative importance of structure and microclimate cannot be assessed.

In addition to using mistletoe as a nest site, there are records from Africa and North America of birds using fresh mistletoe sprigs as nest lining. Whereas the behavior of using green foliage as a nest lining is widespread in birds, at least seven species of raptor have been recorded selecting viscaceous mistletoe as nest lining, often replacing it daily (Brandt 1951, Fry et al. 1988, van Someren 1956). There is evidence that extracts of *Viscum* spp. have antibacterial activity (Grainge & Ahmed 1988), and clinical trials have found some of these chemicals to act as immunostimulants (Fischer et al. 1997, Rentea et al. 1981, Stoffel et al. 1997, Wagner & Proksch 1985). Experimental studies with common starlings (*Sturnus vulgaris*) indicate that fledglings have higher immune function in the presence of various aromatic herbs in the nest; Gwinner et al. (2000) speculated that mistletoe foliage may have a similar effect. Other than isolated records in species accounts, this intriguing behavior has been completely overlooked and merits greater attention. The fact that it has been recorded solely from raptors and scavengers further suggests that viscaceous mistletoe may have a key role in nest hygiene, and that the popularity of mistletoe as a nest site may have a biochemical component in addition to structure and microclimate.

Mistletoe infection has been associated with the formation of hollows and snags (Bennetts 1991, Bennetts et al. 1996, Hawksworth & Wiens 1996), which are limiting resources in many forests (Raphael & White 1984). Mistletoes are considered critical in creating hollows used by a broad range of species for nesting and roosting (Bennetts et al. 1996, Hawksworth & Geils 1996, Mamone 1996, Parks et al. 1996), and influencing larger-scale distributions (Bennetts 1991, Bennetts et al. 1996). A close correlation was found between intensity of mistletoe infection and diversity and abundance of cavity-nesting birds (Bennetts 1991, Bennetts & Hawksworth 1992, Bennetts et al. 1996), suggesting that mistletoe may be important both directly and indirectly in providing nesting and roosting locations for vertebrates.

HYPOTHESIS

Having summarized the life-history of mistletoes and documented the breadth of interactions between mistletoe and vertebrates throughout the world, I propose the following general hypothesis:

Mistletoes function as keystone resources in many forests and woodlands worldwide.

This hypothesis will be evaluated with existing data (both qualitative and quantitative) to allow the first explicit assessment of the importance of mistletoe in forests and woodlands worldwide. Although several researchers have noted the importance of mistletoe in particular regions (Anderson et al. 1978, Bennetts 1991, Reid 1986, Snow & Snow 1988, Turner 1991), the generality of this phenomenon has apparently never been proposed nor evaluated.

This hypothesis builds on work by Terborgh (1986), in which he documented the importance of figs, palm nuts, and nectar to a wide range of vertebrates in neotropical forests. His paper expanded on earlier research by Leighton & Leighton (1983), who proposed that "the reproductive biology of figs makes them uniquely suited to play roles as keystone mutualists for many vertebrates"—one of the earliest examples of applying the "keystone species" concept to a suite of resources. Terborgh (1986) estimated that this resource-base, despite accounting for less than 1% of local plant diversity, supported the majority of frugivorous vertebrates during seasonal periods of scarcity in two neotropical forests. He concluded that figs and other fruit resources were of critical importance to most frugivores, and that fruits function as keystone resources in tropical forests generally.

MISTLETOE AS A KEYSTONE RESOURCE

Since its introduction by Paine (1969), the concept of an ecological keystone has been applied to a wide variety of groups spanning predators, prey, parasitoids, modifiers, links, and resources. Some authors (e.g., Mills et al. 1993) have suggested that the term has become so widely used that it has become too generic and unwieldy. Provided explicit criteria are used to define them, keystones remain useful heuristics both in comparing the structure and function of communities and in identifying priority groups for directed management. Power and associates (1996, p. 609) defined a keystone as a group "whose impact on its community or ecosystem is large and disproportionately large relative to its abundance." Mills and coauthors (1993) stated that an important component of defining keystones was interaction strength, measured by quantifying the community-wide effect of keystone removal that "would likely precipitate loss of obligate and possibly opportunistic users" (p. 220, after Redford 1984).

There have been several small-scale removal experiments to assess the effect of mistletoe infection on various host parameters, but the level of treatment was the tree (Reid et al. 1992, 1994, Shea 1964, Sterba et al. 1993). These data cannot be used to quantify the strength of interactions at the community level, nor address the hypothesis of mistletoe as a keystone resource in forested habitats. This lack of explicit data notwithstanding, the keystone hypothesis can still be assessed. It was primarily on the basis of observational and qualitative data that Terborgh (1986) proposed that figs be considered a keystone resource in tropical forests, with subsequent studies and reviews lending support (Antsett et al. 1997, Nason et al. 1998, but see Gautier-Hion & Michaloud 1989). I examine two well-studied habitats as case studies, summarize relevant data on the role of mistletoes in both systems, and evaluate the validity of the keystone hypothesis.

Case Study 1: Mesquite Woodland of the Southwestern USA

This semiarid habitat is dominated by shrubs and small trees, principally mesquite (*Prosopis* spp.), acacia (*Acacia* spp.), and creosote bush (*Larrea tridentata*), intergrading with different associations in riparian zones and upland areas (Anderson et al. 1979, Blake 1984, Jaeger 1947, Rosenberg et al. 1991, Stamp 1978). Of the two viscaceous genera in the region—*Arceuthobium* and *Phoradendron*—the former is restricted to conifers and is rare to absent in mesquite woodlands. *Phoradendron* species parasitize a broad range of host trees, with *P. californicum* and *P. tomentosum* most common on mesquite trees. Insects (chiefly Diptera and Hymenoptera) are the main pollen vectors (Whittaker 1984), with birds acting as principal fruit dispersers (Rea 1983, Rosenberg et al. 1991, Spooner 1983). The phainopepla (*Phainopepla nitens*) relies on *Phoradendron* berries as its main food source (Walsberg 1975), and several other species (western bluebird *Sialia mexicanus*, cedar waxwing *Bombycilla cedrorum*, northern mockingbird *Mimus polyglottos*, and several species of woodpecker) are partially dependent on the fruits (Rea 1983, Rosenberg et al. 1991, Sutton 1967). Although these species act as primary dispersal agents, many other insectivorous and generalist bird species have been recorded consuming the fruits. Few other fruit resources are available during the winter (Anderson et al. 1979, Austin 1970, Rice et al. 1981), so some birds actively defend mistletoe clumps (Walsberg 1977). During summer, mistletoe fruits are also widely consumed as a source of water (Jaeger 1947, Walsberg 1975). The fruits and leaves are consumed by several mammals (e.g., pronghorn antelope *Antilocapra americana*—Russell 1964; mule deer *Oedocoileus hemionus*—Riney 1951; Great Basin pocket mouse *Perognathus parvus*—Burt 1934), and also feature in the diets of indigenous people as both food and medicine (Curtin 1949, Timbrook 1990). Some mammals appear to be seasonally dependent on mistletoe: *Phoradendron* seeds dominate scats of cacomistles (*Bassariscus astutus*) during peak fruiting (Taylor 1954) and according to Quinton & Horejsi (1977), *P. tomentosum* foliage comprizes up to 65% of the winter diet of white-tailed deer (*Oedocoileus virginianus*). *Phoradendron* clumps are used by many species for nesting (Bleitz 1955, Brandt 1951, Rea 1983, Weathers 1983) and are regarded as preferred nesting sites for several species. A study of phainopeplas in Arizona found 80% of nests in mesquites, of which 80% were in or under mistletoe clumps (Rea 1983); 18% of cactus wren (*Campylorhynchus brunneicappilus*) roosting nests were located in *P. californicum* clumps (Anderson & Anderson 1973). In a study of Abert's towhee (*Pipilo aberti*), 50% of nests were within *P. californicum* clumps, their evergreen habit allowing breeding to commence earlier; the author considered them the most desirable nest site (Finch 1985). Moreover, several birds of prey commonly nest in mistletoe clumps, with other species recorded using fresh sprigs of *Phoradendron* as nest lining (Brandt 1951, Stoner 1932).

Many authors have commented on the interactions between vertebrates and *Phoradendron* in mesquite woodlands, with several spatial and temporal associations between mistletoe and species distributions noted. Austin (1970) reported an increase in abundance of several avian species in response to a major fruiting event of *P. californicum*, and Anderson & Ohmart (1978) and Rice et al. (1981) found *P. californicum* to be a critical habitat component for 15 of 20 species. They noted that *P. californicum* berries constituted the only reliable source of fruit in the area, and thus they determined the distribution of frugivores in mesquite woodlands. This finding was confirmed by Rosenberg et al. (1991), who found the berries to dominate the diets of 11 species and to be of particular importance during winter. Density of *Phoradendron* plants in this habitat varies, with reported values ranging from 1 to 17 plants per hectare in one study (Blake 1984), with another reporting from 0.5% to 8.7% of trees contained mistletoe plants (Stamp 1978).

Case Study 2: Eucalypt Forests of Southeastern Australia

These sclerophyllous forests are found throughout eastern Australia, grading into woodlands and savannas inland. The canopies are dominated by *Eucalyptus* spp., with smaller trees and shrubs (e.g., *Acacia* spp., *Leptospermum* spp., *Casuarina* spp.) often forming a subcanopy, and different associations in gulleys and drier slopes (Smith 1984, Thompson & Owen 1964, Turner 1991). More than 20 loranthaceous species of mistletoe are known from this region, of which seven widespread species commonly parasitize eucalypts [*Amyema bifurcatum, A. miquelii, A. pendulum, Dendrophthoe glabrescens, D. vitellina, Diplatia grandibractea,* and *Muellerina eucalyptoides*; Downey (1998)]. All species are bird-pollinated (Ford et al. 1979, Reid 1986), with insects also playing a role for some species. In addition to nectarivores (primarily in the Meliphagidae, Paton & Ford 1977, Reid 1986), a broad range of birds has been recorded feeding opportunistically on mistletoe flowers (Barker & Vestjens 1989, 1990, Ford et al. 1979, Reid 1986, Turner 1991), especially during summer when nectar availability in these forests is lowest (Paton & Ford 1977). Birds also act as sole fruit dispersers, with two mistletoe-dependent specialists (mistletoebird *Dicaeum hirundaceum* and painted honeyeater *Grantiella picta*) inhabiting these forests. Many other species have been recorded feeding on the fruits (Barker & Vestjens 1989, 1990, Reid 1986) and mistletoe is one of the few reliable fruit sources in this region (Reid 1986). Several mammals regularly consume mistletoe foliage and flowers (Canyon & Hill 1997, Choate et al. 1987, Kavanagh & Lambert 1990, Reid 1986), and may reduce the abundance of mistletoe in some areas. Mistletoes are also used as a foraging substrate by many insectivorous species, and probably represent a concentration of insects (Turner 1991). Mistletoe clumps are noticeably denser than eucalypt foliage (see Figure 2*a*, see color insert) and are frequently used as nest sites by a broad range of species (Beruldson 1980, Campbell 1900, North 1906, Thompson & Owen 1964). While mostly opportunistic, some species appear to prefer nesting in mistletoe clumps—Ford (1999) reported 28% of noisy friarbird and 29% of red wattlebird nests in northern New South Wales were inside mistletoe clumps.

Numerous qualitative datasets have been collected demonstrating a close temporal and spatial relationship between faunal distributions and mistletoe in eucalypt forests, for both particular species (Thompson & Owen 1964, Watson 1997) and entire assemblages (Liddy 1983, Smith 1984, Turner 1991). Thompson & Owen (1964) found ringtail possums (*Pseudocheirus peregrinus*) nested almost exclusively within mistletoe clumps and reported a close correlation in the spatial distribution of the two. Turner (1991) described the importance of mistletoe as a food resource for birds, accounting for the majority of foraging records in terms of both individuals and species. There was a significant relationship between density of mistletoe plants and number of species observed foraging in foliage ($N = 18$, $R = 0.564$, $P \simeq 0.007$ (one-tailed), recalculated from Turner 1991), with almost four times more records in mistletoe than in eucalypts. This involved nectarivores, frugivores, and insectivores, and Turner (1991) proposed that mistletoe be regarded as a critical resource for birds in eucalypt forests. Mistletoe density in eucalypt forests ranges from fewer than three to more than one hundred plants per hectare. These high values are from degraded or highly perturbed habitat (Heather & Griffin 1978, Norton & Stafford Smith 1999); values from intact forest are typically fewer than 10 plants per hectare (Turner 1991).

Inference—Mistletoe as a Keystone Resource

In terms of the qualitative criterion of Power et al. (1996), there is unambiguous support for treating mistletoe as a keystone resource in both habitats. Mistletoe is uncommon in both systems and can be considered a minor to extremely minor vegetational component in terms of abundance, species richness, and biomass. The plants provide a broad range of high-quality nutritional resources that support several obligate and many partially dependent species, especially during periods of seasonal scarcity. They also create habitat used by many species for nesting and roosting. As such, mistletoe is considered a keystone resource in both habitats, having a large and disproportionate impact on both.

The similar role mistletoes play in these two habitats is striking, given the broad differences between the habitats. One is a semiarid, shrub-dominated habitat in which the viscaceous mistletoes are mostly insect-pollinated. The other is temperate mesic forest, with loranthaceous mistletoes pollinated primarily by birds. Ongoing research in the coniferous forests of the Rocky Mountains has also focused on the role of mistletoe (Bennetts 1991, Bennetts & Hawksworth 1992, Bennetts et al. 1996, Hawksworth & Geils 1996) and comprises the most detailed research on the effects of mistletoe density on vertebrate diversity. Bennetts and associates (1996) reported dwarf mistletoe density to be a highly significant predictor of avian richness, positively affecting the abundance of 24 of 28 bird species. Mistletoe density was also correlated with number of snags, cavity nests, and total number of nests and was identified as the single most important variable affecting avian diversity in these boreal forests.

Further case studies supporting the keystone resource hypothesis could have been drawn from studies of mistletoe-vertebrate interactions in savannas of southern Africa (Dean et al. 1994, Fry et al. 1988, Godschalk 1983a,b, 1986, van Someren 1956), acacia shrublands in central Australia (Barker & Vestjens 1989, 1990, Beruldson 1980, North 1906, Reid 1989, 1990), tropical forests in India (Ali & Ripley 1996, 1999, Davidar 1978, 1983, 1985, Mushtaque & Baloch 1979), cactus-dominated deserts in Chile (Martinez del Rio et al. 1995, 1996), or subtropical cloud-forests in Costa Rica (Feinsinger 1978, Sargent 1994, 1995, Skutch 1969, 1980, 1983, Stiles 1985). Thus, given the widespread qualitative and quantitative support, the conclusion that mistletoes function as keystone resources can be provisionally accepted and adopted as a working hypothesis to guide further research.

Beyond Evaluation—Mechanisms and Causation

Although several studies have found greater vertebrate richnesses associated with higher mistletoe densities, few have examined the mechanistic basis of the pattern. In a study of woodland remnants in southern Australia (Watson 1994, Watson et al. 2000), mistletoe density contributed to a vegetation classification that explained the distribution of 12 of 29 common woodland bird species. Avian richnesses increased with mistletoe density until a threshold of approximately 20 plants per ha was reached. Patches with higher densities (up to 93 plants per ha) were characteristically heavily grazed and contained depauperate avian assemblages (Watson 1994, Watson et al. 2000). This pattern has been noted by researchers in other areas (Heather & Griffin 1978, Norton et al. 1995, Norton & Reid 1997, Norton & Stafford Smith 1999), indicating that the mistletoe-diversity association is not a simple linear relationship. In addition to measuring mistletoe density, Watson

(1994) also quantified mistletoe richness and presented the only dataset permitting analysis of the effect of mistletoe richness on vertebrate diversity. The 27 woodland remnants contained between 0 and 4 species of mistletoe, with an associated increase in avian richness (ANOVA, df $= 26; 4, F = 3.875, P \simeq 0.012$, recalculated from Watson 1994). Mean richness of birds in patches with four species of mistletoe was 63% greater than in patches with a single species, suggesting that mistletoe richness may contribute to the mistletoe-diversity patterns noted earlier. As well as increasing net availability of resources, an increase in mistletoe richness may expand the temporal extent of resource availability, given interspecies variation in phenology described earlier.

Although more mistletoes may entail more resources that can potentially support greater richnesses, this assumes no interaction between the consumers. Several researchers have noted birds actively defending fruiting mistletoe plants (Snow & Snow 1984, Walsberg 1977), driving alpha diversities down in the immediate area. Moreover, in many ecosystems the patterns of causal interdependence are unclear. Are areas with more mistletoe more attractive to vertebrates and hence more diverse, or is mistletoe secondarily introduced into diverse areas by seed-dispersing species? For example, data from southern Africa indicate a close relationship between mistletoe distribution and soil type (Dean et al. 1994), revealing mistletoes are more common on more fertile soils. Research in Australia and New Zealand has demonstrated the complex interaction of factors that influence mistletoe distribution, mediated both by environmental factors and interactions with hosts, pollinators, and fruit dispersers (Norton et al. 1995, Norton & Reid 1997). With changes related to European settlement, mistletoe has become more common in eastern Australia (Heather & Griffin 1978, Reid et al. 1994), less common in New Zealand (Ogle & Wilson 1985, Norton 1991), and either absent or superabundant in forest fragments in western Australia (Norton et al. 1995). Associated with these ecosystem-scale changes in vegetation cover have been concomitant changes in the distribution and abundance of animals (Norton & Reid 1997) that may or may not be related to mistletoe distribution.

Although identifying the causal influence of mistletoe on diversity remains elusive, it is clear that mistletoe does have an important role in many habitats throughout the world. As a direct source of nutritional resources, as a provider of nesting and roosting microhabitats, and as an indirect modifier of habitat structure, mistletoes have a pervasive effect in many forests and woodlands. Having identified mistletoes as keystone resources, the next step is to explore the underlying mechanisms and further our understanding of their role in forested ecosystems worldwide.

PRIORITIES FOR FUTURE RESEARCH

Whereas the vast majority of studies that have included mistletoe have found it important in the structure and function of forest and woodland communities, exceptions may emerge as more research is conducted. Tropical regions, in particular, are underrepresented in the mistletoe literature, and it is unclear if mistletoe is as

Figure 2 (*A*) Large *Amyema pendulum* parasitizing eucalyptus tree, New South Wales, Australia. Photograph by author. (*B*) *Psittacanthus* spp. parasitising an oak tree in Oaxaca, Mexico. Note the abundant colorful flowers with short pedicels typical of Loranthaceae, and haustorium at junction with host. Photograph by author. (*C*) Coniferous trees displaying the characteristic dense clumps associated with dwarf mistletoe infection known as witches' brooms. Oregon, USA, photograph by D.L. Nickrent. (*D*) Detail of a dead witch's broom on a pine tree, caused by *Arceuthobium tsugense*. Note the thickened, dense branches associated with the infection, used as a nesting/roosting site by a variety of animals in this habitat. Oregon, USA, photograph by D.L. Nickrent.

important in structuring these highly diverse ecosystems as in less diverse temperate areas. Despite the occurrence of mistletoe on many oceanic islands, the role of mistletoes in these simpler communities is not known. Multitaxon studies are rare, and it is unknown how mistletoe affects diversity patterns of different groups within the same habitat. Many of these gaps can be addressed by modifying ongoing studies. Of the many studies examining diversity patterns of selected faunal groups in forests and woodlands throughout the world, almost all collect a range of habitat metrics, but mistletoe is rarely included. Given the breadth of interactions with pollinators, dispersers, and herbivores, mistletoe density depends on an array of factors and represents a sensitive and accurate assay of many community-level effects (Anderson et al. 1979, Norton et al. 1995, Norton & Reid 1997, Robertson et al. 1999). By incorporating measurements or estimates of mistletoe density into inventory-based field studies, the representation of habitats and faunal groups will be expanded greatly, yielding a broader understanding of mistletoes as keystone resources.

To complement correlational data collected by descriptive studies, manipulative experiments are essential to measure the direct effects of mistletoe on diversity and ecosystem function. Although transplanting is not possible and inoculation difficult, mistletoe plants are relatively easy to remove allowing "replicated" patches of forest to be manipulated and subsequently monitored. As Bennetts et al. (1996) discussed, this is not practical for some habitats where mistletoe infection and resultant habitat change may take centuries. There are other habitats (including the two used here as case studies) in which manipulative experiments would be tractable, with several studies at the individual-tree scale demonstrating the potential feasibility of such an approach (Reid et al. 1992, 1994, Sterba et al. 1993).

To guide future experimental studies, a series of explicit predictions can be made regarding the long-term consequences of removing mistletoes from forested habitats. Compared with control plots (with the usual density and diversity of mistletoe plants), test plots (in which all mistletoe plants have been removed) would be expected to have:

1. lower abundances of mistletoe-obligate frugivores and folivores, with local populations declining toward local extinction;
2. lower abundances of regular mistletoe foragers (folivores, frugivores, and nectarivores);
3. fewer fallen branches, hollows, and snags over the long term;
4. lower abundances of birds and mammals that nest in mistletoe clumps and hollows;
5. lower richness of vertebrates generally; and
6. communities with increased sensitivity to drought and other rare events.

In addition to these research questions, a range of management-oriented applications remain unexplored. Mistletoe density in fragmented habitats is often unusually high or low, revealing the disturbed nature of these landscapes (Heather & Griffin 1978, Norton & Reid 1997, Norton et al. 1995). Could this imbalance be corrected by removing excess plants, or selectively inoculating trees in areas where mistletoe has been extirpated? Some researchers have advocated mistletoe as an integral part of revegetation programs (Anderson et al. 1979). Similar approaches could be extended to commercial plantations, introducing mistletoe to enhance their value for native fauna. Although this may not be applicable to many systems, data from a commercial forest in eastern Australia demonstrate that mistletoe and forestry can coexist (Turner 1991), and the topic merits further exploration.

Finally, the widespread perception of mistletoes as destructive weeds needs to be challenged. Many landholders, managers, and even biologists regard mistletoes as invasive pests, damaging to individual trees and detrimental to forest health. Rather than being viewed as the cause of disturbance or disease, mistletoes need to be promoted as an indicator of habitat health, or in superabundance as a signal of landscape perturbation (Norton & Reid 1997). As demonstrated in this review, mistletoes have a substantial positive role in many forests and woodlands, and should be given appropriate recognition. Effecting such attitudinal changes will take considerable effort, but recognizing mistletoes as a keystone resource is an important first step.

ACKNOWLEDGMENTS

This review would not have been possible without the *Famulus* annotated mistletoe bibliography, an enduring legacy of the late Frank G. Hawksworth. I am indebted to Brian Geils and the other members of The Mistletoe Center for ensuring access to this information; Ralph Mac Nally, Carla Restrepo, and Scott Wing for their insightful comments; Daphne Fautin and Rick Prum for initial encouragement; and Dan Nickrent for photographs and systematic wisdom. Thanks are also due to the countless researchers whom I contacted and who generously provided information about their particular study organism's association with mistletoe.

Visit the Annual Reviews home page at www.AnnualReviews.org

LITERATURE CITED

Alam MK, Rahman MM. 1988. Some observations on the feeding of *Scurrula parasitica* Linn. (*Loranthaceae*) seeds by birds. *Bano Biggyan Patrika* 17:89–90

Ali S. 1980. *Handbook of the Birds of India and Pakistan, Vol. 2: Megapodes to Crab Plover.* Calcutta: Oxford Univ. Press. 2nd ed.

Ali S, Ripley SD. 1996. *Handbook of the Birds of India and Pakistan, Vol. 6: Cuckoo-shrikes to Babaxes.* Bombay: Oxford Univ. Press. 247 pp. 2nd ed.

Ali S, Ripley SD. 1999. *Handbook of the Birds of India and Pakistan, Vol. 10: Flowerpeckers to Buntings.* Calcutta: Oxford Univ. Press. 250 pp. 2nd ed.

Amico G, Aizen MA. 2000. Mistletoe seed dispersal by a marsupial. *Nature* 408:929–30

Anderson AH, Anderson A. 1973. *The Cactus Wren.* Tucson, AZ: Univ. Ariz. Press. 226 pp.

Anderson BW, Ohmart RD, Disano J. 1979. Revegetating the riparian floodplain for wildlife. In *United States For. Serv. Gen. Tech. Rep. WO-12*, pp. 318–31

Andrews JH, Rouse DI. 1982. Plant pathogens

and the theory of *r*- and *K*-selection. *Am. Nat.* 120:283–96

Antsett MC, Hossaert-McKey M, McKey D. 1997. Modelling the persistence of small populations of strongly interdependent species: figs and fig wasps. *Conserv. Biol.* 11: 204–13

Atsatt PR. 1977. The insect herbivore as a predictive model in parasitic seed plant biology. *Am. Nat.* 111:579–86

Atsatt PR. 1983. Mistletoe leaf shape: a host morphogen hypothesis. See Calder & Bernhardt 1983, pp. 259–76

Austin GT. 1970. Breeding birds of desert riparian habitat in southern Nevada. *Condor* 72:432–36

Baker HG, Baker I, Hodges SA. 1998. Sugar compositions of nectars and fruits consumed by birds and bats in the tropics and subtropics. *Biotropica* 30:559–86

Barker RD, Vestjens WJM. 1989. *The Food of Australian Birds. I. Non-passerines.* Melbourne, Aust: CSIRO

Barker RD, Vestjens WJM. 1990: *The Food of Australian Birds. II. Passerines.* Lyneham, Aust.: CSIRO

Barlow BA. 1983. Biogeography of Loranthaceae and Viscaceae. See Calder & Bernhardt 1983, pp. 19–46

Barlow BA, Wiens D. 1971. The cytogeography of the Loranthaceous mistletoes. *Taxon* 20:291–312

Barlow BA, Wiens D. 1977. Host-parasite resemblance in Australian mistletoes: the case for cryptic mimicry. *Evolution* 31:69–84

Bax BN. 1963. Some preliminary observations on the food of elephant in the Tsavo Royal National Park. *East Afr. Wildl. J.* 1:40–53

Bendire CE. 1887. Notes on a collection of birds' nests and eggs from southern Arizona. *Proc. U.S. Nat. Mus.* 10:551–58

Bennetts RE. 1991. *The influence of dwarf mistletoe infestation on bird communities in Colorado ponderosa pine forests.* MS thesis. Colorado State Univ., Fort Collins. 83 pp.

Bennetts RE, Hawksworth FG. 1992. The indirect effects of dwarf mistletoe on bird communities in Colorado ponderosa pine forests.

Proc. 39th West. Int. For. Dis. Work Conf. (1991):14–17

Bennetts RE, White GC, Hawksworth FG, Severs SE. 1996. The influence of dwarf mistletoe on bird communities in Colorado ponderosa pine forests. *Ecol. Appl.* 6:899–909

Bent AC. 1940. *Life Histories of North American Cuckoos, Goatsuckers, Hummingbirds, and Their Allies.* Washington, DC: U.S. Nat. Hist. Mus. Bull. 176. 506 pp.

Bentrupperbaumer JM. 1997. *Reciprocal ecosystem impact and behavioural interactions between cassowaries,* Casuarius casuarius, *and humans,* Homo sapiens: *exploring the natural-human environment interface and its implications for endangered species recovery in North Queensland, Australia.* PhD thesis. James Cook Univ., Townsville. 559 + 144 pp.

Berney FL. 1907. Field notes on birds of the Richmond district. Part V. *Emu* 6:155–58

Beruldson G. 1980. *A Field Guide to Nests and Eggs of Australian Birds.* Adelaide: Rigby. 448 pp.

Blake JG. 1984. A seasonal analysis of bird communities in southern Nevada. *Southwest. Nat.* 29:463–74

Bleitz D. 1955. Adventures with birds in Arizona. *Ariz. Highw.* 31:6–31

Bohl WH. 1970. *A study of the crested tinamou of Argentina. Wildl. No. 131.* US Fish Wildl. Serv. Spec. Sci. Rep. 101 pp.

Boussim IJ. 1991. *Contribution à l'étude des* Tapinanthus *parasites du Karite au Burkina Faso.* PhD diss. (de Troisième Cycle), Univ. Ougadougou, Burkina Faso

Brandt H. 1951. *Arizona and Its Bird Life.* Cleveland, OH: Bird Res. Found. 723 pp.

Bull EL, Wright AL, Hensum MG. 1989. Nesting and diet of long-eared owls in conifer forests, Oregon. *Condor* 91:908–12

Burkhardt F, Smith S, eds. 1990. *The Correspondence of Charles Darwin.* Cambridge: Cambridge Univ. Press. 673 pp.

Burt WH. 1934. The mammals of southern Nevada. *San Diego Soc. Nat. Hist. Trans.* 7:375–428

Bushueva TL, Tonevitsky AG, Maisuryan NA, Kindt A, Franz H. 1990. The effect of pH on the confirmation and stability of the structure of the toxic protein—mistletoe lectin L. *Lectins: Biol. Biochem. Clin. Biochem.* 7:179–85

Calder DM. 1983. Mistletoes in focus: an introduction. See Calder & Bernhardt 1983, pp. 1–18

Calder DM, Bernhardt P, eds. 1983. *The Biology of Mistletoes.* Sydney, Aust: Academic. 348 pp.

Campbell AJ. 1900. *Nests and Eggs of Australian Birds: Including the Geographical Distribution of the Species and Popular Observations Thereon.* Sheffield, UK: Pawson & Brailsford

Canyon DV, Hill CJ. 1997. Mistletoe host-resemblance: a study of herbivory, nitrogen and moisture in two Australian mistletoes and their host trees. *Aust. J. Ecol.* 22:395–403

Chiarlo B, Cajelli E. 1965. Fatty acids and amino acids in the berries of *Loranthus europaeus. Boll. Chim. Farm.* 104:735–43

Choate JH, Andrews RH, Barlow BA. 1987. Herbivory and cryptic mimicry in Australian Loranthaceae. In *Proc. 4th Int. Symp. Parasitic Flowering Plants*, ed. HC Weber, W Forstreuter, pp. 127–34. Marburg, West Ger.

Coleman E. 1949. Menace of the mistletoe. *Vic. Nat.* 66:24–32

Cramp S, ed. 1988. *Handbook of the Birds of Europe, the Middle East and North America: The Birds of the Western Palearctic, Vol. V: Tyrant Flycatchers to Thrushes.* Oxford, UK: Oxford Univ. Press

Cramp S, Perrins CH, eds. 1993. *Handbook of the Birds of Europe, the Middle East and North Africa: The Birds of the Western Palearctic, Vol. VII: Flycatchers to Shrikes.* Oxford, UK: Oxford Univ. Press. 577 pp.

Cramp S, Perrins CM, eds. 1994. *Handbook of the Birds of Europe, the Middle East and North Africa: The Birds of the Western Palearctic, Vol. VIII: Crows to Finches.* Oxford, UK: Oxford Univ. Press

Crome FHJ. 1978. Foraging ecology of an as-semblage of birds in lowland rainforest in northern Queensland. *Aust. J. Ecol.* 3:195–212

Curtin LSM. 1949. *Ethnobotany of the Pima.* Tucson: Univ. Ariz. Press. 156 pp.

Davidar P. 1978. Dispersal in some Loranthaceae of the Nilgiris. *J. Bombay Nat. Hist. Soc.* 75:943–45

Davidar P. 1983. Similarity between flowers and fruits in some flowerpecker pollinated mistletoes. *Biotropica* 15:32–37

Davidar P. 1985. Ecological interactions between mistletoes and their avian pollinators in south India. *J. Bombay Nat. Hist. Soc.* 82:45–60

Davidar P. 1987. Fruit structure in two neotropical mistletoes and its consequences for seed dispersal. *Biotropica* 19:137–39

Davis JN. 1995. Hutton's vireo (*Vireo huttoni*). In *The Birds of North America*, No. 189, ed. A Poole, F Gill, pp. 1–20, Philadelphia, PA: Acad. Nat. Sci./Washington, DC: Am. Ornithol. Union. 20 pp.

Dean WRJ, Midgley JJ, Stock WD. 1994. The distribution of mistletoes in South Africa: patterns of species richness and host choice. *J. Biogeogr.* 21:503–10

De Baar M. 1985. The complex mistletoe-insect community. *Entomol. Soc. Qld. Bull.* 13:100–2

Dickey DR, Van Rossem AJ. 1938. The birds of El Salvador. *Fieldiana: Zool.* 23:1–609

Downey PO. 1998. An inventory of host species for each aerial mistletoe species (Loranthaceae and Viscaceae) in Australia. *Cunninghamia* 5:685–720

Dwiyahreni AA, Kinnaird MF, O'Brien TG, Supriatna J, Andayani N. 1999. Diet and activity of the bear cuscus, *Ailurops ursinus*, in North Sulawesi, Indonesia. *J. Mamm.* 80:905–12

Ehleringer JR, Marshall JD. 1995. Water relations. See Press & Graves 1995, pp. 125–40

Eitniear JC, McGehee S, Waddell W. 1994. Observations on the feeding upon *Psittacanthus calyculatus* by brown–hooded parrots (*Pionopsitta haematotis*). *Ornitol. Neotrop.* 5:119–20

Farentinos RC. 1972. Nests of the tassel-eared squirrel. *J. Mamm.* 53:900–3

Feinsinger P. 1978. Ecological interactions between plants and hummingbirds in a successional tropical community. *Ecol. Monogr.* 48:269–87

Finch DM. 1985. Multivariate analysis of early and late nest sites of Abert's towhees. *Southwest. Nat.* 30:427–32

Fineran BA, Hocking PJ. 1983. Features of parasitism, morphology and haustorial anatomy in Loranthaceous root parasites. See Calder & Bernhardt 1983, pp. 205–29

Fischer S, Scheffler A, Kabeltiz D. 1997. Stimulation of the specific immune system by mistletoe extracts. *Anti-Cancer Drugs* 8(Suppl.):S33–37

Ford HA. 1999. Nest site selection and breeding success in large Austrialian honeyeaters: Are there benefits from being different? *Emu* 99:91–99

Ford HA, Paton DC, Forde N. 1979. Birds as pollinators of Australian plants. *NZ J. Bot.* 17:509–19

Fowler de Neira LE, Johnson MK. 1985. Diets of giant tortoises and feral burros on Volcán Alcedo, Galapagos. *J. Wildl. Manage.* 49:165–69

Fry CH, Keith S, Urban EK. 1988. *The Birds of Africa, Vol. III.* London: Academic. 611 pp.

Gautier-Hion A, Michaloud G. 1989. Are figs always keystone resources for tropical frugivorous vertebrates? A test in Gabon. *Ecology* 70:1826–33

Goddard J. 1968. Food preferences of two black rhinoceros populations. *East Afr. Wildl. J.* 6:1–18

Godschalk SKB. 1983a. Mistletoe dispersal by birds in South Africa. See Calder & Bernhardt 1983, pp. 117–28

Godschalk SKB. 1983b. A biochemical analysis of the fruit of *Tapinanthus leedertziae. S. Afr. J. Bot.* 2:42–54

Godschalk SKB. 1986. To plant a seed on a branch. *Fauna Flora (Pretoria)* 44:28–31

Goodall AG. 1977. Feeding and ranging behaviour of a mountain gorilla group (*Gorilla gorilla beringei*) in the Tshibinda-Kahuzu region (Zaire). *In Primate Ecology: Studies of Feeding and Ranging Behaviour in Lemurs, Monkeys, and Apes,* ed. TH Clutton-Brock, pp. 449–79. London: Academic

Goulding M. 1980. *The Fishes and the Forest: Explorations in Amazonian Natural History.* Berkeley: Univ. Calif. Press

Grainge M, Ahmed S. 1988. *Handbook of Plants with Pest Control Properties.* New York: Wiley. 470 pp.

Guzy MJ, Lowther PE. 1997. Black throated Gray Warbler (*Dendroica nigrescens*). In *The Birds of North America,* No. 319, ed. A Poole, F Gill, pp. 1–20, Philadelphia, PA: Acad. Nat. Sci./Washington, DC.: Am. Ornithol. Union.

Gwinner H, Oltrogge M, Trost L, Nienaber U. 2000. Green plants in starling nests: effects on nestlings. *Anim. Behav.* 59:301–9

Halloran AF, Crandell HB. 1953. Notes on bighorn food in the Sonoran zone. *J. Wildl. Manage.* 17:318–20

Harris T, Franklin K. 2000. *Shrikes and Bush-Shrikes: Including Wood-Shrikes, Helmet-Shrikes, Flycatcher-Shrikes, Philentomas, Batises and Wattle-Eyes,* London: Christopher Helm. 392 pp.

Hawksworth FG. 1969. Ecological aspects of dwarf mistletoe distribution. *Proc. 16th West. Int. For. Disease Work Conf., 1968,* pp. 74–82

Hawksworth FG. 1983. Mistletoes as forest parasites. See Calder & Bernhardt 1983, pp. 317–34

Hawksworth FG, Geils BW. 1996. Biotic associates. See Hawksworth & Wiens 1996, pp. 73–89

Hawksworth FG, Wiens D. 1996. Dwarf mistletoes: biology, pathology, and systematics. *Agric. Handb. 709.* Washington, DC: USDA For. Serv. 410 pp. 2nd ed.

Heather WA, Griffin DM. 1978. The potential for epidemic disease. In *Eucalypts for Wood Production,* ed. WC Hillis, SG Brown, pp. 143–54. Adelaide, Aust: CSIRO

Heine de Balsac H, Mayaud N. 1930. Compléments à l'étude de la propagation du gui (*Viscum album* L.) par les oiseaux. *Alauda* 2:474–93

Hemmingway CA. 1998. Selectivity and

variability in the diet of Milne-Edwards' sifakas (*Propithecus diadema edwardsi*): implications for folivory and seed eating. *J. Primatol.* 19:355–77

Herrera CM. 1985a. Determinants of plant-animal coevolution: the case of mutualistic dispersal of seeds by vertebrates. *Oikos* 44:132–41

Herrera CM. 1985b. Habitat-consumer interactions of frugivorous birds. In *Habitat Selection in Birds*, ed. ML Cody, pp. 341–65. Orlando, FL: Academic

Herzig-Straschil B, Robinson GA. 1978. Ecology of the fruit bat *Rousettus aegyptiacus leachii* in the Tsitsikama Coastal National Park, South Africa. *Koedoe* 0(21):101–10

Howe HF, Estabrook GF. 1977. On intraspecific competition for avian dispersers in tropical trees. *Am. Nat.* 111:817–32

Howe HF, Smallwood J. 1982. Ecology of seed dispersal. *Annu. Rev. Ecol. Syst.* 13:201–28

Howell TR. 1972. Birds of the lowland pine savanna of northeastern Nicaragua. *Condor* 74:316–40

Hudler G, French DW. 1976. Dispersal and survival of seed of eastern dwarf mistletoe. *Can. J. For. Res.* 6:335–40

Hull RJ, Leonard OA. 1964. Physiological aspects of parasitism in mistletoes (*Arceuthobium and Phoradendron*). I. The carbohydrate nutrition of mistletoe. *Plant Physiol.* 39:996–1007

Jaeger EC. 1947. *Desert Wild Flowers.* Stanford, CA: Stanford Univ. Press. 322 pp.

Kavanagh M. 1978. The diet and feeding behaviour of *Cercopithecus aethipos tantalus. Fol. Primatol.* 30:30–63

Kavanagh RP, Lambert MJ. 1990. Food selection by the greater glider *Petauroides volans*: Is foliar nitrogen a determinant of habitat quality? *Aust. Wildl. Res.* 17:285–99

Kingdon J. 1974. *East African Mammals: An Atlas of Evolution in Africa. II(A). Hares and Rodents.* London: Academic. 341 pp.

Kipfmueller KF, Baker WL. 1998. Fires and dwarf mistletoe in a Rocky Mountain lodgepole pine ecosystem. *For. Ecol. Manage.* 108:77–84

Knott J. 1908. The mistletoe (*Viscum album*): a monster in botany: a dryad in mythology: a panacea in therapeutics: a perennial yuletide symbol of seminal survival and reproductive vitality. *NY Med. J.* 88:1159–66

Knutson DM. 1983. Physiology of mistletoe parasitism and disease responses in the host. See Calder & Bernhardt 1983, pp. 295–316

Kuijt J. 1968. Mutual affinities of the Santalalean families. *Brittonia* 20:136–47

Kuijt J. 1969. *The Biology of Parasitic Flowering Plants.* Berkeley, CA: Univ. California Press. 246 pp.

Kuijt J, Mulder D. 1985. Mistletoes parasitic on orchids. *Am. Ochid Soc. Bull.* 54:976–79

Ladley JJ, Kelly D, Robertson AW. 1997. Explosive flowering, nectar production, breeding systems, and pollinators of New Zealand mistletoes (Loranthaceae). *NZ J. Bot.* 35:345–60

Lamont B. 1982. Host range and germination requirements of some South African mistletoes. *S. Afr. J. Sci.* 78:41–42

Lamont B. 1983a. Germination of mistletoes. See Calder & Bernhardt 1983, pp. 129–44

Lamont B. 1983b. Mineral nutrition of mistletoes. See Calder & Bernhardt 1983, pp. 185–204

Lamont B. 1985. Host distribution, potassium content, water relations and control of two co-occurring mistletoe species. *J. R. Soc. West. Aust.* 68:21–25

Landell NE. *Hur Fick Han Idén?* Stockholm: Univ. Stockholm Press. 14 pp.

Leck CF. 1972. Seasonal changes in feeding pressures of fruit- and nectar-eating birds in Panama. *Condor* 74:54–60

Leighton M, Leighton DR. 1983. Vertebrate responses to fruiting seasonality within a Bornean rainforest. In *Tropical Rain Forest: Ecology and Management*, ed. SL Sutton, TC Whitmore, AC Chadwick, pp. 191–96. Oxford, UK: Blackwell Sci.

Liddy J. 1982. The olive-backed oriole: an occasional disseminator of mistletoe. *Corella* 6:93

Liddy J. 1983. Dispersal of Australian mistletoes: the Cowiebank study. See Calder & Bernhardt 1983, pp. 101–16

Lopez de Buen L, Ornelas JF. 1999. Frugivorous birds, host selection and the mistletoe *Psittacanthus schiedianus*, in central Veracruz, Mexico. *J. Trop. Ecol.* 15:329–40

Lord EAR. 1956. The birds of the Murphy's Creek District, southern Queensland. *Emu* 56:100–28

Mack AL, Wright DD. 1996. Notes on occurrence and feeding of birds at Crater Mountain Biological Research Station, Papua New Guinea. *Emu* 96:89–101

Mamone MS. 1996. Wildlife use of Douglas-fir dwarf mistletoe. *Proc. 44th West. Int. For. Dis. Work Conf. 1995*, pp. 75–77

Marchant S, Higgins PJ, eds. 1990. *The Handbook of Australian, New Zealand and Antarctic Birds, Vol. 1, Part B: Australian Pelican to Ducks.* Melbourne: Oxford Univ. Press. 663 pp.

Marshall JT. 1957. Birds of pine-oak woodland in southern Arizona and adjacent Mexico. *Pac. Coast Avifauna* 32:1–125

Martinez del Rio C, Hourdequin M, Silva A, Medel R. 1995. The influence of cactus size and previous infection on bird deposition of mistletoe seeds. *Aust. J. Ecol.* 20:571–76

Martinez del Rio C, Silva A, Medel R, Hourdequin M. 1996. Seed dispersers as disease vectors: bird transmission of mistletoe seeds to plant hosts. *Ecology* 77:912–21

McConkey KR. 1999. *Gibbons as seed dispersers in the rain-forests of central borneo.* PhD thesis. Cambridge Univ., Cambridge

McDonald JD. 1973. *Birds of Australia.* London: Witherby. 552 pp.

McKey D. 1975. The ecology of coevolved seed dispersal systems. In *Coevolution of Plants and Animals*, ed. LE Gilbert, P Raven, pp. 159–91. Austin: Univ. Tex. Press

Mills LS, Soulé ME, Doak DF. 1993. The keystone-species concept in ecology and conservation. *BioScience* 43:219–24

Moermond TC, Denslow JS. 1985. Neotropical avian frugivores: patterns of behavior, morphology, and nutrition, with consequences for fruit selection. In *Neotropical Ornithology*, ed. PA Buckley, MS Foster, ES Morton, RS Ridgely, FG Buckley, pp. 865–97. *Ornithol. Monogr. 36.* Washington, DC: Am. Ornithol. Union

Morgan WTW. 1981. Ethnobotany of the Turkana: use of plants by a pastoral people and their livestock in Kenya. *Econ. Bot.* 35:96–130

Morton ES. 1978. Avian arboreal folivores: why not? In *The Ecology of Arboreal Folivores*, ed. GG Montgomery, pp. 123–30. Washington, DC: Smithson. Inst.

Mushtaque M Baloch GM. 1979. Possibilities of biological control of mistletoes. *Loranthus* spp., using oligophagous insects from Pakistan. *Entomophaga* 24:73–81

Musselman LJ, Press MC. 1995. Introduction to parasitic plants. See Press & Graves 1995, pp. 1–13

Nadkarni NM, Matelson TJ. 1989. Bird use of epiphyte resources in neotropical trees. *Condor* 91:891–907

Nason JD, Herre EA, Hamrick JL. 1998. The breeding structure of a tropical keystone plant resource. *Nature* 391:685–87

Nelson SK. 1997. Marbled murrelet (*Brachyramphus marmoratus*). In *The Birds of North America*, No. 276, ed. A Poole, F Gill, Philadelphia, PA: Acad. Nat. Sci./ Washington, DC: Am. Ornithol. Union

Nicholls TH, Hawksworth FG, Merrill LM. 1984. Animal vectors of dwarf mistletoe, with special reference to *Arcuthobium americanum* on lodgepole pine. In *Biology of Dwarf Mistletoes: Proc. Symp., Gen. Tech. Rep. Rm-111*, ed. FG Hawksworth, RF Scharpf, pp. 102–10. Washington, DC: USDA For. Serv.

Nickrent DL. 2001. Mistletoe phylogenetics: current relationships gained from analysis of DNA sequences. In *Proc. West. Int. For. Dis. Work Conf.*, ed. B Geils, R Mathiasen, USDA For. Serv., Kona, HI

Nickrent DL, Duff RJ, Colwell AE, Wolfe AD, Young ND, et al. 1998. Molecular phylogenetic and evolutionary studies of parasitic plants. In *Molecular Systematics of Plants.*

Vol. 2. *DNA Sequencing*, ed. DE Soltis, PS Soltis, JJ Doyle, pp. 211–41. Boston, MA: Kluwer

Nickrent DL, Franchina CR. 1990. Phylogenetic relationships of the Santales and relatives. *J. Mol. Evol.* 31:294–301

Nickrent DL, Soltis DE. 1995. A comparison of angiosperm phylogenies based upon complete 18S rDNA and rbcL sequences. *Ann. Mo. Bot. Gard.* 82:208–34

North AJ. 1906. *Nests and Eggs of Birds Found Breeding in Australia and Tasmania*, Vol. 2. Sydney: Aust. Mus. 380 pp. 2nd ed.

Norton DA. 1991. *Trilepidea adamsii*: an obituary for a species. *Conserv. Biol.* 5:52–57

Norton DA, Hobbs RJ, Atkins L. 1995. Fragmentation, disturbance, and plant distribution: mistletoes in woodland remnants in the Western Australian wheatbelt. *Conserv. Biol.* 9:426–38

Norton DA, Reid N. 1997. Lessons in ecosystem management from management of threatened and pest loranthaceous mistletoes in New Zealand and Australia. *Conserv. Biol.* 11:759–69

Norton DA, Stafford Smith DM. 1999. Why might roadside mulgas be better mistletoe hosts? *Aust. J. Ecol.* 24:193–98

O'Donnell CFJ, Dilks PJ. 1989. Feeding on fruits and flowers by insectivorous forest birds. *Notornis* 36:72–76

Ogle C, Wilson P. 1985. Where have all the mistletoes gone? *For. Bird* 16:10–14

Ostry ME, Nicholls TH. 1979. Bird vectors of black spruce dwarf mistletoe. *Loon* 51:15–19

Overdorff DJ. 1993. Similarities, differences, and seasonal patterns in the diets of *Eulemur rubriventer* and *Eulemur fulvus rufus* in the Ranomafana National Park, Madagascar. *Int. J. Primatol.* 14:721–53

Overton JM. 1994. Dispersal and infection in mistletoe metapopulations. *J. Ecol.* 82:711–23

Paine RT. 1969. A note on trophic complexity and community stability. *Am. Nat.* 103:91–93

Parks CG, Bull EL. 1997. Technical note— American marten use of rust and dwarf mistletoe brooms in northeastern Oregon. *West. J. Appl. For.* 12:131–33

Parks CG, Bull EL, Tinnin RO. Shepherd JF, Blumton AK. 1999. Wildlife use of dwarf mistletoe brooms in Douglas-fir in northeast Oregon. *West. J. Appl. For.* 14:100–5

Pasteur G. 1982. A classificatory review of mimicry systems. *Annu. Rev. Ecol. Syst.* 13:169–99

Pate JS. 1995. Mineral relationships of parasites and their hosts. See Press & Graves 1995, pp. 80–102

Paton DC, Ford HA. 1977. Pollination by birds of native plants in South Australia. *Emu* 77:73–85

Pekins PJ, Lindzey FG, Gessaman JA. 1991. Physical characteristics of blue grouse winter-use trees and roost sites. *Great Basin Nat.* 51:244–48

Penfield FB, Stevens RE, Hawksworth FG. 1976. Pollination ecology of three Rocky Mountain dwarf mistletoes. *For. Sci.* 22:473–84

Pliny 1960. *Natural History*. Vol. IV; Books 12–16, Book 16. Transl. H. Rackham. Cambridge MA: Harvard Univ. Press

Polhill R, Wiens D. 1998. *Mistletoes of Africa.* Kew, UK: R. Bot. Gard. 370 pp.

Power ME, Tilman D, Estes JA, Menge BA, Bond WJ, et al. 1996. Challenges in the quest for keystones. *BioScience* 46:609–20

Press MC, Graves JD, eds. 1995. *Parasitic Plants*. London: Chapman & Hall. 292 pp.

Prum RO, Razafindratsita VR. 1997. Lek behaviour and natural history of the velvet asity (*Philepitta castanea*: Eurylaimidae). *Wilson Bull.* 109:371–92

Punter D, Gilbert J. 1989. Animal vectors of *Arceuthobium americanum* seed in Manitoba. *Can. J. For. Res.* 19:865–69

Quinton DA, Horejsi RG. 1977. Diets of white-tailed deer on the rolling plains of Texas. *Southwest. Nat.* 22:505–9

Ralph CP. 1975. Life style of *Coccyzus pumilus*, a tropical cuckoo. *Condor* 77:60–72

Rand AS. 1978. Reptile arboreal folivores. In *The Ecology of Arboreal Folivores*, ed. GG

Montgomery, pp. 115–22. Washington, DC: Smithson. Inst.

Raphael MG, White M. 1984. Use of snags by cavity-nesting birds in the Sierra Nevada. *Wildl. Monogr.* 86:1–66

Raven PH, Axelrod DI. 1974. Angiosperm biogeography and past continental movements. *Ann. Mo. Bot. Gard.* 61:539–673

Rea AM. 1983. *Once a River. Bird Life and Habitat Changes on the Middle Gila.* Tucson: Univ. Ariz. Press. 285 pp.

Redford KH. 1984. The termitaria of *Cornitermes cumulans* (Isoptera, Termitidae) and their role in determining a potential keystone species. *Biotropica* 16:112–19

Reid N. 1986. Pollination and seed dispersal of mistletoes (Loranthaceae) by birds in southern Australia. In *The Dynamic Partnership: Birds and Plants in Southern Australia*, ed. HA Ford, DC Paton, pp. 64–84. South Australia: Gov. Printer

Reid N. 1987. The mistletoebird and Australian mistletoes: co-evolution or coincidence? *Emu* 87:130–31

Reid N. 1989. Dispersal of mistletoe by honeyeaters and flowerpeckers: components of seed quality. *Ecology* 70:137–45

Reid N. 1990. Mutualistic interdependence between mistletoes (*Amyema quandang*), and honeyeaters and mistletoebirds in an arid woodland. *Aust. J. Ecol.* 15:175–90

Reid N. 1991. Coevolution of mistletoes and frugivorous birds? *Aust. J. Ecol.* 16:457–69

Reid N, Lange RT. 1988. Host specificity, dispersion and persistence through drought of two arid zone mistletoes. *Aust. J. Bot.* 36:299–313

Reid N, Stafford Smith DM, Venables WN. 1992. Effect of mistletoes (*Amyema preissii*) on host (*Acacia victoriae*) survival. *Aust. J. Ecol.* 17:219–22

Reid N, Yan Z, Fittler J. 1994. Impact of mistletoes (*Amyema miquelii*) on host (*Eucalyptus blakelyi* and *Eucalyptus melliodora*) survival and growth in temperate Australia. *For. Ecol. Manage.* 70:55–65

Rentea R, Lyon E, Hunter R. 1981. Biological properties of iscador: a *Viscum album* preparation. I. Hyperplasia of the thymic cortex and accelerated regeneration of hematopoietic cells following X-irradiation. *Lab. Invest.* 44:43–48

Restrepo C, Sargent S, Levey DJ, Watson DM. 2001. The role of vertebrates in the diversification of New World mistletoes. In *Seed Dispersal and Frugivory: Ecology, Evolution and Conservation*, ed. DJ Levey, WR Silva, M. Galetti. Oxfordshire, UK: CAB Int.

Restrepo C. 1987. Aspectos ecológicos de la diseminación de cinco especies de muérdagos por aves. *Humboldtia* 1:65–116

Reynolds RT, Meslow EC, Wight HM. 1982. Nesting habitat of coexisting Accipiter in Oregon. *J. Wildl. Manage.* 46:124–38

Rice J, Ohmart RD, Anderson B. 1981. Bird community use of riparian habitats: the importance of temporal scale in interpreting discriminant analysis. *USDA For. Serv. Gen. Tech. Rep. RM-* 87, pp. 186–96

Richards PW. 1952. *The Tropical Rainforest: An Ecological Study.* Cambridge: Cambridge Univ. Press, 450 pp.

Richardson KC, Wooler RD. 1988. The alimentary tract of a specialist frugivore, the mistletoebird, *Dicaeum hirundinaceum*, in relation to its diet. *Aust. J. Zool.* 36:378–82

Riney T. 1951. Relationships between birds and deer. *Condor* 53:178–85

Robertson AW, Kelly D, Ladley JJ, Sparrow AD. 1999. Effects of pollinator loss on endemic New Zealand mistletoes (Loranthaceae). *Conserv. Biol.* 13:499–508

Rose AB. 1962. *Reptiles and Amphibians of Southern Africa.* Cape Town: Maskew Miller

Rose WC, Oesterling MJ, Womack M. 1948. Comparative growth on diets containing ten and nineteen amino acids, with further observations upon the role of glutamic and aspartic acids. *J. Biol. Chem.* 176:753–62

Rosenberg KV, Ohmart RD, Hunter WC, Anderson BW. 1991. *Birds of the Lower Colorado River Valley.* Tucson: Univ. Ariz. Press. 416 pp.

Rowe JS. 1983. Concepts of fire effects on plant individuals and species. In *The Role of Fire Effects in Northern Circumpolar*

Ecosystems, ed. RW Wein, DA Maclean, pp. 135–54. Chichester, UK: Wiley

Russell TP. 1964. Antelope of New Mexico. *N. Mex. Dep. Game Fish. Bull.* 12:1–103

Sargent S. 1994. *Seed dispersal of mistletoes by birds in Monteverde, Costa Rica.* PhD thesis. Cornell Univ., Ithaca, NY. 193 pp.

Sargent S. 1995. Seed fate in a tropical mistletoe: the importance of host twig size. *Funct. Ecol.* 9:197–204

Schaffer B, Hawksworth FG, Wullschleger SD. Reid CPP. 1983. Cytokinin-like activity related to host reactions to dwarf mistletoes (*Arceuthobium* spp.). *For. Sci.* 29:66–70

Sharnoff S, Rosentreter R. 1998. *Lichen use by wildlife in North America.* http://www.lichen.com/fauna.html

Shea KR. 1964. Silvicultural control of ponderosa pine dwarf mistletoe in south-central Oregon—a five year study. *J. For.* 62:871–75

Skinner MP. 1928. Yellowstone's winter birds. *Condor* 30:237–42

Skutch AF. 1969. *Life Histories of Central American Birds III: Families Cotingidae, Pipridae, Formicariidae, Furnariidae, Dendrocolaptidae, and Picidae.* Berkeley, CA: Cooper Ornithol. Soc. 580 pp.

Skutch AF. 1980. Arils as a food of tropical American birds. *Condor* 82:31–43

Skutch AF. 1983. *Birds of Tropical America.* Austin: Univ. Texas Press. 305 pp.

Smith GW. 1982. Habitat use by porcupines in a ponderosa pine/douglas-fir forest in northeastern Oregon. *Northwest Sci.* 56:236–40

Smith P. 1984. The forest avifauna near Bega, New South Wales. *I.* Differences between forest types. *Emu* 84:200–10

Smith RB, Wass EF. 1979. Infection trials with three dwarf mistletoe species within and beyond their known ranges in British Columbia. *Can. J. Plant Pathol.* 1:47–57

Snow BK, Snow DW. 1971. The feeding ecology of tanagers and honeycreepers in Trinidad. *Auk* 88:291–322

Snow BK, Snow DW. 1984. Long-term defense of fruit by mistle thrushes *Tudus viscivorus. Ibis* 126:39–49

Snow BK, Snow DW. 1988. *Birds and Berries:*

A Study of An Ecological Interaction. Calton, Staffordshire: T & AD Poyser. 168 pp.

Snow DW. 1971. Evolutionary aspects of fruit eating by birds. *Ibis* 113:194–202

Snow DW. 1982. *The Cotingas: Bellbirds, Umbrellabirds and Other Species.* Ithaca, NY: Cornell Univ. Press

Soini P. 1987. Ecology of the saddle-back tamarin *Saguinus fuscicollis illigeri* on the Rio Pacaya, Northeastern Peru. *Folia Primatol.* 49:11–32

Soper MF. 1976. *New Zealand Birds.* Christchurch, NZ: Whitcoulls. 251 pp. 2nd ed.

Spooner DM. 1983. The northern range of eastern mistletoe. *Phoradendron serotinum* (Viscaceae) and its status in Ohio. *Bull. Torrey Bot. Club* 110:489–93

Stamp NE. 1978. Breeding birds of riparian woodland in south-central Arizona. *Condor* 80:64–71

Sterba H, Andrae F, Pambudhi F. 1993. Crown efficiency of oak standards as affected by mistletoe and coppice removal. *For. Ecol. Manage.* 62:39–49

Stiles FG. 1985. Seasonal patterns and coevolution in the hummingbird-flower community of a Costa Rican subtropical forest. In *Neotropical Ornithology*, ed. PA Buckley, MS Foster, ES Morton, RS Ridgely, FG Buckley, pp.757–87. *Ornith. Monogr. 36.* Washington, DC: Am. Ornithol. Union

Stiles FG, Freeman CE. 1993. Patterns in floral nectar characteristics of some bird-visited plant species from Costa Rica. *Biotropica* 25:191–205

Stoffel B, Kramer K, Mayer H, Beuth J. 1997. Immunomodulating efficacy of combined administration of galactoside-specific lectin standardized mistletoe extract and sodium selenite in BALB/c-mice. *Anticancer Res.* 17:1893–96

Stoner EA. 1932. Some avian use for mistletoe. *Auk* 49:365–66

Stoner K. 1996. Habitat selection and seasonal patterns of activity and foraging of mantled howler monkeys (*Alouatta palliata*) in northeastern Costa Rica. *Int. J. Primatol.* 17:1–30

Sutton GM. 1951. *Mexican Birds—First Impressions.* Norman: Univ. Okla. Press. 282 pp.

Sutton GM. 1967. *Oklahoma Birds: Their Ecology and Distribution with Comments on the Avifauna of the Southern Great Plains.* Norman: Univ. Okla. Press. 674 pp.

Takatsukasa N. 1967. *The Birds of Nippon.* Tokyo: Marnzen. 701 pp.

Taylor WP. 1935. Ecology and life history of the porcupine (*Erethizon epixanthum*) as related to the forests of Arizona and the southwestern United States. *Univ. Ariz. Bull. Biol. Sci.*, Vol. 3. 177 pp.

Taylor WP. 1954. Food habits and notes on life history of the ring-tailed cat in Texas. *J. Mamm.* 35:55–63

Terborgh J. 1986. Keystone plant resources in the tropical forest. In *Conservation Biology: The Science of Scarcity and Diversity*, ed. ME Soulé, pp. 330–44. Sunderland, MA: Sinauer

Thompson JA, Owen WH. 1964. A field study of the Australian ringtail possum *Pseudocheirus peregrinus* (Marsupialia: Phalangeridae). *Ecol. Monogr.* 34:27–52

Timbrook J. 1990. Ethnobotany of Chumash indians, California, based on collections by John P. Harrington. *Econ. Bot.* 44:36–53

Turner RJ. 1991. Mistletoe in eucalypt forest—a resource for birds. *Aust. For.* 54:226–35

van Someren VGL. 1956. Days with birds: studies of habits of some East African species. *Fieldiana Zool.* 38:1–520

Vaughan TA. 1982. Stephen's woodrat, a dietary specialist. *J. Mamm.* 63:53–62

Wagner H, Proksch A. 1985. Immunostimulatory drugs of fungi and higher plants. *Econ. Med. Plant Res.* 1:113–53

Walsberg GE. 1975. The digestive adaptations of *Phainopepla nitens* associated with the eating of mistletoe berries. *Condor* 77:169–74

Walsberg GE. 1977. Ecology and energetics of contrasting social systems in *Phainopepla nitens* (Aves: Ptilogonatidae). *Univ. Calif. Publ. Zool.* 108:1–63

Watson DM. 1994. *The dynamics of bird communities in remnant Buloke* (Allocasuarina luehmanni) *woodlands.* Honours thesis. Monash Univ., Clayton. 39 pp.

Watson DM. 1997. The importance of mistletoe to the white-fronted honeyeater *Phylidonyris albifrons* in western Victoria. *Emu* 97:174–77

Watson DM, Mac Nally R, Bennett AF. 2000. The avifauna of severely fragmented Buloke *Allocasuarina luehmanni* woodland in western Victoria, Australia. *Pac. Conserv. Biol.* 6:46–60

Watson L, Dallwitz MJ. (1992 onwards). *The Families of Flowering Plants: Descriptions, Illustrations, Identification, and Information Retrieval.* Version. 27th Sept. 2000. http://biodiversity.uno.edu/delta/

Weathers WW. 1983. *Birds of Southern California's Deep Canyon.* Berkeley: Univ. Calif. Press. 266 pp.

Wheelwright NT. 1988. Fruit-eating birds and bird-dispersed plants in the tropics and temperate zone. *Trends Ecol. Evol.* 3:270–74

Wheelwright NT, Haber WA, Murray KG, Guindon C. 1984. Tropical fruit-eating birds and their food plants: a survey of a Costa Rican lower montane forest. *Biotropica* 16:173–92

Whittaker PL. 1984. The insect fauna of mistletoe (*Phoradendron tomentosum*, Loranthaceae) in southern Texas. *Southwest. Nat.* 29:435–44

Wiens D. 1978. Mimicry in plants. *Evol. Biol.* 11:365–403

Womack M, Rose WC. 1947. The role of proline, hydroxyproline, and glutamic acid in growth. *J. Biol. Chem.* 171:37–50

Yan Z, Reid N. 1995. Mistletoe (*Amyema miquelii* and *A. pendulum*) seedling establishment on eucalypt hosts in eastern Australia. *J. Appl. Ecol.* 32:778–84

Zona S, Henderson A. 1989. A review of animal mediated seed dispersal of palms. *Selbyana* 11:6–21

Annu. Rev. Ecol. Syst. 2001. 32:251–76

THE ROLE OF DISTURBANCE IN THE ECOLOGY AND CONSERVATION OF BIRDS*

Jeffrey D. Brawn,[1] Scott K. Robinson,[2] and Frank R. Thompson III[3]

[1]*Illinois Natural History Survey and Department of Natural Resources and Environmental Sciences, University of Illinois, Champaign, Illinois 61820; e-mail: j-brawn@uiuc.edu*
[2]*Department of Animal Biology, University of Illinois, Champaign, Illinois 61820; e-mail: skrobins@life.uiuc.edu*
[3]*USDA Forest Service, North Central Research Station, University of Missouri, Columbia, Missouri 65211-7260; e-mail: frthompson@fs.fed.us*

Key Words flood pulse, fire, habitat selection, silviculture, successional habitats

■ **Abstract** Natural ecological disturbance creates habitats that are used by diverse groups of birds. In North America, these habitats or ecosystems include grasslands or prairies, shrublands, savannas, early successional forests, and floodplains. Whereas the extent of all natural habitats has diminished significantly owing to outright loss from agriculture and development, the suppression of disturbance by agents such as fire and flooding has led to further losses. Accordingly, the abundances of many bird species adapted to disturbance-mediated habitats have declined as well. In North America, these declines have been more severe and common than those of species associated with less frequently disturbed habitats such as mature or closed-canopy forests. Field studies consistently reveal the direct role of disturbance and successional processes in structuring avian habitats and communities. Conservation strategies involving the management of disturbance through some combination of flooding, application of fire, or the expression of wildfire, and use of certain types of silviculture have the potential to diversify avian habitats at the local, landscape, and regional scale. Many aspects of the disturbance ecology of birds require further research. Important questions involve associations between the intensity and frequency of disturbance and the viability of bird populations, the scale of disturbance with respect to the spatial structure of populations, and the role of natural vs. anthropogenic disturbance. The effects of disturbance and ensuing successional processes on birds are potentially long-term, and comprehensive monitoring is essential.

---

*The US Government has the right to retain a nonexclusive, royalty-free license in and to any copyright covering this paper.

INTRODUCTION

Disturbance was once viewed largely as an insult to the "balance of nature" and synonymous with habitat destruction (see Botkin 1990). Certain forms of disturbance, however, are now held by ecologists and conservation biologists to play a fundamental and creative role in maintaining the natural heterogeneity in environmental conditions that organisms experience through space, time, or both. Much theory and empirical work has been devoted to understanding how such heterogeneity, or patchiness, affects the evolution of life histories and key ecological processes at the population, community, ecosystem, and landscape levels of organization (Connell 1978, Pickett & White 1985, Southwood 1988, Alverson et al. 1994). The importance of disturbance to the ecology of species and conservation of biodiversity has gained widespread recognition (Connell 1978, Sousa 1984, Pickett & White 1985, Petraitis et al. 1989, DeGraaf & Miller 1996a, Askins 2000) and has been defined by Pickett & White (1985, p. 7) as "any relatively discrete event in time that disrupts ecosystem, community, or population structure and changes resources, substrate availability, or the physical environment."

In the 1970s and 1980s, a series of models appeared that related disturbance to enhanced species diversity at the local (i.e., within-patch) scale. The posited mechanism was generally the prevention of competitive dominance and equilibrial conditions either directly (through density-independent mortality of organisms) or indirectly (through changes in habitat and resource levels). Empirical evaluation of these models has been generally corroborative, but the majority of these studies have been on space-limited, sessile organisms such as barnacles, mussels, and plants (Sousa 1984).

More recent works have considered the role of natural disturbance on species diversity at the landscape or regional scale (Angelstam 1998, Askins 2000). Now, disturbance is viewed as a natural ecological process leading to a mosaic of habitats or successional stages that may enhance both α and β diversity (Anglestam 1998). Indeed, ecological disturbance is an assumption in models that consider spatially structured populations. Depending on the scale of disturbance and the scale at which an organism "perceives" the environment (Vos et al. 2001), natural and, increasingly, anthropogenic disturbance can form the landscape framework for metapopulations and source-sink dynamics (Hanski 1991, Dunning et al. 1992). At a more evolutionary time-scale, the frequency of habitat disturbance has been hypothesized as the source of selective pressure underlying the evolution of dispersal strategies and other key life-history traits (Southwood 1988). In restoration ecology, landscape ecology, and the concept of ecosystem management, natural disturbance is now generally recognized as essential for maintaining biodiversity (Alverson et al. 1994, Askins 2000). Direct application of disturbance (e.g., prescribed burning or silviculture) or allowing the natural agents of disturbance (e.g., wildfire or floods) to proceed, are becoming standard elements of local and regional conservation strategies (Johnson et al. 1998).

This paper reviews how natural and anthropogenic disturbance can affect the population and community ecology of birds, and it emphasizes the conservation implications of these effects. The motivation for this review is twofold. First, the ecology of disturbance with relatively mobile organisms is different, but less well understood, than that for sessile organisms (Sousa 1984). Second, evidence is growing that some form of disturbance is required by the habitat needs of a large segment of the world's terrestrial avifauna (Askins 2000). Much has been written on the loss and fragmentation of closed-canopy forests due to agriculture or urbanization (e.g., Robinson et al. 1995), but less study has been devoted to the role of managing ecosystems and successional processes through the application of disturbance. We limit our review to bird species associated with terrestrial habitats or ecosystems and emphasize species that breed in North America. Finally, we feature the major agents of disturbance that we believe are most relevant to land-use policy or options for avian conservation: fire, silviculture, and floodplain dynamics. Silviculture stands apart as a commercially driven source of habitat disturbance; however, its effects are pervasive and potentially creative. Other important agents of disturbance in terrestrial avian habitats include drought, hurricanes, and herbivory owing to insect outbreaks or mammals (Rotenberry et al. 1995, R. A. Askins, personal communication).

GENERAL BACKGROUND

Habitat Ecology of Birds

The proximate cues used by birds to discriminate among habitats probably include factors that influence the availability of food, risk of predation (to adults or nests), and availability of nest sites. These factors include vegetation structure, floristic composition, densities of conspecifics and heterospecifics, and microclimate (Hildén 1965, James 1971, Hutto 1985, Block & Brennan 1993, Martin 1998). Cues at the landscape or regional scale may include the size, shape, distribution, configuration, and connectedness of different patch types (Wiens et al. 1993). Disturbance likely affects all aspects of avian habitat quality and selection from the microhabitat to regional scale.

Conservation Status of Species in Disturbance-Mediated Habitats and Ecosystems

In a review of the conservation status of birds of eastern North America, Askins (1993) reported a significantly greater proportion of recent declines in estimated abundances for species most often associated with disturbance (grasslands, shrubland, or savannas) than with forest birds (also see Peterjohn & Sauer 1999). Estimates of trends in abundances were derived from the North American Breeding Bird Survey (BBS), a continental avian survey program that was initiated in 1966

(Sauer et al. 2000). We expanded the comparative approach of Askins (1993) to consider all native North American terrestrial species for which abundance trends from 1966 to 1998 could be estimated with reasonable precision. We included 274 species and placed each into one of five habitat categories: grassland ($n = 27$), shrub-scrub ($n = 79$), open woodlands and savanna ($n = 63$), closed-canopy forest (deciduous, coniferous, or mixed, $n = 77$), and generalist ($n = 28$). The last category was used for species associated with two or more categories over all or part of their range and species of urban or suburban habitats. We then consulted Sauer et al. (2000) for estimates of population trends. The significance of these estimated trends are placed by Sauer et al. (2000) into four categories: significant (p < 0.10), upward, or downward, and insignificant (p > 0.10), upward, or downward.

The distribution of species in the four trend categories varied significantly among the different habitats or ecosystems (p = .0004, exact test of proportions). A greater proportion of species experienced significant decreases in all disturbance-mediated habitats than in the forested or generalist categories (Figure 1). The greatest proportion of decreases was within grassland birds where 56% of species in our sample declined significantly, followed by shrub-scrub (39%), and open woodlands (33%). Combining these three categories into a general "disturbance" category revealed that 40% ($n = 169$) of North American species associated with

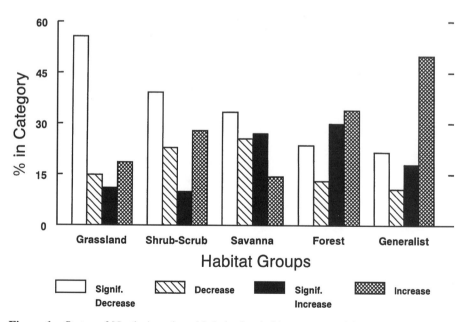

Figure 1 Status of North American birds in five habitat groups with respect to trends in abundance from 1966–1998. Trends from the North American Breeding Bird Survey (see text for explanation of survey, how species were assigned to habitat groups, and how trends were estimated and significance was determined).

some type of disturbance-mediated habitats were significantly decreasing between 1966 and 1998. Conversely, only 17% of the disturbance species experienced significant increases, whereas 34% of the forest species in our sample significantly increased.

EFFECTS OF FIRE

Under natural conditions, fire is the major force that governs the development and function of most deciduous, coniferous, and mixed forest types at temperate latitudes (Wright & Bailey 1982, Attiwill 1994, Frost 1998). Fire, along with other factors, is key in maintaining entire ecosystems such as grasslands and savannas at temperate and tropical latitudes, and Mediterranean-type shrublands (Moreno & Oechel 1994, McPherson 1997). In an extensive analysis of pre–European settlement forests in the United States, Frost (1998) found that fire played a fundamental role in the structure and composition of all but the wettest, most arid, or fire-sheltered forest types. The estimated fire frequency from lightning strikes and Native Americans varied from 1 to 700 y; one half of the continental United States burned every 1 to 12 y. Other historical analyses of forests worldwide have also identified the pervasive role of fire on landscape diversity and the composition, structure, and function of stands (Atiwill 1994, Angelstam 1998).

Suppression of fire and longer intervals between fires have emerged over the past 150 y in nearly all human dominated ecosystems and habitats where fire was a factor historically (Frost 1998, Smith 2000). Prescribed fire as a means of restoring or maintaining biodiversity is becoming commonplace throughout North America and elsewhere (Askins 2000), but long-term studies of periodic fire and birds are few. We review four case histories: coniferous and boreal forests of the western United States and southern Canada, oak savannas of the Midwestern United States, pine savannas of the southeastern United States, and North American grasslands. Other important systems in which fire has known effects on birds include shrubsteppe in the Western United States (Knick & Rotenberry 1999), diverse Australian ecosystems (Reilly 1991), Mediterranen-type ecosystems (Moreno & Oechel 1994), and Cerrado in Brazil (Cavalcanti 1999).

Coniferous and Boreal Forests of the Western United States and Southern Canada

The conifer-dominated forests of the western United States are variable in species composition and include ecoregions (sensu Ricketts et al. 1999) that are geographically widespread. The ecoregions include: Sierra Forests dominated by mixed-conifer stands (Ricketts et al. 1999); Rocky Mountain Forests (split by Ricketts et al. 1999 into several regions) dominated at different elevations by ponderosa pine (*Pinus ponderosa*), spruces (*Picea* spp.), lodgepole pine (*Pinus contorta*), and firs (*Abies* spp.); and mid-elevation forest in the southwest dominated by ponderosa pine. Fire was a common disturbance in these forests; estimates range from 1 per 10 to 300 y (Telfer 2000), with stand replacement fires more common in mixed-conifer forests at relatively high elevations. Historically, replacement fires were uncommon in open ponderosa pine forests (Telfer 2000). Boreal forests are most extensive in southern Canada, with elements found in northern sections of the continental United States. These forests—dominated by spruce, jack pine (*Pinus banksiata*), and quaking aspen (*Populus tremuloides*)—experience relatively frequent stand-replacing fires that occur every 40 to 100 y (Telfer 2000).

The effects of fire on birds in these ecoregions are comparatively well known (Bock et al. 1978, Raphael et al. 1987). The most celebrated case history is the dependence of the Kirtland's Warbler (*Dendroica kirtlandii*) on burned jackpine habitat (Probst & Weinrich 1993). Hutto (1995) examined effects of stand replacing fires on breeding bird communities in coniferous forests of the Northern Rocky Mountain Forests. Fifteen species of birds were markedly more common on recently (i.e., within 1–3 y) burned habitats than in nearly all the other forest types and successional stages sampled in the region. The Black-backed Woodpecker (*Picoides arcticus*) was largely restricted to the recently burned stands (1–2 y after fire). Overall avian community structure was distinctive in burned sites (either 1–2 y or 10–40 y after fire) largely because of greater densities of woodpeckers, flycatchers, or granivorous species. An open stand with many snags (i.e., dead or dying trees) and open ground provides for the nesting and foraging requirements of these species. Other studies in mixed-coniferous forests confirm that cavity-nesting birds in particular benefit from the action of fire disturbance that creates nesting foraging substrate in snags (Caton 1998).

Information on the effects of wildfire and prescribed fire on birds in southwestern ponderosa pine forests, although hampered by a lack of long-term replicated studies (Finch et al. 1997), indicates that several species benefit from disturbance. And in many cases, these are the same species or species groups that respond positively to disturbance in mixed-conifer burns (summarized in Finch et al. 1997). Populations of woodpeckers, certain flycatchers, and certain ground foragers tend to increase after fire.

Boreal forests in southern Canada have provided opportunities to assess the effects of natural burn cycles. As with mixed-conifer sites, stand-replacement burns host distinctive bird communities from the time of initial burns and succession from brush/shrub to relatively mature forest (Helle & Neimi 1996, Imbeau et al. 1999, Drapeau et al. 2000, Schieck & Hobson 2000). Virtually all studies have observed enhanced β diversity with fire disturbance and greater landscape diversity. The Black-backed Woodpecker, in particular, is a specialist in recent burns (Bock & Bock 1974). This species forages on wood-boring beetles that inhabit dying conifers for 2–3 y after exposure to fire (Murphy & Lehnhausen 1998).

Midwestern Oak Savannas and Woodlands

Oak savannas and woodlands were formerly a significant element of the landscapes of the midwestern United States and Canada that covered an estimated 11×10^6 ha

(Nuzzo 1986, Anderson & Bowles 1999; important oak savannas are also found in the southwest United States and California, but are not considered here). Oak savannas and woodlands are distinguished from oak forests by a more open canopy (10% to 70% closure) and a herbaceous understory. Dominant tree species vary according to soil type and geographic location but typically include fire-resistant species such as bur oak (*Quercus macrocarpa*), blackjack oak (*Q. marilandica*), pin oak (*Q. palustris*), and white oak (*Q. alba*) (McPherson 1997). Under natural conditions, the understory is dominated by herbaceous plants that are typically associated with prairies (McPherson 1997). Oak savannas and woodlands were formerly transitional vegetation types between prairies and closed-canopy forests (Anderson & Bowles 1999). Fire was the primary disturbance force in the region, and savannas burned less frequently than prairies, but more often than areas of forest.

Because of development and lack of disturbance, only 1% of the former oak savannas in the Midwest remain, and this habitat type is judged as globally significant (Ricketts et al. 1999). With fire suppression, shaded conditions develop, the herbaceous understory is lost, and encroachment by mesophytic species such as maples (*Acer* spp.) occurs (Packard 1993). Accordingly, prescribed fire, mechanical cutting, and herbiciding for the control of exotics are now common throughout the region in an effort to restore degraded savannas or maintain an open canopy (Packard 1993). Extant savannas in the upper Midwest are characteristically small, isolated remnants (Temple 1998).

Studies of birds and oak savannas have emphasized within-patch effects of prescribed burning. Overall, it is clear that bird communities of oak savanna reflect the transitional status of this vegetation type. Temple (1998) concluded that most bird species characteristic of oak savannas also have affinities with either prairies or forests. In the midwestern United States, this group includes about 40 bird species; of these, about 60% are decreasing significantly in the region (Brawn 1994). Whereas Temple (1998) concluded that there are no true savanna birds in the Midwest, one possible exception may be the Red-headed Woodpecker (*Melanerpes erythocephalus*)—a species that is decreasing throughout much of its range (Sauer et al. 2000). In Illinois' upland habitats, this species is restricted to oak savannas and habitats that mimic savannas such as parks (J. D. Brawn, unpublished data).

Bird communities in oak forest in the upper Midwest change with fire and savanna restoration. Davis et al. (2000) reported that sites with a history of more fires have fewer trees, higher densities of snags, and variable densities of shrubs. Bird species richness over two breeding seasons was markedly greater in the burned sites, and avian community structure varied systematically with the intensity of restoration. Canopy foliage gleaners such as the Red-eyed Vireo (*Vireo olivaceus*) and the Scarlet Tanager (*Piranga olicavea*) decreased with restoration, whereas omnivorous ground or shrub feeders and woodpeckers increased. Species favored by restoration and disturbance included the Baltimore Oriole (*Icterus galbula*), the Red-headed Woodpecker, and the Field Sparrow (*Spizella pusilla*). Many species such as the Indigo Bunting (*Passerina cyanea*), Field Sparrow, and American Goldfinch (*Cardeulis tristis*) breed regularly in oak savannas and nest or forage in shrubs (Sample & Mossman 1994, Davis et al. 2000) and are also characteristic of edge habitats and successional habitats such as abandoned pastures. Frequent and pervasive burns will eliminate shrubs; therefore, the frequency of fire disturbance can significantly change the diversity of bird communities within restored oak savannas and woodlands.

Southeastern Pine Savannas

The upland landscapes of the southeastern United States were once dominated by open pine savannas with an herbaceous understory (Platt 1999). These savannas covered an estimated 200×10^6 ha (Platt 1999) and were dominated by longleaf pine (*Pinus palustris*) (McPherson 1997). At elevations below 75 m, these savannas were nearly continuous along the Atlantic and Gulf coasts from North Carolina to Texas.

Frost (1998) estimated that lightning caused fires every 1–3 y over much of the region. The pine savannas of the southeastern United States, therefore, are one of the most pyrogenic vegetation types known. Not surprisingly, changing land use, fire suppression, commercial logging, and grazing have had immense impact on the vegetation of the region. By 1985, only 4% of the pre–European settlement savannas remained (Noss 1989, Platt 1999) with only 3000 ha in old-growth form.

The avifauna of the region includes three species that are essentially endemic; the Red-cockaded Woodpecker (*Picoides borealis*), the Brown-headed Nuthatch (*Sitta pusilla*), and the Bachman's Sparrow (*Aimophila aestivalis*) (Jackson 1988). The range and abundances of the Red-cockaded Woodpecker have decreased dramatically; this species is classified as federally endangered. The Bachman's Sparrow has also decreased significantly throughout much of its range, although it can use early successional pine fields (Dunning & Watts 1990). Fire-disturbed habitat characterized by large, well-spaced pines and lack of midstory trees (especially encroaching hardwood species) is essential for the Red-cockaded Woodpecker (James et al. 1997, Plentovich et al. 1998). Habitat requirements for the Bachman's Sparrow can vary geographically, but this species was historically also dependent on fire (Dunning & Watts 1990, Plentovich et al. 1998).

The significant historical role of fire in molding the upland landscapes of the southeast, coupled with the endangered status of Red-cockaded Woodpeckers has generated considerable interest in the study of fire and avian biodiversity in the region (Wilson et al. 1995). A 15–y study of fire suppression in a Florida pine woodland documented progressive changes in habitat structure and floristics as the herbaceous understory was lost and mesophytic hardwoods encroached (Engstrom et al. 1984). Avian species composition changed concomitantly, with open-habitat species such as the Bachman's Sparrow and Loggerhead Shrike (*Lanius ludovicianus*) disappearing within 9 y as species characteristic of hardwood forest colonized the site.

Habitat management for Red-cockaded Woodpeckers, such as short intervals between growing season fires and removal of midstory, has provided opportunities to assess how disturbance influences overall community structure. These studies provide evidence that management for an obligate disturbance species can have important collateral benefits for the regional avifauna. Burger et al. (1998) reported that intense management for woodpeckers promotes local increases in seven species that are declining regionally or nationally. These species occur regularly in early successional habitat or mature pine-grassland habitat with an herbaceous understory. Management units for Red-cockaded Woodpeckers in South Carolina also attracted over 30 scrub and successional species soon after burning and thinning of the canopy and midstory (Krementz & Christie 1999).

Grasslands

Before European settlement, about 162×10^6 ha of grasslands occurred in North America in three major ecoregions: tallgrass prairie, shortgrass prairie, and mixed-grass prairie (Samson & Knopf 1996, Herkert & Knopf 1998). Loss of native grassland habitat has been nearly complete in all regions (Samson & Knopf 1996), resulting from conversion to cropland, planting of exotics, overgrazing, and encroachment by shrubs (Vickery et al. 1999). The biogeographic history of North American grassland is complex, but these ecosystems undoubtedly developed with frequent fire from lightning strikes and fires set by Native Americans (see several chapters in Samson & Knopf 1996). The frequency of fire and other disturbance agents such as grazing or mowing (a detrimental form of anthropogenic disturbance depending on when it occurs; see Bollinger et al. 1990) is a major determinant of the structure and plant species composition of most types of prairie. Vickery et al. (1999) identified 32 bird species that are obligate to temperate-zone grasslands in North America, and they listed an additional 52 species that regularly occur in grasslands or ecotonal grasslands such as savannas.

Managing disturbance is the foundation of grassland restoration and conservation (Leopold 1949). Surprisingly few studies have been designed to assess fire effects on grassland birds, but it is evident that the temporal and spatial scales of disturbance have important consequences for habitat suitability. A 23-y study by Johnson (1997) in mixed-grass prairie in North Dakota assessed the effects of a 3–5 y burn rotation on avian abundances. Burning initially rid sites of vegetation and created open ground or short herbaceous vegetation; succession then led to taller vegetation and encroachment of woody vegetation. Avian responses to fire disturbance and subsequent succession fall into three groups: 1. those species that immediately colonized such as the Killdeer (*Charadrius vociferous*), Marbled Godwit (*Limosa fedoa*), or Upland Sandpiper (*Bartramia longicauda*); 2. species that used sites about 2 y after burning, but before woody encroachment, such as the Bobolink (*Dolichonyx oryzivorus*) and Grasshopper Sparrow (*Ammodramus savannarum*); and 3. species that required woody vegetation and relatively long-term protection from disturbance such as Common Yellowthroats (*Geothlypis trichas*). Similar patterns of turnover were found by Madden et al. (1999) in mixed-grass prairies in North Dakota that burn frequently. Therefore, even within the constraints of what is a prairie or grassland, there is an important short-term successional cycle that is disturbance-dependent, with significant turnover in avian community structure. Fire is particularly important in mesic grasslands that are comparatively productive and where woody encroachment proceeds rapidly (Madden et al. 1999).

Time elapsed since fire disturbance also influences bird populations in tallgrass prairie. Local abundances of Henslow's Sparrows (*Ammodramus henslowii*), a species of conservation concern throughout much of its range (Peterjohn & Sauer 1999), are highly sensitive to near-term fire history. In Illinois, newly burned sites are generally avoided and greatest abundances are attained on sites 2–4 y postburn (Herkert & Glass 1999) within a relatively specific range of vegetation heights and litter depths.

EFFECTS OF FLOODPLAIN DISTURBANCE

Floodplains are among the most disturbance-prone ecosystems. Although flood pulses tend to be predictably seasonal in most regions of the earth, the severity and timing of the flood pulse vary greatly (Malanson 1993, Sparks 1995). Floods create oxbow lakes and backwaters, and promote primary successional gradients as plant communities form on newly created soils inside meander loops (point bars) and on islands (Malanson 1993). By depositing silt on the floodplain, floods enhance primary productivity (Johnson et al. 1994). Severe floods can scour most or all vegetation from a floodplain; lack of severe flooding can allow woody encroachment on low-lying sections (Johnson 1994). The "flood pulse concept" (Junk et al. 1989) recognizes the importance of seasonal floods in promoting high biodiversity and productivity in the lateral floodplain. The connection between rivers and their backwaters is crucial for wildlife, especially fishes (Sparks 1995) and waterfowl (Bellrose et al. 1983). The intermediate disturbance hypothesis (Connell 1978) has been implicitly recognized in studies of floodplain ecosystems (Conner & Day 1976). Floodplain disturbances create an extraordinary diversity of habitats in river systems where floods are not constrained or altered and terrestrial bird communities of floodplains are some of the richest in the world (e.g., Terborgh et al. 1990). The river continuum concept (Vannote et al. 1980) relates the magnitude and predictability of seasonal flood pulses to the size of the river or stream with maximal diversity predicted in midsized rivers where levels of disturbance are intermediate.

Perhaps because of their high productivity and rich soils, floodplain habitats are also among the most endangered ecosystems (Dynesius & Nilsson 1994). Rivers have been dammed, channelized, straightjacketed by levees, subjected to increasing agricultural runoff, and lost to agriculture and plantation forestry (Hunt 1988, Sparks 1995). Floodplains have therefore lost much of their fish and wildlife as once-productive backwater swamps, marshes, and lakes have been isolated or subjected to excessive siltation when they do flood (Bellrose et al. 1983, Hunt 1988, Sparks 1995). In North America, three of the four terrestrial bird species that became extinct since European settlement depended largely on floodplain forests, including the Ivory-billed Woodpecker (*Campephilus principalis*),

Carolina Parakeet (*Conuropsis carolinensis*), and Bachman's Warbler (*Vermivora bachmanii*) (Askins 2000). Riparian forest loss in western North America is usually estimated at 95% or greater as a result of conversion to agriculture and alteration of hydrology (Knopf et al. 1988).

In this section, we review the linkages between disturbances and floodplain bird communities in three regions in North America: desert rivers of the southwest United States, big rivers of the central United States, and forested rivers of the southeastern United States.

Riparian Forests of Southwestern North America

Western riparian forests provide important resources for birds (Grinnell & Miller 1944, Rice et al. 1984, Rosenberg et al. 1991). A typical southwestern riparian successional gradient consists of sandbars in the frequently flooded river channel, willows (*Salix* spp.) along the edge of the channel and on islands, cottonwoods (*Populus* spp.) farther back from the river channel, and, in areas less frequently flooded, more diverse hardwoods. Each plant community occurs in a well-defined, often narrow, zone with abrupt borders reflecting the frequency of disturbance (Rosenberg et al. 1991). These zones contain rich breeding bird communities that may be associated with particular plant species for nesting and feeding. Disturbance-dependent willows, for example, are strongly preferred by several species and subspecies that are currently rare and endangered. Examples are Southwestern Willow Flycatchers (*Empidonax traillii extimus*), Arizona Bell's Vireos (*Vireo bellii arizonae*), and Least Bell's Vireos (*V.b. pusillus*) (Askins 2000).

Alteration of the disturbance regimes in western riparian forests has threatened this entire system (Carothers & Johnson 1975, Szaro & DeBano 1985, Knopf et al. 1988, Ohmart 1994). Damming of rivers has reduced the extent of the willow-cottonwood plant communities by reducing the severity and frequency of flood disturbance (Rice et al. 1983, Anderson et al. 1983), with the resulting decline of the many bird species that depend upon these communities (Strong & Bock 1990). Dams reduce seasonal flooding in undammed portions of the river, which dries out backwaters and oxbows. Willows and cottonwoods do not grow along newly created reservoirs because water levels fluctuate too greatly to allow seed germination (Hunter et al. 1987). Invasive introduced salt cedars (*Tamarix chinensis*) have taken over extensive areas of the floodplain in Arizona because damming and irrigation projects have made the soil more salty and hence more favorable to this species (Rosenberg et al. 1991). These dense homogeneous stands of salt cedar have far fewer birds than do native woodlands (Anderson et al. 1977). After much trial and error, some of these riparian zones have been restored successfully (reviewed in Askins 2000), and flood pulses are being reintroduced to the Colorado River in the Grand Canyon.

Large Rivers of Central North America

Large floodplain systems such as those found in the Midwestern United States were historically among the Earth's most productive ecosystems (Bellrose et al. 1983,

Sparks 1995), largely as a result of seasonal flood pulses that enriched soils and created diverse landforms and plant communities including backwaters, oxbow lakes, willow/cottonwood forests, mixed hardwood forests on natural levees, and more open sandbar habitats. In the Midwest, these floodplains supported huge populations of migratory waterfowl (Bellrose et al. 1983) and diverse communities of terrestrial birds (Zimmerman & Tatschl 1975, Knutson et al. 1996).

Disturbance regimes within these systems have been altered by navigation dams, levees, and agriculture (Bellrose et al. 1983, Dynesius & Nilsson 1994, Johnson et al. 1994). Most backwaters are isolated hydrologically or they rapidly fill with silt (Bellrose et al. 1983, Sparks 1995) during the summer. Levees increase the severity of floods in the river channel (Johnson et al. 1994), and flood-intolerant trees are lost in the remaining floodplain (Hosner 1958, Hunt 1988, Knutson & Klaas 1998). As a result, remaining mixed floodplain forests are being replaced by skeletal forests of dead trees and willow thickets (Yin & Nelson 1995, Knutson & Klaas 1997). Before they fall, dead trees temporarily provide habitats for cavity-nesting birds such as Red-headed Woodpeckers and Prothonotary Warblers (*Protonotaria citrea*) (Knutson & Klaas 1997).

With reduced flood severity, many rivers in the Great Plains that formerly lacked woody cover are lined with mixed hardwoods more characteristic of floodplains east of the Mississippi (Knopf & Scott 1990, Johnson 1994). Eastern forest birds use these new wooded corridors to invade western forests, activity that threatens the natural integrity of regional bird communities (Knopf et al. 1988). The loss of sandbar habitat resulting from altered disturbance regimes has been implicated in the declines of the endangered Interior Least Tern (*Sterna antillarum*) and Piping Plover (*Charadrius melodus*) (Schwalbach et al. 1993), which nest most successfully on island sandbars to which predators have little access (Kirsch 1996).

Southeastern North American Floodplain Forests

Southeastern floodplain forests have rich bird communities (Fredrickson 1980, Knutson et al. 1996, Sallabanks et al. 2000) that appear to depend in part on varying levels of disturbance. Carolina Parakeets and Ivory-billed Woodpeckers may have depended upon mature floodplain forest, a habitat that has been almost completely logged and converted to rowcrops (Askins 2000). At least two species, the extinct Bachman's Warbler and the rare Swainson's Warblers (*Limnothlypis swainsonii*) historically inhabited dense stands of cane (*Arundinaria gigantea*) (Meanley 1966, Eddleman et al. 1980, Hamel 1986, Remsen 1986), a plant that thrives at an intermediate level of floodplain disturbance and occasional fire. Human activities have altered disturbance regimes to the extent that cane stands are now rare (Askins 2000). Several bird species depend largely upon tree species associated with varying levels of disturbance (Gabbe et al. 2001). Yellow-throated Warblers (*Dendroica dominica*), for example, depend upon bald cypress (*Taxodium distichum*) growing in backwaters (Gabbe et al. 2001). There is little evidence, however, that bird communities in this region change dramatically along successional gradients as has been documented in rivers with unaltered floodplain dynamics in, for example, Amazonia (Dyrcz 1990, Terborgh et al. 1990, Robinson 1997, Robinson & Terborgh 1997).

Riparian disturbances in southeastern floodplains create natural edges such as the borders of oxbow lakes, natural levees, and the rivers themselves (Hupp & Osterkamp 1985). The natural edges may enhance both bird community diversity and nesting success (Sallabanks et al. 2000, Knutson et al. 2000). The complex vegetation structure along natural disturbances may provide dense cover in which nests may be concealed (Knutson et al. 2000).

EFFECTS OF SILVICULTURAL DISTURBANCE

Nearly one half (1563 million ha) of the area of natural forests worldwide is considered available for timber harvest given current legal and economic restrictions (FAO 1999). In 1991, total world wood production was 4410 million m^3, 37% of which was in the tropics and the balance mostly in northern temperate and boreal conifer forests (Whitmore 1998). Timber harvest is likely a more extensive disturbance agent in developed countries than all forms of natural disturbance. From the early 1980s to the early 1990s, for example, approximately 24% of the timberland in Michigan, Wisconsin, and Minnesota was disturbed by forest management (timber stand improvement, harvesting, planting, etc.) and 13% by natural processes (fire, windthrow, flooding, and pests) (Schmidt et al. 1999).

Silvicultural treatments are typically applied to stands or patches with similar tree composition and structure, often on a definable ecological land type. Regeneration treatments remove existing trees and establish tree reproduction. The clear-cut, shelterwood, and seed-tree methods regenerate even-aged stands and selection methods regenerate uneven-aged stands (Nyland 1996). Intermediate treatments modify existing forest stands (stand improvement) or control its growth and provide early financial returns (thinnings) (Nyland 1996).

Uneven-aged and even-aged methods differ in the scale and intensity of disturbance. Uneven-aged methods maintain a mix of tree sizes or ages within a habitat patch by periodically harvesting individual or small groups of trees. Even-aged methods harvest most or all of the overstory and create a fairly uniform habitat patch dominated by trees of the same age (Nyland 1996). At a landscape level, however, both methods can remove a similar volume of trees and regenerate similar areas of forest but the size and distribution of disturbance patches varies (Shifley et al. 2000). Bird communities differ in their response to regeneration methods as a result of these differences in scale and intensity of disturbance.

For brevity, we review mainly the effects of regeneration practices in natural or semi-natural North American forests where conservation is at least a secondary goal, and do not consider intermediate treatments, intensive culture, or agroforestry.

North American Forests, Within Patch Effects

EVEN-AGED METHODS   Even-age regeneration usually leads to near complete turnover in the breeding bird community (Webb et al. 1977, Conner et al. 1979,

Franzreb & Ohmart 1978). About 50% of neotropical migratory birds in central US hardwood forests prefer disturbance-mediated early successional stands created by harvest and 10–20% prefer these habitats in southern pine forests (Dickson et al. 1995).

In central and northern hardwood forests in the United States, nearly all ($n > 50$) species that previously bred in the mature forest abandon regeneration cuts (Annand & Thompson 1997). Bird communities respond significantly to the disturbance and ensuing successional changes. If residual live and dead trees are left in cut-over stands, some species such as Scarlet and Summer Tanager (*Piranga rubra*), Great Crested Flycatcher (*Myiarchus crinitus*), and Red-bellied Woodpecker (*Melanerpes carolinus*) that use the canopy or tree boles may continue to use the stand. Species that soon colonize and use residual snags, trees, slashpiles, or a rapidly developing herbaceous layer include Eastern Bluebirds (*Sialia* sialis), Northern Flicker (*Colaptes auratus*), Winter Wren (*Troglodytes troglodytes*), Carolina Wren (*Thryothorus ludovicianus*), Field Sparrow, and American Goldfinch. By the second growing season, developing ground vegetation, low shrubs, and young trees attract numerous species, including Common Yellowthroat, Chestnut-sided Warbler (*Dendroica pensylvanica*), Mourning Warbler (*Oporornis philadelphia*), Swainson's Thrush (*Catharus ustulatus*), Yellow-breasted Chat (*Icteria virens*), Prairie Warbler (*Dendroica discolor*), and Blue-winged Warbler (*Vermivora pinus*). By 10 y after timber harvest, tree reproduction forms a closed canopy, most early successional species abandon the site, and mature forest species begin to colonize (DeGraaf & Chadwick 1987, Thompson et al. 1996). Overall, densities of birds in young regenerating forests often are similar to or much greater than those in mature or midsuccessional pole-sized forests (Conner et al. 1979, Yahner 1986, Probst et al. 1992, Hagan et al. 1997, Hobson & Schieck 1999). Moreover, species richness may be greater in regenerating stands in the eastern United States (Conner et al. 1979, Yahner 1986, King et al. 2001).

Regeneration methods influence the future vegetation composition of a stand. For example, use of shelterwood or selection methods as opposed to clear-cutting will favor more shade-tolerant species over intolerant species, such as western hemlock over Douglas fir (*Pseudotsuga menziesii*) in western conifer forests, or sugar maple (*Acer saccharum*) and American beech (*Fagus grandifolia*) over oaks in eastern deciduous forests. The use of the seed tree method can directly control the seed source for regeneration and is often used in southern pine forests. Species composition is sometimes directly manipulated by the planting of improved tree stock and often occurs with Douglas fir in western forests and loblolly (*Pinus taeda*) and shortleaf pine (*Pinus echinata*) in southern forests. These shifts in floristic composition can alter an avian community because many bird species have demonstrated preferences for various tree species as foraging substrates (Franzreb & Ohmart 1978, Holmes & Robinson 1981).

A few studies have reported reproductive rates for songbirds in forest patches managed by even-aged management. In extensively forested northern New England, predation rates on artificial ground and shrub nests were not different among successional stages (DeGraaf & Angelstam 1993). Elsewhere, Morse & Robinson (1999) observed lesser nest predation of Kentucky Warblers in mature forest than in recent clear-cuts. Annand & Thompson (1997) reported comparatively high nest success and low parasitism rates for several shrub nesting birds in stands regenerated by the clear-cut or shelterwood methods in Midwestern US hardwood forests.

Even-aged management regenerates an entire stand, so stand size determines the size of the new habitat patch. Krementz & Christie (1999) found no evidence, however, that species richness or reproductive success varied among stands growing after clear-cuts ranging from 2.8 to 56.7 ha in southeastern longleaf pine forests.

UNEVEN-AGED METHODS   Single-tree and group-selection regeneration methods remove only a portion of the trees in a habitat patch and attempt to maintain a constant tree size/class structure over time. As a result, the turnover in bird species following a regeneration harvest, and during succession between harvests, is less significant than that following even-aged regeneration harvests (Thompson et al. 1995).

In eastern deciduous forests, most mature forest species occur in similar or slightly lower numbers in forests managed by the selection method than in untreated mature forest (Annand & Thompson 1997, Robinson & Robinson 1999). Gap-dependent species such as Hooded Warblers (*Wilsonia citrina*), Indigo Bunting, White-eyed Vireo (*Vireo griseus*), and Carolina Wren, or species dependent on understory density such as the Worm-eating Warbler (*Helmitheros vermivorus*), are more abundant in forests managed by the selection method than in mature even-aged forests. In New England's northern hardwood forests, bird communities in uneven-aged forest managed by selection method are also similar to those in mature forests (DeGraaf & Chadwick 1987). Notwithstanding, there is some evidence that uneven-aged management may selectively favor red maple (*Acer rubrum*), a change that could result in less oak and beech mast for birds and other wildlife (R. A. Askins, personal communication). Selection cutting also appears to have minimal impact on the composition and abundance of bird communities in western conifer forest. Similarly, in mixed-conifer forest in the western Sierra Nevada, there were few consistent differences in bird communities between uneven-aged and mature even-aged forests (Morrison 1992).

Again, there is limited knowledge of reproductive success in forest patches managed by uneven-aged methods. In forests managed by the selection method in southern Illinois, reproductive success of four songbird species was not, on average, greater in logged or nonlogged forest, and cowbird parasitism was only greater for two species in recent cuts (Robinson & Robinson 2001). Nest success of early succession birds does not appear to differ between group selection cuts and clear-cuts in eastern deciduous forests (King et al. 2001, Clawson et al. 2001).

North American Forests, Landscape Level Effects

Landscapes managed by even-aged methods are often a mosaic of different-aged stands, whereas those managed by uneven-aged methods are often less heterogeneous and composed of a range of tree sizes. Thompson et al. (1996) used a simple model to demonstrate how these patterns in composition might affect β diversity in central hardwood forests. The biggest differences among landscapes managed by these silvicultural methods was not in the abundance of species that nest in mature forest, but in early succession species that favored either small gaps (created by selection cutting) or large patches (created by clear-cutting). Thompson et al. (1992) compared the abundance species in landscapes with and without even-aged management (clear-cutting). In general, the abundance of only a few species nesting in mid- or late-successional forest differed significantly between landscapes, and β diversity and abundance of early successional species was greater in managed landscapes. These patterns of diversity vary with rotation time between harvests, but silvicultural disturbance clearly has a potentially beneficial effect on avian diversity.

Nonetheless, timber harvest decreases the continuity of mature or old forest in the landscape and is a source of habitat fragmentation. Forest fragmentation by human dominated land-uses, such as agriculture or urban and suburban development, is hypothesized to reduce reproductive success of forest songbirds by elevating numbers of generalist predators and nest predation in the landscape or by increasing the amount of edge and edge-related nest predation (Robinson et al. 1995). In one of the few controlled landscape-scale studies of the effects of regeneration practices, Clawson et al. (2001) estimated differences in abundance and reproductive success of common forest birds in Midwestern oak forests before and after logging. Densities of some forest species decreased while others increased following logging; however, neither nest predation or cowbird brood parasitism increased following logging.

Syntheses are hampered by differences in methods (especially artificial vs. real nests), the potential effects of other uncontrolled local factors, or effects of landscape context (i.e., Donovan et al. 1997). Manolois (2001) reviewed evidence for edge effects resulting from timber harvest and concluded that many of the studies that reported no edge effects did not have adequate statistical design or power to detect effects if they were present, and 9 of 13 studies with adequate power detected edge effects. General pooling of species may be misleading since lower nesting success near edges may be specific to certain groups such as ground nesters (Flaspohler et al. 2001). Overall, evidence is conflicting for the nature of edge effects between mature forest and recently harvested stands (Yahner & Scott 1988, Morse & Robinson 1999, Manolis et al. 2001).

Conservation Implications and Research Needs

This paper and other analyses (e.g., Alverson et al. 1994, Askins 2000) summarize evidence that ecological disturbance is fundamental to the conservation of birds. Whereas outright habitat loss is the most direct threat to avian biodiversity worldwide, it is clear that significant numbers of bird species are associated with habitats and ecosystems created by diverse forms of disturbance. We do not minimize the conservation significance of species associated with habitats that are rarely subjected to large-scale disturbance (e.g., primary or old growth forests), and we are encouraged that forest birds have benefited by afforestation in the eastern United States (Askins 2000). Rather, we believe that understanding and applying ecological disturbance offers opportunities to conserve large and diverse group species—many of which are declining as a result of habitat loss and successional changes in habitat structure.

The frequency and intensity of ecological disturbance is a determinant of the presence of many terrestrial habitat types (or ecosystems) in North America that support significant components of avian biodiversity (Figure 2). We believe this offers significant management and research opportunities; if conservation strategies for birds are to be effective, some combination of these habitats (along with tracts of mature forest) should be maintained within or across landscapes. Protected habitat should not be protected from processes that maintain or conserve significant elements of biodiversity.

As shown here, the bulk of information about disturbance and birds is on variation in populations and communities resulting from the initial effects of disturbance

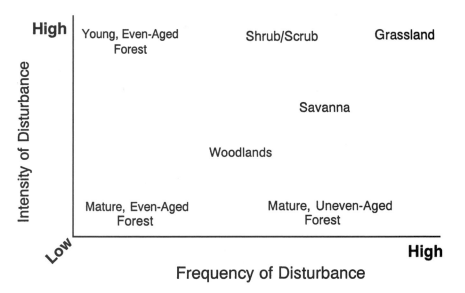

Figure 2   Selected terrestrial habitats used by birds in North America qualitatively ordinated according to frequency and intensity of disturbance. Even-aged forest can result from fire, wind throw, floods, or even-aged silviculture. Uneven-aged forest results from small-scale wind throw or uneven-aged silviculture. Woodlands and savanna result from moderate to light intensity frequent fires. Grasslands generally result from frequent and intense fire, grazing, or both.

or ensuing succession. These studies are highly informative to land managers, but the need for comprehensive understanding of disturbance and birds motivates a complex research agenda. A prevailing need is to design and implement long-term and large-scale studies that monitor the effects of disturbance. Studies to date have generally been short-term, without controls, unreplicated, and unplanned (Finch et al. 1997). Outlined below are the issues and questions that we believe are the most pressing.

Associations Between Habitat Disturbance and Population Viability

Studies of forest birds, habitat fragmentation, and edge effects clearly identify the usefulness of information on annual reproductive success and other demographic traits that affect population viability (Heske et al. 2001). Data on local abundances are crucial but incomplete in the context of estimating viability and assessing land-management options (Brawn & Robinson 1996). With few exceptions (e.g., Johnson & Temple 1990, Krementz & Christie 1999), little is known about the demography of birds in disturbance-mediated or successional habitats. Our expectation is that fecundity will change with successional changes in habitat, especially in grasslands and early successional forests, but details on the time-frame of this dynamic will be highly informative to managers in determining desirable intervals between prescribed fire, silvicultural treatments, or other modes of disturbance. Comparative studies from various geographic locations will be highly informative. In the Midwest, for example, prescribed fire and restoration maintain oak savannas in a specific disturbance-mediated state, but long-term successional changes are common following the less frequent stand-replacing fires in coniferous forests.

The effects of patch size on disturbance-dependent birds also requires close study. Numerous studies confirm that avian diversity and reproductive success is, on average, greater in larger than smaller patches of forest and grasslands (Thompson et al. 2001, Heske et al. 2001). The effects of patch size on birds in more transitional, disturbance-mediated habitats such as early successional forests and oak savannas are less clear. Early results from a study comparing nesting success of forest birds in managed forests found little or no difference in nesting success of disturbance-dependent birds in small selection openings vs. larger clear-cuts (Clawson et al. 2001; see also King et al. 2001).

Disturbance and Landscape Context

Questions about the scale and landscape context of disturbance will be different depending on the mode of disturbance and the ecoregion involved. In the extensive coniferous forests of the western United States and southern Canada, a primary question is how to apply or allow disturbance so that a natural or natural-like mosaic pattern of cover types is maintained (Hunter 1993, Landres et al. 1999). In isolated oak savanna restorations, the question is the effect of surrounding habitat; that is, what are the effects (if any) of nearby grassland, forest, or urban areas on the structure of savanna bird communities (Temple 1998)? Specific issues include the dispersion and arrangement of disturbance patches. Such studies will also serve to address questions from theory on ecological neighborhoods and the scale of ecological patterns or processes (Addicott et al. 1987, Engstrom & Mikushinski 1998). The frequency and extent of natal and breeding dispersal are key to determining the scale at which birds respond demographically to landscape features (Dunning et al. 1992, Vos et al. 2001) and are a persistent uncertainty for most species. Notable exceptions are the model studies on the Bachman's Sparrow and the Red-cockaded Woodpecker that relate ecological disturbance, landscape features, and the spatial structure of their populations (Dunning et al. 1992, Engstrom & Mikusinski 1998, James et al. 1997). Ideally, local conservation decisions regarding ecological disturbance should be made within a multi-scale framework that acknowledges landscape and regional goals based on historical heterogeneity in ecological landtypes (Thompson & DeGraaf 2001).

Effects of Natural vs. Anthropogenic Disturbance

Future land management planning will have to consider a range of approaches to balance economic needs and conservation objectives. Understanding how these different approaches affect the size and viability of birds associated with disturbance is essential. For example, a stand-replacing fire and a clear-cut both remove trees but the long-term implications for avian conservation are likely to be different (Hobson & Schieck 1999, Imbeau et al. 1999). One approach to the conservation of disturbance-dependent birds is to restore natural disturbance regimes, but this is realistic only on large expanses such as those in the western United States. Elsewhere, disturbance-dependent birds are likely to depend on human activities that stem from direct management or commercial activities. Ways to optimize these activities for conservation, renewable resources, and agriculture requires immediate study. In forests with commercial value, for example, the range of options includes (from most to least intensive): short rotation intensive culture, plantation forestry, commercial forestry in natural stands, multiple-use or ecosystem management, and wilderness management through natural disturbance. Another issue is the integrity of floodplain bird communities that depend upon restoration of flood-pulse disturbances (Galat et al. 1998); at present, there is disagreement over whether the flood pulse should be mimicked through moist soil management (Galat et al. 1998) or promoted by directly allowing the river to flow into sections of the floodplain during flood pulses.

Finally, we believe that the social implications of ecological disturbance require exploration. Episodes of opposition to prescribed fire and silviculture, for example, strongly identify the need to educate the public on the positive role that natural and anthropogenic disturbance can play in the conservation of avian biodiversity (Marynowski & Jacobson 1999).

ACKNOWLEDGMENTS

We thank Bob Askins, Dick DeGraff, Jim Herkert, and David King for thoughtful comments. The help of Kathy Jennings was instrumental in preparing the manuscript. Research support to JDB and SKR was provided in part by the National Science Foundation (DEB 99-04058).

Visit the Annual Reviews home page at www.AnnualReviews.org

LITERATURE CITED

Addicott JF, Aho JM, Antolin MF, Padilla DK, Richardson JS, Soluk DA. 1987. Ecological neighborhoods: scaling environmental patterns. *Oikos* 49:340–46

Alverson WS, Kuhlman W, Waller DM. 1994. *Wild Forests, Conservation Biology and Public Policy.* Washington, DC: Island. 300 pp.

Anderson B, Ohmart R, Rice J. 1983. Avian and vegetation community structure and their seasonal relationships in the Lower Colorado River Valley. *Condor* 82:392–405

Anderson RC, Bowles ML. 1999. Deep-soil savannas and barrens of the Midwestern United States. See Anderson et al. 1999, pp. 155–70

Anderson RC, Fralish JS, Baskin JM, eds. 1999. *Savannas, Barrens, and Rock Outcrop Plant Communities of North America.* Cambridge, UK: Cambridge Univ. Press. 470 pp.

Angelstam P. 1998. Maintaining and restoring biodiversity in European boreal forests by developing natural disturbance regimes. *J. Veg. Sci.* 9:593–602

Annand EM, Thompson FR III. 1997. Forest bird response to regeneration practices in central hardwood forests. *J. Wildl. Manage.* 61:159–71

Askins RA. 1993. Population trends in grassland, shrubland, and forest birds in eastern North America. *Curr. Ornithol.* 11:1–34

Askins RA. 2000. *Restoring North America's Birds. Lessons from Landscape Ecology.* New Haven, CT: Yale Univ. Press. 288 pp.

Attiwill PM. 1994. The disturbance of forest ecosystems: the ecological basis for conservative management. *For. Ecol. Manage.* 63:247–300

Bellrose FC, Havera SP, Paveglio FL Jr, Steffeck DW. 1983. *The Fate of Lakes in the Illinois River Valley. Ill. Nat. Hist. Surv. Biol. Notes 119*

Block WM, Brennan LA. 1993. The habitat concept in ornithology: theory and applications. *Curr. Ornithol.* 11:35–91

Bock CE, Bock JB. 1974. On the geographical ecology and evolution of the three-toed woodpeckers, *Picoides tridactylus* and *P. arcticus. Am. Mid. Nat.* 92:397–405

Bock CE, Raphael M, Bock JB. 1978. Changing avian community structure during early post-fire succession in the Sierra Nevada. *Wilson Bull.* 90:119–23

Bollinger EK, Bollinger PB, Gavin TA. 1990. Effects of hay-cropping on eastern populations of the Bobolink. *Wild. Soc. Bull.* 18:142–50

Botkin DB. 1990. *Discordant Harmonies.* Oxford, UK: Oxford Univ. Press. 241 pp.

Brawn JD. 1994. An overview of avian communities in North American savanna habitats: current knowledge and conservation need. *Proc. North Am. Conf. Barrens Savannas,* pp. 145–46. Chicago, IL: US EPA, Great Lakes Nat. Prog. Off.

Brawn JD, Robinson SK. 1996. Source-sink dynamics may complicate interpretation of long-term census data. *Ecology* 77:3–12

Burger LW Jr, Hardy C, Bein J. 1998. Effects of prescribed fire and midstory removal on breeding bird communities in mixed pine-hardwood ecosystems of Southern Mississippi. See Pruden & Brennan 1998, pp. 107–13

Carothers SW, Johnson RR. 1975. Water management practices and their effects on nongame birds in range habitats. In *Proc. Symp. Manage. For. Range Habitats Nongame Birds*, ed. DR Smith, pp. 210–22. Washington, DC: USDA For. Serv. GTR WO-1

Caton EL. 1998. Postfire habitat use by cavity-nesting birds in Northwestern Manitoba. See Pruden & Brennan 1998, p. 364 (Abstr.)

Cavalcanti RB. 1999. Bird species richness and conservation in the Cerrado region of central Brazil. *Stud. Avian Biol.* 19:244–49

Clawson RC, Faaborg J, Gram WK, Porneluzi PA. 2001. Landscape level effects of forest management on bird species in the Ozarks of southeastern Missouri. In *Proc. 2nd Missouri Ozark For. Ecosystem Symp.: Post Treatment Results of the Landscape Experiment*, ed. SR Shifley, JM Kabrick. *USDA For. Serv. Gen. Tech. Rep.* St. Paul, MN. In press

Connell JH. 1978. Diversity in tropical rainforests and coral reefs. *Science* 199:1302–10

Conner RN, Via JW, Prather ID. 1979. Effects of pine-oak clear-cutting on winter and breeding birds in southwestern Virginia. *Wilson Bull.* 91:301–16

Conner WH, Day JW. 1976. Productivity and composition of a bald cypress-water tupelo and bottomland hardwood site in a Louisiana swamp. *Am. J. Bot.* 63:1354–64

Davis MA, Peterson DW, Reich PB, Crozier M, Query T, et al. 2000. Restoring savanna using fire: impact on the breeding bird community. *Rest. Ecol.* 8:30–40

DeGraaf RM, Angelstam P. 1993. Effects of timber size-class on predation of artificial nest in extensive forest. *For. Ecol. Manage.* 61:127–36

DeGraaf RM, Chadwick NL. 1987. Forest type, timber size class and New England breeding birds. *J. Wildl. Manage* 51:212–17

DeGraaf RM, Miller RI, eds. 1996a. *Conservation of Faunal Diversity in Forested Landscapes*. London: Chapman & Hall. 633 pp.

DeGraaf RM, Miller RI. 1996b. The importance of disturbance and land-use history in New England: implications for forested landscapes and wildlife conservation. See DeGraaf & Miller 1996a, pp. 3–36

Dickson JG, Thompson FR III, Conner RN, Franzreb KE. 1995. Silviculture in central and southeastern oak-pine forests. See Martin & Finch 1995, pp. 245–66

Donovan TM, Jones PW, Annand EM, Thompson FR III. 1997. Variation in local-scale edge effects: mechanisms and landscape context. *Ecology* 78:2064–75

Drapeau P, Leduc A, Giroux J-F, Savard J-PL, Bergeron Y, Vickery WL. 2000. Landscape-scale disturbances and changes in bird communities of boreal mixed-wood forests. *Ecol. Monogr.* 70(3):423–44

Dunning JB, Danielson BJ, Pulliam HR. 1992. Ecological processes that affect populations in complex landscapes. *Oikos* 65:169–75

Dunning JB, Watts BD. 1990. Regional differences in habitat occupancy by Bachman's sparrow. *Auk* 107:463–72

Dynesius M, Nilsson C. 1994. Fragmentation and flow regulation of river systems in the northern third of the world. *Science* 266:753–62

Dyrcz A. 1990. Understory bird assemblages in various types of lowland tropical forest in Tambopala Reserve, SE Peru (with faunistic notes). *Acta Zool. Cracov.* 33:215–33

Eddleman WR, Evans KE, Elder WH. 1980. Habitat characteristics and management of Swainson's Warbler in southern Illinois. *Wildl. Soc. Bull.* 8:228–33

Engstrom RT, Crawford RL, Baker WW. 1984. Breeding bird populations in relation to forest structure following fire exclusion: a 15-year study. *Wilson Bull.* 96(3):437–50

Engstrom RT, Mikusinski G. 1998. Ecological neighborhoods in red-cockaded woodpecker populations. *Auk* 115:473–78

FAO. 1999. *State of the World's Forests 1999*. Rome: Food Agric. Organ. UN

Finch DM, Ganey JL, Yong W, Kimball RT, Sallabanks R. 1997. Effects and interactions of fire, logging, and grazing. In *Songbird Ecology in Southwestern Ponderosa Pine Forests: A Literature Review*, ed. WM Block, DM Finch, pp. 103–36. *USDA For. Serv. Gen.*

Tech. Rep. RM-GTR-292. Fort Collins, CO. 152 pp.

Flaspohler DJ, Temple SA, Rosenfield RN. 2001. Species-specific edge effects on nest success and breeding bird density in a forested landscape. Ecol. Appl. 11:32–46

Franzreb KE, Ohmart RD. 1978. The effects of timber harvesting on breeding birds in a mixed-coniferous forest. Condor 80:431–41

Fredrickson LH. 1980. Impact of water management on the resources of lowland hardwood forests. In Integrating Timber and Wildlife Management in Southern Forests, ed. RH Chabreck, RH Hills, pp. 51–64. Baton Rouge: La. State Univ.

Frost CC. 1998. Presettlement fire frequency regimes of the United States: a first approximation. See Pruden & Brennan 1998, pp. 70–81

Gabbe A, Robinson SK, Brawn JD. 2001. Tree species preferences of foraging insectivorous birds: implications for floodplain forest restoration. Conserv. Biol. In press

Galat DL, Fredrickson LH, Humburg DD, Bataille KJ, Bodie JR, et al. 1998. Flooding to restore connectivity of regulated, large-river wetlands. BioScience 48:721–33

Grinnell J, Miller AH. 1944. The distribution of the birds of California. Pacific Coast Avifauna No. 27. Cooper Ornithol. Club, Berkeley, CA

Hagan JM, McKinley PS, Meehan AL, Grove SL. 1997. Diversity and abundance of landbirds in a northeastern industrial forest. J. Wildl. Manage. 61:718–35

Hamel PB. 1986. Bachman's Warbler: A Species in Peril. Washington, DC: Smithson. Inst.

Hanski I. 1991. Single-species metapopulation dynamics: concepts, models, and observation. Biol. J. Linn. Soc. 42:17–38

Helle P, Niemi GJ. 1996. Bird community dynamics in boreal forests. See DeGraaf & Miller 1996a, pp. 209–34

Herkert JR, Knopf FL. 1998. Research needs for grassland bird conservation. See Marzluff & Sallabanks 1998, pp. 273–82

Herkert JR, Glass WD. 1999. Henslow's sparrow response to prescribed fire in an Illinois prairie remnant. Stud. Avian Biol. 19:160–64

Heske EJ, Robinson SK, Brawn JD. 2001. Nest predation and neotropical migrant songbirds: piecing together the fragments. Wildl. Soc. Bull. 29:52–61

Hildén O. 1965. Habitat selection in birds: a review. Ann. Zool. Fenn. 2:53–75

Hobson KA, Schieck J. 1999. Changes in bird communities in boreal mixedwood forest: harvest and wildfire effects over 30 years. Ecol. Appl. 9:849–63

Holmes RT, Robinson SK. 1981. Tree species preferences of foraging insectivorous birds in northern hardwoods forest. Oecologia 48:31–35

Hosner JF. 1958. The effects of complete inundation upon seedlings of six bottomland tree species. Ecology 39:371–73

Hunt CE. 1988. Down By The River: The Impact of Federal Water Projects and Policies On Biological Diversity. Washington, DC: Island. 260 pp.

Hunter ML Jr. 1993. Natural fire regimes as spatial models for managing boreal forests. Biol. Conserv. 65:115–20

Hunter WC, Anderson BW, Ohmart RD. 1987. Avian community structure changes in a mature floodplain forest after extensive flooding. J. Wildl. Manage. 51:495–502

Hupp CR, Osterkamp WR. 1985. Bottomland vegetation distribution along Passage Creek, Virginia, in relation to fluvial landforms. Ecology 66:670–81

Hutto RL. 1985. Habitat selection by nonbreeding migratory land birds. In Habitat Selection in Birds, ed. ML Cody, pp. 455–76. Orlando, FL: Academic. 558 pp.

Hutto RL. 1995. Composition of bird communities following stand-replacement fires in Northern Rocky Mountain (U.S.A.) conifer forests. Conserv. Biol. 9:1041–58

Imbeau L, Savard JPL, Gagnon R. 1999. Comparing bird assemblages in successional black spruce stands originating from fire and logging. Can. J. Zool. 77:1850–60

Jackson JA. 1988. The southeastern pine forest ecosystem and its birds: past, present, and future. *Bird Conserv.* 3:119–59

James FC. 1971. Ordinations of habitat relationships among breeding birds. *Wilson Bull.* 83:215–36

James FC, Hess CA, Kufrin D. 1997. Species-centered environmental analysis: indirect effects of fire history on red-cockaded woodpeckers. *Ecol. Appl.* 71:118–29

Johnson DH. 1997. Effect of fire on bird populations in mixed-grass prairie. In *Ecology and Conservation of Great Plains Vertebrates, Ecol. Stud. No. 125*, ed. FL Knopf, FB Samson, pp. 181–206. New York: Springer-Verlag. 320 pp.

Johnson KN, Sessions J, Franklin J, Gabriel J. 1998. Integrating wildfire into strategic planning for Sierra Nevada forests. *J. For.* 96:42–49

Johnson RG, Temple SA. 1990. Nest predation and brood parasitism of tallgrass prairie birds. *J. Wildl. Manage.* 54:106–11

Johnson RH, Sharitz RR, Dixon PM, Segal DS, Schneider RL. 1994. Woody plant regeneration in four floodplain forests. *Ecol. Monogr.* 64:45–84

Johnson WC. 1994. Woodland expansion in the Platte River, Nebraska: patterns and causes. *Ecol. Monogr.* 64:45–84

Junk WJ, Bayley PB, Sparks RE. 1989. The flood pulse concept in river-floodplain ecosystems. *Can. Spec. Publ. Fish Aquat. Sci.* 106:110–27

King DI, DeGraaf RM, Griffin CR. 2001. Productivity of early-successional shrubland birds in clear-cuts and groupcuts in an eastern deciduous forest. *J. Wildl. Manage.* 65:345–50

Kirsch EM. 1996. Habitat selection and productivity of Least Terns on the Lower Platte River, Nebraska. *Wildl. Monogr.* 60:1–48

Knick ST, Rotenberry JT. 1999. Spatial distribution of breeding passerine bird habitats in a shrubsteppe region of southwestern Idaho. *Stud. Avian Biol.* 19:104–11

Knopf FL, Johnson RR, Rich T, Samson

FB, Szaro RC. 1988. Conservation of riparian ecosystems in the United States. *Wilson Bull.* 100:272–84

Knopf FL, Scott ML. 1990. Altered flows and created landscapes in the Platte River headwaters, 1840–1990. In *Management of Dynamic Ecosystems*, ed. JM Sweeney, pp. 47–70. Lafayette, IN: The Wildl. Soc., North Cent. Sect.

Knutson MG, Gutreuter SJ, Klaas EE. 2000. Patterns of artificial nest depredation in a large floodplain forest. *J. Wildl. Manage.* 64:576–83

Knutson MG, Hoover JP, Klaas EE. 1996. The importance of floodplain forests in the conservation and management of neotropical migratory birds in the Midwest. See Thompson 1996, pp. 168–88

Knutson MG, Klaas EE. 1997. Declines in abundance and species richness of birds following a major flood on the Upper Mississippi River. *Auk* 114:367–80

Knutson MG, Klaas EE. 1998. Floodplain forest loss and changes in forest community composition and structure in the Upper Mississippi River: a wildlife habitat at risk. *Nat. Areas J.* 18:138–50

Krementz DG, Christie JS. 1999. Scrub-successional bird community dynamics in young and mature longleaf pine-wiregrass savannahs. *J. Wildl. Manage.* 63:803–14

Landres PB, Morgan P, Swanson FJ. 1999. Overview of the use of natural variability concepts in managing ecological systems. *Ecol. Appl.* 9:1175–88

Leopold A. 1949. *A Sand County Almanac and Sketches Here and There.* New York: Oxford Univ. Press. 226 pp.

Madden EM, Hanson AJ, Murphy RK. 1999. Influence of prescribed fire history on habitat and abundance of passerine birds in northern mixed-grass prairie. *Can. Field-Nat.* 113:627–40

Malanson GP. 1993. *Riparian Landscapes.* New York: Cambridge Univ. Press. 296 pp.

Manolis JC, Anderson DE, Cuthbert FJ. 2001. Patterns in clear-cut edge and fragmentation effect studies in northern-conifer landscapes: retrospective power analysis and Minnesota results. *Wildl. Soc. Bull.* 28:1088–101

Martin TE. 1998. Are habitat preferences of coexisting species under selection and adaptive? *Ecology* 79:656–70

Martin TE, Finch DM, eds. 1995. *Ecology and Management of Neotropical Migratory Birds.* New York: Oxford Univ. Press

Marynowski SB, Jacobson SK. 1999. Ecosystem management education for public lands. *Wildl. Soc. Bull.* 27:134–145

Marzluff JM, Sallabanks R, eds. 1999. *Avian Conservation: Research and Management.* Washington, DC: Island. 563 pp.

McPherson GR. 1997. *Ecology and Management of North American Savannas.* Tucson: Univ. Ariz. Press. 209 pp.

Meanley B. 1966. Some observations on habitats of Swainson's Warbler. *Living Bird* 5:151–65

Moreno JM, Oechel WC, eds. 1994. *The Role of Fire in Mediterranean-Type Ecosystems. Ecol. Stud. No. 107.* New York: Springer-Verlag. 201 pp.

Morrison ML. 1992. Bird abundances in forests managed for timber and wildlife resources. *Biol. Conserv.* 60:127–34

Morse SF, Robinson SK. 1999. Nesting success of a Neotropical Migrant in a multiple-use, forested landscape. *Conserv. Biol.* 13:327–37

Murphy EC, Lehnhausen WA. 1998. Density and foraging ecology of woodpeckers following a stand-replacement fire. *J. Wildl. Manage.* 62:1359–72

Noss R. 1989. Longleaf pine and wiregrass: keystone components of an endangered ecosystem. *Nat. Areas J.* 9:211–13

Nuzzo VA. 1986. Extent and status of Midwest oak savanna: presettlement and 1985. *Nat. Areas J.* 6:6–36

Nyland RD. 1996. *Silviculture: Concepts and Applications.* St. Louis, MO: McGraw-Hill

Ohmart RD. 1994. The effects of human-induced changes on the avifauna of western riparian habitats. *Stud. Avian Biol.* 15:273–85

Packard S. 1993. Restoring oak ecosystems. *Rest. Manage. Notes* 11:5–16

Peterjohn BG, Sauer JR. 1999. Population status of birds of North American grassland birds from the North American Breeding Bird Survey, 1966–1999. *Stud. Avian Biol.* 19:27–44

Petraitis PS, Latham RE, Nisenbaum RA. 1989. The maintenance of species diversity by disturbance. *Q. Rev. Biol.* 64(4):393–418

Pickett STA, White PS, eds. 1985. *The Ecology of Natural Disturbance and Patch Dynamics.* San Diego: Academic. 472 pp.

Platt WJ. 1999. Southeastern pine savannas. See Anderson et al. 1999, pp. 23–51

Plentovich S, Tucker JW Jr, Holler NR, Hill GE. 1998. Enhancing Bachman's sparrow habitat via management of red-cockaded woodpeckers. *J. Wildl. Manage.* 62(1):347–54

Probst JR, Rakstad DS, Rugg DJ. 1992. Breeding bird communities in regenerating and mature broadleaf forests in the USA Lake States. *For. Ecol. Manage.* 49:43–60

Probst JR, Weinrich J. 1993. Relating Kirtland's Warbler populations to changing landscape composition and structure. *Land. Ecol.* 8:257–71

Pruden TK, Brennan LA, eds. 1998. *Fire in Ecosystem Management: Shifting the Paradigm from Suppression to Prescription. Tall Timbers Fire Ecol. Conf. Proc. No. 20.* Tall Timbers Res. Stn., Tallahassee, FL

Raphael MG, Morrison ML, Yoder-Williams MP. 1987. Breeding bird populations during twenty-five years of postfire succession in the Sierra Nevada. *Condor* 89:614–26

Reilly P. 1991. The effect of wildfire on bird populations in a Victorian Coastal Habitat. *Emu* 91:100–6

Remsen JV Jr. 1986. Was Bachman's Warbler a bamboo specialist? *Auk* 103:216–19

Rice J, Anderson BW, Ohmart RD. 1984. Comparison of the importance of different habitat attributes to avian community organization. *J. Wildl. Manage.* 48:895–911

Ricketts TH, Dinerstein E, Olson DM, Loucks

CJ, Eichbaum W, et al. 1999. *Terrestrial Ecoregions of North America, a Conservation Assessment.* Washington, DC: Island. 485 pp.

Robinson SK. 1997. Birds of a Peruvian oxbow lake: populations, resources, predation, and social behavior. *Ornithol. Monogr.* 48:613–39

Robinson SK, Terborgh J. 1997. Bird community dynamics along primary successional gradients of an Amazonian whitewater river. *Ornithol. Monogr.* 48:641–72

Robinson SK, Thompson FR III, Donovan TM, Whitehead DR, Faaborg J. 1995. Regional forest fragmentation and the nesting success of migratory birds. *Science* 267:1987–90

Robinson WD, Robinson SK. 1999. Effects of selective logging on forest bird populations in a fragmented landscape. *Conserv. Biol.* 13:58–66

Robinson WD, Robinson SK. 2001. Avian nesting success in a selectively harvested north temperate deciduous forest. *Conserv. Biol.* In press

Rosenberg KV, Ohmart RD, Hunter WC, Anderson BW. 1991. *Birds of the Lower Colorado River Valley.* Tucson: Univ. Ariz. Press

Rotenberry JT, Cooper RJ, Wunderle JM, Smith KG. 1995. When and how are populations limited? The roles of insect outbreaks, fire, and other natural perturbations. See Martin & Finch 1995, pp. 55–84

Sallabanks R, Walters JR, Collazo JA. 2000. Breeding bird abundance in bottomland hardwood forests: habitat, edge, and patch size effects. *Condor* 102:748–58

Sample DW, Mossman MJ. 1994. Birds of Wisconsin oak savannas: past, present, and future. *Proc. North Am. Conf. on Barrens Savannas*, pp. 155-61. Chicago, IL: US EPA, Great Lakes-Natl. Prog. Off.

Samson FB, Knopf FL. 1996. *Prairie Conservation: Preserving North America's Most Endangered Ecosystem.* Washington, DC: Island. 339 pp.

Sauer JR, Hines JE, Thomas I, Fallon J, Gough G. 2000. *The North American Breed-*

ing Bird Survey, Results and Analysis 1966–1999. Version 98.1. USGS Patuxent Wildl. Res. Cent., Laurel, Mar.

Schieck J, Hobson KA. 2000. Bird communities associated with live residual tree patches with cut blocks and burned habitat in mixedwood boreal forests. *Can. J. For. Res.* 30:1281–95

Schmidt TL, Miles PD, Hansen MH. 1999. Management and disturbance as agents of change in Lake States timberlands. In *Proc. First North Am. For. Ecol. Workshop*, ed. J Cook, B Oswald, pp. 267–86. Univ. Wis.-Stevens Point Coll. Nat. Res.

Schwalbach MJ, Higgins KF, Dinan J, Dirks BJ, Kruse CD. 1993. Effects of water levels on interior least tern and piping plover nesting along the Missouri River in North Dakota. In *Proc. The Missouri River and Its Tributaries: Piping Plover and Least Tern Symposium*, ed. KF Higgins, MR Brahier, pp. 75–81. Brookings: S. Dak. State Univ.

Shifley SR, Thompson FR III, Larsen DR, Dijak WD. 2000. Modeling forest landscape change in the Missouri Ozarks under alternative management practices. *Comp. Elect. Agric.* 27:7–24

Smith JK, ed. 2000. Wildland fire in ecosystems, effects of fire on fauna. *USDA For. Serv. Gen. Tech. Rep. RMRS-GTR-42,-Vol. 1.* Ogden, UT. 83 pp.

Sousa W. 1984. The role of disturbance in natural communities. *Annu. Rev. Ecol. Syst.* 15:353–91

Southwood TRE. 1988. Tactics, strategies, and templets. *Oikos* 52:3–18

Sparks RE. 1995. Need for ecosystem management of large rivers and their floodplains. *BioScience* 45:168–82

Strong TR, Bock CE. 1990. Bird species distribution patterns in riparian habitats in southeastern Arizona. *Condor* 92:866–85

Szaro RC, DeBano LF. 1985. Changes in hydrologic regime: its impact on riparian system development and wildlife. In *Riparian Ecosystems: Ecology, Conservation, and Management*, ed. RR Johnson, CH Lowe,

PF Foliott. pp. 18–31. Tucson: Univ. Ariz. Press

Telfer EM. 2000. Regional variation in fire regimes. See Smith 2000, pp. 9–15

Temple SA. 1998. Surviving where ecosystems meet: ecotonal animal communities of Midwestern oak savannas and woodlands. *Trans. Wis. Acad. Sci.* 86:207–22

Terborgh J, Robinson SK, Parker TA III, Munn CA, Pierpont N. 1990. Structure and organization of an Amazonian forest bird community. *Ecol. Monogr.* 60:213–38

Thompson FR III, ed. 1996. *Management of Midwestern Landscapes for the Conservation of Neotropical Migratory Birds. USDA For. Serv. Gen. Tech. Rep. NC-187.* North Cent. For. Exp. Stn., St. Paul, MN

Thompson FR III, Brawn JD, Robinson S, Faaborg J, Clawson RL. 2001. Approaches to investigate the effects of forest management on birds in eastern deciduous forests: How reliable is our knowledge? *Wildl. Soc. Bull.* 28:1111–22

Thompson FR III, DeGraaf RM. 2001. Conservation approaches for woody, early successional communities in the eastern United States. *Wildl. Soc. Bull.* In press

Thompson FR III, Dijak WD, Kulowiec TG, Hamilton DA. 1992. Breeding bird populations in Missouri Ozark forests with and without clear-cutting. *J. Wildl. Manage.* 56:23–30

Thompson FR III, Probst JR, Raphael MG. 1995. Impacts of silviculture: overview and management recommendations. See Martin & Finch 1995, pp. 201–19

Thompson FR III, Robinson SK, Whitehead DR, Brawn JD. 1996. Management of central hardwood landscapes for the conservation of migratory. Soc. birds. See Thompson 1996, pp. 117–43

Vannote RL, Minshall GW, Cummins KW, Sedell JR, Cushing CE. 1980. The river continuum concept. *Can. J. Fish Aquat. Sci.* 37:130–37

Vickery PD, Tubaro PL, Cardosa da Silva JM, Peterjohn PG, Herkert JR, Cavalcanti RB. 1999. Conservation of grassland birds in the Western Hemisphere. *Stud. Avian Biol.* 19:2–25

Vos CC, Verboom J, Opdam PFM, Ter Braak CJF. 2001. Toward ecologically scaled landscape indices. *Am. Nat.* 183(1):24–41

Webb WL, Behrend DF, Saisorn B. 1977. Effect of logging on songbird populations in a northern hardwood forest. *Wildl. Monogr.* 55. 35 pp.

Whitmore TC. 1998. *An Introduction to Tropical Rain Forests.* New York: Oxford. 282 pp.

Wiens JA, Stenseth NC, Van Horne B, Ims RA. 1993. Ecological mechanisms and landscape ecology. *Oikos* 66:369–80

Wilson CW, Masters RE, Bukenhofer GA. 1995. Breeding bird response to pine-grassland community restoration for red-cockaded woodpeckers. *J. Wildl. Manage.* 59:56–67

Wright HA, Bailey HW. 1982. *Fire Ecology, United States and Southern Canada.* New York: Wiley. 501 pp.

Yahner RH. 1986. Structure, seasonal dynamics and habitat relationships of avian communities in small even-aged forest stands. *Wilson Bull.* 98:61–82

Yahner RH, Scott DP. 1988. Effects of forest fragmentation on depredation of artificial nests. *J. Wildl. Manage.* 49:508–13

Yin Y, Nelson JC. 1995. Modifications to the Upper Mississippi River and their effects on floodplain forests. *Long Term Res. Monit. Prog., Tech. Rep. 95-T003.* Nat. Biol. Serv., Environ. Manage. Tech. Cent., Onalaska, WI. 17 pp.

Zimmerman JL, Tatschl JL. 1975. Floodplain birds of Weston Bend, Missouri River. *Wilson Bull.* 87:196–206

Annu. Rev. Ecol. Syst. 2001. 32:277–303
Copyright © 2001 by Annual Reviews. All rights reserved

APPROACHES TO THE STUDY OF TERRITORY SIZE AND SHAPE

Eldridge S. Adams

Department of Ecology and Evolutionary Biology, University of Connecticut, Storrs, Connecticut 06269-3043, e-mail: eldridge.adams@uconn.edu

Key Words territoriality, optimality models, game theory, competition, aggression

■ **Abstract** Intraspecific variation in territory size and shape can have strong effects on population structure and dynamics. The traditional theoretical approach to the study of territory size is based on optimality models that analyze decisions of focal residents as responses to the costs and benefits of defense. These models have stimulated numerous empirical studies showing that territory holders adjust their behavior according to rates of intrusion and availability of food. However, models of optimal territory size are applicable only in limited circumstances because they focus on unilateral decisions rather than on interactions. Furthermore, observational and experimental studies often find that territory sizes are insensitive to food supply. Recently, greater emphasis has been placed on two alternative approaches. The first concerns interactions among contiguous neighbors and how these affect use of space. In these models territory size and shape are determined by the balance of pressure exerted at boundaries or arise as the results of local rules of movement and interaction. The second alternative approach views territory size as the outcome of interactions between established residents and potential settlers attempting to acquire territories. By considering the simultaneous actions of multiple competitors, these models allow quantitative prediction of the effects of territory defense on population density and spatial patterns as well as responses to environmental change.

INTRODUCTION

Territoriality can exert strong effects on the population dynamics of aggressive animals (Patterson 1980, Davies & Houston 1984, Łomnicki 1988, Newton 1992, Sutherland 1996). The nature of these effects depends in large part on how territory size is adjusted to ecological circumstances. When territory size is inflexible, local populations may be regulated at stable densities. By contrast, when territory area is readily altered owing to changes in food abundance or the density of competitors, population densities will also vary. Intraspecific variation in territory sizes may also lead to unequal division of resources among competitors, with consequences for differential rates of growth, mortality, and reproduction. The behavioral adjustment

0066-4162/01/1215–0277$14.00

of territory size thus has important consequences for demography, population regulation, and spatial ecology.

My goal is to provide an overview of theoretical studies of the causes of intraspecific variation in territory size and shape, paying particular attention to their potential for predicting population-level phenomena. I argue that the most common approach to the analysis of territory size, which is based on optimality models, is applicable only in limited circumstances. This is because optimality models are constructed to predict the unilateral decisions of focal residents, rather than the outcome of interactions of two or more competitors. In practice, the optimality approach has rarely allowed quantitative prediction of territory sizes or shapes in natural circumstances; even its qualitative predictions can be difficult to test. Alternative approaches are described that explicitly consider the actions and decisions of multiple territory residents.

This review begins with a classification of behavioral models of territory size and shape. Three conceptual approaches are recognized, distinguished by what they view as the essential interaction or decision-making process controlling territory area. These are (*a*) focal resident models, (*b*) models of interactions among neighbors, and (*c*) models of interactions between established residents and potential settlers. Each of these approaches includes both mechanistic models, which demonstrate the effects of particular rules of behavior, and optimality or game theoretical models, which derive the rules that maximize fitness. Mechanistic models are often built on the underlying assumption that behaviors are adaptive, but do not formally solve for optimal behavior. Following this classification, I present a brief review of correlative and experimental studies on how territory size is affected by food availability and intruder pressure, the variables emphasized by the theory of optimal territory size. It turns out that a large fraction of studies show little or no effect of food supply on territory size. This is imperfectly understood and represents an issue for which development of a theory of territory interactions is needed.

Because it is not possible to discuss or even to list all the papers with information on intraspecific variation in territory size in a review of this length, I instead illustrate key points with examples from diverse taxa. Interspecific differences in territory size, the relationship of territory sizes to mating and social systems, and related topics have been reviewed elsewhere (Schoener 1968, Emlen & Oring 1977, Waser & Wiley 1979, Davies 1980, Ydenberg et al. 1988, Newton 1992, Stamps 1994, Temeles 1994, Gordon 1997).

I adopt Wilson's (1975) definition of a territory as "an area occupied more or less exclusively by an animal or group of animals by means of repulsion through overt defense or advertisement." This encompasses a broad range of territory types. At one extreme are territories with sharp limits, often revealed by the locations of distinctive behaviors shown at territory boundaries (e.g., Nursall 1977, Askenmo et al. 1994, Eason et al. 1999). At the other extreme are home ranges that overlap extensively, but that are nonetheless defended to some degree against conspecifics, especially within core areas (e.g., Getty 1981c, Stamps 1990).

FOCAL RESIDENT MODELS

Models of territory size and shape differ according to the number and type of individuals whose actions or decisions are represented. The largest set of models, including the bulk of the optimality models, focuses on the behavior of a single actor, a territorial resident that chooses a region to defend. These can be called focal resident models to distinguish them from models that explicitly represent two or more competitors. The chief question addressed by this body of theory is how optimal territory size is affected by the abundance of resources (especially food) and intruder pressure.

Mechanistic Models

Purely mechanistic models in this category are uncommon, except as points of comparison to models with multiple actors (e.g., White et al. 1996a) (see below). An example of a focal-resident mechanistic model of territory shape is Noakes & McNicol's (1982) description of salmonid fish territories. Territorial salmonids tend to rest at characteristic stations facing upstream. Noakes & McNicol noted that brook charr, *Salvelinus fontinalis*, swim greater distances to attack intruders detected upstream than intruders detected to the sides or downstream. When not distorted by interactions with neighbors, the shapes of defended areas are approximated by cardioids.

Optimality Models: Components and Limitations

The analysis of territory size is often presented as an optimization problem, focusing on a single decision-maker (e.g., Dill 1978, Kodric-Brown & Brown 1978, Pyke 1979, Ebersole 1980, Hixon 1980). This approach has been ably reviewed (Davies 1980, Hölldobler & Lumsden 1980, Schoener 1983, Davies & Houston 1984, Carpenter 1987), with additional examples appearing subsequently (e.g., Grant & Noakes 1987, Schoener 1987). Each of these models assumes that the costs and benefits of holding a territory change with territory area. Graphical or mathematical analysis reveals the range of territory sizes for which benefits exceed costs (economically defendable territories) (Brown 1964) and the particular size that yields maximum net benefits (the optimum territory size). Most of these models concern foraging territories, but a few address other territory functions (e.g., Hunt & Hunt 1976, Stamps 1983). Territory shape is less often considered (but see Eason 1992).

Models of optimal territory size have the following components.

1. *One or more decision variables.* The resident is usually assumed to select the territory area, or to select both territory area and time spent searching for intruders (Schoener 1987).

2. *Optimization criterion and constraints.* Following Schoener (1971), many authors distinguish between "energy maximizers," which seek to maximize

the rate of energy intake, and "time minimizers," which seek to minimize the time spent in foraging and defense. Alternatively, the animal may be assumed to select the maximum territory area for which benefits exceed costs ("area maximizers") (Hixon 1980) or the area that minimizes the probability of starvation (Lima 1984). Time constraints can force a trade-off between defense and foraging, and processing constraints can reduce the rate at which benefits are derived from an increasing food supply (Schoener 1983).

3. A *benefit function* describes how resource acquisition, or other benefits, vary with territory area. For foraging territories, the benefit function is affected by the distribution of food within the territory (often assumed to be uniform), the rate at which food is renewed, the potential for resident satiation, and whether the animal consumes food where it is found or must transport the food to a central site. A few models allow variation in mean rates of food supply (MacLean & Seastedt 1979, Myers et al. 1981) or stochastic variation around the mean (Lima 1984, McNair 1987).

4. A *cost function* describes the relationship between territory area and the costs of defense. Typically, intruders are assumed to arrive randomly on the territory, at a rate proportional to its area or perimeter. The expense of evicting the resident may depend on the distance to the boundary. Intruders may also deplete resources before they are detected and chased away. When the goal is to analyze how optimum territory size varies with food supply and intruder pressure, it is also necessary to specify how the rate of intrusion is affected by food supply (Schoener 1983). To find the optimum territory size, the costs and benefits must be measured in a common currency that estimates effects on the animal's fitness.

Despite the prominence of the optimality approach, its scope and applicability are limited for both practical and conceptual reasons. First, it is usually difficult to make even qualitative predictions regarding the relationship between optimal territory size and the abundance of resources. This was made clear by Schoener's comparisons of alternative optimality models (Schoener 1983, 1987) and by contrasts drawn in other papers (e.g., Hixon 1980, Lima 1984). Whether optimal territory size increases or decreases when food supply increases depends on numerous particulars that can be difficult to determine for any given species and environment. This does not mean that optimality models are inappropriate, but it makes them difficult to test. Rigorous testing requires either detailed information that is usually unavailable, or else judicious choice of animals and environments so that the range of possible predictions can be sharply limited (e.g., Grant & Noakes 1987).

A second set of limitations stem from the assumption that animals can optimize territory area or the behaviors that control territory area. An animal establishing a territory may not be able to predict in advance the level of benefits that the territory will provide (Bollmann et al. 1997), and continual readjustment of the territory may be prohibitively costly. Furthermore, one cannot always expect that phenotypes are optimized with respect to local conditions (Gould & Lewontin 1979). However, the assumption of optimality can be viewed as a way of producing falsifiable predictions about natural systems.

From an ecological standpoint, the greatest limitation of models of optimal territory size arises from the way in which the costs of defense are represented. To infer the nature of the cost function, competitors are essentially treated as particles that rain onto the territory and that can be chased away if the resident chooses to invest the time and energy. There are circumstances in which this is a suitable way to approximate the costs of territory defense, and these are reasonable assumptions for some experimental studies in which territories are widely separated and intruders are easily repelled (e.g., Eberhard & Ewald 1994). However, territory size is often reduced by interactions among neighbors or with potential settlers. These competitors are themselves attempting to select a territory area and may not be so obliging about being removed. For this reason, several theoretical papers explicitly state that the optimality models assume that territories are not contiguous (Hixon 1980, Schoener 1983, Lima 1984). Theoretical representations of more crowded populations require the simultaneous analysis of the actions of multiple animals. When multiple decision-makers are incorporated into a territorial model, the payoffs to alternative strategies are frequency dependent (Davies 1980, Parker & Knowlton 1980) and game theory is needed.

Empirical Tests

Despite these limitations, optimality models are valuable heuristics, especially when territories are not contiguous or when the predicted behaviors do not stimulate strong reactions by opponents. This approach has drawn attention to the costs and benefits of defense; diverse empirical studies confirm that territorial animals respond to these variables. Some studies examine whether costs and benefits determine when an animal is territorial (e.g., Gill & Wolf 1975, Carpenter & MacMillen 1976) or how the mechanisms of defense vary with resource abundance (e.g., Ewald & Orians 1983) rather than making predictions about territory size per se. A few produce quantitative predictions of the optimum size to which measurements of actual territories can be compared (Dill 1978, Hixon et al. 1983). When such a prediction cannot be made with confidence, it may still be possible to test whether the territory sizes selected by experimental subjects yield greater net benefit than alternative choices, or at least whether yield is maximized at an intermediate territory size (Praw & Grant 1999). For example, Carpenter et al. (1983) relied on natural variation in the territory sizes of individual rufous hummingbirds, *Selasphorus rufus*, as the birds adjusted territory areas over a period of days. Some, but not all, of the birds attained the highest rate of weight gain at their final territory size.

In one of the few optimality models to address territory shape, Eason (1992) showed that red-capped cardinals, *Paroaria gularis*, can achieve greater energy intake by defending two short stretches of shoreline on opposite sides of a narrow lake, rather than a single long span along one shore. This is because territory shape affects the speed with which the resident can detect and evict intruders. Eason's analysis also showed that defense costs can have a greater effect than foraging economics on territory shape.

Far more studies test qualitative predictions concerning how territory size is expected to change when resource availability or the costs of defense change. These are considered below in the section on the effects of food availability and intruder pressure.

MODELS OF INTERACTIONS BETWEEN NEIGHBORS

The second major approach to the modeling of territory size and shape focuses on interactions among contiguous neighbors, rather than the unilateral choices of focal residents. These models describe how boundaries are settled and what determines their positions and shapes. Emphasis is on variation among animals, such as asymmetries in strength between adjacent residents, and how this produces unequal division of space. During boundary disputes, competitors assess one another's ability and motivation to fight (e.g., Adams 1990). Asymmetries in these properties affect the persistence and intensity of aggression and hence the size of the territory obtained by each resident. As a result, experimental manipulations of fighting ability, threat displays, or aggressiveness can lead to shifts in territory boundaries (e.g., Peek 1972, Watson & Parr 1981, Adams 1990).

Major variables affecting the sizes of contiguous territories include the degree of crowding and the body sizes, or group sizes, of residents. Field experiments often show that territories expand when neighbors are removed (e.g., Krebs 1971, Nursall 1977, Vines 1979, Hannon 1983, Boutin & Schweiger 1988, Butchart et al. 1999). Exceptions occur when the animals that are removed are replaced by floaters adopting essentially the same area (e.g., Ekman et al. 1981, Price et al. 1986). Some defensive behaviors are density dependent, arising when pressure from neighbors is pronounced (e.g., Clayton 1987). Especially in species with a high variance in body mass, territory area is often correlated with the size of the defending resident. This has been shown in birds (e.g., Petrie 1984, Butchart et al. 1999), fish (e.g., Grant et al. 1989, Tricas 1989, Elliott 1990, Keeley 2000), lizards (e.g., Simon 1975), and invertebrates (e.g., Hart 1985). When territories are defended by social groups, group size may predict territory size, as in social insects (e.g., Mabelis 1979, Adams & Levings 1987, Tschinkel et al. 1995), cooperatively breeding birds (e.g., Woolfenden & Fitzpatrick 1984, Brooker & Rowley 1995, Langen & Vehrencamp 1998), or group-living mammals (Macdonald 1983, Kruuk & MacDonald 1985). There are exceptions among both solitary (Norman & Jones 1984) and group territory holders (Brian et al. 1967, Macdonald 1983).

Both food requirements and fighting ability increase with body size; either may cause changes in the size of the defended area. A few studies show that the relative body size of neighboring residents is a better predictor of territory area than absolute body size of each resident. In the common moorhen, *Gallinula chloropus*, body size predicts territory area, but the ratio of the resident's weight to that of its neighbors is an even better predictor (Petrie 1984). Similarly, in the coot, *Fulica atra*, in which ability to hold territories varies with age, territory size depends on the ratio of a resident male's age to the ages of his neighbors (Cavé et al. 1989). In the fire ant *Solenopsis invicta*, colony size is correlated with territory size, but a tighter correlation is found when the local neighborhood of competitors is taken into account (Adams 1998). These results suggest that territory size in crowded populations is controlled largely by the relative fighting abilities of adjacent competitors.

Mechanistic Models

Two kinds of mechanistic models of neighbor interactions can be distinguished. The first uses geometric techniques to predict boundary positions, thereby dividing the habitat into territories. The second type does not assume the existence of distinct boundaries, but uses local rules of movement and interaction to predict the use of space by adjacent residents.

GEOMETRIC MODELS OF TERRITORY BOUNDARIES   Several models predict where boundaries should be established between adjacent neighbors under simple rules of interaction. These models yield quantitative predictions of the size and shape of each neighbor's territory. Most assume that the residents base their defense from central sites, such as nests or perches, to which they return after foraging bouts. Neighbors are seen as applying pressure against one another; boundaries are formed where this pressure is equal (Maynard Smith 1974, Patterson 1985, Adams 1998). It is often assumed, at least implicitly, that the likelihood or intensity of aggressive defense declines with distance from the central site, as is observed in many territorial animals (e.g., Vines 1979, Patterson 1980, McNicol & Noakes 1981, Melemis & Falls 1982, Giraldeau & Ydenberg 1987, Armstrong 1992, Eason 1992).

Early models suggested that hexagonal territories should be expected under dense packing in homogenous environments (Grant 1968, Maynard Smith 1974). This prediction arises under the assumptions that residents maintain equal spacing and have equivalent fighting abilities. Barlow (1974) showed that the cichlid fish *Tilapia mossambica* produced approximately polygonal territories with visible ridges defining the boundaries, usually with five or six sides. Grant's (1968) analysis of Holmes' data on territories of the pectoral sandpiper, *Calidris melanotos*, showed that adjacent sides of territories meet at an average interior angle of 117°, close to the 120° value expected for hexagons. These studies have been cited as confirming the predictions of hexagonal spacing, but they show considerable variation around the average, including territories with curved sides.

From this starting point, the assumptions underlying the prediction of hexagonal spacing can be relaxed.

1. *Irregular nest spacing.* If nest sites are not evenly spaced, then boundaries may be predicted by Dirichlet tessellation, in which straight line segments are placed midway between adjacent neighbors. This process has been used as a model, or a null model, of territory size or shape (Hasegawa & Tanemura 1976, Buckley & Buckley 1977, Tanemura & Hasegawa 1980, McCleery & Perrins 1985, Doncaster & Woodroffe 1993, Adams 1998). Dirichlet tessellation produces convex polygons, known as Thiessen areas or Voronoi polygons; each encloses the region closer to a particular nest than to any other nest.

2. *Unequal fighting abilities.* Some residents may have advantages owing to head starts in territory acquisition or to greater fighting ability. The resulting territories may be predicted by weighted tessellations, which produce boundary segments that lie farther away from stronger competitors (Getis & Boots 1978, Boots 1980). Models of this kind have been suggested to apply to territories of the badger, *Meles meles* (see Doncaster & Woodroffe 1993), and the fire ant *S. invicta* (see Adams 1998).

3. *Effects of territory compression.* As a territory is compressed, the resident may have a greater incentive to resist encroachment and also a greater ability, because its available time and effort are concentrated in a smaller space. This is one interpretation of Huxley's (1934) elastic disc hypothesis. A consequence is that a territory compressed on one side by a strong neighbor will bulge outward along boundaries with other neighbors (Adams 1998). This phenomenon is also suggested by the increased concentration of scent-marks in smaller territories as seen, for example, in the aardwolf, *Proteles cristatus* (see Richardson 1991).

4. *Uneven topography.* Territory boundaries in some species follow naturally occurring ridges, streams, or discontinuities in substrate or vegetation (e.g., Koford 1957, Watson & Miller 1971, Nursall 1977). Possible causes of this phenomenon have been discussed and tested (e.g., Reid & Weatherhead 1988, Eason et al. 1999). One might expect that placement of boundaries at landmarks would cause territory size to be more resistant to environmental fluctuations.

LOCAL RULES OF MOVEMENT AND INTERACTION   Another set of models views territory size and shape as properties that emerge as a consequence of the way aggressive animals forage, move, and react to competitors. For example, rules of movement and scent-marking by packs of timber wolves, *Canis lupus*, have been modeled to predict how these would affect patterns of spatial organization (Lewis & Murray 1993; White et al. 1996a,b). One set of partial differential equations describes how wolf density varies spatially as a result of a diffusive movement away from the den in search of food, a rate of return back to the den, and reactions to neighbors' scent-marks; a second set of equations describes the production and decay of scent-marks. When neighboring packs interact, the predicted equilibrium behavior is a partitioning of space into territories, with peak concentrations of scent-marks near the border and a trough of low wolf density at the zone of contact. Larger packs are predicted to hold larger territories (White et al. 1996a). All of these features agree with what is seen in natural populations.

When prey populations are also modeled (White et al. 1996b), an additional reason for boundary formation is suggested: prey populations reach peak levels in a buffer zone between wolf packs, reducing the incentive to travel past the boundary into the interior of a neighbor's territory. An advantage of this approach is that it yields predictions about animal density, which can be useful in modeling interactions with other species. However, for many species, it may be difficult and unnecessary to build a description of territorial behavior from rules of movement during foraging. Instead, actions at the boundary may be specifically directed towards competitors and may be shaped by the economics of defense, rather than arising statistically out of movement for other purposes.

Territories may be affected by heterogenous distribution of resources. Several models of foraging area shapes have been developed that take this heterogeneity into account, or that allow its description, but most do not concern territories and thus do not consider the costs of defense (Andersson 1978, Ford & Krumme 1979, Ford 1983). Covich's (1976) graphical model suggests that boundaries will form at zones of contact between adjacent foragers owing to depletion of food. He also discusses how deviations from simple shapes are expected for heterogenous resource distributions, which can occur because space is exploited unequally by the territorial residents. Getty (1981a) developed a space-utilization model that describes how the probability of use of a region of habitat decreases with distance from a central site, increases with the utility of the patch (which is affected by resource abundance), and decreases with use by conspecific animals. However, application of the model to home ranges of the Eastern chipmunk, *Tamias striatus*, produced no evidence that use of patches was affected by the actions of neighboring animals (Getty 1981c).

Stamps & Krishnan (1999) modeled the formation of territories as a learning process governed by sequences of positive and negative experiences occurring in different locations. Positive experiences cause an animal to return to a familiar area, whereas competitive encounters tend to reduce this tendency. As in Getty's model (1981a), the probability of use of a particular region is determined by the attractiveness of that region relative to other regions, from the point of view of the focal animal. In both these models attractiveness can vary as a function of habitat characteristics and according to the degree of use by other animals. However, Stamps & Krishnan's model focuses on the trial-and-error process by which the use of space arises. The learning rule employed is not always adaptive. If it were assumed that the animal learns according to how costs and benefits affect its fitness, the predictions of this model should converge on those of economic models, but Stamps & Krishnan's model shows how stochastic aspects of the learning process can lead to variation in territory size among otherwise equivalent animals.

Cost-Benefit and Game Theoretical Models

A few economic models of contiguous territories consider how differential costs and benefits affect competition among neighbors. Tullock (1983) described territories in which the value of resources declines as a function of distance from the nest site, and in which two or more neighbors may fight for control of locations around their nests. Each resident is willing to expend fighting costs that are equal to the value of the resources that can be obtained from a particular location, discounted by the probability that the resident will lose the fight. The resulting territories can be irregular in shape if resource distribution is heterogeneous or if nests cannot be moved without cost.

Hirata & Seno (1997) modeled a series of territories in a variable habitat, assuming that animals are arranged so that individuals of higher fighting ability are closer to the richest sites. Each animal was assumed to claim its optimum territory size, where benefits vary with resource levels and costs are a simple function of territory size. Analysis shows that the distribution of territory sizes depends on the spatial distribution of the resource. Jones & Krummel (1985) derived the expected sizes of contiguous territories and their dependence on food abundance. Their model incorporates feedback between territory size and population density in that the former determines rates of reproduction and the latter the cost of territory defense. However, food densities are assumed to remain stable long enough for population density and territory sizes to reach stable equilibria.

In these models, as in the single-actor optimality models, the costs of territory defense are not affected by neighbors' behavioral decisions. However, contests among territorial neighbors are interactive, so it is desirable to consider models in which the costs of defending a territory of a particular size depend on competitors' reactions. In this type of model payoffs for a decision rule are frequency dependent; hence, analysis of optimum territory size requires game theory. This is an area in great need of development in the theory of territory size, with only a few previous examples. Formally, one solves for an evolutionarily stable strategy (ESS), for which the expected payoff cannot be exceeded by an individual adopting a rare alternative strategy (Maynard Smith 1974). Some game theoretical models of territory size (e.g., Knowlton & Parker 1979, Parker & Knowlton 1980) are spatially implicit; that is, territories are assumed to be contiguous, but the model does not specify which strategies are adopted by which neighbors. The costs of defense therefore depend on what strategies are being played in the population at large, but not on the particular choices of neighbors. The ESS is to defend spitefully large territories—larger than the size that maximizes the number of surviving offspring—but the degree of spite declines with population size, and may be quite small if defense costs are high. ("Spite" refers to behaviors that lower the fitness of the animal that performs them and also of the animal with which it interacts.)

Maynard Smith (1982) developed a spatially explicit game theoretical model of territory size. A pair of neighbors with fixed territory centers struggles for control of the space between them. Each would like to control more space than the other is willing to concede. The ESS depends on the distance between the territory centers. This model is specifically concerned with signals, which are viewed as bids made during territorial bargaining, with the threat of a mutually injurious escalated fight if the bids do not lead to compromise. This basic framework for examining mutual decision-making and response between contiguous neighbors deserves further development.

There is also a large body of game theoretical models of animal contests for indivisible resources in which the winner takes all (Maynard Smith 1982, Dugatkin & Reeve 1998, Mesterton-Gibbons & Adams 1998). These are relevant to territorial contests in which a nonresident attempts to acquire a territory by displacing its owner; however, these models are not intended to make predictions about territory size. The conclusions of these models cannot be assumed to apply to struggles over the division of space, without explicit confirmation, but major themes may be applicable, including the emphasis on asymmetries between animals as determinants of the outcome of contests (Ydenberg et al. 1988).

MODELS OF INTERACTIONS BETWEEN RESIDENTS AND POTENTIAL SETTLERS

The third approach to territory modeling describes the struggle between residents that have established territories and those that are seeking to acquire them. These models typically lead to predictions about how many animals can acquire territories in a particular habitat patch, which determines average territory size. Three kinds of phenomena are emphasized: the distribution of animals across habitats, the possibility that strategic interactions among neighbors can reduce the risk of losses to new settlers, and the comparative effects of simultaneous and sequential establishment.

In their seminal paper on habitat distribution, Fretwell & Lucas (1970) distinguished alternative ways that territorial behavior might affect animal distributions. If every animal has a perfect ability to assess habitat quality and is able to move to the patch in which its fitness is maximized, an "ideal free distribution" results. Under this hypothesis, territoriality allows animals to assess local density but does not prevent any individual from settling in the most favorable habitat. Alternatively, territorial behavior may limit local densities so that some animals are prevented from settling in their preferred location, producing an "ideal dominance distribution." Average fitness may then vary among habitats, and territory defense may exert negative density-dependent effects on population growth rates. Many variants have been developed (reviewed by Milinski & Parker 1991), including the important extension to cases in which individuals differ in competitive ability (e.g., Parker & Sutherland 1986). These models are game theoretical; they allow determination of ESSs for particular types of competition.

Most of these habitat distribution models assume mechanisms of competition other than territory defense or else do not make explicit predictions about individual territory sizes. An exception is Sutherland's (1996) model, which describes the ESS distribution of territorial animals that vary in competitive ability across habitat patches of differing quality. Each animal is assumed to choose a territory size that optimizes net gain or to become a nonterritorial floater if it cannot obtain a territory of sufficient size and quality. An important result is that the size of the territory obtained by a particular animal depends on the willingness of mobile competitors to compete for space in the same region. This in turns depends on the quality of alternative sites and the density and characteristics of competitors at each location. The resulting relationships can be complex; for example, territory sizes may be either larger or smaller in high quality habitats than in low quality habitats (Sutherland 1996). Models of this kind link territorial decisions to population-level phenomena, including the proportion of floaters and changing densities in preferred and marginal habitats.

Patterson's (1985) laboratory experiments on territory settlement in convict cichlids, *Cichlasoma nigrofasciatum*, illustrate the importance of considering decisions by intruders, as well as by residents, as determinants of local territory size. The fish had the option of establishing territories around preferred nest sites with artificial caves, establishing territories near less-preferred nest sites with flat stones, or remaining in areas with open sand. The equilibrium density of fish was greatest in the preferred habitat; hence, territory size was smaller than in the less-preferred habitat. Detailed behavioral observations showed that territory size was affected more by differential persistence of intruders than by differential defense by residents.

Interactions among neighbors may affect the prospects for additional settlers to acquire territories. As animals establish territories sequentially in open habitat, the first residents may regulate their behavior strategically in order to deny entry by later arrivals. Several possible tactics have been proposed. Established residents may defend areas larger than what would otherwise be favored in order to prevent competitors from establishing themselves in interstitial areas (Burger 1981). Sequentially arriving residents may establish territories so close to adjacent residents that later arrivals are unlikely to find sufficient space between them (Getty 1981b, Stamps & Krishnan 1990). This can result in an early sacrifice of territory area in order to achieve a greater final territory size, but the utility of this tactic requires that early arrivers can predict the number and persistence of later potential settlers (Adams 1994). In addition, adjacent residents may coordinate their defense against competitors that attempt to intrude near their mutual boundary (Getty 1987).

A related question is whether territory sizes are affected by the timing of territory establishment, that is, whether the settlement is synchronous or sequential (e.g., Tanemura & Hasegawa 1980). Maynard Smith (1974) proposed that territory establishment passes through two phases: a phase in which territory centers are not fixed and animals move to positions midway between their boundaries, and a later phase in which territory centers are stationary. Sequential settlement allows animals to space themselves farther from neighbors before fixing their positions, resulting in lower densities and greater territory sizes. Few field experiments have examined the effects of simultaneous arrival during initial territory establishment, and at least some of these provide no support for the hypothesis (Stamps 1992). However, when multiple animals are removed from established territory mosaics, the degree of synchrony of removal may affect the sizes of territories acquired by the replacements (Knapton & Krebs 1974; but see Clayton & Vaughan 1986, Klenner 1991). This appears to be due to the stability of territory boundaries defended by remaining residents. Thus, when only one territory resident is removed or naturally disappears, its replacement inherits the same border positions, but when an entire group of neighbors is removed, new boundaries may be unaffected by the traditions of the previous residents.

FOOD AVAILABILITY AND INTRUDER PRESSURE AS DETERMINANTS OF TERRITORY SIZE

Regardless of modeling approach, a major problem is to understand the effects of resource abundance and intruder pressure on territory size. Many studies examine the effects of only one of these factors. Territory size is often negatively correlated with resource abundance across habitats, years, or individuals within habitats (e.g., Stenger 1958, Cody & Cody 1972, Gass et al. 1976, Salomonson & Balda 1977, Seastedt & MacLean 1979, Village 1982, Smith & Shugart 1987). It is also common to find negative correlations between territory area and some measure of the costs imposed by intruders, such as the local density of competitors (e.g., Turpie 1995), the degree to which the territory boundary is contiguous with neighboring territories (e.g., Norman & Jones 1984), rates of encounters with competitors or intrusion onto the territory (e.g., Dill et al. 1981, Jabłoński 1996), or the degree of visibility, which can affect the ability to detect intruders (Watson & Moss 1972). Experimental manipulation of intruder pressure can lead to changes in territory size (e.g., Norton et al. 1982).

Despite the emphasis in these studies on foraging, it is worth remembering that other environment variables can influence territory size and shape, such as the availability of perch sites (Yosef & Grubb 1994), the visibility of areas that must be defended (Eason & Stamps 1992), and the dispersion of potential mates (e.g., Butchart et al. 1999).

Multivariate Studies

Mutivariate studies allow a more complete understanding than single-factor studies of the interrelationships between resources, intruder pressure, and territory size. Myers et al. (1979) distinguished two possible reasons why territories might shrink when food supply is increased. First, territory residents may respond directly to the abundance of food in deciding how large an area to defend. Second, increased resource density may attract more competitors, causing the resident to lower its territory size. These two possibilities are not mutually exclusive; indeed, models of optimal territory size often consider both effects. Myers et al. (1979) used multivariate statistical methods to evaluate the relationships among rates of intrusion, food abundance, and territory size. The sizes of territories defended by sanderlings, *Calidris alba*, were negatively correlated with both prey density and rates of intrusion by nonterritorial birds. When the relationship between prey density and territory size was accounted for statistically, there was still a significant negative relationship between intruder density and territory size. However, when the effects of intruder density were accounted for, there was no longer a significant relationship between prey density and territory size. Myers et al. (1979) concluded that the costs of defense have a greater effect on territory size variation than direct responses of residents to food supply.

Similar multivariate analyses have been conducted with other animals, with varying results. In some populations, there was no significant effect of food abundance on territory size once the intensity of competition was accounted for (Gauthier 1987, Dunk & Cooper 1994, Tripp & Collazo 1997). In others, food abundance was negatively correlated with territory size even after accounting for competitor density or chase rates (Dill et al. 1981, McFarland 1986, Temeles 1987, Tricas 1989). Occasionally, both prey density and competition by neighbors have significant independent effects (e.g., Temeles 1987).

Caution must be exercised in interpreting these and other correlative studies. Intrusion pressure can be difficult to quantify, and the aspects that are easiest to measure may not be those with the strongest effects on the resident's behavior. Few studies evaluate which measure is the best predictor of territory size (but see Temeles 1987). After territory boundaries have been established, the effect of neighbors may be missed because, in many species, neighbors rarely intrude yet have strong negative effects on territory size. In general, causation can be difficult to discern from correlations. Controlled experiments are the preferred way to determine the direction of causation, and the results of experiments do not always coincide entirely with correlative studies of the same species (e.g., McFarland 1986, Tricas 1989).

Experimental Studies on Food Availability

Table 1 summarizes 40 experimental studies on the effects of food abundance on territory size. The designs and specific intentions of these experiments varied greatly. In some, food was presented at feeding stations in the interior of selected territories, whereas in others food was widely dispersed over several territories. Many researchers followed behavioral adjustments over a period of hours or days; others measured demographic responses over seasons or years. Shortcomings of some studies include small sample sizes, few or no simultaneous controls, and use of nonindependent samples (e.g., contiguous territories). Most were field studies, but some were performed in the laboratory (Symons 1971, Slaney & Northcote 1974, parts of Stamps & Tanaka 1981, Keeley 2000). Boutin (1990) has reviewed additional experiments on the effects of food increases on animal spacing behavior.

TABLE 1 Experimental studies on the effect of food supply on territory size

Taxon	Source	Type[a]	Treatment[b]	Placement[c]	Duration	Sample sizes[d]		Result[f]
						Experimental	Control[e]	
Invertebrates								
Agelenopsis aperta	Riechert 1981	C	V	I	1–4 weeks	1–6 at each level		NS
Leucotrichia pictipes larvae	Hart 1985	NC	R	E	7 days	19	16	Neg
Lottia gigantea	Stimson 1973	NC	R	E	7 months	10	10	Neg
		NC	R	E	2 weeks	3	5	Neg
Fish								
Chaetodon multicinctus	Tricas 1989	C	I & R	D	1–7 days	7	10	Neg[g]
Embiotica jacksoni	Hixon 1980	NC	R	P	4 h	1	1	Neg
Eupomacentrus leucostictus	Ebersole 1980	NC	I	C	1 h–3 days	1	1	Neg
		NC	I	G	2 days	22	0	Females Pos; males NS
Oncorhynchus mykiss	Keeley 2000	C	V	U	2 months	6 at each level		Neg
Parma victoriae	Norman & Jones 1984	NC	R	D	1 week	10	10	NS
Salmo salar	Symons 1971	PC	I & R	D	~1 week	20	10	NS
Salmo gairdneri	Slaney & Northcote 1974	C	V	U	≥65 days	3	0	NS
		C	V	U	8 days	1 at each level		Neg
Lizards								
Anolis aeneus juveniles	Stamps & Tanaka 1981	OV	I	E	1–8 days	5 plots	2 plots	V
		OV	V	E	11 days	3 rooms	0	Neg
		OV	V	E	1 month	2 rooms at each level		NS
Norops humilis	Guyer 1988	OV?	I	G	7 months	3	3	NS
Sceloporus jarrovi	Simon 1975	OV	I	C	10–14 days	17	16	Neg
Sceloporus undulatus	Ferguson et al. 1983	OV	I	I	17 days	28–35	26–37	Neg 1 reps; NS 2 reps
Uta stansburiana	Waldschmidt 1983	OV?	I	I	1 month	8	12	NS

(Continued)

TABLE 1 (*Continued*)

Taxon	Source	Type[a]	Treatment[b]	Placement[c]	Duration	Sample sizes[d] Experimental	Control[e]	Result[f]
Birds								
Anthus petrosus	Askenmo et al. 1994	C	I	C	3 weeks	8	8	NS
	Arvidsson et al. 1997	C	I	C	~1 month	10	10	Pos
Calypte anna	Eberhard & Ewald 1994	NC	V	G	2 days	11	11	V
Lagopus lagopus scoticus	Miller et al. 1970	C	I	P	7 years	1 plot	1 plot	Neg
	Watson et al. 1984	C	I	E	3 years	1 plot	1 plot	Neg
		C	I	E	4 years	3 plots	3 plots	Neg 2 reps; NS 1 rep
Parus major	Krebs 1971	C	I	G	7 months	1 plot	1 plot	NS
Phylidonyris novaehollandiae	McFarland 1994	?	I	D	3 days	3	1	NS
		?	R	D	3 days	3	1	NS
Pipilo erythrophthalmus	Franzblau & Collins 1980	?	I	D?	3 weeks	5	5	NS
Prunella modularis	Davies & Lundberg 1984	C	I	C	3–6 months	28	39	Females Neg; males NS
Prunella collaris	Nakamura 1995	C-PC	I	C	2 months	4–6	4–6	NS; repeated 3 yrs
Psophia leucoptera	Sherman & Eason 1998	C	R	D	6 days	4	0	NS
Selasphorus rufus	Hixon et al. 1983	PC	R	D?	2 days	3	3	Neg
Sitta europaea	Enoksson & Nilsson 1983	C	I	D/C	1–4 months	2 plots	2 plots	Neg
	Enoksson 1990	C	I	G	5 months	3 plot-yrs	3–5 plot-yrs	Neg
Mammals								
Clethrionomys rufocanus	Ims 1987	OV	I	S	25 days	13	11	Females Neg

Madoqua kirkii	Brotherton & Manser 1997	C-PC	I	C	93 days	6	6	NS
Microtus californicus	Ostfeld 1986							
Females		OV	I	P	10 days	2 plots	2 plots	Neg
Males		C	I	P	10 days	2 plots	2 plots	NS
Microtus townsendii	Taitt & Krebs 1981	?	I	G	11 months	3 plots	1 plot	Neg
Peromyscus leucopus	Wolff 1985	OV	I	G	2 years	2 plots	3 plots	Females Neg?; Males NS
Peromyscus maniculatus	Wolff 1986	OV	I	G	105 days	2 plots	2 plots	NS
	Taitt 1981	OV	I	G	1.5 years	1 plot	1 plot	Neg
	Wolff 1985	OV	I	G	2 years	2 plots	3 plots	NS
Sciurus carolinensis	Kenward 1985	OV	I	S	3 weeks	11	9	NS
		OV	I	S	~10 weeks	8	10	Females NS; males Neg
Tamias striatus	Mares et al. 1976	OV	I	C	3 weeks	≥ 9	0	Neg
	Mares et al. 1982	OV	I	C	10 days	?	0	Neg
Tamiasciurus hudsonicus	Klenner 1991	C-PC	I	D	20 days	41–49	0	Neg
				D	up to 2 weeks	2 plots	4	Neg

[a] C, contiguous, largely nonoverlapping territories; PC, partly contiguous, nonoverlapping; C-PC, completeness of contiguity not certain; NC, noncontiguous defended areas; OV, overlapping but partially defended home ranges; ?, degree of contiguity or overlap not described.

[b] I, availability of food increased experimentally; R, availability of food reduced experimentally; V, several levels of food availability created.

[c] C, 1 or 2 central sites in the territory interior; D, dispersed across numerous locations per territories; E, evenly spread throughout territory; G, regular grid of feeding stations; 1, food delivered to individual animal at its location; S, feeding stations used by several animals; P, one habitat patch including parts of several home ranges; U, food delivered upstream, carried downstream by water flow.

[d] Number of individuals, except as noted.

[e] Number of controls followed simultaneously with experimentally altered replicates; most studies also included "before" and "after" measurements of territory size.

[f] Pos, larger territories at higher food availability; Neg, smaller territories at higher food availability; NS, no significant effect of changes in food availability; V, effects varied or complex.

[g] Except at low food densities.

Most studies (23 of 40) found a significant negative relationship between food availability and territory size for at least one comparison, but it was also common to find no significant effect of resource abundance on territory area (14 studies). Positive associations between territory size and food availability are rarely reported and thus deserve special mention. Consistent with the predictions of his model, Ebersole (1980) found that females, but not males, of the reef fish *Eupomacentrus leucostictus* increased territory size when food was added through their territories and beyond. This study has been criticized for lacking controls and because the supplemental food may have been highly attractive compared to what was naturally available (Norman & Jones 1984). However, these supplemental food sources were also placed within the premanipulation boundaries, so the fish could have acquired more food than they originally consumed without defending a larger area. Arvidsson et al. (1997) found that male rock pipits, *Anthus petrosus*, given supplemental food at hidden sites in the interiors of their territories established larger territories than did control males. This was in contrast to an earlier study in which food was offered only after territories had been established (Askenmo et al. 1994). However, the increased size of food-supplemented territories was not due to greater length along the shore of the island at the expense of neighbors, but rather to greater expansion into less-preferred habitat in the interior of the island, which was apparently uncontested.

When territory size decreased, most of the experiments did not estimate the degree to which this was due to increased attraction of competitors. In a study on the Eastern chipmunk, *T. striatus*, Mares et al. (1982) prevented attraction of additional competitors by trapping immigrants. Home range sizes nonetheless declined following experimental supplementation of seeds. In a two-factor experiment Norman & Jones (1984) found that removal of neighbors caused substantial increases in territory size at all food levels, whereas manipulations of food supply had no significant effect. It is also possible to guard against attraction of competitors to food in single-factor experiments by offering the food at a hidden central site, where it is detected only by the resident (Askenmo et al. 1994, Arvidsson et al. 1997).

In many of the experimental studies (35%), there was no significant effect of food abundance on territory size, often in spite of large quantities of food provided and increased attraction of competitors. This parallels the results of many observational studies in which territory size does not vary in response to large fluctuations in the natural supply of food (e.g., Southern 1970). Not all of these negative results can be attributed to low statistical power.

Why do some animals adjust territory size to food abundance, but not others? Several possible reasons have been suggested. First, the territory may serve functions other than protection of food supply (e.g., Waldschmidt 1983, Wolff 1985) or the resident may obtain food off the territory (Myers et al. 1979), lowering its sensitivity to food abundance within the territory limits. Second, some animals may not be able to accurately assess food supply directly (Riechert 1981) and thus may respond more to predictors of expected food supply such as characteristics of the vegetation (Seastedt & MacLean 1979, Franzblau & Collins 1980, Stamps 1994). Third, the nature of the response may depend on the period of observation. For example, Waldschmidt (1983) found that home range estimates for lizards given food supplements rose more slowly with subsequent captures than those of unfed controls, but leveled off at similar values. Too short a period of observation might give the appearance of a reduced territory owing to differences in use of space. Fourth, responses to food supply may not occur where attempts to expand are opposed; thus, short-term adjustments may be less likely when territories are contiguous and nonoverlapping (Patterson 1980, Norman & Jones 1984, Sherman & Eason 1998). Finally, it may be easier to reduce the size of a territory during periods of food abundance than to expand it during periods of food shortage. Territory size may therefore be adjusted to average food availability, or to food availability during periods of shortage (Southern 1970, Myers et al. 1979, von Schantz 1984). This strategy can stabilize population densities in the face of environmental fluctuations. When it is difficult to adjust territory size, residents may respond to changes in food abundance in other ways, such as by allowing more subordinate animals on the territory (von Schantz 1984, McFarland 1994) or by changing defense tactics and the degree of exclusivity (Ewald & Carpenter 1978, Ewald & Bransfield 1987, Sherman & Eason 1998).

Examination of Table 1 suggests that responses to changes in food abundance are less likely for contiguous absolute territories than for noncontiguous territories or overlapping home ranges. Several of the exceptions are long-term studies on populations with contiguous territories in which the food supplementation continued long enough to promote recruitment of juveniles (Miller et al. 1970, Enoksson & Nilsson 1983, Watson et al. 1984, Enoksson 1990). Clearly, other factors contribute to variation in the outcome of these experiments as well, such as the way in which food supplements are distributed.

CONCLUSIONS

The central goal of this review is to clarify conceptual differences among alternative approaches to territory modeling and to indicate the relationship of each approach to empirical studies. The main conclusion is that an expansion of theory beyond the now familiar optimality models will improve our ability to account for population-level effects of territory defense. I conclude by highlighting the strengths and limitations of the three major bodies of theory and by pointing to gaps in our current ability to evaluate these models.

Limitations of optimality models have been discussed in some detail because of the prominence of this approach. Because they focus on decisions by individuals and assume implicitly that territories are not contiguous, simple optimality models are not well suited to predicting the ecological consequences of territory defense in crowded populations. Furthermore, this approach usually does not allow quantitative prediction of territory size in natural circumstances based on information that can be easily gathered. Nevertheless, some of the primary insights of the optimality approach have been confirmed, and the themes identified by this body of work must remain an integral part of territory studies. Optimality models encourage the investigator to examine time and energy budgets in an economic framework. This line of inquiry can reveal the criteria and decision rules used to select territory size (e.g., Hixon 1980). More generally, many empirical studies have shown that the abundance of food and the costs of expelling intruders are important determinants of territory area. However, surprisingly few experiments have been designed to distinguish between direct responses to increased food supply and indirect responses owing to attraction of competitors.

Models of interactions among neighbors are more useful, from the standpoint of population ecology, because they predict the outcome of interactions among multiple competitors in crowded populations. This approach encourages examination of mutual reactions among adjacent residents and how these mold the use of space. Several geometric models produce quantitative predictions of the size and shape of each resident's territory. Although there is the potential for these models to become diverse and complex, it has been possible to account for much of the observed variation in territory size and shape by employing fairly simple assumptions (e.g., Adams 1998). However, geometric models have thus far been tested only by measuring correlations between observed and predicted territories, rather than by experimental manipulation of the factors hypothesized to affect boundary positions. Furthermore, these models are mechanistic, rather than adaptationist, and they largely ignore responses to variable food supply.

An important goal for future studies is to link mechanistic models of boundary formation with the cost-benefit approach. This requires game theory, because the optimum choice by a particular individual depends on what other members of the population are doing. Instead of asking how a focal resident should change its behavior when food supply increases, such a model would ask how the balance of aggression between two or more neighbors can be expected to shift when all individuals experience an increase in food abundance. This may allow an understanding of which territorial systems are sensitive to changes in food availability and which are not.

Territory division among neighbors can also be described as the outcome of local rules of movement and interaction. These models can account for some general features of territorial systems, such as the existence of boundaries, but by and large they have not been subjected to rigorous empirical tests. As a consequence, we do not currently know whether the behavioral rules have been correctly identified or whether models based on these rules yield more robust predictions of population-level phenomena than simpler models lacking this level of detail. A challenge for this body of theory will be to show that it produces more accurate predictions than the alternatives.

Models of contests between established residents and potential settlers focus attention on the decisions of nonterritorial intruders as well as those of territory holders. Because the persistence of an intruder in attempting to acquire a territory should depend on its prospects for obtaining a territory elsewhere, this approach connects the behavior of territory defense to large-scale patterns of distribution across habitat patches. Of the three broad approaches reviewed here, models of resident-settler interactions produce the most complete description of the density-dependent effects of territory defense. They also predict the population responses to habitat loss and degradation (e.g., Sutherland 1996), linking territory behavior to landscape and conservation ecology. Few field studies have been designed specifically to test this set of models, but there will be a considerable payoff for doing so.

ACKNOWLEDGMENTS

I thank Perri K. Eason, Chris Elphick, James W. A. Grant, Michael Mesterton-Gibbons, and Kentwood D. Wells for helpful comments on this manuscript. This work was supported by a grant from the National Science Foundation (IBN-9874451).

Visit the Annual Reviews home page at www.AnnualReviews.org

LITERATURE CITED

Adams ES. 1990. Boundary disputes in the territorial ant *Azteca trigona*: effects of asymmetries in colony size. *Anim. Behav.* 39:321–28

Adams ES. 1994. Settlement tactics in seasonally territorial animals: resolving conflicting predictions. *Am. Nat.* 143:939–43

Adams ES. 1998. Territory size and shape in fire ants: a model based on neighborhood interactions. *Ecology* 79:1125–34

Adams ES, Levings SC. 1987. Territory size and population limits in mangrove termites. *J. Anim. Ecol.* 56:1069–81

Andersson M. 1978. Optimal foraging area: size and allocation of search effort. *Theor. Popul. Biol.* 13:397–409

Armstrong DP. 1992. Correlation between nectar supply and aggression in territorial honeyeaters: causation or coincidence? *Behav. Ecol. Sociobiol.* 30:95–102

Arvidsson B, Askenmo C, Neergaard R. 1997. Food supply for settling male rock pipits affects territory size. *Anim. Behav.* 54:67–72

Askenmo C, Neergaard R, Arvidsson BL. 1994. Food supplementation does not affect territory size in rock pipits. *Anim. Behav.* 47:1235–37

Barlow GW. 1974. Hexagonal territories. *Anim. Behav.* 22:876–78

Bollmann K, Reyer H-U, Brodmann PA. 1997. Territory quality and reproductive success: Can water pipits *Anthus spinoletta* assess the relationship reliably? *Ardea* 85:83–98

Boots BN. 1980. Weighting Thiessen polygons. *Econ. Geog.* 56:248–59

Boutin S. 1990. Food supplementation experiments with terrestrial vertebrates: patterns, problems, and the future. *Can. J. Zool.* 68:203–20

Boutin S, Schweiger S. 1988. Manipulation of intruder pressure in red squirrels (*Tamiasciurus hudsonicus*): effects on territory size and acquisition. *Can. J. Zool.* 66:2270–74

Brian MV, Elmes G, Kelly AF. 1967. Populations of the ant *Tetramorium caespitum* Latreille. *J. Anim. Ecol.* 36:337–42

Brooker M, Rowley I. 1995. The significance of territory size and quality in the mating strategy of the splendid fairy-wren. *J. Anim. Ecol.* 64:614–27

Brotherton PNM, Manser MB. 1997. Female

dispersion and the evolution of monogamy in the dik-dik. *Anim. Behav.* 54:1413–24

Brown JL. 1964. The evolution of diversity in avian territorial systems. *Wilson Bull.* 76:160–69

Buckley PA, Buckley FG. 1977. Hexagonal packing of royal tern nests. *Auk* 94:36–43

Burger J. 1981. Super territories: a comment. *Am. Nat.* 118:578–80

Butchart SHM, Seddon N, Ekstrom JMM. 1999. Polyandry and competition for territories in bronze-winged jacanas. *J. Anim. Ecol.* 68:928–39

Carpenter FL. 1987. The study of territoriality: complexities and future directions. *Am. Zool.* 27:401–9

Carpenter FL, MacMillen RE. 1976. Threshold model of feeding territoriality and test with a Hawaiian honeycreeper. *Science* 194:639–42

Carpenter FL, Paton DC, Hixon MA. 1983. Weight gain and adjustment of feeding territory size in migrant hummingbirds. *Proc. Natl. Acad. Sci. USA* 80:7259–63

Cavé AJ, Visser J, Perdeck AC. 1989. Size and quality of the coot *Fulica atra* territory in relation to age of its tenants and neighbours. *Ardea* 77:87–98

Clayton DA. 1987. Why mudskippers build walls. *Behaviour* 102:185–95

Clayton DA, Vaughan TC. 1986. Territorial acquisition in the mudskipper *Boleophthalmus boddarti* (Teleostei, Gobiidae) on the mudflats of Kuwait. *J. Zool.* 209:501–19

Cody ML, Cody CBJ. 1972. Territory size, clutch size, and food in populations of wrens. *Condor* 74:473–77

Covich AP. 1976. Analyzing shapes of foraging areas: some ecological and economic theories. *Annu. Rev. Ecol. Syst.* 7:235–57

Davies NB. 1980. The economics of territorial behaviour in birds. *Ardea* 68:63–74

Davies NB, Houston AI. 1984. Territory economics. In *Behavioural Ecology: An Evolutionary Approach*, ed. JR Krebs, NB Davies, pp. 148–69. Oxford: Blackwell Sci. 2nd ed.

Davies NB, Lundberg A. 1984. Food distribution and a variable mating system in the dunnock, *Prunella modularis. J. Anim. Ecol.* 53:895–912

Dill LM. 1978. An energy-based model of optimal feeding-territory size. *Theor. Popul. Biol.* 14:396–429

Dill LM, Ydenberg RC, Fraser AHG. 1981. Food abundance and territory size in juvenile coho salmon (*Oncorhynchus kisutch*). *Can. J. Zool.* 59:1801–9

Doncaster CP, Woodroffe R. 1993. Den site can determine shape and size of badger territories: implications for group-living. *Oikos* 66:88–93

Dugatkin LA, Reeve HK, eds. 1998. *Game Theory and the Study of Animal Behaviour.* Oxford: Oxford Univ. Press

Dunk JR, Cooper RJ. 1994. Territory-size regulation in black-shouldered kites. *Auk* 111:588–95

Eason P. 1992. Optimization of territory shape in heterogeneous habitats: a field study of the red–capped cardinal (*Paroaria gularis*). *J. Anim. Ecol.* 61:411–24

Eason PK, Cobbs GA, Trinca KG. 1999. The use of landmarks to define territorial boundaries. *Anim. Behav.* 58:85–91

Eason PK, Stamps JA. 1992. The effect of visibility on territory size and shape. *Behav. Ecol.* 3:166–72

Eberhard JR, Ewald PW. 1994. Food availability, intrusion pressure and territory size: an experimental study of Anna's hummingbirds (*Calypte anna*). *Behav. Ecol. Sociobiol.* 34:11–18

Ebersole JP. 1980. Food density and territory size: an alternative model and a test on the reef fish *Eupomacentrus leucostictus. Am. Nat.* 115:492–509

Ekman J, Cederholm G, Askenmo C. 1981. Spacing and survival in winter groups of willow tit *Parus montanus* and crested tit *P. cristatus*—a removal study. *J. Anim. Ecol.* 50:1–9

Elliott JM. 1990. Mechanisms responsible for population regulation in young migratory trout, *Salmo trutta.* III. The role of territorial behaviour. *J. Anim. Ecol.* 59:803–18

Emlen ST, Oring L. 1977. Ecology, sexual

selection, and the evolution of mating systems. *Science* 197:215–23

Enoksson B. 1990. Autumn territories and population regulation in the nuthatch *Sitta europaea*: an experimental study. *J. Anim. Ecol.* 59:1047–62

Enoksson B, Nilsson SG. 1983. Territory size and population density in relation to food supply in the nuthatch *Sitta europaea* (Aves). *J. Anim. Ecol.* 52:927–35

Ewald PW, Bransfield RJ. 1987. Territory quality and territorial behavior in two sympatric species of hummingbirds. *Behav. Ecol. Sociobiol.* 20:285–93

Ewald PW, Carpenter FL. 1978. Territorial responses to energy manipulations in the Anna hummingbird. *Oecologia* 31:277–92

Ewald PW, Orians GH. 1983. Effects of resource depression on use of inexpensive and escalated aggressive behavior: experimental tests using Anna hummingbirds. *Behav. Ecol. Sociobiol.* 12:95–101

Ferguson GW, Hughes JL, Brown KL. 1983. Food availability and territorial establishment of juvenile *Sceloporus undulatus*. See Huey et al. 1983, pp. 134–48

Ford RG. 1983. Home range in a patchy environment: optimal foraging predictions. *Am. Zool.* 23:315–26

Ford RG, Krumme DW. 1979. The analysis of space use patterns. *J. Theor. Biol.* 76:125–55

Franzblau MA, Collins JP. 1980. Test of a hypothesis of territory regulation in an insectivorous bird by experimentally increasing prey abundance. *Oecologia* 46:164–70

Fretwell SD, Lucas HL Jr. 1970. On territorial behaviour and other factors influencing habitat distribution in birds. I. Theoretical development. *Acta Biotheor.* 19:16–36

Gass CL, Angehr G, Centa J. 1976. Regulation of food supply by feeding territoriality in the rufous hummingbird. *Can. J. Zool.* 54:2046–54

Gauthier G. 1987. Brood territories in buffleheads: determinants and correlates of territory size. *Can. J. Zool.* 65:1402–10

Getis A, Boots B. 1978. *Models of Spatial Processes: An Approach to the Study of Point,* *Line and Area Patterns.* Cambridge: Cambridge Univ. Press

Getty T. 1981a. Analysis of central–place space-use patterns: the elastic disc revisited. *Ecology* 62:907–14

Getty T. 1981b. Competitive collusion: the preemption of competition during the sequential establishment of territories. *Am. Nat.* 118: 426–31

Getty T. 1981c. Territorial behavior of eastern chipmunks (*Tamias striatus*): encounter avoidance and spatial time-sharing. *Ecology* 62:915–21

Getty T. 1987. Dear enemies and the prisoner's dilemma: Why should territorial neighbors form defensive coalitions? *Am. Zool.* 27:327–36

Gill FB, Wolf LL. 1975. Economics of feeding territoriality in the golden-winged sunbird. *Ecology* 56:333–45

Giraldeau L-A, Ydenberg R. 1987. The center-edge effect: the result of a war of attrition between territorial contestants? *Auk* 104:535–38

Gordon DM. 1997. The population consequences of territorial behavior. *Trends Ecol. Evol.* 12:63–66

Gould SJ, Lewontin RC. 1979. The spandrels of San Marco and the Panglossian paradigm: a critique of the adaptationist programme. *Proc. R. Soc. London Ser. B.* 205:581–98

Grant JWA, Noakes DLG. 1987. A simple model of optimal territory size for drift-feeding fish. *Can. J. Zool.* 65:270–76

Grant JWA, Noakes DLG, Jonas KM. 1989. Spatial distribution of defence and foraging in young-of-the-year brook charr, *Salvelinus fontinalis*. *J. Anim. Ecol.* 58:773–84

Grant PR. 1968. Polyhedral territories of animals. *Am. Nat.* 102:75–80

Guyer C. 1988. Food supplementation in a tropical mainland anole, *Norops humilis*: effects on individuals. *Ecology* 69:362–69

Hannon SJ. 1983. Spacing and breeding density of willow ptarmigan in response to an experimental alteration of sex ratio. *J. Anim. Ecol.* 52:807–20

Hart DD. 1985. Causes and consequences of

territoriality in a grazing stream insect. *Ecology* 66:404–14

Hasegawa M, Tanemura M. 1976. On the pattern of space division by territories. *Ann. Inst. Statist. Math* 28:509–19

Hirata H, Seno H. 1997. How does the size distribution of male territories depend on the spatial distribution of females? *Ecol. Model.* 103:193–207

Hixon MA. 1980. Food production and competitor density as the determinants of feeding territory size. *Am. Nat.* 115:510–30

Hixon MA, Carpenter FL, Paton DC. 1983. Territory area, flower density, and time budgeting in hummingbirds: an experimental and theoretical analysis. *Am. Nat.* 122:366–91

Hölldobler B, Lumsden CJ. 1980. Territorial strategies in ants. *Science* 210:732–39

Huey RB, Pianka ER, Schoener TW. eds. 1983. *Lizard Ecology: Studies of a Model Organism.* Cambridge, MA: Harvard Univ. Press

Hunt GL Jr, Hunt MW. 1976. Gull chick survival: the significance of growth rates, timing of breeding and territory size. *Ecology* 57:62–75

Huxley JS. 1934. A natural experiment on the territorial instinct. *Br. Birds* 27:270–77

Ims RA. 1987. Responses in spatial organization and behaviour to manipulations of the food resource in the vole *Clethrionomys rufocanus. J. Anim. Ecol.* 56:585–96

Jabłoński PG. 1996. Intruder pressure affects territory size and foraging success in asymmetric contests in the water strider *Gerris lacustris. Ethology* 102:22–31

Jones DW, Krummel JR. 1985. The location theory of animal populations: the case of a spatially uniform food distribution. *Am. Nat.* 126:392–404

Keeley ER. 2000. An experimental analysis of territory size in juvenile steelhead trout. *Anim. Behav.* 59:477–90

Kenward RE. 1985. Ranging behaviour and population dynamics in grey squirrels. See Sibly & Smith 1985, pp. 319–30

Klenner W. 1991. Red squirrel population dynamics. II. Settlement patterns and the response to removals. *J. Anim. Ecol.* 60:979–93

Knapton RW, Krebs JR. 1974. Settlement patterns, territory size, and breeding density in the song sparrow (*Melospiza melodia*). *Can. J. Zool.* 52:1413–20

Knowlton N, Parker GA. 1979. An evolutionarily stable strategy approach to indiscriminate spite. *Nature* 279:419–21

Kodric-Brown A, Brown JH. 1978. Influence of economics, interspecific competition, and sexual dimorphism on territoriality of migrant rufous hummingbirds. *Ecology* 59:285–96

Koford CB. 1957. The vicuña and the puna. *Ecol. Monogr.* 27:153–219

Krebs JR. 1971. Territory and breeding density in the great tit, *Parus major* L. *Ecology* 52:2–22

Kruuk H, Macdonald D. 1985. Group territories of carnivores: empires and enclaves. See Sibly & Smith 1985, pp. 521–36

Langen TA, Vehrencamp SL. 1998. Ecological factors affecting group and territory size in white-throated magpie-jays. *Auk* 115:327–39

Lewis MA, Murray JD. 1993. Modelling territoriality and wolf-deer interactions. *Nature* 366:738–40

Lima SL. 1984. Territoriality in variable environments: a simple model. *Am. Nat.* 124:641–55

Łomnicki A. 1988. *Population Ecology of Individuals.* Princeton, NJ: Princeton Univ. Press

Mabelis AA. 1979. Wood ant wars. The relationship between aggression and predation in the red wood ant (*Formica polyctena* Först.). *Neth. J. Zool.* 29:451–620

Macdonald DW. 1983. The ecology of carnivore social behaviour. *Nature* 301:379–84

MacLean SF Jr, Seastedt TR. 1979. Avian territoriality: sufficient resources or interference competition. *Am. Nat.* 114:308–12

Mares MA, Lacher TE Jr, Willig MR, Bitar NA, Adams R, et al. 1982. An experimental analysis of social spacing in *Tamias striatus. Ecology* 63:267–73

Mares MA, Watson MD, Lacher TE Jr. 1976.

Home range perturbations in *Tamias striatus*. Food supply as a determinant of home range and density. *Oecologia* 25:1–12

Maynard Smith J. 1974. *Models in Ecology*. Cambridge: Cambridge Univ. Press

Maynard Smith J. 1982. *Evolution and the Theory of Games*. Cambridge: Cambridge Univ. Press

McCleery RH, Perrins CM. 1985. Territory size, reproductive success and population dynamics in the great tit, *Parus major*. See Sibly & Smith 1985, pp. 353–73

McFarland DC. 1986. Determinants of feeding territory size in the New Holland honeyeater *Phylidonyris novaehollandiae*. *Emu* 86:180–85

McFarland DC. 1994. Responses of territorial New Holland honeyeaters *Phylidonyris novaehollandiae* to short-term fluctuations in nectar productivity. *Emu* 94:193–200

McNair JN. 1987. The effect of variability on the optimal size of a feeding territory. *Am. Zool.* 27:249–58

McNicol RE, Noakes DLG. 1981. Territories and territorial defense in juvenile brook charr, *Salvelinus fontinalis* (Pisces: Salmonidae). *Can. J. Zool.* 59:22–28

Melemis SM, Falls JB. 1982. The defense function: a measure of territorial behavior. *Can. J. Zool.* 60:495–501

Mesterton-Gibbons M, Adams ES. 1998. Animal contests as evolutionary games. *Am. Sci.* 86:334–41

Milinski M, Parker GA. 1991. Competition for resources. In *Behavioural Ecology: An Evolutionary Approach*, ed. JR Krebs, NB Davies, pp. 137–68. Oxford: Blackwell Sci. 3rd ed.

Miller GR, Watson A, Jenkins D. 1970. Responses of red grouse populations to experimental improvement of their food. In *Animal Populations in Relation to Their Food Resources*, ed. A Watson, pp. 323–35. Oxford: Blackwell Sci.

Myers JP, Connors PG, Pitelka FA. 1979. Territory size in wintering sanderlings: the effects of prey abundance and intruder density. *Auk* 96:551–61

Myers JP, Connors PG, Pitelka FA. 1981. Optimal territory size and the sanderling: compromises in a variable environment. In *Foraging Behavior: Ecological, Ethological, and Psychological Approaches*, ed. AC Kamil, TD Sargent, pp. 135–58. New York: Garland

Nakamura M. 1995. Responses in spatial organization to manipulations of the food resource in polygynandrous alpine accentors. *Ecol. Res.* 10:281–89

Newton I. 1992. Experiments on the limitation of bird numbers by territorial behaviour. *Biol. Rev.* 67:129–73

Noakes DLG, McNicol RE. 1982. Geometry for the eccentric territory. *Can. J. Zool.* 60:1776–79

Norman MD, Jones GP. 1984. Determinants of territory size in the pomacentrid reef fish, *Parma victoriae*. *Oecologia* 61:60–69

Norton ME, Arcese P, Ewald PW. 1982. Effect of intrusion pressure on territory size in black-chinned hummingbirds (*Archilochus alexandri*). *Auk* 99:761–64

Nursall JR. 1977. Territoriality in redlip blennies (*Ophioblennius atlanticus*—Pisces: Blenniidae). *J. Zool.* 182:205–23

Ostfeld RS. 1986. Territoriality and mating systems of California voles. *J. Anim. Ecol.* 55:691–706

Parker GA, Knowlton N. 1980. The evolution of territory size—some ESS models. *J. Theor. Biol.* 84:445–76

Parker GA, Sutherland WJ. 1986. Ideal free distributions when individuals differ in competitive ability: phenotype-limited ideal free models. *Anim. Behav.* 34:1222–42

Patterson IJ. 1980. Territorial behaviour and the limitation of population density. *Ardea* 68:53–62

Patterson IJ. 1985. Limitation of breeding density through territorial behaviour: experiments with convict cichlids, *Cichlasoma nigrofasciatum*. See Sibly & Smith 1985, pp. 393–405

Peek FW. 1972. An experimental study of the territorial function of vocal and visual display in the male red-winged blackbird (*Agelaius phoeniceus*). *Anim. Behav.* 20:112–18

Petrie M. 1984. Territory size in the moorhen (*Gallinula chloropus*): an outcome of RHP asymmetry between neighbours. *Anim. Behav.* 32:861–70

Praw JC, Grant JWA. 1999. Optimal territory size in the convict cichlid. *Behaviour* 136: 1347–63

Price K, Broughton K, Boutin S, Sinclair ARE. 1986. Territory size and ownership in red squirrels: response to removals. *Can. J. Zool.* 64:1144–47

Pyke GH. 1979. The economics of territory size and time budget in the golden-winged sunbird. *Am. Nat.* 114:131–45

Reid ML, Weatherhead PJ. 1988. Topographical constraints on competition for territories. *Oikos* 51:115–17

Richardson PRK. 1991. Territorial significance of scent marking during the non-mating season in the aardwolf *Proteles cristatus* (Carnivora: Protelidae). *Ethology* 87:9–27

Riechert SE. 1981. The consequences of being territorial: spiders, a case study. *Am. Nat.* 117:871–92

Salomonson MG, Balda RP. 1977. Winter territoriality of Townsend's solitaires (*Myadestes townsendi*) in a piñon-juniper-ponderosa pine ecotone. *Condor* 79:148–61

Schoener TW. 1968. Sizes of feeding territories among birds. *Ecology* 49:123–41

Schoener TW. 1971. Theory of feeding strategies. *Annu. Rev. Ecol. Syst.* 2:369–404

Schoener TW. 1983. Simple models of optimal feeding-territory size: a reconciliation. *Am. Nat.* 121:608–29

Schoener TW. 1987. Time budgets and territory size: some simultaneous optimization models for energy maximizers. *Am. Zool.* 27:259–91

Seastedt TR, MacLean SF. 1979. Territory size and composition in relation to resource abundance in Lapland longspurs breeding in arctic Alaska. *Auk* 96:131–42

Sherman PT, Eason PK. 1998. Size determinants in territories with inflexible boundaries: manipulation experiments on whitewinged trumpeters' territories. *Ecology* 79: 1147–59

Sibly RM, Smith RH, eds. 1985. *Behavioural Ecology: Ecological Consequences of Adaptive Behaviour.* Oxford: Blackwell Sci.

Simon CA. 1975. The influence of food abundance on territory size in the iguanid lizard *Sceloporus jarrovi. Ecology* 56:993–98

Slaney PA, Northcote TG. 1974. Effects of prey abundance on density and territorial behavior of young rainbow trout (*Salmo gairdneri*) in a laboratory stream channels. *J. Fish. Res. Board Can.* 31:1201–9

Smith TM, Shugart HH. 1987. Territory size variation in the ovenbird: the role of habitat structure. *Ecology* 68:695–704

Southern HN. 1970. The natural control of a population of tawny owls (*Strix aluco*). *J. Zool.* 162:197–285

Stamps J. 1994. Territorial behavior: testing the assumptions. *Adv. Study Behav.* 23:173–232

Stamps JA. 1983. Sexual selection, sexual dimorphism, and territoriality. See Huey et al. 1983, pp. 169–204

Stamps JA. 1990. The effect of contender pressure on territory size and overlap in seasonally territorial species. *Am. Nat.* 135:614–32

Stamps JA. 1992. Simultaneous versus sequential settlement in territorial species. *Am. Nat.* 139:1070–88

Stamps JA, Krishnan VV. 1990. The effect of settlement tactics on territory sizes. *Am. Nat.* 135:527–46

Stamps JA, Krishnan VV. 1999. A learning-based model of territory establishment. *Q. Rev. Biol.* 74:291–318

Stamps JA, Tanaka S. 1981. The relationship between food and social behavior in juvenile lizards (*Anolis aeneus*). *Copeia* 1981:422–34

Stenger J. 1958. Food habits and available food of ovenbirds in relation to territory size. *Auk* 75:335–46

Stimson J. 1973. The role of the territory in the ecology of the intertidal limpet *Lottia gigantea* (Gray). *Ecology* 54:1020–30

Sutherland WJ. 1996. *From Individual Behaviour to Population Ecology.* Oxford: Oxford Univ. Press.

Symons PEK. 1971. Behavioural adjustment of population density to available food by juvenile Atlantic salmon. *J. Anim. Ecol.* 40:569–87

Taitt MJ. 1981. The effect of extra food on small rodent populations. I. Deermice (*Peromyscus maniculatus*). *J. Anim. Ecol.* 50:111–24

Taitt MJ, Krebs CJ. 1981. The effect of extra food on small rodent populations. II. Voles (*Microtus townsendii*). *J. Anim. Ecol.* 50:125–37

Tanemura M, Hasegawa M. 1980. Geometrical models of territory. I. Models for synchronous and asynchronous settlement of territories. *J. Theor. Biol.* 82:477–96

Temeles EJ. 1987. The relative importance of prey availability and intruder pressure in feeding territory size regulation by harriers, *Circus cyaneus. Oecologia* 74:286–97

Temeles EJ. 1994. The role of neighbours in territorial systems: When are they "dear enemies"? *Anim. Behav.* 47:339–50

Tricas TC. 1989. Determinants of feeding territory size in the corallivorous butterflyfish, *Chaetodon multicinctus. Anim. Behav.* 37:830–41

Tripp KJ, Collazo JA. 1997. Non-breeding territoriality of semipalmated sandpipers. *Wilson Bull.* 109:630–42

Tschinkel WR, Adams ES, Macom T. 1995. Territory area and colony size in the fire ant *Solenopsis invicta. J. Anim. Ecol.* 64:473–80

Tullock G. 1983. Territorial boundaries: an economic view. *Am. Nat.* 121:440–42

Turpie JK. 1995. Non-breeding territoriality: causes and consequences of seasonal and individual variation in grey plover *Pluvialis squatarola* behaviour. *J. Anim. Ecol.* 64:429–38

Village A. 1982. The home range and density of kestrels in relation to vole abundance. *J. Anim. Ecol.* 51:413–28

Vines G. 1979. Spatial distributions of territorial aggressiveness in oystercatchers, *Haematopus ostralegus* L. *Anim. Behav.* 27:300–8

von Schantz T. 1984. 'Non-breeders' in the red fox *Vulpes vulpes*: a case of resource surplus. *Oikos* 42:59–65

Waldschmidt S. 1983. The effect of supplemental feeding on home range size and activity patterns in the lizard *Uta stansburiana. Oecologia* 57:1–5

Waser PM, Wiley RH. 1979. Mechanisms and evolution of spacing in animals. In *Social Behavior and Communication. Handbook of Behavioral Neurobiology*, ed. P Marler, JG Vandenberg, pp. 159–223. New York: Plenum

Watson A, Miller GR. 1971. Territory size and aggression in a fluctuating red grouse population. *J. Anim. Ecol.* 40:367–83

Watson A, Moss R. 1972. A current model of population dynamics in red grouse. *Proc. Int. Ornothol. Congr., 15th*, pp. 134–49. Leiden: Brill

Watson A, Moss R, Parr R. 1984. Effects of food enrichment on numbers and spacing behaviour of red grouse. *J. Anim. Ecol.* 53:663–78

Watson A, Parr R. 1981. Hormone implants affecting territory size and aggressive and sexual behaviour in red grouse. *Ornis. Scand.* 12:55–61

White KAJ, Lewis MA, Murray JD. 1996a. A model for wolf-pack territory formation and maintenance. *J. Theor. Biol.* 178:29–43

White KAJ, Murray JD, Lewis MA. 1996b. Wolf-deer interactions: a mathematical model. *Proc. R. Soc. London Ser. B.* 263:299–305

Wilson EO. 1975. *Sociobiology: The New Synthesis.* Cambridge, MA: Harvard Univ. Press

Wolff JO. 1985. The effects of density, food, and interspecific interference on home range size in *Peromyscus leucopus* and *Peromyscus maniculatus. Can. J. Zool.* 63:2657–62

Wolff JO. 1986. The effects of food on midsummer demography of white-footed mice, *Peromyscus leucopus. Can. J. Zool.* 64:855–58

Woolfenden GE, Fitzpatrick JW. 1984. *The Florida Scrub Jay.* Princeton, NJ: Princeton Univ. Press

Ydenberg RC, Giraldeau LA, Falls JB. 1988. Neighbours, strangers, and the asymmetric war of attrition. *Anim. Behav.* 36:343–47

Yosef R, Grubb TC. 1994. Resource dependence and territory size in loggerhead shrikes (*Lanius ludovicianus*). *Auk* 111:465–69

Annu. Rev. Ecol. Syst. 2001. 32:305–32
Copyright © 2001 by Annual Reviews. All rights reserved

THE POPULATION BIOLOGY OF INVASIVE SPECIES

Ann K. Sakai,[1] Fred W. Allendorf,[2] Jodie S. Holt,[3] David M. Lodge,[4] Jane Molofsky,[5] Kimberly A. With,[6] Syndallas Baughman,[1] Robert J. Cabin,[7] Joel E. Cohen,[8] Norman C. Ellstrand,[3] David E. McCauley,[9] Pamela O'Neil,[10] Ingrid M. Parker,[11] John N. Thompson,[11] Stephen G. Weller[1]

[1]Department of Ecology and Evolutionary Biology, University of California-Irvine, Irvine, California 92697; e-mail: aksakai@uci.edu, sbaughma@uci.edu, sgweller@uci.edu
[2]Division of Biological Sciences, University of Montana, Missoula, Montana 59812; e-mail: darwin@selway.umt.edu
[3]Department of Botany and Plant Sciences, University of California-Riverside, Riverside, California 92521-0124; e-mail: Jodie.Holt@ucr.edu, ellstrand@pop.ucr.edu
[4]Department of Biological Sciences, University of Notre Dame, Notre Dame, Indiana 46556-0369; e-mail: lodge.1@nd.edu
[5]Department of Botany, University of Vermont, Burlington, Vermont 05405; e-mail: jmolofsk@moose.uvm.edu
[6]Division of Biology, Kansas State University, Manhattan, Kansas 66506; e-mail: kwith@ksu.edu
[7]Department of Biological Sciences, Plattsburgh State University of New York, Plattsburgh, New York 12901; e-mail: bob.cabin@plattsburgh.edu
[8]Laboratory of Populations, Rockefeller University and Columbia University, New York 10021-6399; e-mail: cohen@rockvax.rockefeller.edu
[9]Department of Biological Sciences, Vanderbilt University, Nashville, Tennessee 37235; e-mail: david.e.mccauley@vanderbilt.edu
[10]Department of Biological Sciences, University of New Orleans, New Orleans, Louisiana 70148; e-mail: pgobs@uno.edu
[11]Department of Ecology and Evolutionary Biology, University of California-Santa Cruz, Santa Cruz, California 95064; e-mail: parker@biology.ucsc.edu, thompson@biology.ucsc.edu

Key Words adaptation, alien species, exotic species, rapid evolution, introduced species, nonindigenous species, weeds, invasion resistance, invasibility

■ **Abstract** Contributions from the field of population biology hold promise for understanding and managing invasiveness; invasive species also offer excellent opportunities to study basic processes in population biology. Life history studies and demographic models may be valuable for examining the introduction of invasive species and identifying life history stages where management will be most effective. Evolutionary processes may be key features in determining whether invasive species establish and spread. Studies of genetic diversity and evolutionary changes should be useful for

0066-4162/01/1215-0305$14.00

understanding the potential for colonization and establishment, geographic patterns of invasion and range expansion, lag times, and the potential for evolutionary responses to novel environments, including management practices. The consequences of biological invasions permit study of basic evolutionary processes, as invaders often evolve rapidly in response to novel abiotic and biotic conditions, and native species evolve in response to the invasion.

INTRODUCTION

The impact of invasive species on native species, communities, and ecosystems has been widely recognized for decades (Elton 1958; Lodge 1993a,b; Simberloff 1996), and invasive species are now viewed as a significant component of global change (Vitousek et al. 1996). The severe economic impact of these species is evident; costs of invasive species are estimated to range from millions to billions of dollars annually (U.S. Congr. Off. Technol. Assess. 1993, Pimentel et al. 2000). In addition to economic impacts, invasive species have severe negative consequences for biodiversity. Numerous studies have summarized the impacts of invasive species on native species and community structure (Williamson 1996, Wilcove et al. 1998, Parker et al. 1999, Sala et al. 2000, Stein et al. 2000), and ecosystem-level effects of invasive species are now under study (Vitousek & Walker 1989, Mooney & Hobbs 2000). The impacts of invasive species are eventually expected to be severe throughout all ecosystems, as increasing numbers of nonindigenous (exotic, alien) species become established in new locations (U.S. Congr. Off. Technol. Assess. 1993). In response to the problem, Executive Order #13112 of February 1999 directed several federal agencies "to prevent the introduction of invasive species and provide for their control and to minimize the economic, ecological, and human health impacts that invasive species cause" (Fed. Regist. 64(25):6183–86).

Invasive species have been a target of research in both natural and managed ecosystems as weed scientists, resource managers, conservation biologists, and restoration biologists test various approaches for managing the impacts of these taxa. Although many of these studies are a synthesis of both basic and applied research, there is increasing recognition of the unrealized potential contributions of basic research to the study of invasive species. The purpose of this review is to elucidate the particular role that population biologists can play in studies of invasive species, through life history studies, demographic models, and knowledge of the ecology and evolution of both invasive and native species in a community context (Figure 1). Not only do questions and methods of population biology hold promise for understanding and managing invasiveness, but invasive species offer significant opportunities to study basic processes in population biology. Invasions are like natural experiments, but it may be that the processes are far more rapid than those in purely native systems.

Although the discipline of population biology has already contributed to studies of invasive species biology, untapped potential contributions are even greater,

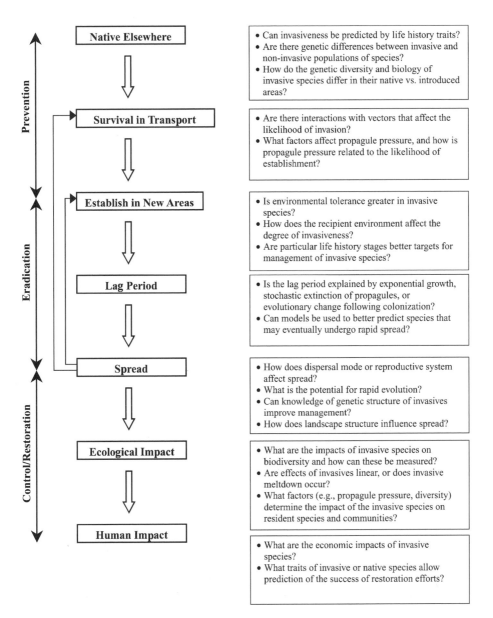

Figure 1   Generalized steps in the invasion process and their relationship to management of invasive species (modified from Lodge 1993b and Kolar & Lodge 2001). The transport, establishment, and spread of many invasive species as well as their effects can be characterized by a series of steps, each with questions that may be relevant to and enhanced by studies in population biology, including studies of life history traits as well as consideration of genetic and evolutionary changes. A few of the questions at each of these stages are highlighted below. Some stages are more relevant to prevention; others are more relevant for issues of control and restoration. Feedback may occur between many of these steps.

and studies from population biology may allow a proactive approach to invasive species. Studies of the phylogeographic structure (Stone & Sunnucks 1993, Bastrop et al. 1998, Pellmyr et al. 1998, Slade & Moritz 1998, Wilson et al. 1999), genetic diversity, and the potential for the rapid evolution of these species may provide novel insights into the colonization dynamics and spread of invasive taxa. The continuing development of new approaches in life-history theory may also lead to predictions of species likely to become serious pests or may identify critical life history stages during which management will be most successful. Current developments in the theory of population dynamics and population genetics may help identify the point at which control rather than eradication efforts would be more effective (Figure 1). Demographic models, including factors influencing dispersal, as well as spatial modeling of populations can be used to examine the spread and management of invasive species. More information about the genetics, evolution, and interactions of invasive species and native species in invaded communities may lead to predictions of the relative susceptibility of ecosystems to invasion, better methods for removal of key alien species, and predictions of the subsequent effects of removal.

Below we focus on the ecological and genetic features of species as well as community properties that may promote invasion, and the ecological and evolutionary effects of invasive species on communities. For each area discussed, we concentrate on how evolution could affect invasiveness, from the perspective of both the invading species and the species residing in the invaded community.

THE ECOLOGICAL BASIS OF INVASIVENESS: LIFE HISTORY CHARACTERISTICS OF INVASIVE SPECIES

Life history traits that make species more invasive have been of continuing interest because of their potential predictive power. Some studies have focused on life history traits that may predispose species to rapid population expansion, whereas others have investigated how the genetic structure of life history traits may provide species with either great phenotypic plasticity or the potential for rapid evolutionary change. In this section we give a historical approach to life history analysis illustrated by several classic botanical and agricultural studies, examine some cross-taxa comparisons, and propose a general approach to the study of life history characteristics of invasive species.

Baker (1965, 1974) discussed several traits associated with weedy plant species and proposed that species with many of these characteristics were more likely to be highly weedy than species with only a few of these traits. Traits promoting weediness included the ability to reproduce sexually and asexually, rapid growth from seedling to sexual maturity, and particularly, adaptation to environmental stress (phenotypic plasticity) and high tolerance to environmental heterogeneity. Although this concept of an "ideal weed" is cited throughout the weed literature and in the literature on exotic species (Baker 1974, Newsome & Noble 1986, Roy

1990, Schiffman 1997), few empirical data exist to support or refute Baker's list of characters (Kolar & Lodge 2001). Species with traits of Baker's ideal weed differ in their invasiveness (Thebaud et al. 1996), and many invasive species have only a subset of the traits described by Baker and others (Williamson & Brown 1986, Roy 1990).

More recently, broad-scale analyses of floras suggest that some plant growth-form and habitat characteristics can be used to predict invasion success. Analysis of exotic species introduced into the Czech Republic since 1492 showed that a species' invasion success was related to plant height, life form, and competitive-ness, and that the sunflower family (Asteraceae) was over-represented in the exotic flora compared with the native flora (Pyšek et al. 1995b). No simple biological predictor of invasion success was found, but some traits were more common in the alien flora than in the native flora and more prevalent for aliens in particular habitats. Rejmanek (1995) found that invasiveness of herbaceous species was best predicted by primary native latitudinal range. Reichard & Hamilton (1997) con-ducted a retrospective analysis of traits of introduced woody plants to distinguish invaders and noninvaders. Discriminant analysis models correctly classified 86% of invaders; high risk of invasiveness was related to vegetative reproduction (see also Daehler 1998), lack of pregermination seed treatment requirements, perfect (hermaphroditic) flowers, and a long period of time in which the fruit was on the plant. In an analysis of global data sets, agricultural weeds tended to be herbaceous, rapidly reproducing, abiotically dispersed species, similar to Baker's ideal weed, whereas plants most likely to become natural area invaders were primarily aquatic or semi-aquatic, grasses, nitrogen-fixers, climbers, and clonal trees (Daehler 1998).

Characteristics common to successful colonists across taxa include r-selected life histories (use of pioneer habit, short generation time, high fecundity, and high growth rates) and the ability to shift between r- and K-selected strategies, but like Baker's characteristics of the ideal weed, many of these ideas have not been tested quantitatively (Kolar & Lodge 2001). Range expansion in birds is related to dis-persal ability, high rate of population increase resulting from large clutch size and production of several clutches per season, ability to compete for resources and habi-tat with native species, repeated introductions (O'Connor 1986), and association with humans (Newsome & Noble 1986). Considering both bird and plant invaders, Newsome & Noble (1986) proposed that successful invaders can be characterized as "gap grabbers" (early germinators with fast initial growth), competitors (for re-sources and nesting sites), survivors (long-lived individuals resistant to mortality), and swampers (mass germinators). Traits that characterize freshwater fish invaders include tolerance to a broad range of environmental conditions, rapid dispersal and colonization, aggressive behavior and competitiveness, and desirability to humans (edibility, sporting qualities, aesthetic characteristics, etc.) (Moyle 1986). Most vertebrate invaders have a close association with humans, as well as high abun-dance in their native range, large size, broad diet, short generation times, ability of females to colonize alone, and ability to function in a wide range of physical conditions (Ehrlich 1989).

Where generalizations fall short, it appears that features specific to the taxonomic group and to the habitat being invaded may be important in determining invasion success. Lack of preadaptation to the new climate, disturbance, competition or predation from native species, and diseases are often cited as reasons for failure of invasions (Lodge 1993b, Moyle 1986, Newsome & Noble 1986). Roy (1990) proposed that the best approach for investigating traits of invaders might be one in which species are classified into functional groups with anticipated similar traits and where the focus is narrowed to particular habitat types.

Studies of the population biology of invasive species may allow a more precise focus on specific characteristics involved in invasiveness (Crawley 1986). Stages that are necessary for successful introduction and subsequent invasion include (*a*) introduction of a species into a new habitat, (*b*) initial colonization and successful establishment of a species, and (*c*) subsequent dispersal and secondary spread into new habitats. During all stages there is great potential for genetic changes to occur through drift or selection. In the following sections we discuss factors related to these stages and identify areas for further exploration.

Introduction of Exotic Species

Most long-distance introductions of nonnative species to new areas are the direct or indirect result of human activities, and social and economic factors are often as critical as biological factors in the introduction of exotic species. Exotic plants have been introduced deliberately as forage, fiber, medicines, or ornamentals; for erosion controls; and for timber plantations (Baker 1974, 1986). Sources of accidental introductions include ballast in ships (Ruiz et al. 2000), impure crop seeds, adhesion to domesticated animals, and soil surrounding roots of nursery stock (Baker 1986). Exotic animals are often introduced deliberately in the pet trade. Activities such as agriculture, logging, and grazing further enhance establishment of exotics by creating disturbed sites for colonization. Agriculture also facilitates invasion when pests in agro-ecosystems are exposed to agricultural practices for many generations, resulting in selection for characteristics that make them persistent and noxious.

Initial Colonization Characteristics and Successful Establishment of Invasive Species

Colonization of new habitats requires that the first arrivals initiate new populations. Many of the traits historically associated with invasive species (e.g., weeds) may be related to initial colonization. For example, species in which isolated individuals can self-fertilize are generally good colonists (Baker 1965). Self-fertility is especially common in plants, but some female insects and vertebrates can store sperm and also colonize from a single introduction (Simberloff 1989, Whittier & Limpus 1996). Species with multiple reproductive strategies (e.g., both vegetative reproduction and seeds) (Huenneke & Vitousek 1990) or plants with multi-seeded fruits may also be good colonists. Phenotypic plasticity has often been cited as a life-history trait needed for colonization of new areas because colonists must be able to cope with a range of environmental conditions (Baker 1965, 1974, Gray 1986). Comparative and experimental studies of invasive species and noninvasive congeners might elucidate the importance of phenotypic plasticity and genetic variation in the colonization by invasive species.

After initial successful colonization, the next stage of invasion is character- ized by establishment of a viable, self-sustaining population. There may be little correlation between traits required for initial colonization and traits needed for establishment. Establishment in a natural community may require different traits than those required upon entering a human-disturbed habitat (Horvitz et al. 1998), and features essential for establishment may not be consistent across taxa. For ex- ample, in a study of introduced insects used as biological control agents, Crawley (1986) found that the species with the highest intrinsic growth rates were more likely to establish successfully. These insects typically had other traits character- istic of r-selected species, including smaller body size and faster time to maturity, resulting in several generations per season (Crawley 1986). In a comparison of insect orders, Lawton & Brown (1986) found that the probability of establishment was positively correlated with smaller body size, and thus possibly correlated with higher rates of population growth and higher carrying capacity. In contrast, when they examined the combined pattern of vertebrates and invertebrates, they found a positive correlation between mean body size and probability of establishment.

Rejmanek & Richardson (1996) found that invasive pine species had smaller seed mass, a shorter juvenile period, and shorter intervals between seed crops. Data on invasive pines were consistent with life-history patterns for invasive insects and birds. In general, smaller seed size was correlated with higher seed production, faster individual growth rate, and the absence of special requirements for germi- nation (Baker 1965). In contrast, Forcella (1985) found that species of agricultural weeds with heavier seeds experienced faster germination rates and were better invaders than species with smaller seeds.

Competitive ability is another trait that may confer an advantage for inva- sive species during establishment. Many studies have documented invaders that show a superior ability to exploit local resources when compared with native resi- dents (Melgoza et al. 1990, Petren & Case 1996, Kupferberg 1997, Holway 1999, Byers 2000) or when compared with noninvading introduced species (Thebaud et al. 1996). Interactions between the invader and the invaded community may be particularly important. Differences between the competing species in the home range and those in the new range may influence an invader's success and ability to dominate a community. For example, *Centaurea diffusa*, a noxious invasive weed in North America, has stronger negative effects on biomass production for North American grasses than for grasses from its native Eurasian communities (Callaway & Aschenhoug 2000). The difference appears to be mediated by al- lelopathy: Plants from the home range are better than those in the introduced range at competing with *Centaurea* in the presence of root exudates. The introduced

Argentine ant (*Linepithema humile*) is competitively superior to the native ant species in both interference and exploitative competition in riparian woodlands of northern California (Holway 1999). Particular life history stages of native species may be especially susceptible to invasive species. In an example of exploitative competition, invasive larval bullfrogs (*Rana catesbeiana*) were able to reduce the survival and growth rates of native larval frogs by depleting benthic algae in a river in northern California (Kupferberg 1997).

Life-history theory predicts a trade-off between fast reproductive rates and competitive ability (MacArthur & Wilson 1967, Pianka 1970, MacArthur 1972, Grime 1979), but this trade-off may not exist for all invasive species. For example, Keddy et al. (1994) found that the invasive wetland species purple loosestrife (*Lythrum salicaria*) has very high fecundity but is also capable of suppressing the biomass of three indicator wetland species when grown in competition. Blossey & Notzold (1995) suggested that invasive species have been released from the pressure of pests in their native habitat and have reallocated biomass used for defense into both reproduction and growth. For example, in *L. salicaria* biomass of plants in the nonnative habitat was greater than biomass in the native habitat (Blossey & Notzold 1995). More comparisons of species in their native and introduced ranges would be useful to test this idea. Experimental manipulations of natural enemies and other presumed selective agents (e.g., Mauricio & Rausher 1997) would provide more direct evidence for their importance in the invasiveness of taxa.

Characteristics Leading to Spread

Once initial colonization and establishment have occurred, invasive species may spread from continuing long-distance dispersal (saltation dispersal) from foreign sources (naturally or aided by humans), and from short-distance dispersal (diffusion dispersal) with lateral expansion of the established population (Smith et al. 1999, Davis & Thompson 2000). Factors influencing the number of propagules, dispersal mode, and vital rates (births, deaths) are critical factors regulating the spread of invasive species.

Continued spread of the established population often occurs because of excellent adaptations for dispersal. Although the route of exotic introductions is usually determined in hindsight, the causes of spread are notably consistent. Wind, water, and animals, particularly birds, are most often the dispersal agents of seeds; these dispersal agents not only move seeds away from parent plants but also may spread the seeds to similar sites, thus increasing the probability of seedling survival (Schiffman 1997). Good dispersal ability is also important in invasions by birds (O'Connor 1986) and fish (Moyle 1986). Despite their importance in the continued range expansion of invasive species, quantifying both the number and distribution of propagules involved in establishment and spread has been very difficult. In an experiment manipulating propagule pressure (seed number) in patches of sedges in a riparian system, Levine (2000) found that propagule pressure was critical in determining which patches were most likely to be invaded. Knowledge of the biology, especially dispersal characteristics, of potential invaders is valuable for developing measures to prevent their spread, which is often easier than controlling large, established populations (Goodell et al. 2000).

GENETICS AND THE EVOLUTIONARY POTENTIAL OF INVASIVE SPECIES

Genetic and evolutionary processes may be key features in determining whether invasive species establish and spread. Invasive species offer an excellent opportunity to study rapid evolution, and some of the best-documented examples of this phenomenon have come from invasive species (e.g., Ellstrand & Schierenbeck 2000, Quinn et al. 2000). Many of the best examples of rapid directional selection in species interactions involve invasive species or native species interacting with invasive species (Thompson 1998). Nevertheless, the genetics and evolution of invasive species have received far less attention than their ecology. Invasive species may evolve both during their initial establishment and during subsequent range expansion, especially in response to selection pressures generated by the novel environment. Hybridization, either interspecific or between previously isolated populations of the same species, may be one important stimulus for the evolution of invasiveness (Ellstrand & Schierenbeck 2000); inbreeding may be another important stimulus (Tsutsui et al. 2000).

It is a fundamental tenet of evolutionary biology that the rate of change in response to natural selection is proportional to the amount of additive genetic variation present (Fisher 1930). If genetic changes, and thus evolution, during and after colonization are characteristic of invasive species, it will be important to understand the role of genetic diversity during this process, and evolutionary analyses may need to be a major focus of work on invasive species biology. Furthermore, studies of genetic variation may help predict the potential for populations of invasive species to evolve in response to management practices (e.g., evolution of resistance to herbicides or biological control agents (Barrett 1992, Van Driesche & Bellows 1996).

Genetic Processes During Colonization

Colonization events may involve a population bottleneck because the number of initial colonists is often small. Genetic drift during colonization may bring about reduced genetic variation in the newly established population. This effect will be especially strong when all colonists are drawn from the same source population. Thus, a newly established population is likely to be much less genetically diverse than the population from which it is derived (Barrett & Kohn 1991). For example, introduced populations of Argentine ants are less diverse genetically than native populations (Tsutsui et al. 2000).

Reduced genetic diversity can have two consequences. First, inbreeding depression may limit population growth and lower the probability that the population will persist (Ellstrand & Elam 1993, Newman & Pilson 1997, Nieminen et al. 2001). The effects of reduced genetic diversity will be especially strong if the population remains small for a number of generations. Second, reduced genetic diversity will limit the ability of the population to evolve. An invading species may be preadapted to some aspects of its new environment, but other aspects will be novel. Although some degree of preadaptation is necessarily a prerequisite for successful invasion, it could well be that adaptive evolution following the initial colonization is equally important.

Lag Times

One common feature of invasions is a lag time between initial colonization and the onset of rapid population growth and range expansion (Mack 1985, Kowarik 1995). This lag time is often interpreted as an ecological phenomenon (the lag phase in an exponential population growth curve). Lag times are also expected if evolutionary change is an important part of the colonization process. This process could include the evolution of adaptations to the new habitat, the evolution of invasive life-history characteristics, or the purging of genetic load responsible for inbreeding depression. It appears likely that in many cases there are genetic constraints on the probability of a successful invasion, and the lag times of successful invasives could be a result of the time required for adaptive evolution to overcome these genetic constraints (Ellstrand & Schierenbeck 2000, Mack et al. 2000).

Multiple introductions are often correlated with the eventual success of non-native species' establishment and invasiveness (Barrett & Husband 1990). Indeed, North America's most successful invasive birds, the European starling (*Sturnus vulgaris*) and the house sparrow (*Passer domesticus*), both became invasive only after repeated introductions (Ehrlich 1989). Migration may be critical not only as a source of continuing propagule pressure, but also as an important source of genetic variation to the colonizing population, if multiple invasions provide the genetic variation necessary for adaptive evolution. Multiple introductions can create invasive populations that are much more genetically diverse than any single source population when the invasive species is highly structured in its native range. Different colonizing populations of the same species are likely to be genetically divergent with different levels of genetic variation and therefore have different capacities to promote invasiveness; characteristics that promote invasiveness might evolve in some populations but not others. Gene flow between populations could result in the spread of invasive genotypes. Alternatively, gene flow between populations that swamps out locally beneficial alleles could prevent evolution of invasiveness (e.g., Kirkpatrick & Barton 1997).

Range Expansion

From a conservation perspective, two problematic features of invasive species are dense local populations and rapid range expansion once they have become established. Some of the genetic constraints likely to influence initial colonization will also influence the rate of spread. For example, dispersal from a point of initial colonization will mean that the invasive species may encounter novel selective regimes. The rate of range expansion, and eventually the boundary of the species, will be influenced by the ability of individuals to survive and reproduce in the new range (Antonovics 1976, Crawley 1986, Hengeveld 1990). Such concerns about the interaction of the environment and genetic changes may become increasingly important with global climate change (e.g., Geber & Dawson 1993, Barrett 2000, Carlton 2000).

The evolution of local adaptation requires genetic variation. Rapid range expansion suggests that the species is highly dispersive, and high dispersal rates are expected to bring about a large amount of gene flow (Barrett & Husband 1990). Whereas high rates of gene flow would help to bring genetic diversity to the edge of the range of a species, they may also act to constrain adaptation to local conditions. In fact, gene flow may be one limit to range expansion (Antonovics 1976, Hoffmann & Blows 1994, Holt 1996, 1997). Recent models by Kirkpatrick & Barton (1997) suggest that gene flow from the center of a species' range may prevent adaptation at the periphery, thereby preventing further range expansion. Hence, the rate and extent of invasive fronts may depend on the degree and pattern of gene flow among populations of an invasive species as the species spreads from its initial sites of colonization. Comparison of the genetic composition of recently established populations with populations in the native range of a species may provide valuable information about the process of invasion.

Protein or DNA genetic markers can be used to measure the amount of genetic diversity in the invasive population. These molecular markers can provide an indication of the amount of genetic variation lost during a colonization bottleneck or provide evidence for multiple population sources. Several studies of this type have already been conducted with invasive species (e.g., Novak et al. 1993; Novak & Mack 1993, 1995; Schierenbeck et al. 1995; Tsutsui et al. 2000; reviewed by Barrett & Husband 1990). Description of the pattern of molecular genetic variation in invasive species might also provide information about the temporal and spatial pattern of invasion (e.g., spread by a simple advancing wave front with extensive gene flow or by a series of long-distance dispersal events and establishment of outlying populations). An evaluation of molecular genetic variation of invasive populations may also allow identification of the source population or populations (e.g., Jousson et al. 2000). These phylogeographic patterns might be viewed as DNA fingerprinting at the level of populations or localities, rather than of individuals. The use of molecular genetic markers has contributed greatly to our understanding of range expansions following the ice ages, both by identifying the refugia from which range expansions have originated and the corridors of spread (Hewitt 2000). It will be interesting to see if these same techniques can be applied to recent invasive events, which presumably occur over a much shorter time frame. This approach has already proven to be valuable in identifying the geographic origins in California of the Medfly (Davies et al. 1999) as well as the origin of California wild oats (Garcia et al. 1989).

Although much information can be gained from molecular markers, characterization of the genetic variation controlling those life-history traits most directly related to establishment and spread is also critical. These traits are likely to be under polygenic control with strong interactions between the genotype and the environment and cannot be analyzed directly with molecular markers, although mapping quantitative traits affecting fitness (QTLs) (Mitchell-Olds 1995), colonizing ability (Barrett 2000), or other traits affecting invasiveness may be possible. For example, variation in the number of rhizomes producing above-ground shoots, a major factor in the spread of the noxious weed johnsongrass, is associated with three QTLs (Paterson et al. 1995). This knowledge may provide opportunities for predicting the location of corresponding genes in other species and for growth regulation of major weeds. Application of the methods of quantitative genetics could be useful for those species in which information can be obtained from a breeding design or from parent-offspring comparisons (Falconer & Mackay 1996, Mazer & LeBuhn 1999). For example, one could compare the additive genetic variance/covariance structure of a set of life history traits of different populations to evaluate the role of genetic constraints on the evolution of invasiveness. Comparisons of the heritability of a trait could be made among different, newly established populations or between invasive populations and the putative source population. Consideration of both the genetic and ecological context of these traits is critical, given the potentially strong interaction of genetic and environmental effects (Barrett 2000).

Management Implications

Consideration of population genetics, with explicit analyses of the genetic structure of invasive species, may allow more effective management of invasive species. Examples of potential applications of population genetics in studies of invasive species include predicting invasiveness to reduce the occurrence of new invasions, predicting the efficacy of alternative control efforts, and improving management of invasive species within native communities.

Successful invasion is often the result of intentional and repeated introduction of many individuals, not only because of increased propagule pressure, but also because these individuals may contain sufficient variation to circumvent the potential genetic problems associated with small population size. One implication of these results is that planned introductions (for horticulture, biological control, etc.) containing lower levels of genetic variation may be safer (but see Tsutsui et al. 2000). Managing for lower invasibility often conflicts with the needs of importers for successful use of these organisms, and more research in this area is clearly needed.

The genetic structure of populations has been shown to affect the efficacy of control of invasives. Burdon & Marshall (1981) noted that asexually reproducing weeds were more often effectively controlled by biological control than sexually reproducing weeds. They attributed the difference to the very different population genetic structures associated with the two reproductive modes. The clonal, genetically more homogeneous, population structure of asexual species makes it easier to match a biological control agent to the host genotype and makes these weeds particularly vulnerable to biological enemies (Van Driesche & Bellows 1996). In sexually reproducing weeds, greater genetic variation apparently allows more rapid adaptive evolution and escape from the biological control agent. Finally, not all populations of an invasive species will necessarily demonstrate invasiveness; the propensity for invasiveness may be a result of genetic factors, ecological factors (e.g., presence or absence of competitors), or a combination of both. For example, there are both stable (noninvasive) and rapidly spreading (invasive) populations of purple loosestrife (*Lythrum salicaria*) in the eastern United States (P. O'Neil, personal observation). Efficient and effective management should focus first on the invasive populations. If eradication is impossible, control strategies might be employed to alter population genetic structure to reduce adaptive variation, to flood populations with maladaptive genes, or to target and destroy invasive genotypes.

THE SUSCEPTIBILITY OF COMMUNITIES TO INVASION: ECOLOGICAL AND GENETIC FACTORS

Invasibility as an Emergent Property of Communities

Few communities are impenetrable to invasion by exotic species (Usher 1988, Lodge 1993a, Gordon 1998), and communities differ in their susceptibility to invasion as well as in their ecological and evolutionary responses to these invasions. Within communities, invasibility is determined by the properties of the invasive species, the native species (e.g., relative competitive abilities, ability to resist disturbance) (Lonsdale 1999), and the community. A species may be invasive either because it shares traits with resident native species or, alternatively, because it possesses traits different from those of native species and thus can occupy "empty niches" (Mack 1996, Levine & D'Antonio 1999).

High current levels of disturbance within communities may increase their invasibility (Horvitz et al. 1998), but recent studies have indicated that the spatial scale of disturbance and local species diversity are as important as the degree of disturbance in understanding invasibility (Levine 2000). Human disturbance of natural communities may have broadened the range of characteristics leading to successful colonization and thus increased the frequency of invasion into existing communities (Vitousek et al. 1996). Hypotheses on the importance of human-mediated introductions, changes in habitat quality, and broadened range of traits for successful establishment could be tested through experimental manipulations of invasive species that are actively colonizing areas.

Species composition, the functional groups present in the community, trophic structure, and the strength of interactions among trophic levels may interact in ways that buffer some communities against invasion more than others. Resistance to invasion may be enhanced in species-rich communities or in communities with diverse functional groups (Elton 1958, Tilman 1997, Lavorel et al. 1999), although this viewpoint has been challenged by May (1972), and empirical studies are mixed in their support (reviewed in Levine & D'Antonio 1999). Empirical approaches have included spatial pattern studies correlating the abundance of invaders and community diversity, invader addition studies adding propagules to natural communities, assembly studies examining community diversity and invasion through time, and direct experimental manipulation of diversity in constructed communities. Levine & D'Antonio (1999) stressed the importance of this latter approach to understanding mechanisms underlying the interaction of species diversity and community resistance. For example, at a broad biogeographic scale, Lonsdale (1999) found a positive relationship between native plant diversity and invasibility, largely because both native and exotic richness were related to the area and habitat diversity of sites included in the analysis. Oceanic islands were more invaded than continents (Elton 1958, Lonsdale 1999), but the cause of this low resistance to invasion was unclear. Native species density was as high on islands as in continental areas, suggesting that perhaps the native island flora was less competitive, particularly when there was grazing pressure from exotic animals (Lonsdale 1999). More experimental tests of this hypothesis could address differences between island and continental areas.

Species Interactions and Invasion Resistance of Communities

Interactions among species may have an adverse effect on the demography of invasive species and prevent them from becoming established in a community. Competition may interact with resource levels to affect invasibility. Tilman (1997, 1999) suggested that the population size of individual species may be lower with lower resource levels, but lowered resource levels may prevent invasions of exotic species if the resource level falls below some threshold required by the invasive species. Reduction in interspecific interactions may also explain why exotic species often flourish in new habitats and become pests. If predators or pathogens of the invasive species are absent in the new community, rapid population growth may occur. Without these negative interactions, the invasive species may have more resources available, thus increasing its competitive abilities and ability to invade (Blossey & Notzold 1995, Van Driesche & Bellows 1996, Tilman 1999).

Native species may also decline relative to invasive species because they are more susceptible to parasites or pathogens. In South Africa the indigenous brown mussel (*Perna perna*) was more susceptible to digenetic trematodes than the invasive Mediterranean mussel (*Mytilus galloprovincialis*). The combined effects of the trematodes gave the invasive species a competitive edge over the native brown mussel (Calvo-Ugarteburu & McQuaid 1998). Occasionally, invasion resistance of the community is enhanced if native predators are able to consume exotic species and decrease their survival (Jaksic 1998). The native muricid snail (*Pteropuzpura festiva*) in San Diego Bay, California, has decimated populations of an introduced mussel (*Musculista senhousia*) because the native snail preferred the introduced mussel over an abundant native clam (*Chione undatella*). Predation thus contributed significantly to invasion resistance in this system (Reusch 1998).

Mutualisms may facilitate invasion or contribute to invasion resistance (Richardson et al. 2000). For example, the absence of ectomycorrhizal fungi, which are spread by red-backed voles (*Clethrionomys gapperi*), limits the invasion of conifers into meadows (Terwilliger & Pastor 1999). Low visitation by pollinators to Scotch broom (*Cytisus scoparius*) slows invasive spread of this exotic shrub (Parker 1997). In contrast, invasion by exotic weeds within the California coastal prairie is facilitated by the native, nitrogen-fixing shrub (bush lupine, *Lupinus arboreus*) because of its mutualism with *Rhizobium* (Maron & Connors 1996).

Genetic Structure of Populations and the Invasibility of Communities

Species interactions contribute to the relationship between diversity, food web connectedness, and stability in communities, thus affecting susceptibility of the community to invasion. Species interactions themselves may evolve rapidly in response to introduced species (Thompson 1998, 1999). Evolutionary host shifts by native species may occur in response to the introduction of an exotic species, as in the case of the use of introduced fennel by the anise butterfly (Thompson 1993). By forging additional links among trophic levels, evolutionary modifications of species interactions after invasion may enhance community stability and resistance to further invasion. Alternatively, postinvasion evolutionary modifications of native species may further disrupt community stability, making these communities even more susceptible to invasion. In a survey of invasive plant species in Florida, Gordon (1998) found that invasive species frequently modified resource availability and competitive interactions, implying that selective pressures on native species might be altered.

Introduction of exotic predators could result in evolution of novel antipredator responses in native prey populations, which might have far-reaching community consequences. For example, brown trout (*Salmo trutta*) in the Taieri River of New Zealand have displaced native fishes, reduced the abundance of grazing invertebrates, and apparently resulted in the evolution of antipredator behaviors in these invertebrates (Townsend 1996). This evolution of antipredator behavior appears to have been responsible for an overall increase in algal biomass (Townsend 1996). Similar evolutionary changes in behavior may facilitate invasions by other species.

Communities may also be more susceptible to invasion when opportunities for hybridization between invasive species and native species are present (Levin et al. 1996). Extinction by hybridization (via either genetic assimilation or outbreeding depression) (Ellstrand & Elam 1993) may occur rapidly (Huxel 1999, Perry et al. 2001b) and lead to loss of diversity within the community. Hybridization may also introduce genes from native into invasive species and thus increase the fitness of the invasive species in the new environment (Ellstrand & Schierenbeck 2000).

Habitat Fragmentation, Population Persistence and the Invasibility of Communities

Spread of invasive species may occur more rapidly in fragmented landscapes. For example, the brown-headed cowbird (*Moluthrus ater*), an avian brood parasite, spread rapidly in response to forest clearing following European settlement of the eastern United States some 200 years ago (Ehrlich 1989; R.D. Holt, D.E. Burhans, S.K. Robinson & S.I. Rothstein, unpublished observations). Brown-headed cowbirds may dramatically reduce the reproductive success of their avian hosts, and this species has been implicated in the decline of many neotropical bird species that breed in the temperate forests of North America. Given the severe selective pressure cowbirds exert on their hosts, the failure of these native bird species to evolve rejection of cowbird eggs may result from the source-sink dynamics of fragmented landscapes. Fragmented landscapes may support sink populations of native host species, but native species may be unable to persist (lambda <1) except for immigration from outside source populations in more contiguous landscapes (where lambda ≥ 1 and cowbird parasitism is less prevalent). Immigrants may therefore "swamp" genetically any adaptations that may emerge in host populations for dealing with cowbird parasitism (R.D. Holt, D.E. Burhans, S.K. Robinson & S.I. Rothstein, unpublished observations). Habitat fragmentation may therefore increase the habitat of the cowbird and at the same time prevent adaptive evolution in response to invasive species, thereby enhancing susceptibility of the community to invasion.

Population Models of Range Expansion

Several basic features of invasions, such as initial lags in population growth, rate of geographic spread, and features of geographic spread (smoothly expanding range with simple advancing wavefronts, or hopscotch jumps with long-distance dispersal), have been addressed using a variety of population modeling approaches. Predictions of the rate and direction of spread of invasive species that have already become established are critical to any management program and represent one of the central challenges in the application of theoretical models of invasive spread. A rich theoretical literature has contributed to our understanding of factors influencing broad-scale spread, such as long-distance dispersal, spatial and temporal heterogeneity, and Allee effects (Skellam 1951, Kot et al. 1996, Lewis 1997, Shigesada & Kawasaki 1997, Neubert et al. 2000). Range expansion models have been applied directly to resource management questions, such as whether barrier zones can effectively be used to slow the spread of the gypsy moth (Sharov & Liebhold 1998) and which lakes or streams are most likely to be invaded by zebra mussels (Horvath et al. 1996, Stoeckel et al. 1997, Bossenbroek et al. 2001). Nonspatial, stage-structured models can also lead to valuable insights into the population biology of invasive species (e.g., Parker 2000, Drayton & Primack 1999, Neubert & Caswell 2000) and have been used to estimate the potential effectiveness of control strategies, such as biological control agents that target specific life history stages (Shea & Kelly 1998, McEvoy & Coombs 1999).

THE ECOLOGICAL AND EVOLUTIONARY CONSEQUENCES OF INVASIONS ON COMMUNITIES

Invasive species have both ecological and genetic impacts on the communities that they invade, and understanding these impacts may aid in reversing them. As shown above, ecological interactions between native and invasive species may be direct (e.g., predation, herbivory, parasitism, competition, mutualism) or indirect (e.g., habitat alteration, apparent predation, cascading trophic interactions) and result in changes in the population biology (births, death, migration) of the native species. Significant genetic and evolutionary changes in both the native and invasive species may also occur. Both genetic drift and natural selection (from biotic interactions and abiotic factors in the new environment) may cause rapid evolution in the invading species. As a consequence, rapid evolutionary changes also may occur in the native species in response to the invading species. In the extreme, hybridization and introgression between invading species and native species may result in extinction of the native species (Levin et al. 1996, Rhymer & Simberloff 1996, Perry et al. 2001b).

Most attention has been focused on a small number of exotic species with major negative impacts. A few aquatic examples offer graphic illustrations of changes in the population biology and community structure of native systems. As zebra mussels continue to spread in the Laurentian Great Lakes and inland lakes and streams of the upper-midwestern United States, native clams are smothered and extirpated and water clarity increases as phytoplankton biomass declines from filtration by zebra mussels. Vascular aquatic plants increase as a result of decreased shading from phytoplankton, and energy flow in general is shunted from the pelagic to the benthic zone (Lodge 2001). The invasive rusty crayfish (*Orconectes rusticus*) has a severe negative impact on the northern crayfish (*Orconectes propinquus*) in the upper-midwestern United States. Several direct and indirect effects of fish predators and competition for food and shelter between these two crayfish species favor the invasive species and lead to the local extirpation of the native northern crayfish in many lakes and streams (Hill & Lodge 1999, Lodge et al. 2000).

In Lake Victoria, East Africa, the predacious Nile perch remained at low population levels for many years after its introduction in the 1950s but more recently has boomed in abundance, inducing behavioral changes in native prey fishes and causing the extinction of perhaps 200 of the about 400 endemic cichlid fishes in the lake (Lodge 2001). In both these cases, invasive species changed both the behavior and ultimately the vital rates of otherwise dominant native species through the threat of predation, actual predation, changes in habitat suitability, and competition for space and food.

In an example of impacts on terrestrial systems, most of the formerly forested lowland ecosystems on the drier, leeward sides of the island of Hawaii are now dominated by invasive species such as fountain grass (*Pennisetum setaceum*). Like many other invasive grasses around the world (D'Antonio & Vitousek 1992), fountain grass has severely altered the ecological and ecosystem dynamics of these arid regions by suppressing native vegetation and promoting fires that have proved devastating to the native flora (Cabin et al. 2000).

The same invasive species can have different impacts in different communities. For example, introductions for biological control range in effectiveness depending on the abiotic conditions, biotic interactions, and genetic structure of the target species (Van Driesche & Bellows 1996). Assessment of the potential economic and ecological damage represented by species introductions would be greatly facilitated by a clear protocol for measuring impacts of different invasive species in different communities. These protocols to compare the impacts of different species are also important if priorities for control of invasive species are motivated in part by removing the most damaging species first (Parker et al. 1999).

Evolutionary Interactions

The consequences of biological invasions represent a powerful opportunity to study basic evolutionary processes. Along with the introduction of a new species into a community comes a chance to observe how the invader evolves in response to novel abiotic and biotic conditions, as well as how resident species evolve in response to the invader (see above).

The use of molecular genetic analyses indicates that hybridization and introgression between invasive and native species may be much more common than is often recognized (Rhymer & Simberloff 1996). Hybridization and introgression can have devastating effects on native species, in some cases contributing to their demise (Levin et al. 1996, Rhymer & Simberloff 1996, Perry et al. 2001a). Invasive species may also cause the evolution of the native species (see previous sections). For example, interactions with the invasive species might lead to natural selection favoring individuals of the native species with traits that were not advantageous prior to the invasion (e.g., by altering competitive interactions, predator-prey interactions, and pollination or seed dispersal). Common garden experiments could be used to compare genetic differences in populations of native species that have been interacting with a particular invader for different lengths of time and for the analysis of genetic changes in native species with removal of invasive species.

Studies of invasive species may provide opportunities to better understand aspects of community dynamics and are crucial for applications of community theory in restoration biology. For example, it is critical to understand the similarities and differences between removing invasive species and reintroducing desirable species back into invaded communities; adding a species into a community may not be the ecological and evolutionary mirror image of taking one away. More study is needed to know if the ecological theory of community assembly rules or community resilience and resistance to invasion can be applied to restoration efforts, where one goal is often to restore invaded systems to a state that more closely resembles its pre-invaded condition. Critical community studies include those that could help managers predict when an invaded community is likely to recover simply by removing invasive species or those that could predict which invasive species should be removed first and which ones may be ignored within a community. Methods of removal may also affect the subsequent ecological and evolutionary response of the invaded community. These same issues may apply equally to the problem of determining which species to add to communities during the process of restoration. Scientists involved in the removal and restoration of invasive species can serve as partners for experiments to measure the population impacts of alternative methodologies, and to create long-term studies of population resilience after removal.

CONCLUSIONS

The synergism arising from combining ecological, genetic, and evolutionary perspectives on invasive species may be essential for developing practical solutions to the economic and environmental losses resulting from these species. As illustrated above, these approaches include an ever expanding array of tools, including molecular techniques, controlled experiments, and mathematical models. Conversely, invasive species offer unique opportunities for population biologists as natural experiments with more rapid processes than occur in most natural systems.

An analysis of the sequence of events associated with the invasion process highlights the need for ecological, genetic, and evolutionary approaches. The invasion sequence can be thought of as a series of steps, each recognizable, and each rich in questions for population biologists (Figure 1). The first question is understanding why some species become invasive and others do not. A growing body of recent literature supports the contention that statistically significant relationships exist between life history characteristics and the potential for invasiveness. This information could be put to immediate practical use in screening potential plant and animal introductions (Reichard & Hamilton 1997, Kolar & Lodge 2000).

More work on the population biology of invasive species may increase the ability to predict invasiveness in a new habitat, including the ecological role of a species in its native habitat (where it may or may not be an aggressive, colonizing species), or the presence of ecological or genetic differences between invasive populations and the populations from which they are derived. During the process of transport to a new environment, interactions between invasive species and vectors may occur, and different vectors may differentially affect birth and death rates in transit. The number of propagules is likely to be a significant element of the establishment process, and yet there are very few quantitative data relating propagule abundance to success during invasion. Despite the obvious practical significance of reducing the frequency of introduction as well as the number of individuals of invasive species involved in introductions, little theoretical or quantitative empirical work has been done to describe the relationship of these factors with the probability of establishment. Development of experimental systems, such as experimental islands to study replicated introductions, could be extremely revealing.

Another question is the extent to which anthropogenic disturbances, including invasions of other species, influence establishment. There is little information on the relative importance of phenotypic plasticity in establishment of a potentially invasive species in a new habitat, the importance of differences between the source and recipient environments, or the extent of genetic variation in invading propagules and its relationship to successful establishment.

The lag period commonly observed during invasion processes has been variously attributed to the exponential growth process, stochastic extinction of propagules, or an evolutionary effect. Evolutionary modification of species following establishment, either through adaptive evolution after a single colonization event or by sorting of adaptive genotypes following multiple colonization events, might cause emergence from the lag period. The relative roles of density dependence and evolutionary change in generating the lag times characteristic of the population growth of many invasive species will also require more research. A greater understanding of how these potential causes of the lag phase interact should provide insights into possible management of the invasion process.

The spread of invasive species is influenced by dispersal mode, landscape structure, and the number of foci of introduction of the invasive species (Moody & Mack 1988). Mode of reproduction also influences the rate of spread of an invasive species. The reproductive system and potential for recombination may determine the rate of spread, particularly if continuous adaptation is a prerequisite for the invasion process. Information about how these factors interact will provide opportunities for management of invasive species, including prevention of introductions and control of established populations.

Many community functions are affected by the population biology of invasive species. The effects of invasive species on recipient communities are unclear; in some cases the effects may be additive, but invasive meltdown (sensu Simberloff & Von Holle 1999) with the presence of invasive species facilitating the invasion of more species is also an alternative. More studies may show when invasive species are likely to cause evolutionary changes in recipient populations that retard or increase the likelihood of further invasions. In addition, they may also help determine the rapidity of an evolutionary response in invasive species and whether this increases the likelihood of further invasion. Population biologists may use invasive species to address basic questions in population biology that at the same time provide data useful for decisions about controlling the spread and minimizing the damage from invasive species.

The stages in the invasion process shown in Figure 1 also suggest a means for evaluating threats from invasive species that directly impact plans for management of invasive species. If a species is known to have potentially severe effects on the communities it invades, then particular attention might be given to preventing the transport of the species (e.g., prevention of invasion of Hawaii by the brown tree snake). Alternatively, if the lag phase is a phenomenon resulting from evolutionary modification related to multiple introductions, there may be a need to be wary of any alien species, even those that show little indication of potential for invasiveness.

Integrating Disciplines

Advances in other academic fields may be relevant to the ideas and approaches we have outlined here. For example, the field of disease epidemiology also addresses processes of colonization and spread and includes an extensive body of theoretical work (Heesterbeek & Zadoks 1986). The applied field of weed science has addressed many general concepts of invasion biology in the specific context of agriculturally important plant invaders, especially with respect to the life history of successful colonizing species (e.g., Forcella 1985, Forcella et al. 1986, Panetta & Mitchell 1991, Holt & Boose 2000, Woolcock & Cousens 2000). Ecotoxicology has developed approaches for measuring biological responses to pollutants, which might be analogous to invasive species (Parks et al. 1991, Bongers & Ferris 1999, Denayer et al. 1999). Links between population biology and these fields could lead to exciting and productive advances in invasion biology.

An important element in the study of invasive species is the necessary link to fields outside of the academic community. For example, federal and state agencies, conservation land trusts (such as The Nature Conservancy), and other conservation managers that detect recently arrived invasive species and monitor their spread can alert researchers so that traits associated with successful colonization can be measured. In turn, research on the ecology, genetics, and evolutionary biology of invasive species may eventually provide the practical information that will be essential for preventing the homogenization of the world's flora and fauna.

ACKNOWLEDGMENTS

This work resulted from a workshop of the Collaboratory on the Population Biology of Invasive Species conducted in October 1999 at the National Science Foundation in Arlington, VA. The Collaboratory is funded by the National Science Foundation (supplement to DEB98-15878). The goal of the Collaboratory is to highlight both the contributions that population biology can make in studies of invasion biology, as well as the opportunities for studies of basic concepts in population biology using invasive species. We thank E. Lyons, S. Scheiner, and M. Courtney for their encouragement and J. Heacock and T. Culley for their technical assistance.

Visit the Annual Reviews home page at www.AnnualReviews.org

LITERATURE CITED

Antonovics J. 1976. The nature of limits to natural selection. *Ann. Mo. Bot. Gard.* 63:224–47

Baker HG. 1965. Characteristics and modes of origin of weeds. In *The Genetics of Colonizing Species*, ed. HG Baker, GL Stebbins, pp. 147–69. New York: Academic. 588 pp.

Baker HG. 1974. The evolution of weeds. *Annu. Rev. Ecol. Syst.* 5:1–24

Baker HG. 1986. Patterns of plant invasion in North America. See Mooney & Drake 1986, pp. 44–57

Barrett SCH. 1992. Genetics of weed invasions. In *Applied Population Biology*, ed. SK Jain, LW Botsford, pp. 91–119. Dordrecht, The Netherlands: Kluwer. 295 pp.

Barrett SCH. 2000. Microevolutionary influences of global changes on plant invasions. See Mooney & Hobbs 2000, pp. 115–39

Barrett SCH, Husband BC. 1990. The genetics

of plant migration and colonization. In *Plant Population Genetics, Breeding and Genetic Resources*, ed. AHD Brown, MT Clegg, AL Kahler, BS Weir, pp. 254–77. Sunderland, MA: Sinauer. 449 pp.

Barrett SCH, Kohn JR. 1991. Genetic and evolutionary consequences of small population size in plants: implications for conservation. In *Genetics and Conservation of Rare Plants*, ed. DA Falk, KE Holsinger, pp. 3–30. New York: Oxford Univ. Press. 283 pp.

Bastrop R, Juerss K, Sturmbauer C. 1998. Cryptic species in a marine polychaete and their independent introduction from North America to Europe. *Mol. Biol. Evol.* 15:97–103

Blossey B, Notzold R. 1995. Evolution of increased competitive ability in invasive nonindigenous plants: a hypothesis. *J. Ecol.* 83:887–89

Bongers T, Ferris H. 1999. Nematode community structure as a bioindicator in environmental monitoring. *Trends Ecol. Evol.* 14:224–28

Bossenbroek JM, Kraft CE, Nekola JC. 2001. Prediction of long-distance dispersal using gravity models: zebra mussel invasion of inland lakes. *Ecol. Appl.* In press

Burdon JJ, Marshall DR. 1981. Biological control and the reproductive mode of weeds. *J. Appl. Ecol.* 18:649–58

Byers JE. 2000. Competition between two estuarine snails: implications for invasions of exotic species. *Ecology* 81(5):1225–39

Cabin RJ, Weller SG, Lorence DH, Flynn TW, Sakai AK, et al. 2000. Effects of long-term ungulate exclusion and recent alien species control on the preservation and restoration of a Hawaiian tropical dry forest. *Conserv. Biol.* 14:439–53

Callaway RM, Aschehoug ET. 2000. Invasive plants versus their new and old neighbors: a mechanism for exotic invasion. *Science* 290:521–23

Calvo-Ugarteburu G, McQuaid CD. 1998. Parasitism and invasive species: effects of digenetic trematodes on mussels. *Mar. Ecol. Program Ser.* 169:149–63

Carlton JT. 2000. Global change and biological invasions in the oceans. See Mooney & Hobbs 2000, pp. 31–53

Crawley MJ. 1986. The population biology of invaders. *Philos. Trans. R. Soc. London Ser. B* 314:711–29

Daehler CC. 1998. The taxonomic distribution of invasive angiosperm plants: ecological insights and comparison to agricultural weeds. *Biol. Conserv.* 84:167–80

D'Antonio CM, Vitousek PM. 1992. Biological invasions by exotic grasses, the grass/fire cycle, and global change. *Annu. Rev. Ecol. Syst.* 23:63–87

Davies N, Villablanca FX, Roderirck GK. 1999. Bioinvasions of the Medfly *Ceratitis capitata*: source estimation using DNA sequences at multiple intron loci. *Genetics* 153:351–60

Davis MA, Thompson K. 2000. Eight ways to be a colonizer; two ways to be an invader: a proposed nomenclature scheme for invasion ecology. *Bull. Ecol. Soc. Am.* 81:226–30

Denayer F-O, Van Haluwyn C, de Foucault B, Schumacker R, Colein P. 1999. Use of bryological communities as a diagnostic tool of heavy metal soil contamination (Cd, Pb, Zn) in northern France. *Plant Ecol.* 140:191–201

Drake JA, Mooney HA, DiCastri HA, Groves HA, Kruger FJ, et al., eds. 1989. *Biological Invasion: a Global Perspective*. New York: Wiley & Sons. 525 pp.

Drayton B, Primack RB. 1999. Experimental extinction of garlic mustard (*Allaria petiolata*) populations: implication for weed science and conservation biology. *Biol. Invasions* 1:159–67

Ehrlich PR. 1989. Attributes of invaders and the invading processes: vertebrates. See Drake et al. 1989, pp. 315–28

Ellstrand NC, Elam DR. 1993. Population genetic consequences of small population size: implications for plant conservation. *Annu. Rev. Ecol. Syst.* 24:217–42

Ellstrand NC, Schierenbeck KA. 2000. Hybridization as a stimulus for the evolution of invasiveness in plants? *Proc. Natl. Acad. Sci. USA* 97:7043–50

Elton CS. 1958. *The Ecology of Invasions by Animals and Plants.* London: Methuen. 181 pp.

Falconer DS, Mackay TFC. 1996. *Introduction to Quantitative Genetics.* Harlow, UK: Prentice Hall. 464 pp. 4th ed.

Fisher RA. 1930. *The Genetical Theory of Natural Selection.* Oxford: Clarendon. 272 pp.

Forcella F. 1985. Final distribution is related to rate of spread in alien weeds. *Weed Res.* 25:181–91

Forcella F, Wood JT, Dillon SP. 1986. Characteristics distinguishing invasive weeds within *Echium* (Bugloss). *Weed Res.* 26:351–64

Garcia P, Vences FJ, Perez de al Vega M, Allard RW. 1989. Allelic and genotypic composition of ancestral Spanish and colonial Californian gene pools of *Avena barbata*: evolutionary implications. *Genetics* 122:687–94

Geber MA, Dawson TE. 1993. Evolutionary responses of plants to global change. In *Biotic Interactions and Global Change*, ed. PM Kareiva, JG Kingsolver, RB Huey, pp. 179–97. Sunderland, MA: Sinauer. 559 pp.

Goodell K, Parker IM, Gilbert GS. 2000. Biological impacts of species invasions: implications for policy makers. In *Incorporating Science, Economics, and Sociology in Developing Sanitary and Phytosanitary Standards in International Trade*, ed. Natl. Res. Counc. US, pp. 87–117. Washington, DC: Natl. Acad. Sci. Press. 275 pp.

Gordon DR. 1998. Effects of invasive, non-indigenous plant species on ecosystem processes: lessons from Florida. *Ecol. Appl.* 8:975–89

Gray AJ. 1986. Do invading species have definable genetic characteristics? *Philos. Trans. R. Soc. London Ser. B* 314:655–72

Grime JP. 1979. *Plant Strategies and Vegetation Processes.* Chichester, UK: Wiley & Sons. 222 pp.

Heesterbeek JAP, Zadoks JC. 1986. Modelling pandemics of quarantine pests and diseases: problems and perspectives. *Crop Prot.* 6:211–21

Hengeveld R. 1990. *Dynamic Biogeography.* London: Cambridge Univ. Press. 249 pp.

Hewitt G. 2000. The genetic legacy of the quaternary ice ages. *Nature* 405:907–13

Hill AM, Lodge DM. 1999. Replacement of resident crayfishes by an exotic crayfish: the roles of competition and predation. *Ecol. Appl.* 9:678–90

Hoffmann AA, Blows MW. 1994. Species borders: ecological and evolutionary perspectives. *Trends Ecol. Evol.* 9:223–27

Holt JS, Boose AB. 2000. Potential for spread of *Abutilon theophrasti* in California. *Weed Sci.* 48:43–52

Holt RD, Gomulkiewicz R. 1997. How does immigration influence local adaptation? A reexamination of a familiar paradigm. *Am. Nat.* 149:563–72

Holt RD. 1996. Demographic constraints in evolution: toward unifying ecological theories of senescence and niche conservatism. *Evol. Ecol.* 10:1–11

Holway DA. 1999. Competitive mechanisms underlying the displacement of native ants by the invasive argentine ant. *Ecology* 80:238–51

Horvath TG, Lamberti GA, Lodge DM, Perry WL. 1996. Zebra mussels in flowing waters: role of headwater lakes in downstream dispersal. *J. North Am. Benthol. Soc.* 15:564–75

Horvitz C, Pascarella J, McMann S, Freedman A, Hofsetter RH. 1998. Functional roles of invasive non-indigenous plants in hurricane-affected subtropical hardwood. *Ecol. Appl.* 8:947–74

Huenneke LF, Vitousek PM. 1990. Seedling and clonal recruitment of the invasive tree *Psidium cattleianum*: implications for management of native Hawaiian forests. *Biol. Conserv.* 53:199–211

Huxel GR. 1999. Rapid displacement of native species by invasive species: effects of hybridization. *Biol. Conserv.* 89:143–52

Invasive Species. 1999. Executive Order 13112 of Feb. 3. *Fed. Regist.* 64(25):6183–86

Jaksic FM. 1998. Vertebrate invaders and their

ecological impacts in Chile. *Biodivers. Conserv.* 7:1427–45

Jousson O, Pawlowski J, Zaninetti L, Zechman FW, Dini F, et al. 2000. Invasive alga reaches California. *Nature* 408:157–58

Keddy PA, Twolan-Strutt L, Wisheu IC. 1994. Competitive effect and response ranking in 20 wetland plants: Are they consistent across three environments? *J. Ecol.* 82:635–43

Kirkpatrick M, Barton NH. 1997. Evolution of a species' range. *Am. Nat.* 150:1–23

Kolar C, Lodge DM. 2001. Progress in invasion biology: predicting invaders. *Trends Ecol. Evol.* 16:199–204

Kolar C, Lodge DM. 2000. Freshwater nonindigenous species: interactions with other global changes. See Mooney & Hobbs 2000, pp. 3–30

Kot M, Lewis MA, van den Driesshe P. 1996. Dispersal data and the spread of invading organisms. *Ecology* 77:2027–42

Kowarik I. 1995. Time lags in biological invasions with regard to the success and failure of alien species. See Pyšek et al. 1995a, pp. 15–38

Kupferberg SJ. 1997. Bullfrog (*Rana catesbeiana*) invasion of a California river: the role of larval competition. *Ecology* 78:1736–51

Lavorel S, Priur-Richard AH, Grigulis K. 1999. Invasibility and diversity of plant communities: from patterns to processes. *Divers. Distrib.* 5:41–49

Lawton JH, Brown KC. 1986. The population and community ecology of invading insects. *Philos. Trans. R. Soc. London Ser. B* 314:607–16

Levin DA, Francisco-Ortega J, Jansen RK. 1996. Hybridization and the extinction of rare plant species. *Conserv. Biol.* 10:10–16

Levine JM. 2000. Species diversity and biological invasions: relating local process to community pattern. *Science* 288:852–54

Levine JM, D'Antonio CM. 1999. Elton revisited: a review of evidence linking diversity and invasibility. *Oikos* 87:15–26

Lewis MA. 1997. Variability, patchiness and jump dispersal in the spread of an invading population. In *Spatial Ecology: The Role of Space in Population Dynamics and Interspecific Interactions*, ed. D Tilman, P Karieva, pp. 46–69. Princeton, NJ: Princeton Univ. Press. 368 pp.

Lodge DM. 1993a. Biological invasions: lessons for ecology. *Trends Ecol. Evol.* 8:133–37

Lodge DM. 1993b. Species invasions and deletions: community effects and responses to climate and habitat change. In *Biotic Interactions and Global Change*, ed. PM Karieva, JG Kingsolver, RB Huey, pp. 367–87. Sunderland, MA: Sinauer. 559 pp.

Lodge DM. 2001. Lakes. In *Future Scenarios of Global Biodiversity*, ed. FS Chapin III, OE Sala, E Huber-Sannwald. New York: Springer-Verlag. In press

Lodge DM, Taylor CA, Holdich DM, Skurdal J. 2000. Nonindigenous crayfishes threaten North American freshwater biodiversity: lessons from Europe. *Fisheries* 25:7–20

Lonsdale WM. 1999. Global patterns of plant invasions and the concept of invasibility. *Ecology* 80:1522–36

MacArthur RH. 1972. *Geographical Ecology: Patterns in the Distribution of Species.* New York: Harper & Row. 269 pp.

MacArthur RH, Wilson EO. 1967. *The Theory of Island Biogeography.* Princeton, NJ: Princeton Univ. Press. 203 pp.

Mack RN. 1985. Invading plants: their potential contribution to population biology. In *Studies on Plant Demography: A Festschrift for John L. Harper*, ed. J White. pp. 127–42. London: Academic. 393 pp.

Mack RN. 1996. Predicting the identity and fate of plant invaders: emergent and emerging approaches. *Biol. Conserv.* 78:107–21

Mack RN, Simberloff D, Lonsdale WM, Evans H, Clout M, Bazzaz FA. 2000. Biotic invasions: causes, epidemiology, global consequences, and control. *Ecol. Applic.* 10:689–710

Maron JL, Connors PG. 1996. A native nitrogen-fixing shrub facilitates weed invasion. *Oecologia* 105:302–12

Mauricio R, Rausher MD. 1997. Experimental manipulation of putative selective agents provides evidence for the role of natural enemies in the evolution of plant defense. *Evolution* 51:1435–44

May RM. 1972. Will a large complex system be stable? *Nature* 238:413–14

Mazer SJ, LeBuhn G. 1999. Genetic variation in life-history traits: heritability estimates within and genetic differentiation among populations. In *Life History Evolution in Plants*, ed TO Vuorisalo, PK Mutikainen, pp. 85–135. Dordrecht: Kluwer Academic. 348 pp.

McEvoy PB, Coombs EM. 1999. Biological control of plant invaders: regional patterns, field experiments, and structured population models. *Ecol. Appl.* 9:387–401

Melgoza G, Nowak RS, Tausch RJ. 1990. Soil water exploitation after fire: competition between *Bromus tectorum* (cheatgrass) and two native species. *Oecologia* 83:7–13

Mitchell-Olds T. 1995. The molecular basis of quantitative genetic variation in natural populations. *Trends Ecol. Evol.* 10:324–28

Moody ME, Mack RN. 1988. Controlling the spread of plant invasions: the importance of nascent foci. *J. Appl. Ecol.* 25:1009–21

Mooney HA, Drake JA, eds. 1986. *Ecology of Biological Invasions of North America and Hawaii*. New York: Springer-Verlag. 321 pp.

Mooney HA, Hobbs RJ, eds. 2000. *Invasive Species in a Changing World*. Washington, DC: Island. 457 pp.

Moyle PB. 1986. Fish introductions into North America: patterns and ecological impact. See Mooney & Drake 1986, pp. 27–43

Neubert MG, Caswell H. 2000. Demography and dispersal: calculation and sensitivity analysis of invasion speed for structured populations. *Ecology* 81:1613–28

Neubert MG, Kot M, Lewis MA. 2000. Invasion speeds in fluctuating environments. *Proc. R. Soc. Biol. Sci. Ser. B* 267:1603–10

Newman D, Pilson D. 1997. Increased probability of extinction due to decreased genetic effective population size: experimental populations of *Clarkia pulchella*. *Evolution* 51:354–62

Newsome AE, Noble IR. 1986. Ecological and physiological characters of invading species. In *Ecology of Biological Invasions*, ed. RH Groves, JJ Burdon, pp. 1–20. Cambridge: Cambridge Univ. Press. 166 pp.

Nieminen M, Singer MC, Fortelius W, Schops K, Hanski I. 2001. Experimental confirmation that inbreeding depression increases extinction risk in butterfly populations. *Am. Nat.* 157:237–44

Novak SJ, Mack RN. 1993. Genetic variation in *Bromus tectorum* (Poaceae): comparison between native and introduced populations. *Heredity* 71:167–76

Novak SJ, Mack RN. 1995. Allozyme diversity in the apomictic vine, *Bryonia alba* (Cucurbitaceae): potential consequences of multiple introductions. *Am. J. Bot.* 82:1153–62

Novak SJ, Mack RN, Soltis PS. 1993. Genetic variation in *Bromus tectorum* (Poaceae): introduction dynamics in North America. *Can. J. Bot.* 71:1441–48

O'Connor RJ. 1986. Biological characteristics of invaders among bird species in Britain. *Philos. Trans. R. Soc. London Ser. B* 314:583–98

Panetta FD, Mitchell ND. 1991. Homoclime analysis and the prediction of weediness. *Weed Res.* 31:273–84

Parker IM. 1997. Pollinator limitation of *Cytisus scoparius*, an invasive exotic shrub. *Ecology* 788:1457–70

Parker IM. 2000. Invasion dynamics of *Cytisus scoparius*: a matrix model approach. *Ecol. Appl.* 10:726–43

Parker IM, Simberloff D, Lonsdale WM, Goodell K, Wonham M, et al. 1999. Impact: toward a framework for understanding the ecological effects of invaders. *Biol. Invasions* 1:3–19

Parks JW, Craig PJ, Neary BP, Ozburn G, Romani D. 1991. Biomonitoring in the mercury-contaminated Wabigoon-English-Winnipeg River (Canada) system: selection of the best available bioindicator. *Appl. Organomet. Chem.* 5:487–95

Paterson AH, Schertz KF, Lin Y-R, Liu S-C, Chang Y-L. 1995. The weediness of wild plants: molecular analysis of genes influencing dispersal and persistence of johnsongrass, *Sorghum halepense* (L.) Pers. *Proc. Natl. Acad. Sci. USA* 92:6127–31

Pellmyr O, Leebens-Mack M, Thompson JN. 1998. Herbivores and molecular clocks as tools in plant biogeography. *Biol. J. Linn. Soc.* 63:367–78.

Perry WL, Feder JL, Lodge DM. 2001a. Hybridization and introgression between introduced and resident *Orconectes* crayfishes in northern Wisconsin. *Conserv. Biol.* In press

Perry WL, Feder JL, Lodge DM. 2001b. Hybrid zone dynamics and species replacement between *Orconectes* crayfishes in a northern Wisconsin lake. *Evolution.* In press

Petren K, Case TJ. 1996. An experimental demonstration of exploitation competition in an ongoing invasion. *Ecology* 77:118–32

Pianka ER. 1970. On r- and K-selection. *Am. Nat.* 104:592–97

Pimentel D, Lach L, Zuniga R, Morrison D. 2000. Environmental and economic costs of nonindigenous species in the United States. *BioScience* 50:53–65

Pyšek P, Prach K, Rejmanek M, Wade M, eds. 1995a. *Plant Invasions—General Aspects and Special Problems.* Amsterdam, The Netherlands: SPB Academic. 263 pp.

Pyšek P, Prach K, Smilauer P. 1995b. Relating invasion success to plant traits: an analysis of the Czech alien flora. See Pyšek et al. 1995a, pp. 39–60

Quinn TP, Unwin MJ, Kinnison MT. 2000. Evolution of temporal isolation in the wild: genetic divergence in timing of migration and breeding by introduced chinook salmon populations. *Evolution* 54:1372–85

Reichard SH, Hamilton CW. 1997. Predicting invasions of woody plants introduced into North America. *Conserv. Biol.* 11:193–203

Rejmanek M. 1995. What makes a species invasive? See Pyšek et al. 1995a, pp. 3–13

Rejmanek M, Richardson DM. 1996. What attributes make some plant species more invasive? *Ecology* 77:1655–61

Reusch TBH. 1998. Native predators contribute to invasion resistance to the nonindigenous bivalve *Musculista senhousia* in southern California. *Mar. Ecol. Program Ser.* 170:159–68

Rhymer JM, Simberloff D. 1996. Extinction by hybridization and introgression. *Annu. Rev. Ecol. Syst.* 27:83–109

Richardson DM, Allsopp N, D'Antonio CM, Milton SJ, Rejmanek M. 2000. Plant invasions: the role of mutalisms. *Biol. Rev.* 75:65–93

Roy J. 1990. In search of the characteristics of plant invaders. In *Biological Invasions in Europe and the Mediterranean Basin,* ed. F di Castri, AJ Hansen, M Debussche, pp. 335–52. Dordrecht, The Netherlands: Kluwer Academic. 463 pp.

Ruiz GM, Rawlings TK, Dobbs FC, Drake LA, Mulladay T, et al. 2000. Global spread of microorganisms by ships. *Nature* 408:49–50

Sala OE, Chapin FS III, Armesto JJ, Berlow E, Bloomfield J, et al. 2000. Global biodiversity scenarios for the year 2100. *Science* 287:1770–74

Schierenbeck KA, Mamrick JL, Mack RN. 1995. Comparison of allozyme variability in a native and an introduced species of *Lonicera. Heredity* 75:1–9

Schiffman PM. 1997. Animal-mediated dispersal and disturbance: driving forces behind alien plant naturalization. In *Assessment and Management of Plant Invasions,* ed. JO Luken, JW Thieret, pp. 87–94. New York: Springer-Verlag. 324 pp.

Sharov AA, Liebhold AM. 1998. Model of slowing the spread of gypsy moth (Lepidoptera: Lymantriidae) with a barrier zone. *Ecol. Appl.* 8:1170–79

Shea K, Kelly D. 1998. Estimating biocontrol agent impact with matrix models: *Carduus nutans* in New Zealand. *Ecol. Appl.* 8:824–32

Shigesada N, Kawasaki K. 1997. *Biological Invasions: Theory and Practice.* London: Oxford Univ. Press. 205 pp.

Simberloff D. 1989. Which insect introductions

succeed and which fail? See Drake et al. 1989, pp. 61–76

Simberloff D. 1996. Impacts of introduced species in the United States. *Consequences: Nat. Implic. Environ. Change* 2:13–22

Simberloff D, Von Holle M. 1999. Synergistic interactions of nonindigenous species: invasional meltdown? *Biol. Invasions* 1:21–32

Skellam JB. 1951. Random dispersal in theoretical populations. *Biometrika* 38:196–218

Slade RW, Moritz C. 1998. Phylogeography of *Bufo marinus* from its natural and introduced ranges. *Proc. R. Soc. London Ser. B* 265:769–77

Smith HA, Johnson WS, Shonkwiler JS. 1999. The implications of variable or constant expansion rates in invasive weed infestations. *Weed Sci.* 47:62–66

Stein B, Kutner LS, Adams JS. 2000. *Precious Heritage: The Status of Biodiversity in the United States.* Oxford: Oxford Univ. Press. 399 pp.

Stoeckel JA, Schneider DW, Soeken LA, Blodgett KD, Sparks RE. 1997. Larval dynamics of a riverine metapopulation: implications for zebra mussel recruitment, dispersal, and control in a large-river system. *J. North Am. Benthol. Soc.* 16:586–601

Stone GN, Sunnucks P. 1993. Genetic consequences of an invasion through a patchy environment—the cynipid gallwasp *Andricus queruscalicis* (Hymenoptera: Cynipidae). *Mol. Ecol.* 2:251–68

Terwilliger J, Pastor J. 1999. Small mammals, ectomycorrhizae, and conifer succession in beaver meadows. *Oikos* 85:83–94

Thebaud CA, Finzi C, Affre L, Debusscche M, Escarre J. 1996. Assessing why two introduced *Conyza* differ in their ability to invade Mediterranean old fields. *Ecology* 77:791–804

Thompson JN. 1993. Oviposition preference and the origins of geographic variation in specialization in swallowtail butterflies. *Evolution* 47:1585–94

Thompson JN. 1998. Rapid evolution as an ecological process. *Trends Ecol. Evol.* 13:329–32

Thompson JN. 1999. The evolution of species interactions. *Science* 284:2116–18

Tilman D. 1997. Community invasibility, recruitment limitation, and grassland biodiversity. *Ecology* 78:81–92

Tilman D. 1999. The ecological consequences of changes in biodiversity: a search for general principles. *Ecology* 80:1455–74

Townsend CR. 1996. Invasion biology and ecological impacts of brown trout *Salmo trutta* in New Zealand. *Biol. Conserv.* 78:13–22

Tsutsui ND, Suarez AV, Holway DA, Case TJ. 2000. Reduced genetic variation and the success of an invasive species. *Proc. Natl. Acad. Sci. USA* 97:5948–53

U.S. Congr. Off. Technol. Assess. 1993. *Harmful non-indigenous species in the United States, OTF-F-565.* Washington, DC: US GPO

Usher MB. 1988. Biological invasions of nature reserves: a search for generalizations. *Biol. Conserv.* 44:119–35

Van Driesche RG, Bellows TS Jr. 1996. *Biological Control.* New York: Chapman & Hall. 539 pp.

Vitousek PM, D'Antonio CM, Loope LL, Westbrooks R. 1996. Biological invasions as global environmental change. *Am. Sci.* 84:218–28

Vitousek PM, Walker LR. 1989. Biological invasion by *Myrica faya* in Hawaii: plant demography, nitrogen fixation, ecosystem effects. *Ecol. Monogr.* 59:247–65

Whittier JM, Limpus D. 1996. Reproductive patterns of a biologically invasive species: the brown tree snake (*Boiga irregularis*) in eastern Australia. *J. Zool.* 238:591–97

Wilcove DS, Rothstein D, Dubow J, Phillips A, Losos E. 1998. Quantifying threats to imperiled species in the United States. *BioScience* 48:607–15

Williamson M. 1996. *Biological Invasions.* New York: Chapman & Hall. 244 pp.

Williamson M, Brown K. 1986. The analysis

and modelling of British invasions. *Philos. Trans. R. Soc. London Ser. B* 314:505–22

Wilson AB, Nalsh K-A, Boulding EG. 1999. Multiple dispersal strategies of the invasive quagga mussel (*Dreissena bugensis*) as re-vealed by microsatellite analysis. *Can. J. Fish. Aquat. Sci.* 56:2248–61

Woolcock JL, Cousens R. 2000. A mathematical analysis of factors affecting the rate of spread of patches of annual weeds in an arable field. *Weed Sci.* 48:27–34

Annu. Rev. Ecol. Syst. 2001. 32:333–65
Copyright © 2001 by Annual Reviews. All rights reserved

STREAMS IN THE URBAN LANDSCAPE

Michael J. Paul[1] and Judy L. Meyer

Institute of Ecology, University of Georgia, Athens, Georgia 30602;
e-mail: mike@sparc.ecology.uga.edu, meyer@sparc.ecology.uga.edu

Key Words   impervious surface cover, hydrology, fluvial geomorphology, contaminants, biological assessment

■ **Abstract**   The world's population is concentrated in urban areas. This change in demography has brought landscape transformations that have a number of documented effects on stream ecosystems. The most consistent and pervasive effect is an increase in impervious surface cover within urban catchments, which alters the hydrology and geomorphology of streams. This results in predictable changes in stream habitat. In addition to imperviousness, runoff from urbanized surfaces as well as municipal and industrial discharges result in increased loading of nutrients, metals, pesticides, and other contaminants to streams. These changes result in consistent declines in the richness of algal, invertebrate, and fish communities in urban streams. Although understudied in urban streams, ecosystem processes are also affected by urbanization. Urban streams represent opportunities for ecologists interested in studying disturbance and contributing to more effective landscape management.

INTRODUCTION

Urbanization is a pervasive and rapidly growing form of land use change. More than 75% of the U. S. population lives in urban areas, and it is expected that more than 60% of the world's population will live in urban areas by the year 2030, much of this growth occurring in developing nations (UN Population Division 1997, US Census Bureau 2001). Whereas the overall land area covered by urban growth remains small (2% of earth's land surface), its ecological footprint can be large (Folke et al. 1997). For example, it is estimated that urban centers produce more than 78% of global greenhouse gases (Grimm et al. 2000) and that some cities in the Baltic region claim ecosystem support areas 500 to 1000 times their size (Boland & Hanhammer 1999).

This extensive and ever-increasing urbanization represents a threat to stream ecosystems. Over 130,000 km of streams and rivers in the United States are im-

---

[1]Present address: Tetra Tech, Inc., 10045 Red Run Blvd., Suite 110, Owings Mills, Maryland 21117.

0066-4162/01/1215-0333$14.00   **333**

paired by urbanization (USEPA 2000). This makes urbanization second only to agriculture as the major cause of stream impairment, even though the total area covered by urban land in the United States is minor in comparison to agricultural area. Urbanization has had similarly devastating effects on stream quality in Europe (House et al. 1993).

Despite the dramatic threat urbanization poses to stream ecosystems, there has not been a thorough synthesis of the ecological effects of urbanization on streams. There are reviews discussing the impacts of a few aspects of urbanization [biology of pollution (Hynes 1960), physical factors associated with drainage (Butler & Davies 2000), urban stream management (Baer & Pringle 2000)] and a few general reviews aimed at engineers and invertebrate biologists (House et al. 1993, Ellis & Marsalek 1996, Suren 2000), but the ecological effects of urban growth on stream ecosystems have received less attention (Duda et al. 1982, Porcella & Sorenson 1980).

An absolute definition of urban is elusive. *Webster's New Collegiate Dictionary* defines urban as "of, relating to, characteristic of, or constituting a city," where the definition of city is anything greater than a village or town. In human population terms, the U. S. Census Bureau defines urban as "comprising all territory, population, and housing units in urbanized areas and in places of 2,500 or more persons outside urbanized areas," where urbanized areas are defined as places with at least 50,000 people and a periurban or suburban fringe with at least 600 people per square mile. The field of urban studies, within sociology, has a variety of definitions, which all include elements of concentrated populations, living in large settlements and involving some specialization of labor, alteration of family structure, and change in political attitudes (Danielson & Keles 1985). In this review, we rely on the census-based definition, as it includes suburban areas surrounding cities, which are an integral part of many urban ecological studies and represent, in many cases, areas that will develop into more densely populated centers. However, many industrial/commercial/transportation areas that are integral parts of urban and urbanizing areas have low resident population densities, but are certainly contained within our view of urban areas.

Ecological studies of urban ecosystems are growing (McDonnell & Pickett 1990, USGS 1999, Grimm et al. 2000). A valuable distinction has been drawn between ecology in cities versus ecology of cities (Grimm et al. 2000). The former refers to the application of ecological techniques to study ecological systems within cities, whereas the latter explores the interaction of human and ecological systems as a single ecosystem. Although our review focuses on stream ecology in cities, it is our hope that it will provide information of value to the development of an ecology of cities. The goal of this review is to provide a synthesis of the diverse array of studies from many different fields related to the ecology of urban streams, to stimulate incorporation of urban streams in ecological studies, and to explore ecological findings relevant to future policy development. This review is a companion to the review of terrestrial urban ecosystems by Pickett et al. (2001). The review is structured in three parts that focus on the physical, chemical, and biological/ecological effects of urbanization on streams.

PHYSICAL EFFECTS OF URBANIZATION

Hydrology

A dominant feature of urbanization is a decrease in the perviousness of the catchment to precipitation, leading to a decrease in infiltration and an increase in surface runoff (Dunne & Leopold 1978). As the percent catchment impervious surface cover (ISC) increases to 10–20%, runoff increases twofold; 35–50% ISC increases runoff threefold; and 75–100% ISC increases surface runoff more than fivefold over forested catchments (Figure 1) (Arnold & Gibbons 1996). Imperviousness has become an accurate predictor of urbanization and urban impacts on streams (McMahon & Cuffney 2000), and many thresholds of degradation in streams are associated with an ISC of 10–20% (Table 1) [hydrologic and geomorphic (Booth & Jackson 1997), biological (Klein 1979, Yoder et al. 1999)].

Various characteristics of stream hydrography are altered by a change in ISC. Lag time, the time difference between the center of precipitation volume to the center of runoff volume, is shortened in urban catchments, resulting in floods that peak more rapidly (Espey et al. 1965, Hirsch et al. 1990). Decreases in flood peak widths from 28–38% over forested catchments are also observed, meaning floods are of shorter duration (Seaburn 1969). However, peak discharges are higher in urban catchments (Leopold 1968). Flood discharges increase in proportion to ISC and were at least 250% higher in urban catchments than forested catchments in Texas and New York after similar storms (Espey et al. 1965, Seaburn 1969). Flood discharges with long-term recurrence intervals are less affected by urbanization than more frequent floods, primarily because elevated soil moisture associated with large storms results in greater surface runoff in forested catchments (Espey et al. 1965, Hirsch et al. 1990). Some exceptions to these observations have been noticed, largely depending on the location of urbanization within a catchment. If the ISC occurs lower in a catchment, flooding from that portion can drain faster than stormflow from forested areas higher in the catchment, leading to lower overall peak flood discharge and increased flood duration (Hirsch et al. 1990). In addition, blocked culverts and drains, swales, etc. may also detain water and lower peak flood discharges (Hirsch et al. 1990).

A further result of increased runoff is a reduction in the unit water yield: a greater proportion of precipitation leaves urban catchments as surface runoff (Figure 1) (Espey et al. 1965, Seaburn 1969). This reduces groundwater recharge and results in a reduction of baseflow discharge in urban streams (Klein 1979, Barringer et al. 1994). However, this phenomenon has been less intensively studied than

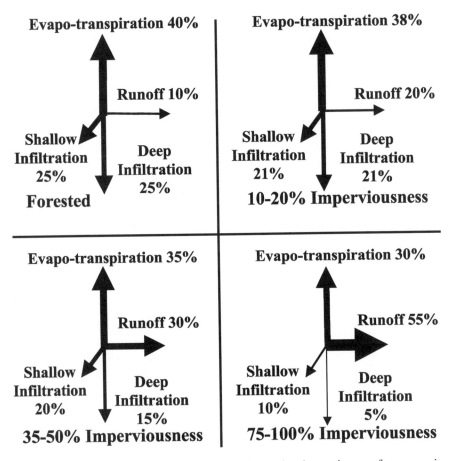

Figure 1 Changes in hydrologic flows with increasing impervious surface cover in urbanizing catchments (after Arnold & Gibbons 1996).

flooding, and the effects of irrigation, septic drainage, and interbasin transfers may mitigate the effects of reduced groundwater recharge on baseflow (Hirsch et al. 1990). Baseflow may also be augmented by wastewater treatment plant (WWTP) effluent. The Acheres (Seine Aval) treatment plant, which serves 8.1 million people, discharges 75 km west of Paris and releases 25,000 liters/s during low flow periods (Horowitz et al. 1999), increasing baseflow discharge in the Seine by up to 40% during low flow periods. More strikingly, wastewater effluent constitutes 69% annually and at times 100% of discharge in the South Platte River below Denver, Colorado (Dennehy et al. 1998). In our experience, high percentage contributions of wastewater discharge to urban rivers are not uncommon.

TABLE 1   Effects of impervious surface cover (ISC) resulting from urbanization on various physical and biological stream variables[a]

Study subject	Findings	Reference
Physical responses: hydrology		
Streams in Texas	Peak discharge increases and lag time decreases with ISC.	Espey et al. 1965
Streams in Pennsylvania	Bankfull discharge increases and lag time decreases with catchment ISC.	Leopold 1968
Review	Surface runoff increases and lag time decreases with increasing ISC (see Figure 1).	Arnold & Gibbons 1996
Streams in Washington	Increase in bankfull discharge with increasing ISC. At 10%, 2 y urban flood equals a 10 y forested flood.	Booth & Jackson 1997
Physical responses: geomorphology		
Streams in Pennsylvania	Channel enlargement increases with increasing ISC.	Hammer 1972
Streams in New York	Channel enlargement begins at 2% ISC.	Morisawa & LaFlure 1979
Streams in New Mexico	Dramatic changes in channel dimensions at 4% ISC	Dunne & Leopold 1978
Streams in Washington	Channels begin widening at 6% ISC; channels universally unstable above 10% ISC	Booth & Jackson 1997
Physical responses: temperature		
Streams in Washington, DC	Stream temperatures increase with increasing ISC.	Galli 1991
Biological responses: fish		
Streams in Maryland	Fish diversity decreased dramatically above 12–15% ISC and fish were absent above 30–50% ISC.	Klein 1979
Streams in Ontario, Canada	Fish IBI decreased sharply above 10% ISC, but streams with high riparian forest cover were less affected.	Steedman 1988
Streams in New York	Resident and anadromous fish eggs and larvae densities decreased to 10% urban land use and then were essentially absent.	Limburg & Schmidt 1990

(Continued)

TABLE 1   (*Continued*)

Study subject	Findings	Reference
Streams in Maryland	Fish diversity decreased dramatically above 10–12% ISC.	Schueler & Galli 1992
Streams in Wisconsin	Fish IBI decreased rapidly at 10% ISC.	Wang et al. 1997
Streams in Ohio	Fish IBI decreased rapidly between 8% and 33% urban land use.	Yoder et al. 1999
Biological responses: invertebrates		
Streams in Maryland	Invertebrate diversity decreased sharply from 1% to 17% ISC.	Klein 1979
Streams in Northern Virginia	Insect diversity decreased between 15% and 25% ISC.	Jones & Clark 1987
Streams in Maryland	Insect diversity metrics moved from good to poor at 15% ISC.	Schueler & Galli 1992
Streams in Washington	Insect IBI decreased sharply between 1% and 6% ISC, except where streams had intact riparian zones.	Horner et al. 1997
Streams in Ohio	Insect diversity, biotic integrity decreased between 8% and 33% ISC.	Yoder et al. 1999

[a]IBI, index of biotic integrity.

Geomorphology

The major impact of urbanization on basin morphometry is an alteration of drainage density, which is a measure of stream length per catchment area (km/km^2). Natural channel densities decrease dramatically in urban catchments as small streams are filled in, paved over, or placed in culverts (Dunne & Leopold 1978, Hirsch et al. 1990, Meyer & Wallace 2001). However, artificial channels (including road culverts) may actually increase overall drainage densities, leading to greater internal links or nodes that contribute to increased flood velocity (Graf 1977, Meyer & Wallace 2001).

A dominant paradigm in fluvial geomorphology holds that streams adjust their channel dimensions (width and depth) in response to long-term changes in sediment supply and bankfull discharge (recurrence interval average $= 1.5$ years) (Dunne & Leopold 1978, Roberts 1989). Urbanization affects both sediment supply and bankfull discharge. During the construction phase erosion of exposed soils increases catchment sediment yields by 10^2–10^4 over forested catchments and can be more exaggerated in steeply sloped catchments (Wolman 1967, Leopold 1968, Fusillo et al. 1977). Most of this export occurs during a few large, episodic floods (Wolman 1967). This increased sediment supply leads to an aggradation phase

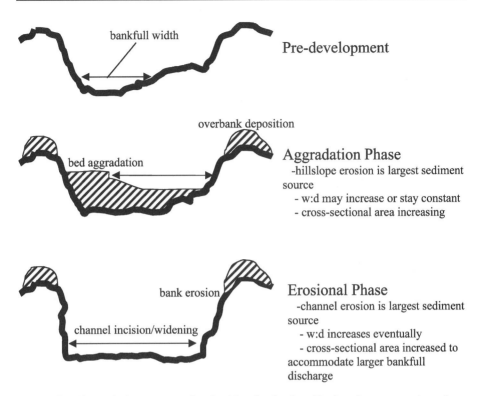

bankfull width

Pre-development

overbank deposition

bed aggradation

Aggradation Phase
-hillslope erosion is largest sediment
source
- w:d may increase or stay constant
- cross-sectional area increasing

bank erosion

channel incision/widening

Erosional Phase
-channel erosion is largest sediment
source
- w:d increases eventually
- cross-sectional area increased to
accommodate larger bankfull
discharge

Figure 2   Channel changes associated with urbanization. During the construction phase
of urbanization, hillslope erosion increases sediment supply leading to bed aggradation and
overbank deposition. After construction ceases hillslope sediment supply is reduced, but
bankfull flows are increased owing to increases in imperviousness. This leads to increased
channel erosion as channel incision and widening occur to accommodate increased bankfull
discharge.

as sediments fill urban channels (Figure 2). During this phase stream depths may
decrease as sediment fills the channel, and the decreased channel capacity leads to
greater flooding and overbank sediment deposition, raising bank heights (Wolman
1967). Therefore, overall channel cross-sections stay the same or even decrease
slightly (Robinson 1976). Ironically, the flooding associated with aggradation may
help attenuate increased flows resulting from increased imperviousness by stor-
ing water in the floodplain, temporarily mitigating urban effects on hydrography
(Hirsch et al. 1990).

After the aggradation phase sediment supply is reduced and geomorphic readjustment initiates a second, erosional phase (Figure 2). High ISC associated
with urbanization increases the frequency of bankfull floods, frequently by an order
of magnitude or, conversely, increases the volume of the bankfull flood (Leopold
1973, Dunne & Leopold 1978, Arnold et al. 1982, Booth & Jackson 1997). As

a result, increased flows begin eroding the channel and a general deepening and widening of the channel (channel incision) occurs to accommodate the increased bankfull discharge (Hammer 1972, Douglas 1974, Roberts 1989, Booth 1990). Increased channel water velocities exceed minimum entrainment velocities for transporting bed materials, and readily moveable sediment is lost first as channels generally deepen (Leopold 1973, Morisawa & LaFlure 1979). Channels may actually narrow during this phase as entrained sediment from incision is deposited laterally in the channel (Dunne & Leopold 1978). After incision channels begin to migrate laterally, bank erosion begins, which leads to general channel widening (Booth 1990, Booth & Jackson 1997, Trimble 1997).

During the erosional phase channel enlargement can occur gradually if increases in width and depth keep pace with increases in discharge associated with increasing ISC. In this case the channel enlargement may be barely noticeable (Booth 1990). However, erosion more commonly occurs disproportionately to discharge changes, often leading to bank failure and catastrophic erosion in urban streams (Neller 1988, Booth 1990). In developed urban catchments, as a result of this erosional readjustment phase, the majority of sediment leaving the catchment comes from within-channel erosion as opposed to hillslope erosion (Trimble 1997). The magnitude of this generalized geomorphic response will vary longitudinally along a stream network as well as with the age of development, catchment slope, geology, sediment characteristics, type of urbanization, and land use history (Gregory et al. 1992).

Urban streams differ in other geomorphic characteristics from forested catchments as well. The spacing between pool-riffle sequences (distance between riffles) is generally constant at 5–7 times channel width in forested catchments (Gregory et al. 1994). Generally, this ratio stays constant in urban channels as they widen, which means the absolute distance between pool-riffle units increases, although there is some evidence that this spacing may decrease to 3–5 times channel width (Gregory et al. 1994).

Changes in sediment supply may also alter channel pattern. Increased sediment supply during construction has converted some meandering streams to braided patterns or to straighter, more channelized patterns (Arnold et al. 1982). In the latter case, channelizing leads to increased slope and therefore higher in-stream velocities, especially where artificial channel alteration is carried out to increase the efficiency of the channel in transporting flows (Pizzuto et al. 2000).

Urbanization can also alter sediment texture. Less fine sediment, increased coarse sand fractions, and decreased gravel classes have been observed in urban channels as a result of alteration of sediment supply and altered velocities (Finkenbine et al. 2000, Pizzuto et al. 2000). In addition to sediment changes, large woody debris is also reduced in urban channels. Catchments in Vancouver, British Columbia with greater than 20% ISC generally have very little large woody debris, a structural element important in both the geomorphology and ecology of Pacific Northwest stream ecosystems (Finkenbine et al. 2000).

Other geomorphic changes of note in urban channels include erosion around bridges, which are generally more abundant as a result of increased road densities in urban channels (Douglas 1974). Bridges have both upstream and downstream effects, including plunge pools created below bridge culverts that may serve as barriers to fish movement. Knickpoints are another common feature of urban channels. These readily erodeable points of sudden change in depth are created by channel erosion, dredging, or bridge construction and are transmitted throughout the catchment, causing channel destabilization (Neller 1988). Other features include increased tree collapse, hanging tributary junctions as a result of variable incision rates, and erosion around artificial structures (e.g., utility support pilings) (Roberts 1989).

Changes in the hydrology and geomorphology of streams likely affect the hydraulic environment of streams, altering, among other things, the velocity profiles and hyporheic/parafluvial dynamics of channels. Such changes would affect many ecological processes, from filter-feeding organisms (Hart & Finelli 1999) to carbon processing and nutrient cycling (Jones & Mulholland 2000).

Temperature

Stream temperature is an important variable affecting many stream processes such as leaf decomposition (Webster & Benfield 1986) and invertebrate life history (Sweeney 1984). Urbanization affects many elements of importance to stream heat budgets. Removal of riparian vegetation, decreased groundwater recharge, and the "heat island" effect associated with urbanization, covered more fully in a companion review (Pickett et al. 2001), all affect stream temperature (Pluhowski 1970), yet very little published data exists on temperature responses of streams to urbanization. In one study on Long Island urban streams had mean summer temperatures 5–8°C warmer and winter temperatures 1.5–3°C cooler than forested streams. Seasonal diurnal fluctuations were also greater in urban streams, and summertime storms resulted in increased temperature pulses 10–15°C warmer than forested streams, a result of runoff from heated impervious surface (Pluhowski 1970). Similar effects on summer temperatures and daily fluctuations have also been observed elsewhere (Table 1) (Galli 1991, Leblanc et al. 1997).

CHEMICAL EFFECTS OF URBANIZATION

Chemical effects of urbanization are far more variable than hydrologic or geomorphic effects and depend on the extent and type of urbanization (residential versus commercial/industrial), presence of wastewater treatment plant (WWTP) effluent and/or combined sewer overflows (CSOs), and the extent of stormwater drainage. Overall, there are more data on water and sediment chemistry in urban streams than any other aspect of their ecology. This is aided by several very large national datasets of stream chemistry that focus in whole or in part on urbanization [e.g., National Urban Runoff Program (United States), National Water Quality Assessment Program (USGS 2001), Land-Ocean Interaction Study (UK) (Neal & Robson 2000)].

In general, there is an increase in almost all constituents, but consistently in oxygen demand, conductivity, suspended solids, ammonium, hydrocarbons, and metals, in urban streams (Porcella & Sorenson 1980, Lenat & Crawford 1994, Latimer & Quinn 1998, USGS 1999). These increases can be attributed to both WWTP effluent and non–point source (NPS) runoff. Many countries have accomplished significant reductions in chemical constituents as a result of adopting better WWTP technologies (e.g., Krug 1993, Litke 1999). However, treatment cannot remove all constituents from wastewater, treatment systems fail, and permitted discharge limits are exceeded. There are more than 200,000 discharges subject to permitting in the United States (USEPA 2001), and of 248 urban centers studied, 84% discharge into rivers (40% of those into rivers with mean annual discharges less than 28 m³/s) (Heaney & Huber 1984). In addition, CSO systems are still common, in which stormwater and untreated sewage are combined and diverted to streams and rivers during storms. At least 28% of the urban centers mentioned above contained CSOs, and in the United Kingdom 35% of the annual pollutant discharge comes from CSOs and storm drains during less than 3% of the time (Heaney & Huber 1984, Faulkner et al. 2000). In addition, illicit discharge connections, leaking sewer systems, and failing septic systems are a large and persistent contributor of pollutants to urban streams (Faulkner et al. 2000). In the Rouge River catchment in Detroit, Michigan, the focus of an intense federal NPS management program, septic failure rates between 17% and 55% were reported from different subcatchments, and it was estimated that illicit untreated sewage discharge volume at more than 193,000 m³/yr (Johnson et al. 1999). The ubiquitous nature of small, NPS problems in urban catchments has led some to suggest that the cumulative effect of these small problems may be the dominant source of biological degradation in urban catchments (Duda et al. 1982).

Nutrients and Other Ions

Urbanization generally leads to higher phosphorus concentrations in urban catchments (Omernik 1976, Meybeck 1998, USGS 1999, Winter & Duthie 2000). An urban effect is most often seen in total phosphorus as a result of increased particle-associated phosphorus, but dissolved phosphorus levels are also increased (Smart et al. 1985). In some cases increases in phosphorus can even rival those seen in agricultural catchments both in terms of concentration and yield (Omernik 1976). Even an attempt to understand the agricultural contribution to catchment phosphorus dynamics in a midwestern catchment discovered that urbanization was a dominant factor (Osborne & Wiley 1988). Even though urban areas constituted only 5% of the catchment area and contributed only a small part to the total annual yield of dissolved phosphorus, urban land use controlled dissolved phosphorus concentration throughout the year.

Sources of phosphorus in urban catchments include wastewater and fertilizers (LaValle 1975). Lawns and streets were the primary source of phosphorus to urban streams in Madison, Wisconsin as a result of fertilizer application (Waschbusch et al. 1999). Soils are important in phosphorus dynamics, and the retention of groundwater phosphorus from septic fields affects stream phosphorus concentrations (Hoare 1984, Gerritse et al. 1995). Phosphorus stored in soils as a result of fertilization, however, can be mobilized by soil erosion and contribute to eutrophication of receiving waters. This effect has been called the "chemical time bomb" and is of particular concern when previously agricultural land is cleared for urban growth (Bennett et al. 1999).

Although phosphorus concentrations are elevated in urban streams, the effective increase is not as great as that observed for nitrogen. Urban centers have been shown to increase the nitrogen concentration in rivers for hundreds of kilometers (Meybeck 1998, USGS 1999). Increases have been observed for ammonium as well as nitrate (McConnell 1980, Hoare 1984, Zampella 1994, Wernick et al. 1998). The extent of the increase depends on wastewater treatment technology, degree of illicit discharge and leaky sewer lines, and fertilizer use. As with phosphorus, nitrogen concentrations in streams draining agricultural catchments are usually much higher (USGS 1999), but some have noticed similar or even greater levels of nitrogen loading from urbanization (Omernik 1976, Nagumo & Hatano 2000). Soil characteristics also affect the degree of nitrogen retention, of importance when on-site septic systems are prevalent (Hoare 1984, Gerritse et al. 1995).

Other ions are also generally elevated in urban streams, including calcium, sodium, potassium, and magnesium (McConnell 1980, Smart et al. 1985, Zampella 1994, Ometo et al. 2000). Chloride ions are elevated in urban streams, especially where sodium chloride is still used as the principal road deicing salt. A significant portion of the more than 100,000 tons of sodium chloride applied in metropolitan Toronto annually for deicing enters long-turnover groundwater pools and is released slowly, raising stream chloride concentrations throughout the year (Howard & Haynes 1993). The combined effect of heightened ion concentrations in streams is the elevated conductivity observed in most urban streams. The effect is so common that some have suggested using chloride concentration or conductivity as general urban impact indicators (Wang & Yin 1997, Herlihy et al. 1998).

Metals

Another common feature of urban streams is elevated water column and sediment metal concentrations (Bryan 1974, Wilber & Hunter 1977, Neal et al. 1997, Horowitz et al. 1999, Neal & Robson 2000). The most common metals found include lead, zinc, chromium, copper, manganese, nickel, and cadmium (Wilber & Hunter 1979), although lead has declined in some urban river systems since its elimination as a gas additive (Frick et al. 1998). Mercury is also elevated in some urban streams, and particle-bound methyl-mercury can be high during stormflow (Mason & Sullivan 1998, Horowitz et al. 1999). In addition to industrial discharges, there are many NPSs of these metals in urban catchments: brake linings contain nickel, chromium, lead, and copper; tires contain zinc, lead, chromium, copper, and nickel; and metal alloys used for engine parts contain nickel, chromium, copper, and manganese among others (Muschak 1990, Mielke et al. 2000). All of these metals accumulate on roads and parking lots (Sartor et al. 1974, Forman & Alexander 1998). Many other metals have been found in elevated concentrations in urban stream sediments including arsenic, iron, boron, cobalt, silver, strontium, rubidium, antimony, scandium, molybdenum, lithium, and tin (Khamer et al. 2000, Neal & Robson 2000). Not surprisingly, it appears that NPSs of metals are more important than point sources in urban streams (Wilber & Hunter 1977, Mason & Sullivan 1998).

The concentration, storage, and transport of metals in urban streams is connected to particulate organic matter content and sediment characteristics (Tada & Suzuki 1982, Rhoads & Cahill 1999). Organic matter has a high binding capacity for metals, and both bed and suspended sediments with high organic matter content frequently exhibit 50–7500 times higher concentrations of zinc, lead, chromium, copper, mercury, and cadmium than sediments with lower organic matter content (Warren & Zimmerman 1994, Mason & Sullivan 1998, Gonzales et al. 2000). Sediment texture is also important, and metal concentration in sediments was inversely correlated to sediment particle size in several urban New Jersey streams (Wilber & Hunter 1979). In addition, geomorphic features have been shown to influence metal accumulations. Higher sediment metal concentrations were found in areas of low velocity (stagnant zones, bars, etc.) where fine sediments and organic particles accumulate, whereas areas of intermediate velocities promoted the accumulation of sand-sized metal particles, which can also be common in urban streams (Rhoads & Cahill 1999).

Several organisms (including algae, mollusks, arthropods, and annelids) have exhibited elevated metal concentrations in urban streams (Davis & George 1987, Rauch & Morrison 1999, Gundacker 2000), and ecological responses to metals include reduced abundances and altered community structure (Rauch & Morrison 1999). It is important to note that the route of entry appears to be both direct exposure to dissolved metals and ingestion of metals associated with fine sediments and organic matter. This has led a few researchers to suggest that metal toxicity is most strongly exerted through the riverbed rather than the overlying water (Medeiros et al. 1983, House et al. 1993), although only dissolved metal concentrations in the water column are regulated in the United States.

Pesticides

Pesticide detection frequency is high in urban streams and at concentrations frequently exceeding guidelines for the protection of aquatic biota (USGS 1999, Hoffman et al. 2000). These pesticides include insecticides, herbicides, and fungicides (Daniels et al. 2000). In addition, the frequent detection of banned substances such as DDT and other organochlorine pesticides (chlordane and dieldrin) in urban streams remains a concern (USGS 1999). Most surprising is that many organochlorine pesticide concentrations in urban sediments and biota frequently exceed those observed in intensive agricultural areas in the United States (USGS 1999), a phenomenon observed in France as well (Chevreuiel et al. 1999). Additionally, it is estimated that the mass of insecticides contributed by urban areas is similar to that from agricultural areas in the United States (Hoffman et al. 2000).

There are many sources of pesticides in urban catchments. Urban use accounts for more than 136,000 kg, which is a third of U.S. pesticide use (LeVeen & Willey 1983). They are frequently applied around homes (70–97% of U.S. homes use pesticides) and commercial/industrial buildings and are intensively used in lawn and golf course management (LeVeen & Willey 1983, USGS 1999). Areal application rates in urban environments frequently exceed those in agricultural applications by nearly an order of magnitude (Schueler 1994b). For example, pesticide application rates on golf courses (including herbicides, insecticides, and fungicides) exceed 35 pounds/acre/year, whereas corn/soybean rotations receive less than 6 pounds/acre/year (Schueler 1994b). However, unlike agricultural use, urban pesticide application rates are generally not well documented (LeVeen & Willey 1983, Coupe et al. 2000).

As with metals, the main vector of transport of pesticides into urban streams appears to be through NPS runoff rather than WWTP effluent (Foster et al. 2000). A strong correlation between particle concentration and pesticide concentration was found in the Anacostia River basin in Maryland and the San Joaquin River in California, suggesting NPS inputs are most important (Pereira et al. 1996, Foster et al. 2000). Volatilization and aerosol formation contributed to higher pesticide concentrations, including atrazine, diazinon, chlorpyrifos, p,p'-DDE (a DDT metabolite), and other organochlorines, in precipitation in urban areas and may contribute directly to greater pesticide concentrations and yields in urban areas (Weibel et al. 1966, Coupe et al. 2000).

Other Organic Contaminants

A whole suite of other organic contaminants are frequently detected in urban streams, including polychlorinated biphenyls (PCBs), polycyclic aromatic hydrocarbons (PAHs), and petroleum-based aliphatic hydrocarbons (Whipple & Hunter 1979, Moring & Rose 1997, Frick et al. 1998). PCBs are still frequently detected in urban areas of the United States, even though their use in manufacturing was outlawed because of their carcinogenic effects. These compounds are very stable and are still found in fish at concentrations exceeding consumption-level guidelines in urban rivers such as the Chattahoochee River below Atlanta, Georgia (Frick et al. 1998). PCB concentrations were highly correlated with urban land use in the Willamette Basin in Oregon as well (Black et al. 2000). As with metals and pesticides, PCBs are primarily particle associated, and in the absence of industrial point sources, it is assumed that stormwater runoff is the major route of entry (Foster et al. 2000).

PAHs are a large class of organic compounds that include natural aromatic hydrocarbons but also many synthetic hydrocarbons including organic solvents with different industrial uses (Yamamoto et al. 1997). For this reason, the unnatural PAHs are probably derived from industrial effluent or episodic spills. Very little is known about these compounds in urban streams. In Dallas–Fort Worth, Texas streams, 24 different industrial PAHs were detected, including 4 of the top 10 U. S. Environmental Protection Agency (EPA) most hazardous substances, and at concentrations exceeding human health criteria (Moring & Rose 1997). In Osaka, Japan streams, 55 PAHs were detected, including 40 EPA target compounds. Organic solvents (e.g., toluene, trichloroethane, and dichlorethane) were most common (Yamamoto et al. 1997).

It is difficult to find automobile parking spaces without oil stains in any city. The result of these leaky crankcases is a cornucopia of different petroleum-based aliphatic hydrocarbons in storm runoff associated primarily with particles (Whipple & Hunter 1979). Although there are natural aliphatic hydrocarbons in streams, these are generally overwhelmed by petroleum-based compounds in urban stream bed and water-column sediments (Hunter et al. 1979, Mackenzie & Hunter 1979, Eganhouse et al. 1981). Evidence suggests that these are frequently at concentrations that are stressful to sensitive stream organisms (Latimer & Quinn 1998). Most striking is the yield of these compounds from urban catchments. An estimated 485,000 liters of oil enters the Narragansett Bay each year, a volume equal to nearly 50% of the disastrous 1989 *World Prodigy* oil spill in that same bay (Hoffman et al. 1982, Latimer & Quinn 1998). Similarly, it is estimated that the Los Angeles River alone contributes about 1% of the annual world petroleum hydrocarbon input to the ocean (Eganhouse et al. 1981).

Lastly, recent data suggest pharmaceutical substances from hospital effluent may contribute an array of different chemical compounds into streams. Detectable levels of antiobiotics, genotoxic chemotherapeutic drugs, analgesics, narcotics, and psychotherapeutic drugs have been reported from effluent and/or surface waters (Halling-Sorensen et al. 1998). Although there is some information on the toxicity of these different compounds from laboratory studies, there are insufficient data on the nature or extent of the threat they pose to urban stream biota.

BIOLOGICAL AND ECOLOGICAL EFFECTS OF URBANIZATION

The ecological implications of urbanization are far less studied than the chemical effects, an absence noted in several studies (Porcella & Sorenson 1980, Duda et al. 1982, Medeiros et al. 1983). Nevertheless, much is known about the response of stream organisms, especially invertebrates, to urbanization; far less is known about urban effects on fish (Mulholland & Lenat 1992). Of even greater concern is the lack of mechanistic studies; few studies analyze whether physical habitat, water quality, or food web disturbances (either resource effects or altered community interactions) are the cause of biological degradation in urban streams (Suren 2000). Grossly underrepresented are studies of population dynamics, community interactions, and ecosystem ecology of urban streams, which is surprising given the level of knowledge within the field (Allan 1995). Lastly, very little information has been gathered on biological monitoring of restoration or best management practice implementation in urban catchments (Riley 1998). Most studies assess performance based on stream channel condition or pollutant reduction; few, if any, monitor biological response (Benke et al. 1981, Center for Watershed Protection 2000). In this section, we discuss the effects of urbanization on microbes, algae, macrophytes, invertebrates, and fish.

Microbes

Bacterial densities are usually higher in urban streams, especially after storms (Porcella & Sorenson 1980, Duda et al. 1982). Much of this is attributable to increased coliform bacteria, especially in catchments with wastewater treatment plant (WWTP) and combined sewer overflow (CSO) effluent (Gibson et al. 1998, Young & Thackston 1999). In Saw Mill Run, an urban stream near Pittsburgh, Pennsylvania, fecal coliform colony–forming units (CFU) increased from 170–13,300 CFU/100 ml during dry weather to 6,100–127,000 CFU/100 ml during wet weather (Gibson et al. 1998). CSOs contributed 3,000–85,000 CFU/100 ml during wet weather. These data indicate that non–point sources (NPSs) as well as point sources contribute to fecal coliform loads in urban streams. High values during dry weather are not uncommon in urban streams and may indicate chronic sewer leakage or illicit discharges. Storm sewers were also a significant source of coliform bacteria in Vancouver, British Columbia; stormwater there contained both human and nonhuman fecal coliform bacteria (Nix et al. 1994). Other pathogens, including *Cryptosporidum* and *Giardia*, have also been associated with CSOs (Gibson et al. 1998).

Increased antibiotic resistance has been seen in some urban bacterial populations (Goni-Urriza et al. 2000). Increased resistance to several antiobiotics, including nalidixic acid, tetracycline, beta-lactam, and co-trimoxazole, has been observed from several enteric as well as native stream species isolated from a river downstream of a WWTP discharge in Spain. It may be that resistant bacteria are passing through the treatment process and conferring resistance to native bacteria. Recent evidence suggests that metal toxicity may also be indirectly involved in increasing antibiotic resistance in stream bacteria. Bacterial resistance to streptomycin and kanamycin were positively correlated with sediment mercury concentration in streams below nuclear reactors and industrial facilities, a result of indirect selection for metal tolerance (McArthur & Tuckfield 2000). Metals may also affect bacterial enzyme activity in urban streams. Enzyme levels were inversely correlated to sediment metal concentration in an urban stream, and this was especially pronounced below an industrial effluent (Wei & Morrison 1992).

Nitrifying bacteria, responsible for the oxidation of reduced nitrogen, are also influenced by urbanization. WWTP effluent can represent a significant source of nitrifying bacteria to urban streams (Brion & Billen 2000). These bacteria are used to oxidize ammonium during the treatment process, but escape into streams in effluent and contribute to the high nitrifier activity observed below some WWTP discharges (Jancarkova et al. 1997). Nitrification rates were as much as six times higher in treated effluent entering the Seine than in receiving river water upstream (Brion & Billen 2000). Ironically, because so many nitrifiers entered the Seine River in France via untreated sewage historically, the reduction in untreated sewage via improved sewage design contributed to a reduction in ammonium oxidation rates in the river from 1.5 μmol/liter/h in 1976 to 1.0 μmol/liter/h in 1993 (Brion & Billen 2000). In addition to nitrifiers, iron-oxidizing bacteria are often abundant in urban streams, especially where reduced metals emerge from anoxic urban groundwater or storm sewers (Dickman & Rygiel 1998).

Algae

The use of algae to indicate water quality in Europe and the United States has a long history (Kolkwitz & Marsson 1908, Patrick 1973). As a result, information exists on algal species and community responses to organic pollution; however, the response of algae to all aspects of urbanization is far less studied. The increasing proportion of urban land use in a catchment generally decreases algal species diversity, and this change has been attributed to many factors including water chemistry (Chessman et al. 1999). Elevated nutrients and light levels typically favor greater algal biomass, which has been observed in many urban streams, where algae do not appear to be nutrient limited (Chessman et al. 1992, Richards & Host 1994). However, the shifting nature of bed sediment in urban streams, frequent bed disturbance, and high turbidity may limit algal accumulation (Burkholder 1996, Dodds & Welch 2000). In addition, several algal species are sensitive to metals, and stream sediment metal accumulation can result in reduced algal biomass (Olguin et al. 2000). Lastly, the frequent detection of herbicides in streams, some with known effects on algae (Davies et al. 1994), will undoubtedly affect stream algal communities

Macrophytes

Little has been written on macrophyte response to urbanization. Most of the work has been done in New Zealand and Australia, where bed sediment changes, nutrient enrichment, and turbidity all contribute to reduced diversity of stream macrophytes (Suren 2000). Exotic species introductions in urban streams have also resulted in highly reduced native macrophyte diversity (Arthington 1985, Suren 2000). Excessive macrophyte growth as a result of urbanization has not been observed in New Zealand, even though nutrient and light levels are higher (Suren 2000).

Invertebrates

Literature searches revealed more studies of urban effects on aquatic invertebrates than on any other group, and the available data are being expanded by groups biomonitoring urban systems (e.g., USGS National Water Quality Assessment,

U.S. EPA, state agencies, and others). All aspects of aquatic invertebrate habitat are altered by urbanization. One of the historically well-studied aspects has been the effects of organic pollutants (especially WWTP effluent) on invertebrates. Organic pollution generally reduces invertebrate diversity dramatically, resulting in a community dominated by Chironomidae (Diptera) and oligochaetes (Campbell 1978, Seager & Abrahams 1990, Wright et al. 1995). However, general effects of urbanization on stream invertebrates have also been studied and general invertebrate responses can be summarized as follows: decreased diversity in response to toxins, temperature change, siltation, and organic nutrients; decreased abundances in response to toxins and siltation; and increased abundances in response to inorganic and organic nutrients (Resh & Grodhaus 1983, Wiederholm 1984).

Studies of the effects of urban land use on invertebrates can be divided into three types: those looking along a gradient of increasing urbanization in one catchment, those looking at an urbanized versus a reference catchment, and large studies considering urban gradients and invertebrate response in several catchments. All single catchment gradient studies find a decrease in invertebrate diversity as urban land use increases, regardless of the size of the catchment (Pratt et al. 1981, Whiting & Clifford 1983, Shutes 1984, Hachmoller et al. 1991, Thorne et al. 2000). Decreases were especially evident in the sensitive orders—Ephemeroptera, Plecoptera, and Trichoptera (Pratt et al. 1981, Hachmoller et al. 1991). Most of these studies observed decreases in overall invertebrate abundance, whereas the relative abundance of Chironomidae, oligochaetes, and even tolerant gastropods increased (Pratt et al. 1981, Thorne et al. 2000). Comparative catchment studies show the same trends with increasing urbanization as those observed in single catchment studies: decreased diversity and overall abundance and increased relative abundance of tolerant Chironomidae and oligochaetes (Medeiros et al. 1983, Garie & McIntosh 1986, Pederson & Perkins 1986, Lenat & Crawford 1994).

The multi-catchment studies attempt to relate differing amounts of urbanization in many catchments to particular invertebrate community responses, often using a gradient analysis approach. As discussed above, all find decreases in diversity and overall invertebrate abundance with increased urbanization. This response is correlated with impervious surface cover, housing density, human population density, and total effluent discharge (Klein 1979, Benke et al. 1981, Jones & Clark 1987, Tate & Heiny 1995, Kennen 1999). Klein (1979) studied 27 small catchments on the Maryland Piedmont and was among the first to identify impervious surface cover (ISC) as an important indicator of degradation. Invertebrate measures declined significantly with increasing ISC until they indicated maximum degradation at 17% ISC (Table 1). Degradation thresholds at ISC between 10 and 20% have been supported by numerous other studies for many different response variables (see Schueler 1994a). Residential urbanization in Atlanta, Georgia had dramatic effects on invertebrate diversity, but there were very few clues as to the mechanisms responsible, although leaky sewers were implicated in these and other urban residential catchments (Benke et al. 1981, Johnson et al. 1999).

Few studies have considered specific mechanisms leading to the observed effects of urbanization. This is a difficult task because of the multivariate nature of urban disturbance. Increased turbidity has been associated with higher drift densities of insects (Doeg & Milledge 1991), but more work has focused on the instability of smaller and more mobile bed sediments associated with urban sedimentation. In general, the change in bed sediments favors species adapted to unstable habitats, such as the chironomid dipterans and oligochaete annelids (Pedersen & Perkins 1986, Collier 1995). Where slopes are steeper, and smaller sediments are removed by increased water velocities, localized areas of higher invertebrate diversity are observed within the coarser sediments (Collier 1995). Pools are particularly affected by sediment accumulation in urban streams, and invertebrate communities within these habitats are degraded (Hogg & Norriss 1991). Lastly, sedimentation associated with urban streams reduces available refugial space, and invertebrates are more susceptible to drift when refugial space is limited during the frequent floods characteristic of urban environments (Borchardt & Statzner 1990). Storm-flows in urban streams introduce the majority of pollutants and also move the bed sediment frequently. The mortality of *Pteronarcys dorsata* (Plecoptera) in cages in urban streams was attributed to sedimentation associated with storms (Pesacreta 1997).

Sediment toxicity has also been explored. As mentioned above, benthic organic matter binds many toxins and is also a major food resource for many stream invertebrates (Benke & Wallace 1997). Mortality of aquatic invertebrates remains high in many urban streams even during low flow periods, suggesting that toxicity associated with either exposure in the bed or ingestion of toxins associated with organic matter contributes to invertebrate loss (Pratt et al. 1981, Medeiros et al. 1983).

Riparian deforestation associated with urbanization reduces food availability, affects stream temperature, and disrupts sediment, nutrient, and toxin uptake from surface runoff. Invertebrate bioassessment metrics decreased sharply in Puget Sound, Washington tributaries with increasing ISC (Horner et al. 1997). However, streams that had higher benthic index of biotic integrity scores for a given level of ISC were always associated with greater riparian forest cover in their catchment, suggesting that riparian zones in some urban catchments may buffer streams from urban impacts. Above 45% ISC, all streams were degraded, regardless of riparian status. The value of riparian forests is also reduced if the stormwater system is designed to bypass them and discharge directly into the stream.

Road construction associated with urbanization impacts stream invertebrates. Long-term reductions (>6 y) in invertebrate diversity and abundances were observed in association with a road construction project in Ontario (Taylor & Roff 1986). General effects of roads on streams has been reviewed recently (Forman & Alexander 1998).

Very little ecological data beyond presence/absence or abundance data have been reported for urban stream invertebrates. Aquatic insect colonization potential was reported to be high in some urban streams, suggesting restoration efforts would not be limited in this regard (Pedersen & Perkins 1986), but little is known about colonization or adult aquatic insect ecology in urban streams. Urban stream restoration work focuses largely on channel geomorphological stability, with relatively little attention given to biological restoration (Riley 1998), although restoration of Strawberry Creek on the campus of the University of California at Berkeley has resulted in detectable increases in invertebrate diversity and abundance (Charbonneau & Resh 1992). Drift of aquatic invertebrates is a well studied phenomenon in streams, but with one exception (Borchardt & Statzner 1990), little has been published on insect drift in urban streams. We found no published work regarding life cycle ecology (e.g., voltinism or emergence timing), population dynamics, behavioral ecology, community interactions, or production of aquatic invertebrates in urban streams.

Fish

Less is known about fish responses to urbanization than about invertebrates, and a general response model does not exist. However, the Ohio Environmental Protection Agency has a very large database of land use and fish abundance from around their state and has suggested three levels of general fish response to increasing urbanization: from 0 to 5% urban land use, sensitive species are lost; from 5 to 15%, habitat degradation occurs and functional feeding groups (e.g., benthic invertivores) are lost; and above 15% urban land use, toxicity and organic enrichment result in severe degradation of the fish fauna (Table 1) (Yoder et al. 1999). This model has not been verified for other regions of the country, where studies have focused on various aspects of urbanization. Here we consider three types of urban land use studies with regards to fish: gradients of increasing urbanization within a single catchment, comparing an urban and reference catchment, and large, multi-catchment urban gradient studies.

Along urban gradients within single catchments, fish diversity and abundances decline, and the relative abundance of tolerant taxa increases with increasing urbanization (Table 1) (Onorato et al. 2000, Boet et al. 1999, Gafny et al. 2000). Invasive species were also observed to increase in more urbanized reaches of the Seine River, France, and this effect extended more than 100 km below Paris (Boet et al. 1999). Summer storms in that river were associated with large fish kills as a result of dissolved oxygen deficits, an effect also observed for winter floods in Yargon Stream, the largest urban stream in Israel (Gafny et al. 2000). Comparisons with historical collections, an approach used commonly with fish studies, revealed that several sensitive species were extirpated from the Upper Cahaba River system in Alabama between 1954 and 1995, a period coinciding with the rapid growth of Birmingham, Alabama (Onorato et al. 2000). Extirpation of fish species is not uncommon in urban river systems (Ragan & Dietmann 1976, Weaver & Garman 1994, Wolter et al. 2000).

Comparative catchment studies also find dramatic declines in fish diversity and abundances in urban catchments compared with forested references (Scott et al.

1986, Weaver & Garman 1994, Lenat & Crawford 1994). Kelsey Creek, a well-studied urban stream in Washington, is unusual in that it has sustained salmonid populations, especially cutthroat trout (*Oncorhynchus clarki*), even though coho salmon (*Oncorhyncus kisutch*) and many nonsalmonid species have disappeared (Scott et al. 1986). Salmonids in the urban stream actually grow more rapidly and to larger sizes, increasing fish production up to three times that in the forested reference site, presumably a result of warmer temperatures and greater invertebrate biomass in the urban stream. However, the population size structure is different in the two streams, with year 0 and 1 cutthroat underrepresented in the urban stream (Scott et al. 1986).

Large multi-site studies of fish responses to urban gradients also find dramatic decreases in diversity or fish multimetric indices [index of biotic integrity (IBI)] with increasing ISC or other urban land use indicators (Table 1) (Klein 1979, Steedman 1988, Wang et al. 1997, Frick et al. 1998, Yoder et al. 1999). Similar to effects observed for invertebrates, these studies also find precipitous declines in fish metrics between 0 and 15% ISC or urban land use, beyond which fish communities remain degraded (Klein 1979, Yoder et al. 1999). The effect of urbanization on fish appears at lower percent land area disturbed than effects associated with agriculture. In Wisconsin and Michigan few fish community effects were observed in agricultural catchments up to 50% agricultural land use in the catchment (Roth et al. 1996, Wang et al. 1997), and mixed agriculture and urban catchments had significantly lower IBI scores than strictly agricultural catchments (Wang et al. 2000). This suggests that although total urban land use occupies a smaller area globally, it is having disproportionately large effects on biota when compared with agriculture. However, it is crucial to recognize that all urban growth does not have the same effects. Extensive fish surveys in Ohio suggest that residential development, especially large-lot residential development, has less of an effect on stream fishes than high-density residential or commercial/industrial development (Yoder et al. 1999). They hypothesize that riparian protection and less channel habitat degradation are responsible for protecting the fauna in these streams, even up to 15% urban land use. Similar benefits of riparian forests to fish in urban streams were observed in the Pacific Northwest (Horner et al. 1997).

Few studies have explored specific mechanisms causing changes in fish assemblages with urbanization. Sediment is presumably having effects on fish in urban streams similar to those observed in other systems although toxin-mediated impacts may be greater (Wood & Armitage 1997). Road construction results in an increase in the relative abundance of water-column feeders as opposed to benthic feeders, likely a response to a decrease in benthic invertebrate densities (Taylor & Roff 1986). Benthic feeders quickly reappeared as sedimentation rates declined after construction. Flow modification associated with urbanization also affects stream fish. In the Seine, modification of flow for flood protection and water availability has affected pike (*Esox lucius*) by reducing the number of flows providing suitable spawning habitat. With urbanization, the river contains enough suitable spawning habitat in only 1 out of 5 years as opposed to 1 out of every 2 years historically (Boet et al. 1999). Last, WWTP effluent clearly affects fishes. Reductions in WWTP effluent have been associated with the recovery of the fish community in a River Trent tributary near Birmingham, England (Harkness 1982). After nearly 250 years of degradation, effluent reductions, improved treatment, and construction of run-of-the river purification have resulted in an increase in fish diversity and abundances.

A few studies have actually examined ecological factors regulating stream fish populations and communities in urban streams. Recruitment of anadromous fish in the Hudson River Basin in New York was limited by suitable spawning habitat as a result of urbanization (Limburg & Schmidt 1990). Numbers of alewife (*Alosa pseudoharengus*) eggs and larvae in tributary streams decreased sharply between 0 and 15% urban land use. Beyond 15%, no eggs or larvae were found. The Kelsey Creek study discussed above showed impacts on salmonid population structure associated with urbanization, suggesting that urban streams may serve as population sinks for cutthroat, and that fish populations in those streams are dependent on recruitment from source populations with normal population age structures (Scott et al. 1986). Few data on the diet of fish in urban streams have been published, although a shift in diet was observed for fish along an urban gradient in Virginia (Weaver & Garman 1994).

Introduced fish species are also a common feature of urban streams. As a result of channelization, other river transportation modifications, and voluntary fisheries efforts in the Seine around Paris, 19 exotic species have been introduced, while 7 of 27 native species have been extirpated (Boet et al. 1999). The red shiner (*Cyprinella lutrensis*), a Mississippi drainage species commonly used as a bait fish, has invaded urban tributaries of the Chattahoochee River in Atlanta, Georgia where it has displaced native species and now comprises up to 90% of the fish community (DeVivo 1995).

As observed above for invertebrates, real gaps exist in our understanding of fish ecology in urban streams. The effects of urbanization on fishes have focused primarily on patterns of species presence, absence, or relative abundance. We found no published information on behavioral ecology, community interactions, or the biomass and production of nonsalmonid fishes in urban streams.

Ecosystem Processes

Ecosystem processes such as primary productivity, leaf decomposition, or nutrient cycling have been overlooked in urban streams, although they have been extensively studied in other types of stream ecosystems (Allan 1995). A few studies have considered organic matter in streams. WWTP effluent and CSO discharges can dramatically increase dissolved and particulate organic carbon concentrations, especially during storms (McConnell 1980). However, much less is known about baseflow concentrations of particulate and dissolved carbon in urban streams—natural or anthropogenic. The carbon inputs associated with sewage are generally more labile than natural transported organic matter and they affect dissolved oxygen in streams. Oxygen deficits associated with high biological oxygen demand during and after storms are common (McConnell 1980, Faulkner et al. 2000, Ometo et al. 2000). In addition, nonrespiratory oxygen demands associated with chemical oxidation reactions are also elevated in urban streams and can be much higher than biological oxygen demand in stormwater runoff (Bryan 1972). These inputs explain in part why more than 40% of 104 urban streams studied in the United States showed a high probability of greater than average oxygen deficits, with dissolved oxygen concentrations below 2 mg/liter and daily fluctuations up to 7 mg/liter not uncommon (Keefer et al. 1979). In a comparison of 2 forested and 4 urban catchments, average organic matter standing stocks were significantly lower in urban streams near Atlanta, Georgia (Paul 1999). This was attributed to greater scouring of the highly mobile sandy substrates in urban channels as a result of more severe flows.

Organic matter quality has been characterized in a few urban streams. In Kelsey Creek, particulate organic matter (POM) carbohydrate concentrations were higher than in POM in a nearby forested reference stream, suggesting that urbanization affects the nature of transported organic matter as well (Sloane-Richey et al. 1981). In addition to differences in organic matter quantity and quality, urban streams also differ in organic matter retention. Coarse and fine particles released to measure organic matter transport in Atlanta, Georgia streams traveled much farther before leaving the water column in urban streams than in forested streams (Paul 1999). Combined with the data from benthic organic matter (BOM) storage, these data indicate that these urban streams retain less organic matter, a fact that could limit secondary production in these urban streams (Paul 1999).

Ecosystem metabolism has also been measured in a few urban streams. In a comparison of three rivers in Michigan the urban river had higher gross primary production and community respiration than the forested river (Ball et al. 1973). In addition, the gross primary productivity to community respiration (P/R) ratio in the urban river without municipal effluent was greater than the forested stream and greater than 1.0, indicating that autotrophy dominated organic matter metabolism. However, in a downstream reach of the urban river receiving effluent, respiration was higher and the P/R ratio less than the forested river and far less than 1.0, indicating that heterotrophic metabolism predominated. Similar results were observed for urban streams in Atlanta, where gross primary production and community respiration were higher in urban streams than forested streams, and urban streams had more negative net ecosystem metabolism (gross primary production–community respiration), indicating greater heterotrophy (Paul 1999). However, because carbon storage was far less in the urban streams, carbon turnover was faster, supporting the hypothesis that respiration in urban streams was driven by more labile sources of carbon, such as sewage effluent.

Decomposition of organic matter has been measured in a few urban streams. Willow leaves decayed much faster in two suburban New Zealand streams than ever reported for any other stream; this occurred regardless of whether shredding insects were present or absent (Collier & Winterbourn 1986). The same results were observed for chalk maple (*Acer barbatum*) decay in urban streams in Atlanta, where rates were far faster in urban streams than rates observed for any woody leaf species in any stream (Paul 1999). Fungal colonization of leaves was only slightly lower in the urban streams, but there were no shredding insects associated with packs. These results suggested that higher stormflow was responsible for greater fragmentation of leaves in the urban streams, resulting in faster decay rates (Paul 1999).

Removal of added nutrients and contaminants is an ecological service provided by streams and relied upon by society. Although nutrient uptake in flowing waters has been extensively studied in forested ecosystems (Meyer et al. 1988, Stream Solute Workshop 1990, Marti & Sabater 1996), urban settings have been largely ignored. Studies in enriched reaches of river below the effluent from wastewater treatment plants have provided opportunities to examine patterns of denitrification in rivers (e.g., Hill 1979) and seasonal patterns of phosphorus removal and retention in a eutrophic river (e.g., Meals et al. 1999). Recently, ecologists have used the nutrients added by a wastewater treatment plant to measure nutrient uptake length, which is the average distance downstream traveled by a nutrient molecule before it is removed from the water column (Marti et al. 2001, Pollock & Meyer 2001). Uptake lengths in these rivers are much longer than in nonurban rivers of similar size, suggesting that not only is nutrient loading elevated in urban streams, but also nutrient removal efficiency is greatly reduced. The net result of these alterations in urban streams is increased nutrient loading to downstream lakes, reservoirs, and estuaries.

OPPORTUNITIES AND IMPERATIVES FOR AN ECOLOGY OF URBAN STREAMS

Urban streams are common features of the modern landscape that have received inadequate ecological attention. That is unfortunate because they offer a fertile testing ground for ecological concepts. For example, hydrologic regime is a master variable in streams (Minshall 1988), influencing channel form, biological assemblages, and ecosystem processes. As discussed in this review, impervious surfaces result in characteristically altered and often extreme hydrologic conditions that provide an endpoint on a disturbance gradient and that offer opportunities to quantify the relationships between channel form, biological communities, and ecosystem processes (Meyer et al. 1988). Does a continuous gradient of impervious surface cover result in a similar gradient of ecological pattern and process or are there thresholds? Answering that question is of both theoretical and practical interest. Developing a mechanistic understanding of the linkages between urbanization and stream ecosystem degradation is elusive but essential if ecologists hope to understand the nature of ecological response to disturbance and if they want to contribute to the development of scenarios that can guide planning decisions.

Many urban centers developed around rivers, which were the lifeblood of commerce. These commercial uses of rivers ignored and degraded the ecological services rivers provide, a phenomenon continuing today as urban sprawl accelerates.

Despite widespread degradation, urban rivers and streams offer local communities an easily accessible piece of nature. Most people live in urban areas, and many children first encounter nature playing in urban streams. Hence, urban streams offer opportunities for ecological outreach and education that ecologists are only beginning to explore. The meteoric rise in numbers of local catchment associations and adopt-a-stream monitoring groups is testimony to an audience eager for ecological insights.

Urban streams also offer ecologists an opportunity to test concepts of system organization through restoration projects. The field of urban stream restoration is dominated by physical scientists and engineers and rarely extends beyond stormwater management and bank stabilization with a goal of reestablishing a channel geomorphology in dynamic equilibrium with the landscape (e.g., Riley 1998). Little attention is given to restoration of a native stream biota or the ecological services streams provide. Urban stream restoration offers challenges not only in integrating physical, chemical, and biological processes to rehabilitate impaired ecosystems, but also requires an attention to esthetics and human attitudes toward the landscape. This offers an opportunity for the integration of ecological and social sciences with landscape design, which if successful will provide an avenue for ecologists to participate in the creation of the sustainable metropolitan centers of the future.

Cities have been a part of human history for millenia, and projections suggest most humans will live in cities in the future. Hence, urban areas lie at the intersection of human and ecological systems. If we are to succeed in that often-stated goal of incorporating humans as components of ecosystems, cities and their streams can no longer be ignored.

ACKNOWLEDGMENTS

This work is dedicated to those who have braved the urban stream. We apologize to those many the restrictions in length prohibited us from including. Our research on urban streams in Atlanta has been supported by the EPA/NSF Waters and Watersheds Program (EPA R 824777-01-0) for work on the Chattahoochee River and by the EPA Ecological Indicators program (EPA R 826597-01-0) for work on the Etowah River.

Visit the Annual Reviews home page at www.AnnualReviews.org

LITERATURE CITED

Allan JD. 1995. *Stream Ecology: Structure and Function of Running Waters*. New York: Chapman & Hall

Arnold CL, Boison PJ, Patton PC. 1982. Sawmill Brook: an example of rapid geomorphic change related to urbanization. *J. Geol.* 90:155–66

Arnold CL, Gibbons CJ. 1996. Impervious surface coverage: the emergence of a key environmental indicator. *Am. Planners Assoc. J.* 62:243–58

Arthington AH. 1985. The biological resources of urban creeks. *Aust. Soc. Limnol. Bull.* 10:33–39

Baer KE, Pringle CM. 2000. Special problems of urban river conservation: the encroaching megalopolis. In *Global Perspectives on River Conservation: Science, Policy, and Practice*, ed. PJ Boon, BR Davies, GE Petts, pp. 385–402. New York: Wiley

Ball RC, Kevern NR, Haines TA. 1973. *An ecological evaluation of stream eutrophication. Tech. Rep. 36*, Inst. Water Res., Mich. State Univ., E. Lansing

Barringer TH, Reiser RG, Price CV. 1994. Potential effects of development on flow characteristics of two New Jersey streams. *Water Resour. Bull.* 30:283–95

Benke AC, Wallace JB. 1997. Trophic basis of production among riverine caddisflies: implications for food web analysis. *Ecology* 78:1132–45

Benke AC, Willeke GE, Parrish FK, Stites DL. 1981. *Effects of urbanization on stream ecosystems. Environ. Resour. Ctr. Rep. 07–81*, Georgia Inst. Tech., Atlanta

Bennett EM, Reed-Andersen T, Hauser JN, Gabriel JR, Carpenter SR. 1999. A phosphorus budget for the Lake Mendota watershed. *Ecosystems* 2:69–75

Black RW, Haggland AL, Voss FD. 2000. Predicting the probability of detecting organochlorine pesticides and polychlorinated biphenyls in stream systems on the basis of land use in the Pacific Northwest, USA. *Environ. Toxicol. Chem.* 19:1044–54

Boet P, Belliard J, Berrebi-dit-Thomas R, Tales E. 1999. Multiple human impacts by the City of Paris on fish communities in the Seine River basin, France. *Hydrobiologia* 410:59–68

Boland P, Hanhammer S. 1999. Ecosystem services in urban areas. *Ecol. Econ.* 29:293–301

Booth DB. 1990. Stream channel incision following drainage-basin urbanization. *Water Resour. Bull.* 26:407–17

Booth DB, Jackson CR. 1997. Urbanization of aquatic systems: degradation thresholds, stormwater detection, and the limits of mitigation. *J. Am. Water Resour. Assoc.* 33:1077–90

Borchardt D, Statzner B. 1990. Ecological impact of urban stormwater runoff studied in experimental flumes: population loss by drift and availability of refugial space. *Aquat. Sci.* 52:299–314

Brion N, Billen G. 2000. Wastewater as a source of nitrifying bacteria in river systems: the case of the River Seine downstream from Paris. *Water Resour. Res.* 34:3213–21

Bryan EH. 1972. Quality of stormwater drainage from urban land. *Water Resour. Bull.* 8:578–88

Bryan EH. 1974. Concentrations of lead in urban stormwater. *J. Water Pollut. Control Fed.* 46:2419–21

Burkholder JM. 1996. Interactions of benthic algae with their substrata. In *Algal Ecology: Freshwater Benthic Ecosystems*, ed. RJ Stevenson, ML Bothwell, RL Lowe, pp. 253–97. San Diego, CA: Academic. 753 pp.

Butler D, Davies JW. 2000. *Urban Drainage*. New York: E & FN Spon

Campbell IC. 1978. Biological investigation of an organically polluted urban stream in Victoria. *Aust. J. Mar. Freshw. Res.* 29:275–91

Center for Watershed Protection. 2000. *Urban Stream Restoration Practices: An Initial Assessment*. Ellicott City, MD: Cent. Watershed Prot.

Charbonneau R, Resh VH. 1992. Strawberry Creek on the University of California, Berkeley campus—a case-history of urban stream restoration. *Aquat. Conserv. Mar. Freshw. Ecosyst.* 2:293–307

Chessman BC, Growns I, Currey J, Plunkett-Cole N. 1999. Predicting diatom communities at the genus level for the rapid biological assessment of rivers. *Freshw. Biol.* 41:317–31

Chessman BC, Hutton PE, Burch JM. 1992. Limiting nutrients for periphyton growth in sub-alpine, forest, agricultural, and urban streams. *Freshw. Biol.* 28:349–61

Chevreuiel M, Garmouma M, Fauchon N. 1999. Variability of herbicides (triazines, phenylureas) and tentative mass balance as a function of stream order in the River Marne basin (France). *Hydrobiologia* 410:349–55

Collier KJ. 1995. Environmental factors affecting the taxonomic composition of aquatic macro-invertebrate communities in lowland waterways of Northland, New Zealand. *N. Z. J. Mar. Freshw. Res.* 29:453–65

Collier KJ, Winterbourn MJ. 1986. Processing of willow leaves in two suburban streams in Christchurch, New Zealand. *N. Z. J. Mar. Freshw. Res.* 20:575–82

Coupe RH, Manning MA, Foreman WT, Goolsby DA, Majewski MS. 2000. Occurrence of pesticides in rain and air in urban and agricultural areas of Mississippi, April-September 1995. *Sci. Total Environ.* 248:227–40

Daniels WM, House WA, Rae JE, Parker A. 2000. The distribution of micro-organic contaminants in river bed-sediment cores. *Sci. Total Environ.* 253:81–92

Danielson MN, Keles R. 1985. *The Politics of Rapid Urbanization: Government and Growth in Modern Turkey.* New York: Holmes & Meier

Davies PE, Cook SJ, Barton JL. 1994. Triazine herbicide contamination of Tasmanian streams: sources, concentrations, and effects on biota. *Aust. J. Mar. Freshw. Res.* 45:209–26

Davis JB, George JJ. 1987. Benthic invertebrates as indicators of urban and motorway discharges. *Sci. Total Environ.* 59:291–302

Dennehy KF, Litke DW, Tate CM, Qi SL, McMahon PB, et al. 1998. *Water quality in the South Platte River Basin, Colorado, Nebraska, and Wyoming, 1992–1995. USGS Circular 1167*

DeVivo JC. 1995. Impact of introduced red shiners (*Cyprinella lutrensis*) on stream fishes near Atlanta, Georgia. In *Proc. 1995 Georgia Water Resources Conf.*, ed. K Hatcher, pp. 95–98. Athens: Carl Vinson Sch. Gov., Univ. Georgia

Dickman M, Rygiel G. 1998. Municipal landfill impacts on a natural stream located in an urban wetland in regional Niagara, Ontario. *Can. Field Nat.* 112:619–30

Dodds WK, Welch EB. 2000. Establishing nutrient criteria in streams. *J. N. Am. Benthol. Soc.* 19:186–96

Doeg TJ, Milledge GA. 1991. Effect of experimentally increasing concentrations of suspended sediment on macroinvertebrate drift. *Aust. J. Mar. Freshw. Res.* 42:519–26

Douglas I. 1974. The impact of urbanization on river systems. In *Proc. Int. Geogr. Union Reg. Conf.*, pp. 307–17. N. Z. Geograph. Soc.

Duda AM, Lenat DR, Penrose DL. 1982. Water quality in urban streams—what we can expect. *J. Water Pollut. Control Fed.* 54:1139–47

Dunne T, Leopold LB. 1978. *Water in Environmental Planning.* New York: Freeman. 818 pp.

Eganhouse RP, Simoneit BRT, Kaplan IR. 1981. Extractable organic matter in urban stormwater runoff. 2. Molecular characterization. *Environ. Sci. Technol.* 15:315–26

Ellis JB, Marsalek J. 1996. Overview of urban drainage: environmental impacts and concerns, means of mitigation and implementation policies. *J. Hydraulic Res.* 34:723–31

Espey WH Jr, Morgan CW, Masch FD. 1965. *A study of some effects of urbanization on storm runoff from a small watershed. Tech. Rep. 44D 07–6501 CRWR-2*, Ctr. for Res. in Water Resour., Univ. Texas, Austin

Faulkner H, Edmonds-Brown V, Green A. 2000. Problems of quality designation in diffusely populated urban streams—the case of Pymme's Brook, North London. *Environ. Pollut.* 109:91–107

Finkenbine JK, Atwater DS, Mavinic DS. 2000. Stream health after urbanization. *J. Am. Water Resour. Assoc.* 36:1149–60

Folke C, Jansson A, Larsson J, Costanza R. 1997. Ecosystem appropriation by cities. *Ambio* 26:167–72

Forman RTT, Alexander LE. 1998. Roads and their major ecological effects. *Annu. Rev. Ecol. Syst.* 29:207–31

Foster GD, Roberts EC Jr, Gruessner B, Velinsky DJ. 2000. Hydrogeochemistry and transport of organic contaminants in an urban watershed of Chesapeake Bay (USA). *Appl. Geochem.* 15:901–16

Frankie GW, Kohler CS, eds. 1983. *Urban Entomology: Interdisciplinary Perspectives.* New York: Praeger

Frick EA, Hippe DJ, Buell GR, Couch CA, Hopkins EE, et al. 1998. *Water quality in the Appalachicola-Chattahoochee-Flint River Basin, Georgia, Alabama, and Florida, 1992–1995. USGS Circular 1164*

Fusillo TV, Nieswand GH, Shelton TB. 1977. Sediment yields in a small watershed under suburban development. In *Proc. Int. Symp. Urban Hydrology, Hydraulics, and Sediment Control.* Lexington: Univ. Kentucky

Gafny S, Goren M, Gasith A. 2000. Habitat condition and fish assemblage structure in a coastal Mediterranean stream (Yargon, Israel) receiving domestic effluent. *Hydrobiologia* 422/423:319–30

Galli FJ. 1991. *Thermal Impacts Associated with Urbanization and Stormwater Management Best Management Practices.* Washington, DC: Washington Council of Governments

Garie HL, McIntosh A. 1986. Distribution of benthic macroinvertebrates in a stream exposed to urban runoff. *Water Resour. Bull.* 22:447–55

Gerritse RG, Adeney JA, Dommock GM, Oliver YM. 1995. Retention of nitrate and phosphate in soils of the Darling Plateau in Western Australia: implications for domestic septic tank systems. *Aust. J. Soil Res.* 33:353–67

Gibson CJ, Stadterman KL, States S, Sykora J. 1998. Combined sewer overflows: a source of *Cryptosporidium* and *Giardia*? *Water Sci. Technol.* 38:67–72

Goni-Urriza M, Capdepuy M, Arpin C, Raymond N, Caumette P, et al. 2000. Impact of an urban effluent on antibiotic resistance of riverine *Enterobacteriaceae* and *Aeromonas* spp. *Appl. Environ. Microbiol.* 66:125–32

Gonzales AE, Rodriguez MT, Sanchez JCJ, Espinoza AJF, de la Rosa FJB. 2000. Assessment of metals in sediments in a tributary of Guadalquiver River (Spain): heavy metal partitioning and relation between water

and sediment system. *Water Air Soil Pollut.* 121:11–29

Graf WL. 1977. Network characteristics in suburbanizing streams. *Water Resour. Res.* 13:459–63

Gregory KJ, Davis RJ, Downs PW. 1992. Identification of river channel change due to urbanization. *Appl. Geogr.* 12:299–318

Gregory KJ, Gurnell AM, Hill CT, Tooth S. 1994. Stability of the pool-riffle sequence in changing river channels. *Regul. Rivers: Res. Manage.* 9:35–43

Grimm NB, Grove MJ, Pickett STA, Redman CL. 2000. Integrated approaches to long-term studies of urban ecological systems. *Bioscience* 50:571–84

Gundacker C. 2000. Comparison of heavy metal bioaccumulation in freshwater molluscs of urban river habitats in Vienna. *Environ. Pollut.* 110:61–71

Hachmoller B, Matthews RA, Brakke DF. 1991. Effects of riparian community structure, sediment size, and water quality on the macroinvertebrate communties in a small, suburban stream. *Northwest Sci.* 65:125–32

Halling-Sorensen B, Nielsen SN, Lanzky PF, Ingerslev F, Holten-Lutzhoft HC, et al. 1998. Occurrence, fate, and effects of pharmaceutical substances on the environment—a review. *Chemosphere* 36:357–93

Hammer TR. 1972. Stream channel enlargement due to urbanization. *Water Resour. Res.* 8:1530–40

Harkness N. 1982. The River Tame—a short history of water pollution and control within an industrial river basin. *Water Sci. Technol.* 14:153–65

Hart DD, Finelli CM. 1999. Physical-biological coupling in streams: the pervasive effects of flow on benthic organisms. *Annu. Rev. Ecol. Syst.* 30:363–95

Heaney JP, Huber WC. 1984. Nationwide assessment of urban runoff on receiving water quality. *Water Resour. Bull.* 20:35–42

Herlihy AT, Stoddard JL, Johnson CB. 1998. The relationship between stream chemistry and watershed land cover data in the

Mid-Atlantic region, US. *Water Air Soil Pollut.* 105:377–86

Hill AR. 1979. Denitrification in the nitrogen budget of a river ecosystem. *Nature* 281:291

Hirsch RM, Walker JF, Day JC, Kallio R. 1990. The influence of man on hydrologic systems. In *Surface Water Hydrology (The Geology of America, Vol. O-1)*, ed. MG Wolman, HC Riggs, pp. 329–59. Boulder, CO: Geol. Soc. Am.

Hoare RA. 1984. Nitrogen and phosphorus in Rotorua urban streams. *N. Z. J. Mar. Freshw. Res.* 18:451–54

Hoffman EJ, Latimer JS, Mills GL, Quinn JG. 1982. Petroleum hydrocarbons in urban runoff from a commercial land use area. *J. Water. Pollut. Control Fed.* 54:1517–25

Hoffman RS, Capel PD, Larson SJ. 2000. Comparison of pesticides in eight U.S. urban streams. *Env. Toxicol. Chem.* 19:2249–58

Hogg ID, Norris RH. 1991. Effects of runoff from land clearing and urban development on the distribution and abundance of macroinvertebrates in pool areas of a river. *Aust. J. Mar. Freshw. Res.* 42:507–18

Horner RR, Booth DB, Azous A, May CW. 1997. Watershed determinants of ecosystem functioning. In *Effects of Watershed Development and Management on Aquatic Ecosystems*. ed. C Roessner, pp. 251–74. New York: Am. Soc. Civil Eng.

Horowitz AJ, Meybeck M, Idlafkih Z, Biger E. 1999. Variations in trace element geochemistry in the Seine River Basin based on floodplain deposits and bed sediments. *Hydrol. Process.* 13:1329–40

House MA, Ellise JB, Herricks EE, Huitved-Jacobsen T, Seager J, et al. 1993. Urban drainage—impacts on receiving water quality. *Water Sci. Technol.* 27:117–58

Howard KWF, Haynes J. 1993. Urban geology. 3. Groundwater contamination due to road deicing chemicals—salt balance implications. *Geosci. Can.* 20:1–8

Hunter JV, Sabatino T, Gomperts R, Mackenzie MJ. 1979. Contribution of urban runoff to hydrocarbon pollution. *J. Water Pollut. Control Fed.* 51:2129–38

Hynes HBN. 1960. *The Biology of Polluted Waters.* Liverpool, UK: Liverpool Univ. Press

Jancarkova I, Larsen TA, Gujer W. 1997. Distribution of nitrifying bacteria in a shallow stream. *Water Sci. Technol.* 36:161–66

Johnson B, Tuomori D, Sinha R. 1999. Impacts of on-site sewage systems and illicit discharges on the Rouge River. In *Proc. Natl. Conf. Retrofit Opportunities for Water Resour. Prot. Urban Environ.*, pp. 132–35. EPA/625/R-99/002. Washington, DC: EPA

Jones JB, Mulholland PJ. 2000. *Streams and Ground Waters.* San Diego, CA: Academic

Jones RC, Clark CC. 1987. Impact of watershed urbanization on stream insect communities. *Water Resour. Bull.* 23:1047–55

Keefer TN, Simons RK, McQuivey RS. 1979. *Dissolved oxygen impact from urban storm runoff. EPA-600/2–79–156.* Washington, DC: EPA

Kennen JG. 1999. Relation of macroinvertebrate community impairment to catchment characteristics in New Jersey streams. *J. Am. Water Resour. Assoc.* 35:939–55

Khamer M, Bouya D, Ronneau C. 2000. Metallic and organic pollutants associated with urban wastewater in the waters and sediments of a Moroccan river. *Water Qual. Res. J. Can.* 35:147–61

Klein RD. 1979. Urbanization and stream quality impairment. *Water Resour. Bull.* 15:948–63

Kolkwitz R, Marsson M. 1908. Okologie der pflanzlichen saprobien. *Ber. Deutsche Bot. Ges.* 26a:505–19

Krug A. 1993. Drainage history and land use pattern of a Swedish river system—their importance for understanding nitrogen and phosphorus load. *Hydrobiologia* 251:285–96

Latimer JS, Quinn JG. 1998. Aliphatic petroleum and biogenic hydrocarbons entering Narragansett Bay from tributaries under dry weather conditions. *Estuaries* 21:91–107

LaValle PD. 1975. Domestic sources of stream phosphates in urban streams. *Water Res.* 9:913–15

Leblanc RT, Brown RD, Fitzgibbon JE. 1997. Modeling the effects of land use change on water temperature in unregulated urban streams. *J. Environ. Manage.* 49:445–69

Lenat DR, Crawford JK. 1994. Effects of land use on water quality and aquatic biota of three North Carolina Piedmont streams. *Hydrobiologia* 294:185–99

Leopold LB. 1968. *Hydrology for Urban Land Planning—A Guidebook on the Hydrologic Effects of Urban Land Use. USGS Circular 554*

Leopold LB. 1973. River channel change with time—an example. *Bull. Geol. Soc. Am.* 88:1845–60

LeVeen EP, Willey WRZ. 1983. A political economic analysis of urban pest management. See Frankie & Kohler 1983, pp. 19–40

Limburg KE, Schmidt RE. 1990. Patterns of fish spawning in Hudson River tributaries: response to an urban gradient? *Ecology* 71:1238–45

Litke DW. 1999. *Review of phosphorus control measures in the United States and their effects on water quality. USGS Water Resourc. Invest. Rep. 99–4007*

Mackenzie MJ, Hunter JV. 1979. Sources and fates of aromatic compounds in urban stormwater runoff. *Environ. Sci. Technol.* 13:179–83

Marti E, Aumatell J, Gode L, Poch M, Sabater F. 2001. Effects of wastewater treatment plant inputs on stream nutrient retention. *Water Resour. Res.* In press

Marti E, Sabater F. 1996. High variability in temporal and spatial nutrient retention in Mediterranean streams. *Ecology* 77:854–69

Mason RP, Sullivan KA. 1998. Mercury and methyl-mercury transport through an urban watershed. *Water Res.* 32:321–30

McArthur JV, Tuckfield RC. 2000. Spatial patterns in antibiotic resistance among stream bacteria: effects of industrial pollution. *Appl. Environ. Microbiol.* 66:3722–26

McConnell JB. 1980. *Impact of urban storm runoff on stream quality near Atlanta, Georgia. EPA-600/2–80–094.* Washington, DC: EPA

McDonnell MJ, Pickett STA. 1990. Ecosystem structure and function along urban-rural gradients: an unexploited opportunity for ecology. *Ecology* 71:1232–37

McMahon G, Cuffney TF. 2000. Quantifying urban intensity in drainage basins for assessing stream ecological conditions. *J. Am. Water Resour. Assoc.* 36:1247–62

Meals DW, Levine SN, Wang D, Hoffmann JP, Cassell EA, et al. 1999. Retention of spike additions of soluble phosphorus in a northern eutrophic stream. *J. N. Am. Benthol. Soc.* 18:185–98

Medeiros C, Leblanc R, Coler RA. 1983. An *in situ* assessment of the acute toxicity of urban runoff to benthic macroinvertebrates. *Environ. Toxicol. Chem.* 2:119–26

Meybeck M. 1998. Man and river interface: multiple impacts on water and particulates chemistry illustrated in the Seine River Basin. *Hydrobiologia* 373/374:1–20

Meyer JL, McDowell WH, Bott TL, Elwood JW, Ishizaki C, et al. 1988. Elemental dynamics in streams. *J. N. Am. Benthol. Soc.* 7:410–32

Meyer JL, Wallace JB. 2001. Lost linkages in lotic ecology: rediscovering small streams. In *Ecology: Achievement and Challenge.* ed. M Press, N Huntly, S Levin, pp. 295–317. Boston: Blackwell Sci. In press

Mielke HW, Gonzales CR, Smith MK, Mielke PW. 2000. Quantities and associations of lead, zinc, cadmium, manganese, chromium, nickel, vanadium, and copper in fresh Mississippi Delta alluvium and New Orleans alluvial soils. *Sci. Total Environ.* 246:249–59

Minshall GW. 1988. Stream ecosystem theory: a global perspective. *J. N. Am. Benthol. Soc.* 8:263–88

Moring JB, Rose DR. 1997. Occurrence and concentrations of polycyclic aromatic hydrocarbons in semipermeable membrane devices and clams in three urban streams of the Dallas-Fort Worth Metropolitan Area, Texas. *Chemosphere* 34:551–66

Morisawa M, LaFlure E. 1979. Hydraulic geometry, stream equilibrium, and urbanization. In *Adjustments of the Fluvial System,* ed. DD Rhodes, GP Williams, pp. 333–50. Dubuque, IA: Kendall-Hunt

Mulholland PJ, Lenat DR. 1992. Streams of the southeastern Piedmont, Atlantic Drainage. In *Biodiversity of the Southeastern United States—Aquatic Communities*, ed. CT Hackney, SM Adams, WA Martin, pp. 193–232. New York: Wiley

Muschak W. 1990. Pollution of street runoff by traffic and local conditions. *Sci. Total Environ.* 93:419–31

Nagumo T, Hatano R. 2000. Impact of nitrogen cycling associated with production and consumption of food on nitrogen pollution of stream water. *Soil Sci. Plant Nutr.* 46:325–42

Neal C, Robson AJ. 2000. A summary of river water quality data collected within the Land-Ocean Interaction Study: core data for eastern UK rivers draining to the North Sea. *Sci. Total Environ.* 251/252:585–665

Neal C, Robson AJ, Jeffery HA, Harrow ML, Neal M, et al. 1997. Trace element interrelationships for hydrological and chemical controls. *Sci. Total Environ.* 194:321–43

Neller RJ. 1988. A comparison of channel erosion in small urban and rural catchments, Armidale, New South Wales. *Earth Surf. Process.* 13:1–7

Nix PG, Daykin MM, Vilkas KL. 1994. Fecal pollution events reconstructed and sources identified using a sediment bag grid. *Water Environ. Res.* 66:814–18

Olguin HF, Salibian A, Puig A. 2000. Comparative sensitivity of *Scenedesmus acutus* and *Chlorella pyrenoidosa* as sentinel organisms for aquatic ecotoxicity assessments: studies on a highly polluted urban river. *Environ. Toxicol.* 15:14–22

Omernik JM. 1976. *The influence of land use on stream nutrient levels. EPA-600/2–76–014.* Washington, DC: EPA

Ometo JPHB, Martinelli LA, Ballester MV, Gessner A, Krusche A, et al. 2000. Effects of land use on water chemistry and macroinvertebrates in two streams of the Piracicaba River Basin, southeast Brazil. *Freshw. Biol.* 44:327–37

Onorato D, Angus RA, Marion KR. 2000. Historical changes in the ichthyofaunal assemblages of the Upper Cahaba River in Alabama

associated with extensive urban development in the watershed. *J. Freshw. Ecol.* 15:47–63

Osborne LL, Wiley MJ. 1988. Empirical relationships between land use/cover and stream water quality in an agricultural watershed. *J. Environ. Manage.* 26:9–27

Patrick R. 1973. Use of algae, especially diatoms, in the assessment of water quality. In *Biological Methods for the Assessment of Water Quality*, ed. J Cairns, KL Dickson, pp. 76–95. Philadelphia: Am. Soc. Testing & Mat.

Paul MJ. 1999. *Stream ecosystem function along a land use gradient.* PhD thesis, Univ. Georgia, Athens

Pedersen ER, Perkins MA. 1986. The use of benthic macroinvertebrate data for evaluating impacts of urban runoff. *Hydrobiologia* 139:13–22

Pereira WE, Domagalski JL, Hostettler FD, Brown LR, Rapp JB. 1996. Occurrence and accumulation of pesticides and organic contaminants in river sediment, water, and clam tissue from the San Joaquin River and tributaries, California. *Environ. Toxicol. Chem.* 15:172–80

Pesacreta GJ. 1997. Response of the stonefly *Pteronarcys dorsata* in enclosures from an urban North Carolina stream. *Bull. Environ. Contam. Toxicol.* 59:948–55

Pickett STA, Cadenasso ML, Grove JM, Nilon CH, Pouyat RV, et al. 2001. Urban ecological systems: linking terrestrial ecological, physical, and socio-economic components of metropolitan areas. *Annu. Rev. Ecol. Syst.* 32:127–57

Pizzuto JE, Hession WC, McBride M. 2000. Comparing gravel-bed rivers in paired urban and rural catchments of southeastern Pennsylvania. *Geology* 28:79–82

Pluhowski EJ. 1970. *Urbanization and its effect on the temperature of streams in Long Island, New York. USGS Prof. Paper 627–D*

Pollock JB, Meyer JL. 2001. Phosphorus assimilation below a point source in Big Creek. In *Proc. 2001 Georgia Water Resour. Conf.*, ed. KJ Hatcher, pp. 509–9. Athens: Univ. Georgia

Porcella DB, Sorensen DL. 1980. *Characteristics of non-point source urban runoff and its effects on stream ecosystems. EPA-600/3–80–032*. Washington, DC: EPA

Pratt JM, Coler RA, Godfrey PJ. 1981. Ecological effects of urban stormwater runoff on benthic macroinvertebrates inhabiting the Green River, Massachusetts. *Hydrobiologia* 83:29–42

Ragan RM, Dietmann AJ. 1976. *Characteristics of urban runoff in the Maryland suburbs of Washington, DC*. College Park, MD: Water Resourc. Res. Cent., Univ. Maryland

Rauch S, Morrison GM. 1999. Platinum uptake by the freshwater isopod *Asellus aquaticus* in urban rivers. *Sci. Total Environ.* 235:261–68

Resh VH, Grodhaus G. 1983. Aquatic insects in urban environments. See Frankie & Kohler 1983, pp. 247–76

Resh VH, Rosenberg DM, eds. 1984. *The Ecology of Aquatic Insects*. New York: Praeger

Rhoads BL, Cahill RA. 1999. Geomorphological assessment of sediment contamination in an urban stream system. *Appl. Geochem.* 14:459–83

Richards C, Host G. 1994. Examining land use influences on stream habitats and macroinvertebrates: a GIS approach. *Water Resour. Bull.* 30:729–38

Riley AC. 1998. *Restoring Streams in Cities: A Guide for Planners, Policymakers, and Citizens*. Washington, DC: Island Press

Roberts CR. 1989. Flood frequency and urban-induced change: some British examples. In *Floods: Hydrological, Sedimentological, and Geomorphological Implications*, ed. K Beven, P Carling, pp. 57–82. New York: Wiley

Robinson AM. 1976. Effects of urbanization on stream channel morphology. In *Proc. Natl. Symp. Urban Hydrology, Hydraulics, and Sediment Control*. Univ. Ky. Coll. Eng. Publ. III, Lexington

Roth NE, Allan JD, Erickson DL. 1996. Landscape influences on stream biotic integrity assessed at multiple spatial scales. *Landsc. Ecol.* 11:141–56

Sartor JD, Boyd GB, Agardy FJ. 1974. Water pollution aspects of street surface contaminants. *J. Water Pollut. Control Fed.* 46:458–67

Schueler TR. 1994a. The importance of imperviousness. *Watershed Prot. Tech.* 1:100–11

Schueler TR. 1994b. Minimizing the impact of golf courses on streams. *Watershed Prot. Tech.* 1:73–75

Schueler TR, Galli J. 1992. Environmental impacts of stormwater ponds. In *Watershed Restoration Source Book*. ed. P Kumble, T Schueler, Washington, DC: Metropol. Wash. Counc. Gov.

Scott JB, Steward CR, Stober QJ. 1986. Effects of urban development on fish population dynamics in Kelsey Creek, Washington. *Trans. Am. Fish. Soc.* 115:555–67

Seaburn GE. 1969. *Effects of urban development on direct runoff to East Meadow Brook, Nassau County, New York. USGS Prof. Paper 627–B*

Seager J, Abrahams RG. 1990. The impact of storm sewage discharges on the ecology of a small urban river. *Water Sci. Technol.* 22:163–71

Shutes RBE. 1984. The influence of surface runoff on the macro-invertebrate fauna of an urban stream. *Sci. Total Environ.* 33:271–82

Sloane-Richey JS, Perkins MA, Malueg KW. 1981. The effects of urbanization and stormwater runoff on the food quality in two salmonid streams. *Verh. Int. Ver. Theor. Ang. Limnol.* 21:812–18

Smart MM, Jones JR, Sebaugh JL. 1985. Stream-watershed relations in the Missouri Ozark Plateau Province. *J. Environ. Qual.* 14:77–82

Steedman RJ. 1988. Modification and assessment of an index of biotic integrity to quantify stream quality in southern Ontario. *Can. J. Fish. Aquat. Sci.* 45:492–501

Stream Solute Workshop. 1990. Concepts and methods for assessing solute dynamics in stream ecosystems. *J. N. Am. Benthol. Soc.* 9:95–119

Suren AM. 2000. Effects of urbanisation. In *New Zealand Stream Invertebrates: Ecology and Implications for Management*. ed. KJ

Collier, MJ Winterbourn, pp. 260–88. Hamilton: N.Z. Limnol. Soc.

Sweeney BW. 1984. Factors influencing life history patterns of aquatic insects. See Resh & Rosenberg 1984, pp. 56–100

Tada F, Suzuki S. 1982. Adsorption and desorption of heavy metals in bottom mud of urban rivers. *Water Res.* 16:1489–94

Tate CM, Heiny JS. 1995. The ordination of benthic invertebrate communities in the South Platte River Basin in relation to environmental factors. *Freshw. Biol.* 33:439–54

Taylor BR, Roff JC. 1986. Long-term effects of highway construction on the ecology of a Southern Ontario stream. *Environ. Pollut. Ser. A* 40:317–44

Thorne RSJ, Williams WP, Gordon C. 2000. The macroinvertebrates of a polluted stream in Ghana. *J. Freshw. Ecol.* 15:209–17

Trimble SJ. 1997. Contribution of stream channel erosion to sediment yield from an urbanizing watershed. *Science* 278:1442–44

US Census Bureau. 2001. http://www.census.gov

UN Population Division. 1997. *Urban and Rural Areas, 1950–2030 (The 1996 Revision).* New York: United Nations

US Environ. Prot. Agency (USEPA). 2000. *The quality of our nation's waters. EPA 841–S-00–001*

US Environ. Prot. Agency (USEPA). 2001. www.epa.gov/owm/gen2.htm

US Geol. Surv. (USGS). 1999. *The quality of our nation's waters—nutrients and pesticides. USGS Circular 1225*

US Geol. Surv. (USGS). 2001. http://water.usgs.gov/nawqa/

Wang L, Lyons J, Kanehl P, Bannerman R, Emmons E. 2000. R. Watershed urbanization and changes in fish communities in southeastern Wisconsin streams. *J. Am. Water Resour. Assoc.* 36:1173–89

Wang L, Lyons J, Kanehl P, Gatti R. 1997. Influences of watershed land use on habitat quality and biotic integrity in Wisconsin streams. *Fisheries* 22:6–12

Wang X, Yin Z. 1997. Using GIS to assess the relationship between land use and water quality at a watershed level. *Environ. Int.* 23:103–14

Warren LA, Zimmerman AP. 1994. Suspended particulate oxides and organic matter interactions in trace metal sorption reactions in a small urban river. *Biogeochemistry* 23:21–34

Waschbusch RJ, Selbig WR, Bannerman RT. 1999. *Sources of phosphorus in stormwater and street dirt from two urban residential basins in Madison, Wisconsin. USGS Water Resour. Invest. Rep. 99–4021*

Weaver LA, Garman GC. 1994. Urbanization of a watershed and historical changes in a stream fish assemblage. *Trans. Am. Fish. Soc.* 123:162–72

Webster JR, Benfield EF. 1986. Vascular plant breakdown in freshwater ecosystems. *Annu. Rev. Ecol. Syst.* 17:567–94

Wei C, Morrison G. 1992. Bacterial enzyme activity and metal speciation in urban river sediments. *Hydrobiologia* 235/236:597–603

Weibel SR, Weidner RB, Cohen JM, Christianson AG. 1966. Pesticides and other contaminants in rainfall and runoff. *J. Am. Water Works Assoc.* 58:1075–84

Wernick BG, Cook KE, Schreier H. 1998. Land use and streamwater nitrate-N dynamics in an urban-rural fringe watershed. *J. Am. Water Resour. Assoc.* 34:639–50

Whipple W Jr, Hunter JV. 1979. Petroleum hydrocarbons in urban runoff. *Water Resour. Bull.* 15:1096–104

Whiting ER, Clifford HF. 1983. Invertebrates and urban runoff in a small northern stream, Edmonton, Alberta, Canada. *Hydrobiologia* 102:73–80

Wiederholm T. 1984. Responses of aquatic insects to environmental pollution. See Resh & Rosenberg 1984, pp. 508–57

Wilber WG, Hunter JV. 1977. Aquatic transport of heavy metals in the urban environment. *Water Resour. Bull.* 13:721–34

Wilber WG, Hunter JV. 1979. The impact of urbanization on the distribution of heavy metals in bottom sediments of the Saddle River. *Water Resour. Bull.* 15:790–800

Winger JG, Duthie HC. 2000. Export coefficient modeling to assess phosphorus loading in an urban watershed. *J. Am. Water Resour. Assoc.* 36:1053–61

Wolman MG. 1967. A cycle of sedimentation and erosion in urban river channels. *Geogr. Ann.* 49a:385–95

Wolter C, Minow J, Vilcinskas A, Grosch UA. 2000. Long-term effects of human influence on fish community structure and fisheries in Berlin waters: an urban water system. *Fish. Manage. Ecol.* 7:97–104

Wood PJ, Armitage PD. 1997. Biological effects of fine sediment in the lotic environment. *Environ. Manage.* 21:203–17

Wright IA, Chessman BC, Fairweather PG, Benson LJ. 1995. Measuring the impact of sewage effluent on the macroinvertebrate community of an upland stream. The effect of different levels of taxonomic resolution and quantification. *Aust. J. Ecol.* 20:142–49

Yamamoto K, Fukushima M, Kakatani N, Kuroda K. 1997. Volatile organic compounds in urban rivers and their estuaries in Osaka, Japan. *Environ. Pollut.* 95:135–43

Yoder CO, Miltner RJ, White D. 1999. Assessing the status of aquatic life designated uses in urban and suburban watersheds. In *Proc. Natl. Conf. Retrofit Opportunities for Water Resour. Prot. Urban Environ.*, pp. 16–28. *EPA/625/R-99/002*

Young KD, Thackston EL. 1999. Housing density and bacterial loading in urban streams. *J. Environ. Eng.* 125:1177–80

Zampella RA. 1994. Characterization of surface water quality along a watershed disturbance gradient. *Water Resour. Bull.* 30:605–11

Annu. Rev. Ecol. Syst. 2001. 32:367–96
Copyright © 2001 by Annual Reviews. All rights reserved

INTEGRATING FUNCTION AND ECOLOGY IN STUDIES OF ADAPTATION: Investigations of Locomotor Capacity as a Model System

Duncan J. Irschick[1] and Theodore Garland, Jr.[2]

[1]*Department of Ecology and Evolutionary Biology, 310 Dinwiddie Hall, Tulane University, New Orleans, Louisiana 70118; e-mail: irschick@tulane.edu*
[2]*Department of Zoology, University of Wisconsin, 430 Lincoln Drive, Madison, Wisconsin 53706; e-mail: tgarland@facstaff.wisc.edu*

Key Words performance, evolution, field, lizards, marine mammals

■ **Abstract** Understanding adaptation in morphological and physiological traits requires elucidation of how traits relate to whole-organism performance and how performance relates to fitness. A common assumption is that performance capacities are utilized by and important to organisms. For some systems, it is assumed that high levels of physical fitness, as indexed by measures of locomotor performance, lead to high fitness levels. Although biologists have appreciated this, little attention has been paid to quantifying how organisms use their performance capacities in nature. We argue that for the study of adaptation to proceed, greater integration of laboratory studies of performance and behavioral/ecological studies is needed, and we illustrate this approach by examining two questions. First, how does the environment affect locomotor function in nature? Second, what percentage of locomotor capacities do animals use in nature? A review of studies in several animal groups shows widespread effects of the environment on measures of locomotor function.

INTRODUCTION

Studies of locomotor function have contributed to our understanding of critical physiological issues, such as whether structure matches function and whether natural selection favors individuals with enhanced performance capacities (reviewed in Bennett & Huey 1990, Garland & Losos 1994, Gans et al. 1997, Boggs & Frappell 2000). Such studies have shown applications for the fields of conservation biology, evolutionary biology, and ecology (Miles 1994, Wainwright 1994, Turchin 1998). This synergy among fields is not surprising, as locomotor capacity is intimately associated with the ecology of animals, in part because the ability to move in a particular environment affects which portions of the habitat are accessible. Further, the modulation of locomotor speed determines the temporal component of how animals exploit space. Thus, understanding the locomotor capacities of organisms sheds light on numerous issues relating to traditional ecological and evolutionary ideas.

Locomotor capacities have long intrigued biologists because of the assumption that they are critical to Darwinian fitness, although few studies have provided empirical evidence to support this assumption (Garland & Losos 1994). Arnold (1983) first codified a theoretical framework for relating variation among individuals in morphology, performance, and fitness. This "performance paradigm" assumes that variation in lower level morphological, physiological, and biochemical traits (subordinate traits) determines variation in some ecologically relevant performance capacity (e.g., sprint running speed). In turn, the performance measure may be correlated with Darwinian fitness (some measure of reproductive success in the wild, although precise empirical and theoretical definitions are complicated (e.g., see McGraw & Caswell 1996). In this manner, a direct link is created between morphology and fitness, but organismal performance forms the critical intermediate step between them. This paradigm has inspired numerous studies relating morphology to performance but a smaller number that relate performance to fitness (Pough 1989, Bennett & Huey 1990, Garland & Losos 1994).

Because of their presumed importance for Darwinian fitness, locomotor capacities have played a significant role in discussions of adaptation (Pough 1989, Bennett & Huey 1990, Garland & Losos 1994, Dickinson et al. 2000). However, relatively few studies have measured locomotor function in nature, particularly in comparison to laboratory studies (Wainwright 1994, Irschick & Losos 1998, 1999). Our goal is to review field studies that directly measure locomotor function in nature or that measure characteristics related to locomotion. We argue that any understanding of locomotor adaptation will be incomplete without data on how organisms function in nature.

We review four bodies of literature to demonstrate the diverse applicability of field studies for addressing physiological hypotheses: speed of birds, energetics and performance of lizards, diving physiology of marine mammals, and performance and kinematics of flying insects. Our criteria for inclusion in this review are that a paper (subsequent to 1950) provides quantitative data on locomotor performance in nature or contains quantitative data on some physiological function that is intimately related to locomotion in nature. We do not exclude papers using relatively simple methods of measurement. We include some studies that used trained marine mammals that were allowed to move unfettered in the ocean.

Although energetic efficiency is one aspect of locomotor performance, we do not consider the ecological cost of transport, i.e., the percentage of the total daily energy budget that is attributable to locomotor costs (Garland 1983). This measure is typically calculated by estimating the actual distance that animals walk in nature, then using laboratory data on the incremental cost of locomotion to estimate costs attributable to locomotion. This quantity can then be divided by an estimate of the total daily field energy expenditure. Recent work in this area has shown that larger animals tend to spend relatively more on locomotor costs, and lizards spend more than mammals (Goszczynski 1986, Kenagy & Hoyt 1989, Baudinette 1991,

Karasov 1992, Christian et al. 1997, Altmann 1998, Gorman et al. 1998, Drent et al. 1999, Steudel 2000, Girard 2001, T. Garland, unpublished observations).

Ecological Function

Central to any concept of ecologically relevant locomotor capacity is what we term "ecological function." A hypothetical example showcases the distinction between ecological function and traditional laboratory measures of function. Consider a biologist who studies the maximal sprinting capacities of two rodent species and finds that the maximal speed of both species under optimal laboratory conditions is 1 m/s. By optimal, we mean that the substrate provides good traction, etc., and that attempts have been made to motivate the animal to achieve maximal performance. In nature, the biologist finds that species A runs at about 0.9 m/s (90% of maximum) when escaping from a natural predator (a hawk), 0.7 m/s (70% maximum) when chasing insect prey, and 0.5 m/s (50% maximum) when chasing rival males from its territory. By contrast, species B runs at speeds of 1.5 m/s when escaping a hawk, 1 m/s when chasing prey, and 0.7 m/s when chasing rival males.

Thus, the speeds achieved by each species in the field form the ecological performance capacities of each species. This example is instructive in several ways. First, if the biologist found a positive correlation between laboratory maximal speed and survivorship in either species, then without quantification of ecological performance, one would have no understanding of why fitness is related to locomotor performance (see also Pough 1989). For the example given, though the positive correlation between Darwinian fitness and maximal speed consists of three components (escape from a predator, feeding, and chasing rival males), the field data suggest that the first is likely to be the most important because only during this activity do the animals run near their maximal abilities. Thus, laboratory studies alone, and correlational studies of natural selection alone, reveal little about the ecological reasons why selection favors high levels of performance. In addition, the above example illuminates that ecological performance may be greater than laboratory performance (species B in our example) if the animal is not actually maximally motivated in the laboratory, or if the animal's performance is constrained by the laboratory set-up (Irschick & Jayne 1999, Bonine & Garland 1999). Finally, studies of ecological function can be a rich source of information about organisms and may yield insights into their behavior that were not predicted by theory.

Two questions pervade this review. First, how does the environment affect ecological function? Second, what proportion of their maximal capacities do species use in nature during different behaviors? Understanding how the environment affects function has been a central goal of physiologists for many years (e.g., see Garland & Carter 1994, Feder et al. 1987, Feder et al. 2000), but most studies have examined this question under controlled circumstances in the laboratory. Understanding whether, and to what extent, species use their maximal locomotor capacities in nature is also important (Hertz et al. 1988), because if species never use their maximal capacities, then selection cannot act directly on such traits as maximal sprint speed (Irschick & Losos 1998). Studies of ecological function can reveal the extent to which performance capacities are used in nature during different behaviors, and thus provide information on the potential for selection to act on a trait. Thus, Arnold's (1983) original paradigm has been expanded to include consideration of behavior as a potential "filter" between whole-organism performance and the direct effects of natural (or sexual) selection (Garland & Carter 1994, Garland & Losos 1994).

REVIEW OF LITERATURE

Speed of Birds

A substantial body of work has investigated how fast bird species fly in nature, using a variety of techniques (e.g., radar). The goals of these studies typically fall into several categories. Optimization tests compare competing hypotheses about which factors limit speeds used during migration and other activities. These studies test some optimization criterion under the assumption that birds either minimize or maximize some physiological quantity. Several reviews present details of these and other hypotheses (Welham 1994, Hedenstrom & Alerstam 1995, Pennycuick 1997). A smaller set of studies correlates variation in speeds to variation in habitat (e.g., movement over sea vs. land), variation in the technique of flying (e.g., cruising, soaring), or simply to describe the speeds that birds use. A third set of studies addresses whether species of special interest (e.g., the peregrine falcon) achieve the high levels of performance predicted by popular belief.

OPTIMIZATION TESTS OF FLIGHT   One optimization hypothesis posits that birds will maximize the distance traveled per unit energy expended (termed the maximum range speed, V_{mr}), resulting in high ratios of speed to power. Several equations have been proposed for this optimization criterion (Welham 1994). Another hypothesis posits that birds will fly at speeds that minimize the amount of power they expend during flight (termed the minimum power speed, V_{mp}). In contrast to V_{mr}, V_{mp} should minimize the ratio of power to speed. Importantly, whether a bird is predicted to use V_{mr} or V_{mp} depends on both the ecological context (e.g., duration of flight, time of day) and the species involved (Norberg 1981). Because power or energy expenditure cannot be easily measured in free-flying birds, researchers have tested these hypotheses indirectly by measuring speeds of a variety of species during different activities (e.g., long-distance migration, feeding young). In addition to empirical studies, theoretical models of how fast birds should fly during different activities have been generated, usually based on V_{mr} and V_{mp}.

Birds may be engaged in multiple activities when speed is measured, and the predicted speeds can vary depending on the behavioral context (Hedenstrom &

Alerstam 1995). Several recent papers have reviewed studies of whether birds fly at predicted speeds (Hedenstrom & Alerstam 1995, Pennycuick 1997). We do not summarize how many studies support a particular hypothesis, but review studies of speed in birds by considering them in an ecological and behavioral context.

EFFECTS OF ENVIRONMENT ON SPEED

Activity   Typically, researchers have examined flight performance during several activities: migration, movement between feeding sites and nests, and song flight. In addition, a number of studies have examined birds when their behavioral activity is unclear. Migration speeds are frequently double those during other activities (Hedenstrom & Alerstam 1992, 1994a, 1995), which may result in part from differences in flight behavior between migration and other activities (e.g., song flight). During migration, birds ascend to a preferred height and then fly level for long periods. By contrast, during song flight in skylarks (*Alauda arvensis*), individuals first climb to a culminating level flight where the bird flies against the wind and finally descends by parachuting, gliding, flapping flight, or occasionally a rapid dive (Hedenstrom & Alerstam 1992).

Migration speeds are not constant: Speeds over water are typically higher than over land (Alerstam 1975). This effect may be caused by rising thermals over land, resulting in soaring flight behavior, which is slower than steady-state flapping (Alerstam 1975, Pennycuick 1982a). Speeds of nonsoaring birds over land were similar to those observed over open water (Alerstam 1975). Although slope soaring had little overall effect on speed in the wandering albatross (*Diomedea exulans*) (Pennycuick 1982b), this species had a greater variance in speeds over land than sea. In a study of several species migrating across the English Channel, Parslow (1969) found that birds migrating in larger groups had higher speeds than did those in smaller groups, although Parslow (1969) attributed this effect to more favorable winds on nights with large groups of migrants. For 11 species of seabirds, airspeeds during nonforaging flights were higher than for foraging flights, although each was variable (Alerstam et al. 1993). Behavioral interactions also affect flight speed; when a pair of imperial shags (*Phalacrocorax alriceps*) were pursued and attacked by a south polar skua (*Catharacta maccormicki*), the skua accelerated rapidly from a speed of about 6 m/s to over 16 m/s, while the speeds of the Imperial shags remained almost constant (Alerstam et al. 1993).

Orientation   The relationship between the orientation of flying (ascending, descending) and speed is unclear. For some bird species (e.g., skylarks), descending speeds tend to be the fastest and climbing speeds the slowest (Hedenstrom 1995), although for the Little blue heron, speeds are lower during descent than level flight (Tucker & Schmidt-Koeing 1971). However, in a variety of other species, speed and incline are unrelated (Tucker & Schmidt-Koeing 1971). Air speeds and climb rates were negatively related within migrating bird species (Piersma et al.

1997). Speeds can also vary dramatically as a function of the type of flight used. For instance, skylarks (*A. arvensis*) most often descend by parachuting at a mean sinking speed of 1.5 m/s, but occasionally dive 8.4 m/s, and flapping descent results in speeds of about 1.6 m/s (Hedenstrom 1995). Alerstam (1987a) found that when several bird species migrated across a strong magnetic analomy (iron ore deposits), they tended to descend for about 2 minutes before leveling off, thus changing both their altitude and vertical speed.

Wind   The available data support the notion that airspeeds are higher in headwinds than with tailwinds (Bellrose 1967, Able 1977, Larkin 1980, Larkin & Thompson 1980, Bloch & Bruderer 1982, Wege & Raveling 1984, McLaughlin & Montgomerie 1985, Gudmundsson et al. 1992, Wakeling & Hodgson 1992; see also Schnell 1965, Bruderer & Steidinger 1972, Tucker 1974, Schnell & Hellack 1979), although some studies are inconclusive (Blokpoel 1974, Tucker & Schmidt-Koeing 1971). Hedenstrom & Alerstam (1994b) argued that the climb rates of knots (*Calidris canutus*) and turnstones (*Arenaria interpes*) were enhanced by turbulent wind, such that the mean climb rate under conditions of no wind was 1.0 m/s, whereas under windy conditions it was 1.2 m/s (but see Piersma et al. 1997). At least one study (Able 1977) has also documented that birds alter their flight posture when flying into strong headwinds by flying at steep angles.

Body size   Several studies have investigated both how body mass relates to observed speed and how closely observed speed matches predicted speeds based on theoretical models of how much power should be produced (i.e., V_{mp} or V_{mr}). Smaller birds tend to fly either consistently faster than, or equal to V_{mp} or V_{mr}, whereas larger birds tend to fly at speeds that are consistently lower than predicted estimates (Pennycuick 1982b, 1997, Hedenstrom & Alerstam 1992, 1994a, Welham 1994). Some authors have argued that the relatively slow speeds of larger birds result from the ability of small birds to extract power from their flight muscles to fly at any characteristic speed during level flight (Pennycuick 1997). Behavior may also be important: Some large species appear to have relatively slow speeds because they more frequently slope-soar than do smaller birds (Pennycuick 1997).

Diurnal and seasonal variation   Relatively few studies have investigated systematically whether bird speeds change with either the time of year or time of day. In a survey of 12 species, undisturbed evening flight speeds were greater, although not substantially so, than midday flights (Evans & Drickamer 1994). In a massive survey of about 3500 individual birds (representing an unknown number of species) flying at night over southeastern New York, Larkin & Thompson (1980) found little seasonal variation in airspeed, with the exception that slightly more slow birds were detected during the fall. During the spring fast birds generally oriented northeast, whereas during the fall they oriented due west. By contrast, slow birds showed no obvious heading in either the spring or fall. These findings show that performance can be affected by environmental factors not easily studied in the laboratory.

Type of flight   Birds exhibit several kinds of flight, including gliding, flapping (Pennycuick 1987) and diving behavior in which the wings are folded into the body. Further, birds fly in a zig-zag pattern, which is associated with slope soaring along waves, (Pennycuick 1982a). In skylarks (Hedenstrom 1995), flapping flight is typically faster than parachuting or diving flight. Birds that rely on high-speed dives to capture prey modify their wing and body posture to alter their speed. For instance, gyrfalcons (*Falco rusticolus*) fold their wings as they accelerate during high-speed dives (Tucker et al. 1998). The kind of flight used, and consequently the airspeed, can be affected by wind and other conditions. For example, during calm conditions, albatrosses use swell soaring and turning within a width of about 300–500 m to travel about 10 m/s. Under windy conditions, they travel faster (about 22.5 m/s) by a combination of wave soaring and dynamic soaring (Alerstam et al. 1993). One group of barn swallows (*Hirundo rustica*) had low and straight flight paths (mean speed = 8.8 m/s), whereas another group used high and erratic flight behavior (mean speed = 6.8 m/s) during undisturbed activity (Blake et al. 1990).

Loading   Pennycuick et al. (1994) trained falcons and hawks (six species) to fly 500 m to a lure and showed that only the gyrfalcon (*F. rusticolus*) had a lower air speed with a load (a transmitter) than without. Although these studies were not completed under natural conditions, they are a good first step for understanding the effects of loading on flight performance in unconstrained surroundings.

Effects of food availability   Variation in speed relative to food availability tests the idea that birds are either maximizing the distance traveled per unit energy expended (foraging efficiency) or their overall rate of energy delivery. Lapland longspurs (*Calcarius lapponicus*) do not fly faster when food is more available, which is consistent with maximizing the distance traveled per unit energy expended (foraging efficiency) (McLaughlin & Montgomerie 1990).

WHAT PERCENTAGE OF MAXIMUM CAPACITY DO BIRDS USE DURING VARIOUS ACTIVITIES?   Few studies have examined this issue, primarily because researchers have been more interested in testing whether birds moved at preferred speeds that matched theoretical optima. We address this issue in two ways. First, we review a small number of studies that have tested whether birds reach maximal speeds based on theoretical models of flight. Second, we describe some of the mean speeds that species use during migration (when speeds appear to be fastest) and consider these data in the absence of information on maximum capacities.

Because falcons are considered to be among the fastest animals, much effort has centered on both modeling and measuring the speeds they can achieve in nature. Mathematical models of an ideal peregrine falcon (*Falco peregrinus*) predict that it should reach top speeds of 89–112 m/s for a vertical dive, the higher speed representing larger birds (Tucker 1998). If drag is considered, then top speeds could range from 138 to 174 m/s. This model further predicts that an ideal falcon diving at angles between 15 and 90°, with a body mass of 1 kg, reaches 95% of top speed after traveling about 1200 m (Tucker 1998).

In trials with a closely related species (a 1.02 kg gyrfalcon, *F. rusticolus*), during the first (acceleration) phase of the dive maximal speeds ranged from 52 to 58 m/s, which closely matched predicted speeds assuming minimum drag (Tucker et al. 1998). Falcons then began a constant-speed phase, which lasted no more than a few seconds, during which they increased the amount of drag and thus slowed considerably. Whereas these speeds are among the fastest reported, predicted speeds were much higher than actual speeds. This difference is attributable in part to the behavior of increasing drag after the initial acceleration phase, whereas Tucker's (1998) model assumes continued acceleration.

Alerstam (1987b) used tracking radar to detail diving behavior of two species known for their high-speed hunting dives, the peregrine falcon (*F. peregrinus*), and the goshawk (*Accipiter gentilis*). The initial portion of their dives consisted of flapping flight, during which birds accelerated along a level path. The next stage consisted of flapping flight and a gentle descent, followed quickly by diving at steep angles. After the strike animals leveled off. The falcons attained speeds ranging from 19 to 23 m/s during the initial period of vigorous flapping flight, which was nearly horizontal, but when they dived without wing beats, speeds approached 30 m/s. These observed speeds for shallow dives were similar to maximal speeds based on theoretical predictions (Pennycuick 1975). Theory predicts a positive curvilinear relationship between the angle of dive and velocity, but the observed relationship between these variables is nearly flat, suggesting that either the model is flawed or birds are adopting behaviors that affect diving speeds. One possibility is that birds use moderate stooping speeds to gain in hunting precision (Alerstam 1987b). This finding is intriguing in that it highlights how high levels of locomotor performance may not be desirable: Accuracy in locating prey may be more important than speed for some species.

In a review of undisturbed speeds of 36 species, Pennycuick (1997) listed mean speeds of 8.8–19.1 m/s. Given the maximal speeds of diving falcons and goshawks, which are presumably among the fastest of birds (Alerstam 1987b, Tucker et al. 1998), these estimates suggest that birds are using well over 50% of the maximal speeds during migration and likely are moving at speeds that are 70–90% of maximum capacity. For 48 species of birds (most of which were different from those in Pennycuick 1997), mean speeds ranged from 8.0–30.6 m/s during migration, although the average speeds for most species were approximately 11–13 m/s (Welham 1994). Thompson (1961) estimated the speed of a flying red-breasted merganser at 44 m/s, which would make it one of the fastest animals in the world. In short, the available data show that birds regularly move at extremely fast speeds during migration, suggesting that they are able to fly long distances at close to or maximal speeds.

Energetics of Lizards

Field studies of energetics in lizards are few and typically involve direct measurements of metabolic rates by use of doubly labeled water. Most studies have not directly measured speed during locomotor activity, but the papers reviewed here have attempted to quantify movement in relation to metabolic rates. Christian et al. (1997) and Nagy et al. (1999) have extensively reviewed the field energetics of lizards, and we examine a subset of those studies in the context of locomotion.

INTERSEXUAL DIFFERENCES   Because males and females frequently differ in their energetic requirements, primarily owing to reproduction, one might expect sex differences in locomotor behavior that would also affect metabolic rates. Lichtenbelt et al. (1993) found that differences between male and female green iguanas (*Iguana iguana*) in average daily energy expenditure (DEE) were small, although males tended to have higher metabolic rates during the mating season, which is related to their higher levels of activity during this period.

In the temperate-zone lizard *Sceloporus virgatus* Merker & Nagy (1984) revealed an interesting interaction among activity levels, season, and sex. In the spring males had higher metabolic rates than females and were active for twice as long as females (7.6 vs. 3.6 h/day). Nevertheless, females had higher rates of energy intake than males (8.3 vs. 3.6 times resting metabolic rate), resulting in a weight loss for males. In the summer both sexes maintained constant body masses and similar energy budgets, but females were active for longer periods. This example shows how locomotor behavior is intimately related to many other factors in the environment, and further studies that quantify how fast and/or how often these lizards move would shed even more insight on the factors that affect field energetics.

TYPE OF ACTIVITY   Behaviors often differ in their energetic costs. Lichtenbelt et al. (1993) found that climbing was about six times more energetically costly than movement on horizontal surfaces for green iguanas (*I. iguana*), which is consistent with laboratory studies in lizards that show movement up steep inclines is energetically expensive (Farley & Emshwiller 1996). Marine iguanas (*Amblyrhynchus cristatus*) forage in both the intertidal and subtidal zones, but total DEE apparently does not differ significantly between habitats (Drent et al. 1999).

Predatory lizards often differ in foraging behavior, with some lizards relying on infrequent, fast movements for capturing prey (sit-and-wait) and others moving actively in search of prey (actively-foraging) (see Perry 1999 for a recent review). A widely foraging lizard *Eremias lugubris* has a substantially higher metabolic rate than its close relative, *E. lineoocellata*, despite the forager being active for shorter periods [2.75 h/day compared with 10.25 h/day, respectively (Nagy et al. 1984)]. A recent review of DEEs (Nagy et al. 1999) did not address whether such a correlation holds across all lizard species that have been studied, perhaps because of the scarcity of quantitative data on movement rates in the field (but see Garland 1993, Perry 1999).

INTERSEASONAL VARIATION   Seasonal variation is typically associated with two key variables that influence locomotor behavior, temperature and reproduction. Varanid lizards are frequently inactive during the dry season, although this pattern can change depending on habitat type. During the dry season *Varanus panoptes* was largely inactive in woodland areas, whereas on floodplains, it often walked for 3.5 h/day, which is remarkably high for a reptile, and consequently led to high rates of energy expenditure (Christian et al. 1995). In the Galapagos islands marine iguanas, *A. cristatus*, were about three times as active during the warm than the cold season, resulting in substantially higher values of DEE for the former season (Drent et al. 1999).

Annual variation in energy allocation differed between male and female *I. iguana*, such that males primarily devoted energy on locomotion for social activities, whereas females devoted about 15% of their annual energy budget to the production of eggs (Lichtenbelt et al. 1993). Seasonal variation also affected activity levels, and hence the metabolic rates of male and female striped plateau lizards (*Sceloporus virgatus*), such that females were more active in the summer, whereas males were more active in the spring (Merker & Nagy 1984).

HABITAT DIFFERENCES   The lizard *Cnemidophorus hyperythrus* exhibited higher DEE in woodland areas than in thorn scrub sites (330 J/g/day vs. 219 J/g/day), likely because woodland lizards were active most of the day (about 9 h), whereas lizards in thorn scrub habitat were active mostly in the morning (about 3.5 h/day) (Karasov & Anderson 1984).

Performance of Lizards

Lizards have long been used as a model for studies of both maximal loco-motor performance (e.g., Garland & Losos 1994, Bonine & Garland 1999, Van Hooydonck et al. 2001) and ecology (Vitt & Pianka 1994). Until recently, however, relationships between performance and ecology were poorly understood (but see Aerts et al. 2000). A recent body of work on both terrestrial and arboreal lizards has shed considerable light on how ecological performance is affected by morphology, habitat use, and behavior (Irschick & Losos 1998, Jayne & Ellis 1998, Irschick 2000a,b, Irschick & Jayne 1998, 1999a,b, Jayne & Irschick 2000).

HOW DOES THE ENVIRONMENT AFFECT LOCOMOTION?   Footprints of two species of lizard (*Callisaurus draconoides, Uma scoparia*) were studied in the soft sand of the Kelso dune system in southern California to examine how incline, vegetative cover, and other factors affected locomotion when lizards were escaping a threat (approach of a human) and when they moved undisturbed through the habitat (Jayne & Ellis 1998, Irschick & Jayne 1999a, Jayne & Irschick 2000). These two lizards are closely related, are morphologically and behaviorally different, yet occur sympatrically in various sand dune systems. From tracks left in the sand, one can gain accurate values of stride length, as well as determine the orientation of paths relative to landmarks in the environment. Studies in the laboratory on sand show that both lizards exhibit a linear relationship between stride length and speed (Irschick & Jayne 1998, Jayne & Irschick 2000), which can be used to estimate speeds of movement in nature.

The fringe-toed lizard (*U. scoparia*) has several specializations for movement on sandy surfaces, such as laterally oriented toe fringes, a countersunk jaw, and smooth skin that facilitates burrowing into sand (Stebbins 1944). This lizard has a relatively stout body, short limbs, and a short tail and moves predominantly by quadrupedal locomotion. By contrast, the closely related zebra-tailed lizard (*C. draconoides*) is more specialized for movement on firm substrates and is considered a bipedal specialist (Irschick & Jayne 1998). Consequently, *Callisaurus* has several specializations for high-speed bipedal locomotion, including long hindlimbs, a long tail, and long distal elements (Snyder 1962, Bonine & Garland 1999).

Because moving up inclines is energetically more expensive than moving on level, or near-level surfaces (Taylor et al. 1972, Farley & Emshwiller 1996), one might predict that lizards will avoid moving on steep inclines, which on sand dunes reach as high as 32°. On the other hand, because incline does not affect maximal speed or acceleration as greatly in small animals as large animals (Huey & Hertz 1982, 1984), one might expect that lizards will preferentially flee uphill when escaping from a predator. Jayne & Ellis (1998) elicited escape locomotion of fringe-toed lizards by approaching the animals and measuring how speed (based on a stride-by-stride basis) and the orientation of escape paths were related to incline, angle of turning, and the location of nearby landmarks (vegetative cover, burrows, the steepest available incline). By comparing the frequency distributions of inclines used during escape with the inclines available in the habitat, they determined that *Uma* escaped randomly with respect to incline, but that maximal speeds were negatively affected both by running up steep hills and by large turn angles. Thus, running in a straight line on a level surface enhances maximal speed. Further, most *Uma* appeared to use a predetermined escape route (e.g., Stamps 1995), by which they ran toward and then down burrows (Jayne & Ellis 1998).

Bipedal locomotion has been cited as an important behavior that allows some lizards to move at faster speeds than strictly quadrupedal lizards, but few studies have examined how often lizards run bipedally in nature and whether speeds during bipedal locomotion are actually faster than for quadrupedal locomotion. Irschick & Jayne (1999a) examined the effects of incline and other habitat variables on the escape locomotion of *C. draconoides* and also measured whether each stride was bipedal or quadrupedal. Bipedal strides were, on average, 12% longer than quadrupedal strides, so if one assumes a similar stride frequency between the two modes, bipedal locomotion is significantly faster. Laboratory studies in *Callisaurus* for locomotion on a high-speed treadmill show a similar ratio of longer stride lengths for bipedal locomotion, suggesting that this result is robust (Irschick & Jayne 1999b). Another result in common between *U. scoparia* and *C. draconoides* was that maximal stride lengths were typically not achieved until several meters into the escape path, some of which were 30 m long (Irschick & Jayne 1999a).

Therefore, racetrack estimates of speed likely underestimate the maximal speeds of which both *Uma* and *Callisaurus* are capable, a problem that may be overcome by use of high-speed treadmills (Bonine & Garland 1999, Irschick & Jayne 1999b).

Callisaurus lizards exhibited an interesting threshold effect in regards to incline use. On shallow slopes (<15°), over which most of the locomotion occurred, lizards moved randomly with respect to the incline. By contrast, lizards avoided moving directly up or down steep hills (>15°), preferring to run horizontally across the hill (this cut-off value was found by evaluating movements on a variety of inclines). Thus, animals may have complex and often unpredictable behaviors regarding usage of habitats that pose functional challenges.

A field experiment was used to examine the effects of incline and vegetative cover on the undisturbed locomotion of *U. scoparia* (Jayne & Irschick 2000). Three 40 × 100-m plots that differed in both incline and amount of vegetative cover were established. Before lizards were active each morning, all tracks were erased on the sand dunes, and paths were examined several hours later. Similar to escape locomotion, *U. scoparia* moved around the habitat randomly with respect to incline, but at two preferred speeds. Slow locomotion occurred near burrows and mounds of vegetation, which *Uma* used as retreats from both predators and high temperatures, whereas the high-speed movements (typically >2 m/s) generally occurred in open areas of the dune [undisturbed desert iguanas, *Dipsosaurus dorsalis*, have been observed to engage in a similar bimodality of speeds near Dale Dry Lake, San Bernardino County, California (T. Garland Jr., personal observations)].

Uma also tended to move more slowly on steep inclines and near vegetation, resulting in different locomotor behavior in different parts of the sand dune. On steep slopes with little vegetation the average speed was only 1.54 m/s, whereas on shallow slopes with more vegetation the average speed was 1.76 m/s. In addition, incline, speed, and orientation interacted for the undisturbed locomotion of *Uma*. On shallow inclines locomotion tended to be relatively fast and was distributed approximately equally on different inclines. On steep surfaces lizards favored direct uphill over downhill locomotion, and most locomotion was relatively slow. This research underscores several themes: (*a*) Environmental effects on locomotor performance are complex and interactive, (*b*) racetrack speeds may underestimate the true maximal speeds of lizards, and (*c*) species may exhibit threshold effects in terms of habitat use.

THE EVOLUTION OF ECOLOGICAL PERFORMANCE IN CARIBBEAN ANOLIS LIZARDS
An important, yet rarely addressed, issue in ecological and evolutionary physiology is whether performance capacity, as measured in the laboratory, and performance levels exhibited during natural behaviors have co-evolved. Several unresolved questions bear on this issue. First, is laboratory performance always maximal? Second, do species with low maximal capacities compensate for them by using a greater percentage of their performance capacities than species with high maximal capacities? These issues have been examined both across species and ontogenetically in Caribbean *Anolis* lizards (Irschick & Losos 1998, Irschick 2000a,b).

The ecology, morphology, and behavior of Caribbean *Anolis* lizards have been studied extensively (Williams 1972, Losos 1990a,b, Roughgarden 1995), but until recently the relationship between locomotion and habitat use has not been well understood. Within each of the Greater Antillean islands (e.g., Cuba, Hispaniola), largely independent radiations (Losos et al. 1998, Jackman et al. 1999) of *Anolis* have resulted in a series of ecologically and morphologically distinct forms, termed "ecomorphs" (trunk-ground, trunk-crown, twig, crown-giant, trunk, and grass-bush, named for the portion of the habitat they prefer) (Losos 1990a,b). For example, the trunk-ground ecomorph has long hindlimbs, a long tail, and tends to occupy broad tree trunks close to the ground (<1 m). By contrast, twig anoles have short hindlimbs, a short tail, and tend to occupy narrow surfaces higher in the canopy. Irschick & Losos (1998) and Irschick (2000a) focused on eight anole species that are similar in size yet represent three ecomorph types (Figure 1). The convergent evolution of ecomorphs provides enhanced statistical power for examining the evolution of ecological performance (e.g., see Van Hooydonck & Van Damme 1999).

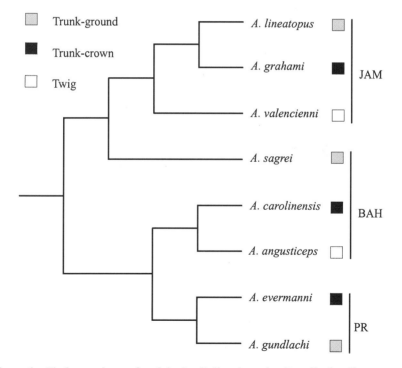

Figure 1 Phylogenetic tree for eight *Anolis* lizard species. Locality headings next to species names represent the Caribbean islands on which these species occur, although *A. carolinensis* was studied in Louisiana. JAM, Jamaica; PR, Puerto Rico; BAH, Bahamas. From Irschick & Losos (1998).

Irschick & Losos (1998) measured sprinting and jumping of all eight anoles during four behaviors, three in the field (escape from a threat, feeding, undisturbed locomotion) and one in the laboratory (maximal speed on a broad surface with sure footing). An initial test revealed that laboratory performance for both jumping and sprinting was significantly greater than for all three natural behaviors, with anoles using an average of about 90% of their maximal sprinting capacities during escape, 70% of maximum during feeding, and 33% during undisturbed locomotion. By contrast, the maximal jump distances in the field during feeding and undisturbed movements (the two activities in which they were most often used) were <40% of the maximum performance elicited in the lab. Therefore, two key factors, the ecological context and the kind of performance, both markedly affect the levels of performance. These results also show that if one were to conduct a study of natural selection on maximal jumping ability as measured in the laboratory, one likely would find no correlation with Darwinian fitness, because anoles do not use their maximum jumping capacities in nature. Phylogenetically informed, cross-species comparisons showed that for sprinting, performance capacity has evolved in positive correlation with both escape and feeding behavior; hence, performance capacity is a good predictor of performance in nature.

However, this finding does not address whether species with low performance capacities compensate by using a greater fraction of their capacities in the field. One possibility is that each species will use a similar fraction of its maximal abilities for a particular behavioral task. This possibility might be correct if successful escape from a predator is a direct function of absolute speed (but see Van Damme & Van Dooren 1999). Alternatively, a species with a lower capacity could compensate by using a greater fraction of its performance capacity for a given task as compared with species with a higher capacities. This possibility would be more likely if a particular absolute speed facilitated escape from predators. In other words, animals may perform only to the lowest level needed for successfully escaping a predator. This idea was tested both among species that vary in performance capacity (Irschick & Losos 1998) and for ontogenetic classes that also vary in performance capacity (Irschick 2000b).

For escape performance (elicited by human approach) among species, the hypothesis of compensation is supported (Irschick & Losos 1998), as species with low sprinting capacities, such as twig anoles, tended to use nearly all of their capacities, whereas the speedier trunk-ground anoles escaped by using as little of their performance capacities as possible (Figure 2). Irschick (2000b) studied maximal laboratory and field speeds during escape, feeding, and undisturbed locomotion for juveniles, adult females, and adult males of the trunk-ground anole *Anolis lineatopus*. Maximal speed is generally correlated with size in *Anolis* lizards, such that adult males are significantly faster than adult females, which, in turn, are significantly faster than juveniles (Macrini & Irschick 1998). In support of the hypothesis of compensation, juveniles used a greater fraction of their sprinting capacities in comparison with adult males and females during both escape and feeding. However, adult females did not use a significantly greater fraction of their sprinting

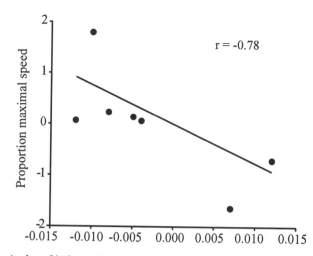

Figure 2   A plot of independent contrasts between maximal speed (x-axis) and proportion maximal speed used during escape (y-axis) for eight anole species P < 0.05. From Irschick & Losos (1998).

capacities than adult males during either escape or feeding, despite the females having, on average, an 11% lower sprinting capacity. Overall, the available data seem to support the notion that performance-limited lizard species or ontogenetic classes compensate by using a greater fraction of their performance capacities in nature (see also Clobert et al. 2000), but the hypothesis does not explain all the observed variation. One means of rigorously testing the hypothesis of compensation, and also determining which of the above metrics of performance [absolute speed, relative speed (see also Van Damme & Van Dooren 1999, Packard & Boardman 1999), percent maximal speed] is under selection would be to carry out a field study relating sprinting performance during different behaviors to Darwinian fitness among individuals of different sexes and sizes.

EVOLUTION OF PREFERRED SPEED IN ANOLIS LIZARDS   Few studies have related the preferred speeds of animals in nature to their habitat conditions, particularly in a phylogenetic context. Arboreal animals, such as some lizards, are ideal for addressing this issue because they must move through environments that vary in incline and surface diameter, both of which affect locomotion (Taylor et al. 1972, Cartmilll 1985, Losos & Sinervo 1989, Farley & Emshwiller 1996). For example, because of the negative effect of surface diameter on speed (Losos & Sinervo 1989, Macrini & Irschick 1998), arboreal lizards that frequently utilize narrow surfaces may move more slowly than lizards that typically utilize broad surfaces. Alternatively, if surface diameter does not pose a strong constraint on movement at slow speeds, then lizards that use broad surfaces may move at similar preferred speeds to those lizards that use narrow surfaces.

Irschick (2000a) examined the evolution of preferred speeds in the eight anoles mentioned above and tested the hypothesis that preferred speeds have co-adapted with habitat use. In contrast to *Uma*, the distributions of speed for all eight anoles were unimodal, with a preponderance of slow locomotion. Indeed, for all anoles, at least 75% of their movements were slower than 40% of each species' maximum speed. Comparative methods revealed that anole species that use narrow perches move at slower preferred speeds than species that use broader surfaces. Further, species with slow preferred speeds use a wider variety of surface diameters than species with high preferred speeds. This research demonstrates the usefulness of integrating ecological studies of habitat use and information on physiological data, such as preferred speed.

ARBOREAL VERSUS TERRESTRIAL HABITATS   Arboreal habitats are characterized by discrete perches that vary in both incline and diameter and thus pose functional challenges for locomotion. Terrestrial habitats pose fewer constraints for locomotion. The habitat of *Uma* and *Callisaurus* consists of clumps of vegetation and burrows in an open terrestrial environment. Both lizards use these patches as retreats from predators and extreme temperatures. If one assumes that the threat of predation or overheating in open areas is higher than in retreats, then the high-speed movements in open areas may occur because the lizards perceive the open areas as dangerous (Lima & Dill 1990). The golden-mantled ground squirrel (*Spermophilus marmoratus*) primarily uses high-speed movements between retreats but slower movements when closer to retreats (Kenagy & Hoyt 1989). It also lives in a relatively open habitat interspersed with patches of vegetation.

At the other extreme, much of the undisturbed locomotion (jumping and running) of arboreal *Anolis* lizards is well below maximum capacity. Because of the characteristics of the discrete perches of arboreal anoles, much of their locomotion consists of short movements and is highly intermittent, which is an obvious constraint on achieving high speeds. Therefore, maximal speeds as measured under laboratory settings (broad diameter, long surface) are good approximations of their performance capacity. One difficulty is that *Uma* and *Callisaurus* are closely related, but anoles are distantly related to these species, and *Uma* and *Callisaurus* share many differences in morphology and behavior, which could confound terrestrial-arboreal comparisons. Thus, future comparisons would be most fruitful if they were to compare closely related species that differed in their degree of arboreality.

Diving Physiology of Marine Mammals

Common to all of the studies of ecological locomotor function in marine mammals is some test of actual versus predicted performance based on models of aerobic and anaerobic physiology. For example, the aerobic dive limit is the amount of time that an animal should be able to dive on a single breath of air and yet not suffer a substantial cost because of buildup of anaerobic waste products (e.g., lactate).

Consequently, the goal of many studies has been to test whether marine mammals exceed their aerobic dive limit, and if so, how often. Unlike terrestrial and arboreal habitats, aquatic systems are ideal for studies relating changes in physiology to locomotion. This is because most physiological variables will change in a regular manner with two related variables: dive depth and dive duration. Some of the relevant functional variables measured include speed, breathing and heart rates, levels of lactate before and after a dive, and levels of gases (O_2 and CO_2) in the blood. A few researchers (e.g., Kooyman et al. 1980) also measured additional aspects of blood chemistry (e.g., arterial hemoglobin concentrations), which will not be discussed here.

Some generalities have emerged in regard to the effects of depth on ecological locomotor function. As marine mammals dive, heart rate typically decreases (Fedak et al. 1988, Elsner et al. 1989, Andrews et al. 1997, Ponganis et al. 1997, Williams et al. 1999), in some cases as low as four beats per minute (Thompson & Fedak 1993). In many marine mammals heart rate decreases during the descent and subsequently increases during the ascent, with a noticeable tachycardia just prior to surfacing (Fedak et al. 1988, Elsner et al. 1989, Thompson & Fedak 1993, Andrews et al. 1997, Williams et al. 1999). Once the animal surfaces, heart rate increases, presumably to increase blood flow to oxygen-starved tissues (Thompson & Fedak 1993, Andrews et al. 1997, Williams et al. 1999). The amount of lactate in tissues increases with dive time (Kooyman et al. 1983, Ponganis et al. 1997, Shaffer et al. 1997, Williams et al. 1999), but for some marine mammals, lactic acid concentrations were not different from resting levels unless the dive was quite long (e.g., >20 min) [Weddell seals (Kooyman et al. 1983)].

How deep marine mammals dive, and the characteristic speeds and behaviors during dives, also shows some trends. The typical pattern is that marine mammals dive to a certain depth, spend a brief time moving horizontally, and then ascend to the surface [Weddell seals (Kooyman et al. 1980), sea lions (Feldkamp et al. 1989), narwhals (Martin et al. 1994), white whales (Shaffer et al. 1997)]. Some marine mammals, however, move for relatively long periods both slowly and horizontally at the bottom of their dives, resulting in some cases in a dramatically decreased metabolic rate [some gray seals (Thompson & Fedak 1993), elephant seals (Andrews et al. 1997)]. In elephant seals the amount of time spent moving horizontally along the bottom is substantial (Hindell et al. 1992), although males appear to have more flat-bottomed dives than females (Le Boeuf et al. 1988, 1996). For gray seals, Thompson & Fedak (1993) hypothesized that this horizontal movement is part of a sit-and-wait strategy in which the animal moves slowly along the bottom in hopes of finding passing prey. More research into what behaviors (e.g., feeding) are used during diving (possibly through the use of video cameras) could provide more information as to the reasons for long dives.

Average transit times are generally higher during deep than shallow dives [humpback whales during ascent (Dolphin 1987a,b), elephant seals (Hindell et al. 1992), harp seals (Lydersen & Kovacs 1993), narwhals (Martin et al. 1994), white whales (Shaffer et al. 1997)], which is consistent with the notion that marine mammals can spend only a limited amount of time under water. In bottlenose dolphins, however, speeds for both ascent and descent are similar for dives to 60 m and 210 m (Williams et al. 1999). Speeds tend to be faster during ascent (when presumably the cost of lactate build-up is at its greatest) than descent [humpback whales (Dolphin 1987a,b), elephant seals (Le Boeuf et al. 1988), white whales (Shaffer et al. 1997), bottlenose dolphins (Williams et al. 1999)], although in some marine mammals there appear to be few differences in speed between ascent and descent [harp seals (Lydersen & Kovacs 1993), narwhals (Martin et al. 1994), white whales (but see below) (Ridgway et al. 1984)]. The former pattern suggests that marine mammals are conserving oxygen at the beginning of a dive. In a study of four species of otariids, Ponganis et al. (1990) found that speeds when moving along the surface were similar to diving speeds for species that ranged in size from 30 to 130 kg.

Descent and ascent rates in some marine mammals are related to the depth of the dive. The relationship between descent and ascent rate and depth is curvilinear in some cases, such that on deep dives, the speeds of both ascent and descent are relatively slow compared with shallower dives [narwhals (Martin et al. 1994)]. For greater dives ($\geq$20 m in depth), both the average and maximum rates of ascent and descent were positively related to the maximum depth of diving in narwhals. However, for dives $\geq$20 m in depth and lasting >10 minutes, there were negative relationships between the average rates of descent and ascent, and the durations of diving. In brief, during extremely long dives, narwhals moved more slowly as the duration of the dive increased. In white whales speeds during ascent and descent were similar during shallow dives (<200 m), but for deep dives ascent speeds were significantly faster than descent speeds (Ridgway et al. 1984). These findings suggest complicated relationships between the environment and performance and also show the value of a comparative approach, as not all species react to the environment in the same manner.

Do marine mammals travel at the predicted speeds that would minimize their energetic cost of transport? Gray whales that migrate southwards move at speeds predicted to minimize the cost of transport (Rice & Wolman 1971, Sumich 1983), as do fur seals, Galapagos sea lions, and Galapagos fur seals (Ponganis et al. 1990), but such is not the case for Hooker's sea lions (Ponganis et al. 1990), narwhals (Martin et al. 1994), and bottlenose dolphins (Williams et al. 1999). The direction of migration may also affect average speed, as northbound gray whales migrate about half as fast as southbound gray whales (Pike 1962). More data on the frequency distributions of speed are needed to evaluate this question to provide a point of comparison with birds, in which migration also occurs.

Respiratory rate decreases with speed in white whales (Shaffer et al. 1997), in contrast to the pattern in other animal groups (Taylor et al. 1987). This pattern may be related to surface/submergence patterns, rather than coupling between respiratory patterns and locomotor movements. In some marine mammals, such as gray whales (*Eschrichitus robustus*) (Sumich 1983), this pattern is reversed. Some studies have measured blood gas levels in diving marine mammals and have generally found a substantial decrease in the partial pressure of oxygen after animals have been diving for extended periods [weddell seals (Kooyman et al. 1980), white whales (Shaffer et al. 1997), bottlenose dolphins (Williams et al. 1999)].

Seasonal and diel variation may affect how long animals dive, potentially because of variation in prey availability. For instance, in 1983 the Southern Californian Bight experienced a major influx of warm water, which correlated with a 20-fold reduction in zooplankton biomass and a marked decrease in commercial catches of squid, salmon, and other fish (McGowan 1984). Sea lions dove more often and for significantly longer periods than prior to the reduction in food, which Feldkamp et al. (1989) suggested was compensation for a reduced prey encounter rate. It would be interesting to know whether the speeds and durations of dives during periods of low food availability differ from those during periods of greater food abundance.

Numerous studies have addressed whether marine mammals dive for time periods that break the predicted aerobic dive limit (ADL), and if so, why [e.g., Weddell seals (Kooyman et al. 1980, 1983), humpback whales (Dolphin 1987a,b), sea lions (Feldkamp et al. 1989), harp seals (Lydersen & Kovacs 1993), elephant seals (Hindell et al. 1992), bottlenose dolphins (Williams et al. 1999)]. ADLs are calculated from estimated speeds of movement, metabolic rates, and the amount of oxygen stores in the animal prior to diving. From these values, aerobic dive limits are calculated as the amount of time an animal can spend underwater before oxygen stores are depleted and anaerobic waste products begin to increase. Consequently, estimated ADLs can change depending on the values input to the equation. The presumed cost of such a lactate build-up is that the marine mammal would have to spend additional surface time to rid the body of the anaerobic waste products. Most marine mammals dive for periods shorter than their ADLs [Weddell seals (Kooyman et al. 1983), humpback whales (Dolphin 1987a,b), sea lions (Feldkamp et al. 1989), harp seals (Lydersen & Kovacs 1993), narwhals (Martin et al. 1994), bottlenose dolphins (Williams et al. 1999)]. However, for some marine mammals, some individuals substantially exceeded the estimated ADLs [gray seals (Thompson & Fedak 1993), elephant seals (Hindell et al. 1992, Andrews et al. 1997, Williams et al. 1999)]. For instance, some bottlenose dolphins dive for over 243 s, which exceeds their predicted oxygen stores by 28%, yet they do not incur the predicted anaerobic costs (Williams et al. 1999).

Why can some of these marine mammals break the rules? Part of the answer may lie in their locomotor behavior. An implicit assumption in calculations of ADL is that marine mammals are moving continuously, but the intermittent locomotion of some marine mammals appears to violate this assumption. During ascent, when oxygen stores are presumably at their lowest, dolphins and other marine mammals use burst-and-glide locomotion in which they glide for extended periods following active propulsion (Williams et al. 1999). This behavior is especially interesting because it is rarely observed in marine mammals in captivity, in part because tanks are too small. Other factors also may play a role in increasing dive times without incurring substantial costs. Ridgway et al. (1969) found that bottlenose dolphins swimming horizontally at depths <25 m had lower pulmonary oxygen reserves (and therefore greater metabolic demands) than the same individuals diving to 200 m, and lung compression and changes in the mode of swimming may have enhanced these energy savings. Thus, calculations of aerobic time limits may, in some cases, overestimate the energetic costs of locomotion under water (Williams et al. 1999). Finally, marine mammals such as elephant seals may substantially decrease their metabolic rates when diving, which would lead to incorrect values of ADL (see Hindell et al. 1992). The common thread to all three of these factors is that animals can alter their behavior in such a way as to violate the basic assumptions of physiological models, such as calculations of ADL. ADL is also an example of a physiological hypothesis that can only be tested in the field.

Performance and Kinematics of Flying Insects

A small body of work has measured ecological locomotor function in flying insects, typically their speeds of movement and kinematics. Two kinds of studies are apparent. First, studies on migratory locusts (*Locusta*) have yielded important insights into differences in performance and kinematics between field and laboratory settings. Second, work on Neotropical butterflies has documented interesting relationships between morphology, behavior, and performance in nature.

Locust flight has been studied extensively in the laboratory (see Baker et al. 1981 for references), but only a few studies have measured kinematics or flight speeds in nature. By filming swarms of migrating locusts (*Locusta migratoria*), Baker et al. (1981) found that both wing-beat frequencies and flight speeds were higher in the field than in the laboratory. For instance, the mean wing-beat frequency of locusts under tethered flight was 19.8 Hz, compared with 22.9 Hz during natural free flight. Similarly, mean speeds were 40% greater in the field than for tethered flight (Baker et al. 1981), but the linear relationship between flight speed and wing-beat-frequency was similar between free and tethered flight. Kinematic analyses of the wing motions of individuals within swarms show that a variety of parameters (e.g., ratio of upstroke to downstroke, stroke angles) are more variable in nature than during tethered laboratory flight (Baker & Cooter 1979). Other studies have shown that natural flight speed in locusts increases with air temperature (Waloff 1972, range 22–33°C), a result that also differs from laboratory studies, which have found flight speed to be temperature independent (Weis-Fogh 1956), despite a similar range of temperatures tested in the laboratory (range 25–35°C). These studies support the notion that certain aspects of tethered flight may be ecologically unrealistic, although such laboratory studies have clearly contributed to our understanding of insect flight.

In an elegant group of studies Dudley and his colleagues examined the natural (undisturbed) airspeeds of dozens of Neotropical butterfly species by following them across Lake Gutan in Panama. They were able to correlate these airspeeds to the morphological and behavioral characteristics of different species. In a study of 27 genera of Neotropical butterflies Srygley & Dudley (1993) found that palatable butterflies fly at faster natural speeds and are better able to avoid predators in a small cage, primarily owing to their center of mass being positioned near their wing base. By contrast, unpalatable butterflies fly at relatively slow natural speeds and are less effective at eluding predators, likely as a consequence of their posterior position of their center of mass. The natural airspeeds of the butterflies that were studied were 2–6 times faster than ground speeds obtained from the same genera in insectaries (Srygley & Dudley 1993). Data on natural airspeeds in habitats other than a large lake are needed to verify the generality of these findings.

In another comparative study of 62 Neotropical butterfly species, natural airspeed was positively correlated with body mass, thoracic mass, and wing loading (Dudley & Srygley 1994). Once body mass was controlled statistically, higher wing loadings were correlated with increased flight speed, but flight speed and wing aspect ratio were negatively correlated. These results suggest that butterfly airspeeds under natural conditions can reasonably be predicted from morphometric measurements (Dudley & Srygley 1994).

Studies of natural flight within butterfly species have also generated important insights into the biomechanics of insect flight. Dudley & DeVries (1990) found that certain aspects of their flight for the moth *Urania fulgens* consisted of unsteady aerodynamic movements because of the relatively large lift coefficients (between 2 and 3). For most flight sequences, however, lift coefficients were closer to 1.0 or were substantially less, suggesting quasi-steady aerodynamic mechanisms (Dudley & DeVries 1990). Verification of whether organisms use fundamentally similar flight mechanics in the field and in nature is important for understanding the generality of basic models of insect flight.

Intersexual differences in flight physiology were also studied for *U. fulgens*. DeVries & Dudley (1990) studied the effect of internal (thoracic) and ambient temperature, as well as morphology, on the airspeeds of both sexes. Airspeeds generally increased with increased thoracic temperature but did not change with ambient temperatures. Males and females differed significantly in external morphology but not airspeeds or thoracic temperatures, suggesting that neither ambient temperatures nor morphology greatly affect the natural flight speeds of this moth.

Srygley et al. (1996) examined natural airspeeds, wind speeds, and headings of two migrating butterflies and one moth over Lake Gutan in Panama. The Pierid *Aphrissa statira* and the nymphalid *Marpesia chiron* were capable of wind-drift compensation during migration, whereas the moth *U. fulgens* was not. Thus, information on how fast free-living animals move in nature provides interesting clues to many aspects of their biology, including migration behavior, which could have important implications for the conservation of these species.

CONCLUSIONS AND PROSPECTUS FOR THE FUTURE

An emerging theme from this review is the complex manner in which behavior and habitat interact to affect ecological locomotor function. This theme reinforces the suggestion that the "performance paradigm" (Arnold 1983, Wainwright 1994) must be expanded to include behavior (see Garland & Losos 1994, Garland & Carter 1994).

Several trends are noteworthy. First, levels of performance or relationships between environmental variables and function can be dissimilar between the field and laboratory. Laboratory studies indicate that flight speed is independent of temperature in locusts, whereas field studies show that flight speeds increase with temperature (Weis-Fogh 1956, Waloff 1972). In some cases the difference may be methodological; the slower speeds of the fringe-toed lizard (*U. scoparia*) in the laboratory relative to the field may be a consequence of using a racetrack (Jayne & Ellis 1998, Bonine & Garland 1999). Second, many species show seasonal variation in locomotor behavior. Studies of the field metabolic rates of lizards show that they vary their activity levels seasonally, resulting in different energetic requirements. Especially interesting would be studies that combine energetic information with more exact data on speeds.

Birds and marine mammals dominate the literature on ecological locomotor function; more data are needed for terrestrial organisms. The different results from studies of preferred speed in terrestrial *Uma* lizards and arboreal *Anolis* lizards show that generalities about ecological function are difficult to establish when there are substantial differences in habitat and behavior. For instance, do arboreal animals always move at slow preferred speeds? Do animals that occur in open habitats with patchily distributed retreats typically exhibit bimodal distributions of speed? Further, despite the impressive amount of data for some animal groups, certain key issues remain unresolved. First, few studies have used an evolutionary approach for investigating ecological function. The amount of data required for evolutionary comparisons would be prohibitive for certain physiological measurements, but comparative data can yield insights that would not be possible from studies on single species. For instance, comparative studies of ecological performance in *Anolis* lizards revealed that low-performance species compensate for their poor sprinting capacities by using a greater fraction of their capacities when fleeing from a threat. Other questions pertaining to the co-adaptation of physiology and behavior could also be addressed. Has maximum speed (Djawdan & Garland 1988) or evasiveness (Djawdan 1993) of rodents co-adapted with how fast they move when escaping a natural predator? Do species that gain an energetic advantage from intermittent locomotion also use the mode of locomotion more often in nature?

Few studies have developed methods for quantifying intermittent locomotion in nature. Recent laboratory studies have shown that terrestrial animals experience enhanced distance-running capacity if they move intermittently (Full & Weinstein 1992, Weinstein & Full 1992, 2000, Adamczewska & Morris 1998), but quantifying intermittent locomotion is a challenge. Unlike steady-speed locomotion, intermittent locomotion is characterized by three variables: the durations of movements, the durations of the pauses of movements, and the speed of movements. Several questions remain unanswered. Do the lengths of pauses among movements in nature correspond to pauses that minimize the energetic cost of transport, or are species using pauses that are not energetically efficient but instead enhance some other physiological or behavioral function, such as lactate clearance or vigilance? Do species switch between intermittent and steady-speed locomotion in nature depending on the ecological context? A recent experiment with laboratory mice found that lines selectively bred for high levels of voluntary wheel running (Koteja et al. 1999) also exhibit more intermittent locomotion, as compared with unselected (random bred) control lines (Girard et al. 2001).

Most glaringly, no studies have examined the relationship between fitness and ecological function, despite the importance of this issue for evolutionary theory. For example, one could examine the relationships of laboratory performance, field performance, and fitness. The difficulties for these studies are measuring performance on the same individuals over time and accurately determining the fitness of free-living animals. It would be interesting to examine sympatric species that vary in the degree to which they use a performance capacity for a given task. One might predict that in the species that uses the performance capacity to the greatest extent, fitness is most closely correlated with performance, whereas in the species in which only a small fraction of their capacity is used, fitness will be unrelated to performance. However, empirical complications may occur if individuals with low performance abilities compensate for those low abilities, or if high performance has both positive and negative consequences. For example, a recent study of the lizard *Lacerta vivipara* found that individuals with low stamina, as measured in the lab at birth, tended to exhibit reduced activity in the field, lower growth rate, and higher parasite load, but apparently lower predation risk as assessed by tail losses (Clobert et al. 2000). Individuals with high stamina showed higher rates of activity in the field, higher growth rates, and lower parasite loads, but higher incidence of broken tails. Across all individuals, stamina at birth did not predict survivorship to the age of sexual maturity.

With the exception of those on diving physiology, few studies have followed the movement of individual animals over time and published complete histograms for whole-organism performance variables. Such histograms are invaluable because they provide a baseline for comparison to less common kinds of performance, such as when animals use locomotion to capture prey. Most studies provide only point estimates in which a performance variable is measured at a single point in time. This limitation results in part from the methods used to measure performance, such as radar guns for measuring speeds.

Finally, a potentially fruitful source of data is artificial enclosures, which provide much of the complexity of natural habitats, yet are simple enough that certain variables can be controlled (Watkins 1996, Losos et al. 2000). For instance, one might design an enclosure with artificial dowels of different diameters and inclines to investigate how these variables affect the speed of arboreal lizards. A crucial issue for such experiments is accurate replication of the scale and complexity of natural habitats, which is especially problematic for large animals that move long distances (e.g., many marine mammals). Consequently, these techniques will probably be most useful for small organisms that move short distances. Especially useful would be studies that combine data from artificial manipulations with field data on how natural environmental variation affects locomotor function.

ACKNOWLEDGMENTS

The authors' research is supported in part by NSF grants to D. J. I. (IBN-9983003) and T. G. (IBN-9723758, IBN-9728434). We thank I. A. Girard for assistance with literature.

Visit the Annual Reviews home page at www.AnnualReviews.org

LITERATURE CITED

Able KP. 1977. The flight behaviour of individual passerine nocturnal migrants: a tracking radar study. *Anim. Behav.* 25:924–35

Adamczewska AM, Morris S. 1998. Strategies for migration in the terrestrial Christmas Island red crab *Gecarcoidea natalis*: intermittent versus continuous locomotion. *J. Exp. Biol.* 201:3221–31

Aerts P, Van Damme R, Van Hooydonck B, Zaaf A, Herrel A. 2000. Lizard locomotion: how morphology meets ecology. *Neth. J. Zool.* 50:261–77

Alerstam T. 1975. Crane *Grus grus* migration over sea and land. *Ibis* 117:489–95

Alerstam T. 1987a. Bird migration across a strong magnetic analomy. *J. Exp. Biol.* 130:63–86

Alerstam T. 1987b. Radar observations of the stoop of the Peregrine Falcon *Falco peregrinus* and the Goshawk *Accipiter gentilis*. *Ibis* 129:267–73

Alerstam T, Gudmundsson GA, Larsson B. 1993. Flight tracks and speeds of Antarctic and Atlantic seabirds: radar and optical measurements. *Proc. R. Soc. London* B340:55–67

Altmann SA. 1998. *Foraging for Survival: Yearling Baboons in Africa*. Univ. Chicago Press. 608 pp.

Andrews RD, Jones DR, Williams JD, Thorson PH, Oliver GW, et al. 1997. Heart rates of northern elephant seals diving at sea and resting on the beach. *J. Exp. Biol.* 200:2083–95

Arnold SJ. 1983. Morphology, performance and fitness. *Am. Zool.* 23:347–61

Baker PS, Cooter RJ. 1979. The natural flight of the migratory locust, *Locusta migratoria* L. I. Wing movements. *J. Comp. Physiol.* 131:79–87

Baker PS, Gewecke M, Cooter RJ. 1981. The natural flight of the migratory locust, *Locusta migratoria* L. III. Wing-beat frequency, flight speed and altitude. *J. Comp. Physiol.* 141:233–37

Baudinette RV. 1991. The energetics and cardio respiratory correlates of mammalian terrestrial locomotion. *J. Exp. Biol.* 160:209–31

Bellrose FC. 1967. Radar in orientation research. *Proc. XIV Intern. Ornith. Congr.*, pp. 281–309

Bennett AF, Huey RB. 1990. Studying the evolution of physiological performance. *Oxford Surv. Evol. Biol.* 7:251–84

Blake RW, Kolotylo R, de la Cueva H. 1990. Flight speeds of the barn swallow, *Hirundo rustica*. *Can. J. Zool.* 68:1–5

Bloch R, Bruderer B. 1982. The air speed of migrating birds and its relationship to the wind. *Behav. Ecol. Soc.* 11:19–24

Blokpoel H. 1974. Migration of lesser snow and blue geese in spring across southern Manitoba. Part 1. Distribution, chronology, directions, numbers, heights, and speeds. *Can. Wildl. Serv. Rep. Ser.* No. 28

Boggs DF, Frappell PB. 2000. Unifying principles of locomotion: foreward. *Physiol. Biochem. Zool.* 73:647–50

Bonine KE, Garland T Jr. 1999. Sprint performance of phrynosomatid lizards, measured on a high-speed treadmill, correlates with hindlimb length. *J. Zool. London* 248:255–65

Bruderer B, Steidinger P. 1972. Methods of quantitative and qualitative analysis of bird migration with a tracking radar. *NASA Spec. Publ.* NASA SP-262, 151–67

Cartmill M. 1985. Climbing. In *Functional Vertebrate Morphology*, ed. M Hildebrand, DM Bramble, KF Liem, DB Wake, pp. 73–88. Cambridge: Belknap

Christian KA, Baudinette RV, Pamula Y. 1997. Energetic costs of activity by lizards in the field. *Funct. Ecol.* 11:392–97

Christian KA, Corbett L, Green B, Weavers B. 1995. Seasonal activity and energetics of two species of varanid lizards in tropical Australia. *Oecologia* 103:349–57

Clobert JA, Oppliger G, Sorci B, Ernande HG, Swallow, Garland T Jr. 2000. Trade-offs in phenotypic traits: endurance at birth, growth, survival, predation, and susceptibility to parasitism in a lizard, *Lacerta vivipara*. *Funct. Ecol.* 4:675–84

DeVries PJ, Dudley R. 1990. Morphometrics, airspeed, thermoregulation, and lipid reserves of migrating *Urania fulgens* (Uraniidae) moths in natural free flight. *Physiol. Zool.* 63:235–51

Dickinson MH, Farley CT, Full RJ, Koehl MAR, Kram R, Lehman S. 2000. How animals move: an integrative view. *Science* 288:100–6

Djawdan M. 1993. Locomotor performance of bipedal and quadrupedal heteromyid rodents. *Funct. Ecol.* 7:195–202

Djawdan M, Garland T Jr. 1988. Maximal running speeds of bipedal and quadrupedal rodents. *J. Mammal.* 69:765–72

Dolphin WF. 1987a. Dive behavior and estimated energy expenditure of foraging humpback whales in southeast Alaska. *Can. J. Zool.* 65:354–62

Dolphin WF. 1987b. Ventilation and dive patterns of humpback whales, *Megaptera novaeangliae*, on their Alaskan feeding grounds. *Can J. Zool.* 65:83–90

Drent J, van Marken Lichtenbelt WD, Wikelski M. 1999. Effects of foraging mode and season on the energetics of the Marine Iguana, *Amblyrhynchus cristatus*. *Funct. Ecol.* 13:493–99

Dudley R, DeVries PJ. 1990. Flight physiology of migrating *Urania fulgens* (Uraniidae) moths: kinematics and aerodynamics of natural free flight. *J. Comp. Physiol. A* 167:145–54

Dudley R, Srygley RB. 1994. Flight physiology of Neotropical butterflies: allometry of airspeeds during natural free flight. *J. Exp. Biol.* 191:125–39

Elsner R, Wartzok D, Sonafrank NB, Kelly BP. 1989. Behavioral and physiological reactions of arctic seals during under-ice pilotage. *Can. J. Zool.* 67:2506–13

Evans TR, Drickamer LC. 1994. Flight speeds of birds determined using Doppler radar. *Wilson Bull.* 106:154–56

Farley CT, Emshwiller M. 1996. Efficiency of uphill locomotion in nocturnal and diurnal lizards. *J. Exp. Biol.* 199:587–92

Fedak MA, Pullen MR, Kanwisher J. 1988. Circulatory responses of seals to periodic breathing: heart rate and breathing during exercise and diving in the laboratory and open sea. *Can. J. Zool.* 66:53–60

Feder ME, Bennett AF, Burggren WW, Huey RB. 1987. *New Directions in Evolutionary Physiology*. New York: Cambridge Univ. Press. 364 pp.

Feder ME, Bennett AF, Huey RB. 2000. Evolutionary physiology. *Annu. Rev. Ecol. Syst.* 31:315–41

Feldkamp SD, DeLong RL, Antonelis GA. 1989. Diving patterns of California sea lions, *Zalophus californianus*. *Can. J. Zool.* 67:872–83

Full RJ, Weinstein RB. 1992. Integrating the physiology, mechanics and behavior of rapid running ghost crabs: slow and steady doesn't always win the race. *Am. Zool.* 32:382–95

Gans C, Gaunt AS, Webb PW. 1997. Vertebrate locomotion. In *Handbook of Physiology*. Section 13. *Comparative Physiology*, ed. WH Dantzler, 1:55–213. New York: Oxford Univ. Press

Garland T Jr. 1983. Scaling the ecological cost of transport to body mass in terrestrial mammals. *Am. Nat.* 121:571–87

Garland T Jr. 1993. Laboratory endurance capacity predicts variation in field locomotor behaviour among lizard species. *Anim. Behav.* 57:77–83

Garland T Jr, Carter PA. 1994. Evolutionary physiology. *Annu. Rev. Physiol.* 56:579–621

Garland T Jr, Losos JB. 1994. Ecological morphology of locomotor performance in squamate reptiles. See Wainwright & Reilly 1994, pp. 240–302

Girard I. 2001. Field cost of activity in the kit fox, *Vulpes macrotis. Physiol. Biochem. Zool.* 74:191–202

Girard I, McAleer MA, Rhodes JS, Garland T Jr. 2001. Increased intermittency of locomotion in house mice selectively bred for high voluntary wheel running. *Am. Zool.* In press

Gorman ML, Mills MG, Raath JP, Speakman JR. 1998. High hunting costs make African wild dogs vulnerable to kleptoparasitism by hyaenas. *Nature* 391:479–81

Goszczynski J. 1986. Locomotor activity of terrestrial predators and its consequences. *Acta Theriol.* 31:79–95

Gudmundsson SA, Alerstam T, Larsson B. 1992. Radar observations of northbound migration of the arctic tern, *Sterna paradisaea*, at the Antarctic Peninsula. *Antarctic Sci.* 4:163–70

Hedenstrom A. 1995. Song flight performance in the skylark *Alauda arvensis. J. Avian. Biol.* 26:337–42

Hedenstrom A, Alerstam T. 1992. Climbing performance of migrating birds as a basis for estimating limits for fuel-carrying capacity and muscle work. *J. Exp. Biol.* 164:19–38

Hedenstrom A, Alerstam T. 1994a. Skylark optimal flight speeds for flying nowhere and somewhere. *Behav. Ecol.* 7:121–26

Hedenstrom A, Alerstam T. 1994b. Optimal climbing flight in migrating birds: predictions and observations of knots and turnstones. *Anim. Behav.* 48:47–54

Hedenstrom A, Alerstam T. 1995. Optimal flight speed of birds. *Philos. Trans. R. Soc. London Ser. B* 348:471–87

Hertz PE, Huey RB, Garland T Jr. 1988. Time budgets, thermoregulation, and maximal locomotor performance: Are reptiles Olympians or boy scouts? *Am. Zool.* 28:927–38

Hindell MA, Slip DJ, Burton HR, Bryden MM. 1992. Physiological implications of continuous prolonged, and deep dives of the Southern elephant seal (*Mirounga leonine*). *Can. J. Zool.* 70:370–79

Huey RB, Hertz PE. 1982. Effects of body size and slope on sprint speed of a lizard (*Stellio (Agama) stellio*). *J. Exp. Biol.* 97:401–9

Huey RB, Hertz PE. 1984. Effects of body size and slope on acceleration of a lizard (*Stellio stellio*). *J. Exp. Biol.* 110:113–23

Irschick DJ. 2000a. Comparative and behavioral analyses of preferred speed: *Anolis* lizards as a model system. *Physiol. Biochem. Zool.* 73:428–37

Irschick DJ. 2000b. Effects of behavior and ontogeny on the locomotor performance of a West Indian lizard *Anolis lineatopus. Funct. Ecol.* 14:438–44

Irschick DJ, Jayne BC. 1998. Effects of incline on acceleration, body posture, and hindlimb kinematics in two species of lizard, *Callisaurus draconoides* and *Uma scoparia. J. Exp. Biol.* 201:273–87

Irschick DJ, Jayne BC. 1999a. A field study of the effects of inclines on the escape locomotion of a bipedal lizard. *Physiol. Biochem. Zool.* 72:44–56

Irschick DJ, Jayne BC. 1999b. Comparative three-dimensional kinematics of the hindlimb for high-speed bipedal and quadrupedal locomotion of lizards. *J. Exp. Biol.* 202: 1047–65

Irschick DJ, Losos JB. 1998. A comparative analysis of the ecological significance of locomotor performance in Caribbean *Anolis* Lizards. *Evolution* 52:219–26

Irschick DJ, Losos JB. 1999. Do lizards avoid habitats in which performance is submaximal? The relationship between sprinting capabilities and structural habitat use in Caribbean anoles. *Am. Nat.* 154:293–305

Jackman TR, Larson A, de Queiroz K, Losos JB. 1999. Phylogenetic relationships and tempo of early diversification in *Anolis* lizards. *Syst. Biol.* 48:254–85

Jayne BC, Ellis RV. 1998. How inclines affect the escape behaviour of a dune dwelling lizard, Uma scoparia. *Anim. Behav.* 55:1115–30

Jayne BC, Irschick DJ. 2000. A field study of incline use and preferred speeds for the locomotion of lizards. *Ecology* 81:2969–83

Karasov WH. 1992. Daily energy expenditure and the cost of activity in mammals. *Am. Zool.* 32:238–48

Karasov WH, Anderson RA. 1984. Interhabitat differences in energy acquisition and expenditure in a lizard. *Ecology* 65:235–47

Kenagy GJ, Hoyt DF. 1989. Speed and time-energy budget for locomotion in golden-mantled ground squirrels. *Ecology* 70:1834–39

Kooyman GL, Castellini MA, Davis RW, Maue RA. 1983. Aerobic diving limits of immature Weddell seals. *J. Comp. Phys.* 151:171–74

Kooyman GL, Wahrenbrock EA, Castellini MA, Davis RW, Sinnett EE. 1980. Aerobic and anaroebic metabolism during voluntary diving in Weddell seals: evidence of preferred pathways from blood chemistry and behavior. *J. Comp. Physiol.* 138:335–46

Koteja P, Swallow JG, Carter PA, Garland T Jr. 1999. Energy cost of wheel running in house mice: implications for coadaptation of locomotion and energy budgets. *Physiol. Biochem. Zool.* 72:238–49

Larkin RP. 1980. Transoceanic bird migration: evidence for detection of wind direction. *Behav. Ecol. Sociobiol.* 6:229–32

Larkin RP, Thompson D. 1980. Flight speeds of birds observed with radar: evidence of two phases of migratory flight. *Behav. Ecol. Sociobiol.* 7:301–17

Le Boeuf BJ, Costa DP, Huntley AC, Fedlkamp SD. 1988. Continuous, deep diving in female northern elephant seals, *Mirounga angusitrostris. Can. J. Zool.* 66:446–58

Le Boeuf BJ, Morris PA, Blackwell SB, Crocker DE, Costa DP. 1996. Diving behavior of juvenile northern elephant seals. *Can. J. Zool.* 74:1632–44

Lichtenbelt WDVM, Wesselingh RA, Vogel JT, Albers KBM. 1993. Energy budgets in free-living green iguanas in a seasonal environment. *Ecology* 74:1157–72

Lima SL, Dill LM. 1990. Behavioral decisions made under the risk of predation: a review and prospectus. *Can. J. Zool.* 68:619–39

Losos JB. 1990a. Concordant evolution of locomotor behavior, display rate and morphology in *Anolis* lizards. *Anim. Behav.* 39:879–90

Losos JB. 1990b. Ecomorphology, performance capability, and scaling of West Indian Anolis lizards: an evolutionary analysis. *Ecol. Monogr.* 60:369–88

Losos JB, Creer DA, Glossip D, Goellner R, Hampton A, et al. 2000. Evolutionary implications of phenotypic plasticity in the hindlimb of the lizard *Anolis sagrei. Evolution* 54:301–5

Losos JB, Jackman TR, Larson A, de Queiroz K, Rodríguez-Schettino L. 1998. Historical contingency and determinism in replicated adaptive radiations of island lizards. *Science* 279:2115–18

Losos JB, Sinervo B. 1989. The effect of morphology and perch diameter on sprint performance of *Anolis* lizards. *J. Exp. Biol.* 145:23–30

Lydersen C, Kovacs KM. 1993. Diving behaviour of lactating harp seal, *Phoca groenlandica*, females from the Gulf of St. Lawrence, Canada. *Anim. Behav.* 46:1213–21

Macrini TE, Irschick DJ. 1998. An intraspecific analysis of trade-offs in sprinting performance in a West Indian lizard (*Anolis lineatopus*). *Biol. J. Linn. Soc.* 63:579–91

Martin AR, Kingsley MCS, Ramsay MA. 1994. Diving behaviour of narwhals (*Monodon monoceros*) on their summer grounds. *Can. J. Zool.* 72:118–25

McGowan JA. 1984. The California El Niño, 1983. *Oceanus* 27:48–51

McGraw JB, Caswell H. 1996. Estimation of individual fitness from life-history data. *Am. Nat.* 147:47–64

McLaughlin RL, Montgomerie RD. 1985. Flight speeds of central place foragers:

female Lapland longspurs feeding nestlings. *Anim. Behav.* 33:810–16

McLaughlin RL, Montgomerie RD. 1990. Flight speeds of parent birds feeding nestlings: maximization of foraging efficiency or food delivery rate? *Can. J. Zool.* 68:2269–74

Merker GP, Nagy KA. 1984. Energy utilization by free-ranging *Sceloporus virgatus* lizards. *Ecology* 65:575–81

Miles DB. 1994. Population differentiation in locomotor performance and the potential response of a terrestrial organism to global environmental change. *Am. Zool.* 34:422–36

Nagy KA, Girard IA, Brown TK. 1999. Energetics of free-ranging mammals, reptiles, and birds. *Annu. Rev. Nutr.* 19:247–77

Nagy KA, Huey RB, Bennett AF. 1984. Field energetics and foraging mode of Kalahari lacertid lizards. *Ecology* 65:588–96

Norberg RA. 1981. Optimal flight speeds in birds when feeding young. *J. Anim. Ecol.* 50:473–77

Packard GC, Boardman TJ. 1999. The use of percentages and size-specific indices to normalize physiological data for variation in body size: wasted time, wasted effort? *Comp. Biochem. Physiol.* 122A:37–44

Parslow JLF. 1969. The migration of passerine night migrants across the English channel studied by radar. *Ibis* 111:48–79

Pennycuick CJ. 1975. Mechanics of flight. In *Avian Biology*, ed. DS Farner, JR King, pp. 1–75. New York: Academic

Pennycuick CJ. 1982a. The ornithodolite: an instrument for collecting large samples of bird speed measurements. *Philos. Trans. R. Soc. London Ser. B* 300:61–73

Pennycuick CJ. 1982b. The flight of petrels and albatrosses (Procellariiformes), observed in South Georgia and its vicinity. *Philos. Trans. R. Soc. London Ser. B* 300:75–106

Pennycuick CJ. 1987. Flight of auks (Alcidae) and other northern seabirds compared with southern Prolellariiformes: Orithodolite observations. *J. Exp. Biol.* 128:335–47

Pennycuick CJ. 1997. Actual and 'optimum' flight speeds: field data reassessed. *J. Exp. Biol.* 200:2355–61

Pennycuick CJ, Fuller MR, Oar JJ, Kirkpatrick SJ. 1994. Falcon versus grouse: flight adaptations of a predator and its prey. *J. Avian Biol.* 25:39–49

Perry G. 1999. The evolution of search modes: ecological versus phylogenetic perspectives. *Am. Nat.* 153:98–109

Piersma T, Hedenstrom A, Bruggemann HJ. 1997. Climb and flight speeds of shorebirds embarking on an intercontinental flight: Do they achieve the predicted optimal behaviour? *Ibis* 139:299–304

Pike GC. 1962. Migration and feeding of the Gray whale (*Eschrichtius robustus*). *J. Fish. Res. Board Can.* 19:815–38

Ponganis PJ, Kooyman GL, Winter LM, Starke LN. 1997. Heart rate and plasma lactate responses during submerged swimming and trained diving in California sea lions, *Zalophus californianus. J. Comp. Physiol.* 167:9–16

Ponganis PJ, Ponganis EP, Ponganis KV, Kooyman GL, Gentry RL, Trillmich F. 1990. Swimming velocities in otariids. *Can. J. Zool.* 68:2105–12

Pough FH. 1989. Organismal performance and Darwinian fitness: approaches and interpretations. *Physiol. Zool.* 62:199–236

Rice DW, Wolman AA. 1971. The life history and ecology of the Gray whale (*Eschrichtius robustus*). *Am. Soc. Mamm. Sp. Pub. No.* 3:1–21

Ridgway SH, Bowers CA, Miller D, Schultz ML, Jacobs CA, Dooley CA. 1984. Diving and blood oxygen in the white whale. *Can. J. Zool.* 62:2349–51

Ridgway SH, Scronce BL, Kanwisher J. 1969. Respiration and deep diving in the bottlenose dolphin. *Science* 166:1651–54

Roughgarden J. 1995. *Anolis Lizards of the Caribbean: Ecology, Evolution, and Plate Tectonics.* New York: Oxford Univ. Press

Schnell GD. 1965. Recording the flight speeds of birds by Doppler radar. *Living Bird* 4:79–87

Schnell GD, Hellack JJ. 1979. Bird flight

speeds in nature: optimized or a compromise? *Am. Nat.* 113:53–66

Shaffer SA, Costa DP, Williams TM, Ridgway SH. 1997. Diving and swimming performance of white whales, *Delphinapterus leucas*: an assessment of plasma lactate and blood gas levels and respiratory rates. *J. Exp. Biol.* 200:3091–99

Snyder RC. 1962. Adaptations for bipedal locomotion of lizards. *Am. Zool.* 2:191–203

Srygley RB, Dudley R. 1993. Correlations with the position of center of body mass with butterfly escape tactics. *J. Exp. Biol.* 174:155–66

Srygley RB, Oliveira EG, Dudley R. 1996. Wind drift compensation, flyways, and conservation of diurnal, migrant neotropical Lepidoptera. *Proc. R. Soc. London* 263: 1351–57

Stamps J. 1995. Motor learning and the value of familiar space. *Am. Nat.* 146:41–58

Stebbins RC. 1944. Some aspects of the ecology of the iguana genus *Uma. Ecol. Monogr.* 14:313–32

Steudel K. 2000. The physiology and energetics of movement: effects on individuals and groups. In *On the Move: How and Why Animals Travel in Groups*, ed. S Boinski, PA Garber, pp. 9–23. Chicago: Univ. Chicago Press

Sumich JL. 1983. Swimming velocities, breathing patterns, and estimated costs of locomotion in migrating gray whales, *Eschrichtius robustus. Can. J. Zool.* 61:647–52

Taylor CR, Caldwell SL, Rowntree VJ. 1972. Running up and down hills: some consequences of size. *Science* 178:1096–97

Taylor CR, Weibel ER, Karas RH, Hoppeler H. 1987. Adaptive variation in the mammalian respiratory structure in relation to energetic demand. VIII. Structural and functional design principles determining the limits to oxidative metabolism. *Respir. Physiol.* 69:117–27

Thompson D, Fedak MA. 1993. Cardiac responses of gray seals during diving at sea. *J. Exp. Biol.* 174:139–64

Thompson MC. 1961. The flight speed of a red-breasted merganser. *Condor* 63:265

Tucker VA. 1974. Energetics of natural avian flight. In *Avian Energetics*, ed. RA Paynter, pp. 298–333. Cambridge: Nuttall Ornithol. Club

Tucker VA. 1998. Gliding flight: speed and acceleration of ideal falcons during diving and pull out. *J. Exp. Biol.* 201:403–14

Tucker VA, Cade T, Tucker AE. 1998. Diving speeds and angles of a gyrfalcon (*Falco rusticolus*). *J. Exp. Biol.* 201:2061–70

Tucker VA, Schmidt-Koeing K. 1971. Flight speeds of birds in relation to energetics and wind directions. *Auk* 88:97–107

Turchin P. 1998. *Quantitative Analysis of Movement: Measuring and Modeling Population Redistribution in Animals and Plants.* Sunderland, MA: Sinauer

Van Damme R, Van Dooren TJM. 1999. Absolute versus per unit body length speed of prey as an estimator of vulnerability to predation. *Anim. Behav.* 57:347–52

Van Hooydonck B, Van Damme R, Aerts P. 2001. Speed and stamina trade off in lacertid lizards. *Evolution.* In press

Van Hooydonck B, Van Damme R. 1999. Evolutionary relationships between body shape and habitat use in lacertid lizards. *Evol. Ecol. Res.* 1:785–805

Vitt LJ, Pianka ER. 1994. *Lizard Ecology: Historical and Experimental Perspectives.* Princeton, NJ: Princeton Univ. Press

Wainwright PC. 1994. Functional morphology as a tool in ecological research. In *Ecological Morphology: Integrative Organismal Biology*, ed. PC Wainwright, SM Reilly, pp. 42–59. Chicago: Univ. Chicago Press

Waloff Z. 1972. Observations on the airspeeds of freely flying locusts. *Anim. Behav.* 20:367–72

Watkins TB. 1996. Predator-mediated selection on burst swimming performance in tadpoles of the Pacific tree frog, *Pseudacris regilla. Physiol. Zool.* 69:154–67

Wege ML, Raveling DG. 1984. Flight speed and directional responses to wind by migrating Canada Geese. *Auk* 101:342–48

Weinstein RB, Full RJ. 1992. Intermittent exercise alters endurance in an eight-legged ectotherm. *Am. J. Physiol.* 262 (*Reg. Int. Comp. Physiol.* 31):R852–59

Weinstein RB, Full RJ. 2000. Intermittent locomotion increases endurance in a gecko. *Physiol. Biochem. Zool.* 72:732–39

Weis-Fogh T. 1956. Biology and physics of locust flight. II. Flight performance of the desert locust (*Schistocerca gregaria*). *Philos. Trans. R. Soc. London. Ser. B* 239:459–510

Welham CVJ. 1994. Flight speeds of migrating birds: a test of maximum range speed predictions from three aerodynamic equations. *Behav. Ecol.* 5:1–8

Williams EE. 1972. The origin of faunas. Evolution of lizard congeners in a complex island fauna: a trial analysis. *Evol. Biol.* 6:47–89

Williams TM, Haun JE, Friedl WA. 1999. The diving physiology of bottlenose dolphins (*Tursiops truncates*). I. Balancing the demands of exercise for energy conservation at depth. *J. Exp. Biol.* 202:2739–48

Annu. Rev. Ecol. Syst. 2001. 32:397–414
Copyright © 2001 by Annual Reviews. All rights reserved

THE SOUTHERN CONIFER FAMILY ARAUCARIACEAE: History, Status, and Value for Paleoenvironmental Reconstruction

Peter Kershaw and Barbara Wagstaff

Centre for Palynology and Palaeoecology, School of Geography and Environmental Science, PO Box 11a, Monash University, Victoria 3800, Australia;
e-mail: peter.kershaw@arts.monash.edu.au, barbara.wagstaff@arts.monash.edu.au

Key Words biogeography, vegetation history, climate change, environmental variability

■ **Abstract** The Araucariaceae are important to biogeography because they have an ancient origin and are a distinctive and sometimes dominant component of southern hemisphere forest communities. This paper examines recent information on ecology and phylogeny and on pollen and macrofossil assemblages to assess the history and present-day status of the family and its potential for refinement of past environmental, particularly climatic, conditions. From an origin in the Triassic, the family expanded and diversified in both hemispheres in the Jurassic and Early Cretaceous and remained a significant component of Gondwanan vegetation until the latter part of the Cenozoic. The development of angiosperms in the Middle Cretaceous probably assisted in the demise of some araucarian components but there was also evolution of new genera. Recorded diversity in the early Cenozoic of Australia is as high as it was in the Early Cretaceous. Continental separation and associated climatic drying, cooling, and increased variability progressively reduced the ranges of conifers to moist, predominantly mesothermal climates on continents. However, tectonic and volcanic activity, partially associated with Australia's collision with Southeast Asia, provided new opportunities for some araucarian components on Asia-Pacific islands. Araucarians provide information on climatic conditions suitable for rainforest vegetation throughout their recorded period, even prior to the recognition or even existence of these forests in the fossil record. High pollen abundance is also indicative of marginal rainforest environments where these canopy emergents can compete effectively with angiosperm forest taxa. Despite their apparent relictual status in many areas, they provide precise paleoclimatic estimates in late Quaternary pollen records and have particular value in providing evidence of climatic variability that has otherwise been difficult to detect.

INTRODUCTION

van Steenis (1971) established *Nothofagus* as the focus of study of southern hemisphere biogeography and paleoenvironments at a time when continental drift, revitalized as plate tectonics, was becoming rapidly accepted as a basic theory in the earth sciences (Le Grand 1988). Since then, a plethora of papers has appeared, not only on the ecology and biogeography of the genus, but also on its role in cool temperate rainforest dynamics, in quantitative paleoclimatic reconstructions, and in the development of methods in biogeographic study. Much of this research is presented in the recent compilation of Veblin et al. (1996). However, there is still no consensus over the place and time of origin of the genus, its ancestral source, or the subsequent biogeographic histories of its subgenera. These uncertainties, combined with its restricted climatic range and relatively recent history in geological terms means that *Nothofagus* has limited scope for elucidation of past climates and changing biogeographies over much of the hemisphere.

Araucariaceae has potential to rival or at least complement *Nothofagus* in elucidating patterns of vegetation, climate, and tectonic change in the southern hemisphere. Araucariaceae, together with Podocarpaceae and some genera of Cupressaceae, make up the southern conifers that have a history extending back to the middle or early Mesozoic Period. Both Araucariaceae and Podocarpaceae have good fossil records. Although the latter has a broader geographic range and a greater number of genera with often distinctive morphological characters that preserve in fossil material, Araucariaceae possesses a variety of features that make it an appealing subject for study.

Most members of the family have an impressive emergent habit or distinctive form that ensures their distribution is relatively well known. The family also has a fossil record composed of both macrofossils and pollen, with pollen being abundant in association with parent plants but, unlike many podocarp and *Nothofagus* species, generally not widely dispersed beyond source vegetation, except perhaps by water.

Like *Nothofagus*, araucarians are almost entirely restricted to rainforest, apart from occurrences in the unique sclerophyllous maquis vegetation on ultramafic substrates in New Caledonia (Jaffré 1995), and tend to be most common at the margins of more complex forest types. Consequently, they indicate the presence or extent of rainforest vegetation in fossil records where a majority of taxa, being angiosperms, have limited pollen representation or uncertain affinities. Being ancient, there is also the possibility that they can be useful in the identification of environments suitable for rainforest before rainforests, broadly recognized today as communities with a continuous tree canopy generally dominated by angiosperms, evolved.

Although distributed in association with rainforest, their emergent habit sets araucarians apart from the rainforest canopy to the extent that some vegetation classifications (e.g., Webb 1959) have typed communities with common araucarian emergents as woodlands rather than forests. The distinction between some araucarian emergent species and associated rainforest has also been recognized in studies on forest basal area, where it has been found that the forest canopy trees have a constant basal area regardless of the presence or absence of araucarian emergents (Enright 1982, Ogden 1985). This "additive basal area" phenomenon (Enright & Ogden 1995) could suggest that some araucarians are distributed independently of general forest composition or structure and provide different or additional information on environmental change. Consequently, this attribute, combined with their exposure to the atmosphere above the forest canopy, may make araucarians ideal indicators of regional climatic conditions. On the other hand, their frequent inability to regenerate under a dense canopy, in the absence of disturbance, means they may not reflect their full potential climate range (Enright & Ogden 1995).

More general concern about the status of Araucaricaeae, along with other conifers, has been expressed from a consideration of both present-day age class distributions and fossil studies. A number of studies have revealed that many southern conifers do not tend to conform to the reverse J-curve characteristic of continuous population recruitment but have an over-representation of larger size classes. This feature suggests a general demise that has variously been attributed to disequilibrium with the present-day climate, alteration of angiosperm/conifer balance with the extinction of major herbivores (New Zealand), and the continuation of a long-term trend towards angiosperm dominance (Enright & Ogden 1995). However, recent research (much of it summarized in Enright & Hill 1995) has shown that the predominant regeneration strategy of southern conifers is intermittent recruitment resulting from occasional disturbances, a strategy consistent with the general shade tolerance and longevity of araucarians. Consequently, fossil or tree ring records, extending beyond the time-span of ecological studies, are required to test the stability and viability of many present-day conifer populations.

The long-term fossil record shows that there has been a relative decrease in conifers generally since the angiosperms evolved (e.g., Lidgard & Crane 1990). Regal (1977) provided a comprehensive and convincing explanation for this decline based on the greater flexibility of angiosperms with respect to reproductive strategies, including the use of animals for pollination and seed dispersal, and on their high growth rates and chemical defenses against herbivory. Populations of conifers are seen as progressively isolated, a situation common in the Araucariaceae. However, the fossil record also indicates that the decline in abundance does not seem to be matched by a decrease in diversity (Lidgard & Crane 1990).

Despite the uncertainty about the degree to which araucarian distributions relate to their potential climate ranges, component taxa have been used to reconstruct or refine paleoclimatic estimates from pollen data, especially for the late Quaternary. We examine the present-day distributions and fossil history of the Araucariaceae in relation to aspects of their ecology and phylogeny to assess the validity of climatic estimates made and to examine the value of these conifers for paleoenvironmental, particularly paleoclimatic, reconstruction.

PRESENT-DAY DISTRIBUTIONS

Members of the Araucariaceae are now restricted to the South American and Southwest Asia–Western Pacific region (Figure 1) despite their extensive distribution in both hemispheres during the Mesozoic (Stockey 1990). A relictual status is also suggested by generally low species diversity and disjunct distributions on continental masses. However, relatively high diversity is recorded on islands in the Pacific and the Southeast Asian region, indicating conditions suitable for survival and continuous evolution through a long period of geological time or more recent dispersal and rapid evolution from continental sources. A long residence time may be inferred for New Caledonia and New Guinea, which were part of the southern megacontinent Gondwana during the Mesozoic Period. Recent dispersal is more likely on some relatively recently formed Southeast Asian and Western Pacific islands.

Greater insights into present distributions can be gleaned from inferred taxonomic relationships within the Araucariaceae (Setoguchi et al. 1998) (Figure 2). The family contains two major genera, *Araucaria* and *Agathis*, together with the recently discovered monotypic *Wollemia*, recorded only from a few small patches in protected gorges in southeastern Australia (Jones et al. 1995). *Wollemia nobilis* is clearly relictual and, phylogenetically, may be considered the most basal extant taxon in the family. *Araucaria* and *Agathis* are sister taxa on the basis of phylogenies derived from *rbcL* data, but extant representatives of the former show

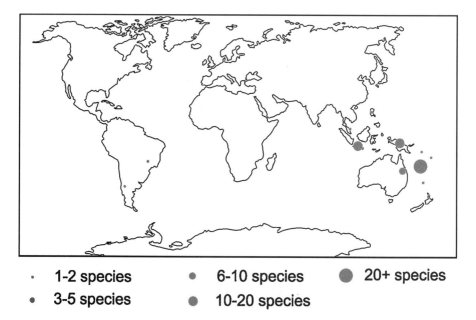

| · | **1-2 species** | ● | **6-10 species** | ⬤ | **20+ species** |
| ● | **3-5 species** | ● | **10-20 species** | | |

Figure 1 A generalized, global representation of species of Araucariaceae (data from Enright & Hill 1995).

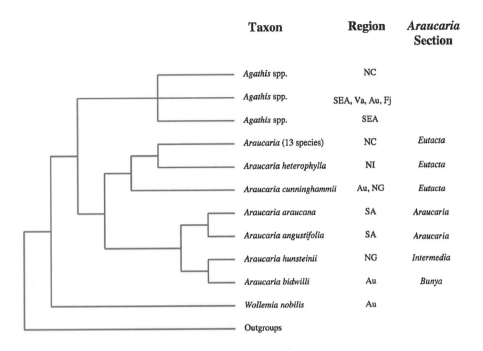

Figure 2   Phylogenetic relationships within the Araucariaceae inferred from *rbcL* gene sequences (adapted from Setoguchi et al. 1998). Geographical region abbreviations: NC, New Caledonia; SEA, Southeast Asia; Va, Vanuatu; Au, Australia; Fj, Fiji; NI, Norfolk Island; NG, New Guinea; SA, South America).

greater genetic and morphological differentiation at the species and section levels and greater geographic spread.

Within *Araucaria*, *Eutacta* has a broad distribution on Gondwanan terranes, including Australia, New Guinea, Norfolk Island, and New Caledonia, suggesting considerable geological antiquity for this section. The high diversity of species on New Caledonia has been presumed to be the result of diversification from ancestral Gondwanan stock during the Eocene Period when much of the island was covered in ultramafic rocks (Jaffré 1995). However, there is some geological evidence that the island was below sea level in the early Cenozoic (R. Hall, personal communication), and long-distance transport from Australia following re-emergence, uplift, and emplacement of ultramafic rocks during the Eocene (Aitchison et al. 1995) may have to be invoked to explain its presence. The isolation of the section *Araucaria* in South America may indicate divergence at this level since the effective environmental and physical separation of the South American and Australasian regions, probably in the early Cenozoic. The monotypic sections *Bunya* and *Intermedia* represented in Australia and New Guinea, respectively, are considered to have diverged relatively recently, perhaps in the later Cenozoic, when the development of a dry corridor effectively separated the two islands.

Agathis, unlike *Araucaria*, extends into the Southeast Asian islands and Malay Peninsula, leaving open the possibility that it had a Laurasian source, or was re-cycled into Gondwanan territory by India as it collided with this region in the Eocene Period (Morley 1998). However, the confinement of *Agathis* to the south-west Pacific region, combined with small morphological and genetic differentia-tion compared with *Araucaria*, strongly favors an Australasian source. Dispersal and diversification may have taken place since the Eocene in the case of New Caledonia, and perhaps also New Zealand if Pole's (1994) hypothesis is correct (i.e., that New Zealand became submerged in the Cenozoic and all extant taxa are derived from long-distance dispersal). Dispersal into Southeast Asia would most likely have occurred since the Early Miocene as the northward-moving Australian continent came into contact with this region.

Overall, the evidence is not consistent with a gradual decline in Araucariaceae as a result of competition with evolving angiosperms. Most taxa appear to have ori-gins postdating the Mesozoic and early Cenozoic Period of continental drift, with long-distance dispersal and some probable evolutionary radiations in the middle to late Cenozoic. The islands, rather than the major continental masses, demonstrate this latest phase of activity, presumably because of the ability of Araucariaceae to colonize new substrates and to compete effectively under conditions of regular disturbance. Such disturbances characterize islands, especially those that experi-ence volcanism and tectonism. The islands may also have been buffered against climatic variability that affected the continents.

Despite the unusual present-day distribution pattern of the Araucariaceae, there is some consistency in representation in relation to climatic conditions, with a concentration in lower mid-latitudes. On the continental masses of South America and Australia, *Araucaria* occurs predominantly in the subtropics, extending into the marginal tropics (Figure 1). *Agathis australis*, the only extant araucarian in New Zealand, is restricted to the "subtropical" tip of the North Island. The majority of taxa, occurring on New Caledonia, are within the tropics. A mesothermal climate is also indicated for many taxa within the equatorial region of New Guinea and Southeast Asia, because these are distributed in the lower montane zone. Major exceptions to this distribution are the *Agathis* species that occur in equatorial peat swamps, where perhaps substrates with low nutrients and poor drainage allow them to be competitive with angiosperms, and *Araucaria araucana* in Chile that extends to the treeline. However, *A. araucana* does not extend into the cool temperate forests of South America that support the majority of southern conifers on this continent.

THE FOSSIL RECORD

Overview

The representation of fossil data is patchy and sparse for most parts of the world. Only in Australia is there a substantial record of both macrofossils and pollen; as this is the only landmass to have extant representatives of all three genera, discussion focuses on this continent. Figure 3 (see color insert) summarizes the evidence from Australia, in the form of stratigraphic ranges, macrofossil species records from the Cretaceous and Cenozoic periods, and averaged pollen percentages, shown against major environmental variables and events. Figure 4 (see color insert) shows more detailed pollen site records.

The early part of the record is characterized by uncertainty, both in what constitutes a species or even an araucarian. The earliest unequivocal Araucariaceae date from the Jurassic (Stockey 1982), and macrofossils include *Araucaria* sections Eutacta and Bunya as well as indeterminate *Araucaria* types (Hill & Brodribb 1999) and possibly other genera, although pollen grains putatively attributable to the family are recorded from the Early Triassic in Australia (De Jersey 1968).

During the Jurassic and Cretaceous *Araucaria* sections *Eutacta* and *Bunya* appear to have been widespread in both hemispheres. *Araucaria* section *Araucaria* has been recorded occasionally from the Early Cretaceous of southeastern Australia as well as from southern South America, where it currently lives (Hill & Brodribb 1999), whereas section *Intermedia* has been recorded from the Early Cenozoic of New Zealand (Pole 1994). Macrofossils attributed to *Agathis* have been recorded from the Jurassic and Cretaceous periods, although these lack sufficient diagnostic characters to allow confident identification (Hill & Brodribb 1999), and the oldest substantiated *Agathis* fossils date only from the Middle Eocene of southeastern Australia (Carpenter & Pole 1995). Unfortunately, the pollen record does not assist in determination of the history of *Agathis*, as the grains are morphologically similar to those of *Araucaria* and it has proved impossible, at least with older taxa, to separate the two genera. By contrast, *Wollemia* has distinctive pollen that is first recorded in the Turonian of southeastern Australia and subsequently in Antarctica and perhaps New Zealand (Macphail et al. 1995). However, Chambers et al. (1998) have suggested that some Early to Middle Cretaceous macrofossils may also be related to *Wollemia*.

The Araucariaceae appear to be one group of plants that evolved after the major extinction phase at the end of the Permian, prompted by arid conditions throughout much of Gondwana (Hill et al. 2000). The earliest pollen records of Araucariaceae, in the Triassic, are found in coastal Queensland, perhaps only coincidentally within the present limited geographical range of the family in Australia. They increased in abundance and diversity during the Jurassic, with high pollen values in both eastern and particularly western Australia peaking in the Late Jurassic. It is likely that rising sea levels through the Jurassic corresponded to both increased precipitation and temperature that facilitated the development of forest vegetation. However, throughout the period of evolution of the bulk of Gondwanan vegetation (the Jurassic and particularly the Cretaceous and early Cenozoic), Australia was at much higher latitudes than it is today. Consequently, light would have been a major limiting factor, placing constraints on forest development. The emergent habit could have imparted a great advantage on conifers through interception of high angle solar radiation. Another pertinent feature of Australia during the Jurassic and Early Cretaceous was that the continent was rotated with the west at lowest latitudes. The high araucarian pollen percentages through this period within the western part of the continent may indicate an early preference of the family for these lower latitudes where independent evidence suggests climates were relatively warm and dry (McLoughlin & Hill 1996).

The initiation of continental rifting toward the end of the Jurassic resulted in the exclusion of India and southern Africa from subsequent changes in Gondwanan vegetation that included a major phase of modernization. The beginning of the Cretaceous saw perhaps the first appearance of *Wollemia*-related taxa and at least four times the number of *Araucaria* species currently represented on the continent. By contrast, there are few araucarian macrofossils recorded from the Middle to Late Cretaceous, and there is a corresponding fall in araucarian pollen abundance. However, Dettmann (1994) suggested that both Araucariaceae and Podocarpaceae were major canopy dominants through most of the Cretaceous, with a reduction in Araucariaceae toward the very end of this period. Owing to the emergence of most of the continent above sea level, fossil evidence is largely restricted to southeastern Australia. It is possible that araucarians flourished through the Middle-Late Cretaceous in the northern part of Australia, as there are high pollen percentages in the Albian and Cenomanian when fossils are preserved.

Relatively low pollen percentages for Araucariaceae in the southern part of Australia may have been a result of competition with evolving angiosperms. However, high precipitation may have been the critical factor, at least until the end of the Cretaceous when *Nothofagus*, which produces copious pollen, would have had an additional affect on reducing araucarian pollen representation. It is unfortunate that there is little quantitative information from sites in southeastern Australia within the unstable area of continental rifting, for it is considered to have been a major center of evolution of the extant Australian flora (Dettmann 1989). This period of angiosperm development appears to have included the evolution of *Wollemia* species closely related to *W. nobilis* and also to many extant podocarp genera, questioning the idea that the conifers were outcompeted by angiosperms. However, there were extinctions within the Araucariaceae, as indicated on Figure 3 (see color insert), that would have contributed to reduced pollen percentages.

There is some increase in Araucariaceae pollen percentages in the Palaeocene and early Eocene, with the relatively high percentage in the latter epoch derived from *Wollemia* (Macphail et al. 1994). However, macrofossil evidence is slight. There is then an increase in macrofossil records, demonstrating high diversity in the Middle Eocene and Early Oligocene. These peaks in diversity are perhaps surprising, as diversity of angiosperms was also high, with complex rainforest covering much of the continent. Pollen percentages remain low. It is likely that many Araucariaceae were restricted to unstable environments such as river margins suitable for macrofossil preservation. In addition, much of the diversity was contained within *Agathis*, which seldom achieves high population densities and the pollen of which is seldom recorded in abundance. Conditions were clearly unsuitable for the development of *Araucaria* forests.

Diversity, initially in *Agathis* spp., decreased from the Late Oligocene and especially the Early Miocene owing, according to Hill & Brodribb (1999), to the onset of continental drying. However, from studies of pollen and lithotype variation in coal deposits of southeastern Australia, increased climatic variability may have been the critical factor (Kershaw et al. 1991). There was also a global trend toward cooler conditions at high latitudes that may have contributed to the elimination of *Agathis* from southeastern Australia, where all macrofossil records are located, and its contraction to the northeastern part of the continent. Although the movement of Australia northward into lower latitudes moderated this temperature decrease, as illustrated by temperature estimates derived from fossil assemblage data (Figure 3, color insert), *Agathis* figures prominently in determination of the lower limit of a number of temperature parameter estimates derived for the Latrobe Valley in southeastern Australia during the Early Miocene period. Consequently, any subsequent reduction in temperature is likely to have eliminated the genus. By contrast, there is an increase in representation of Araucariaceae pollen. This increase is consistent with drier or more variable climates, as regeneration in some species of *Araucaria* is facilitated by a more open rainforest canopy that is promoted by such conditions. In fact, it is *Araucaria* that defines the upper limit of the estimates of annual, wettest month and driest month rainfall in the Latrobe Valley during the Early Miocene, consistent with both a reduction in rainfall and an increase in seasonal rainfall distribution from the Cenozoic climatic "optimum." We propose that a widespread emergent *Araucaria* layer, analogous to that in the Middle Jurassic to Early Cretaceous, re-formed, perhaps suggesting a return to similar climatic conditions.

The macrofossil record ends in the Late Miocene. It is likely that expansion of the Antarctic ice sheet to close to present dimensions around the end of the Miocene and the intensification of oceanic and atmospheric circulation patterns (Bowler 1982) produced cooler and drier conditions over southeastern Australia that caused a decline in conifer diversity. However, the lack of macrofossil records is as much related to unfavorable conditions for preservation as it is to conifer representation, as the Araucariaceae achieve their highest Cenozoic pollen percentages in the Early Pliocene, although these high percentages are virtually restricted to coastal and subcoastal locations along the eastern seaboard. In southeastern Australia the distributional change in high pollen percentages from more inland to more coastal locations from the Miocene to the Early Pliocene plots the drier margin of rainforest as it contracted to present high rainfall areas along the highland coastal fringe.

The subsequent contraction of the Araucariaceae to the northeastern part of the continent in the Late Pliocene to Pleistocene could be a result of the intensification of the westerly wind system that extended over southeastern Australia (Bowler 1982). This system subjected the conifers, possibly for the first time, to a winter rainfall regime, or to the onset of Quaternary glacial/interglacial scale climatic oscillations (Shackleton et al. 1995). Increased global climatic variability in itself was probably not the major factor, as high araucarian pollen abundance values are maintained in the northeastern part of the continent.

The ability of species of Araucariaceae to survive for long periods under adverse climatic conditions in southeastern Australia is demonstrated by continued low percentages of *Araucaria* pollen in a record from the western plains of Victoria through the whole of the Early Pleistocene (Wagstaff et al. 2001) and by the extant small stands of *Wollemia* within protected gorges in New South Wales 2 million years after the last appearance of pollen in fossil records (Macphail et al. 1995). However, it is likely that the increased amplitude of climatic oscillations that became established around the Early-Middle Pleistocene boundary (Shackleton et al. 1995) resulted in the final demise of *Araucaria* within southeastern Australia.

In general terms, this Late Cenozoic pattern of decline in Araucariaceae is also registered in New Zealand and South America (Kershaw & McGlone 1995), but there is little evidence for changing abundance or distribution in other areas that support members of the family.

Late Quaternary patterns

It is primarily within the later part of the Quaternary that araucarian fossils have been used as an indicator of environmental change. There have been substantial changes in both distribution and abundance, especially in *Araucaria*. In northeastern Queensland the record from Lynch's Crater (Figure 5, color insert) illustrates alternation of high levels of complex rainforest, dominated by angiosperm taxa, during high rainfall interglacials, with araucarian forest and open sclerophyll forest during drier glacial stages. This pattern changed during the latter part of the last glacial period (isotope stages 4–2) when *Araucaria*, together with other southern conifers, were largely eliminated from the region and sclerophyll vegetation dominated the pollen record until complex rainforest returned under the high rainfall conditions of the Holocene. As there was no evidence for any change in the pattern of global climate cyclicity, the destruction of araucarian forest was most likely caused by burning activities of people, who, from archaeological evidence, arrived in Australia by around 40,000 years BP (Kershaw 1986). The sharp and sustained increase in charcoal abundance at this time adds support to this hypothesis.

However, this interpretation of araucarian forest decline has been questioned by evidence from a similar long pollen record constructed from marine core ODP 820 taken off the coast of northeastern Australia adjacent to the Atherton Tableland (Moss 1999, Moss & Kershaw 2000) (Figures 5 & 6, color insert). The record demonstrates that the regional decline in *Araucaria* commenced much earlier than around Lynch's Crater and well before any archaeological evidence of people, whose arrival is now considered to have been between around 50,000 to 60,000 years ago (Roberts et al. 1993). It is also apparent from the charcoal curve that fire has been a continuous feature of the region through at least the past 250,000 years. The first substantial decline in araucarian forest occurred around 130,000 years ago, at the height of the penultimate glacial period, and was associated with a peak in fire activity and an initial increase in eucalypt-dominated sclerophyll vegetation. A further sustained decline occurred about 35,000 years ago; it was also associated with a major peak in burning and a further increase

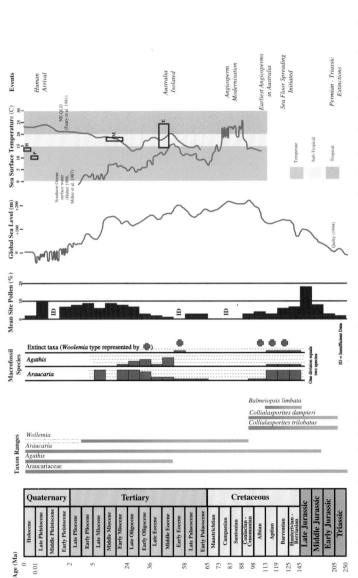

Figure 3 Australian records of Araucariaceae including taxon stratigraphic ranges, representation of predominantly southeastern Australian macrofossil species (data from Hill & Brodribb 1999) and average Araucariaceae percentage from pollen records shown on Figure 4, for individual geological periods, shown in relation to major environmental changes and events. Land temperature estimates from southeastern Australia: E — from leaf size index of Greenwood (1994), M — from overlapping bioclimatic ranges of pollen and macrofossil taxa from the brown coals of the Latrobe Valley (Kershaw 1997), P — from Last Glacial Maximum pollen data (Kershaw 1998), H — from bioclimatic profile of an early Holocene occurrence of the aquatic *Brasenia* (Lloyd & Kershaw 1997).

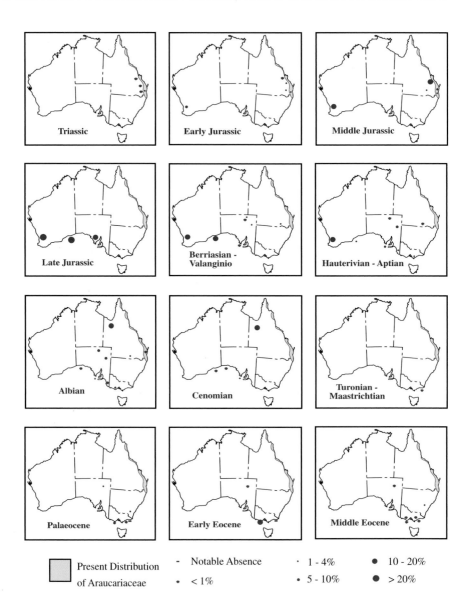

Present Distribution of Araucariaceae

- Notable Absence
- < 1%
· 1 - 4%
• 5 - 10%
● 10 - 20%
● > 20%

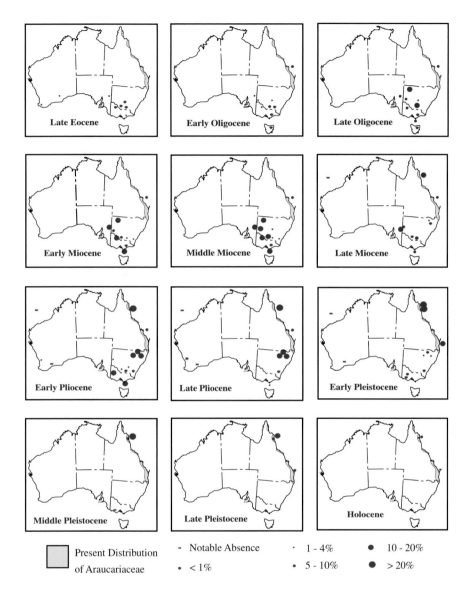

Figure 4 Average percentages of Araucariaceae pollen for geological time periods in Australian fossil sites; (*opposite*) Triassic - Middle Eocene; (*above*) Late Eocene-Holocene (from a variety of sources).

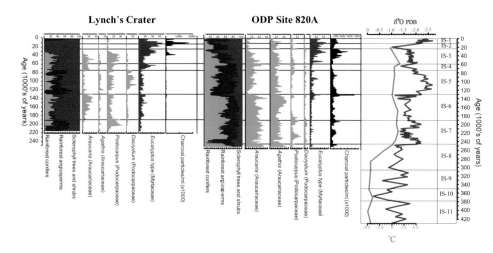

Figure 5 Selected attributes of the pollen records of Lynch's Crater (Kershaw 1986) and ODP 820 (Moss 1999) in relation to the oxygen isotope and inferred sea surface temperature change records from ODP 820 (Peerdeman et al. 1993). All pollen values are expressed as percentages of the sum of total dryland forest taxa for each sample.

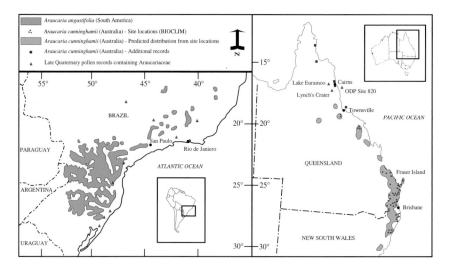

Figure 6 Distribution of *Araucaria angustifolia* in southeastern South America (from Veblin et al. 1995) and location of late Quaternary records containing *Araucaria* (from Berling 1998, Ledru et al. 1998), site records and predicted bioclimatic distribution of *Araucaria cunninghamii* in Australia (after Nix 1991), and location of major late Quaternary pollen records containing Araucariaceae.

in abundance of *Eucalyptus*. This decline can be correlated with that at Lynch's Crater.

It could be argued that the earlier change was climatically induced, the fires promoted by dry glacial conditions, whereas the second *Araucaria* decline was the result of increased burning owing to human activity. However, this would not explain why araucarian forests remained intact through earlier, dry glacial periods. The key to understanding this geologically late vegetation transformation may lie in evidence provided by the extended oxygen isotope record from the ODP 820 core (Peerdeman et al. 1993). A systematic shift in isotope values, superimposed on those attributed to glacial-interglacial cyclicity, between about 350,000 and 250,000 years BP (Figure 5, color insert) has been interpreted as evidence of an increase in sea surface temperatures of some 4°C. Such an increase, not recorded beyond this region, could relate to the development of the West Pacific Warm Pool, which is centered off eastern New Guinea, to the north of this area (Isern et al. 1996).

Although a temperature increase may be expected to have led to an increase in precipitation within the region, and consequently to an expansion of rainforest vegetation, the West Pacific Warm Pool is part of a temperature gradient across the Pacific that is a prerequisite for El Niño-Southern Oscillation (ENSO) climatic variability. Consequently, the decline in araucarian forest might have been a result of increased burning during dry El Niño events, regardless of mean precipitation values. ENSO is noted for its activity on scales of a few years, but it is predicted that activity may have varied on scales of thousands of years (Clement et al. 1999). Spectral analysis has identified a preferred frequency of occurrence of 30,000 years in both the charcoal and southern conifer curves from the ODP record (Moss & Kershaw 1999). This is not a frequency of orbital solar variation that controls glacial-interglacial cyclicity, but could be an ENSO frequency that, over the past 250,000 years, has been the major influence on the dynamics of the araucarian forests. Its major impacts were around 130,000 years ago, when high ENSO activity corresponded with a dry glacial period, and between 40,000 and 35,000 years ago, when high ENSO activity coincided with the presence of people on the continent.

Sea surface temperatures could have increased within the Middle Pleistocene owing to the continued movement of Australia into the Southeast Asian region and consequent alteration of land-sea configurations. Any blockage of the major ocean current that transports warm water from the Pacific to the Indian Ocean through the Indonsian region as a component of the global oceanic circulation system would have caused the build-up of warm water in the equatorial west Pacific and influenced climate in this region and perhaps also over other parts of the globe.

The continued movement of Australia northward might also have had a direct influence on the decline in *Araucaria* through the attainment of critical temperature levels, as this movement brought northern Australia into tropical latitudes for the first time in the history of *Araucaria* on the continent. There is also evidence for a recent decline of *Araucaria* in the subtropics (Longmore 1997), although araucarian forests have remained much more abundant in these latitudes (Figure 6, color insert).

Although climatic variability may have had a significant influence on the decline of araucarian forest, it may also have had a positive influence on conifers within wetter forest systems. At Lake Euramoo, increased representation of *Agathis* in the more complex rainforests of the Atherton Tableland over the past 5000 years may have been due to increased levels of instability associated with the onset of the latest phase of high ENSO activity (McGlone et al. 1992). This activity may have led to a similar expansion of *Agathis* in the warm-temperate to subtropical forests of New Zealand (Ogden et al. 1992).

A clearer indication of distributional changes in *Araucaria* within the late Quaternary is provided by southeastern South America (Figure 6, color insert). Limited data suggest that *A. angustifolia* may have been widely distributed during the last glacial period (prior to 20,000 years BP) but was restricted to small, moist, isolated patches during the latest Pleistocene and early Holocene. Recently analyzed records do not support a substantial shift northward of araucarian forest during cooler conditions of the last glacial period, as suggested by Colinvaux et al. (1996).

However, some northern locations may have experienced local expansion of *Araucaria* during the Pleistocene-Holocene transition. This expansion could be explained by increased precipitation and reduced temperatures from frequent southerly incursions of polar air masses facilitated by the location of the intertropical convergence zone much further north than it is today (Ledru et al. 1998). After about 8000 years BP, the frequency of polar incursions decreased within this region, and araucarian forest was replaced by semi-deciduous forest. Seasonal conditions in the early to mid-Holocene restricted the distribution of rainforest generally, although rainforest expanded gradually with a subsequent increase in effective rainfall. Araucarian forests expanded only within the past 4000 years or so, owing to the wettest and least seasonal conditions in the Holocene period. Ledru et al. (1998) suggested that the spread of *Araucaria* was also facilitated by a return to a high frequency of polar incursions, as there was no response in *Araucaria* forests north of 25°S, where the southern boundary of the intertropical convergence zone is presently located. Ledru et al. (1998) did not address the issue of ENSO variability, except to note that southern Brazil is influenced by this phenomenon. However, high Holocene levels of burning, determined from one site with a charcoal record (Behling 1997), are related to the late Holocene *Araucaria* expansion, and it is possible that this was a result of the global increase in ENSO at this time.

The pattern of increased burning through the Holocene may also relate to increased human activity within the past few thousand years. Intensification of occupation is certainly a feature of the late Holocene of Australia (Lourandos & David 2001) and may confound interpretation of late Holocene vegetation changes there. No such explanation is valid for New Zealand, though, where people did not arrive until within the past 1000 years (Anderson & McGlone 1992).

Synthesis

The Arauciaceae possibly evolved in response to ameliorating climatic conditions during or after the Permian-Triassic extinction event and Early Triassic global dry phase, and have been an important component of forest vegetation, especially in the southern hemisphere, since the Jurassic. *Araucaria* is identified in the fossil record as the oldest genus, dating back to the Jurassic, although the recently discovered genus *Wollemia* is considered, on the basis of phylogenetic relationships, to have derived prior to differentiation of *Araucaria* and *Agathis*. However, macrofossils comparable to *Wollemia* date back only to the Early Cretaceous, and the distinctive pollen is recorded from only the Late Cretaceous. From the limited range of *Wollemia* pollen in southern Australia, Antarctica, and perhaps New Zealand at its maximum during the Late Cretaceous and early Cenozoic, it is unlikely that this genus was ever extensively distributed over the southern hemisphere (Dettmann & Jarzen 1999). However, the discovery of this new genus is leading to a re-examination of taxonomic relationships of early Araucariaceae.

Within the genus *Araucaria*, sections *Eutacta* and *Bunya* have the oldest fossil records, both dating back to the Jurassic, where they were widely distributed in both hemispheres (Hill & Brodribb 1999). However, their subsequent history has been very different, with *Eutacta* still broadly distributed and diverse, apparently having been able to take advantage of mid- to late-Cenozoic opportunities for colonization, whereas *Bunya* is reduced to one species restricted to northeastern Australia and without a clear macrofossil record during the Cenozoic. Setoguchi et al. (1998) suggested, from genetic analysis, that older and younger populations attributed to *Bunya* may not be related. The other two sections of *Araucaria* (*Araucaria* and *Intermedia*) indicate wider distributions in the past, and the fossil record demonstrates their greater connection with South America. All extant sections of *Araucaria* appear to have evolved well before the final break-up of Gondwana. The extant representatives of *Agathis* show relatively minor genetic and morphological differences compared with the variation evident in *Araucaria*, and it is likely that species of the former represent the products of a relatively recent phase of evolutionary radiation in the southwestern Pacific region.

The Araucariaceae declined in abundance and range from the Early Cretaceous. Although the evolution of angiosperms is likely to have been influencial in this decline, many araucarians may have been disadvantaged by the Middle Cretaceous peak in temperatures and humidity. However, *Agathis* and *Wollemia* may have evolved in association with early angiosperms and been part of the proposed center of biotic radiation within the unstable rift valley between southern Australia and Antarctica.

The more substantial fossil record for the Cenozoic illustrates marked differences between pollen and macrofossil representation. High macrofossil diversity is recorded within the period of peak temperature and rainfall conditions of the Eocene period. Many of these araucarian taxa are attributable to *Agathis*, which is able to survive in more complex rainforest than *Araucaria*. The decline in macrofossil diversity and representation is accompanied by higher pollen relative abundances, suggesting the development of dense, species-poor araucarian forests under drier and more variable climatic conditions of the later Cenozoic. The gradual restriction of rainforest toward the eastern coastal margins of Australia that accompanied continental drying, and subsequent contraction of complex rainforest to the northeastern part of the continent under temperature decline and changing atmospheric circulation patterns are well tracked by *Araucaria* pollen.

The final restriction of *Araucaria* to small, isolated pockets, at least in northeastern Australia, was, we suggest, caused by increased climatic variability and biomass burning resulting from atmospheric and oceanic circulation changes associated with Australia's continued movement into the Southeast Asian region. However, this ENSO variability may have simultaneously facilitated the regeneration of *Agathis* in northeastern Australia and also araucarian species in angiosperm-dominated rainforests of other ENSO-affected areas such as New Zealand and southeastern South America during the late-Holocene phase of high ENSO activity.

The potential of the araucarians for palaeoenvironmental reconstruction is demonstrated by both past and present distributions, in combination with a knowledge of their ecology. In general terms, the expansion and subsequent contraction of the Araucariaceae, at least on the southern hemisphere continents, corresponds with a major climatic cycle from the Triassic to Late Quaternary. Both the development and contraction phases show high pollen abundances that suggest a dense emergent cover of *Araucaria*, regardless of the understorey, under relatively dry climatic conditions. Optimal conditions for plant growth through much of the middle of this period are characterized by high macrofossil diversity but relatively low pollen abundance of araucarians, consistent with dense understorey or canopy coverage.

The araucarians also appear to have maintained a preference for subtropical or mesothermal conditions. This is well illustrated by the present latitudinal distributions of the major continental *Araucaria* species (*A. angustifolia* in southeastern South America and *A. cunninghamii* and *A. bidwillii* in Australia) that are remarkably similar, despite an obviously long period of continental separation and representation in different sections of the genus (Figure 6, color insert). Mean annual precipitation is not a significant factor, as *A. angustifolia* requires a minimum of 1400 mm per annum, greater than the mean of either Australian species. All species have a minimum requirement of about 10°C in winter months that, perhaps together with a requirement for significant summer rainfall, could explain the southern extent of the species.

The lower latitude limit of the *Araucaria* species in Australia may relate to a sensitivity to high temperatures that made the taxon vulnerable in the tropics to increased climatic variability. The length of record in South America is insufficient to determine whether there is a trend toward reduced araucarian representation. However, this is unlikely considering that the continent has not moved northward to the same degree as Australia, and *Araucaria* appears to have been advantaged by increased climatic variability and burning in the late Holocene. Perhaps the position of the intertropical convergence zone, and hence disturbances resulting from polar air incursions, have been critical to the survival of the species at low latitudes. This example illustrates that, although patterns of representation of many araucarians on Gondwanic land masses appear relictual and current ranges may not reflect the full potential climatic range, araucarians can be useful for refinement and probably quantification of climate from high-resolution pollen records in particular areas. Their response to climatic variability also suggests that they can provide data in addition to mean values that characterize most climate reconstructions from pollen records.

In contrast to a general decline in representation of Araucariaceae on major continental masses, there have been major expansions and evolution of *Agathis* and *Araucaria* section *Eutacta* in Southeast Asia–Pacific islands. This proliferation illustrates significant dispersal ability, an attribute not expected from the groups' continental history or current distributions. Colonization and establishment appear to have commenced by the Eocene and have been facilitated by a number of factors. These factors include emergence of islands and increased proximity to the Australian continent (via tectonic convergence and development of island arc corridors); the production of environments within humid climates not conducive to the development of a dense rainforest cover, such as extensive peat swamps, ultramafic soils, and steep slopes; and frequent disturbances resulting from tectonic and volcanic activity. There are insufficient data on the history of these island conifers derived from recent radiations to determine their response to past climate change.

ACKNOWLEDGMENTS

We thank Simon Haberle and Stephen McLoughlin for their very helpful comments on a draft of this paper; Daphne Fautin for substantial improvement of English expression; Hermann Behling, Mary Dettmann, Neil Enright, Bob Hill, Marie-Pierre Ledru, Mike Macphail, Helene Martin, John Ogden, and Liz Truswell for valuable discussion and information on this topic; and Jonathan Brown and Gary Swinton for drafting the text figures.

Visit the Annual Reviews home page at www.AnnualReviews.org

LITERATURE CITED

Aitchison J, Clarke GL, Meffre S, Cluzel D. 1995. Eocene arc-continent collision in New Caledonia and implications for regional Southwest Pacific tectonic evolution. *Geology* 23:161–64

Anderson A, McGlone M. 1992. Living on the edge—prehistoric land and people in New Zealand. In *The Naive Lands: Prehistory and Environmental Change in Australia and the Southwest Pacific*, ed. J. Dodson, pp. 199–241. Melbourne, Aust.: Longman Cheshire

Behling H. 1997. Late Quaternary vegetation,

climate and fire history of the *Araucaria* forest and campos region from Serra Campos Gerais, Paranà State (South Brazil). *Rev. Palaeobot. Palynol.* 97:109–21

Behling H. 1998. Late Quaternary vegetational and climate changes in Brazil. *Rev. Palaeobot. Palynol.* 99:143–56

Bowler JM. 1982. Aridity in the Tertiary and Quaternary of Australia. In *Evolution of the Flora and Fauna of Arid Australia*, ed. WR Barker, PJM Greenslade, pp. 35–45. Frewville, S. Aust.: Peacock

Carpenter RJ, Pole MS. 1995. Eocene plant fossil from the Lefroy and Cowan palaeodrainages, Western Australia. *Aust. Syst. Bot.* 8:1107–54

Chambers TC, Drinnan AN, McLoughlin S. 1998. Some morphological features of Wollemi pine (*Wollemia nobilis*: Araucariaceae) and their comparison to Cretaceous plant fossils. *Int. J. Plant Sci.* 159:160–71

Clement AC, Seager R, Cane MA. 1999. Orbital controls on the El Niño/Southern Oscillation and the tropical climate. *Paleoceanography* 14:441–56

Colinvaux PA, Liu K-B, De Oliveira P, Bush MB, Miller MC, Kamman MS. 1996. Temperature depression in the lowland tropics in glacial times. *Clim. Change* 32:19–33

De Jersey N. 1968. *Triassic spores and pollen grains from the Clematis Sandstone. Geol. Surv. Queensland Publ. No. 338, Palaeontol. Pap. No. 14.* Brisbane, Aust.: Geol. Surv. Queensland. 44 pp.

Dettmann ME. 1989. Antarctica: Cretaceous cradle of austral temperate rainforests? In *Origins and Evolution of the Antarctic Biota*, ed. JA Crane, pp. 89–105. London: Geol. Soc. Spec. Publ. No. 47

Dettmann ME. 1994. Cretaceous vegetation: the microfossil record. See Hill 1994, pp. 143–70

Dettmann ME, Jarzen DM. 2000. Pollen of extant *Wollemia* (Wollemi pine) and comparisons with pollen of other extant and fossil Araucariaceae. In *Pollen and Spores Morphology and Biology*, ed. MM Harley, CM

Morton, S Blackmore, pp. 187–203. Kew: R. Bot. Gard.

Enright NJ. 1982. Does *Araucaria hunsteinii* compete with its neighbours? *Aust J. Ecol.* 7:97–99

Enright NJ, Hill RS. 1995. *Ecology of the Southern Conifers.* Melbourne, Aust.: Melbourne Univ. Press. 342 pp.

Enright NJ, Ogden J. 1995. The southern conifers—a synthesis. See Enright & Hill 1995, pp. 271–87

Greenwood DR. 1994. Palaeobotanical evidence for Tertiary climates. See Hill 1994, pp. 44–59

Hill RS, ed. 1994. *History of the Australian Vegetation: Cretaceous to Recent.* Cambridge: Cambridge Univ. Press

Hill RS, Brodribb TJ. 1999. Southern conifers in time and space. *Aust. J. Bot.* 47:639–96

Hill RS, Truswell EM, McLoughlin S, Dettmann ME. 2000. Evolution of the Australian flora: fossil evidence. In *Flora of Australia, Vol. 1 (Introduction)*, ed. AE Orchard, pp. 251–320. Melbourne, Aust.: CSIRO. 2nd ed.

Huber BT. 1998. Enhanced: tropical paradise at the Cretaceous poles. *Science* 282:2199–200

Isern AR, McKenzie JA, Feary DA. 1996. The role of sea-surface temperature as a control on carbonate platform development in the western Coral Sea. *Palaeogeogr. Palaeoclimatol. Palaeoecol.* 124:247–72

Jaffré T. 1995. Distribution and ecology of the conifers of New Caledonia. See Enright & Hill 1995, pp. 171–96

Jones WG, Hill KD, Allen JM. 1995. *Wollemi nobilis*, a new living Australian genus and species in the Araucariaceae. *Telopea* 6:173–76

Kershaw AP. 1986. Climate change and Aboriginal burning through the last two glacial-interglacial cycles from northeastern Queensland. *Nature* 322:47–49

Kershaw AP. 1997. A bioclimatic analysis of Early to Middle Miocene brown coal floras, Latrobe Valley, southeastern Australia. *Aust. J. Bot.* 45:373–87

Kershaw AP. 1998. Estimates of regional climatic variation within southeastern mainland

Australia since the Last Glacial Maximum from pollen data. *Palaeoclimates* 3:107–34

Kershaw AP, Bolger P, Sluiter IRK, Baird J, Whitelaw M. 1991. The origin and evolution of brown coal lithotypes in the Latrobe Valley, Victoria, Australia. *Int. J. Coal Geol.* 18:233–49

Kershaw AP, McGlone MS. 1995. The Quaternary history of the southern conifers. See Enright & Hill 1995, pp. 30–63

Ledru M-P, Salgaro-Labouriau ML, Lorscheitter ML. 1998. Vegetation dynamics in southern and central Brazil during the last 10,000 yr B.P. *Rev. Palaeobot. Palynol.* 99:131–42

Le Grand HE. 1988. *Drifting Continents and Shifting Theories.* Cambridge: Cambridge Univ. Press. 313 pp.

Lidgard S, Crane PR. 1990. Angiosperm diversification and Cretaceous floristic trends: a comparison of palynofloras and leaf macrofloras. *Palaeobiol.* 16:77–93

Lloyd PJ, Kershaw AP. 1997. Late Quaternary vegetation and early Holocene quantitative climatic estimates from Morwell Swamp, Latrobe Valley, south-eastern Australia. *Aust. J. Bot.* 45:549–63

Longmore ME. 1997. Quaternary palynological records from the perched lake sediments of Fraser Island, Queensland, Australia: rainforest, forest history and climatic control. *Aust. J. Bot.* 45:507–26

Lourandos H, David B. 2001. Long-term archaeological and environmental trends: a comparison from late Pleistocene-Holocene Australia. In *The Environmental and Cultural History and Dynamics of the Australian-Southeast Asian Region,* ed. AP Kershaw, B David, NJ Tapper, D Penny, J. Brown. Reiskirchen, Ger.: Catena. In press

Macphail M, Alley NF, Truswell EM, Sluiter IRK. 1994. Early Tertiary vegetation: evidence from pollen and spores. See Hill 1994, pp. 189–261

Macphail M, Hill K, Partridge A, Truswell E, Foster C. 1995. Australia: Wollemi pine: old pollen records for a newly discovered genus of gymnosperm. *Geol. Today* March-April:42–44

McGlone MS, Kershaw AP, Markgraf V. 1992. El Niño/Southern Oscillation climatic variability in Australasian and South American palaeoenvironmental records. In *El Nino: Historical and Palaeoclimatic Aspects of the Southern Oscillation,* ed. HF Diaz, V Markgraf, pp. 435–62. Cambridge: Cambridge Univ. Press

McLoughlin S, Hill RS. 1996. The succession of Western Australian Phanerozoic terrestrial floras. In *Gondwana Heritage: Past, Present and Future of the Western Australian Biota,* ed. SD Hopper, JA Chappell, MS Harvey, AS George, pp. 61–80. Chipping-Norton, UK: Beatty & Sons

Miller KG, Fairbanks RG, Mountain GS. 1987. Tertiary oxygen isotope synthesis, sea level history and continental margin erosion. *Paleoceanography* 2:1–19

Morley RJ. 1998. Palynological evidence for Tertiary plant dispersals in the SE Asian region in relation to plate tectonics and climate. In *Biogeography and Geological Evolution of SE Asia,* ed. R Hall, JD Holloway, pp. 211–34. Leiden, The Netherlands: Backbuys

Moss PT. 1999. *Late Quaternary environments of the humid tropics of northeastern Australia.* PhD thesis. Monash Univ., Melbourne. Aust. 269 pp.

Moss PT, Kershaw AP. 1999. Evidence from marine ODP Site 820 of fire/vegetation/climate patterns in the humid tropics of Australia over the last 250,000 years. In *Bushfire 99, Proc. Aust. Bushfire Conf., Albury, Aust., July,* pp. 269–79. Albury: Charles Sturt Univ.

Moss PT, Kershaw AP. 2000. The last glacial cycle from the humid tropics of northeastern Australia: comparison of a terrestrial and a marine record. *Palaeogeogr. Palaeoclimatol. Palaeoecol.* 155:155–76

Nix HA. 1991. An environmental analysis of Australian rainforests. In *The Rainforest Legacy: Australian National Rainforests Study. Vol. 2, Flora and Fauna of the Rainforests,* ed. G Werren, AP Kershaw, pp. 1–26.

Canberra: Aust. Heritage Comm. Publ. Ser. No. 7 (2)

Ogden J. 1985. An introduction to plant demography with special reference to New Zealand trees. *NZ J. Bot.* 23:751–72

Ogden J, Wilson A, Hendy C, Newnham RM. 1992. The late Quaternary history of kauri (*Agathis australis*) in New Zealand and its climatic significance. *J. Biogeogr.* 19:611–22

Peerdeman FM, Davies PJ, Chivas AR. 1993. The stable oxygen isotope signal in shallow-water, upper-slope sediments off the Great Barrier Reef (Hole 820A). *Proc. Ocean Drilling Program Sci. Res.* 133:163–73

Pole MS. 1994. The New Zealand flora—entirely long-distance dispersal? *J. Biogeogr.* 21:625–35

Quilty PG. 1994. The background: 144 million years of Australian palaeoclimate and palaeogeography. See Hill 1994, pp. 14–43

Regal P. 1977. Ecology and the evolution of flowering plant dominance. *Science* 196:622–29

Roberts R, Jones R, Smith MA. 1993. Optical dating of Deaf Adder Gorge, Northern Territory indicates human occupation between 53,000 and 60,000 years ago. *Aust. Arch.* 37:58–59

Setoguchi H, Osawa TA, Oinaud J-C, Jaffré T, Veillon J-M. 1998. Phylogenetic relationships within Araucariaceae based on *rbcL* gene sequences. *Am. J. Bot.* 85:1507–16

Shackleton NJ, Crowhurst S, Hagelberg T, Pisias N, Schnieder DA. 1995. A new late Neogene timescale: applications to leg 138 sites. *Proc. Ocean Drilling Program Sci. Res.* 138:73–101

Stockey RA. 1982. The Araucariaceae: an evolutionary perspective. *Rev. Palaeobot. Palynol.* 37:133–54

Stockey RA. 1990. Antarctic and Gondwana conifers. In *Antarctic Paleobiology*, ed. TN Taylor, EL Taylor, pp. 179–91. New York: Springer-Verlag

van Steenis CGGJ. 1971. *Nothofagus*, a key genus to plant geography, in time and space, living and fossil, ecology and phylogeny. *Blumea* 19:65–98

Veblin TT, Burns BR, Kitzberger T, Lara A, Villalba R. 1995. The ecology of the conifers of southern South America. See Enright & Hill 1995, pp. 120–70

Veblin TT, Hill RS, Read J. 1996. *The Ecology and Biogeography of Nothofagus Forests*. New Haven, CT: Yale Univ. Press. 403 pp.

Wagstaff BE, Kershaw AP, O'Sullivan P, Harle KJ, Edwards J. 2001. An Early to Middle Pleistocene palynological record from Pejark Marsh, Western Plains of Victoria, southeastern Australia. *Palaeogeogr. Palaeoclimatol. Palaeoecol.* In press

Webb LJ. 1959. A physiognomic classification of Australian rainforests. *J. Ecol.* 8:118–29

Annu. Rev. Ecol. Syst. 2001. 32:415–48
Copyright © 2001 by Annual Reviews. All rights reserved

THE UNITS OF SELECTION ON MITOCHONDRIAL DNA

David M. Rand

Department of Ecology and Evolutionary Biology, Brown University, Box G-W, 69 Brown Street, Providence, Rhode Island 02912; e-mail: David_Rand@brown.edu

Key Words   heteroplasmy, natural selection, genetic drift, neutrality test, population genetics

■ **Abstract**   Mitochondrial DNA (mtDNA) exists in a nested hierarchy of populations. There are multiple mtDNAs within each mitochondrion, a population of mitochondria in each cell, multiple oocytes within each reproductive female, multiple females in each population, and so on up through species and higher clades. The metabolic properties of mitochondria make them highly mutagenic environments for the naked, circular mtDNAs that lie within them. This mutational pressure introduces mtDNA variation (i.e., heteroplasmy) into the cytoplasmic population of cell lineages that are particularly prone to mutational decay and Muller's ratchet owing to the asexual, maternal inheritance of mtDNA. Neutrality tests show that deleterious mutations are common in mtDNA evolution. Population cage experiments further show that mtDNA fitnesses are influenced by nuclear-mitochondrial interactions. These selective processes are pervasive despite the long-standing use of mtDNA as a neutral marker in population and evolutionary biology. These evolutionary dynamics are also unique in the nested hierarchy of mtDNA populations because mutation, selection, and drift can act—and interact—at multiple levels. Multi-level selection can facilitate the escape from Muller's ratchet and help resolve intragenomic conflicts. This review addresses recent advances in the transmission genetics, population genetics, and evolution of mtDNA. A primary goal of the review is to motivate additional empirical studies that might clarify the many units of selection acting on mtDNA.

INTRODUCTION

Mitochondrial DNA (mtDNA) has been the most widely studied region of eukaryotic genomes and has played a critical role in the development of population and evolutionary genetics (Brown 1985, Avise et al. 1987, Moritz et al. 1987, Harrison 1989, Avise 1991, Rand 1994). During its first decade of use in evolutionary biology, it was an implicit—if not explicit—assumption that mtDNA was a neutral genetic marker. More recently, there has been growing interest in how selection might act on mtDNA, and two different approaches have been taken to address this question. Direct experiments with laboratory populations have attempted to identify

0066-4162/01/1215-0415$14.00

fitness effects of distinct mtDNA haplotypes (e.g., MacRae & Anderson 1988, Clark & Lyckegaard 1988, Nigro & Prout 1990, Fos et al. 1990, Kambhampati et al. 1992, Hutter & Rand 1995, Kilpatrick & Rand 1995, Garcia-Martinez et al. 1998, Rand et al. 2001). Alternatively, evidence for selection on mtDNA has been sought using statistical tests of neutral models of molecular evolution based on static samples of DNA sequences from natural populations (Whittam et al. 1986; Excoffier 1990; Ballard & Kreitman 1994; Ballard 2000a,b; Nachman et al. 1994, 1996; Nachman 1998; Quesada et al. 1999; Rand et al. 1994, 2000; Rand & Kann 1996, 1998; Templeton 1996; Wise et al. 1998; Weinreich & Rand 2000; Blouin 2000). Both of these approaches have identified nonneutral evolution of mtDNA that has implicated selection at the level of the individual organism.

However, a large and rapidly growing literature in cell biology and molecular genetics has demonstrated the importance of selection and genetic drift within and among cytoplasms as primary mechanisms affecting the phenotypic expression of mtDNA mutations. It is clear from comparative and medical studies that deleterious mutations define much of the selection on mtDNA. The answer to how organelle genomes purge these deleterious mutations and escape Muller's ratchet in the absence of recombination and sex may lie in the dynamics afforded by multi-level selection (Birky 1995, Bergstrom & Pritchard 1998, Jacobs et al. 2000). The classic examples of multi-level selection are the *petite* mutants in yeast. These mtDNA mutants lack genes for oxidative phosphorylation but retain functional origins of replication. They can out-compete wild type mtDNAs by "selfish" replication in mixed (i.e., heteroplasmic) cytoplasms but are ultimately selected against at the level of the cell owing to reduced growth rates from deficiencies in oxidative phosphorylation (e.g., Reid 1980, Birky 1983). Analogous systems of multi-level selection involving mtDNA deletion mutants may well have operated among some of the retinal and brain cells that are processing the words that you are now reading (Wallace 1999).

At a macroevolutionary scale, a similar multi-level selection has operated. The origin and evolution of mtDNA—and the eukaryotic cell itself—has involved extensive gene transfer from a proto-mitochondrion to the nuclear genome, as well as outright gene loss (Lang et al. 1999, Gray et al. 1999). Deleted mtDNAs have undoubtedly won many cytoplasmic "replication races" during the history of life, but selection at the level of the integrated nuclear-mitochondrial system has undoubtedly produced the lineages that carry streamlined mtDNAs today. Because there are clear parallels between both micro- and macro-evolutionary (as well as clinical) models of the units of selection on mtDNA, this view provides an opportunity for a synthesis across a broad spectrum of the biological sciences.

The connection between organelle evolution and the units of selection has been discussed in other reviews (Lewontin 1970, Eberhard 1980, Cosmides & Tooby 1981, Buss 1987, Hastings 1992, Birky 1995, Blackstone 1995, Maynard Smith & Szathmáry 1995). Whereas these reviews have defined many fundamental questions, the derived state of current mitochondria and their free-living sister taxa (Lang et al. 1999) present problems for historical inference, and certainly for direct experimental studies of phenotypes. The primary aim of this review is to survey recent studies of selection and drift acting on mtDNA in the hierarchy of populations from organelles, germ cells, and tissues through the individual and population levels. The review is biased towards the population genetics of animal mtDNA. However, certain plant and fungal systems offer clear advantages for some aspects of multi-level selection analysis and are mentioned (e.g., Albert et al. 1996). A second aim of the review is to motivate additional manipulative experiments that might reveal the importance of evolutionary forces acting across the hierarchy of organization inherent in organelle genomes.

THE CYTOPLASM AS A POPULATION

As Lewontin (1970) pointed out, selection will act among units at any level of organization that exhibits three basic principles: 1. phenotypic variation, 2. differential fitness of phenotypes, and 3. transmission of fitness phenotypes. The hierarchical nesting of mtDNAs within organelles, cells, tissues, and individuals allows for each of these principles to apply at more than one level. For example, the phenotypic variation at the level of cells can be an emergent property of the heteroplasmic composition of individual cells; the phenotypic variation at the level of individual organisms can be an emergent property of the heteroplasmic composition of various tissues, each of which might have distinct effects on the life history of the organism. Moreover, the maternal inheritance and general lack of recombination in (animal) mtDNA provides a clear context in which to apply the basic population genetics of mutation, selection, and drift across this hierarchy of populations (see Figure 1).

Consider a cell that is fixed (homoplasmic) for one mtDNA type (see Figure 2). Mutation will introduce variation into this population, providing phenotypic variation (principle 1). Mutant mtDNAs will replicate at different rates, either by kinetic advantages in replication, or by differential rates of organelle or cell proliferation (principle 2). The frequencies of the two (or more) mtDNA variants within a cell will be subject to random genetic drift during the sampling event imposed by mitosis and cytokinesis. Organelles or cells that enjoy a fitness benefit from the novel frequencies of mtDNA will transmit these fitness states to their offspring (principle 3). However, random drift at each of these levels can undermine principle 3 and provides a clear population genetic distinction between the units of selection and the units of evolution (cf. Vrba & Gould 1986).

Mutation

MUTATION VERSUS SUBSTITUTION   A hallmark of mtDNA is rapid sequence evolution, at least in mammals (Upholt & Dawid 1977, Brown et al. 1979). Given the functional importance of oxidative phosphorylation (OXPHOS) in eukaryotic metabolism, this was surprising. The rapid rate of evolution was explained

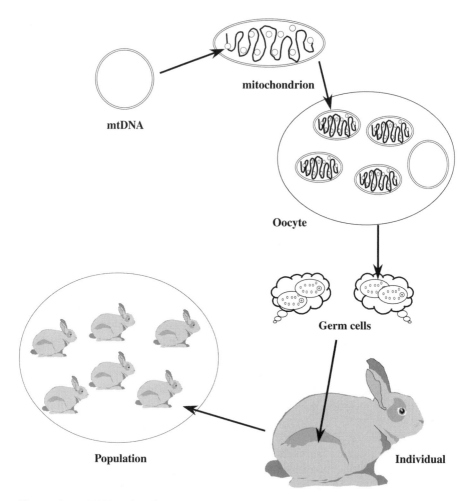

Figure 1   mtDNA exists in a nested hierarchy of populations spanning the mtDNA molecule, the organelle, the cell, the individual organism, and on up through natural populations and species.

by a high mutation rate, relaxed selective constraints, or limited mtDNA repair mechanisms (Brown 1985, Avise 1991). Mutation and substitution (or fixation) need to be distinguished. The rate of DNA sequence evolution (i.e., substitution) between species is a function of the mutation rate (u) and the probability of fixation (p_{fix}) (Kimura 1983). Many more mutations arise than ultimately become fixed, because most mutations are deleterious. The substitution rate inferred from comparative sequence data fails to record mutations that never enter populations as polymorphisms and polymorphisms that never reach fixation. This distinction is particularly important in the context of a units-of-evolution view of mtDNA

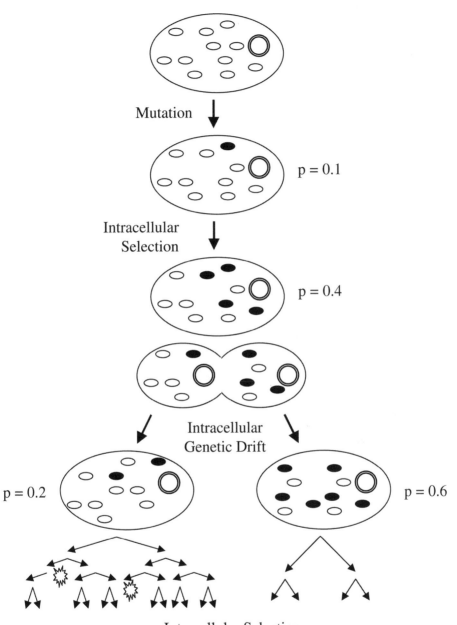

Figure 2 Population genetics of the cytoplasm. Mutation introduces variation at the lowest level of the hierarchy. Natural selection and random genetic drift act to change mtDNA haplotype frequencies within and among cytoplasms. Selection and drift can act at multiple levels in the hierarchy of populations: within organelles, within cells, or among maternal lineages.

because the cytoplasmic population is an additional level of the biological hierarchy that can affect the distinction between mutation and substitution. Hence, heteroplasmy is an obligatory, if transient, phase in the evolution of mtDNA. The hierarchical population genetics of mutation, drift, and selection thus defines the spectrum of phenotypic variation upon which selection can act at each level from the mtDNA molecule to the organismal population.

DIRECT MEASUREMENTS OF MUTATION RATES   Mitochondrial mutation rates have been measured by sequencing partial or complete mtDNAs in defined pedigrees. Parsons et al. (1997) sequenced the control region (CR or D-loop) from grandparent- or parent-offspring pairs in 134 human lineages spanning 327 generational events. They detected 10 mutations, resulting in a rate estimate of 1 mutation per 33 generations, or 2.5 mutations per site per million years. This is 20 times higher than the mutation rates estimated using phylogenetic methods of sequence comparisons between humans and chimps (Parsons et al. 1997). In the nematode *Caenorhabditis elegans*, sequences from 75% of the mtDNA genome were obtained from 74 mutation accumulation lines maintained by single-worm transmission for an average of 214 generations (Denver et al. 2000). This protocol reduces natural selection to a minimum. Of the 26 mutations observed, 16 were single base changes and 10 were insertions or deletions (indels). The majority of the point mutations were nonsynonymous, and some of the deletions disrupted reading frames. The estimated mutation rate (14.3 mutations per site per million years) is two orders of magnitude higher than the phylogenetic approach for humans discussed above, and about five times higher than the mutation rate estimate in human pedigrees. This discrepancy between mutation and substitution rates illustrates the clear deleterious nature of the spontaneous mutations and underscores the important qualitative distinction between mutational processes operating at these different time scales.

HOMOPLASMY AND HETEROPLASMY   Initial studies of mtDNA variation in nature showed that all mtDNA molecules within an individual were identical (homoplasmy). The preponderance of homoplasmy, coupled with high levels of variation within and between species, dictates that the necessary heteroplasmic phase is transient during mtDNA evolution. The first studies that uncovered high levels of heteroplasmy involved insertion and deletion of repeated sequences with high mutation rates (Solignac et al. 1983, Hauswirth et al. 1984, Harrison et al. 1985; reviewed in Rand 1993). In the pedigree analyses discussed above, both Parsons et al. (1997) and Denver et al. (2000) detected individuals heteroplasmic for point mutations or indels. Because complete shifts in mtDNA genotypes were also observed between lineages in the pedigrees, germ line mutations had clearly arisen and gone through the heteroplasmic phase to fixation. These studies illustrate the timescales that are relevant for the transition between distinct homoplasmic states.

The biochemical processes of the mitochondrion itself are believed to be a major source of mutations. Most of the cell's energy production and oxygen consumption occur during OXPHOS, which is performed by five enzyme complexes embedded in the inner mitochondrial membrane. Complex I (NADH dehydrogenase) and particularly complex III (cytochrome bc$_1$) are major sources of reactive oxygen species, which are potent mutagens (Finkel & Holbrook 2000). The proximity of mtDNA to the sites of OHPHOS and reactive oxygen species production, and the fact that mtDNA is not associated with histones, are two primary explanations for high mutation pressure on mtDNA (Shigenaga et al. 1994, de Gray 1999).

Coupled with the lack of recombination (Ballard 2000a, Ingman et al. 2000), the high rate of deleterious mutation in mtDNA should result in an accumulation of mutations leading to a "mutational meltdown" and extinction of lineages (Lynch 1997, Lynch & Blanchard 1998). A solution to this problem is likely provided by both drift and selection within and among cytoplasms. Replicative purging of the cytoplasmic population can occur if mtDNAs with the most lesions are replicated less efficiently (Birky 1983, Lightowlers et al. 1997, Marchington et al. 1997). Indeed, age-dependent accumulation of heteroplasmies is greater in nondividing tissue than in dividing tissue (Cortopassi et al. 1992). Random drift of mtDNA mutants during germ line divisions leads to the loss of variation within cells but increases the variance among oocytes, somatic tissues, or maternal lineages. This will increase the efficacy of selection at the level of the cell or individual, and possibly counter the "ratchet" effects of reactive oxygen species–induced mutations (Bergstrom & Pritchard 1998). Nevertheless, mtDNA evolution may be characterized by a mutation-drift-selection balance in which most cells carry transient levels of heteroplasmy for point mutations, deletions, and rearrangements.

We need additional studies of the accumulation of mtDNA mutations in a diverse array of organisms (e.g., Parsons et al. 1997, Denver et al. 2000). Such studies done in organisms with very different patterns of germ line differentiation will provide important comparative data on the fate of mtDNA mutations through the mitochondrial population booms and bottlenecks of extended germ lines. Plants offer excellent material for these studies as well. Because the germ line is dispersed, the opportunity to study somatic mosaicism and potential intra-individual selection on chloroplast or mtDNA is great (Albert et al. 1996, Klekowski 1998).

Genetic Drift Within and Among Cytoplasms

BEAN-BAG POPULATION GENETICS AND BOTTLENECKS IN THE CYTOPLASM   Some of the earliest analyses of the transmission of a population of organelles through the cytoplasm came from studies of mutant and wild-type chloroplasts in variegated plants (reviewed in Gillham 1978, Beale & Knowles 1979). Studies of the segregation of mtDNAs in heteroplasmic yeast cultures lead to the application of population genetic theory to mitochondrial genetics (Thrailkill & Birky 1980, Birky 1983). The ability to construct experimental heteroplasmies by mating yeast strains with different mutant forms of mtDNA has obvious advantages over relying on naturally occurring heteroplasmies. However, these single-cell studies are limited in their ability to characterize the genetics of mitochondria in organisms with clearly defined germ lines. Transmission of mitochondria from one organismal generation to the next involves repeated sampling events during the mitotic divisions that occur during the formation of the germ line and subsequent production of competent oocytes.

The evolution of this cytoplasmic population can be modeled using a drift equation for the evolution of the variance among demes (Wright 1968, p. 346; Solignac et al. 1984, Rand & Harrison 1986):

$$V_n = p(1-p)[1-(1-1/N_{e\text{-}mt})^{gn}], \qquad\qquad 1.$$

where V_n is the variance among cells (i.e., demes) after n organismal generations, p and $(1-p)$ are the initial frequencies of the two mtDNA types in the founding mother (or oocyte), $N_{e\text{-}mt}$ is the effective population size of mitochondria during cell division, and g is the number of germ line cell divisions per organismal generation. V_n and p are relatively straightforward to estimate using molecular methods. Recent advances in competitive PCR have made the "field work" of this population genetic problem accurate and repeatable (Chen et al. 1995; Barritt et al. 2000). The main limitation to this approach is the number of oocytes or offspring that can be sampled from any one mother, because estimates of variance are sensitive to sample size. This is not a problem in insects but can be in mammals, and especially in human pedigrees.

The number of germ cell generations per organismal generation has been estimated to be 7 in *Drosophila* (Solignac et al. 1987) and about 15 in mice (Jenuth et al. 1996) and humans (Howell et al. 1992, Poulton et al. 1998; although g = 24 was used by Jenuth et al. 1996), and possibly as high as 50 in cows (Upholt & Dawid 1977, Hauswirth & Laipis 1982). Empirical studies of the evolution of the variance across organismal generations can be used to estimate the effective population size for mtDNA through the germ line. This sampling-effective population size may involve mtDNAs, organelles, or aggregations of organelles (see below). As illustrated in Figure 3, the demography and population structure of the germ line, as defined by the drift equation, can have important consequences for the evolution of mtDNA.

STUDIES OF MITOCHONDRIAL TRANSMISSION   Transmission studies of mtDNA in yeast, insects, and mammals indicate that this "bean bag" population genetics provides a relatively good fit with the number of mitochondria sampled from a cytoplasmic population. Ultrastructural observations suggest that there are about 500–1000 mitochondria per cell (Gillham 1978). Assuming that this population is sampled during cell division, one would expect $N_{e\text{-}mt}$ to be lower than the census size, or on the order of a few hundred mitochondria.

Using the drift equation above, $N_{e\text{-}mt}$ is approximately 50 in the budding yeast *Saccharomyces* and somewhat higher in the fission yeast *Schizzosaccharomyces* (Birky 1983, Thrailkill & Birky 1980). In *Drosophila* $N_{e\text{-}mt}$ ranges from 545 to 700 (Solignac et al. 1984, 1987; Kann et al. 1998). Crickets have a somewhat lower estimate of $N_{e\text{-}mt}$ [87–395, (Rand & Harrison 1986)]. A transmission study

Evolution of the variance due to drift

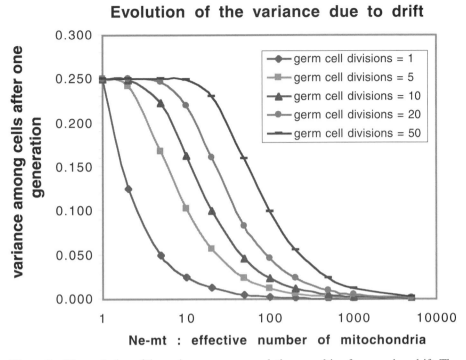

Figure 3   The evolution of the variance among populations resulting from random drift. The amount of variation among germ cells (or F1 offspring) from a single heteroplasmic female is shown on the y-axis, as a function of the effective number of segregating mitochondria (or other units such as mtDNAs or clusters of organelles) on the x-axis. Separate curves are generated from Equation 1 based on the number of germ cell generations that span a single organismal generation. The heteroplasmic mother is assumed to have two mtDNAs at equal frequency ($p = q = 0.5$); hence, from Equation 1, fixation occurs at a variance of $V = 0.25$. Note that a single-celled organism with two mitochondria has comparable transmission dynamics to that of a multicellular organism with 50 germ cell generations per organismal generation, but with $N_{e\text{-}mt} \sim 100$ (variances of about 0.125).

in an experimental strain of heteroplasmic mice carrying wild-type mtDNAs reported estimates of $N_{e\text{-}mt}$ that ranged from 76 to 867. Importantly, analyses of primary oocytes, mature oocytes, and F_1 offspring from these experimental females produced statistically indistinguishable estimates of the effective number of mitochondria, leading to a general estimate of $N_{e\text{-}mt} = \sim 200$ (Jenuth et al. 1996). Similar results have been reported for a distinct strain of heteroplasmic mice carrying wild-type and deleted mtDNAs (Inoue et al. 2000). This is surprising given the pathological nature of the deleted mtDNA, but the close fit to neutral drift suggests there is no major effect of OXPHOS deficiency on transmission through germ cells in this strain.

Very different, and conflicting, results have been obtained from cows and humans. In cows, complete shifts from one mtDNA haplotype to another have been observed in one or two generations (Ashley et al. 1989, Koehler et al. 1991). These observations suggest a strong bottleneck in the effective population size of mtDNA through the germ line, possibly as small as a single mtDNA template. Rapid shifts from one homoplasmic state to another have also been observed in human pedigrees (Parsons et al. 1997, Blok et al. 1997), but this is not a general rule; Howell et al. (1992) obtained estimates of $N_{e\text{-mt}}$ ranging from 36 to 2400 (see also Poulton et al. 1998, Chinnery et al. 2000 for results similar to mice). Moreover, very different estimates of $N_{e\text{-mt}}$ have been reported when transmission was studied using blood samples ($N_{e\text{-mt}}$ from 800 to 4000) versus oocyte ($N_{e\text{-mt}}$ from 15 to 70; Blok et al. 1997). This is surprising given the many more cell generations that intervene between blood samples of two children and two oocytes. During the differentiation of mature oocytes from primary oocytes in mammals, the mitochondrial population is amplified at least 50-fold and the number of mtDNAs per mitochondrion is reduced from $\sim$10 to an average of 2 (Hauswirth & Laipis 1982, Blok et al. 1997). However, little mtDNA replication occurs during the early divisions following fertilization (Piko & Matsumoto 1976, Meirelles & Smith 1998). If alternative mtDNA haplotypes or mitochondria are sampled at random to serve as a "master copy" during the amplification phase, a population bottleneck would result.

WHAT ARE THE SEGREGATING UNITS?   The spatial distribution of mitochondria is very different in different tissues and changes dramatically during development, suggesting considerable control by nuclear genes (reviewed in Yaffe 1999). Moreover, mitochondria are associated with a number of cytoskeletal proteins involved in cytoplasmic movement (kinesin, cytoplasmic dynein, and dynamin homologs of GTP binding proteins) (Yaffe 1999). Whereas mtDNAs appear to be sampled randomly for replication during the cell cycle (Bogenhagen & Clayton 1977), this does not mean that subsequent transmission is a passive process. The dynamic associations between cytoskeletal proteins, nuclear signaling, and mitochondria suggest that a regulated mitochondrial "mitosis" may be involved in the partitioning of mitochondria prior to cytokinesis (Yaffe 1999, Jacobs et al. 2000). Rapid shifts in mtDNA haplotypes between generations could easily be achieved with pulses of replication, targeted mitochondrial trafficking, and a random distribution of mutant mtDNAs in the cell.

It is not clear whether clusters of distinct mitochondria (Jacobs et al. 2000) or individual heteroplasmic mitochondria (Howell et al. 1992) are the units sampled for transmission. Nor is it clear how this differs between somatic and germ cells. We need additional transmission studies in organisms with different patterns of germ line development and somatic partitioning.

Drift at More than One Level

GERMLINE VS. MATRILINE   It is intuitive that genetic drift within germ cells should occur faster than drift among maternal lineages. Jenuth et al. (1996) have quantified this difference. Using a mean value of $N_{e-mt} = 187$, taken from several studies of human mtDNA inheritance, the drift model indicates that the mean number of cell generations to reach fixation in homoplasmic cells is 370. At 24 germ cell generations per human generation, a mutant mtDNA will drift to fixation in a human matriline in 15 generations. Under neutrality the sojourn time to fixation for a novel mtDNA mutation in an organismal population is $2N_e$ generations. Based on an accepted estimate of the human population size ($N_e = 10,000$ individuals, or $\sim$5,000 females) (Takahata 1993), a neutral mtDNA mutation will drift to fixation in $\sim$10,000 generations, vastly longer than the fixation time of any given heteroplasmy (Jenuth et al. 1996).

This difference in the sorting time for drift within germ lines and drift among maternal lineages justifies the assumption that heteroplasmy can be ignored in some theoretical treatments of the evolutionary dynamics of mtDNA (Avise et al. 1984, Takahata & Palumbi 1985). From a units-of-evolution perspective, the fact that drift is operating at two (or more) levels of a hierarchy offers a unique opportunity to dissect the interaction among levels. Standing amounts of polymorphism for neutral mutations are determined by an equilibrium between mutational input and loss by genetic drift. Drift operating within germ cells leads to the loss of variability at the level of the individual organism. Thus, the action of an evolutionary force at one level affects the uniqueness of the higher-level individuals. In the terminology of hierarchy theory, there is "upward causation" (Vrba & Gould 1986). Because drift and selection can operate in different directions at the level of tissues (see below), the individual organism can acquire a uniqueness from the infinity of ways an organism can exhibit somatic mosaicism. However, drift at the level of individual organisms has a downward effect on variation at the cytoplasmic level and will tend to reduce it. The development of mature oocytes from a heteroplasmic stem cell in the germ line sorts variation from within to among populations. If sampling at the level of the individual is added on top of this, the approach to homoplasmy will be accelerated. In the terminology of hierarchy theory, drift at the individual level has "downward causation" effects on drift at the cytoplasmic level (Vrba & Gould 1986).

APPORTIONMENT OF DIVERSITY AMONG LEVELS   Mutation will introduce variation only at the lowest level, i.e., among mtDNAs within organelles. However, the relative contributions of drift at the multiple levels in the hierarchy will lead to some equilibrium apportionment of gene diversity within and among these levels. At very high mutation rates, most of the gene diversity will lie within cells because drift will not be "fast" enough to eliminate it. At lower mutation rates, the levels of the hierarchy carrying the greatest diversity will characterize the net effect of drift at multiple levels. The apportionment of diversity across this hierarchy can be quantified in a manner analogous to population structure analyses using F-statistics (with the caveat that Mendelian expectations do not apply to cytoplasmic systems).

Gene diversity is defined as $K = 1 - \Sigma x_i^2$, where x_i is the frequency of the ith variant at the focal level of the hierarchy (Birky et al. 1989, Rand & Harrison 1989). Following the notation of Birky et al. (1989), K_a is the diversity within a cell, K_b

is the diversity within an individual organism, K_c is the diversity within a deme (or collecting locality), and K_t is the diversity within the total sample. Following Lewontin (1972), these diversity measures can be apportioned into different hierarchical components: the within-individual component of the total diversity, the among-individual-within-deme component of the total diversity, and the among-deme component of the total diversity. With the appropriate data, apportionment of diversity across additional levels of a hierarchy are easily incorporated into these statistics (Birky et al. 1989, Rand & Harrison 1989).

Apportionment of mtDNA variability across this hierarchy has been done for mitochondrial VNTR (variable number of tandem repeat) systems in which the mutation rate is sufficiently high that heteroplasmy is easily observed (Rand & Harrison 1989, Arnason & Rand 1992, Wilkinson & Chapman 1991, Lunt et al. 1998). These studies showed that differences in the mutation-drift balance lead to distinct apportionment of diversity for different organisms (e.g., codfish have a higher within-individual component than evening bats, owing to a highly mutable VNTR). This approach has not been applied to the many heteroplasmies reported in mice and humans, but such studies would provide important contributions to the units of evolution view of mtDNA. With the advent of high-resolution methods for detecting heteroplasmy (Michikawa et al. 1999) and for quantifying the mtDNA in individual mitochondria (Cavelier et al. 2000), the apportionment of variation could be quantified for multiple levels in a variety of organisms.

Selection Within and Among Cytoplasms

MATERNAL INHERITANCE   With biparental inheritance of organelles, the evolutionary interests of the organelle may be in conflict with the evolutionary interests of the host cell or organism. It has been argued that uniparental inheritance is one means by which this conflict can be resolved (e.g., Cosmides & Tooby 1981, Hastings 1992). Some insight into the nature of these events can be provided by comparative analysis of the mechanisms that ensure uniparental inheritance of organelle DNA (Birky 1995). Studies indicating that "male" mitochondria are actively destroyed by the egg cytoplasm or early embryo provide compelling proximate evidence for interference competition in the cytoplasm. That these mechanisms are under control of multiple nuclear genes further suggests an ultimate mechanism for the continued avoidance of the presumed intracellular genomic conflict.

In *Drosophila*, sperm tail structures composed of proteins and mitochondrial derivatives persist throughout embryogenesis but are eventually enveloped by the developing midgut and defecated after larval hatching (Pitnick & Karr 1998). In mice, mitochondria from sperm that are microinjected into eggs get eliminated by the time of birth, but mitochondria from liver do not (Shitara et al. 2000), indicating that mitochondria from different tissues may be differentially tagged and recognized by the host cell. A study of the domestic cow has shown that sperm mitochondria are tagged with ubiquitin and targeted for destruction in the early embryo (Sutovsky et al. 1999).

The doubly uniparental inheritance of mtDNA in mussels is a special case. In *Mytilus* two distinct types of mtDNAs are found in males and females. Females transmit their mtDNA (F type) to female and male offspring, whereas males transmit their mtDNA (M type) only to males (Zouros et al. 1994, Skibinski et al. 1994). At fertilization, both male and female embryos receive M mtDNA from the sperm, but within 24 hours the M mtDNA is eliminated in female embryos and maintained in male embryos (Sutherland et al. 1998). In male germline the M mtDNA is preferentially amplified. These M and F mtDNA types predate the divergence of *Mytilus* species (Rawson & Hilbish 1995), but independent M and F mtDNA types have evolved in the history of marine mussels (Hoeh et al. 1997). A similar system exists in freshwater mussels that diverged from *Mytilus* more than 450 mya (Liu et al. 1996). Thus, cells of different tissue type and sex can detect and eliminate alternative mitochondrial types. Although direct homology of these pathways has yet to be determined, they may provide a means to dissect the mechanisms by which the nuclear host genome avoids intragenomic conflict by ensuring the uniparental inheritance of mtDNA.

Additional insight into the mechanisms of maternal inheritance comes from cases in which paternal leakage has been detected. Paternal leakage tends to be detected in crosses between distinct subspecies or species of animals, rather than between strains of the same species. Mechanisms that eliminate "male" mtDNA may thus be species specific and break down in disrupted, hybrid nuclear genomes. Kondo et al. (1990) observed a paternal leakage rate of 0.1% in crosses between *Drosophila simulans* females and *D. mauritiana* males. Gyllensten et al. (1991) detected paternal mtDNA in all tissues in crosses between *Mus musculus* and *M. spretus*, albeit at low levels (0.1%). Their analyses further indicated increases in proportion of paternal mtDNA through additional backcross generations. Shitara et al. (1998) detected paternal leakage in mice, but paternal mtDNA was restricted to the F_1 generation animals and was not found in all tissues, unlike the results of Gyllensten et al. (1991). In contrast to domestic cattle, ubiquitin tagging was not evident on sperm mitochondria in crosses between domestic cow and the Asian wild gaur, and male mitochondria persist beyond the third embryonic division (Sutovsky et al. 1999, 2000).

In *Mytilus* a breakdown of the typical doubly uniparental pattern of inheritance is also observed in hybrid zones between *M. galloprovincialis* and *M. trossulus* but not between the former and *M. edulis* (see Rawson et al. 1996). Moreover, where hybridization has occurred, the F type of *M. edulis* has invaded the M route of transmission in *M. trossulus* (see Quesada et al. 1999), suggesting occasional breakdown of the transmission barrier. Breakdown of the elimination of male mitochondria in hybrids could provide a powerful means to dissect the underlying nuclear control of this presumed mechanism for the avoidance of intragenomic conflict.

A RACE FOR REPLICATION IN THE CYTOPLASM   Organelle genomes with functional origins of replication should out-compete full-length wild-type molecules during replication (see Selosse et al. 2001). When these small genomes are not beneficial to the cell, their frequency within a heteroplasmic cytoplasm is inversely proportional to fitness at the cell level, providing a clear example of conflicting levels of selection (Reid 1980, Birky 1983, Hurst et al. 1996). Similar systems exhibiting conflicting selection at two levels have been known in *Aspergillus* and *Nerurospora* for some time (Lewontin 1970, p. 4). However, a simple kinetic race for replication does not fully account for the selective advantage of deleted mtDNAs. A complex signaling network regulates the replication and transcription of mtDNA, with ATP balance and OXPHOS activity of individual mitochondria serving as potential signals (Allen 1993). The nucleus of a cell carrying a heteroplasmic mixture of wild type and defective mtDNAs may receive signals that some mitochondria are respiring at reduced levels and send signals to those mitochondria. Hence, active mechanisms under nuclear control may play important roles in the over-replication of defective mtDNA in heteroplasmic cells (de Gray 1999, p. 91).

In both *Drosophila* (Solignac et al. 1984) and crickets (Rand & Harrison 1986), in which mtDNA heteroplasmy is due to VNTR length variation in the origin of replication region, the smaller mtDNA showed significant increases in frequency across animal generations. These results were interpreted as evidence for a race for replication favoring smaller molecules. However, subsequent studies in both *D. mauritiana* (Solignac et al. 1987) and *D. melanogaster* (Kann et al. 1998) showed that as a heteroplasmic female ages, the longer mtDNA shows a transmission advantage from mother to offspring, indicating that the selection coefficient for the short and long mtDNAs changes sign as heteroplasmic females age. Because these heteroplasmies involve indels of ~400 bp repeats in the origin of replication region that has secondary structure potential (Lewis et al. 1994), the length of the mtDNAs may be less important than the ability of these repeat arrays to initiate and complete mtDNA replication.

The construction of heteroplasmic lines using microinjection of intact mitochondria provides a powerful means with which to study selection between competing mtDNAs (see King & Attardi 1988 for a study in cell culture). Mitochondria from different strains and species of *Drosophila* have been transplanted in various combinations to construct experimental heteroplasmic lines, revealing clear evidence for selection in the cytoplasm (Matsuura et al. 1989, 1991; de Stordeur et al. 1989, 1997). However, manipulation of temperature (Matsuura et al. 1993) and nuclear background (Doi et al. 1999) can have considerable effects on the selection coefficients of mtDNAs in heteroplasmic germlines.

All pairwise combinations of microinjections were performed among the divergent strains of *D. simulans* (*si*I, *si*II, and *si*III) and *D. mauritiana* (*ma*I and *ma*II) (de Stordeur et al. 1989, de Stordeur 1997). When the donor was *si*I or *ma*II, no successful heteroplasmic lines were established. In all remaining heteroplasmies, *si*II showed a consistent advantage over other mtDNAs in most nuclear backgrounds. Overall, a fitness hierarchy among heteroplasmic mtDNAs could be defined as *si*II > *si*III = *ma*I > *si*I = *ma*II (de Stordeur 1997). Manipulative experiments of multi-level selection can be addressed in this system by competing these same mtDNA haplotypes against one another in population cages of homoplasmic flies. The available data suggest that the fitness hierarchy at the cytoplasmic level is maintained at the organismal level, but additional experiments are needed to determine the generality of this pattern in different nuclear backgrounds (see below).

TISSUE-SPECIFIC SELECTION   The two mouse models for mtDNA heteroplasmy discussed above provide contrasting pictures of tissue-specific selection for mtDNA haplotypes. In one study the two mtDNAs were essentially wild-type genomes from different lab strains (BALB and NZB) and showed random transmission when individual oocytes or whole F_1 offspring were assayed (Jenuth et al. 1996). Analyses of heteroplasmies in blood, kidney, liver, and spleen showed distinct tissue-specific selection for the alternative mtDNAs as the mice aged from 2 to 17 months. Because the two mtDNAs had identical sequences in origin of replication, selection may have been acting among mitochondria based on the proportion of the two mtDNAs within each organelle (Jenuth et al. 1997).

Different results were obtained for the heteroplasmic strain of mice carrying wild type and deleted mtDNAs (Inoue et al. 2000). Transmission between generations was essentially random, but in contrast to the BALB/NZB study, the deletion strain showed very little tissue-specific variation in heteroplasmy (Inoue et al. 2000). Low variation among tissues is in contrast with human studies, which show marked tissue-specific variation in heteroplasmies when pathological deletions are present with wild-type mtDNAs (Shoubridge 2000a). A comparable system of heteroplasmy has been studied in *Drosophila subobscura*, with a wild strain carrying a large deletion (Volz-Lingenhohl et al. 1992, Petit et al. 1998). Germ tissue and oocytes show the lowest levels of the deletion mutant (60%), which increases to over 80% in the third larval stage and remains stable in the adult.

Barritt et al. (2000) determined that the frequency of the "common" mtDNA deletion was 33% within heteroplasmic human oocytes. However, in later stage embryos the frequency of the mtDNA mutant was 8%, suggesting that oocytes with higher frequencies of the mtDNA mutant were less likely to be implanted as embryos. A replication race is not believed to cause this pattern because little mtDNA replication takes place from the mature oocyte through the blastocyst stage in mammals (Meirelles & Smith 1998). Furthermore, because ≥ 1000 mtDNAs are present in each cell, pathological mutations must be present at frequencies between 60 and 95% to have an effect, depending on the severity of the mutation (Blok et al. 1997, Jacobs et al. 2000). Thus, there is some disagreement in the literature about the evidence for selection among oocytes owing to heteroplasmy (Shoubridge 2000b).

Experimental work is needed to determine the relative importance of (*a*) selection between two mtDNAs in replication, (*b*) different rates of organelle replication, and (*c*) selection at the level of the cell. Such studies will help clarify the role that multi-level selection could play in stopping the advance of Muller's ratchet by modulating the efficacy of among-lineage purging of deleterious mutations (see Bergstrom & Pritchard 1998).

SELECTION AT THE LEVEL OF THE
INDIVIDUAL ORGANISM

Fitness Assays

The possibility of selection on mtDNA was mentioned in a seminal paper by Upholt & Dawid (1977) and raised by D. Ellis in the symposium questions following a paper presented by John Avise (see Avise 1986). The first studies that explicitly addressed selection on mtDNA were theoretical (Clark 1984, Gregorius & Ross 1984). These were followed by empirical studies seeking to detect fitness differences between mtDNA haplotypes collected from nature.

SELECTION ON HAPLOIDS   Clark (1984) reasoned that the extensive mtDNA polymorphism in natural populations might have three explanations: (*a*) neutrality with a high mutation rate, (*b*) multiple-niche polymorphism or frequency dependence, or (*c*) nuclear-mitochondrial fitness interactions. For haploid genomes, the maintenance of polymorphism through selection in multiple niches, or by frequency-dependent selection, requires a very restrictive set of reciprocally balanced fitness values. Clark (1984) and Gregorius & Ross (1984) showed that the addition of various cytonuclear fitness interactions did little to promote the stable maintenance of haploid variation. By inference, most of the nonneutral mtDNA variation within a panmictic population would be eliminated by selection. A consequence of this, however, is that mtDNA fitness variation might accumulate between populations.

Empirical support for this theory has been obtained using chromosomal fitness assays in *Drosophila* (Clark 1985, Clark & Lyckegaard 1988). Pairwise combinations of wild second chromosomes and cytoplasms were constructed such that the fitness of each chromosome could be tested with each cytoplasm and vice versa. No significant nuclear-cytoplasmic fitness interactions were detected within geographic populations of *D. melanogaster*, but when chromosomes and cytoplasm derived from Old World and New World populations were combined, significant nuclear x cytoplasmic interactions were detected. Statistical analyses of the mtDNA haplotypes in these experiments revealed that the fitness of individual chromosomes was influenced by mtDNA haplotypes, but only when divergent populations were involved. These experimental designs identify the unit of selection as individual nuclear chromosomes in transient heterozygous combinations with distinct cytoplasms (and mtDNA).

These results have been extended to ask if cytonuclear fitness effects are distinct for sex chromosomes and autosomes (Rand et al. 2001). The patterns of chromosomal transmission for males and females, coupled with maternal inheritance of mtDNA, establishes distinct patterns of cotransmission for autosomal-mtDNA versus sex chromosome–mtDNA combinations. In mammals and insects, for example, females are diploid at all chromosomes, whereas males are effectively haploid for the X-chromosome. In a random mating population the probability of cotransmission of an autosome with the cytoplasm is 0.5, whereas for an X-chromosome the probability is 0.67. By modifying earlier simulation studies,

A. Clark was able to show that X-chromosome-cytoplasm fitness interactions can maintain joint nuclear and cytoplasmic polymorphisms, unlike previous autosomal models (Rand et al. 2001). Chromosome segregation studies further showed that X-chromosome-cytoplasm fitness interactions could be detected within geographic populations of *D. melanogaster*, and that these interactions were sexually antagonistic: The high-fitness mtDNAs, cytoplasms, and X-chromosomes in females tended to have low-fitness states in males (Rand et al. 2001). These studies show that selection on cytonuclear interactions are very distinct for X-chromosomes and autosomes.

POPULATION CAGE STUDIES   Most population cage experiments have shown that one mtDNA variant increases in frequency over another across several generations (MacRae & Anderson 1988, Nigro & Prout 1990, Fos et al. 1990, Kambhampati et al. 1992, Nigro 1994, Hutter & Rand 1995, Kilpatrick & Rand 1995, Datta et al. 1996, Garcia-Martinez et al. 1998). In some cases, the apparent mtDNA fitness effect could be attributed to cytoplasmic incompatibility resulting from *Wolbachia* infections (e.g., Nigro & Prout 1990, Kambhampati et al. 1992). Cytoplasmic incompatibility occurs when an uninfected female mates with an infected male, resulting in reduced egg hatch rates. In the reciprocal cross normal egg hatch rates are observed. Thus, cytoplasms carrying *Wolbachia* have a fitness advantage over cytoplasms free of *Wolbachia*, resulting in the "hitchhiking" of other cytoplasmic factors, such as mtDNA. Nigro & Prout (1990) showed that this incompatibility was sufficient to account for the observed changes in mtDNA frequencies in their experimental populations, and that the nonneutral trajectories could not be attributed to any fitness differences between the mtDNAs themselves.

NUCLEAR-CYTOPLASMIC INTERACTIONS   In *Drosophila* population cage studies in which *Wolbachia* was eliminated by tetracycline treatment, repeatable changes in mtDNA frequencies were detected, but the fate of competing mtDNA variants was influenced by nuclear genetic background. Using distinct strains of *D. melanogaster* from Argentina and Central Africa, Kilpatrick & Rand (1995) showed that mtDNA haplotypes behaved as neutral variants when competed on either the Argentinian or the Central African nuclear background, but the Argentinian mtDNA had a transient advantage on a mixed (hybrid) nuclear background. In reciprocally backcrossed lines of *D. pseudoobscura* and *D. persimilis* the *D. pseudoobscura* mtDNA showed a strong and repeatable home team advantage on its own nuclear background, owing almost entirely to greater egg production. On the *D. persimilis* nuclear background, however, the two mtDNAs were effectively neutral variants (Hutter & Rand 1995). Because mtDNA haplotypes behave as unlinked loci with respect to nuclear chromosomes, the repeatable association of a mtDNA haplotype with fecundity differences across multiple generations and a frequency perturbation suggests a causal role for mtDNA. Datta et al. (1996) developed a neutrality test of change in joint nuclear and mtDNA frequencies applicable to these sorts of population cage studies. The test derives expectations for the decay of the cytonuclear disequilibrium measures defined by Asmussen et al. (1987). Applied to population cage data from *D. melanogaster*, the decay of cytonuclear disequilibria was significantly slower than expected, suggesting cytonuclear fitness interactions.

In *D. subobscura*, two distinct mtDNA haplotypes (I and II) are found throughout its range in Europe. In replicated population cages, haplotype II showed significant increases in frequency (Garcia-Martinez et al. 1998). Whereas the experimental results are clearly nonneutral, the data are difficult to reconcile with the observation that the haplotypes remain polymorphic in natural populations. If the natural polymorphism is due to selection, exactly reciprocal cytonuclear fitness interactions, habitat fitnesses, or frequency-dependent fitnesses are required (Clark 1984, Gregorius & Ross 1984). The difference between the experimental and natural populations is likely due to cytonuclear fitness interactions that were introduced in establishing the population cages.

These population cage experiments do suggest that selection at the level of the individual organism can be detected experimentally and that nuclear-cytoplasmic interactions are a fundamental context of this selection. Whereas there are limitations in applying the results of laboratory experiments to natural populations, it is clear that mtDNA markers can behave as nonneutral variants that could compromise evolutionary inference in studies of population structure, gene flow, and hybrid zones.

SELECTION AT THE CYTOPLASM AND ORGANISMAL LEVELS   Two-level selection on distinct mtDNAs can be addressed by comparing the data from heteroplasmic transmission of microinjected strains of *Drosophila* (Matsuura et al. 1989, de Stordeur 1997) and population cage studies (Aubert & Solignac 1990, Nigro 1994). Aubert & Solignac (1990) established replicate bottles of a single *D. simulans si*III female and 32 *D. mauritiana ma*II females; the *si*III mtDNA increased markedly in three generations. Whereas this experiment involved mixing of nuclear genomes as well, the selection is in the same direction as the heteroplasmic competition between *si*III and *ma*II at the cytoplasmic level (de Stordeur 1997). Using microinjected strains of *D. simulans* in which the *si*II haplotype had gone to fixation on the *si*III background, the *si*II haplotype was selected against when competed at the individual level in population cages on the *si*III background (Nigro 1994). This selection is opposite in direction from that reported for heteroplasmy (de Stordeur 1997). When the *si*II and *si*III mtDNAs were competed on a mixed nuclear background (from reciprocal F$_1$ crosses), the *si*II mtDNA increased in frequency. Additional population cage experiments are needed in each of the nuclear backgrounds to complement the network of pairwise combinations of selection at the cytoplasm level.

Neutrality Tests of mtDNA Sequences

MILDLY DELETERIOUS MTDNA EVOLUTION   The first suggestion that mtDNA polymorphisms depart from a neutral model were based on the frequency spectrum of polymorphic restriction sites in human mtDNA (Whittam et al. 1986). A test of the Ewens-Watterson model of neutrality using mtDNA restriction map data showed that high-frequency alleles (restriction sites) were more common than expected, intermediate frequency alleles were observed less often than expected, and singletons were more common than expected under the neutral assumptions (Whittam et al. 1986). Whereas demographic issues and population structure may have played a role in these patterns (Excoffier 1990, Merriwether et al. 1991), the action of purifying selection was supported by the observation that amino-acid altering restriction sites were observed as singletons more frequently than expected (Whittam et al. 1986).

An excess of amino acid polymorphism is the most consistent pattern of non-neutral evolution that has emerged from efforts to detect selection on mtDNA (Kaneko et al. 1993; Ballard & Kreitman 1994; Nachman et al. 1994, 1996; Rand et al. 1994, 2000; Rand & Kann 1996, 1998; Hasegawa et al. 1998; Nachman 1998; Templeton 1996; Wise et al. 1998; Quesada et al. 1999, Weinreich & Rand 2000, Ballard 2000a). This "excess" is defined in the context of the McDonald-Kreitman, or MK, test (McDonald & Kreitman 1991): The ratio of polymorphism to divergence at nonsynonymous sites ($r_{pd-nonsynonymous}$) tends to be greater than the ratio of polymorphism to divergence at synonymous "silent" sites ($r_{pd-synonymous}$). These data are generally interpreted as evidence for mildly deleterious selection on amino acid polymorphisms (Hasegawa et al. 1998, Nachman 1998, Rand & Kann 1998). Mutations that change the amino acid sequence of a protein are presumed to experience weak negative selection, which should reduce their rate of fixation. Relative to a strictly neutral mutation, a mildly deleterious mutation will contribute more to polymorphism than to divergence (Figure 3.7 in Kimura 1983). Because selection on synonymous sites is presumed to be weaker than for amino acid–altering sites, ($r_{pd-nonsynonymous}$) > ($r_{pd-synonymous}$) (see Akashi 1995, Rand & Kann 1996, 1998, Nachman 1998).

The data applied to an MK test can be used to estimate the historical selection coefficient (e.g., Mcdonald & Kreitman 1991, Sawyer & Hartl 1992; Akashi 1995). An index that correlates well with this selection coefficient can be derived from the basic data of an MK test [referred to as a neutrality index (NI)] (Rand & Kann 1996):

$$NI = (r_{pd-nonsynonymous})/(r_{pd-synonymous}). \qquad 2.$$

NI values >1 indicate negative selection and NI values <1 indicate positive selection. Note that a ratio of (dN/dS within species) to (dN/dS between species) is comparable to NI (where dN and dS are rates of nonsynonymous and synonymous substitutions, respectively) (see Hasegawa et al. 1998). With one degree of freedom, similar NI values can result from very different departures of neutrality. NI >1 can result from an excess of mildly deleterious nonsynonymous polymorphism, diversifying selection on nonsynonymous sites within species, or an excess of synonymous fixed differences. Several additional lines of evidence point to weak negative selection. At the ND3 gene in mice, polymorphic amino acid sites are not found at deep nodes in the genealogy of mtDNA haplotypes, which might be expected under balancing selection (Nachman et al. 1994). At the ND5 gene in *Drosophila*, amino acid polymorphisms have significantly negative Tajima's D values, whereas synonymous polymorphisms do not (Rand & Kann 1996). Nielsen & Weinreich (1999) showed that the mean age of nonsynonymous polymorphisms was less than that for synonymous polymorphisms, consistent with the mildly deleterious hypothesis. Nematodes provide a striking example in which mtDNA shows NI<1 (Blouin 2000). Although indicative of positive selection, this result stems from saturation of synonymous changes between species owing to a strongly biased base composition. An underestimation of the number of fixed synonymous differences will lead to an underestimation of NI values.

DELETERIOUS MUTATIONS VS. POLYMORPHISMS   Studies of mutation rates in mtDNAs provide further direct evidence for the deleterious effects of mtDNA mutations (Parsons et al. 1997, Denver et al. 2000). In the mutation accumulation lines in *C. elegans* (Denver et al. 2000), 6 synonymous and 8 nonsynonymous mutations were detected. Between 2 wild isolates of *C. elegans*, 26 synonymous and 3 nonsynonymous polymorphisms were detected (Denver et al. 2000). This is a significant (P<0.001) "excess" of amino acid mutations relative to amino acid polymorphisms. This comparison is essentially an MK test conducted at a lower level of organization and gives an NI value of 11.5. This is considerably higher than the average NI of 4.41 ± 4.5 for 31 mtDNA data sets tabulated from the literature (Weinreich & Rand 2000), indicating somewhat stronger negative selection than that inferred from polymorphism and divergence data.

Using the nonsynonymous/synonymous ratio for the mutation accumulation lines ($8/6 = 1.333$), Denver et al. (2000) estimated that 34.7 nonsynonymous polymorphisms would be detected between the two strains of *C. elegans* under strict neutrality (with 26 observed synonymous polymorphisms, $1.33 \times 26 = 34.7$). The efficiency of selection on amino acid changes is thus $1-(3$ observed polymorphisms/34.7 expected) $= 91.4\%$, implying the width of the "selective sieve" of 8.6%. Using 31 data sets on polymorphism and divergence reported for mtDNA in Weinreich & Rand (2000), a similar calculation provides a mean estimate of the efficiency of selection of $73.7\% \pm 26.8\%$, or a selective sieve of 26.3%. The mutation data from *C. elegans* are on the high end of the selective spectrum for these types of data. The selective filtering that prevents a spontaneous nonsynonymous mutation from entering the population as a polymorphism appears to be slightly stronger than the selective filtering that prevents existing nonsynonymous polymorphisms from reaching fixation.

MILDLY DELETERIOUS SYNONYMOUS EVOLUTION   If the mildly deleterious story is general, then this pattern should hold for any two classes of nucleotide change that show different levels of functional constraint. Insects tend to have highly AT-rich mtDNAs [78.6% in *Drosophila* (Clary & Wolstenholme 1985), 84.9% in honeybee (Crozier & Crozier 1993)], and the third codon position shows even greater skew in base composition than the average for the entire genome. Thus, A- and T-ending codons are used much more frequently than C- or G-ending codons. Mutations that change an A- or T-ending codon to a C- or G-ending codon should be deleterious or "unpreferred" relative to a change from C- or G-ending codon to A- or T-ending codons (Ballard & Kreitman 1994; see also Akashi 1995). Consistent with the standard MK tests, there is an excess of "unpreferred" synonymous changes from A- or T-codons to C- or G-codons within *Drosophila* mtDNA (Ballard & Kreitman 1994, Rand & Kann 1998). This result is also evident in a reciprocal pattern in the genes encoded on opposite strands in *Drosophila* mtDNA (Rand & Kann 1998, Ballard 2000a).

If synonymous sites show a pattern of excess "unpreferred" polymorphism in mtDNA, it would follow that functionally distinct classes of amino acid changes should exhibit a similar pattern, but this does not appear to be the case. When amino acids are classified as conservative or radical based on charge, polarity, or volume, there is no excess of radical polymorphisms (comparable to nonsynonymous polymorphism at the codon level) (Rand et al. 2000). It may be that the amino acid polymorphisms that are detectable in sequencing surveys have been filtered by purifying selection to a degree that they are effectively neutral by the criteria of an MK test. However, the partitioning of amino acid polymorphism into distinct classes reduces the sample size, and hence the power, of the MK tests. Additional studies of this sort are needed.

MILDLY DELETERIOUS EVOLUTION AND MITOCHONDRIAL GENETICS   To address why mtDNA might be predisposed to a mildly deleterious evolution, Weinreich & Rand (2000) compared 36 nuclear data sets from *Drosophila* with 31 mtDNA data sets from diverse species. The mean of NI values for the nuclear loci was significantly smaller than for the mitochondrial data sets, and this difference between genomes was even greater when only those genes showing significant MK tests were considered (Figure 4).

Among the several ways that the genetics of nuclear and mitochondria genomes differ, recombination is a likely source of the difference in average NI values (Weinreich & Rand 2000). Linkage does not prevent genes or nucleotide sites with different levels of functional constraint from evolving at different rates (e.g., Birky & Walsh 1988). However, linkage has a considerable effect on the levels of polymorphism, as a consequence of hitchhiking (Maynard Smith & Haigh 1974) or background selection (Charlesworth et al. 1993). For mtDNA this discrepancy emerges as a nonneutral MK test, which is a consequence of the relevant units of selection. For divergence the relevant units of selection are the different classes of functionally constrained sites, but for polymorphism the unit of selection is the entire mtDNA haplotype. An alternative explanation for the nuclear-mtDNA difference in NI values is that the genes in modern mitochondria have cellular functions more likely to evolve under purifying selection. Many nuclear genes have functions in which diversity is key: pathogen recognition and avoidance, sex peptides, etc. No such genes are carried in mtDNA, and hence are less likely to evolve by positive selection (Weinreich & Rand 2000).

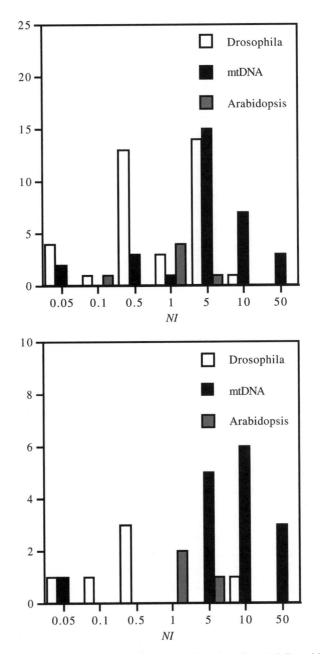

Figure 4 The distribution of neutrality index (NI) scores from McDonald-Kreitman (MK) tests for Drosophila nuclear, animal mtDNA, and Arabidopsis nuclear data sets. mtDNA and Arabidopsis tend to show NI values >1, indicative of negative selection. Top, all data sets; bottom, data sets in which the MK tests reject neutrality at the 5% level. Data from Weinreich & Rand (2000) and Rand et al. (2000).

SELECTIVE SWEEPS   Evidence for selective sweeps in mtDNA was first revealed by "unusually low" levels of restriction site polymorphism in geographic samples of *D. simulans* (Baba-Aïssa et al. 1988, Nigro 1988). Subsequent Hudson, Kreitman, Aguadé (HKA) tests (Hudson et al. 1987) of mtDNA sequences and neutral contrast loci showed that this low polymorphism was below that predicted by a neutral model (Ballard & Kreitman 1994, Rand et al. 1994). A whole-mtDNA sequence analysis by Ballard (2000b) showed that three independent selective sweeps have occurred in each of the three distinct mtDNA haplotypes of *D. simulans* (*si*I, *si*II, and *si*III). Because each of these strains carries a distinct type of *Wolbachia*, it is currently believed that these selective sweeps are driven by the *Wolbachia*-mediated cytoplasmic incompatibility systems, not by adaptive mutations in mtDNA (Ballard 2000b, Rand 2001).

Recently, data sets sufficient to perform tests of selective sweeps in mtDNA have accumulated in a diversity of taxa. The virtual absence of studies documenting selective sweeps targeted to mtDNA is surprising given the general pattern of low recombination regions of nuclear genomes showing reduced variation. Either mtDNA provides a small target for adaptive mutations, or selective sweeps may occur with normal frequency in mtDNA and their footprint is rapidly erased by the replenishment of polymorphism owing to a high mutation rate.

EVOLUTION ABOVE THE SPECIES LEVEL

Rate Heterogeneity

Rate heterogeneity exists both among nucleotide sites within mtDNA and among lineages of organisms (reviewed in Mindell & Thacker 1996, Xia 1998, Gissi et al. 2000). Considerable attention has been focused on potential mutation-based explanations for evolutionary rate variation among lineages, such as the generation time and metabolic rate hypotheses (Martin & Palumbi 1993, Rand 1994, Nunn & Stanley 1998, Gissi et al. 2000, Weinreich 2001). From the perspective of selection on mtDNA, however, evolutionary rate analyses in species with very different population sizes provide a means of studying the modulation of effective selection.

RATE ACCELERATION IN SMALL POPULATIONS   As predicted by the nearly neutral theory, the efficacy of selection should be reduced in small populations (Ohta 1992). One of the first studies to suggest this for mtDNA was a comparison of substitution rates in lineages of Hawaiian *Drosophila* based on restriction site maps (DeSalle & Templeton 1988). In the lineage that had experienced repeated founder events, mtDNA evolution was accelerated, consistent with a relaxation of purifying selection (see also Lynch 1997). Primates have a higher ratio of nonsynonymous to synonymous divergence (dN/dS ratio) for samples of nuclear genes than rodents, owing to presumed smaller effective population size in primates

(Ohta 1995). A parallel pattern has been documented for mtDNAs of primates and rodents, also consistent with the mild-deleterious model (Weinreich 2001). However, this study showed that changes in mutation rate—and hence dS values—in rodents were partly responsible for the reduced dN/dS ratio. Rates of divergence for nonsynonymous and synonymous sites in mitochondrial genes have been described for island-endemic lineages with small populations that should lead to greater lineage-specific accelerations for nonsynonymous sites than for synonymous sites. This prediction is upheld in island birds compared with their mainland sister taxa (Johnson & Seger 2001).

In *Drosophila* the situation has proved more complex (Ballard 2000a). *D. melanogaster* is believed to have a lower effective population size than its sister species, *D. simulans* (see Akashi 1995). Indeed, there is an excess of amino acid substitutions on the branch leading to *D. melanogaster* (Ballard 2000a). However, among synonymous sites, the more constrained "unpreferred" G- or C-ending codons do not show an excess of substitutions along the *D. melanogaster* branch. In island endemic lineages of the *D. simulans* clade, excess substitution for the functionally more-constrained classes of nucleotide sites is not consistently observed (Ballard 2000a). Hence, mtDNA evolution in this group of flies rejects both the neutral theory and simple predictions of the nearly neutral theory (Rand 2001). While lineage-specific changes in mutation rates are a possibility, fluctuating selection associated with changing population sizes seems more parsimonious.

Whole genome comparisons are providing new approaches for studying selection on mtDNA. Relaxed purifying selection in small effective population sizes should affect the entire mtDNA, but changes in the texture of selection could have different effects on specific mitochondrial genes. This can be detected statistically using a contingency table approach in which lineages are considered "rows" and genes are considered "columns." Lineage x locus interactions provide a powerful approach for discovering how selection acts on specific gene loci in nonrecombining mtDNA. Different genes show different patterns of rate acceleration in distinct lineages of the *D. melanogaster* subgroup (Ballard 2000a) and mammals (Gissi et al. 2000, Weinreich 2001).

A particularly promising variation on this theme of comparative genomics will be to explore the coevolution of nuclear and mitochondrial genomes. Lineage-specific covariation in evolutionary rates for nuclear- and mtDNA-encoded components of mitochondrial enzyme complexes is one means of taking a predictive approach to the lineage x locus tests of neutrality (Schmidt et al. 2001). If covariation in rate changes is detected for a nuclear and mitochondrial protein pair, the null hypothesis is that any random nuclear protein would show the same pattern of covariation. However, when the comparative sequence analysis can be done with nuclear and mitochondrial subunits that are known to interact, one has a set of candidate genes for the presumed coevolution (Weinreich 1998). As the biochemistry of OXPHOS proteins progresses, this analysis can be extended to individual amino acid sites. Moreover, comparisons of patterns of sequence divergence with functional assays of mitochondrial enzyme complexes is a promising approach to the coevolution of nuclear and mitochondrial genes (Kenyon & Moraes 1997, Edmands & Burton 1999).

PURIFYING SELECTION IS PERVASIVE   Despite clear modulation of dN/dS ratios, all mitochondrial proteins that have been examined to date have dN/dS ratios well below 1.0. Among hominoid mtDNAs, specific codons with dN/dS ratios >1 have been detected, but these represent <1% of the coding sites (Yang et al. 2000). This pattern is consistent with all of the data that have been presented in this review: mtDNA evolves under the scrutiny of purifying selection at the level of the individual organism.

Macroevolutionary Patterns of Genome Selection

If selection on mtDNA acts among genomes within cells, tissues, individuals, and populations, what is the important level of selection for mtDNA? The answer depends on the timescale of inquiry, but will always involve an interaction among levels. At the macroevolutionary scale the relevant units of evolution have been the dynamic interactions of the cyto-nuclear genome. These interactions have governed the two major evolutionary themes for organelle DNA: the reduction of genome size and the evolution of uniparental inheritance. Both of these transitions require the transfer of considerable control from the organelle "parasite" to the nuclear "host" genome.

GENOME SIZE   The extensive reduction in genome size during the evolution of mtDNA from a protomitochondrial genome (Gray et al. 1999, Lang et al. 1999) and the extreme genetic economy of modern animal mtDNA (Attardi 1985) indicate a long history of selection. The reduction in size is a combination of the transfer of genes to the nuclear host genome and the outright loss of genes that may be redundant or not necessary (Sogin 1997, Selosse et al. 2001). These patterns support the notion of selection at the level of the integrated cytonuclear genotype. A pattern of sequential gene transfer and loss occurs, with genes of central metabolism being the first to go and rRNA genes, cytochrome b, and subunits of cytochrome oxidase generally being retained (Gray et al. 1999, Lang et al. 1999). In plastid genomes the same set of genes has been retained in different lineages, indicating parallel gene loss or transfer for a diversity of other proto-plastid genes. Typically, organelle genome size has been reduced beyond the "standard" set of genes only in cases in which the organism—the cytonuclear genomic unit–has become parasitic (e.g., nonphotosynthetic, saprophytic plants) (Wolfe et al. 1992, Palmer et al. 2000). Thus, selection for specific biochemical functions appear to have dictated the outcome of genomic evolution (Race et al. 1999). These patterns of parallel evolution are as suggestive of adaptive evolution as parallel dog or cat ecomorphs in placental and marsupial mammals.

The transitions to reduced organelle genomes must have involved a heteroplasmic state, however brief, in which the smaller genome out-competed the resident wild-type genome. The establishment of the reduced organelle genome as a new wild-type requires selection at the level of the integrated nuclear-cytoplasmic unit because such a transition seems unlikely without some level of gene redundancy in the nuclear host genome (Selosse et al. 2001). These large-scale genome reorganizations are relatively rare in animal mtDNAs. Given the rate of organelle deletion and rearrangement in plants, species with major organelle genome rearrangements may be discovered in which wide crosses allow direct genetic experimentation of nuclear and cytoplasmic genomes. This could provide material for direct manipulation of the units of selection.

UNIPARENTAL INHERITANCE   Cytonuclear coevolution has also been a driving force in the evolution of uniparental inheritance. The best explanation for this focuses on the units-of-selection problem associated with competition in the cytoplasm for variant genomes. The conflict inherent in the opportunity for selfish organelle genomes to spread under biparental transmission can be resolved by the evolution of asexual, uniparental inheritance (Eberhard 1980, Cosmides & Tooby 1981, Hastings 1992). Again, this is not achieved by the organelle or the nuclear genome, but by an intricate coevolution of the two. The sequestering of organelles in sex-specific germ cells dictates the presence of distinct surface proteins of the organelles from sperm versus egg and the recognition of these epitopes by different tissues (see above) Sutovsky et al. 1999, Shitara et al. 2000). These transitions have largely been completed among extant organisms, and polymorphisms for degree of uniparental inheritance are not common. For direct experimentation, the doubly uniparental inheritance in mussels provides the most accessible system. However, with increased knowledge of the proteins that tag—and recognize— male and female mitochondria in a diversity of species, we may be able to disentangle the complex selective events that have lead to the common pattern of uniparental inheritance (Birky 1995).

SUMMARY AND PROSPECTS FOR FUTURE STUDIES

The hierarchical organization of organelle genomes (nested populations of DNAs within organelles, cells, germlines, populations, species, and higher clades) dictates a units-of-evolution perspective for organelle evolution. The reduction of genome size, uniparental inheritance, and the purging of deleterious mutations represent important themes in the macroevolution of organelle genomes. These themes have played out under the pressures of mutation, selection and drift operating at multiple levels across this nested hierarchy. Understanding the relative contributions of these forces at each level remains the biggest challenge for a mechanistic view of organelle evolution. We need additional analyses of mtDNA mutation accumulation, transmission, tissue-level variation, and whole-organism phenotype. A promising area of research will be the cytoplasmic population genetics of stem cells and the formation of mature oocytes. Comparisons of the mtDNA populations of oocytes with those that survive to emerge as F_1 offspring could help establish if this transition between levels is an especially important phase in the purging of deleterious mutations. Some of the most promising systems are those in which experimental heteroplasmies can be constructed and mtDNAs tested for fitness effects at more than one level. In these systems, the context of conflict and cooperation among mtDNAs can be manipulated, allowing true units-of-selection experiments to be done. Much of the statistical machinery for cytoplasmic population genetics is in place, as is much of the theoretical underpinning of what cooperation and conflict might look like (Buss 1987, Keller 1999, Michod 1999).

As the nuclear machinery for mitochondrial function becomes elucidated, comparative genomics will provide rich opportunities for empirical studies of evolutionary questions. Population samples of whole mitochondrial genomes are the new standard for neutrality tests (Ballard 2000a,b) and will provide crucial haplotype information for fitness studies in distinct nuclear backgrounds. Systems in which direct genetic manipulations are not tractable will still play an important role in the dissection of selection on mtDNA. Particular attention should be focused on comparative cytonuclear genomics of organisms in which the standard rules of mitochondrial genetics may not hold (e.g., maternal inheritance, lack of recombination, atypical gene sets or orders). When sister groups exhibit variation in these properties, strong inferences about the mechanisms leading to these transitions can be made.

Visit the Annual Reviews home page at www.AnnualReviews.org

LITERATURE CITED

Akashi H. 1995. Inferring weak selection from patterns of polymorphism and divergence at "silent" sites in Drosophila DNA. *Genetics* 139:1067–76

Albert B, Godelle B, Atlan A, De Paepe R, Gouyon PH. 1996. Dynamics of plant mitochondrial genome: model of a three-level selection process. *Genetics* 144:369–82

Allen JF. 1993. Control of gene expression by redox potential and the requirement for chloroplast and mitochondrial genomes. *J. Theor. Biol.* 165:609–31

Arnason E, Rand DM. 1992. Heteroplasmy of short tandem repeats in mitochondrial DNA of atlantic cod, *Gadus morhua. Genetics* 132:211–20

Ashley MV, Laipis PJ, Hauswirth WW. 1989. Rapid segregation of heteroplasmic bovine

mitochondria. *Nucleic Acids Res.* 17:7325–31

Asmussen MA, Arnold J, Avise JC. 1987. Definition and properties of disequilibrium statistics for associations between nuclear and cytoplasmic genotypes. *Genetics* 115:755–68

Attardi G. 1985. Animal mitochondrial DNA: an extreme example of genetic economy. *Int. Rev. Cytol.* 93:93–145

Aubert J, Solignac M. 1990. Experimental evidence for mitochondrial DNA introgression between *Drosophila* species. *Evolution* 44:1272–82

Avise JC. 1986. Mitochondrial DNA and the evolutionary genetics of higher animals. *Philos. Trans. R. Soc. London Ser.* B 312:325–42

Avise JC. 1991. Ten unorthodox perspectives on

evolution prompted by comparative population genetic findings on mitochondrial DNA. *Annu. Rev. Genet.* 25:45–69

Avise JC, Arnold J, Ball RM, Bermingham E, Lamb T, et al. 1987. Intraspecific phylogeography: the mitochondrial DNA bridge between population genetics and systematics. *Annu. Rev. Ecol. Syst.* 18:489–522

Avise JC, Neigel JE, Arnold J. 1984. Demographic influences on mitochondrial DNA lineage survivorship in animal populations. *J. Mol. Evol.* 20:99–105

Baba-Aïssa F, Solignac M, Dennebouy N, David JR. 1988. Mitochondrial DNA variability in *Drosophila* simulans: quasi absence of polymorphism within each of the three cytoplasmic races. *Heredity* 61:419–26

Ballard JWO. 2000a. Comparative genomics of mitochondrial DNA in members of the *Drosophila* melanogaster subgroup. *J. Mol. Evol.* 51:48–63

Ballard JWO. 2000b. Comparative genomics of mitochondrial DNA in *Drosophila simulans*. *J. Mol. Evol.* 51:64–75

Ballard JWO, Kreitman M. 1994. Unraveling selection in the mitochondrial genome of Drosophila. *Genetics* 138:757–72

Barritt JA, Brenner CA, Willadsen S, Cohen J. 2000. Spontaneous and artificial changes in human ooplasmic mitochondria. *Hum. Reprod.* 15(Suppl. 2):207–17

Beale G, Knowles J. 1979. *Extranuclear Genetics*. Baltimore: Univ. Park Press. 142 pp.

Bergstrom CT, Pritchard J. 1998. Germline bottlenecks and the evolutionary maintenance of mitochondrial genomes. *Genetics* 149:2135–46

Birky CW Jr. 1983. Relaxed cellular controls and organelle heredity. *Science* 222:468–75

Birky CW Jr. 1995. Uniparental inheritance of mitochondrial and chloroplast genes: mechanisms and evolution. *Proc. Natl. Acad. Sci. USA* 92:11331–38

Birky CW Jr, Fuerst P, Maruyama T. 1989. Organelle gene diversity under migration, mutation, and drift: equilibrium expectations, approach to equilibrium, effects of hetero-

plasmic cells, and comparison to nuclear genes. *Genetics* 121:613–27

Birky CW Jr, Walsh JB. 1988. Effects of linkage on rates of molecular evolution. *Proc. Natl. Acad. Sci. USA* 85:6414–18

Blackstone NW. 1995. A units-of-evolution perspective on the endosymbiont theory of the origin of the mitochondrion. *Evolution* 49:785–96

Blok RB, Gook DA, Thorburn DR, Dahl H-HM. 1997. Skewed segregation of the mtDNA nt 8993 (T→G) mutation in human oocytes. *Am. J. Hum. Genet.* 60:1495–501

Blouin MS. 2000. Neutrality tests on mtDNA: unusual results from nematodes. *J. Hered.* 91:156–58

Bogenhagen D, Clayton DA. 1977. Mouse L cell mitochondrial DNA molecules are selected randomly for replication throughout the cell cycle. *Cell* 11:719–27

Brown WM. 1985. The mitochondrial genome of animals. In *Molecular Evolutionary Genetics*, ed. RJ Macintyre, pp. 95–130. New York: Plenum

Brown WM, George M Jr, Wilson AC. 1979. Rapid evolution of animal mitochondrial DNA. *Proc. Natl. Acad. Sci. USA* 76:1967–71

Buss LW. 1987. *The Evolution of Individuality*. Princeton, NJ: Princeton Univ. Press

Cavelier L, Johannisson A, Gyllensten U. 2000. Analysis of mtDNA copy number and composition of single mitochondrial particles using flow cytometry and PCR. *Exp. Cell. Res.* 259:79–85

Charlesworth B, Morgan MT, Charlesworth D. 1993. The effect of deleterious mutations on neutral molecular variation. *Genetics* 134:1289–303

Chen X, Prosser R, Simonetti S, Sadlock J, Jagiello G, et al. 1995. Rearranged mitochondrial genomes are present in human oocytes. *Am. J. Hum. Genet.* 57:239–47

Chinnery PF, Thorburn DR, Samuels DC, White SL, Dahl HM, et al. 2000. The inheritance of mitochondrial DNA heteroplasmy: random drift, selection or both? *Trends Genet.* 16:500–5

Clark AG. 1984. Natural selection with nuclear and cytoplasmic transmission. I. A deterministic model. *Genetics* 107:679–701

Clark AG. 1985. Natural selection with nuclear and cytoplasmic transmission. II. Tests with *Drosophila* from diverse populations. *Genetics* 111:97–112

Clark AG, Lyckegaard EM. 1988. Natural selection with nuclear and cytoplasmic transmission. III. Joint analysis of segregation and mtDNA in Drosophila melanogaster. *Genetics* 118:471–81

Clary DO, Wolstenholme DR. 1985. The mitochondrial DNA molecule of *Drosophila yakuba:* nucleotide sequence, gene organization, and genetic code. *J. Mol. Evol.* 22:252–71

Cortopassi G, Shibata D, Soong NW, Arnheim N. 1992. A pattern of accumulation of a somatic deletion of mitochondrial DNA in aging human tissues. *Proc. Natl. Acad. Sci. USA* 89:7370–74

Cosmides LM, Tooby J. 1981. Cytoplasmic inheritance and intragenomic conflict. *J. Theor. Biol.* 89:83–129

Crozier RH, Erozier YC. 1993. The mitochondrial genome of the honeybee *Apis mellifera*: complete sequence and genome organization. *Genetics* 133:97–117

Datta S, Kiparsky M, Rand DM, Arnold J. 1996. A statistical test of a neutral model using the dynamics of cytonuclear disequilibria. *Genetics* 144:1985–92

de Gray ADNJ. 1999. *The Mitochondrial Free Radical Theory of Aging.* Georgetown, TX: Landes

Denver DR, Morris K, Lynch M, Vassilieva L, Thomas WK. 2000. High direct estimate of mutation rate in the mitochondrial genome of *Caenorhabditis elegans. Science* 289:2342–44

DeSalle R, Templeton AR. 1988. Founder effects and the rate of mitochondrial DNA evolution in Hawaiian *Drosophila. Evolution* 42:1076–84

de Stordeur E. 1997. Nonrandom partition of mitochondria in heteroplasmic *Drosophila. Heredity* 79:615–23

de Stordeur E, Solignac M, Monnerot M, Mounolou JC. 1989. The generation of transplasmic *Drosophila simulans* by cytoplasmic injection: effects of segregation and selection on the perpetuation of mitochondrial DNA heteroplasmy. *Mol. Gen. Genet.* 220:127–32

Doi A, Suzuki H, Matsuura ET. 1999. Genetic analysis of temperature-dependent transmission of mitochondrial DNA in Drosophila. *Heredity* 82:555–60

Eberhard WG. 1980. Evolutionary consequences of intracellular organelle competition. *Q. Rev. Biol.* 55:231–49

Edmands S, Burton RS. 1999. Cytochrome *c* oxidase activity in interpopulation hybrids of a marine copepod: a test for nuclear-nuclear or nuclear-cytoplasmic coadapatation. *Evolution* 53:1972–78

Excoffier L. 1990. Evolution of human mitochondrial DNA: evidence for departure from a pure neutral model of populations at equilibrium. *J. Mol. Evol.* 30:125–39

Finkel T, Holbrook NJ. 2000. Oxidants, oxidative stress and the biology of aging. *Nature* 408:239–47

Fos M, Dominguez MA, Latorre A, Moya A. 1990. Mitochondrial DNA evolution in experimental populations of *Drosophila subobscura. Proc. Natl. Acad. Sci. USA* 87:4198–201

Garcia-Martinez J, Castro JA, Ramon M, Latorre A, Moya A. 1998. Mitochondrial DNA haplotype frequencies in natural and experimental populations of *Drosophila subobscura. Genetics* 149:1377–82

Gillham NW. 1978. *Organelle Heredity.* New York: Raven. 650 pp.

Gissi C, Reyes A, Pesole G, Saccone C. 2000. Lineage-specific evolutionary rate in mammalian mtDNA. *Mol. Biol. Evol.* 17:1022–31

Gray M, Burger G, Land BF. 1999. Mitochondrial evolution. *Science* 283:1476–81

Gregorius H-R, Ross MD. 1984. Selection with gene-cytoplasm interactions. I. Maintenance of cytoplasmic polymorphisms. *Genetics* 107:165–78

Gyllensten U, Wharton D, Josefsson A, Wilson AC. 1991. Paternal inheritance of mitochondrial DNA in mice. *Nature* 352:255–57

Harrison RG. 1989. Mitochondrial DNA as a genetic marker in population and evolutionary genetics. *Trends. Ecol. Evol.* 4:6–11

Harrison RG, Rand DM, Wheeler WC. 1985. Mitocondrial DNA size variation within individual crickets. *Science* 228:1446–48

Hasegawa M, Cao Y, Yang Z. 1998. Preponderance of slightly deleterious polymorphism in mitochondrial DNA: nonsynonymous/synonymous rate ratio is much higher within species than between species. *Mol. Biol. Evol.* 15:1499–505

Hastings IM. 1992. Population genetic aspects of deleterious cytoplasmic genomes and their effect on the evolution of sexual reproduction. *Genet. Res. Camb.* 59:215–25

Hauswirth WW, Laipis PJ. 1982. Mitochondrial DNA polymorphism in a maternal lineage of Holstein cows. *Proc. Natl. Acad. Sci. USA* 79:4686–90

Hauswirth WW, Van De Walle MJ, Laipis PJ, Olivo PD. 1984. Heterogeneous mitochondrial DNA D-loop sequences in bovine tissue. *Cell* 37:1001–7

Hoeh WR, Stewart DT, Saavedra C, Sutherland BW, Zouros E. 1997. Phylogenetic evidence for role-reversals of gender-associated mitochondrial DNA in Mytilus (Bivalvia: Mytilidae). *Mol. Biol. Evol.* 14:959–67

Howell N, Halvorson S, Kubacka I, McCullough DA, Bindoff LA, et al. 1992. Mitochondrial gene segregation in mammals: is the bottleneck always narrow? *Hum. Genet.* 90:117–20

Hudson RR, Kreitman M, Aguadé M. 1987. A test of neutral molecular evolution based on nucleotide data. *Genetics* 116:153–59

Hurst LD, Atlan A, Bengtsson BO. 1996. Genetic conflicts. *Q. Rev. Biol.* 71:317–64

Hutter CM, Rand DM. 1995. Competition between mitochondrial haplotypes in distinct nuclear genetic environments: *Drosophila pseudoobscura* vs. *D. persimilis*. *Genetics* 140:537–48

Ingman M, Kaessmann H, Paabo S, Gyllensten U. 2000. Mitochondrial genome variation and the origin of modern humans. *Nature* 408:708–13

Inoue K, Nakada K, Ogura A, Isobe K, Goto Y, et al. 2000. Generation of mice with mitochondrial dysfunction by introducing mouse mtDNA carrying a deletion into zygotes. *Nat. Genet.* 26:176–81

Jacobs HT, Lehtinen SK, Spelbrink JN. 2000. No sex please, we're mitochondria: a hypothesis on the somatic unit of inheritance of mammalian mtDNA. *BioEssays* 22:564–72

Jenuth JP, Peterson AC, Fu K, Shoubridge EA. 1996. Random genetic drift in female germline explains the rapid segregation of mammalian mitochondrial DNA. *Nat. Genet.* 14:146–51

Jenuth JP, Peterson AC, Shoubridge EA. 1997. Tissue-specific selection for different mtDNA genotypes in heteroplasmic mice. *Nat. Genet.* 16:93–95

Johnson K, Seger J. 2001. Elevated rates of nonsynonymous evolution in island birds. *Mol. Biol. Evol.* 18:874–81

Kambhampati S, Rai KS, Verleye DM. 1992. Frequencies of mitochondrial DNA haplotypes in laboratory cage populations of the mosquito, *Aedes albopictus*. *Genetics* 132:205–9

Kaneko M, Satta Y, Matsuura ET, Chigusa SI. 1993. Evolution of the mitochondrial ATPase 6 gene in *Drosophila*: unusually high level of polymorphism in *D. melanogaster*. *Genet. Res.* 61:195–204

Kann LM, Rosenblum EB, Rand DM. 1998. Aging, mating, and the evolution of mtDNA heteroplasmy in *Drosophila melanogaster*. *Proc. Natl. Acad. Sci. USA* 5:2372–77

Keller L. 1999. *Levels of Selection in Evolution*. Princeton, NJ: Priunceton Univ. Press

Kenyon L, Moraes CT. 1997. Expanding the functional human mitochondrial DNA database by the establishment of primate xenomitochondrial cybrids. *Proc. Natl. Acad. Sci. USA* 94:9131–35

Kilpatrick ST, Rand DM. 1995. Conditional hitchhiking of mitochondrial DNA: frequency shifts of *Drosophila melanogaster* mtDNA variants depend on nuclear genetic background. *Genetics* 141:1113–24

Kimura M. 1983. *The Neutral Theory of Molecular Evolution.* Cambridge: Cambridge Univ. Press

King MP, Attardi G. 1988. Injection of mitochondria into human cells leads to a rapid replacement of the endogenous mitochondrial DNA. *Cell* 52:811–19

Klekowski EJ. 1998. Mutation rates in mangroves and other plants. *Genetica* 102/103: 325–31

Koehler CM, Lindberg GL, Brown DR, Beitz DC, Freeman AE, et al. 1991. Replacement of bovine mitochondrial DNA by a sequence variant within one generation. *Genetics* 29:247–55

Kondo R, Satta Y, Matsuura ET, Ishiwa H, Takahata N, et al. 1990. Incomplete maternal transmission of mitochondrial DNA in *Drosophila. Genetics* 126:657–63

Lang BF, Gray MW, Burger G. 1999. Mitochondrial genome evolution and the origin of the eukaryotes. *Annu. Rev. Genet.* 33:351–97

Lewis DL, Farr CL, Farquhar AL, Kaguni LS. 1994. Sequence, organization, and evolution of the A+T region of *Drosophila melanogaster* mitochondrial DNA. *Mol. Biol. Evol.* 11:523–38

Lewontin RC. 1970. The units of selection. *Annu. Rev. Ecol. Syst.* 1:1–18

Lewontin RC. 1972. The apportionment of human diversity. *Evol. Biol.* 6:381–98

Lightowlers RN, Chinnery PF, Turnbull DM, Howell N. 1997. Mammalian mitochondrial genetics: heredity, heteroplasmy and disease. *Trends Genet* 13:450–55

Liu H-P, Mitton JB, Wu S-K. 1996. Paternal mitochondrial DNA differentiation far exceeeds maternal mitochondrial DNA and allozyme differentiation in the freshwater mussel, *Anodonta grandis grandis. Evolution* 50:952–57

Lunt DH, Whipple LE, Hyman BC. 1998. Mitochondrial DNA variable number of tandem repeats (VNTRs): utility and problems in molecular ecology. *Mol. Ecol.* 7:1441–55

Lynch M. 1997. Mutation accumulation in nuclear, organelle, and prokaryotic transfer RNA genes. *Mol. Biol. Evol.* 14:914–25

Lynch M, Blanchard JL. 1998. Deleterious mutation accumulation in organelle genomes. *Genetica* 102/103:29–39

MacRae AF, Anderson WW. 1988. Evidence for non-neutrality of mitochondrial DNA haplotypes in *Drosophila pseudoobscura. Genetics* 120:485–94

Marchington DR, Hartshorne GM, Barlow D, Poulton J. 1997. Homopolymeric tract heteroplasmy in mtDNA from tissues SS and single oocytes: support for a genetic bottleneck. *Am. J. Hum. Genet.* 60:408–16

Martin AP, Palumbi SR. 1993. Body size, metabolic rate, generation time, and the molecular clock. *Proc. Natl. Acad. Sci. USA* 90:4087–91

Matsuura ET, Chigusa SI, Niki Y. 1989. Induction of mitochondrial DNA heteroplasmy by intra- and interspecific transplantation of germ plasm in *Drosophila. Genetics* 122:663–67

Matsuura ET, Niki Y, Chigusa SI. 1991. Selective trasmission of mitochondrial DNA in heteroplasmic lines for intra- and interspecific combinations in *Drosophila melanogaster. Jpn. J. Genet.* 66:197–207

Matsuura ET, Niki Y, Chigusa SI. 1993. Temperature-dependent selection in the transmission of mitochondrial DNA in *Drosophila. Jpn. J. Genet.* 8:127–35

Maynard Smith J, Haigh J. 1974. The hitchhiking effect of a favourable gene. *Genet. Res.* 23:23–35

Maynard Smith J, Szathmáry E. 1995. *The Major Transitions in Evolution.* Oxford: Freeman. 346 pp.

McDonald JH, Kreitman M. 1991. Adaptive protein evolution at the *Adh* locus in *Drosophila. Nature* 351:652–54

Meirelles FV, Smith LC. 1998. Mitochondrial genotype segregation during preimplantation development in mouse heteroplasmic embryos. *Genetics* 148:877–83

Merriwether DA, Clark AG, Ballinger SW, Schurr TG, Soodyall H, et al. 1991. The structure of human mitochondrial DNA variation. *J. Mol. Evol.* 33:543–55

Michikawa Y, Mazzucchelli F, Bresolin N, Scarlato G, Attardi G. 1999. Aging-dependent large accumulation of point mutations in the human mtDNA control region for replication. *Science* 286:774–79

Michod RE. 1999. *Darwinian Dynamics: Evolutionary Transitions in Fitness and Individuality.* Princeton, NJ: Princeton Univ. Press

Mindell DP, Thacker CE. 1996. Rates of molecular evolution: phylogenetic issues and applications. *Annu. Rev. Ecol. Syst.* 27:279–303

Moritz C, Dowling TE, Brown WM. 1987. Evolution of animal mitochondrial DNA: relevance for population biology and systematics. *Annu. Rev. Ecol. Syst.* 18:269–92

Nachman MW. 1998. Deleterious mutations in animal mitochondrial DNA. *Genetica* 102/103:61–69

Nachman MW, Boyer SN, Aquadro CF. 1994. Non-neutral evolution at the mitochondrial ND3 gene in mice. *Proc. Natl. Acad. Sci. USA* 91:6364–68

Nachman MW, Brown WM, Stoneking M, Aquadro CF. 1996. Nonneutral mitochondrial DNA variation in humans and chimpanzees. *Genetics* 142:953–63

Nielsen R, Weinreich DM. 1999. The age of nonsynonymous and synonymous mutations in animal mtDNA and implications for the mildly deleterious theory. *Genetics* 153:497–506

Nigro L. 1988. Natural populations of *Drosophila simulans* show great uniformity of the mitochondrial DNA restriction map. *Genetica* 77:133–36

Nigro L. 1994. Nuclear background affects frequency dynamics of mitochondrial DNA variants in *Drosophila simulans. Heredity* 72:582–86

Nigro L, Prout T. 1990. Is there selection on RFLP differences in mitochondrial DNA? *Genetics* 125:551–55

Nunn GB, Stanley SE. 1998. Body size effects and rates of cytochrome b evolution in Tubenosed seabirds. *Mol. Biol. Evol.* 15:1360–71

Ohta T. 1992. The nearly neutral theory of molecular evolution. *Annu. Rev. Ecol. Syst.* 23:263–86

Ohta T. 1995. Synonymous and nonsynonymous substitutions in mammalian genes and the nearly neutral theory. *J. Mol. Biol.* 40:56–63

Palmer JD, Adams KL, Cho Y, Parkinson CL, Qiu YL, et al. 2000. Dynamic evolution of plant mitochondrial genomes: mobile genes and introns and highly variable mutation rates. *Proc. Natl. Acad. Sci. USA* 97:6960–66

Parsons TJ, Muniec DS, Sullivan K, Woodyatt N, Alliston-Greiner R, et al. 1997. A high observed substitution rate in the human mitochondrial DNA control region. *Nat. Genet.* 15:363–68

Petit N, Touraille S, Debise R, Morel F, Renoux M, et al. 1998. Developmental changes in heteroplasmy level and mitochondrial gene expression in a *Drosophila subobscura* mitochondrial deletion mutant. *Curr. Genet.* 33:330–39

Piko L, Matsumoto L. 1976. Number of mitochondria and some properties of mitochondrial DNA in the mouse egg. *Dev. Biol.* 49:1–10

Pitnick S, Karr TL. 1998. Paternal products and by-products in *Drosophila* development. *Proc. R. Soc. London Ser.* B 265:821–26

Poulton J, Macauay V, Marchington DR. 1998. Mitochondrial genetics '98: is the bottleneck cracked? *Am. J. Hum. Genet.* 62:752–57

Quesada H, Wenne R, Skibinski DO. 1999. Interspecies transfer of female mitochondrial DNA is coupled with role- reversals and departure from neutrality in the mussel *Mytilus trossulus. Mol. Biol. Evol.* 16:655–65

Race HL, Herrmann RG, Martin W. 1999. Why have organelles retained genomes? *Trends. Genet.* 15:364–70

Rand DM. 1993. Endotherms, ectotherms, and mitochondrial genome-size variation. *J. Mol. Evol.* 37:281–95

Rand DM. 1994. Thermal habit, metabolic rate and the evolution of mitochondrial DNA. *Trends Ecol. Evol.* 9:125–31

Rand DM. 2001. Mitochondrial genomics flies high. *Trends Ecol. Evol.* 16:2–4

Rand DM, Clark AG, Kann LM. 2001. Sexually antagonistic cytonuclear fitness effects in *Drosophila melanogaster*. *Genetics*. In press

Rand DM, Dorfsman M, Kann LM. 1994. Neutral and non-neutral evolution of *Drosophila* mitochondrial DNA. *Genetics* 138:741–56

Rand DM, Harrison RG. 1986. Mitochondrial DNA transmission genetics in crickets. *Genetics* 114:955–70

Rand DM, Harrison RG. 1989. Molecular population genetics of mtDNA size variation in crickets. *Genetics* 121:551–69

Rand DM, Kann LM. 1996. Excess amino acid polymorphism in mitochondrial DNA: contrasts among genes from *Drosophila*, mice, and humans. *Mol. Biol. Evol.* 13:735–48

Rand DM, Kann LM. 1998. Mutation and selection at silent and replacement sites in the evolution of animal mitochondrial DNA. *Genetica* 102/103:393–407

Rand DM, Weinreich DM, Cezairliyan BO. 2000. Neutrality tests of conservative-radical amino acid changes in nuclear- and mitochondrially-encoded proteins. *Gene* 261:115–25

Rawson PD, Hilbish TJ. 1995. Evolutionary relationships among the male and female mitochondrial DNA lineages in the *Mytilus edulis* species complex. *Mol. Biol. Evol.* 12:893–901

Rawson PD, Secor CL, Hilbish TJ. 1996. The effects of natural hybridization on the regulation of doubly uniparental mtDNA inheritance in blue mussels (*Mytilus* spp.). *Genetics* 144:241–48

Reid RA. 1980. Selfish DNA in "petite" mutants. *Nature* 285:620

Sawyer SA, Hartl DL. 1992. Population genetics of polymorphism and divergence. *Genetics* 132:1161–76

Schmidt TR, Wu W, Goodman M, Grossman LI. 2001. Evolution of nuclear- and mitochondrial-encoded subunit interaction in cytochrome c oxidase. *Mol. Biol. Evol.* 18:563–69

Selosse M-A, Albert B, Godelle B. 2001. Reducing the genome size of organelles favours gene transfer to the nucleus. *Trends. Ecol. Evol.* 16:135–41

Shigenaga MK, Hagen TM, Ames BN. 1994. Oxidative damage and mitochondrial decay in aging. *Proc. Natl. Acad. Sci. USA* 91:10771–78

Shitara H, Hayashi JI, Takahama S, Kaneda H, Yonekawa H. 1998. Maternal inheritance of mouse mtDNA in interspecific hybrids: segregation of the leaked paternal mtDNA followed by the prevention of subsequent paternal leakage. *Genetics* 148:851–57

Shitara H, Kaneda H, Sato A, Inoue K, Ogura A, et al. 2000. Selective and continuous elimination of mitochondria microinjected into mouse eggs from spermatids, but not from liver cells, occurs throughout embryogenesis. *Genetics* 156:1277–84

Shoubridge EA. 2000a. A debut for mitomouse. *Nat. Genet.* 26:132–34

Shoubridge EA. 2000b. Mitochondrial DNA segregation in the developing embryo. *Hum. Reprod.* 15(Suppl. 2):229–34

Skibinski DO, Gallagher C, Beynon CM. 1994. Mitochondrial DNA inheritance. *Nature* 368:817–18

Sogin M. 1997. History assignment: When was the mitochondrion founded? *Curr. Opin. Genet. Dev.* 7:792–99

Solignac M, Genermont J, Monnerot M, Mounolou J-C. 1984. Genetics of mitochondria in *Drosophila*: mtDNA inheritance in heteroplasmic strains of *D. mauritiana*. *Mol. Gen. Genet.* 197:183–88

Solignac M, Genermont J, Monnerot M, Mounolou J-C. 1987. *Drosophila* mitochondrial genetics: evolution of heteroplasmy through germ line cell divisions. *Genetics* 117:687–96

Solignac M, Monnerot M, Mounolou JC. 1983. Mitochondrial DNA heteroplasmy in *Drosophila* mauritiana. *Proc. Natl. Acad. Sci. USA* 80:6942–46

Sutherland B, Stewart D, Kenchington ER, Zouros E. 1998. The fate of paternal mitochondrial DNA in developing female mussels, *Mytilus edulis*: implications for the mechanism of doubly uniparental inheritance of mitochondrial DNA. *Genetics* 148:341–47

Sutovsky P, Moreno RD, Ramalho-Santos J, Dominko T, Simerly C, et al. 1999. Ubiquitin tag for sperm mitochondria. *Nature* 402:371–72

Sutovsky P, Moreno RD, Ramalho-Santos J, Dominko T, Simerly C, et al. 2000. Ubiquitinated sperm mitochondria, selective proteolysis, and the regulation of mitochondrial inheritance in mammalian embryos. *Biol. Reprod.* 63:582–90

Takahata N. 1993. Allelic genealogy and human evolution. *Mol. Biol. Evol.* 10:2–22

Takahata N, Palumbi SR. 1985. Extranuclear differentiation and gene flow in the finite island model. *Genetics* 109:441–57

Templeton AR. 1996. Contingency tests of neutrality using intra/interspecific gene trees: the rejection of neutrality for the evolution of mitochondrial cytochrome oxidase II gene in the hominoid primates. *Genetics* 144:1263–70

Thrailkill KM, Birky CW Jr. 1980. Intracellular population genetics: evidence for random drift of mitochondrial allele frequencies in *Saccharomyces cerevisiae* and *Schizosaccharomyces pombe*. *Genetics* 96:237–62

Upholt WB, Dawid IB. 1977. Mapping of mitochondrial DNA of individual sheep and goats: rapid evolution in the D loop region. *Cell* 11:571–83

Volz-Lingenhöhl A, Solignac M, Sperlich D. 1992. Stable heteroplasmy for a large-scale deletion in the coding region of *Drosophila subobscura* mitochondrial DNA. *Proc. Natl. Acad. Sci. USA* 89:11528–32

Vrba ES, Gould SJ. 1986. The hierarchical expansion of sorting and selection: Sorting and selection cannot be equated. *Paleobiology* 12:217–38

Wallace D. 1999. Mitochondrial diseases in man and mouse. *Science* 283:1482–88

Weinreich DM. 1998. *OXPHOS enzymes and the molecular clock*. PhD thesis, Harvard Univ., Cambridge, MA

Weinreich DM. 2001. The rates of molecular evolution in rodent and primate mitochondrial DNA. *J. Mol. Evol.* 52:40–50

Weinreich DM, Rand DM. 2000. Contrasting patterns of non-neutral evolution in proteins encoded in nuclear and mitochondrial genomes. *Genetics* 156:385–99

Whittam TS, Clark AG, Stoneking M, Cann RL, Wilson AC. 1986. Allelic variation in human mitochondrial genes based on patterns of restriction site polymorphism. *Proc. Natl. Acad. Sci. USA* 83:9611–15

Wilkinson GS, Chapman AM. 1991. Length and sequence variation in evening bat D-loop mtDNA. *Genetics* 128:607–17

Wise CA, Sraml M, Easteal S. 1998. Departure from neutrality at the mitochondrial NADH dehydrogenase subunit 2 gene in humans, but not in chimpanzees. *Genetics* 148:409–21

Wolfe KH, Morden CW, Ems SC, Palmer JD. 1992. Rapid evolution of the plastid translational apparatus in a nonphotosynthetic plant: loss or accelerated sequence evolution of tRNA and ribosomal protein genes. *J. Mol. Biol.* 35:304–17

Wright S. 1968. *Evolution and the Genetics of Populations,Vol. 2. The Theory of Gene Frequencies*. Chicago: Univ. Chicago Press. 511 pp.

Xia X. 1998. The rate heterogeneity of nonsynonymous substitutions in mammalian mitochondrial genes. *Mol. Biol.Evol.* 15:336–44

Yaffe MP. 1999. The machinery of mitochondrial inheritance and behavior. *Science* 283:1493–97

Yang Z, Nielsen R, Goldman N, Pedersen AM. 2000. Codon-substitution models for heterogeneous selection pressure at amino acid sites. *Genetics* 155:431–49

Zouros E, Ball AO, Saavedra C, Freeman KR. 1994. Mitochondrial DNA inheritance. *Nature* 368:818

Annu. Rev. Ecol. Syst. 2001. 32:449–80

Evolutionary Patterns Among Permo-Triassic Therapsids[*]

Bruce S. Rubidge[1] and Christian A. Sidor[2]

[1]*Bernard Price Institute for Palaeontological Research, University of Witwatersrand, Johannesburg, South Africa; e-mail: 106gar@cosmos.wits.ac.za*
[2]*Department of Paleobiology, National Museum of Natural History, Smithsonian Institution, Washington, DC 20560; e-mail: sidor.christian@nmnh.si.edu*

Key Words   Synapsida, Therapsida, mammal-like reptile, Permian, phylogeny

■ **Abstract**   A rich fossil record documents nonmammalian evolution. In recent years, the application of cladistic methodology has shed valuable light on the relationships within the therapsid clades Biarmosuchia, Dinocephalia, Anomodontia, and Cynodontia. Recent discoveries from South Africa suggest that Gondwana, rather than Laurasia, was the center of origin and radiation for many early therapsids. Because of their relative abundance and global distribution, therapsids have enjoyed widespread use in biostratigraphy, basin analysis, and paleo-environmental and -continental reconstructions. Synapsids (including therapsids) form the bulk of tetrapod diversity (in terms of both number of species and abundance) from Early Permian to Middle Triassic times and thus can provide critical information on the nature of the Permo-Triassic extinction in the terrestrial realm. Quantitative techniques have produced headway into understanding the relative importance of homoplasy and convergent evolution in the origin of mammals.

INTRODUCTION

Central to an understanding of mammalian origins has been an assemblage of primarily Permian and Triassic fossils termed "nonmammalian therapsids" or "mammal-like reptiles" (Hopson & Crompton 1969, Crompton & Jenkins 1973, Kemp 1982). Although the curious blend of reptilian and mammalian features preserved in these fossils puzzled the nineteenth century paleontologists who first described them (e.g., Owen 1844), the morphological stages they preserve are now recognized as documenting in exceptional detail the acquisition of mammalian features within an evolving lineage. More generally, nonmammalian therapsids were the dominant terrestrial vertebrates for much of the late Paleozoic and early

---

[*]The US Government has the right to retain a nonexclusive, royalty-free license in and to any copyright covering this paper.

Mesozoic eras and thus figure prominently in the diversification of vertebrates on land and the establishment of modern terrestrial ecosystems (Sumida & Martin 1997). In the past few decades an accelerating number of fossil discoveries, the widespread application of cladistic methodology, and the increased use of quantitative techniques have combined to fuel a renaissance in therapsid research.

We review the current state of knowledge of Permo-Triassic therapsid evolution with an emphasis on new fossil discoveries and the increased systematic research that they have engendered. We also highlight the utility of fossil therapsids to areas of geological research such as the relative dating of rocks and paleogeography. Finally, we consider how the therapsid fossil record has been used to shed light on macroevolutionary questions such as the dynamics of large-scale evolutionary trends, rates of evolution, and the prevalence of morphological convergence.

EARLY SYNAPSID EVOLUTION: THE "PELYCOSAURS"

Synapsida is the clade including all taxa, extinct and extant, that share a closer relationship with mammals than they do with reptiles (Figure 1). The fundamental split between the mammal and reptile lineages is remarkably ancient, occurring

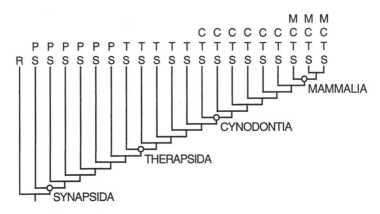

Figure 1   Diagrammatic representation of the cladistic terminology used in this paper. Synapsida (S) includes all fossil and Recent taxa that are more closely related to living mammals (M) than they are to living reptiles (R). Major synapsid subclades include Therapsida (T) and Cynodontia (C), both of which include mammals as their extant representatives. Synapsids that are not therapsids (i.e., nontherapsid synapsids) are a paraphyletic group traditionally termed "pelycosaurs" (P). Because fossil synapsids show a blend of mammalian and reptilian characteristics, the latter actually being retentions of the primitive amniote condition, the term "mammal-like reptile" was often employed. In phylogenetic phraseology, the term mammal-like reptiles corresponds to nonmammalian synapsids. Features diagnosing the clades Synapsida, Therapsida, and Cynodontia are discussed in the text.

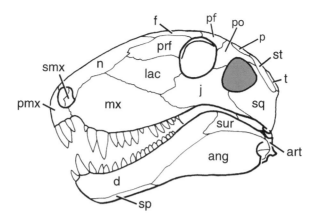

Figure 2 Skull of the pelycosaur-grade synapsid *Dimetrodon* in lateral view. The shaded region represents the lateral temporal fenestra, a diagnostic feature of all synapsids, including modern mammals. Abbreviations: ang, angular; art, articular; d, dentary; f, frontal; j, jugal; lac, lacrimal; mx, maxilla; n, nasal; p, parietal; pf, postfrontal; pmx, premaxilla; po, postorbital; prf, prefrontal; smx, septomaxilla; sp., splenial; sq, squamosal; st, supratemporal; sur, surangular; t, tabular. Reconstruction modified from Sidor & Hopson (1998).

at least 300 Mya (Laurin & Reisz 1995). At the time of their first appearance in the fossil record, however, early synapsids are distinguished from their reptilian contemporaries by only a few features, chief among which is their possession of a lateral temporal fenestra (Figure 2). Although explanations for this feature's origin remain speculative, a temporal opening is retained in all synapsids, including modern mammals.

The early stages of synapsid diversification are primarily recorded in the Upper Pennsylvanian and Lower Permian rocks of North America (Romer & Price 1940). Nontherapsid synapsids, "pelycosaurs" of traditional usage, occupied a wide variety of ecological niches and taxonomically dominated most paleofaunal assemblages in which they are found. Six family-level clades encompassing ~30 genera are currently recognized (Reisz 1986).

Two points are of special ecological note. First, large-bodied sphenacodontids such as the familiar sailback *Dimetrodon* (Figure 2) likely represent the earliest terrestrial carnivores to prey on similar-sized animals. Second, phylogenetic studies have postulated that herbivory originated twice within the early synapsids—once in the edaphosaurid *Edaphosaurus* (Modesto 1995) and once in the Caseidae (Olson 1968). Indeed, the successful diversification of early amniotes in general, and synapsids in particular, has been suggested by at least some workers to be tied to the advent of herbivory in these groups (Modesto 1992, Reisz & Sues 2000).

Pelycosaur-grade synapsids and their descendents, the therapsids, have long been known to have disjunct geographic and temporal distributions. In general,

Early Permian pelycosaurs are recorded from paleoequatorial regions (e.g., Europe and North America), whereas therapsids were, until very recently, predominantly known from the Late Permian of Russia and South Africa (i.e., high latitude regions). The apparent lack of overlap between these two groups has been considered by some to indicate a physiological advance of the latter permitting their geographic dispersal (McNab 1978, Parrish et al. 1986). More recent discoveries, however, have increased the degree of overlap between these two groups. Finds of early therapsids alongside the pelycosaur genera *Elliotsmithia* (Dilkes & Reisz 1996, Reisz et al. 1998, Modesto et al. 2001), *Mesenosaurus* (Reisz & Berman 2001), and *Ennatosaurus* (Ivachnenko 1990) clearly demonstrate the ability of pelycosaur-grade taxa to persist into the Late Permian in presumably more temperate climates.

THE ORIGIN OF THERAPSIDS

All synapsids phylogenetically more derived than sphenacodontids form the clade Therapsida (Laurin & Reisz 1996). When compared with their pelycosaur-grade precursors, the members of this clade are characterized by a greatly increased number of mammal-like features in both the cranial and postcranial skeleton (Kemp 1982, Sidor & Hopson 1998). For example, the loss of several skull elements [e.g., supratemporal (Figure 2) anterior coronoid] foreshadows the drastically reduced complement of skull bones characteristic of mammals (Sidor 2001), and the acquisition of an inflected femoral head has been thought to signal a more upright hindlimb posture in this group (Kemp 1978; Blob 1998, 2001). In addition, carnivorous adaptations initiated within sphenacodont-grade taxa are carried further within therapsids: The temporal fenestra is increased in size such that jaw adductor muscles have a larger area of origination (contrast Figure 2 with Figures 4 and 5), the upper and lower canines are enlarged such that the maxilla (mx in Figure 4) gains a new contact with the prefrontal (prf in Figure 4) and the dentary is expanded anteriorly, and fusion of the basicranial articulation reinforces the skull against the forces of subduing larger prey.

Therapsid remains are first recovered unequivocally from Upper Permian strata in Russia (Ivachnenko et al. 1997) and South Africa (Rubidge 1990a, 1995). Although the Russian forms have traditionally been thought to represent an earlier and more primitive radiation (e.g., Olson 1962), the discovery of a new paleo-fauna from the base of the Beaufort Group in South Africa has cast doubt on this theory (Rubidge 1990a). Moreover, recent phylogenetic work has postulated a Gondwanan origin for several therapsid subgroups (e.g., Modesto et al. 1999). A recently discovered therapsid fauna from the Permian of China (Li & Cheng 1995) is of uncertain temporal relationship to those of Russia and South Africa, but based on the primitive cast of its constituent taxa, it might represent the oldest therapsids yet recovered.

THE POSITION OF *TETRACERATOPS*   On the basis of several derived features, Laurin & Reisz (1996) recently suggested that *Tetraceratops insignis* represents the most primitive therapsid known (i.e., the sister taxon to all other therapsids). Because of its relatively late appearance (middle Early Permian), Laurin & Reisz's phylogenetic hypothesis implies that the greater than 30 characters diagnosing more advanced therapsids probably evolved in a remarkably short window of time (~8–10 myr) (Sidor & Hopson 1998). Although not impossible, the damaged and poorly preserved nature of the single *Tetraceratops* specimen makes a confident assessment of its phylogenetic placement difficult. For the remainder of our discussion, the terms "Therapsida" and "therapsid" will be used to denote post-*Tetraceratops* taxa.

HIGHER-LEVEL THERAPSID RELATIONSHIPS

Higher-level therapsid relationships have been the subject of cladistic attention for over 15 years, and although some areas of consensus have emerged, other areas remain contentious. As noted by Hopson (1991b), these differing perspectives on therapsid phylogeny inevitably produce contrasting views on the prevalence of evolutionary phenomena such as convergent evolution. In this section we review recent work on the relationships among the higher groups of therapsids (i.e., biarmosuchians, dinocephalians, anomodonts, gorgonopsians, therocephalians, and cynodonts). The cladogram depicted in Figure 3 and outlined below represents what we consider to be the best-supported hypothesis, although we note differing viewpoints where appropriate.

Therapsida

Romer & Price (1940) recognized that all therapsids arose from a single sphenacodont lineage, but Olson (1959, 1962) later advocated a polyphyletic origin of Therapsida, with anomodonts being independently derived from caseid ancestors. A similar view was held by Boonstra (1972), who argued for three independent originations of the therapsid grade of organization. Cladistic analyses have shown, however, that diverse features unite all therapsids in a clade, and that the features shared between anomodonts and caseids are more easily interpreted as convergent adaptations to an herbivorous diet.

As briefly discussed above, numerous morphological changes distinguish therapsids from their Early Permian pelycosaur forebears. For example, Sidor & Hopson (1998) suggested that between 36 and 48 characters can be used to identify Therapsida, depending on character optimization (see also Rowe 1986, Laurin & Reisz 1996). In addition to those previously discussed, modifications of the therapsid skull include an elongation of the dorsal process of the premaxilla (pmx in Figure 2), the pineal foramen being raised on a prominent boss or chimney (pin b in Figure 4), the posterior coronoid shifting ventrally so that it fails to form the dorsal margin of the lower jaw in medial view, and the reflected lamina adopting a characteristic pattern of ridges on its lateral surface and becoming more deeply

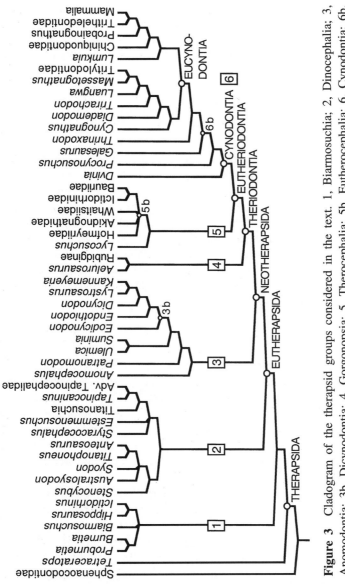

Figure 3 Cladogram of the therapsid groups considered in the text. 1, Biarmosuchia; 2, Dinocephalia; 3, Anomodontia; 3b, Dicynodontia; 4, Gorgonopsia; 5, Therocephalia; 5b, Eutherocephalia; 6, Cynodontia; 6b, Epicynodontia. This cladogram is by no means exhaustive. For example, only the most derived pelycosaur family is shown here (e.g., Sphenacodontidae). Relationships depicted here are based on those proposed by Hopson (1991, 1994), Rubidge & van den Heever (1997), Sidor & Hopson (1998), Modesto et al. (1999) and B. S. Rubidge & C. A. Sidor (unpublished data). See text for further details.

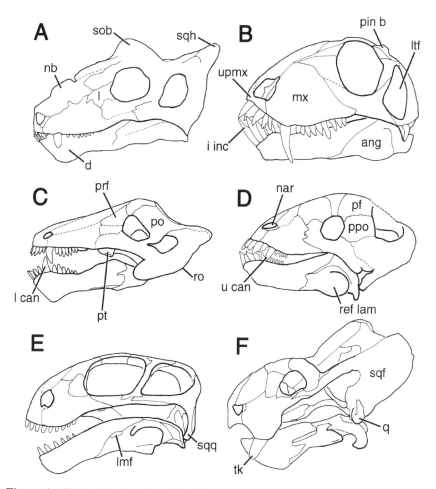

Figure 4 Skulls of selected nontheriodont therapsids in lateral view (not to scale).
(*A*) the burnetiamorph biarmosuchian *Proburnetia*; (*B*) the anteosaurid dinocephalian
Stenocybus; (*C*) the tapinocephalian dinocephalian *Styracocephalus*; (*D*) the basal
tapinocephalid dinocephalian *Tapinocaninus*; (*E*) the basal anomodont *Patranomodon*;
(*F*) the advanced dicynodont *Kannemeyeria*. Anatomical abbreviations: ang, angular;
d, dentary; i inc, intermeshing incisors; l, lacrimal; l can, lower canine; lmf, lateral
mandibular fenestra; ltf, lateral temporal fenestra; mx, maxilla; nar, external naris;
nb, nasal boss; pf, postfrontal; pin b, pineal boss; po, postorbital; ppo, pachyostosed
postorbital bar; pt, transverse flange of the pterygoid bone; ref lam, reflected lamina
of the angular; ro, anteriorly rotated occiput; sob, supraorbital boss; sqf, squamoal
fossa for jaw adductor musculature; sqh, squamosal horn; sqq, quadrate process of
squamosal; tk, tusk-like upper canine; u can, upper canine; upmx, upturned premaxilla.
Reconstructions modified from the following original sources: (*B*) Cheng & Li (1997);
(*C*) Rubidge & van den Heever (1997); (*D*) Rubidge (1991); (*E*) Hopson (1994); (*F*)
Renaut; 2000.

incised dorsally and, thus, more separated from the body of the angular (ang, ref lam in Figures 2, 4). Postcranial synapomorphies are also abundant and are suggestive of improved locomotor ability in therapsids. These include a deepening and rounding of both the pectoral glenoid and pelvic acetabulum, the replacement of the primitively screw-shaped humeral articular head with a rounded surface, the inflection of the femoral head to permit the hindlimbs to adopt a more adducted configuration, a narrowing of the scapular blade, and loss of intercentra in the trunk region.

Eutherapsida

Hopson & Barghusen (1986) and Rowe (1986) initially recognized a clade uniting dinocephalians, anomodonts, and theriodonts, which Hopson (1994) later termed Eutherapsida. However, because of the divergent morphologies of its subgroups, unambiguous synapomorphies uniting this proposed clade have been difficult to identify. Among the best candidates are that (*a*) the zygomatic arches are bowed such that the temporal fenestra is expanded laterally, (*b*) the ulna lacks a distinct, ossified olecranon process, and (*c*) the fifth pedal digit has only three phalanges. Rowe & van den Heever (1988) have suggested that the mammalian phalangeal formula (i.e. 2-3-3-3-3) should be considered a eutherapsid synapomorphy, but therapsid phalangeal evolution is extremely complex and recent workers have rejected the use of the mammalian phalangeal formula as a synapomorphy at this hierarchical level (Hopson 1995, Sidor 2000). Kemp (1982), Battail (1992), and Sidor (2000) have put forward alternatives to the recognition of a eutherapsid clade, but these are also supported by only a few characters.

Neotherapsida

The clade including all therapsids more advanced than biarmosuchians and dino-cephalians (i.e., anomodonts and theriodonts) has recently been termed Neother-apsida by Hopson (1999) (see Figure 3). Cranial synapomorphies for this group include a ventrally expanded squamosal (sqq in Figure 4; sq in Figure 5) that hides most of the quadrate (q in Figures 4, 5) and quadratojugal in posterior view and an epipterygoid (epi in Figure 5) that broadly contacts the underside of the parietal (p in Figure 5). Postcranial synapomorphies include the presence of at-lantal epipophyses and an enlarged obturator foramen on the puboischiadic suture. Although most workers accept the monophyly of a clade including anomodonts, gorgonopsians, therocephalians, and cynodonts, the branching sequence among these taxa has been strongly disputed (discussed below).

Theriodontia

Theriodonts are advanced carnivorous therapsids. This clade comprises three main subgroups, Gorgonopsia, Therocephalia, and Cynodontia, that are united by their common possession of a flat, low snout (i.e., the dorsal surface of the nasals is

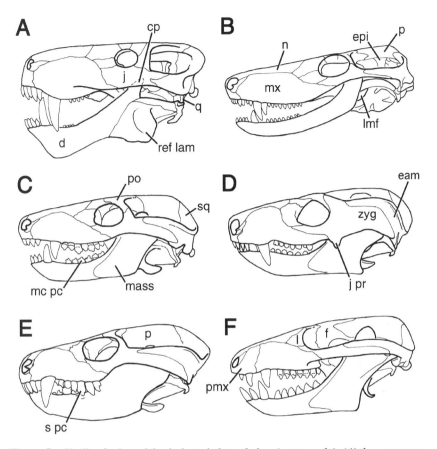

Figure 5   Skulls of selected theriodonts in lateral view (not to scale). (*A*) the gorgonopsian *Leontocephalus*; (*B*) the eutherocephalian *Ictidosuchoides*; (*C*) the basal cynodont *Thrinaxodon*; (*D*) the basal cynognathian *Trirachodon*; (*E*) the basal probainognathian *Lumkuia*; (*F*) the tritheledontid *Pachygenelus*. Anatomical abbreviations: cp, coronoid process of the dentary; d, dentary; eam, fossa on the squamosal termed the external auditory meatus; epi, epipterygoid; f, frontal; j, jugal; j pr, jugal suborbital process; l, lacrimal; lmf, lateral mandibular fenestra; mass, masseteric fossa; mc pc, multicusped postcanine teeth; mx, maxilla; n, nasal; p, parietal; pmx, premaxilla; po, postorbital; q, quadrate; ref lam, reflected lamina of the angular; s pc, sectorial postcanine teeth; sq, squamosal; zyg, zygomatic arch (composed of the jugal and squamosal bones). Reconstructions modified from the following original sources: (*A*) Sidor & Hopson (1998); (*B*) (*C*) (*D*) (*F*) Hopson (1994); (*E*) Hopson & Kitching (2001).

horizontal), a narrow temporal roof (such that the intertemporal width is equal to or less than the interorbital width), and most importantly, a free-standing coronoid process (for the localized insertion of jaw adductor musculature) (cp in Figure 5).

Kemp (1988) contested the features used by Hopson & Barghusen (1986) to support the grouping of gorgonopsians with therocephalians and cynodonts (to form a monophyletic Theriodontia), but gave little indication of where gorgonopsians should be placed. In contrast, Rowe (1986; see also Gauthier et al. 1988) suggested that anomodonts are more closely related to eutheriodonts (therocephalians + cynodonts) than are gorgonopsians, thereby rendering Theriodontia paraphyletic. After assessing these proposals, Hopson (1991b, p. 659) suggested that "the alternatives to accepting an association of Gorgonopsia with Therocephalia and Cynodontia are not sufficiently convincing to require abandonment of the Theriodontia." In a final permutation, Modesto et al. (1999) suggested that therocephalians and anomodonts form a clade to the exclusion of dinocephalians and gorgonopsians. The paucity of phylogenetic research on gorgonopsians and therocephalians has led to little progress in understanding theriodont monophyly or its alternatives.

Eutheriodontia

A sister-group relationship between therocephalians and cynodonts is currently accepted by most, if not all, workers (e.g., Kemp 1982, 1988; Hopson & Barghusen 1986, Rowe 1986, Hopson 1991b, 1994; Sidor & Hopson 1998). Eutheriodontia is a well-supported clade with synapomorphies that can be easily traced to their mammalian homologues. For example, eutheriodonts lose palatine teeth, anteroposteriorly expand the epipterygoid (epi in Figure 5), which is the homologue of the alisphenoid in mammals, reduce the temporal roof to a narrow sagittal crest, and shorten the posterior ramus of the postorbital (po in Figure 5) so that it fails to contact the squamosal (sq in Figure 5).

Hillenius (1992, 1994) noted the presence of ridges on the inner surface of the nasal and maxilla bones in the basal therocephalian *Glanosuchus* and several primitive cynodonts and argued that they indicate the presence of respiratory turbinate bones. This type of turbinate is found exclusively in mammals and birds and is thought to be an osteological correlate of endothermy. As such, this apparent synapomorphy may be indicative of increased oxygen consumption rates and an eutheriodont beginning for mammal-like metabolism.

Cynodontia

Cynodonts, in the form of *Procynosuchus* and *Cynosaurus* from southern Africa and *Dvinia* from Russia, first appear near the end of the Late Permian and include mammals as their extant subgroup. Many of the morphological hallmarks of mammals first appear as apomorphies in the premammalian cynodonts, and as noted by Hopson (1991b), most of these can be related to obtaining and orally processing food. In addition, some of these features have been suggested to be adaptations indicative of some degree of endothermy (McNab 1978, Hillenius 1994). Cynodontia is the best-supported therapsid subclade, with at least 25 characters diagnosing it (Hopson & Barghusen 1986, Rowe 1986, Sidor & Hopson 1998). We restrict our discussion to only a few features of special evolutionary significance.

The appearance of several apomorphic features early in cynodont evolution signal a fundamental reorganization of the jaw closing musculature. In *Dvinia* and *Procynosuchus*, the two earliest cynodonts known, the coronoid process of the dentary bears a shallow fossa on its dorsolateral surface (Tatarinov 1968, Kemp 1979). This fossa, in combination with a flaring of the zygomatic arch that situates the coronoid process near the middle of the temporal fenestra, has been taken as evidence for the differentiation of the primitively unipartite *m. adductor mandibulae* into the mammalian *m. temporalis* and *m. masseter* (Barghusen 1968). The possession of these two muscles is a uniquely mammalian feature that underlies the muscular "sling" that permits the complex jaw movements characteristic of this group (Crompton & Parker 1978). In cynodonts more derived than *Dvinia* and *Procynosuchus* (Epicynodontia sensu Hopson & Kitching 2001) (clade 6b in Figure 3), the masseteric fossa is enlarged such that it approaches the ventral margin of the mandible (mass in Figure 5).

Several other cynodont apomorphies can be interpreted as consequences of a more refined set of jaw-closing muscles. The parietal bone extends ventrally to more extensively contribute to the sidewall of the braincase, and the epipterygoid is expanded to alisphenoid-like proportions and makes a new contact with the frontal (f in Figure 5). In addition, the origination site of the *m. temporalis* on the parietal crest is elongated anteriorly so as to incorporate the pineal foramen.

Two more features have been considered to represent adaptations to the increased oral processing (i.e., chewing) of food. First, the postcanine dentition is elaborated early in cynodont evolution by the addition of mesial and distal accessory cusps (mc pc in Figure 5) and a lingual cingulum (a small shelf on the medial side of the tooth's crown). Second, all cynodonts possess at least a rudimentary secondary palate. In the most basal forms (e.g., *Dvinia, Procynosuchus, Galesaurus*), the bony plates extending medially from the maxilla and palatine bones fail to contact one another. In *Thrinaxodon* and eucynodonts, however, the plates suturally connect along the midline and thereby create a complete secondary palate (Fourie 1974). As in modern mammals, the secondary palate serves to separate the airway from the food-processing system.

Based on phylogenetic evidence, at least two cynodont lineages survived the Permo-Triassic mass extinction. The cynodonts *Galesaurus* and *Thrinaxodon* dominated the Early Triassic, but by the beginning of the Middle Triassic, cynodonts were diversifying and had diverged phylogenetically into two main subgroups: Probainognathia and Cynognathia. The latter group, principally distinguished by their transversely expanded cheek teeth, was much more numerous than the former, but both groups persisted beyond the Triassic. Advanced cynodonts are considered in more detail below.

SYSTEMATIC REVIEW OF THERAPSID DIVERSITY

In this section we review the six major clades of Permo-Triassic therapsids in greater detail. It should be noted that in contrast to the amount of attention that higher-level therapsid relationships have attracted, lower-level systematics have lagged behind. As a result, some of the results we present are based on very recent, or as yet unpublished, research.

Biarmosuchia

The higher-taxon Biarmosuchia (clade 1 in Figure 3) was the most recently recognized higher-level therapsid clade, with Hopson & Barghusen (1986) being the first to determine that taxa such as *Biarmosuchus*, *Hipposaurus*, *Burnetia*, and *Ictidorhinus* were the most primitive of the post-*sphenacodont* synapsids. More recently, Hopson (1991b) produced several characters in support of biarmosuchian monophyly, and Sidor (2000) has provided a preliminary lower-level analysis of the group.

In contrast to those of most other therapsids, biarmosuchian fossil remains are exceedingly rare. Most genera are known from only one or two poorly preserved skulls and, as a consequence, the diagnostic characters common to the group have been difficult to ascertain with confidence.

Biarmosuchians were typically small-to-medium sized carnivores that combine several primitive pelycosaur-grade features (e.g., the configuration of the temporal region), with several unique specializations of their own. The latter include a squamosal with a long ventral quadrate ramus and an elongate zygomatic process that extends under the orbit. In addition, the ventral edge of the occiput is moderately rotated anteriorly, and there is a pronounced difference in the height between the canine and postcanine regions of the dentary. Although few taxa possess postcranial remains, two possible biarmosuchian postcranial synapomorphies are the elongation of the cervical vertebrae (Sidor 2000) and the fusion of distal tarsals IV and V (Hopson 1991b). Although not all taxa have been included in formal cladistic analysis, biarmosuchians can be divided into a (probably) paraphyletic grade including most of the "ictidorhinids" (e.g., *Ictidorhinus*, *Hipposaurus*) and a derived subclade, Burnetiamorpha.

Burnetiamorph skulls are adorned with numerous protuberances and horn-like outgrowths (nb, sob, sqh in Figure 4). For much of this group's history, however, it was known from only two taxa, *Burnetia* and *Proburnetia* (Figure 4A). Two recently discovered taxa are in the process of being described (Sidor 2000, B. S. Rubidge & J. W. Kitching, unpublished data). These new taxa, in addition to the recognition that *Niuksenitia sukhonensis* is a burnetiamorph (Ivachnenko et al. 1997), have prompted the first cladistic appraisals of burnetiamorph relationships (Sidor 2000, B. S. Rubidge & J. W. Kitching, unpublished data). Biogeographically, the phylogenetic conclusions of these workers suggest that burnetiamorphs had a Gondwanan origin and that this clade displays no evidence of Late Permian endemism, as has been suggested for basal anomodonts (e.g., Modesto et al. 1999).

Dinocephalia

Dinocephalians (clade 2 in Figure 3) are a diverse group of basal therapsids primarily known from Upper Permian rocks of South Africa (Boonstra 1963; Rubidge 1991, 1994; Rubidge & van den Heever 1997) and Russia (Ivachnenko 1995, 1996a, 2000), although new finds have been recorded from Zimbabwe (Lepper et al. 2000, Munyikwa 2001), Brazil (Langer 2000), and China (Li & Cheng 1995; Cheng & Li 1996, 1997; Li et al. 1996). Dinocephalia is the earliest group of therapsids for which a significant radiation can be identified, but for all their early success, they became extinct by the middle Late Permian (Boonstra 1971). Important new discoveries have been made since King's (1988) review, and as such, are emphasized here.

Dinocephalians, or the "terrible-headed" therapsids, can be divided into two main groups: the carnivorous anteosaurs and the herbivorous tapinocephalians. Derived characters uniting these two groups have been debated but can be shown to include nonterminal external nares (nar in Figure 4) (Grine 1997), the loss of a vomerine process of the premaxilla so that vomer abuts the body of the premaxilla, a preorbitally positioned transverse flange of the pterygoid (pt in Figure 4), the lack of ridges and fossae on the lateral surface of the reflected lamina (ref lam in Figure 4), and the presence of a foramen on the inner surface of the lower jaw between the prearticular and angular (Sidor 2000). The group's traditional defining feature, intermeshing upper and lower incisors (i inc in Figure 4), has recently been identified in several biarmosuchians (Sigogneau-Russell 1989; CA Sidor & BS Rubidge, unpublished data) and thus requires a more detailed appraisal of its phylogenetic distribution. Postcranially, the fifth metapodial of all dinocephalians in which it is known is more robust and at least as long as the fourth metapodial.

The most basal dinocephalians are thought to be the Russian anteosaurs (Boonstra 1971, Hopson & Barghusen 1986), although some workers consider the estemmenosuchids to be more primitive (Kemp 1982, King 1988). The massive South African genus *Anteosaurus* is considered the most derived anteosaur, prompting the conclusion that dinocephalians originated in Russia and arrived in southern Africa via overland migrations. The description of *Australosyodon* from the *Eodicynodon* Assemblage Zone, however, has shown that primitive anteosaurs are also present in the southern hemisphere (Rubidge 1994). *Sinophoneus* (Cheng & Li 1996) and *Stenocybus* (Figure 4B) (Cheng & Li 1997), from the Gansu Province of China (Li & Cheng 1995, Li et al. 1996), also appear to be basal anteosaurian dinocephalians.

The Tapinocephalia has been defined as all dinocephalians more closely related to *Tapinocephalus* than to *Anteosaurus* (Sidor 2000). Morphological variation related to three main characteristics (viz. adaptations to herbivory, body size, and skull pachyostosis) has formed the basis for most tapinocephalian classifications.

The most primitive members of the Tapinocephalia are *Estemmenosuchus*, from Russia, and *Styracocephalus* (Figure 4C), from South Africa. Both species of *Estemmenosuchus*, *E. uralensis* and *E. mirabilis* (a slightly smaller form), have strange bony protuberances on the skull roof and large temporal fenestrae. As in other dinocephalians, they have enlarged, intermeshing incisors (but these lack heels), and they also have prominent canines, as are present in anteosaurids and titanosuchians. *Styracocephalus* was previously known only from its very distorted holotype, but several recently discovered specimens have permitted a redescription of the skull (Rubidge & van den Heever 1997). The genus is considered to be characterized by a large posteriorly projected horn made up of the postorbital and squamosal bones. Despite the appearance of a close relationship to *Estemmenosuchus* suggested by the presence of horns in both genera, these structures develop from different bones of the skull roof. In the absence of clear synapomorphies, these two genera are best considered as basal tapinocephalians forming a trichotomy with the remaining members of the clade (Rubidge & van den Heever 1997).

The next most derived taxa in the Tapinocephalia are *Jonkeria* and *Titanosuchus*, the Titanosuchia of Boonstra (1969). Although titanosuchians retain a prominent canine tooth (as is present in anteosaurians and more primitive tapinocephalians), the incisors have a large talon and a prominent crushing heel on their lingual surface. The upper incisors intermesh with the lowers in such a way that when the jaws occlude, the heels meet to form a crushing surface, used to grind plant material. *Estemmenosuchus* and *Jonkeria/Titanosuchus* share several dental features not seen in other dinocephalians (e.g., the lower postcanine teeth are inset from the lateral margin of the dentary, and the upper postcanines are continuous with the incisors medial to the canine).

Tapinocephalians more derived than the titanosuchians form the Tapinocephalidae. A plethora of tapinocephalid genera have been described from the *Tapinocephalus* Assemblage Zone of South Africa (see Boonstra 1969, King 1988). Other tapinocephalids include one species of *Criocephalus* from the Madumabisa mudstones of Zimbabwe (Boonstra 1969) and *Ulemosaurus* from the Esheevo complex of Russia (Efremov 1940).

Tapinocephalids carry even further the specializations toward herbivory initiated in the titanosuchians. In tapinocephalids, however, the postcanines as well as the incisors have a lingual heel, and interdigitation of the entire dentition occurs instead of just the anterior teeth. The basal form *Tapinocaninus* (Figure 4D) (Rubidge 1991) has canine teeth in both the upper and lower jaws (e.g., u can in Figure 4), a feature that is lost in all other tapinocephalids. *Tapinocaninus* also retains a large temporal fenestra compared with other, more derived, tapinocephalids, in which it is impinged upon by the increasingly pachyostotic bones of the skull roof (particularly the postorbital) (ppo in Figure 4). Barghusen (1975) makes convincing arguments that the heavily thickened skulls of tapinocephalids were used for head butting.

Anomodontia

Anomodontia (clade 3 in Figure 3) and their highly diverse subclade, Dicynodontia (clade 3b), were arguably the single most successful group of Permian therapsids (King 1990). Appearing early in the Late Permian and surviving until the Late Triassic, anomodonts are known from all continental landmasses. Only a few genera lie phylogenetically outside the dicynodont clade, but it is within this pre-dicynodont grade that most investigations into the adaptive radiation of dicynodonts have been focused (King et al. 1989, King 1994). Most recent workers restrict the term Anomodontia to dicynodonts and their near relatives, although in the past this name was considered to encompass dinocephalians as well (e.g., Watson & Romer 1956). Recent cladistic propositions of a sister-group relationship between dinocephalians and anomodonts (sensu stricto) have been refuted (Hopson 1991b, Grine 1997; contra King 1988).

Dimacrodon hottoni, from the San Angelo Formation of Texas, has been suggested to be the oldest anomodont (Olson 1962). Restudy of the material, which includes a partial lower jaw and very fragmentary remains of the skull, by Sidor & Hopson (1995) has shown that this specimen consists of misidentified pelycosaur remains and displays none of the diagnostic features of anomodonts.

The most basal member of the Anomodontia is the newly described *Anomocephalus africanus* from the Beaufort Group of South Africa (Modesto & Rubidge 2000). The position of this genus is supported by the relatively long snout compared with other anomodonts, a vertically aligned and blade-like zygomatic process of the squamosal, and the fact that the squamosal does not contact the ventral tip of the postorbital. Other primitive members of the Anomodontia include *Patranomodon* (Figure 4E), which is known from a well-preserved skull and lower jaw as well as several postcranial elements including a beautifully preserved manus (Rubidge & Hopson 1996). Additional early anomodonts include *Otsheria*, *Ulemica*, and *Suminia*, which have been proposed to form a clade of Russian endemic genera (Rybczynski 2000; contra Ivachnenko 1994, 1996b), and the South African taxa *Galepus*, *Galechirus*, and *Galeops* (Brinkman 1981, Rubidge & Hopson 1996).

Dicynodontia

Dicynodonts form the bulk of anomodont taxonomic diversity and dominated terrestrial herbivorous niches for most of the Late Permian. It is estimated that in South Africa's Karoo Basin their remains are 10 times more plentiful than those of contemporaneous carnivores. Indeed, their abundance has been used as ecological evidence that dicynodonts were primary consumers and that modern trophic structure had appeared by this time. Until relatively recently, the vast number of named dicynodont species clouded our understanding of this group's evolutionary history. However, as a result of the efforts of several workers (e.g., Cluver & Hotton 1981, Cluver & King 1983, King 1988, King & Rubidge 1993, Keyser 1993, Keyser & Cruickshank 1979, Renaut 2000), the number of valid taxa has been greatly reduced and it has become possible to resolve relationships within the Dicynodontia (Cox 1998, Angielczyk 2001).

The oldest and most primitive dicynodont is *Eodicynodon*, from the base of the Beaufort Group in South Africa (Rubidge 1990b). As in other dicynodonts, the premaxilla and anterior portion of the dentary are edentulous in *Eodicynodon*. Both of these areas are pock-marked with numerous foramina that suggest the presence of a horny covering in life. Adding to their unusual appearance, the dicynodont dentition consists of paired, tusk-like upper caniniforms (tk in Figure 4) and a vestigal complement of small postcanine teeth. These features, in addition to their characteristic sliding jaw joint, have been traditionally regarded as underlying the success of dicynodonts (Crompton & Hotton 1967, Barghusen 1976, King 1994).

Until the recent discovery of a new species from India (Ray 2000), the aberrant dicynodont *Endothiodon* had been known only from rather fragmentary material (e.g., Cox 1964). The most striking feature of this genus, which contrasts sharply to the condition present in most dicynodonts, is its numerous obliquely arranged rows of teeth in both the upper and lower jaws. New work by Latimer et al. (1995) has demonstrated that these tooth rows correspond to the concept of Zahnreihen, as in several other Paleozoic genera with multiple tooth rows.

The end of the Permian brought a catastrophic drop in dicynodont numbers so that by the beginning of the Triassic ($\sim$251 Ma), only *Lystrosaurus* was common. This genus was extremely widespread, however, as its remains have been found in Western Europe, Russia, India, China, Africa, Antarctica, and possibly Australia (King 1988). *Lystrosaurus* was traditionally considered to have first appeared in the Early Triassic, but the recent discovery of an overlap in the stratigraphic ranges of *Dicynodon* and *Lystrosaurus* has demonstrated that the latter also occurs in the Late Permian and thus it was the only dicynodont known to cross the Permo-Triassic boundary (Cheng 1993, Smith 1995, King & Jenkins 1997).

Dicynodonts underwent a second, relatively minor, radiation during the Middle Triassic, with taxa such a *Kannemeyeria* (Figure 4F) becoming common. Using only cranial characters, King (1988) provided the first cladistic analysis of the Triassic dicynodonts. More recently, Surkov (2000) provided a cladogram based on postcranial features and has reiterated the traditional viewpoint (e.g., Cluver 1971) that Triassic dicynodonts probably originated from an advanced Late Permian form similar to *Dicynodon*. The latest surviving dicynodonts are preserved in the Middle and Upper Triassic sediments of North and South America and tend to be massively built, with genera such as *Stahleckeria*, *Ischigualastia*, and *Placerias* being the largest dicynodonts on record. Compared with their Permian antecedents, Triassic dicynodonts were scarce and faced competition with a host of newly evolved herbivores, including bauriid therocephalians, gomphodont cynodonts, and rhynchosaurs.

Postcranial studies have suggested that dicynodonts were powerful animals and that the forelimb retained a sprawling posture, whereas the hindlimb could adopt both sprawling and upright positions (King 1988, 1990). Histological studies on eight dicynodont genera spanning all stratigraphic levels of the Beaufort Group have established the presence of fibro-lamellar alternating with lamellated bone tissue in all taxa except *Diictodon*, which had only the former type of tissue (Chinsamy & Rubidge 1993). This suggests that in most dicynodonts, bone was deposited rapidly, but with intermittent periods of slow growth. The absence of annuli in *Diictodon* suggests that it was capable of sustained rapid growth, as are extant endotherms.

Gorgonopsia

Gorgonopsians (clade 4 in Figure 3; Figure 5A) were the dominant large-bodied carnivores for much of the Late Permian. In the rocks of South Africa's Lower Beaufort Group, from where they are most abundantly known, hundreds of specimens have been collected and over 25 genera are currently recognized (Sigogneau-Russell 1989). All gorgonopsians possess a distinctive palatal morphology. The unpaired vomer is broad anteriorly but tapers rapidly posteriorly and, when well-preserved, can be seen to possess a system of three ridges on its ventral surface: a weak median ridge and two lateral ridges that parallel the choanal margin. In addition, the palatines form an extensive sutural contact on the ventral midline of the palate, thereby separating the vomer from the pterygoids. On the skull roof the preparietal (when discernible) is diamond-shaped and positioned between the parietals and frontal (i.e., it does not contact the pineal foramen as in some biarmosuchians and anomodonts), and the jugal (j in Figure 5) lacks a distinct postorbital process. The lower jaw of gorgonopsians is also highly diagnostic; the reflected lamina of the angular (ref lam in Figure 5) has an attached dorsal margin and bears a unique system of lateral ridges including a near-vertical, anterior ridge in front of a deep, pocket-like depression.

Studies of gorgonopsian functional morphology have principally focused on the jaw opening and closing mechanism. Although there remains room for dispute, two features are clear: (*a*) Gorgonopsians possessed a quadrate (q in Figure 5) that was only loosely attached to the skull (i.e., they lacked a sutural connection between the quadrate and squamosal), and (*b*) when the jaw was fully opened, the shape of the jaw joint forced the articular to be translated relative to the quadrate (i.e., as the articular rotated to open the jaw, it was compelled to slide laterally as well) (Parrington 1955). Kemp (1969) and Tatarinov (2000) have proposed alternative interpretations of quadrate movement, whereas Laurin (1998) argued against any type of motion because of the quadrate's firm connection with the epipterygoid and stapes.

Gorgonopsians flourished during the Late Permian despite a rather unsophisticated masticatory apparatus. Although both the canines and large incisors of gorgonopsians were armed with serrations, a lack of regular wear facets and the reduced postcanine dentition in some forms indicate that little processing of food was done. Instead, it appears that gorgonopsians simply tore hunks of flesh out of their prey. Despite their common preservation in Upper Permian sediments, gorgonopsians were victims of the end-Permian extinction.

The unsatisfactory state of gorgonopsian taxonomy has been the single largest impediment to a broader understanding of this group's evolution. Indeed, gorgonopsians possess such a stereotyped cranial morphology that ontogenetic changes appear to have been used to identify species and even genera. Since the precladistic work of Sigogneau (1970; see also Sigogneau-Russell 1989), little progress has been made in determining the interrelationships within Gorgonopsia. However, one group recognized by Sigogneau (1970) that does appear to represent a valid clade is the subfamily Rubidgeinae. This group includes most of the latest appearing, large gorgonopsians that possess a thickened postorbital bar and a very deep zygomatic arch whose ventral margin descends to a level lower than that of the quadrate.

Therocephalia

In contrast to the morphological conservatism seen in gorgonopsians, therocephalians (clade 5 in Figure 3) evolved a wide range of morphologies and presumed ecologies in the course of their evolutionary history. The earliest therocephalians occur as low in the Permian sediments of South Africa as any of the other major therapsid groups (Rubidge et al. 1983), and they survived until early Middle Triassic times (Rubidge 1995). Geographically, therocephalians are known not only from southern and eastern Africa (Keyser 1973, Kitching 1977, Gay & Cruickshank 1999), but from China (Li & Cheng 1995), Russia (Ivachnenko et al. 1997), and Antarctica (Colbert & Kitching 1981) as well.

By their first appearance in the fossil record, early taxa such as *Lycosuchus* and *Scylacosaurus* share several synapomorphies with all other therocephalians that indicate the monophyly of this group. Most notably, bilateral fenestrae (termed suborbital vacuities) are developed in the palate between the pterygoid, ectopterygoid, and palatine bones. In addition, the stapes is rod-like, lacking a stapedial foramen. Postcranially, the ilium of all therocephalians possesses a distinct, finger-like process that emanates from the bone's anterior margin.

Therocephalians more advanced than lycosuchids and scylacosaurids have been termed eutherocephalians (clade 5b in Figure 3; Figure 5B) (Hopson & Barghusen 1986, van den Heever 1994). Although this clade has strong character support, the interrelationships of its subordinate taxa (viz. Hofmeyriidae, Akidnognathidae, Whaitsiidae, and Baurioidea) remain poorly understood (Hopson 1991b).

Therocephalians increased in taxonomic diversity throughout the Late Permian and are inferred to have been exclusively carnivores during this time. Of particular note in this regard is *Euchambersia*, a short-snouted form possessing specializations that are suggestive of a snake-like venomous bite (e.g., a deep fossa on its cheek communicates with a groove on the upper canine) (Hotton 1991). Beginning in the Early Triassic, however, bauriid therocephalians adopted a herbivorous lifestyle. In this clade the postcanine teeth were transversely expanded and became tightly packed into a curved tooth row. In addition, regular wear facets indicate that bauriids gained precise postcanine occlusion similar to that of advanced cynodonts. Although therocephalians parallel the development of several other cynodont features (e.g., the acquisition of a secondary palate and loss of the postfrontal bone, among others), they retained many other primitive characters and became extinct in the early Middle Triassic.

Cynodontia

Among therapsids, cynodonts (clade 6 in Figure 3) have garnered the greatest amount of attention because of their intimate relationship with the origin of mammals (Kemp 1983, Rowe 1988, Hopson 1994). The early evolution of cynodonts was discussed in a previous section; here we focus on the derived, post-Early Triassic forms.

Cynodonts more derived than *Thrinaxodon* (Figure 5C) have been termed the Eucynodontia (Kemp 1982, 1988) and are characterized by a number of features that produce an increasingly mammal-like skull and postcranial skeleton. Important modifications include: (*a*) an enlarged dentary and reduced, more rod-like, postdentary bones, (*b*) the fusion of the dentaries at the mandibular symphysis, (*c*) a supplementary contact between the surangular and squamosal lateral to the quadrate-articular jaw joint, (*d*) the formation of an acromion process on the leading edge of the scapula, and (*e*) the acquisition of the mammalian phalangeal formula.

Advanced nonmammalian cynodont systematics still lack consensus, but most phylogenetic hypotheses fall into one of three camps (Figure 6). The first supports the traditional view that tritylodontids are derived from a lineage of gomphodont cynodonts with transversely expanded cheek teeth, whereas mammals arose from a lineage with a persistently sectorial (i.e., blade-like) dentition (Crompton & Ellenberger 1957; Crompton 1972; Hopson & Kitching 1972, 2001; Sues 1985; Hopson 1991b; Luo 1994). The second camp contends that most gomphodonts (i.e., diademodontids, trirachodontids, and traversodontids) form a clade, but that tritylodontids are distinct and more closely related to mammals (and possibly tritheledonts) (Kemp 1982, 1983). The final permutation dissolves the tooth-type dichotomy completely, and intersperses gomphodont with sectorial taxa as successive mammal outgroups (Gauthier et al. 1988; Rowe 1988, 1993). Battail (1991) groups gomphodonts and tritylodontids but suggests that mammals evolved from a *Thrinaxodon*-grade ancestor, a hypothesis unlike that of any of the previous workers.

Hopson & Kitching (2001) have provided the most recent investigation of the higher cynodont problem. Their results support the traditional hypothesis (i.e., Figure 6A) and also have the benefit of the most extensive taxon sampling—an important factor in accurately reconstructing phylogenies (Wagner 2000). For the purposes of the following discussion, we follow Hopson (e.g., Hopson & Barghusen 1986, Hopson 1991b, 1994, Sidor & Hopson 1998) in dividing eucynodonts into two major clades, Cynognathia and Probainognathia.

Cynognathia

A rich fossil record of early cynognathians is known from the Triassic sediments of South Africa and South America. Early in this sequence the most primitive and only carnivorous member of this group, *Cynognathus*, is found alongside its herbivorous relatives, *Diademodon* and *Trirachodon* (Figure 5D). The latter genus is

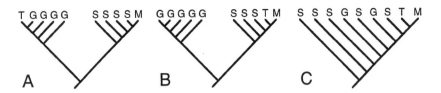

Figure 6 Diagrammatic representation of three hypotheses of advanced cynodont phylogeny. (*A*) The traditional hypothesis (cf. Crompton & Ellenberger 1957, Hopson & Kitching 1972, Hopson 1991); (*B*) that of Kemp (1982, 1983); (*C*) that of Rowe (1988, 1993; see also Gauthier et al. 1988). G, herbivorous, gomphodont cynodonts; T, tritylodontids (the teeth of which are morphologically gomphodont); S, cynodonts with sectorial postcanine teeth; M, early mammals (which have sectorial teeth). The following taxa are characterized by transversely-expanded (gomphodont) postcanine teeth in Figure 3: *Diademodon, Trirachodon, Luangwa, Massetognathus,* and *Tritylodontidae*; all other cynodonts have sectorial (slicing) postcanine teeth. Although we favor the hypothesis depicted in (*A*), advanced cynodont systematics lacks consensus, and further work is required.

of special interest in that it is one of the only therapsid genera in which qualitative morphologic change can be demonstrated within a taxon's stratigraphic range. The earliest-occurring form, known only from recently collected material, maintains sectorial teeth along much of its tooth row (Neveling 1998). The succeeding morphospecies, *Trirachodon kannemeyeri*, has transversely widened (gomphodont) teeth anteriorly in its tooth row but retains sectorial teeth more posteriorly. The final species, *Trirachodon berryi*, by contrast, has a full complement of gomphodont teeth. The stratigraphic and ontogenetic sequences preserved by fossils of this genus may shed light on the origin and evolution of the gomphodont type of dentition.

Cynognathians can be diagnosed by several synapomorphies including a very deep zygomatic arch that extends above the middle of the orbit (zyg in Figure 5), a suborbital process on the jugal (j pr in Figure 5) for the origination of the masseter jaw closing muscle, and a very deep groove on the lateral surface of the squamosal that connected the middle ear with the outside world (i.e., the "external auditory meatus") (Allin & Hopson 1992) (eam in Figure 5). Gomphodont cynodonts more derived than *Diademodon* and *Trirachodon* typically lack sectorial teeth and have traditionally been termed traversodontids. Members of this group include *Exaeretodon* (Bonaparte 1962), *Luangwa* (Kemp 1980), *Massetognathus* (Abdala & Giannini 2000), and *Pascualgnathus* (Bonaparte 1966), all of which share lower postcanines with two anterior, transversely aligned cusps in front of a posterior basin. Recent fossil finds in Madagascar have prompted Flynn et al. (1999, 2000) to phylogenetically define Traversodontidae as a stem-based group. This type of definition is unfortunate in this case because it ignores the traditional division between pre-traversodontid genera (viz. *Diademodon, Trirachodon,* and even *Cynognathus*) and the more advanced forms. In addition, under most workers' relationships, this definition would be supplanted by Cynognathia (Hopson 1991b, 1994).

According to the phylogenetic hypothesis followed here, tritylodontids represent the most highly derived members of the cynognathian lineage. This group's name is derived from its members' highly modified postcanine dentition, in which the lower molariforms possess three longitudinal rows of cusps. These three rows occlude against two rows in the upper dentition and produce wear facets interpreted as the result of strong, bilateral, backwards jaw movement (Clark & Hopson 1985, Sues 1986). The lower incisors of tritylodontids are enlarged and fit between the uppers to yield a rodent-like appearance. The postcranial skeleton of tritylodontids is mammal-like. For example, the greater trochanter is separated from the femoral head by a notch and, in the pelvis the reduced pubis is positioned posteroventral to the acetabulum and the ilium is rod-like. Hopson (1991a, 1995), however, has suggested that homoplasy is relatively common in the therapsid postcranial skeleton. Tritylodontids are the latest-occurring cynognathians, known primarily from the Jurassic, though recent discoveries have now extended their range to the Early Cretaceous (Tatarinov & Matchenko 1999).

Probainognathia

Compared with the diversity and abundance of basal cynognathians, the fossil record of early probainognathians is relatively poor. Indeed, the earliest-appearing member of this group has only very recently been described (Hopson & Kitching 2001). This taxon, *Lumkuia fuzzi* (Figure 5E), shares with other probainognathians the lack of a pineal foramen and expanded ribs, as well as a posteriorly elongated secondary palate. Prior to their description, probainognathians were known exclusively from younger sediments, principally the upper Middle and Upper Triassic of Argentina (Romer 1969b, 1970; Martinez & Forster 1996; Martinez et al. 1996) and Brazil (Romer 1969a, Hopson 1985).

Advanced probainognathians can be divided into two subgroups, the chiniquodontids and a lineage including *Probainognathus*, tritheledontids, and mammals. The former subgroup includes taxa such as *Probelesodon*, *Aleodon*, and *Chiniquodon* that uniformly possess a characteristic right-angle bend at the rear of the maxilla, but span a wide range of dental morphologies. For example, *Probelesodon* has mitten-shaped postcanine teeth that completely lack cingula (i.e., the sides of the teeth are smooth), whereas *Aleodon* possesses transversely expanded, multicusped postcanines that led to its early classification as a gomphodont (Crompton 1955). *Probainognathus* coexisted with *Probelesodon*, but is advanced in a mammalian direction in aspects of its postcanine dentition.

Tritheledontids (Figure 5F) are poorly understood as a group, although they have played an important role in understanding the origin of mammals (Crompton 1963, Luo 1994). Although its homology with that of mammals remains controversial, the dentary of the derived tritheledontid, *Pachygenelus*, appears to have made a contact with the squamosal, although an articular fossa is not present (Hopson 1994). This contact foreshadows the condition in early mammals such as *Morganucodon* and *Sinoconodon*, in which a well-developed dentary-squamosal jaw joint functions beside the primitive quadrate-articular jaw joint (Crompton & Luo 1993).

GEOLOGIC UTILITY OF THERAPSIDS

The therapsid fossil record enjoys a long pedigree of use in the geological sciences. The rocks of the Beaufort Group, in which there are a paucity of basin-wide lithostratigraphic markers, are divided into eight therapsid-defined "assemblage zones" (Rubidge 1995). Based on their well-known faunal content and the global distribution of some therapsid genera, these zones are now the international standard for global correlation of Permian-Jurassic nonmarine deposits (e.g., Ochev & Shishkin 1989, Lucas 1998).

Therapsid fossils have also been used to interpret paleoenvironments (e.g., Kitching 1977), and more recent high-resolution taphonomic studies have played an important role in delineating the spatial aspects of floodplain subenvironments within the Lower Beaufort Group (Smith 1993, 1995; Smith et al. 1993), and regional base level changes in the Upper Triassic/Lower Jurassic Elliot Formation (Smith & Kitching 1997). Therapsid fossils have also been utilized in defining formational contacts, such as the long disputed Ecca-Beaufort contact, which can now be defined using both lithological and paleontological criteria (Hancox & Rubidge 1997; Rubidge et al. 1999, 2000). Similarly, subdivision of the *Cynognathus* Assemblage Zone has permitted the diachronous (i.e., time-transgressive) nature of the Beaufort-"Stormberg" boundary to be recognized (Hancox et al. 1995, Hancox 2000).

Biodiversity patterns in the Lower Beaufort Group point to major faunal turnovers between the *Tapinocephalus* and *Pristerognathus* Assemblage Zones and between the *Dicynodon* and *Lystrosaurus* Assemblage Zones. The latter turnover traditionally has been considered to coincide with the Permo-Triassic boundary (Smith 1995), a time of global mass extinction (Erwin 1993). Correlating marine and nonmarine strata has been a recalcitrant problem when assessing the various mechanisms proposed to underlie the end-Permian event (Olson 1989). Further research into the Permo-Triassic boundary in the Karoo Basin (i.e., radiometric dating, paleomagnetism, therapsid diversity patterns) will allow for better correlation with the international marine sequence and may help unravel extinction mechanisms (MacLeod et al. 2000, Ward et al. 2000).

MAJOR PATTERNS OF THERAPSID EVOLUTION

Documenting the Acquisition of Mammalian Features

The synapsid fossil record has been considered of special importance because it chronicles the origin of mammals in exceptional detail. Indeed, Kemp (1982,

1985) has stressed the importance of dense fossil sampling in interpreting the macroevolutionary process by which higher taxa emerge. Although the evolution of mammals represents only one branch on the larger synapsid evolutionary tree, a major research goal has been to understand the evolutionary questions of when, how, and why the features we recognize as mammalian first appeared. The numerous cladistic studies published to date have afforded a rigorous basis to answering the first of these questions. The phylogenetic first appearance of many features is now well known. By contrast, our understanding of how and why these features appeared is, and will doubtless continue to be, much more difficult to pin down.

Progress has been made on the evolutionary origins of mammalian limb posture. Experimental approaches using extant reptiles and mammals have helped to identify plausible and implausible loading regimes for hindlimb elements (Blob 1998, 2001), and recent anatomical work has detailed postcranial evolution within therocephalians (Fourie 2001). In addition, preliminary data suggest that discriminating between sprawling and nonsprawling locomotion may be possible on a quantitative basis (Beck et al. 2000).

The evolution of endothermy in the mammalian stem-lineage has also attracted recent work. The formation of a bony secondary palate (Maier et al. 1996), possession of nasal turbinates (Hillenius 1992, 1994), specific bone histologies (de Ricqlès 1974, Chinsamy & Rubidge 1993, Botha & Chinsamy 2001), loss of a pineal foramen, and modern predator/prey ratios (Bakker 1975) have all been suggested as correlates of mammal-like endothermy in therapsids. Although the acquisition of endothermy was probably piecemeal (Ruben 1995), and none of the aforementioned characters alone provides conclusive evidence, taken together they strongly suggest that increased metabolic activity probably characterized eutheriodonts, and that some form of endothermy was present in advanced cynodonts.

The acquisition of mammalian features has also been studied in a temporal framework. Sidor & Hopson (1998) have shown a good correlation between the duration of gaps in the synapsid fossil record and the number of mammalian features inferred to have been gained during those gaps. They argued that this relationship supports the hypothesis of a gradual appearance of mammalian features within the synapsid lineage.

Evolutionary Trends

The synapsid fossil record has also been used to provide insight into broad-scale patterns of evolution. Olson (1959) advocated a polyphyletic origin of mammals, with multiple therapsid lineages independently attaining the mammalian grade of organization. Although this view has now fallen into disfavor among paleontologists, establishing the prevalence of morphological parallelism in nonmammalian therapsids remains an important avenue of research.

Based on phylogenetic arguments, Hopson (1995) has shown that the mammalian phalangeal formula appeared multiple times within therapsid evolution. Hopson argued that this feature, along with fusions in the distal carpals and tarsals and increasing symmetry among the metapodials, is related to an increasingly forward-facing hand and foot, brought about by the requirements of a less-sprawling posture.

Sidor (2000, 2001) addressed evolutionary parallelism in the synapsid skull. When skull complexity was defined in terms of the number of distinct elements, Sidor was able to show that simplification (in terms of bone loss or fusion) was a widespread feature of synapsid evolution. This pattern was prevalent not only at the broadest scale possible (i.e., all synapsids), but it typified nearly all synapsid subgroups as well. Stratigraphic and phylogenetic tests indicate that skull bone loss in synapsids corresponds to what McShea (1994) termed a driven trend.

SUMMARY AND FUTURE DIRECTIONS

The past decade has produced fundamental advances in synapsid research. Fossil discoveries have offered new taxa with suprising combinations of primitive and derived features and have prompted a new wave of detailed phylogenetic research. The availability of large numbers of finely prepared specimens has helped to refine the taxonomy of many therapsid groups and permits the use of quantitative approaches to understanding their evolutionary history.

The therapsid fossil record also makes valuable contributions to the geological sciences. Therapsid fossils are increasingly utilized as paleoenvironmental indicators and are also routinely employed for both inter- and intracontinental correlation of stratigraphic sequences. Although many of these advances are at an early stage of development, the use of therapsids as age indicators has led to the establishment of global biochrons.

Enhanced paleocontinental reconstructions, together with phylogenetic analyses, have improved our understanding of therapsid biogeography. Whereas therapsids were previously thought to have originated in North America or Russia, it now appears that southern Africa was home to many therapsid groups. Africa has many Permo-Triassic basins that have received only cursory geological and paleontological attention. We predict that these areas are likely to produce important new therapsid faunas in the future.

Systematic research has progressed for anomodonts, biarmosuchians, dinocephalians, basal therocephalians, and higher cynodonts, but almost nothing has appeared on gorgonopsians or derived therocephalians. These and tapinocephalid dinocephalians are obvious targets for future studies. Similarly, although a steady stream of therapsid cranial descriptions have appeared in the past decade, comparable postcranial descriptions are sorely needed. This disparity has resulted in cladistic studies being overly reliant on cranial features, which might lead to biased phylogenetic results if the cranial and postcranial skeletons preserve different qualities of phylogenetic information. An accurate assessment of Permo-Triassic therapsid biodiversity is an important long-term goal. Meeting this goal will require a combination of research on older, previously described taxa, and the continuing discovery of new therapsid faunas.

ACKNOWLEDGMENTS

We thank Jim Hopson, Robert Reisz, and an anonymous referee for kindly reviewing the manuscript, and Allison Beck, Ross Damiani, and John Hancox for reading earlier drafts. Research leading to this review has been supported by the National Research Foundation of South Africa, the University of the Witwatersrand, the National Science Foundation, the University of Chicago Hinds Fund, the Richard Estes Award (Society of Vertebrate Paleontology), and a fellowship from the Smithsonian Institution.

Visit the Annual Reviews home page at www.AnnualReviews.org

LITERATURE CITED

Abdala F, Giannini NP. 2000. Gomphodont cynodonts of the Chañares Formation: the analysis of an ontogenetic sequence. *J. Vertebr. Paleontol.* 20:501–6

Allin EF, Hopson JA. 1992. Evolution of the auditory system in Synapsida ("mammal-like reptiles" and primitive mammals) as seen in the fossil record. In *The Evolutionary Biology of Hearing*, ed. DB Webster, RR Fay, AN Popper, pp. 587–614. New York: Springer-Verlag

Angielczyk KD. 2001. Preliminary phylogenetic analysis and stratigraphic congruence of the dicynodont anomodonts (Synapsida: Therapsida). *Palaeontol. Afr.* 37:In press

Bakker RT. 1975. Experimental and fossil evidence for the evolution of tetrapod bioenergetics. In *Perspectives of Biophysical Ecology*, ed. DM Gates, RB Schmerl, pp. 365–99. New York: Springer-Verlag

Barghusen HR. 1968. The lower jaw of cynodonts (Reptilia, Therapsida) and the evolutionary origin of mammal-like adductor jaw musculature. *Postilla* 116:1–49

Barghusen HR. 1975. A review of fighting adaptation in dinocephalians (Reptilia, Therapsida). *Paleobiology* 12:95–311

Barghusen HR. 1976. Notes on the adductor jaw musculature of *Venjukovia*, a primitive anomodont therapsid from the Permian of the U.S.S.R.. *Ann. S. Afr. Mus.* 69:249–60

Battail B. 1991. Les Cynodontes (Reptilia,

Therapsida): une phylogenie. *Bull. Mus. Natl. Hist. Nat.* 13:17–105

Battail B. 1992. Sur la phylogénie des thérapsides (Reptilia, Therapsida). *Ann. Paléontol.* 78:83–42

Beck AL, Blob RW, Hopson JA. 2000. Interpreting limb posture in fossil tetrapods: morphological indicators of sprawling and non-sprawling locomotion. *J. Vertebr. Paleontol.* 20:29A

Blob RW. 1998. *Mechanics of non-parasagittal locomotion in* alligator and iguana: *functional implications for the evolution of non-sprawling posture in the Therapsida*. PhD thesis, Univ. Chicago. 324 pp.

Blob RW. 2001. Evolution of hindlimb posture in non-mammalian therapsids: biomechanical tests of paleontological hypotheses. *Paleobiology* 27:14–38

Bonaparte JF. 1962. Descripción del cráneo y mandibula de *Exaeretodon frenguellii*, Cabrera, y su comparación con Diademodontidae, Tritylodontidae y los cinodontes sudamericanos. *Publ. Mus. Munic. Nat. Tradic, Mar Plata* 1:135–202

Bonaparte JF. 1966. Una nueva "fauna" Triasica de Argentina (Therapsida: Cynodontia; Dicynodontia); considaciones filogeneticas y paleobiogeográficas. *Ameghiniana* 4:243–96

Boonstra LD. 1963. Diversity within the South African Dinocephalia. *S. Afr. J. Sci.* 59:196–206

Boonstra LD. 1969. The fauna of the *Tapinocephalus* Zone (Beaufort beds of the Karoo). *Ann. S. Afr. Mus.* 56:1–73

Boonstra LD. 1971. The early therapsids. *Ann. S. Afr. Mus.* 59:17–46

Boonstra LD. 1972. Discard the names Theriodontia and Anomodontia: a new classification of the Therapsida. *Ann. S. Afr. Mus.* 59:315–38

Botha J, Chinsamy A. 2001. Growth patterns deduced from the bone histology of the cynodonts *Diademodon* and *Cynognathus. J. Vertebr. Paleontol.* 20:705–11

Brinkman D. 1981. The structure and relationships of the dromasaurs (Reptilia: Therapsida). *Breviora* 465:1–34

Cheng Z, Li J. 1996. First record of a primitive anteosaurid dinocephalian from the Upper Permian of Gansu, China. *Vertebr. Pal. Asiat.* 34:123–34

Cheng Z, Li J. 1997. A new genus of primitive dinocephalian: the third report on Late Permian Dashankou lower tetrapod fauna. *Vertebr. PalAsiat.* 35:35–43

Cheng Z-W. 1993. On the discovery and significance of the nonmarine Permo-Triassic transition zone at Dalongkou in Jimusar, Xinjiang, China. In *The Nonmarine Triassic*, ed. SG Lucas, M Morales, pp. 65–67. Albuquerque: New Mexico Mus. Nat. Hist. Sci. Bull. No. 3

Chinsamy A, Rubidge BS. 1993. Dicynodont (Therapsida) bone histology: phylogenetic and physiological implications. *Palaeontol. Afr.* 30:97–102

Clark JM, Hopson JA. 1985. Distinctive mammal-like reptile from Mexico and its bearing on the phylogeny of the Tritylodontidae. *Nature* 315:398–400

Cluver MA. 1971. The cranial morphology of the dicynodont genus *Lystrosaurus. Ann. S. Afr. Mus.* 56:155–274

Cluver MA, Hotton N III. 1981. The genera *Dicynodon* and *Diictodon* and their bearing on the classification of the Dicynodontia (Reptilia, Therapsida). *Ann. S. Afr. Mus.* 83:99–146

Cluver MA, King GM. 1983. A reassessment of the relationships of the Permian Dicynodontia (Reptilia, Therapsida) and a new classification of dicynodonts. *Ann. S. Afr. Mus.* 91:195–273

Colbert EH, Kitching JW. 1981. Scalopsaurian reptiles from the Triassic of Antarctica. *Am. Mus. Novit.* 1–22

Cox CB. 1964. On the palate, dentition, and classification of the fossil reptile *Endothiodon* and related genera. *Am. Mus. Novit.* 2171:1–25

Cox CB. 1998. The jaw function and adaptive radiation of the dicynodont mammal-like reptiles of the Karoo Basin of South Africa. *Zool. J. Linn. Soc.* 122:349–84

Crompton AW. 1955. On some Triassic cynodonts from Tanganyika. *Proc. Zool. Soc. Lond.* 125:617–69

Crompton AW. 1963. The evolution of the mammalian jaw. *Evolution* 17:431–39

Crompton AW. 1972. Postcanine occlusion in cynodonts and tritylodontids. *Bull. Br. Mus. (Nat. Hist.) Geol.* 21:30–71

Crompton AW, Ellenberger F. 1957. On a new cynodont from the Molteno Beds and the origin of the tritylodontids. *Ann. S. Afr. Mus.* 44:1–14

Crompton AW, Hotton N. 1967. Functional morphology of the masticatory apparatus of two dicynodonts (Reptilia, Therapsida). *Postilla* 109:1–51

Crompton AW, Jenkins FA Jr. 1973. Mammals from reptiles: a review of mammalian origins. *Annu. Rev. Earth Planet. Sci.* 1:131–55

Crompton AW, Luo Z. 1993. Relationship of the Liassic mammals *Sinoconodon, Morganucodon oehleri*, and *Dinnetherium*. See Szalay et al. 1993, pp. 30–44

Crompton AW, Parker P. 1978. Evolution of the mammalian masticatory apparatus. *Am. Sci.* 66:192–201

de Ricqlès A. 1974. Evolution of endothermy. *Evol. Theory* 1:51–80

Dilkes DW, Reisz RR. 1996. First record of a basal synapsid ('mammal-like reptile') in Gondwana. *Proc. R. Soc. London Ser. B* 263:1165–70

Efremov IA. 1940. *Ulemosaurus svijagensis*

Riab.—ein Dincephale aus den Ablagerungen des Perm der UdSSR. *Nove Acta Leopoldina* 9:155–205

Erwin DH. 1993. *The Great Paleozoic Crisis: Life and Death in the Permian.* New York: Columbia Univ. Press

Flynn JJ, Parrish JM, Rakotosamimanana B, Ranivoharimanana L, Simpson WF, Wyss AR. 2000. New traversodontids (Synapsida: Eucynodontia) from the Triassic of Madagascar. *J. Vertebr. Paleontol.* 20:422–27

Flynn JJ, Parrish JM, Rakotosamimanana B, Simpson WF, Whatley RL, Wyss AR. 1999. A Triassic fauna from Madagascar, including early dinosaurs. *Science* 286:763–65

Fourie H. 2001. *Morphology and function of the postcrania of selected genera of Therocephalia (Amniota: Therapsida).* PhD thesis, Univ. Witwatersrand, Johannesburg, S. Afr. 185 pp.

Fourie S. 1974. The cranial morphology of *Thrinaxodon liorhinus* Seeley. *Ann. S. Afr. Mus.* 65:337–400

Gauthier J, Kluge A, Rowe T. 1988. Amniote phylogeny and the importance of fossils. *Cladistics* 4:105–209

Gay SA, Cruickshank ARI. 1999. Biostratigraphy of the Permian tetrapod faunas from the Ruhuhu Valley, Tanzania. *J. Afr. Earth Sci.* 29:195–210

Grine FE. 1997. Dinocephalians are not anomodonts. *J. Vert. Paleontol.* 17:177–83

Hancox PJ. 2000. The continental Triassic of South Africa. *Z. Geol. Paläontol.* 1998: 1285–324

Hancox PJ, Rubidge BS. 1997. The role of fossils in interpreting the development of the Karoo Basin. *Palaeontol. Afr.* 33:41–54

Hancox PJ, Shishkin MA, Rubidge BS, Kitching JW. 1995. A threefold subdivision of the *Cynognathus* Assemblage Zone (Beaufort Group, South Africa) and its palaeogeographical implications. *S. Afr. J. Sci.* 91:143–44

Hillenius WJ. 1992. The evolution of nasal turbinates and mammalian endothermy. *Paleobiology* 18:17–29

Hillenius WJ. 1994. Turbinates in therapsids: evidence for Late Permian origins of mammalian endothermy. *Evolution* 48:207–29

Hopson JA. 1985. Morphology and relationships of *Gomphodontosuchus brasiliensis* von Huene (Synapsida, Cynodontia, Tritylodontoidea) from the Triassic of Brazil. *Neues Jahrb. Geol. Paläontol. Monatsch.* 5:285–99

Hopson JA. 1991a. Convergence in mammals, tritheledonts, and tritylodonts. *J. Vertebr. Paleontol.* 11:36A

Hopson JA. 1991b. Systematics of the nonmammalian Synapsida and implications for patterns of evolution in synapsids. See Schultze & Trueb 1991, pp. 635–93

Hopson JA. 1994. Synapsid evolution and the radiation of non-eutherian mammals. In *Major Features of Vertebrate Evolution.*, ed. DB Prothero, RM Schoch, pp. 190–219. Knoxville, TN: Paleontol. Soc.

Hopson JA. 1995. Patterns of evolution in the manus and pes of non-mammalian therapsids. *J. Vertebr. Paleontol.* 15:615–39

Hopson JA. 1999. Therapsids. In *Encyclopedia of Paleontology*, ed. R Singer, pp. 1256–66. Chicago: Fitzroy Dearborn

Hopson JA, Barghusen H. 1986. An analysis of therapsid relationships. See Hotton et al. 1986, pp. 83–106

Hopson JA, Crompton AW. 1969. Origin of mammals. In *Evolutionary Biology*, ed. T Dobzhansky, MK Hecht, WC Steere, pp. 15–71. East Norwalk, CT: Appleton-Century-Croft

Hopson JA, Kitching JW. 1972. A revised classification of cynodonts (Reptilia; Therapsida). *Palaeontol. Afr.* 14:71–85

Hopson JA, Kitching JW. 2001. A probainognathian cynodont from South Africa and the phylogeny of non-mammalian cynodonts. *Bull. Mus. Comp. Zool.* 156: In press

Hotton N. 1991. The nature and diversity of synapsids: prologue to the origin of mammals. See Schultze & Trueb 1991, pp. 598–634

Hotton N, MacLean PD, Roth JJ, Roth EC, eds. 1986 *The Ecology and Biology of the Mammal-Like Reptiles*. Washington, DC: Smithsonian Inst. Press

Ivachnenko MF. 1990. Elements of the Early Permian tetrapod faunal assemblages of Eastern Europe. *Paleontol. J.* 1990:102–11

Ivachnenko MF. 1994. A new Late Permian dromasaurian (Anomodontia) from Eastern Europe. *Paleontol. J.* 1994:77–84

Ivachnenko MF. 1995. Primitive Late Permian dinocephalian-titanosuchids of Eastern Europe. *Paleontol. J.* 29:120–29

Ivachnenko MF. 1996a. New primitive therapsids from the Permian of Eastern Europe. *Paleontol. J.* 30:337–43

Ivachnenko MF. 1996b. Primitive anomodonts, venyukoviids, from the Late Permian of Eastern Europe. *Paleontol. J.* 30:575–82

Ivachnenko MF. 2000. Estemmenosuches and primitive theriodonts from the Late Permian. *Paleontol. J.* 34:184–92

Ivachnenko MF, Golubev VK, Gubin YM, Kalandadze NN, Novikov IV, et al. 1997. *Permian and Triassic Tetrapods of Eastern Europe*. Moscow: GEOS

Kemp TS. 1969. On the functional morphology of the gorgonopsid skull. *Philos. Trans. R. Soc. London Ser. B* 256:1–83

Kemp TS. 1978. Stance and gait in the hindlimb of a therocephalian mammal-like reptile. *J. Zool. London* 186:143–61

Kemp TS. 1979. The primitive cynodont *Procynosuchus*: functional anatomy of the skull and relationships. *Philos. Trans. R. Soc. London Ser. B* 285:73–122

Kemp TS. 1980. Aspects of the structure and functional anatomy of the Middle Triassic cynodont *Luangwa*. *J. Zool. London* 191:193–239

Kemp TS. 1982. *Mammal-Like Reptiles and the Origin of Mammals*. New York: Academic

Kemp TS. 1983. The relationships of mammals. *Zool. J. Linn. Soc.* 77:353–84

Kemp TS. 1985. Synapsid reptiles and the origin of higher taxa. *Spec. Pap. Palaeontol.* 33:175–84

Kemp TS. 1988. Interrelationships of the Synapsida. In *The Phylogeny and Classification of the Tetrapods: Mammals*, ed. MJ

Benton, 2:1–22. Oxford: Systematics Assoc. Spec. Vol. No. 35B

Keyser AW. 1973. A new Triassic vertebrate fauna from South West Africa. *Palaeontol. Afr.* 13:1–15

Keyser AW. 1993. A re-evaluation of the smaller Endothiodontidae. *Mem. Geol. Surv. S. Afr.* 82:1–53

Keyser AW, Cruickshank ARI. 1979. The origins and classifications of Triassic dicynodonts. *Trans. Geol. Soc. S. Afr.* 82:81–108

King G. 1990. *The Dicynodonts: A Study in Palaeobiology*. New York: Chapman & Hall

King GM. 1988. Anomodontia. In *Encyclopedia of Paleoherpetology*, ed. P Wellnhofer, 17C:1–174. Stuttgart: Fischer

King GM. 1994. The early anomodont *Venjukovia* and the evolution of the anomodont skull. *J. Zool. London* 232:651–73

King GM, Jenkins I. 1997. The dicynodont *Lystrosaurus* from the Upper Permian of Zambia: evolutionary and stratigraphic implications. *Palaeontology* 40:149–56

King GM, Oelofsen BW, Rubidge BS. 1989. The evolution to the dicynodont feeding system. *Zool. J. Linn. Soc.* 96:185–211

King GM, Rubidge BS. 1993. A taxonomic revision of small dicynodonts with postcanine teeth. *Zool. J. Linn. Soc.* 107:131–54

Kitching JW. 1977. *The distribution of the Karoo vertebrate fauna. Bernard Price Inst. for Palaeontol. Res.*, 1:1–131. Memoir

Langer MC. 2000. The first record of dinocephalians in South America: Late Permian (Rio do Rasto Formation) of the Paraná Basin, Brazil. *Neues Jahrb. Geol. Paläontol. Abhandlug.* 215:69–95

Latimer EM, Gow CE, Rubidge BS. 1995. Dentition and feeding niche of *Endothiodon* (Synapsida; Anomodontia). *Palaeontol. Afr.* 32:75–82

Laurin M. 1998. New data on the cranial morphology of *Lycaenops* (Synapsida, Gorgonopsidae) and reflections on the possible presence of streptostyly in gorgonopsians. *J. Vertebr. Paleontol.* 18:765–76

Laurin M, Reisz RR. 1995. A reevaluation of

early amniote phylogeny. *Zool. J. Linn. Soc.* 113:165–223

Laurin M, Reisz RR. 1996. The osteology and relationships of *Tetraceratops insignis*, the oldest known therapsid. *J. Vertebr. Paleontol.* 16:95–102

Lepper J, Raath MA, Rubidge BS. 2000. A diverse dinocephalian fauna from Zimbabwe. *S. Afr. J. Sci.* 96:403–5

Li J, Cheng Z. 1995. A new Late Permian vertebrate fauna from Dashankou, Gansu with comments on Permian and Triassic vertebrate assemblage zones of China. In *Short Papers of Sixth Symposium on Mesozoic Terrestrial Ecosystems and Biota*, ed. AL Sun, YQ Wang, pp. 33–37. Beijing: China Ocean Press

Li J, Rubidge BS, Cheng Z. 1996. A primitive anteosaurid dinocephalian from China—implications for the distribution of earliest therapsid faunas. *S. Afr. J. Sci.* 92:252–53

Lucas SG. 1998. Global Triassic tetrapod biostratigraphy and biochronology. *Palaeogeogr. Palaeoclimat. Palaeoecol.* 143:347–84

Luo Z. 1994. Sister-group relationships of mammals and transformations of diagnostic mammalian characters. In *In the Shadow of the Dinosaurs: Early Mesozoic Tetrapods*, ed. NC Fraser, H-D Sues, pp. 98–128. New York: Cambridge Univ. Press

MacLeod KG, Smith RMH, Koch PL, Ward PL. 2000. Timing of mammal-like reptile extinctions across the Permian-Triassic boundary in South Africa. *Geology* 28:227–30

Maier W, van den Heever J, Durand F. 1996. New therapsid specimens and the origin of the secondary hard and soft palate of mammals. *J. Zool. Syst. Evol. Res.* 34:9–19

Martinez RN, Forster CA. 1996. The skull of *Probelesodon sanjuanensis*, Sp. Nov., from the Late Triassic Ischigualasto Formation of Argentina. *J. Vertebr. Paleontol.* 16:285–91

Martinez RN, May CL, Forster CA. 1996. A new carnivorous cynodont from the Ischigualasto Formation (Late Triassic, Argentina), with comments on eucynodont phylogeny. *J. Vertebr. Paleontol.* 16:271–84

McNab BK. 1978. The evolution of endothermy in the phylogeny of mammals. *Am. Nat.* 112:1–21

McShea DW. 1994. Mechanisms of large-scale evolutionary trends. *Evolution* 48:1747–63

Modesto SP. 1992. Did herbivory foster early amniote diversification? *J. Vertebr. Paleontol.* 11:49A

Modesto SP. 1995. The skull of the herbivorous synapsid *Edaphosaurus boanerges* from the Lower Permian of Texas. *Palaeontology* 38:213–39

Modesto SP, Rubidge BS, Welman J. 1999. The most basal anomodont therapsid and the primacy of Gondwana in the evolution of the anomodonts. *Proc. R. Soc. London Ser. B* 266:331–37

Modesto S, Rubidge B. 2000. A basal anomodont therapsid from the Lower Beaufort Group, Upper Permian of South Africa. *J. Vertebr. Paleontol.* 20:515–21

Modesto S, Sidor C, Rubidge B, Welman J. 2001. A second varanopseid skull from the Upper Permian of South Africa: implications for Late Permian 'pelycosaur' evolution. *Lethaia* 34:In press

Munyikwa D. 2001. *Cranial morphology of a primitive dinocephalian (Amniota: Therapsida) from the Madumabisa Mudstone Formation, Zimbabwe.* MSc thesis, Univ. Witwatersrand, Johannesburg, S. Afr. 75 pp.

Neveling J. 1998. *The biostratigraphy and sedimentology of the contact area between the* Lystrosaurus *and* Cynognathus *Assemblage Zones (Beaufort Group: Karoo Supergroup).* MSc Thesis. Univ. Witwatersrand, Johannesburg, S. Afr. 126 pp.

Ochev VG, Shishkin MA. 1989. On the principles of global correlation of the continental Triassic on the tetrapods. *Acta Palaeontol. Polon.* 34:149–73

Olson EC. 1959. The evolution of mammalian characters. *Evolution* 13:344–53

Olson EC. 1962. Late Permian terrestrial vertebrates, U.S.A. and U.S.S.R.. *Trans. Am. Philos. Soc.* 52:1–224

Olson EC. 1968. The family Caseidae. *Fieldiana Geol.* 17:224–349

Olson EC. 1989. Problems of Permo-Triassic

terrestrial vertebrate extinctions. *Hist. Biol.* 2:17–35

Owen R. 1844. Description of certain fossil crania, discovered by A. G. Bain, Esq., in sandstone rocks at the south-eastern extremity of Africa, referable to different species of an extinct genus of Reptilia (*Dicynodon*), and indicative of a new tribe or suborder of Sauria. *Proc. Geol. Soc. London* 4:500–4

Parrington FR. 1955. On the cranial anatomy of some gorgonopsids and the synapsid middle ear. *Proc. Zool. Soc. London* 125:1–40

Parrish JMJ, Parrish JT, Ziegler AM. 1986. Permian-Triassic paleogeography and paleoclimatology and implications for therapsid distribution. See Hotton et al. 1986, pp. 109–31

Ray S. 2000. Endothiodont dicynodonts from the Late Permian Kundaram Formation, India. *Palaeontology* 43:375–404

Reisz RR. 1986. Pelycosauria. In *Encyclopedia of Paleoherpetology*, ed. P Wellnhofer, 17A:1–102. Stuttgart: Fischer

Reisz RR, Berman DS. 2001. The skull of *Mesenosaurus romeri*, a small varanopseid (Synapsida: Eupelycosauria) from the Upper Permian of the Mesen River Basin, northern Russia. *Ann. Carnegie Mus.* 70:113–132

Reisz RR, Dilkes DW, Berman DS. 1998. Anatomy and relationships of *Elliotsmithia longiceps* Broom, a small synapsid (Eupelycosauria: Varanopseidae) from the Late Permian of South Africa. *J. Vertebr. Paleontol.* 18:602–11

Reisz RR, Sues H-D. 2000. Herbivory in Late Paleozoic and Triassic terrestrial vertebrates. In *Evolution of Herbivory in Terrestrial Vertebrates*, ed. H-D Sues, pp. 9–41. New York: Cambridge Univ. Press

Renaut AJ. 2000. *A re-evaluation of the cranial morphology and taxonomy of the Triassic dicynodont genus* Kannemeyeria. PhD thesis, Univ. Witwatersrand, Johannesburg, S. Afr. 266 pp

Romer AS. 1969a. The Brazilian Triassic cynodont reptiles *Belesodon* and *Chiniquodon*. *Breviora* 332:1–16

Romer AS. 1969b. The Chañares (Argentina) Triassic reptile fauna V. A new chiniquodontid cynodont, *Probelesodon lewisi*—cynodont ancestry. *Breviora* 333:1–24

Romer AS. 1970. The Chañares (Argentina) Triassic reptile fauna VI. A chiniquodontid cynodont with an incipient squamosal-dentary jaw articulation. *Breviora* 344:1–18

Romer AS, Price LI. 1940. Review of the Pelycosauria. *Geol. Soc. Am. Spec. Pap.* 28:1–538

Rowe T. 1986. *Osteological diagnosis of Mammalia, L. 1758, and its relationship to extinct Synapsida.* PhD thesis, Univ. Calif. 446 pp.

Rowe T. 1988. Definition, diagnosis, and the origin of Mammalia. *J. Vertebr. Paleontol.* 8:241–64

Rowe T. 1993. Phylogenetic systematics and the early history of mammals. See Szalay et al. 1993, pp. 129–45

Rowe T, van den Heever JA. 1988. The hand of *Anteosaurus magnificus* (Dinocephalia: Therapsida) and its bearing on the origin of the mammalian manual phalangeal formula. *S. Afr. J. Sci.* 82:641–45

Ruben J. 1995. The evolution of endothermy in mammals and birds: from physiology to fossils. *Annu. Rev. Physiol.* 57:69–95

Rubidge BS. 1990a. A new vertebrate biozone at the base of the Beaufort Group, Karoo Sequence (South Africa). *Palaeontol. Afr.* 27:17–20

Rubidge BS. 1990b. Redescription of the cranial morphology of *Eodicynodon oosthuizeni* (Therapsida: Dicynodontia). *Navors. Nas. Mus. Bloemfontein* 7:1–25

Rubidge BS. 1991. A new primitive dinocephalian mammal-like reptile from the Permian of southern Africa. *Palaeontology* 34:547–59

Rubidge BS. 1994. *Australosyodon*, the first primitive anteosaurid dinocephalian from the Upper Permian of Gondwana. *Palaeontology* 37:579–94

Rubidge BS. 1995. *Biostratigraphy of the Beaufort Group (Karoo Supergroup).* Pretoria, S. Afr.: Gov. Printer

Rubidge BS, Hancox PJ, Catuneanu O. 2000. Sequence analysis of the Ecca-Beaufort contact in the southern Karoo of South Africa. *S. Afr. J. Geol.* 103:81–96

Rubidge BS, Hopson JA. 1996. A primitive anomodont therapsid from the base of the Beaufort Group (Upper Permian) of South Africa. *Zool. J. Linn. Soc.* 117:117–39

Rubidge BS, Kitching JW, van den Heever JA. 1983. First record of a therocephalian (Therapsida: Pristerognathidae) from the Ecca of South Africa. *Navors. Nas. Mus. Bloemfontein* 4:229–35

Rubidge BS, Modesto SP, Sidor CA, Welman J. 1999. *Eunotosaurus* from the Ecca-Beaufort contact in the Northern Cape: implications for Karoo Basin development. *S. Afr. J. Sci.* 95:553–55

Rubidge BS, van den Heever JA. 1997. Morphology and systematic position of the dinocephalian *Styracocephalus platyrhinus. Lethaia* 30:157–68

Rybczynski N. 2000. Cranial anatomy and phylogenetic position of *Suminia getmanovi*, a basal anomodont (Amniota: Therapsida) from the Late Permian of Eastern Europe. *Zool. J. Linn. Soc.* 130:329–73

Schultze H-P, Trueb L, eds. 1991. *Origins of the Higher Groups of Tetrapods: Controversy and Consensus.* Ithaca, NY: Comstock

Sidor CA. 2000. *Evolutionary trends and relationships within the Synapsida.* PhD thesis, Univ. Chicago. 370 pp.

Sidor CA. 2001. Simplification as a trend in synapsid cranial evolution. *Evolution* 55:1419–42

Sidor CA, Hopson JA. 1995. The taxonomic status of the Upper Permian eotheriodont therapsids of the San Angelo Formation (Guadalupian), Texas. *J. Vertebr. Paleontol.* 15:53A

Sidor CA, Hopson JA. 1998. Ghost lineages and "mammalness": assessing the temporal pattern of character acquisition in the Synapsida. *Paleobiology* 24:254–73

Sigogneau D. 1970. *Révision Systématique des Gorgonopsiens Sud-Africains.* Paris: Cahiers Paléontol.

Sigogneau-Russell D. 1989. Theriodontia I. In *Encyclopedia of Paleoherpetology, Part 17B*, ed. P Wellnhofer, pp. 1–127. Stuttgart, Ger.: Fischer

Smith RMH. 1993. Vertebrate taphonomy of Late Permian floodplain deposits in the southwestern Karoo Basin of South Africa. *Palaios* 8:45–67

Smith RMH. 1995. Changing fluvial environments across the Permian-Triassic boundary in the Karoo Basin, South Africa and possible causes of tetrapod extinctions. *Palaeogeogr. Palaeoclimat. Palaeoecol.* 117:81–104

Smith RMH, Eriksson PG, Botha WJ. 1993. A review of the stratigraphy and sedimentary environments of the Karoo-aged basins of southern Africa. *J. Afr. Earth Sci.* 16:143–69

Smith RMH, Kitching JW. 1997. Sedimentology and vertebrate taphonomy of the *Tritylodon* Acme Zone: a reworked palaeosol in the Lower Jurassic Elliot Formation, Karoo Supergroup, South Africa. *Palaeogeogr. Palaeoclimat. Palaeoecol.* 131:29–50

Sues H-D. 1985. The relationships of the Tritylodontidae (Synapsida). *Zool. J. Linn. Soc.* 85:205–17

Sues H-D. 1986. The skull and dentition of two tritylodontid synapsids from the Lower Jurassic of western North America. *Bull. Mus. Comp. Zool.* 151:217–68

Sumida SS, Martin KLM, eds. 1997. *Amniote Origins: Completing the Transition to Land.* New York: Academic. 510 pp.

Surkov MV. 2000. On the historical bigeography of Middle Triassic anomodonts. *Paleontol. J.* 34:79–83

Szalay FS, Novacek MJ, McKenna MC, eds. 1993. *Mammal Phylogeny: Mesozoic Differentiation, Multituberculates, Monotremes, Early Therians, and Marsupials.* New York: Springer-Verlag

Tatarinov LP. 1968. Morphology and systematics of the Northern Dvinia cynodonts (Reptilia, Therapsida; Upper Permian). *Postilla* 126:1–15

Tatarinov LP. 2000. A new gorgonopid (Reptilia, Theriodontia) from the Upper Permian of the Vologda region. *Paleontol. J.* 34:75–83

Tatarinov LP, Matchenko EN. 1999. A find of

an aberrant tritylodont (Reptilia, Cynodontia) in the Lower Cretaceous of the Kemerovo region. *Paleontol. J.* 33:422–28

van den Heever JA. 1994. The cranial anatomy of the early Therocephalia (Amniota: Therapsida). *Univ. van Stellenbosch Ann.* 1: 1–59

Wagner PJ. 2000. The quality of the fossil record and the accuracy of phylogenetic inferences about sampling and diversity. *Syst. Biol.* 49:65–86

Ward PD, Montgomery DR, Smith RMH. 2000. Altered river morphology in South Africa related to the Permian-Triassic extinction. *Science* 289:1740–43

Watson DMS, Romer AS. 1956. A classification of therapsid reptiles. *Bull. Mus. Comp. Zool.* 114:37–89

Annu. Rev. Ecol. Syst. 2001. 32:481–517
Copyright © 2001 by Annual Reviews. All rights reserved

ECOLOGY, CONSERVATION, AND PUBLIC POLICY

Donald Ludwig,[1] Marc Mangel,[2] and Brent Haddad[2]

[1]*Departments of Mathematics and Zoology, University of British Columbia, Vancouver, British Columbia V6T 1Z2, Canada; e-mail: ludwig@math.ubc.ca*

[2]*Department of Environmental Studies, University of California, Santa Cruz, California 95064; e-mail: msmangel@cats.ucsc.edu; bhaddad@cats.ucsc.edu*

Key Words scientific uncertainty, Bayesian inference, regulatory decision-making, scientism, social construction

■ **Abstract** A new sense of urgency about environmental problems has changed the relationship between ecology, other disciplines, and public policy. Issues of uncertainty and scientific inference now influence public debate and public policy. Considerations that formerly may have appeared to be mere technicalities now may have decisive influence. It is time to re-examine our methods to ensure that they are adequate for these new requirements. When science is used in support of policy-making, it cannot be separated from issues of values and equity. In such a context, the role of specialists diminishes, because nobody can be an expert in all the aspects of complicated environmental, social, ethical, and economic issues. The disciplinary boundaries that have served science so well in the past are not very helpful in coping with the complex problems that face us today, and ecology now finds itself in intense interaction with a host of other disciplines. The next generation of ecologists must be prepared to interact with such disciplines as history, religion, philosophy, geography, economics, and political science. The requisite training must involve not only words, but core skills in these disciplines. A sense of urgency has affected not only ecology but other disciplines that influence environmental problems: they are undergoing a similar transformation of their outlook and objectives.

INTRODUCTION

More than any previous generation, the current generation of ecologists is concerned about applications of their work to society's problems. Since the first Earth Day about 30 years ago there has been a change in the public perception of environmental problems and a corresponding growth in the sensitivity of ecologists to issues of public policy (Brown 2000, Kaiser 2000, Padilla & Gibson 2000). Berkes (1999) reiterates the question of Roszak (1972, p. 404), "Which will ecology be: the last of the old sciences or the first of the new?" Such hopes and expectations place a heavy burden on ecologists. Furthermore, the scope and setting of ecology has changed in the last three decades. Mangel et al. (1996) summarized these changes as developments in (*a*) ecological and biological understanding,

0066-4162/01/1215-0481$14.00

481

(*b*) the combination of economics and ecology, (*c*) institutions and policy, and (*d*) technology and methodologies.

There are two themes in our review. The first is scientific uncertainty in ecological problems and its influence on relations between ecologists and policy makers. Uncertainty in ecological knowledge causes difficulties when this knowledge is used for regulatory purposes. This effect may be compounded when regulatory agencies adopt policies advocated by industries that are being regulated. Many scientists experience such phenomena when they work for government agencies or when they attempt to influence public policy. The second theme is an appreciation of insights from other domains of knowledge; here we cannot be exhaustive, but hope to provide entry into the relevant literature. Clearly scientific understanding, and particularly ecological understanding, is an important tool for dealing with environmental problems. Now it is also clear that scientific knowledge will never be enough. Somehow that knowledge must be integrated with political, economic, social, ethical, and religious insight, and tempered with respect for human dignity and for the biosphere. For example, although the modern world has characterized indigenous and traditional peoples as "primitive," some of them have (or had) highly evolved social and religious systems that promoted sustainable use of natural resources for hundreds and thousands of years (Bodley 1990). We believe that ecologists can be effective in conservation efforts only if they understand and appreciate the social and ethical aspects of conservation. Because this wider understanding is at the heart of an effective approach to conservation, we first provide examples illustrating the issues.

Environmental Problems are "Wicked"

Problems such as conservation of world forest resources, conservation of endangered and threatened species, and global climate change are not merely ecological or scientific. They involve a host of traditional academic disciplines that cannot be separated from issues of values, equity, and social justice. Whether our progeny live in a world at war or at peace may hinge on equitable resolutions of environmental problems (Homer-Dixon 1991, 1994; Homer-Dixon et al. 1993). Rittel & Webber (1973) defined "wicked problems" as those with no definitive formulation, no stopping rule, and no test for a solution. Solutions are judged good or bad instead of true or false, and there will likely never be a final resolution of any of these problems. Each wicked problem is unique and defies classification. Roe (1998) calls these problems "truly complex" or "complex all the way down." Funtowicz et al. (1999) call such problems "post normal," characterized by "radical uncertainty" and a "plurality of legitimate perspectives." They characterize analysis of complex systems as "feeling the elephant," in analogy with the classic story of the blind men and the elephant. The same analogy is used in the context of koala conservation in Clark et al. (2000, p. 699). Maddox (2000) notes that "the best environmental policy depends upon how you frame the question." These problems will not be dealt with by optimizing or managing anything (who manages climate change?); there are no experts for such problems, nor can there be. Dealing with these problems involves much more than ecology and science; it requires an understanding of how economic and social factors interweave with ecological and evolutionary science. The construction of effective approaches may require specification of mechanisms, processes, and parameters that may not be accessible. Here are some examples of such problems.

Forests

The World Commission on Forests and Sustainable Development was established by a group of distinguished world leaders following the Earth Summit in 1992. The Commission worked for five years, holding hearings around the world. Their report (WCFSD 1999) emphasizes that problems of forest conservation cannot be separated from problems of equity and of governance:

> [p. 59] . . . forest dwelling communities can be found in all types of forests, in every geographic region of the world. Everywhere they are beset by similar forces: loggers, ranchers, and others moving onto their lands; erosion of their traditional rights of access and use; displacement of their homes; erosion of their livelihoods; ignorance of their values, their historical custodial functions, their accumulated intellectual property; disregard of the authorities; often persecuted by the powers that be and the politically strong. These forces are likely to become intensified as the world gets more crowded, as the demands on forests increase, and as forest capital further declines. Yet there was no evidence in any region of a constructive and compassionate movement on the part of political leaders to prevent these abuses and protect the poor and politically weak, despite the fact that we are in the United Nations Decade for Indigenous Peoples. [p. 60] It is a travesty of justice that the rights of such communities—communities which via their custodial services subsidize the world—are not secured, while financial fortunes from exploitation of a country's forest capital accrue to private corporate interests and government treasuries, to merchants and middlemen.

Very similar messages with abundant detail appear in Bodley (1990). Applications to rangeland appear in Busby & Coz (1994). Robinson (1993) notes that by not recognizing the conflicts and contradictions inherent in conservation and development, one can reach simplistic and inappropriate conclusions; that even "sustainable development" will lead to loss of biological diversity; and that "carrying capacity is not an ecosystem characteristic, but is defined for the population of a given species" (p. 23). See Cohen (1995) for a general discussion of the meaning of human carrying capacity.

Endangered Species

Various institutions have attempted to ensure the survival of endangered species by establishing protected areas. These are generally far too small for many species, and they cannot cover more than a small fraction of species under threat. Furthermore, the areas that are established often cannot be protected because of incursions by people living nearby and extensive forest removal and road building in adjoining areas. In response, community-based conservation has been attempted. Many of these efforts have also failed: the local communities that must be involved in conservation efforts are often powerless to resist exploitation by industries, governments, and the military. They may see their traditional use-rights curtailed and their legal status altered to "poacher." Although at first sight the problem of conservation seems to be a scientific one, it may primarily be a problem of governance and of equity (Clad 1988, Goodland 1988, Wells et al. 1999, Wells & Brandon 1992, Brandon 1997, van Schaik et al. 1997, Sanderson & Redford 1997) or population (Mangel et al. 1996). Sometimes it is possible to involve local populations in conservation successfully but "to devise policies that lead to resource conservation through indigenous management, we believe it will be necessary to understand the distinctive features of the foraging *economy*, as a matter of behavior with practical consequences" (Winterhalder & Lu 1997, p. 1362). It is worth noting that, in some cases, individuals of the species being protected cannot even be easily identified (McElroy et al. 1997).

Climate Change

The problem of climate change is intractable by our traditional scientific methods (Hamilton 1999). Bretherton (1994) notes that different ways in which individuals respond to uncertainty in everyday life leads to different perspectives on global environmental change that includes different views of the agenda, the analyses needed, and the scientific implications. Morgan & Dowlatabadi (1996) point out that, in addition to all the other uncertainties about climate change, we are uncertain about the identities of the decision makers. They conclude (p. 363):

> ... the more we work on integrated assessment of climate change, the more we realize that the biggest challenges are philosophical and methodological.... We have never worked on a problem in which the labile and adaptive nature of values, or the number of different actors with different values, is as central as it is in climate policy. Finally, we have been doing analysis for the entire world, using basic ideas of causation, probability, and rational expectation: ideas that are probably not shared by 80% of the world's peoples. The available tools of policy analysis are simply not up to the challenges we face, so we are busy inventing new tools.

Uncertainty and Ecological Science

Complex problems such as those described in the previous section involve varying degrees of knowledge and ignorance. It is helpful to characterize the aspects that are well understood, so that they can be differentiated from those that are more problematical. There is a large literature on scientific uncertainty, but uncertainty per se in ecology has not been surveyed. Some aspects of uncertainty are considered in Levins (1966), Beck (1987), Simberloff (1988), Dovers & Handmer (1995), Hilborn et al. (1995), Lélé & Norgaard (1995), Talbot (1996), Mangel et al. (1996), and Hilborn et al. (1999). Smithson (1989) and Wynne (1992) differentiate kinds of ignorance. For example, we face risk determined by the significance level if we reject a null hypothesis. We may be uncertain about the value of a parameter that measures the strength of competition between two species with similar requirements. A few well-designed experiments may remove such uncertainty. We may be ignorant about the form of a predation response: Is it Holling type I, II, or III? Reduction of such ignorance may require a long series of well-controlled experiments. It may be undetermined whether a community is structured by competition, predation, or some other interaction. It may be impossible to give a general solution of such indeterminate problems. The answer may vary from species to species, from place to place, and from time to time. The implications of these different kinds of uncertainty are profound for both sustainability and application of a precautionary approach (Dovers & Handmer 1995). Although we may often use stochastic models, fundamental ignorance may dominate fluctuations. Harcourt (1995) notes, in work on the population viability of the Virunga gorilla (*Gorilla gorilla*): (p. 134) "Deterministic change in habitat is a greater threat than stochastic demographic variation, and yet our ecological ignorance is such that we could not begin to model the consequences of removal of even the main food plant." Similarly, in view of recent amphibian declines (Pounds et al. 1997, Alford & Richards 1992), action may be required before the full story is known. The different types of uncertainty may mandate different types of actions. The fundamental law in biology—evolution by natural selection—is not a predictive law in the sense of Newton's equations (Maddox 1999). The more detailed implications of this fundamental law, such as the rules that characterize population growth and interaction or energy flows, are typically unknown or may vary from one system to another (Shrader-Frechette & McCoy 1993). This forces ecologists to create and use a variety of models of uncertain validity (Crowder et al. 1994, Halley & Dempster 1996, Berryman 1997, Starfield 1997).

Data and Their Interpretation: Hypothesis Testing and Bayesian Inference

In spite of the lack of sweeping general laws such as those in physics, ecologists can make statistical inferences from their data. These are useful provided that the chain of inference is well documented. The most commonly used method is hypothesis testing. Although application of the technique is straightforward, there are many pitfalls in interpretation. Imagine that we are interested in an ecological hypothesis H and have collected some data D in the course of investigating the hypothesis. Ideally, we wish to know the probability that H is true, given the data D. Symbolically we write

$$\Pr\{H|D\} = \text{Probability that } H \text{ is true, given the data.} \qquad 1.$$

However, in using the approach of hypothesis testing, we construct a null hypothesis H_0, which is the complement of H in the sense that

$$\Pr\{H\} + \Pr\{H_0\} = 1. \qquad 2.$$

We then evaluate the probability of observing the data D (or results more extreme) on the assumption that H_0 is true:

$$\Pr\{D|H_0\} = \text{Probability of observing data } D \text{ (or more extreme results)}$$
$$\text{when } H_0 \text{ is true.} \qquad 3.$$

The problem is this: it is rarely true that the quantities in Equations 1 and 3 are equal; in symbols

$$\Pr\{D|H_0\} \neq \Pr\{H|D\} \qquad 4.$$

The quantities on either side of relationship {4} refer to incompatible ways of viewing the data and the hypotheses. For the left-hand side of {4}, the hypothesis is either true or false; there is no probability associated with its truth. The data are random, drawn from the infinite set of data possible under conditions where H_0 is true. For the right-hand side of {4}, the truth of the hypothesis is random, and we wish to calculate its probability in view of the single set of data at hand.

Relationship {4} aside, the most common error in interpretation is to draw an inference from failure to reject a null hypothesis. Failure to reject is often taken as evidence in favor of the null hypothesis; some even believe that the truth of the null hypothesis is thereby established (Brook et al. 2000). However, the significance level addresses only the issue of false rejection of the null hypothesis, assuming its truth. If the null hypothesis is not rejected, the quantity of interest is the probability of accepting the null hypothesis when it is, in fact, false. The complement of this quantity is termed the power of the test, and it depends upon which alternative to the null hypothesis is, in fact, true (Peterman 1990, Peterman & M'Gonigle 1992, Osenberg et al. 1994, Steidl et al. 1997). Management based on hypothesis testing without consideration of the power of the test may be disastrous, as we show below. A second error in interpretation of hypothesis testing is to interpret the significance level as the probability that the null hypothesis is true. As we have pointed out above, such an inference is nonsensical in standard (frequentist) statistics, because hypotheses are either true or false in that framework: They do not have probabilities attached to them.

On the other hand, Bayesian statistics assigns probabilities to hypotheses (Apostolakis 1990, Howson & Urbach 1993, Hilborn & Mangel 1997, Press 1997, Malakoff 1999, Wade 2000). A Bayesian approach proceeds as follows (Hilborn & Mangel 1997). Suppose that p is the probability that the hypothesis of interest H is true; then $1 - p$ is the probability that the null hypothesis H_0 is true. If we apply the definition of conditional probability, with $\Pr\{H,D\}$ denoting the probability that H is true and of observing the data D, the result is

$$\Pr\{H|D\} = \Pr\{D|H\}\,\Pr\{H\}/\Pr\{D\}, \qquad 5.$$

where $\Pr\{D\}$ is the probability of observing the data. Since

$$\Pr\{H,D\} = \Pr\{D|H\}\Pr\{H\} = \Pr\{D|H\}p, \qquad 6.$$

and

$$\Pr\{D\} = \Pr\{D|H\}p + \Pr\{D|H_0\}(1-p), \qquad 7.$$

we obtain

$$\Pr\{H|D\} = \Pr\{D|H\}p/[\Pr\{D|H\}p + \Pr\{D|H_0\}(1-p)]. \qquad 8.$$

Equation 8 gives us what we want, and generalizes in a natural manner for cases in which multiple hypotheses are possible explanations.

There are various objections to Bayesian inference. Some concern technical difficulties in implementing it, and about the manner in which the value of p (called the prior probability of hypothesis H) is chosen. Dennis (1996) claims that Bayesian methods are not useful for ecological research. He objects (rightly in our view) to Bayesian neglect of methods such as randomization, examination of residuals, and design of sample surveys; these should also be part of the toolkit. Anderson (1998) presents psychological evidence that people find it difficult to reason about probabilities attached to hypotheses. She recommends standardization and improved methods of presentation to overcome some of these difficulties. It is fair to say that Bayesian methods avoid some common pitfalls of scientific inference and interpretation, but the limitations of Bayesian methods should also be recognized; they are best used as part of a comprehensive framework of analysis. Methods of choosing and implementing appropriate statistical methods are undergoing vigorous development; see Mayo (1996) and references therein.

Uncertainty

Uncertainty about the value of a parameter can be addressed by ordinary statistical methods. A common omission in interpretation of data is to fail to draw the full implications when the null hypothesis is rejected. An effect has been detected; is it of biological (as opposed to statistical) significance? If the data are sufficient to justify rejection of the null hypothesis, they are generally sufficient to supply a quantitative estimate of the size of the effect; one should not terminate the analysis with rejection. Confidence intervals are a common tool for such assessment, but their interpretation is problematical. A common error is to interpret a 95% confidence interval by stating that the probability that the quantity lies within the interval is 95%. This is nonsense in frequentist statistics because (as described above) hypotheses don't have probabilities. Bayesian credibility intervals do have the desired property; they are determined by finding the smallest set of hypotheses that have a given total probability. A Bayesian posterior distribution assigns probabilities to all combinations of hypotheses. Thus it conveys more information than a few credibility intervals, since all credibility intervals can be computed from it. Ironically, the calculations for the Bayesian analysis are often identical to those for a conventional analysis, but the conventional result is interpreted as Bayesian (attaching a probability to a hypothesis) even though frequentists may object vigorously to Bayesian approaches.

Ignorance About Model Structure

Although the method of multiple hypotheses (Chamberlain 1897) was clearly enunciated more than one hundred years ago, scientists still tend to use one model rather than a variety of models and tend to force problems into a format that their one model can handle (Hilborn & Mangel 1997). For some recent exceptions that deal with dispersal and invasion, see Kot et al. (1995), Rejmanek & Richardson (1996), or Clark et al. (1998); population trajectories, see Kot et al. (1995), Rejmanek & Richardson (1996), or Clark et al. (1998); trophic cascades, see Strong et al. (1999); abundance range relationships, see Gaston et al. (1997). These are all recent applications.

The over-reliance on single models has two consequences. The first is even the best ecological journals generally require that analysis of data be conducted in the context of a hypothesis test. As we indicated above, it is preferable to estimate parameters that describe how a response is related to a putative cause. A second consequence of adherence to a single model is that one may end up like Ptolemy, "who tinkered endlessly with his cosmological theory to preserve the fiction that the earth was at the center of the universe. When the heavenly lights failed to move in perfect circles around the earth, he proposed that their orbits included curlicues called epicycles. And when observation and theory still wouldn't mesh, he added epicycles to the epicycles" (Johnson 1999, p. 235). Ignorance about model structure is best addressed within the context of multiple hypotheses, with results adjudicated by the data. Bayesian methods are well suited to this purpose, but a variety of approaches based upon consideration of likelihood will also serve (Edwards 1972).

Response to Various Degrees of Uncertainty

The manner in which humans respond to uncertainty has received considerable attention in psychological literature (Wallsten 1980, Hogarth 1983), in political literature (Linnerooth 1984), and in some ecological literature (Funtowicz & Ravetz 1991; Morgan & Dowlatabati 1996; Hilborn et al. 1999; Lempert et al. 1996; Murphy & Noon 1991, 1992; Smithson 1989; Van Valen 1982). One approach is to ignore uncertainty in all of its forms. In fact, there may be times when a deterministic model suffices. Caughley (1994) notes that conservation biology was driven by two different threads—what causes populations to become small, and what happens to populations once they are small. He states: "The declining population paradigm ... is urgently in need of more theory. The small population paradigm ... needs more practice. Each has much to learn from the other. A cautious inter-mixing of the two might well lead to a reduction in the rate at which species are presently going extinct" (p. 215). Caughley's dichotomy was not well received by the establishment of conservation biology (Hedrick et al. 1996). Ecologists have recognized the necessity of stochastic approaches ever since the classic experiments of Park on flour beetles. Nevertheless, classic fisheries, forest, and wildlife management typically use deterministic models with known parameters, whereas the systems typically show large random fluctuations, and the model parameters are poorly determined from available data (Hilborn & Walters 1992, Quinn & Deriso 1999). Scientists may be asked to make deterministic predictions in situations that are poorly understood, or where environmental and population fluctuations are important (Shrader-Frechette 1993, Oreskes et al. 1994, Lauck et al. 1998, Taylor et al. 2000).

There is always an element of risk in adopting a course of action. Even carefully planned and well thought-out management plans may fail. The public should not be led to expect success in every ecologically based plan. The sort of risk that is analogous to flipping a coin does not present major difficulties for planning. The situation is more problematical for other types of uncertainty. Because ecological data may be expensive to collect, sparse, or of poor quality, critical parameters and processes may be poorly determined (Walters & Ludwig 1981, Ludwig & Walters 1981). There is a temptation to make recommendations on the basis of best available data in situations that are of critical importance (Ruckelshaus et al. 1997, Slade et al. 1998). This often means adopting a single best value for a parameter and a best hypothesis about the structure of the system. Such an approach may be misleading because it ignores the range of consequences that are plausible but not excluded on the basis of available data (Mangel et al. 1996).

A particularly harmful practice is to apply hypothesis testing methods to management situations. If one insists upon significant results at a 95% level before taking action, timely action may never be taken. Doak (1995) uses simple source sink models to analyze habitat degradation and grizzly bear (*Ursus arctos horribilis*) persistence. The models allowed a power analysis of habitat degradation and showed that if habitat loss is slow, more than a decade may pass between when the critical amount of habitat is lost and when it can be detected under best circumstances; also see Mangel & Hofman (1999). Johnston et al. (2000) apply power analysis to a situation where stocks of harp seal (*Pagophilus groenlandicus*) are judged not to be at risk unless a statistically significant decline is observed. They point out that a trend of 1.5% could be detected with a probability of 95% only after 19 annual surveys; detection of a trend of 0.5% would require 39 annual surveys. That is, after 10–15 years of a 1.5% decline per year, one would not be able to determine with 95% probability that the stock had declined, even though the stock decline would be considerable. Thus, the burden of proof for those who wish to assert that human intervention in an ecosystem shows lack of effect is considerably more complicated than applying a hypothesis test (Hilborn & Ludwig 1993, Mangel 1993).

How can we make decisions when there is ignorance about the model structure? One approach is to analyze the consequences of various types of error. McElroy et al. (1997) discuss the trade-off between type I and type II errors (which they call the "producer's versus consumer's risk") in the analysis of an endangered species problem. Within the context of classical statistics, an objective might be to minimize the chance of a type I error while maximizing power. Another approach might be to attempt to minimize the probability of the worst possible scenario. This last approach is not very stable, because there is no limit to possible scenarios. A better procedure would be one in which the differing scenarios are weighted by their plausibility (Lindley 1985). Decision theory (e.g., Ralls & Starfield 1994) provides a framework in which one can be pragmatic, defined by Farber (1999) as:

> Being pragmatic does not mean the rejection of rules or principles in favor of ad hoc decision making or raw intuition. Rather, it means a rejection of the view that rules, in and of themselves, dictate outcomes. ... Hard policy decisions can't be programmed into a spreadsheet. ... But we also need an analytic framework to help structure the process of making environmental decisions. Intuition is often an unhelpful guide because environmental law concerns issues outside our normal, everyday experience. ... Rather than rigid rules or mechanical techniques, we need a framework that leaves us open to the unique attributes of each case, without losing track of our more general normative commitments. (pp. 10, 11).

Instead of making isolated decisions on the basis of a fixed body of knowledge, adaptive management (Walters & Hilborn 1978, Walters 1986, Williams 1996, Parma et al. 1998, Shea et al. 1998) regards each decision as part of an ongoing series; our knowledge of the system may change in response to these decisions. Adaptive management is commonly stated as a goal of management, but is widely misunderstood. Adaptive management does not consist of trial and error, i.e., adapting after something untoward happens; it requires a prior experimental design and generally long-term studies. Such long-term study may be unwarranted or difficult to justify because of economic discounting of uncertain returns in the far future (Ludwig & Hilborn 1983, Walters 1986, Halbert 1993, Taylor et al. 1997, Walters 1997). We can learn from careful comparative studies where management has been adaptive or tried to be adaptive (Gunderson et al. 1995).

Radical Uncertainty

The preceding approaches can be effective as long as the structure of the system is reasonably well understood. None of them is sufficient to make important decisions because, as we indicated above, a host of considerations (not all of which are scientific) must be examined. In some cases information about natural history may be essential, but not yet available. In such cases of overwhelming ignorance it is important not to be swayed by expert opinion that is mere guesswork, or by databases that are lacking in empirical foundations. There are many important cases in which prediction (even in the probabilistic sense) may not be possible. Structural uncertainties such as natural catastrophes or unforeseen consequences of human interventions are important, but inherently unpredictable (Simberloff 1988, Menges 1990, Schoener & Spiller 1992, Mangel & Tier 1994, Young 1994, Root 1998, Ludwig 1999).

Even when the system is more or less understood, it may be that objectives are fundamentally in conflict. Jackson (1997) discusses the situation in which one wishes to minimize PCBs in fish and maximize the chance of sustainable fisheries. These goals are in conflict because minimizing PCBs requires increased stocking rates of chinook salmon (the predator), which then may increase the probability that the prey species (alewives) will decline with an associated reduction in the chance of a sustainable fishery. As pointed out above, for the problem of climate change even the identities of the decision-makers are not known with certainty.

Science in Support of Policy Making

Although ecologists are often consulted in deciding a great variety of issues, they may feel that their results are distorted or their advice ignored. What causes this lack of trust, lack of proper weighting of scientific opinion and advice, or, stated broadly, the inability of those seeking the consultations to effectively utilize scientific input in the public policy process? The issue is complicated by misunderstanding or misstatements of the reliability of scientific inferences on both sides of the technical divide. As we have indicated above, there are many obstacles in the path to firm conclusions. Scientists are usually aware of these difficulties at some level, but they may also exaggerate the extent of their adherence to general principles of scientific inference. In their zeal to communicate clearly and forcefully—or perhaps in an attempt to achieve unique authority—they may neglect the more troublesome aspects of particular inferences. For instance, many scientists advocate the campaign for 20% of the world's oceans to be marine reserves by 2020. However, a 20% reserve will improve sustainability of the stock only if the biological characteristics of the stock and fishing pressure outside of the reserve are at appropriate levels (Mangel 2000). When pressed, proponents of 20% by 2020 acknowledge that their advocacy is based on political realities rather than a scientific analysis. Media and political figures have their own reasons to exaggerate the reliability of scientific inference, because it is often advantageous to pretend that important political and social issues can be reduced to technical ones and put in the hands of technicians. Examples are given below under the heading Equity, Values, and the Public Policy Process. The public response may vacillate between unquestioning acceptance of dicta supposedly backed by scientific fact, and wholesale rejection of the process as politically and ideologically motivated (Butler 2000, Dickson 2000).

Where scientific input to policy discussions is potentially significant, both the science and policy tend to be distorted (Bolin 1994, Lélé & Norgaard 1996). The entanglement of political and social objectives may not be apparent to the scientists involved. As a case in point, Sagoff (1985) cites the hypothesis that ecological diversity promotes stability. This is an important hypothesis that may be true in some circumstances, but it is by no means firmly established as a general principle (Grime 1997, Hooper & Vitousek 1997, Tilman et al. 1997, McCann 2000). Yet it has become entangled with calls for preservation of biodiversity that are rallying cries for those concerned about widespread destruction of habitats and other human assaults upon our life support system (Kaiser 2000); see the remarkable exchange in *The Bulletin of the Ecological Society of America* (Wardle et al. 2000, Tilman 2000, Naeem 2000). It is ironic that when stability and diversity relationships are discussed in introductory ecology, they are appropriately introduced as a scientific debate, yet when they enter the policy arena it becomes a fact that diversity promotes stability. The scientist acting as an advisor "must recognize, not necessarily accept, political or other constraints under which a decision-maker is acting" (Bolin 1994, p. 25) and be careful not to distort the science.

The Process of Policy-Making

Ecologists provide input to a complex, evolving, political process out of which policy emerges. The policy process involves the preparation, expression, and implementation of a government opinion or course of action. Policies can be expressed through: 1. goals articulated by political leaders; 2. points of view expressed by representatives of government agencies; 3. formal statutes, rules, or regulations; and 4. practices of administrative agencies and courts charged with implementing or overseeing programs. A policy process can occur anywhere in the government, even the judicial bench. Broad descriptions of this process start with an agenda-setting process, proceed through intermediate steps of policy formation, legitimation, and implementation, and then conclude with review and revision of existing policies (Kraft 1996, Hempel 1996). Ecologists may participate at all stages in the policy process. Here we focus on scientific input to regulatory decision-making.

From the policy-process perspective, one reason for disregard of scientific advice may be political influence. Regulatory decisions cannot escape such influence because of the necessity of accommodating various stakeholders. The behavior of regulatory agencies may sometimes seem to conflict with their mandate to protect the public interest. This can be understood as follows. Most public agencies depend upon steady support from the legislatures that established them. This support is best achieved if a group of constituents (typically those who are most affected by the actions of the agency) influence the legislature to support the agency. Influence may be exerted by mustering votes, or campaign contributions, or both. Such legislative support is easiest to mobilize if a small group derives substantial benefits from a sympathetic stance by the agency. These principles apply, in the United States, to such agencies as the Army Corps of Engineers, the Bureau of Land Management, the National Forest Service, and the National Park Service (Maas 1951, Clarke & McCool 1985, Culhane 1981). The Army Corps of Engineers and Bureau of Land Management are examples of successful bureaucracies because their staff and appropriations rise steadily. These agencies have well-organized supporters who benefit substantially from their activities. In contrast, the National

Park Service has had a continual struggle to obtain support. Many thousands of people enjoy services from national parks, but they are not readily organized into a lobby to protect the interests of the National Park Service. In cases where industries are regulated, the industry has a strong interest in supporting agencies that protect the interest of the industry, but much less interest in supporting an agency that looks after the public interest. This frequently results in "capture" of regulatory agencies by the industries they are supposed to regulate (McConnell 1966, Trebilcock 1978, Rourke 1984, Schrecker, 1990).

Another reason for discounting the input of ecologists to public policy concerns the form of the data and recommendations that ecologists provide. Scientists often have difficulty in communicating the implications of technical analyses of complex issues. The policy process (especially in industrialized nations) has come to expect relatively simple, precise (often interpreted as "correct") answers from scientists. If ecologists (or any scientists) fail to provide firm predictions, they may be less influential in policy discussions (Glantz 1979, Saetersdahl 1979, Saville 1979, Walters & Maguire 1996, Finlayson and McCay 1997). On the other hand, if scientists overstate the reliability of their results, they may eventually cause a lack of public trust in scientific advice (Haerlin & Parr 1999). Perhaps the most outstanding example of overstatement of reliability has been the selling of nuclear power (Shrader-Frechette 1993). Scientists may overestimate the impact of phenomena because of their concern about the organisms involved. Garshelis (1997) concludes that the previous estimates of sea otter mortality in the *Exxon Valdez* oil spill had been too high because scientists used inadequate data and then notes (pp. 913–914):

> Scientifically unsound death tolls may have even more profound implications in terms of their effect on public opinion . . . the lower estimate may seem low only by comparison . . . Once numbers are made public, however, they become a benchmark by which to judge all subsequent estimates, including those from future environmental catastrophes. Conservation hinges on public opinion, and public opinion is shaped by the information provided by scientists. If conservation biologists are to be regarded as rigorous and objective scientists . . . they must adhere to high scientific standards . . . and must admit to the limits and uncertainties in their work . . . Biologists involved in assessing the impacts of such events must keep in mind that their work will be intensely scrutinized and that their errors will affect the credibility and hence the effectiveness of conservation biologists in general.

These issues will become especially important if the Endangered Species Act is successful and species are considered for delisting (Pagel et al. 1996), because "There are no easy genetic or demographic answers to the questions of 'when is recovery complete', or 'how should fiscal triage be applied to provide the best in situ conservation of a listed species?'" (p. 434). Great care must be taken to avoid averaging different scientific positions as a mechanism for reaching consensus. Similarly, when predictions do not come true, as in the mineral resource shortages predicted in the 1950s–1980s (Hodges 1995), we must ask why this is the case.

Equity, Values, and The Public Policy Process

Because regulatory decisions typically involve conflict between opposing interests and values, an immediate issue is the relative power and resources available to various groups. For example, public interest groups rarely attend routine advisory committee meetings, but industry is generally well represented (Jasanoff 1990, p. 247). Hence significant policy decisions, particularly decisions not to act, may be reached after deliberations inordinately influenced by only one set of interests. Jasanoff (1990) attributes this difference in attendance to the differing resources available to the two groups. But if decisions are perceived as arbitrary or excessively favoring one group over another, they tend to be discredited, and they invite litigation.

Most governmental regulations impose costs on one segment of the population, and provide benefits to a different segment. Hence the evaluation of regulations involves weighing of disparate sets of values and interests. This is manifestly not a scientific exercise. Governmental officials may attempt to convert a difficult political issue into a technical problem by involving expert scientific advice. Recent advice to the British government on bovine spongiform encephalopathy (BSE) is an example (Loder 1999; Haerlin & Parr 1999). The administrative dilemma was whether to expose the whole population to a risk of infection or to destroy an industry. The administrative solution was to convert this impossible choice into a technical evaluation of the risk of infection from eating British beef (Ridley 1999). No amount of scientific evidence can prove that there is no risk, because that is beyond the capability of scientific inference. Nevertheless, the government contrived to produce such a verdict from an eminent scientific panel. The French government, acting for a different set of constituents, decided to ban British beef. The final resolution, reached in the fall of 1999, was arbitrated by a scientific panel of the European Union, but the scientific conclusions had little effect on the political process.

Hutchings et al. (1997) write, "We contend that political and bureaucratic interference in government fisheries science compromises the DFO's [Department of Fisheries and Oceans of the Government of Canada] efforts to sustain fish stocks and, thereby, the socioeconomic well-being of fishing people and fishing communities." They go on to document how the DFO suppressed and ignored evidence that declines in northern cod (*Gadus morhua*) stocks might be caused by overfishing. They also report instructions to DFO scientific staff that they must support the Minister's position on adequate stream flows for conservation of salmon stocks (*Oncorhynchus* spp.) while adhering to the scientific advice. Those who were unable to do so were advised to find other employment. See also Charles (1995), Kurlansky (1997), Taylor (1995), and Myers et al. (1997).

The role of values in formulating regulations is not always appreciated. Whittemore (1983) has shown how values enter into risk assessment for environmental toxicants. Brunk et al. (1991) describe a situation in which scientific panels employed by industry and the unions arrived at wildly disparate estimates of acceptable exposures for Alachlor, based upon the same scientific evidence. An analogous incident involved the California sardine, in which action was not taken to reduce harvests because the scientific evidence was inconclusive, is described by Radovich (1981); also see McEvoy (1986). Costanza & Folke (1997) note that managers of the Patuxent River drainage basin in Maryland are attempting to embed three sets of values—surrounding efficiency, fairness, and sustainability—in ecological-economic modeling and analysis. Haddad (2000) argues that policy analysis should be organized around the categories of feasibility and appropriateness, the latter of which is concerned with values.

Drawing Boundaries Between Science and Policy

Experience with regulatory science shows that boundaries between science and policy are difficult to draw and that they may require a long period of trial and negotiation (Salter 1988, Brunk et al. 1991). Jasanoff (1990) provides a history of several decisions made by the Environmental Protection Agency (EPA) and the Food and Drug Administration (FDA) in the United States. According to Jasonoff, the Scientific Advisory Board (SAB) of the EPA was created in 1974 in response to a requirement by the U.S. Congress that the EPA consult with external scientific advisors. The EPA's procedures had been under attack for its use of unpublished and unreviewed data. Its methods had frequently been challenged in the courts, and a National Academy of Sciences (NAS) study suggested that the adversarial process might better be replaced by one involving technical rationality. The composition of the SAB immediately became a political issue. Funding for the scientific activities of the EPA was cut, and a hit list of EPA's scientific advisors, compiled by the Reagan administration, was uncovered in 1983. Political involvement seems inevitable, because questions to be decided in regulatory science generally involve subjective judgment. Eventually an accommodation was reached involving an agreement between the EPA and the SAB on the boundaries between science and policy. This process included early decisions that were challenged because of lack of competent scientific advice, later decisions in which the composition of the scientific panel was stacked to favor one set of values over another, and still later ones in which there was substantial cooperation between the agency and scientific advisory committees. The latter process involved negotiations and compromise concerning the jurisdictions of the scientific advisory committee and the agency staff. It was impossible to make a complete separation between the science and the policy, for the reasons given above. Nevertheless, in cases where a compromise could be reached, it led to a much smoother process and greater acceptance of the final results by all parties.

How Can We Provide Sound and Influential Scientific Advice?

If we accept that decisions about environmental problems inevitably combine scientific and political aspects, how can scientific input be both sound and influential?

One means may be to broaden the composition of the panel providing advice. Natural scientists are accustomed to thinking of themselves as objective, but they are influenced by cultural and social differences between one group and another. For example, scientists involved in fishery management are sometimes charged with a tendency to discount traditional knowledge and anecdotal evidence in favor of the more structured forms of information available from industrial sources (Finlayson & McCay 1997, Berkes & Folke 1998). Funtowicz et al. (1999) point out that systems of management of environmental problems that do not involve science, and which cannot immediately be explained on scientific principles, are commonly dismissed as the products of blind tradition or chance. When people with no formal qualifications attempt to participate in the processes of innovation, evaluation, or decision, their efforts may be viewed with scorn or suspicion. Such attitudes do not arise from malevolence; they are inevitable products of scientific training, which presupposes and then indoctrinates, the assumption that all problems are amenable to scientific methods.

These observations suggest that scientific panels would benefit from more diverse composition and input. In contrast to natural scientists, sociologists and anthropologists who study resource systems tend to emphasize the value of multiple forms, methods, and means of acquiring and validating knowledge. This difference in outlook between natural and social scientists may help to balance what otherwise might be a very lopsided set of deliberations (Finlayson & McCay 1997; Berkes & Folke 1998). There are risks as well as benefits associated with broadening advisory panels, because results of natural science may not be understood or trusted by all the participants. In order to mitigate such effects, it may be well to change the format in which advice is presented. Ozawa (1991) describes how consensus-based methods may help to lessen the gap between expert and the laity, and improve the quality of expert advice. Consensus-based methods are aimed at obtaining approval from all participants. A preliminary series of meetings is held to discuss proposed regulation. A special master convenes discussions on scientific bases for the decision. Scientists participate in intensive question-and-answer sessions, and present alternative interpretations or alternative analysis. The objective of consensus-based methods is to bring all individuals up to a common plane of technical competency. When experts are aware that they must explain the logic of their arguments rather than simply ride on their reputations to win concurrence, they make more serious efforts to educate the stakeholders. The division between experts and nonexperts narrows, and science benefits by having a more knowledgeable public. A similar process is envisaged by Clark et al. (2000). The result of a consensus-based approach may be very different from a method in which uncertainties are neglected by concocting a single best estimate of the state of the system. A common understanding of the scientific issues and uncertainties enables a variety of stakeholders to support their interests effectively. The politics of environmental mediation (a quite different process) is dealt with in Amy (1983, 1987) and Weber (1998).

THE NEXT GENERATION OF ECOLOGISTS

It is clear from the preceding discussion that ecologists face unprecedented challenges and opportunities when dealing with environmental issues. How can we prepare ourselves to deal with these issues? As we pointed out in the Introduction, traditional disciplines and training are inadequate for wicked problems involving the interaction of humans with their environment. This has crucial implications for the future training of ecologists. The important problems that we face can be dealt with only by combining insights from a variety of sources. There is much to be learned from history, religion, philosophy, geography, economics, political science, and other disciplines. The training for those interested in solving environmental problems must be broader and deeper than the training of a disciplinary scholar. To be successful, interdisciplinary training must be more—not less—rigorous than disciplinary training. That is, interdisciplinary training requires mastering the core skills, not just the key words, in disciplines of the natural and social sciences.

Disciplines are Essential, but Disciplinary Boundaries are an Impediment

Great progress has been made by defining the natural scientific disciplines and by setting up disciplinary boundaries. Science progressed as it shed metaphysical and religious dogma from its scope. Following Francis Bacon, scientists have attempted to exclude internal bias from our interpretation of empirical observation (Roszak 1972, Gould 2000). Biology progressed by rejecting vitalism (Mayr 1982). Ecology progressed by excluding human interventions from its scope: we generally consider that an experiment has been hopelessly compromised if there are unplanned applications of pesticides or other disruptions of the study site.

This progress has imposed costs. Attached to the dogma and biases was context: an implicit recognition that what scientists of one era considered rational in terms of methods and proof might be different from the beliefs of earlier or later scientists (Toulmin 1990). The Austrian economist von Hayek (1945) extolled the importance of "the knowledge of particular circumstances of time and place" in operating economic systems. Declining respect for context in science reduces the value scientists place on knowledge of long-standing, effective institutional regimes for managing ecosystems. Many perceive a conflict between science and religion, but Benjamin & Mangel (1999) note a connection between parts of the Hebrew Bible and modern statistical thinking; they note that much of the challenge raised by Darwin was to social structure rather than to deep religious thinking.

Our relationship with nature has suffered from the Newtonian concept of biological organisms (including humans) as machines (Callicott 1994, Callicott 1999, Hodgson 1999) rather than as complex adaptive systems (Waldrop 1992, Lewin 2000) and has led to oxymorons such as "ecosystem management" or "maximum sustainable yield" (Grumbine 1994, Wood 1994, Ludwig 1994, Stanley 1995). The artificial distinction between natural and social sciences has produced many anomalies. Ecologists produced a manifestly political "Sustainable Biosphere Initiative" (Lubchenko et al. 1991) that ignores human population dynamics and patterns of resource use (Grumbine 1994, Wood 1994, Ludwig 1994, Stanley 1995). It is futile to attempt to conserve species through protected areas if conditions leading to human encroachments are ignored. Forests cannot be conserved without consideration of their inhabitants. Problems resulting from climate change cannot be addressed without appealing to conceptions of social justice. Wicked problems cannot be addressed within the old disciplinary boundaries.

Insights from Geography, History, and Anthropology

Although we are still struggling to understand interactions between humans and their environment, there is already a rich literature. Geographers have long been concerned with human impacts and their consequences. Marsh's work of 1864 is a classic (Marsh 1965), and it has inspired more recent updates (Thomas 1956, Turner et al. 1990, Dasmann 1984). It is clear that many regional conservation problems require a geographic perspective, as Brawn and Robinson (1996 p. 10) note:

> In fragmented landscape, a regional approach to conserving migrant species appears warranted. Conservation strategies designed to reduce nest predation and parasitism in Illinois' forest birds may be ineffective without plans to maintain potential source populations within extensive forests in surrounding states . . . Lack of data on dispersal is the major gap in understanding the population dynamics of neotropical migrants and prescribing effective conservation measures.

Historians provide additional scope to the vision of man's relation to nature (Crosby 1986, Worster 1993, Cronon 1996, Diamond 1997, McNeill 2000). Worster (1993, p. 24) advises that scientists acknowledge the connection between the nature they study and thousands or millions of years of human history, its ideas and social forces:

> Environmental historians would argue that scientists need them to answer a very big question that the latter have themselves raised but are unequipped to answer: Why are we in a state of crisis with the global environment? . . . natural science cannot by itself fathom the sources of the crisis it has identified, for the sources lie not in the nature that scientists study but in the *human nature* and, especially, in the *human culture* that historians and other humanists have made their study. We are facing a global crisis today not because of how ecosystems function, but rather because of how our ethical systems function.

Redman (1999) points out that many environmental problems that we perceive today are similar to those faced by ancient societies. He suggests that the collapse of many such societies can be attributed to destruction of natural resources, perhaps prompted by institutions that promote ever higher production in order to satisfy the demands of an elite. He makes a persuasive case for the contribution of archaeology to the discussion of environmental problems. Bodley (1990) claims that environmental problems must be viewed in long-term perspective as a struggle between two incompatible cultural systems—tribes and states. The most critical features of tribal groups are their political independence, reliance on local natural resources, and relative internal social equality. In comparison with states, and especially industrial states, tribal systems tend to expand more slowly and have been environmentally less destructive. These differences explain why territories still controlled by tribal groups are so attractive to developing nations because tribal territories contain "underutilized" resources. Bodley goes on to explain how the ensuing pattern of exploitation is justified by an ethnocentric attitude that nullifies the interests and rights of indigenous people (WCFSD 1999, Clad 1988, Goodland 1988, Wells et al. 1999, Wells & Brandon 1992, Brandon 1997, van Schaik et al. 1997, Sanderson & Redford 1997). There is an interesting debate in an issue of *Conservation Biology* about the relationship between indigenous peoples and conservation. Redford & Stearman (1993a, p. 252) point out the great diversity of indigenous groups, and the severe pressures that have left many of them destitute and estranged from traditional ways. They state "To expect indigenous people to retain traditional low-impact patterns of resource use is to deny them the right to grow and change in ways compatible with the rest of humanity." The discussion continues with Alcorn (1993), and Redford & Stearman (1993b). Also see Ruttan & Borgerhoff-Mulder (1999) and more recent issues of *Conservation Biology*. Ponting (1992) provides many examples of tribal groups causing severe environmental problems.

If we adopt a more sympathetic attitude towards traditional peoples and their ways, perhaps we can learn something of value for our present problems. Ostrom (1990) provides a well-articulated account of things to be learned. Ostrom is concerned with "Common Pool Resources" such as fisheries, in which there is joint management of the stock but the proceeds of the harvest accrue to individuals. She lists properties that tend to promote successful management, such as clearly defined rights and norms of behavior, collective choice in devising and application of such rules, monitoring of compliance by the participants, graduated sanctions for offenses, a mechanism for conflict resolution, and protection of local rights to organize from challenges by external authorities. She (Ostrom 1992) gives a series of injunctions for those contemplating modification of local institutions. In essence, these involve knowledge of and respect for existing rules and institutions. Perhaps most important, she advises that one should propose new rules only if there are no existing rules, or if existing rules are ineffective, and if one is thoroughly familiar with the configuration of existing institutions and their function. Also see Berkes (1999), Berkes & Folke (1998), Gadgil et al. (1993), Gadgil & Malhotra (1994), Hanna et al. (1996), and Ostrom et al. (1999).

Insights from Economics

The relationship between economics and ecology has been a vexing one. Both fields of study share the Greek root oikos, which means "home," and both address questions of scarcity and competition. By the early twentieth century, mainstream economics had adopted a mechanistic perspective closely linked to Newtonian mechanics (Hodgson 1999) and elaborated an economic interpretation of utilitarian ethics. Both of these century-long themes have had profound implications for relations between economics and ecology. Arguments in favor of economically-efficient resource use have played roles in the passage of major U.S. environmental legislation (Cropper & Oates 1992). Gillis (1991) uses economics to critique policies with direct and indirect environmental impacts. He cites artificially low grazing fees that led to overgrazing and desertification, near zero prices charged for irrigation water, and maldistribution of income as major causes of ecological problems. The development of nonmarket valuation techniques, including contingent valuation, has enabled policy-makers to consider the value of forests and wetlands on other terms than simply sources of raw-material inputs to the economy (NOAA 1993, Hanemann 1994). Within the mainstream, economic inquiry into environmental questions is known as environmental economics.

The century-old commitment of economics to mechanics and utilitarianism has imposed costs. Economists' regular use of linear dynamics, quadratic costs, and Gaussian random variables stands in contrast to ecological situations that are often unique, nonlinear, and nonGaussian. Common economic modeling assumptions such as complete reversibility of transactions and complete substitutability of goods and services are incompatible with ecologists' conceptions of evolution and uniqueness of ecosystems. The doctrinal domination of utilitarianism mutes the discussion of environmental policies that could emerge from alternative ethical bases, such as sustainability or equity (Haddad & Howarth 2001). Ecologists place much greater emphasis on experiment (as opposed to theory) than do economists. McCloskey (1994), writing for fellow economists, points out that part of professional training is inclusion of certain aspects of experience, and exclusion of others, which results in "learned incapacity." McCloskey claims that neoclassical economists' disregard for history, philosophy, geography, psychology, anthropology, sociology, law, and political science leads others to write economics off. Heilbroner (1999) also challenges the narrowness of focus of neoclassical economics. Similar criticisms can be leveled at ecology and ecologists when we attempt to advise on important policy issues but ignore other viewpoints that may be pertinent (e.g., Bodmer et al. 1994).

Growing numbers of economists are challenging mechanistic metaphors and strict adherence to utilitarianism. Evolutionary economists are attempting to restore biological metaphors and insights when explaining economic systems (Nelson & Winter 1992, Nelson 1995, Hodgson 1999, England 1994). Ecological economists are attempting to employ both evolutionary and coevolutionary perspectives (Norgaard 1984, Norgaard 1994, Gowdy 1994, Hodgson 1999) as well as to explore practical and theoretical implications of the sustainability ethic [Constanza

1991, Daly (1991, 1996), Howarth 1995, Howarth & Norgaard 1990]. A seminal work that places the human economy squarely within the confines of physical and ecological systems is Georgescu-Roegen (1971).

Amartya Sen is an exceptional economist who has been concerned with distributional issues as well as the usual criterion of economic efficiency. Sen (1999, p. 6) notes that ... no famine has ever taken place in the history of the world in a functioning democracy, be it economically rich (as in contemporary Western Europe or North America) or relatively poor (as in post-independence India, or Botswana, or Zimbabwe). Famines have tended to occur in colonial territories governed by rulers from elsewhere (as in British India or in an Ireland administered by alienated English rulers), or in one party states (as in the Ukraine in the 1930s, or China during 1958–1961, or Cambodia in the 1970s), or in military dictatorships (as in Ethiopia, or Somalia, or some of the Sahel countries in the recent past).

This is elaborated later in the work Sen (1999, pp. 160–188). Some philosophical aspects of ecological economics are discussed in Goulder & Kennedy (1997). Thoughtful introductions are given by Common (1995) and Cooper (1981). Funtowicz et al. (1999) propose multicriteria evaluation as a tool to deal with incommensurability and complexity, and they describe a variety of promising approaches. The case for an institutional economic approach (which would involve insights from a variety of social sciences) is made by Jacobs (1994). Haddad (1997, 2000) combines perspectives from institutional and ecological economics in evaluating the role of markets in environmental protection. Farber (1999) is guided by legal precedents and the history of legislative enactments of environmental protection rather than abstract arguments in favor of protection. He includes a critical discussion of cost-benefit calculations and discounting of future costs and benefits.

Insights from Political Science

Greenbaum et al. (1995) provide an excellent selection of papers dealing with environmental law and ethics. Issues of the environment and social justice are treated in Faber (1998), Sachs (1993), Westra & Wenz (1995), Low & Gleeson (1998), Mason (1999), and Coward et al. (2000). Some political scientists have been concerned with the challenges that environmental issues pose to our systems of governance. Ophuls (1977), Ophuls & Boyan (1992), and Heilbroner (1974) are pessimistic about the prospects for democratic resolution of ecological conflicts, but more recent thought has concluded that democratic institutions offer the best prospects (Paehlke 1996, Paehlke & Torgerson 1990a, John 1994, Lafferty & Meadowcroft 1996, Dryzek 1999, Faber 1998). Cortner & Moote (1999) provide a comprehensive survey of political aspects of resource management, with emphasis on newer approaches that emphasize citizen participation.

On the international level, Homer-Dixon (1991, 1994) and Homer-Dixon et al. (1993) have alerted us to possible violent conflict over resources. Some of the troubles in the Middle East are driven by a combination of conflict over oil and conflict over water (Hubbard 1991, Odell 1986, Yergin 1991, Clark 1993, Gleick 1993, Ohlsson 1995, Postel 1992, Postel 1996).

Paehlke (1989) sees environmentalism as "a political movement that seeks to impose upon the physical sciences and engineering restraints based upon the findings and judgments of the social and life sciences." Therefore, "science can never again be an activity solely devoted to removing humanity from nature, lifting us out of natural limits—for centuries, if not millennia, its implicit goal." Hence there may be a gulf between environmentalism and liberalism, and indeed between environmentalism and all ordinary, moderate politics (Berlin 1990, Ignatieff 1998). This poses a challenge for all who are concerned about the future of the earth.

Insights from Religion, Philosophy, Psychology, and Ethics

Surveys of the relationship between religion and ecology are provided in Rockefeller & Elder (1992), Callicott (1994), Anderson (1996), Berkes (1999) and Waskow (2000). From the Confucian period in China, we have sayings such as "Mountains empty, rivers gorged." Referring to cattle and goats, Meng Tse said, "To these things is owing the bare and stripped appearance of the mountain, and when people now see it, they think it was never finely wooded. But is this the nature of the mountain?" (Glacken 1956, p. 70). According to Plato, "Long ago there were abundant forests in the mountains, which provided fodder for the animals and storage for water, which could then issue forth in springs and rivers. The water was not lost, as it is today, by running off a bare ground into the sea" (Glacken 1956, p. 70). The book of *Exodus* of the Hebrew Bible contains an injunction to leave land fallow one year in seven (Dunham & Coward 2000, Waskow 1995). Religious restraints and taboos have served to protect many of the world's forests until recently. The most profound change wrought by Western influence in India was the onslaught on the Hindu concept of the forest as sacred, which had in effect made the forest an ecological reserve (Ashton 1988, Gadgil & Guha 1992, Gadgil et al. 1997). Shalomi (1993) has called for the development of a "Gaia theology."

The classic statement of the ethical consequences of an ecological and evolutionary perspective is Leopold (1966 [1949]). Further literature and interpretation is in Callicott (1999). The contributions in Coward et al. (2000) center around the concept of "ecosystem justice." These ideas have an ancient history. Callicott (1999) gives an interesting exposition of the various versions of creation in the book of *Genesis*. The passage about God creating man in God's own image, and giving him dominion over the earth appears first in the standard version of the text. White (1967) claims that this passage provided the ethical sanction for unrestricted exploitation of the earth and all that live in it. This interpretation, rooted in nineteenth-century Protestant thinking, did great damage to the relationship between deep religious thinking and the environment. But it is problematical, in view of the complexities of interpretation of the Hebrew Bible, written without vowels or punctuation (Herczeg 1995). The second account of the creation in *Genesis* is actually older than the first one. It refers to Adam being "formed of the dust of the ground." This image is reinforced in the name "Adam," which is derived from the Hebrew word for earth; "Eve" is connected to the Hebrew root for "life." In an interesting parallel, Weaver (1996, p. 12) quotes Salvador Palomino as follows: "The Earth, our Mother Earth, has always been part of our collectivity. We belong to her, she does not belong to us. Land and community are the souls of our peoples." A similar sentiment is expressed by Leakey (1996 p. 253), based upon the common origins of all life forms on Earth.

Roszak (1992 pp. 68, 78) explores connections between ecology and psychology, noting that "the species that destroys its own habitat in pursuit of false values, in willful ignorance of what it does, is 'mad' if the word means anything." To the extent the natural realm possesses a sacramental quality, traditional psychotherapy should insist that people remain vitally connected to nature.

Science and Values

As indicated above, scientific inferences and theories cannot be separated from the values of those who conduct research. This must be clearly understood if ecologists are to be effective in formulating and executing policies (Lélé & Norgaard 1996, Matthews 1995, Fischer 1990, Meffe & Viederman 1995, Orians et al. 1986, Soulé 1986). The relationship between science and values has been generally glossed over in training (Norton 1998) but is now generally recognized. The implications of the values of scientists can be understood through examination of case histories. See Maguire (1994) for an account of the "Wildlands Project." The connection between the mission of the management agency and scientific conclusions is particularly clear in Radovich (1981), in which scientists employed by different agencies came to differing conclusions from the same data. Also see Brunk et al. (1991) for the issue of setting standards for exposure of workers to hazardous materials. A poignant case in which scientific advice was tempered by political concerns with disastrous results for a fishery is described by Saville (1979); also see Hutchings et al. 1997.

Once we admit that environmental problems may reflect our own culture and attitudes as much as a scientific or technical problem, we have greater scope for possible responses. There is a rich literature on this topic. Caldwell (1990) identifies some dominant modern ideologies: Economism means placing an exceptional and inordinate emphasis upon utilitarian economic values in contradistinction to all others. This approach gives little or no value to nature apart from its immediate utility for economic purposes. Scientism is the belief that science, in its several meanings, is inherently capable of solving almost all human problems. It is an extrapolation from the unquestioned achievements of science, and reflects an oversimplification of the ways in which science relates to the social and political issues of human society. Technocratism is an effort to achieve policy solutions by recourse to technological innovation or through what is sometimes called a technological fix (Ehrenfeld 1978, 1993). Cotgrove (1982) points out that technocracy seeks to depoliticize value clashes by reducing them to an economic or technical calculus, as the British government did for the BSE crisis, and as all governments have done for nuclear power. The technocratic mode dismisses values that are not easily quantified or monetarized; see also Callicott (1994). The philosophical underpinnings of the contemporary approach to resource management are explained by Cortner & Moote (1999). They point out that many of our attitudes date from the period of the Enlightenment: ideals of rationality, social equality, and progress. Along with these goes an attitude of domination and mastery over nature, which is still the dominant approach to natural resources.

Perhaps the most approachable account of postmodern ideas is Norgaard (1994), who points out that although a few have attained material abundance, resource depletion and environmental degradation now endanger many and threaten the hopes of all. Modern values, knowledge, organization, and technological systems reflect the availability of fossil hydrocarbons rather than the features needed to interact, and continue to coevolve effectively, with ecosystems. Norgaard (1994) describes some consequences of an evolutionary approach. Because prediction is very difficult or impossible, experimentation should be done cautiously. Experiments that can be undone quickly are preferred. Diversity is necessary for evolution; hence it is inherently good. Most changes are likely to be selected out, but small, compatible changes can eventually change the coevolutionary course.

The idea that our attitudes toward natural systems are a social construction is part of a wide-ranging inquiry in the social sciences that has caused much conflict among natural scientists (Gould 2000, Hacking 1999). We have given some references in the context of fisheries management. Charles (1994) and Matthews (1985) have inquired into attitudes behind the recent collapse of the Northern Cod. There are three competing metaphors: biological/ecological, economic, and socio-cultural. The socio-cultural fishery has four aspects: 1. it preserves a way of life, 2. it is the employer of last resort, 3. it provides economic benefits that enable more people to survive in a mixed economy, and 4. it supports a community. During the 1970s, Canadian fishery officials moved from a biologically-driven understanding of the fishery to one that considered it governed by an open-access property regime, a perspective heavily influenced by the economic model. The social view of the fishery was championed by the Province of Newfoundland, which for many years has been faced with economic decline and emigration of its young citizens. In the end, there was no way of satisfying all of these competing interests. In fact, none was satisfied.

On the other hand, because ecologists may differ in the ways that they describe mortality, reproduction, or competition does not mean these are social constructions. They are real and we need words to describe the reality. Isaiah Berlin (1990 p. 42), writing about Comte, notes that.

> He understood the role of natural science and the true reasons for its prestige better than most contemporary thinkers. He saw no depth in mere darkness; he demanded evidence; he exposed shams; he denounced intellectual impressionism ... he provided weapons in the war against the enemies of reason, many of which are far from obsolete today. Above all he grasped the central issue of all philosophy—the distinction between words (or thoughts) that are about

words, and words (or thoughts) that are about things, and thereby helped to lay the foundation of what is best and most illuminating in modern empiricism.

In the context of our main discussion, words about words (e.g., Keller & Lloyd 1992) may contribute little to the solution of wicked problems. Indeed, not all aspects of the social sciences (any more than all aspects of the natural sciences) will be helpful in finding solutions, and we need to think carefully about which aspects of the social sciences can contribute (or at least not make things worse).

CONCLUDING REMARKS

We began this review by pointing out changes in concerns and attitude of ecologists in response to environmental problems. It is apparent at the end of this review that the challenges are broad and deep. They apply not only to ecology, but to a host of other disciplines and ways of thought, some of which might first appear to be esoteric and unrelated to major human concerns. Environmental problems challenge our concepts of experts and of rational optimizing approaches, because there frequently is no consensus on a final goal or set of objectives. As with global climate change, there may be no clearly defined set of decision-makers. Environmental problems challenge policy-makers to deal effectively with scientific data that are accurate but not precise.

Because of the great demand for reliable knowledge and prediction in the face of a host of unknowns and unknowable aspects of conservation, great demands are made on our methods of data analysis and statistical inference. Philosophical differences such as those that divide frequentists and Bayesians become pressing issues, because important decisions may hinge on the choice of adequate and reliable statistical procedures and communication of their implications. Blunders in statistical inference may have only minor consequences for research programs in well-established fields, but the potential for damage is enormous when such blunders influence environmental decisions.

Experience has shown that when science is used to support policy-making, both the science and the policy are altered in fundamental ways. The idea that science can be completely objective and value-free cannot be supported in such a context. The closer the issue is to fundamental human goals and aspirations, the more difficult it is to separate scientific conclusions from other influence. Scientists cannot expect to be granted a privileged position in environmental deliberations; they will have to devise ways of communicating their insights to a variety of people, some of whom may have quite different values and ways of knowing, and opposing interests.

Our work also has implications for policy. The coevolutionary relationship between the practices of science and policy-making suggests that the policy process also needs change. We encourage policy-makers, some of whom are themselves ecologists, to find new ways to integrate ecological knowledge into the policy process. This includes not only integrating different forms of data and conclusions based on data into decision processes, but also integrating multiple and complex expressions of values. Wicked environmental problems require not only innovative policy responses but also innovative methods of arriving at responses.

The disciplinary structure that has served science so well over past generations is inadequate to address contemporary problems. This realization is quite general over many academic disciplines. One of the most striking things we have noticed in examining material from the humanities and social sciences is the widespread feeling that, although the respective discipline has much to offer to understand and cope with environmental challenges, it will require a fundamental realignment to achieve those goals. This applies to history, geography, anthropology, economics, political science, and ethics. The science wars that emerged from social deconstruction of natural science (Hacking 1999) must be resolved in ways that facilitate exchange of insights across the disciplinary divide. The issues are no longer esoteric because we cannot progress in other ways until we progress in this one.

This is an exciting time in which to work because the challenges are new and fundamental and the opportunities are correspondingly great. We may imagine that a great bell has tolled. It makes not a single sound but a great wash of sounds, both high and low, some of them discordant and harsh. It resonates in every object within its reach. It tolls for all of us.

ACKNOWLEDGMENTS

We thank Peter Kareiva, Richard Norgaard, and Samuel Scheiner for helpful comments on the manuscript. This work was partly supported by NSERC of Canada under grant #A-9209.

Visit the Annual Reviews home page at www.AnnualReviews.org

LITERATURE CITED

Alcorn J. 1993. Indigenous peoples and conservation. *Conserv. Biol.* 7:424–26

Alford RA, Richards SJ. 1999. Global amphibian declines: a problem in applied ecology. *Annu. Rev. Ecol. Syst.* 30:133–64

Amy DJ. 1983. The politics of environmental mediation. *Ecol. Law Q.* 11:1–19

Amy DJ. 1987. *The Politics of Environmental Mediation.* New York: Columbia Univ. Press. 255 pp.

Anderson EN. 1996. *Ecologies of the Heart: Emotion, Belief, and the Environment.* Oxford: Oxford Univ. Press. 256 pp.

Anderson JL. 1998. Embracing uncertainty: the interface of Bayesian statistics and cognitive psychology. *Conserv. Ecol.* 2(1). Available on the internet at http://www.consecol.org/vol2/iss1/art1

Apostolakis G. 1990. The concept of probability in safety assessments of technological systems. *Science* 250:1359–64

Ashton PS. 1988. A question of sustainable use. In *People of the Tropical Rainforest,* ed. J Denslow, C Padoch, pp. 185–96. Berkeley, CA: Univ. California Press

Beck MB. 1987. Water quality modeling: a review of the analysis of uncertainty. *Water Resourc. Res.* 23:1393–42

Bender DJ, Conteras TA, Fahrig L. 1998. Habitat loss and population decline: a meta-analysis of the patch size effect. *Ecology* 79:517–33

Benjamin T, Mangel M. 1999. The ten plagues and statistical sciences as a way of knowing. *Judaism* 48:17–34

Berkes F. 1999. *Sacred Ecology. Traditional Ecological Knowledge and Management Systems.* London: Taylor & Frances. 209 pp.

Berkes F, Folke C. 1998. *Linking Social and Ecological Systems.* Cambridge: Cambridge Univ. Press. 459 pp.

Berlin I. 1990. *Four Essays on Liberty.* New York: Oxford Univ. Press. 213 pp.

Berryman AA. 1997. On the principles of population dynamics and theoretical models. *Am. Entomol.* 43(3):147–51

Bodley J. 1990. *Victims of Progress.* Mountain View, CA: Mayfield. 261 pp. 3rd ed.

Bodmer RE, Fang TG, Moya L, Gill R. 1994. Managing wildlife to conserve Amazonian forests: Population biology and economic considerations of game hunting. *Biol. Conserv.* 67:29–35

Bolin B. 1994. Science and policy making. *Ambio* 23:25–29

Brandon K. 1997. Policy and practical considerations in land use strategies for biodiversity conservation. See Kramer et al. 1997, pp. 90–114

Brawn JD, Robinson SK. 1996. Source-sink population dynamics may complicate the interpretation of long-term census data. *Ecology* 77:3–12

Bretherton F. 1994. Perspectives on policy. *Ambio* 23:96–97

Brook BW, O'Grady JJ, Chapman AP, Burgman MA, Akçakaya HR, Frankham R. 2000. Predictive accuracy of population viability analysis in conservation biology. *Nature* 404:385–87

Brown KS. 2000. A new breed of scientist-advocate emerges. *Science* 287:1192–95

Brunk C, Haworth L, Lee B. 1991. Is a scientific assessment of risk possible? Value assumptions in the Canadian Alachlor controversy. *Dialogue* 30:235–45. Reprinted in Greenbaum A, Wellington A, Baar E. eds. 1995. *Social Conflict and Environmental Law.* North York, Ontario: Captus. 2 vols.

Burnham KP, Anderson DR. 1998. *Model Selection and Inference. A Practical Information-Theoretic Approach.* New York: Springer-Verlag

Busby FE, Coz CA. 1994. Rangeland health: new methods to classify, inventory and monitor rangelands. *Ren. Res. J.* 12(1):13–19

Butler D. 2000. The role of science is to illuminate political choices, not enforce them. *Nature* 403:6–7

Butzer K. 2000. The human role in environmental history. *Science* 287:2427–28

Caldwell LK. 1990. *Between Two Worlds: Science, the Environmental Movement, and Policy Choice.* Cambridge: Cambridge Univ. Press. 285 pp.

Callicott JB. 1994. *Earth's Insights: a Multicultural Survey of Ecological Ethics from the Mediterranean Basin to the Australian Outback.* Berkeley, CA: Univ. California Press. 285 pp.

Callicott JB. 1999. *Beyond the Land Ethic: More Essays in Environmental Philosophy.* Albany, NY: State Univ. New York Press. 427 pp.

Caughley G. 1994. Directions in conservation biology. *J. Anim. Ecol.* 63:215–44

Chamberlain TC. 1897. The method of multiple working hypotheses. *J. Geol.* 5:837–48

Charles A. 1994. Towards sustainability: the fishery experience. *Ecol. Econ.* 11:201–11

Charles A. 1995. The Atlantic Canadian groundfishery: roots of a collapse. *Dalhousie Law J.* 18:65–83

Charles A. 1998. Living with uncertainty in fisheries: analytical methods, management priorities and the Canadian groundfishery experience. *Fish. Res.* 37:37–50

Clad JR. 1988. Conservation and indigenous peoples: a study of convergent interests. See McNeely & Pitt 1988, pp. 45–62

Clark JS, Fastie C, Hurtt G, Jackson ST, Johnson C, et al. 1998. Reid's paradox of rapid plant migration. *BioScience* 48:13–24

Clark TW. 1993. Creating and using knowledge for species and ecosystem conservation: science, organizations, and policy. *Persp. Biol. Med.* 36:3. Reprinted in 1994. *Environmental*

Policy and Biodiversity, ed. RE Grumbine. Washington, DC: Island

Clark TW, Mazur N, Begg RJ, Cook SJ. 2000. Interdisciplinary guidelines for developing effective Koala conservation policy. *Conserv. Biol.* 14:69–701

Clarke JN, McCool D. 1985. *Staking Out the Terrain: Power Differentials among Natural Resource Management Agencies*. Albany: State Univ. New York Press

Cohen J. 1995. *How Many People Can the Earth Support?* New York: Norton. 532 pp.

Common M. 1995. *Sustainability and Policy: Limits to Economics*. Cambridge: Cambridge Univ. Press. 348 pp.

Cooper C. 1981. *Economic Evaluation and the Environment*. London: Hodder & Staughton. 161 pp.

Cortner HJ, Moote MA. 1999. *The Politics of Ecosystem Management*. Washington, DC: Island. 179 pp.

Costanza R, ed. 1991. *Ecological Economics: The Science and Management of Sustainability*. New York: Columbia Univ. Press. 525 pp.

Costanza R, Folke C. 1997. Valuing ecosystems services with efficiency, fairness, and sustainability as goals. In *Nature's Services: Societal Dependence on Natural Ecosystems*, ed. G Daily, pp. 49–68. Washington, DC: Island

Cotgrove S. 1982. *Catastrophe or Cornucupia*. New York: Wiley

Coward H, Ommer R, Pitcher T. 2000. *JustFish: Ethics and Canadian Marine Fisheries*. St. Johns, Newfoundland: Inst. Soc. Econ. Res. Memorial Univ. Newfoundland. 304 pp.

Cronon W. 1995. *Uncommon Ground: Toward Reinventing Nature*. New York: Norton. 561 pp.

Cropper ML, Oates WE. 1992. Environmental Economics: A Survey, *J. Econ. Lit.* 30: 675–740

Crosby AW. 1986. *Ecological Imperialism: the Biological Expansion of Europe, 900–1900*. Cambridge: Cambridge Univ. Press. 368 pp.

Crowder LB, Crouse DT, Heppell SS, Martin TH. 1994. Predicting the impact of turtle extruder devices on loggerhead sea turtle populations. *Ecol. Appl.* 4:437–45

Crowley PH. 1992. Resampling methods for computation-intensive data analysis in ecology and evolution. *Annu. Rev. Ecol. Syst.* 23:405–47

Culhane P. 1981. *Public Lands Politics: Interest Group Influence on the Forest Service and the Bureau of Land Management*. Baltimore: Johns Hopkins Univ. Press. 398 pp.

Daly H. 1991. *Steady-State Economics*. Washington, DC: Island. 302 pp. 2nd ed.

Daly H. 1996. *Beyond Growth: The Economics of Sustainable Development*. Boston: Beacon. 253 pp.

Dasmann R. 1984. *Environmental Conservation*. New York: Wiley

DeGroot MH. 1972. *Optimal Statistical Decisions*. New York: McGraw-Hill. 489 pp.

Dennis B. 1996. Discussion: should ecologists become Bayesians? *Ecol. Appl.* 6(4):1095–103

Dennis B, Taper ML. 1994. Density dependence in time series observations of natural populations: estimation and testing. *Ecol. Monogr.* 64:205–24

Diamond J. 1997. *Guns, Germs, and Steel: the Fates of Human Societies*. New York: Norton. 480 pp.

Dickson D. 2000. If knowledge is king, we may need a revolution. *Nature* 403:9

Doak DF. 1995. Source-sink models and the problem of habitat degradation: general models and applications to the Yellowstone Grizzly. *Conserv. Biol.* 9:1370–79

Dovers SR, Handmer JW. 1995. Ignorance, the precautionary principle and sustainability. *Ambio* 24:92–97

Dryzek JS. 1999. Global ecological democracy. See Low 1999, pp. 264–82

Dunham S, Coward H. 2000. World religions and ecosystem justice in the marine fishery. See Coward et al. 2000, pp. 47–66

Edwards AWF. 1972. *Likelihood*. Cambridge: Cambridge Univ. Press. 235 pp.

Ehrenfeld D. 1978. *The Arrogance of Humanism*. New York: Oxford Univ. Press. 286 pp.

Ehrenfeld D. 1993. *Beginning Again.* New York: Oxford University Press. 216 pp.

Engen S, Lande R, Saether B-E. 1997. Harvesting strategies for fluctuating populations based on uncertain population estimates. *J. Theor. Biol.* 186:201–12

England RW, ed. 1994. *Evolutionary Concepts in Economics.* Ann Arbor, MI: Univ. Michigan Press. 255 pp.

Faber DJ. 1998. *The Struggle for Ecological Democracy. Environmental Justice Movements in the United States.* New York: Guilford. 366 pp.

Farber DA. 1999. *Eco-Pragmatism. Making Sensible Environmental Decisions in an Uncertain World.* Chicago: Univ. Chicago Press. 210 pp.

Finlayson AC, McCay B. 1998. Crossing the threshold of ecosystem resilience: the commercial extinction of northern cod. See Berkes & Folke 1998, pp. 11–37

Fischer R. 1990. *Technocracy and the Politics of Expertise.* Newbury Park, CA: Sage. 387 pp.

Funtowicz SO, Ravetz JR. 1991. A new scientific methodology for global environmental issues. See Costanza 1991, pp. 137–52

Funtowicz SO, Martinez-Alier J, Munda G, Ravetz JR. 1999. Information tools for environmental policy under conditions of complexity. *Environ. Iss. Ser.: Eur. Environ. Ag..* Available at http://www.eea.eu.int. 54 pp.

Gadgil M, Berkes F, Folke C. 1993. Indigenous knowledge for biodiversity conservation. *Ambio* 22:151–56

Gadgil M, Guha R. 1992. *This Fissured Land: an Ecological History of India.* Berkeley: Univ Calif. Press

Gadgil M, Hemam N, Reddy B. 1998. People, refugia and resilience. See Berkes & Folke 1998, pp. 30–47

Gadgil M, Malhotra KC. 1994. The ecological significance of caste. In *Social Ecology,* ed. R Guha, pp. 27–41. New Delhi: Oxford Univ. Press

Garshelis DL. 1997. Sea otter mortality estimated from carcasses collected after the Exxon Valdez oil spill. *Conserv. Biol.* 11:905–16

Gaston KJ, Blackburn TM, Lawton JH. 1997. Interspecific abundance-range size relationships: an appraisal of mechanisms. *J. Anim. Ecol.* 66:579–601

Georgescu-Roegen N. 1971. *The Entropy Law and the Economic Process.* Cambridge, MA: Harvard Univ. Press. 457 pp.

Gillis M. 1991. Economics, ecology and ethics: mending the broken circle for tropical forests. In *Ecology, Economics, Ethics,* ed. FH Bormann, SR Kellert, pp. 155–79. New Haven, CT: Yale Univ. Press

Glacken CJ. 1956. Changing ideas of the habitable world. See Thomas 1956, pp. 70–92

Glantz MH. 1979. Science, politics and economics of the Peruvian anchoveta fishery. *Mar. Policy* 3(3):201–10

Gleick P, ed. 1993. *Water in Crisis: a Guide to the World's Fresh Water Resources.* New York: Oxford Univ. Press. 473 pp.

Goodland R. 1988. Tribal peoples and economic development: the human ecological dimension. See McNeely & Pitt 1988, pp. 13–32

Gould SJ. 2000. Deconstructing the "science wars" by reconstructing an old mold. *Science* 287:253–60

Goulder LH, Kennedy D. 1997. Valuing ecosystem services: philosophical bases and empirical methods. In *Nature's Services,* ed. G Dailey, pp. 23–47. Washington, DC: Island

Gowdy J. 1994. *Coevolutionary Economics: The Economy, Society, and the Environment.* Boston: Kluwer Academic. 246 pp.

Greenbaum A, Wellington A, Baar E, eds. 1995. *Social Conflict and Environmental Law.* North York, Ontario: Captus. 2 vols.

Greene C, Umbanhowar J, Mangel M, Caro T. 1998. Animal breeding systems, hunter selectivity, and consumptive use in wildlife conservation. In *Behavioral Ecology and Conservation Biology,* ed. T Caro, pp. 271–305, New York: Oxford Univ. Press. 582 pp.

Grime JP. 1997. Biodiversity and ecosystem function: the debate deepens. *Science* 277:1260–61

Grumbine RE. 1994. *Environmental Policy and Biodiversity.* Washington, DC: Island. 416 pp.

Gunderson LH, Holling CS, Light SS, eds. *Barriers and Bridges to the Renewal of Ecosystems and Institutions.* New York: Columbia Univ. Press. 593 pp.

Hacking I. 1999. *The Social Construction of What?* Cambridge, MA: Harvard Univ. Press. 261 pp.

Haddad B. 1997. Putting markets to work: the design and use of marketable permits and obligations. *Paris: Org. Econ. Coop. Dev. Occ. Pap. #19*

Haddad B. 2000. *Rivers of Gold: Designing Markets to Allocate Water in California.* Washington, DC: Island

Haddad B, Howarth R. 2001. Commensurability, substitution, and protest bids: contingent valuation and ecological economics. In *Handbook of Contingent Valuation,* ed. J Kahn, A Alberini, D Bjornstad. London: Edward Elgar. In press

Haerlin B, Parr D. 1999. How to restore public trust in science. *Nature* 400:499

Hakanson L, Peters RH. 1995. *Predictive Limnology: Methods for Predictive Modelling.* Amsterdam: SPB Academic. 464 pp.

Halbert CL. 1993. How adaptive is adaptive management? *Rev. Fish Biol. Fish.* 1:261–83

Halley JM, Dempster JP. 1996. The spatial population dynamics of insects exploiting a patchy food source: a model study of local persistence. *J. Appl. Ecol.* 33:439–54

Hamilton C. 1999. Justice, the market and climate change. See Low 1999, pp. 90–105

Hanemann WM. 1994. Valuing the environment through contingent valuation. *J. Econ. Persp.* 8:19–43

Hanna S, Folke C, Mäler K-G, eds. 1996. *Rights to Nature. Ecological, Economic, Cultural, and Political Principles of Institutions for the Environment.* Washington, DC: Island. 298 pp.

Harcourt AH. 1995. Population viability estimates: theory and practice for a wild gorilla population. *Conserv. Biol.* 8:134–42

Hedrick PW, Lacy RC, Allendorf FW, Soule ME. 1996. Directions in conservation biology: comments on Caughley. *Conserv. Biol.* 10:1312–20

Heilbroner RL. 1974. *An Inquiry into the Human Prospect.* New York: Norton. 180 pp.

Heilbroner RL. 1999. *The Worldly Philosophers.* New York: Touchstone

Hempel LC. 1996. *Environmental Governance: The Global Challenge.* Washington, DC: Island. 291 pp.

Herczeg YIZ. 1995. *Rashi. The Torah with Rashi's Commentary Translated, Annotated, and Elucidated.* Brooklyn, NY: Mesorah

Hilborn R, Ludwig D. 1993. The limits of applied ecological research. *Ecol. Appl.* 3:550–52

Hilborn R, Mangel M. 1997. *The Ecological Detective: Confronting Models with Data.* Princeton: Princeton Univ. Press. 315 pp.

Hilborn R, Pikitch EK, Francis RC. 1999. *Current trends in including risk and uncertainty in stock assessment and management decisions.* Sch. Fish., Rep. Univ. Washington. 24 pp.

Hilborn R, Walters CJ. 1992. *Quantitative Fisheries Stock Assessment. Choice, Dynamics and Uncertainty.* New York: Chapman & Hall. 570 pp.

Hilborn R, Walters CJ, Ludwig D. 1995. Sustainable exploitation of renewable resources. *Annu. Rev. Ecol. Syst.* 26:45–67

Hodges CA. 1995. Mineral resources, environmental issues and land use. *Science* 268:1305–12

Hodgson GM. 1999. *Evolution and Institutions: on Evolutionary Economics and the Evolution of Economics.* Cheltenham, UK: Edward Elgar. 345 pp.

Hogarth RM. 1983. *Judgment and Choice. The Psychology of Decision.* Chichester, UK: Wiley Intersci.

Homer-Dixon TF. 1991. On the threshold. Environmental changes as causes of acute conflict. *Inter. Secur.* 16:76–116

Homer-Dixon TF. 1994. Environmental scarcities and violent conflict. *Inter. Secur.* 19:5–40

Homer-Dixon TF, Boutwell JF, Rathjens GW. 1993. Environmental change and violent conflict. *Sci. Am.* 268:38–45

Hooper D, Vitousek P. 1997. The effects of

plant composition and diversity on ecosystem processes. *Science* 277:1302–5

Howarth R. 1995. Sustainability under uncertainty: a deontological approach. *Land Econ.* 71:417–27

Howarth RB, Norgaard RB. 1990. Intergenerational resource rights, efficiency, and social optimality. *Land Econ.* 66:1–11

Howson C, Urbach P. 1993. *Scientific Reasoning: The Bayesian Approach.* La Salle, IL: Open Court

Hubbard HM. 1991. The real cost of energy. *Sci. Am.* 264:36–42

Hutchings J, Walters CJ, Haedrich RL. 1997. Is scientific inquiry incompatible with government information control? *Can. J. Fish. Aquat. Sci.* 54:1198–210

Ignatieff M. 1998. *Isaiah Berlin. A Life.* New York: Owl. 356 pp.

Jackson LJ. 1997. Piscivores, predation, and PCBs in Lake Ontario's pelagic food web. *Ecol. Appl.* 7:991–1001

Jacobs M. 1994. The limits to neoclassicism. Towards an institutional environmental economics. See Redclift & Benton eds. 1994, pp. 67–91

Jacobs M. 1997. Crossing the threshold of ecosystem resilience: the commercial extinction of northern cod. See Berkes & Folke 1998, pp. 311–37

Jasanoff S. 1990. *The Fifth Branch: Scientists as Policy-Makers.* Cambridge MA: Harvard Univ. Press. 302 pp.

John D. 1994. *Civic Environmentalism.* Washington DC: CQ. 347 pp.

Johnson G. 1999. *Strange Beauty. Murray Gell-Mann and the Revolution in Twentieth-Century Physics.* New York: Knopf

Johnston DW, Meisenheimer P, Lavigne D. 2000. An evaluation of management objectives for Canada's commercial Harp Seal hunt, 1996–1998. *Conserv. Biol.* 14:729–37

Jonsson B, Andersen R, Hansen LP, Fleming IA, Bjorge A. 1993. Sustainable use of biodiversity. *NINA Utredning* 48:1–22. Nor. Inst. Nat. Res., Trondheim, Norway

Kaiser J. 2000. Ecologists on a mission to save the world. *Science* 287:1188–92

Karban R, Agrawal AA, Mangel M. 1997. The benefits of induced defenses against herbivores. *Ecology* 78:1351–55

Keller EF, Lloyd EA. 1992. *Key Words in Evolutionary Biology.* Cambridge, MA: Harvard Univ. Press

Kot M, Lewis MA, Van den Driessche P. 1996. Dispersal data and the spread of invading organisms. *Ecology* 77:2027–42

Kraft ME. 1996. *Environmental Policy and Politics.* New York: Harper-Collins

Kramer R, ed. 1997. *Last Stand: Protected Areas and the Defense of Tropical Biodiversity.* New York: Oxford Univ. Press. 242 pp.

Kurlansky M. 1997. *Cod.* New York: Penguin. 294 pp.

Lafferty W, Meadowcroft J, eds. 1996. *Democracy and the Environment: Problems and Prospects.* Cheltenham, UK: Edward Elgar. 276 pp.

Lauck T, Clark CW, Mangel M, Munro GR. 1998. Implementing the precautionary principle in fisheries management through marine reserves. *Ecol. Appl.* 8(Suppl.): S72–78

Leakey R. 1996. *The Sixth Extinction.* London: Weidenfeld & Nicolson. 271 pp.

Lélé S, Norgaard RB. 1996. Sustainability and the scientist's burden. *Conserv. Biol.* 10:354–65

Lempert RJ, Schlesinger ME, Bankes SC. 1996. When we don't know the costs or the benefits: adaptive strategies for abating climate change. *Clim. Change* 33:235–74

Leopold A. 1966 [1949]. *A Sand County Almanac.* New York: Oxford Univ. Press

Levins R. 1966. The strategy of model building in population biology. *Am. Sci.* 54:421

Lewin R. 2000. *Complexity. Life at the Edge of Chaos.* Chicago: Univ. Chicago Press. 2nd ed.

Lindley DV. 1985. *Making Decisions.* New York: Wiley. 207 pp.

Linnerooth J. 1984. The political processing of uncertainty. *Acta Psychol.* 56:219–31

Loder N. 1999. BSE advisers admit giving up a purely scientific role. *Nature* 400:490

Low N, ed. 1999. *Global Ethics and the Environment.* London: Routledge. 320 pp.

Low N, Gleeson B. 1998. *Justice, Society, and*

Nature: an Exploration of Political Ecology. London: Routledge. 256 pp.

Lubchenko J, Olson AM, Brubaker LB, Carpenter SR, Holland MM, et al. 1991. The sustainable biosphere initiative: an ecological research agenda. *Ecology* 72:371–412

Ludwig D. 1993. Environmental sustainability: magic, science, and religion in natural resource management. *Ecol. Appl.* 3: 555–58

Ludwig D. 1994. Uncertainty and fisheries management. In *Frontiers in Mathematical Biology.* ed. S Levin. *Lecture Notes in Biomathematics.* 100:516–28. Berlin: Springer-Verlag

Ludwig D. 1999. Is it meaningful to estimate a probability of extinction? *Ecology* 80:298–310

Ludwig D, Hilborn R. 1983. Adaptive probing strategies for age-structured fish stocks. *Can. J. Fish. Aquat. Sci.* 40:559–69

Ludwig D, Walters CJ. 1981. Measurement errors and uncertainty in parameter estimates for stock and recruitment. *Can. J. Fish. Aquat. Sci.* 38:711–20

Maas A. 1951. *Muddy Waters: the Army Engineers and our Nations Rivers.* Cambridge, MA: Harvard Univ. Press

Maddox J. 1999. *What Remains to Be Discovered? Mapping the Secrets of the Universe, the Origins of Life, and the Future of the Human Race.* New York: Martin Kessler. 434 pp.

Maddox J. 2000. Positioning the goal posts. *Nature* 403:139

Maguire L. 1994. Science, values, and uncertainty: a critique of the wildlands project. In *Environmental Policy and Biodiversity*, ed. RE Grumbine, pp. 267–72. Washington, DC: Island

Malacoff D. 1999. Bayes offers a 'new' way to make sense of numbers. *Science* 286:1460–64

Mangel M. 1993. Comparative analyses of the effects of high seas driftnets on the Northern Right Whale Dolphin *Lissodelphus Borealis. Ecol. Appl.* 3:221–29

Mangel M. 2000. On the fraction of habitat allocated to marine reserves. *Ecol. Lett.* 3:15–22

Mangel M, Hofman RJ. 1999. Ecosystems.

Patterns, processes and and paradigms. In *Conservation and Management of Marine Mammals*, ed. JR Twiss, RR Reeves, pp. 87–98. Washington, DC: Smithsonian

Mangel M, Talbot LM, Meffe GK, Agardy M, Alverson DL, et al. 1996. Principles for the conservation of wild living resources. *Ecol. Appl.* 6:338–62

Mangel M, Tier C. 1994. Four facts every population biologist should know about persistence. *Ecology* 75:607–14

Marsh GP. 1965 [1864]. *Man and Nature; or The Earth as Modified by Human Action.* Cambridge, MA: Belnap Press of Harvard Univ. Press. 472 pp.

Mason M. 1999. *Environmental Democracy.* London: Earthscan. 266 pp.

Matthews DR. 1995. 'Constructing' fisheries management: a values perspective. *Dalhousie Law J.* 18:44–57

Mayo DG. 1996. *Error and the Growth of Experimental Knowledge.* Chicago: Univ. Chicago Press. 493 pp.

Mayr E. 1982. *The Growth of Biological Thought.* Cambridge, MA: Harvard Univ. Press. 974 pp.

McCann KS. 2000. The diversity-stability debate. *Nature* 405:228–33

McCloskey DN. 1994. *Knowledge and Persuasion in Economics.* Cambridge: Cambridge Univ. Press. 445 pp.

McConnell G. 1966. *Private Power and American Democracy.* New York: Knopf. 397 pp.

McElroy DM, Shoemaker JA, Douglas ME. 1997. Discriminating Gila robusta and Gila cypha: risk assessment and the Endangered Species Act. *Ecol. Appl.* 7:958–67

McEvoy A. 1986. *The Fisherman's Problem.* Cambridge: Cambridge Univ. Press. 368 pp.

McNeely JA, Pitt D, eds. 1988. *Economics and Biological Diversity: Developing and Using Economic Incentives to Conserve Biological Resources.* Gland, Switzerland: Int. Union Conserv. Nat. Nat. Resourc. 236 pp.

McNeill JR. 2000. *Something New Under the Sun: An Environmental History of the Twentieth-Century World.* New York: Norton. 421 pp.

Meffe G, Viederman S. 1995. Combining science and policy in conservation biology. *Wildl. Soc. Bull.* 23:327–32

Menges ES. 1990. Population viability analysis for an endangered plant. *Conserv. Biol.* 4: 52–62

Morgan MG, Dowlatabadi H. 1996. Learning from integrated assessment of climate change. *Clim. Change* 34:337–68

Murphy DD, Noon BR. 1991. Coping with uncertainty in wildlife management. *J. Wildl. Man.* 55:773–82

Murphy DD, Noon BR. 1992. Integrating scientific methods with habitat conservation planning: reserve design for Northern Spotted Owls. *Ecol. Appl.* 2:3–17

Myers R, Hutchings J, Barrowman N. 1997. Why do fish stocks collapse? The example of cod in Atlantic Canada. *Ecol. Appl.* 7:91–106

Naeem S. 2000. Reply to Wardle et al. *Bull. Ecol. Soc. Am.* 81:241–46

Nakaoka M. 1996. Dynamics of age- and size-structured populations in fluctuating environments: applications of stochastic matrix models to natural populations. *Res. Popul. Ecol.* 38:141–52

National Atmospheric and Oceanic Administration. 1993. Report of the NOAA Panel on Contingent Valuation. *Fed. Regist.* 58:4601–14

Nelson RR. 1995. Recent Evolutionary Theorizing about Economic Change. *J. Econ. Lit.* 33:48–90

Nelson RR, Winter SG. 1982. *An Evolutionary Theory of Economic Change*. Cambridge, MA: Belknap

Norgaard R. 1984. Coevolutionary development potential. *Land Econ.* 60:160–73

Norgaard R. 1994. *Development Betrayed: the End of Progress and a Coevolutionary Revisioning of the Future*. London: Routledge. 280 pp.

Norton BG. 1998. Improving ecological communication: the role of ecologists in environmental policy formation. *Ecol. Appl.* 8:350–64

Odell P. 1986. *Oil and World Power*. London: Pelican. 8th ed.

Ohlsson L. 1995. *Hydropolitics*. London: Zed. 230 pp.

Ophuls W. 1977. *Ecology and the Politics of Scarcity*. San Francisco: Freeman. 303 pp.

Ophuls W, Boyan AS. 1992. *Ecology and the Politics of Scarcity Revisited*. New York: Freeman. 379 pp.

Oreskes N, Shrader-Frechette K, Belitz K. 1994. Verification, validation, and confirmation of numerical models in the earth sciences. *Science* 263:641–46

Orians G, et al. *Ecological Knowledge and Environmental Problem Solving*. Washington, DC: National Academy. 388 pp.

Osenberg CW, Schmitt RJ, Holbrook SJ, Abu-Saba KE, Flegal AR. 1994. Detection of environmental impacts: natural variability, effect size, and power analysis. *Ecol. Appl.* 4:16–30

Ostrom E. 1990. *Governing the Commons: the Evolution of Institutions for Collective Action*. New York: Cambridge Univ. Press. 280 pp.

Ostrom E. 1992. Rudiments of a theory of common property institutions. In *Making the Commons Work*, ed. D Bromley, pp. 293–318. San Francisco: ICS

Ostrom E, Burger J, Field CB, Norgaard R, Polikansky D. 1999. Revisiting the commons: local lessons, global challenges. *Science* 284:278–82

Ozawa CP. 1991. *Recasting Science: Consensual Procedures in Public Policy Making*. Boulder, CO: Westview. 143 pp.

Padilla A, Gibson I. 2000. Science moves to centre stage. *Nature* 403:357–59

Paehlke RC. 1989. *Environmentalism and the Future of Progressive Politics*. New Haven, CT: Yale Univ. Press. 325 pp.

Paehlke RC. 1996. Environmental challenges to democratic practice. In *Democracy and the Environment: Problems and Prospects*, ed. W Lafferty, J Meadowcroft, pp. 18–38. Cheltenham, UK: Edward Elgar

Paehlke R, Torgerson D, eds. 1990a. *Managing Leviathan: Environmental Politics and the Administrative State*. Peterborough Ontario: Broadview. 310 pp.

Paehlke R, Torgerson D. 1990b. Environmental politics and the administrative state. See. Paelke & Torgerson 1990a, pp. 285–301

Pagel JE, Bell DA, Norton BE. 1996. De-listing the American peregrine falcon: is it premature. *Wildl. Soc. Bull.* 24:429–35

Parma A, Amarasekare P, Mangel M, Moore J, Murdoch WW, et al. 1998. What can adaptive management do for our fish, forests, food and biodiversity? *Int. Biol.* 1:16–26

Peterman RM. 1990. Statistical power analysis can improve fisheries research and management. *Can. J. Fish. Aquat. Sci.* 47:2–15

Peterman RM, M'Gonigle M. 1992. Statistical power analysis and the precautionary principle. *Mar. Pollut. Bull.* 24:231–34

Platt JR. 1964. Strong inference. *Science* 146:347–53

Ponting C. 1992. *A Green History of the World, The Environment and the Collapse of Great Civilizations.* New York: St. Martin's

Postel S. 1992. *Last Oasis: Facing Water Scarcity.* New York: Norton. 239 pp.

Postel S. 1996. *Dividing the Waters: Food Security, Ecosystem Health, and the New Politics of Scarcity.* Washington, DC: Worldwatch. 76 pp.

Pounds JA, Fogden MP, Savage JM, Gorman GC. 1997. Tests of null models for amphibian declines on a tropical mountain. *Conserv. Biol.* 11:1307–22

Press WH. 1997. Understanding data better with Bayesian and global statistical methods. In *Unsolved Problems in Astrophysics*, ed. JN Bahcall, JP Ostriker, pp. 49–60. Princeton, NJ: Princeton Univ. Press

Quinn TJ, Deriso RB. 1999. *Quantitative Fish Dynamics.* New York: Oxford Univ. Press. 542 pp.

Radovich J. 1981. The collapse of the California Sardine fishery. What have we learned? In *Resource Managment and Environmental Uncertainty: Lessons from Coastal Upwelling Fisheries*, ed. M Glantz, JD Thompson, pp. 107–36. New York: Wiley

Ralls K, Starfield AM. 1995. Choosing management strategy: Two structured decision-making methods for evaluating the predic-

tions of stochastic simulation models. *Conserv. Biol.* 9:175–81

Redclift M, Benton T, eds. 1994. *Social Theory and the Global Environment.* London: Routledge. 271 pp.

Redford KH, Stearman AM. 1993a. Forest-dwelling native Amazonians and the conservation of biodiversity. *Conserv. Biol.* 7:248–55

Redford KH, Stearman AM. 1993b. On common ground? Response to Alcorn. *Conserv. Biol.* 7:427–28

Redman CL. 1999. *Human Impact on Ancient Environments.* Tucson, AZ: Univ. Arizona Press. 239 pp.

Rejmanek M, Richardson DM. 1996. What attributes make some plant species more invasive? *Ecology* 77:1655–61

Ridley M. 1999. *Genome. The Autobiography of a Species in 23 Chapters.* New York: Harper-Collins.

Rittel H, Webber M. 1973. Dilemmas in a general theory of planning. *Polit. Sci.* 4:155–69

Robinson JG. 1993. The limits to caring: sustainable living and the loss of biodiversity. *Conserv. Biol.* 7:20–28

Rockefeller S, Elder J, eds. 1992. *Spirit and Nature: Why the Environment is a Religious Issue.* Boston: Beacon. 236 pp.

Roe E. 1998. *Taking Complexity Seriously: Policy Analysis, Triangulation and Sustainable Development.* Boston: Kluwer Academic. 138 pp.

Root KV. 1998. Evaluating the effects of habitat quality, connectivity, and catastrophes on a threatened species. *Ecol. Appl.* 8:854–65

Roszak T. 1972. *Where the Wasteland Ends.* Garden City, NY: Doubleday. 492 pp.

Roszak T. 1992. *The Voice of the Earth.* New York: Touchstone. 365 pp.

Rourke F. 1984. *Bureaucracy, Politics and Public Policy.* Glenview IL: Foresman. 3rd ed.

Ruckelshaus M, Hartway C, Kareiva P. 1997. Assessing the data requirements of spatially explicit dispersal models. *Conserv. Biol.* 11:1298–306

Ruttan L, Bogerhoff-Mulder M. 1999. Are East

African pastoralists truly conservationists? *Curr. Anthropol.* 40:621–52

Sachs W. 1993. *Global Ecology.* Atlantic Highlands: Zed. 262 pp.

Saetersdahl G. 1979. A review of past management of some pelagic stocks and its effectiveness. *Rapp. P.-v. Reun. Conserv. Int. Explor. Mer.* 177:505–12

Sagoff M. 1995. Fact and value in environmental science. See Greenbaum, et al. 1995, pp. 136–42

Salter L. 1988. *Mandated Science.* Dordrecht: Kluwer. 221 pp.

Sanderson S, Redford KH. 1997. Biodiversity politics and the contest for ownership of the world's biota. See Kramer 1997, pp. 115–32

Sanford C. 1999. *The Hunting Apes.* Princeton: Princeton Univ. Press. 253 pp.

Saville A. 1979. discussion and conclusions of the symposium on the biological basis of pelagic fish stock management. *Rapp. P.-V. Reun. Conserv. Int. Explor. Mer.* 177:513–17

Schacher-Shalomi Z. 1993. *Paradigm Shift.* New York: Jason Aronson

Schoener TW, Spiller DA. 1992. Is extinction related to temporal variability in population size? An empirical answer for orb spiders. *Am. Nat.* 139:1176–207

Schrecker T. 1990. Resisting environmental regulation: the cryptic pattern of business-government relations. See Paehlke & Torgerson 1990a, pp. 165–99

Sen A. 1999. *Development as Freedom.* New York: Knopf. 366 pp.

Shea K. 1998. Management of populations in conservation, harvesting and control. *Trends Ecol. Evol.* 13:371–75

Shrader-Frechette K. 1993. *Burying Uncertainty.* Berkeley: Univ. Calif. Press. 346 pp.

Shrader-Frechette K, McCoy E. 1993. *Method in Ecology: Strategies for Conservation.* Cambridge: Cambridge Univ. Press. 328 pp.

Simberloff D. 1988. The contribution of population and community biology to conservation science. *Annu. Rev. Ecol. Syst.* 19:473–511

Slade NA, Gomulkiewicz R, Alexander HM. 1998. Alternatives to Robinson and Redford's method of assessing overharvest from incomplete demographic data. *Conserv. Biol.* 12:148–55

Smithson M. 1989. *Ignorance and Uncertainty.* New York: Springer-Verlag. 393 pp.

Soulé ME. 1986. *Conservation Biology. The Science of Scarcity and Diversity.* Sunderland, MA: Sinauer. 584 pp.

Stanley TR. 1995. Ecosystem management and the arrogance of humanism. *Conserv. Biol.* 9:255–62

Starfield AM. 1997. A pragmatic approach to modeling for wildlife management. *J. Wildl. Man.* 61:261–70

Steidl RJ, Hayes JP, Schauber S. 1997. Statistical power analysis in wildlife research. *J. Wildl. Man.* 612:270–79

Sterman JD. 1988. A skeptic's guide to computer models. In *Foresight and National Decisions,* ed. L Grant, pp. 133–69. Lanham, MD: University

Stevens TH, Belkner R, Dennis D, Kittredge D, Willis C. 2000. Comparison of contingent valuation and conjoint analysis in ecosystem management. *Ecol. Econ.* 32:63–74

Strong DR, Whipple AV, Child AL, Dennis B. 1999. Model selection for a subterranean trophic cascade: root-feeding caterpillars and entomopathogenic nematodes. *Ecology* 80:2750–61

Talbot L. 1996. *Living Resource Conservation: An International Overview.* Marine Mammal Commission, Washington, DC

Taylor B, Kremsater L, Ellis R. 1997. *Adaptive Management of Forests in British Columbia.* British Columbia Ministry of Forests–Forest Practices Branch. Victoria, BC, Canada. 93 pp.

Taylor BL, Wade PR, De Master DP, Barlow J. 2000. Incorporating uncertainty into management models for marine mammals. *Conserv. Biol.* 14:1243–52

Taylor G. 1995. The collapse of the Northern Cod fishery: a historical perspective. *Dalhousie Law J.* 18:13–22

Thomas WL. 1956. *Man's Role in Changing*

the Face of the Earth. Chicago: Univ. Chicago Press

Tilman D. 2000. What Issues in Ecology is, and isn't. Bull. Ecol. Soc. Am. 81:240

Tilman D, Knops J, Wedin D, Reich P, Ritchie M, Siemann E. 1997. The influence of functional diversity and composition on ecosystem processes. Science 277:1300–2

Toulmin S. 1990. Cosmopolis: The Hidden Agenda of Modernity. Chicago: Univ. Chicago Press. 228 pp.

Trebilcock MJ. 1978. The consumer interest and regulatory reform. In The Regulatory Process in Canada, ed. GB Doern, pp. 94–127. Toronto: Macmillan

Turner BL, Clark WC, Cates RW, Richards JF, Mathews JT, Meyer WB, eds. 1990. The Earth as Transformed by Human Action. Cambridge: Cambridge Univ. Press. 713 pp.

van Schaik CP, Terborgh J, Dugelby B. 1997. The silent crisis: the state of rain forest nature preserves. See Kramer 1997, pp. 64–89

Van Valen L. 1982. Why misunderstand the evolutionary half of biology? In Conceptual Issues in Ecology, ed. E Saarinen, pp. 323–42. Boston: Reidel

von Hayek FA. 1945. The use of knowledge in society. Amer. Econ. Rev. 35:519–30

Vucetich JA, Peterson RO, Waite TA. 1996. Effects of social structure and prey dynamics on extinction risk in gray wolves. Cons. Biol. 11:957–65

Wade PR. 2000. Bayesian methods in conservation biology. Cons. Biol. 14:1308–5

Waldrop MM. 1992. Complexity. The Emerging Science at the Edge of Order and Chaos. New York: Simon & Schuster. 380 pp.

Wallsten TS. 1980. Cognitive Processes in Choice and Decision Behavior. Hillsdale, NJ: Lawrence Erlbaum. 285 pp.

Walters CJ. 1986. Adaptive Management of Renewable Resources. New York: Macmillan. 374 pp.

Walters CJ. 1997. Challenges in adaptive management of riparian and coastal ecosystems.

Cons. Ecol. 1:1. Available on the internet at http://www.consecol.org/vol1/iss2/art1

Walters CJ, Hilborn R. 1978. Ecological optimization and adaptive management. Annu. Rev. Ecol. Syst. 9:157–88

Walters CJ, Ludwig D. 1981. Effects of measurement errors on the assessment of stock-recruitment relationships. Can. J. Fish. Aquat. Sci. 38:704–10

Walters CJ, Maguire J-J. 1996. Lessons for stock assessment from the northern cod collapse. Rev. Fish Biol. Fish. 6:125–37

Wardle DA. 2000. Biodiversity and function: an issue on ecology. Bull. Ecol. Soc. Am. 81:235–39

Waskow A. 1995. Down to Earth Judaism. New York: Morrow

Weaver J. 1996. Defending Mother Earth: Native American Perspectives on Environmental Justice. Maryknoll, NY: Orbis. 205 pp.

Weber EP. 1998. Pluralism by the Rules: Conflict and Cooperation in Environmental Regulation. Washington, DC: Georgetown. 308 pp.

Wells MS, Brandon K. 1992. People and Parks. Linking Protected Area Managment with Local Communities. Washington, DC: World Bank. 99 pp.

Wells MS, Guggenheim S, Khan A, Wardjo W, Jepson P. 1999. Investing in Biodiversity. A Review of Indonesia's Integrated Conservation and Development Projects. Washington, DC: World Bank

Westra L, Wenz P. 1995. Faces of Environmental Racism. Confronting Issues of Global Justice. Lanham, MD: Rowman & Littlefield. 246 pp.

White L. 1967. The historical roots of our ecological crisis. Science 155:1203–7. Reprinted in 1969. The Subversive Science, ed. P. Shepard. Boston: Houghton Mifflin

Whittemore AS. 1983. Facts and values in risk analysis for environmental toxicants. Risk Analysis 3:23–33

Williams BK. 1996. Adaptive optimization of renewable natural resources: Solution algorithms and a computer program. Ecol. Mod. 93:101–11

Wilmott P. 1998. *Derivatives*. New York: Wiley

Winterhalder B, Lu F. 1997. A forager-resource population ecology model and implications for indigenous conservation. *Conserv. Biol.* 11:1354–64

Wood CA. 1994. Ecosystem management: Achieving the new land ethic. *Ren. Res. J.* 12(1):6–12

Woodroffe R, Ginsberg JR. 1998. Edge effects and the extinction of populations inside protected areas. *Science* 280:2126–28

World Commission on Forestry and Sustainable Development. 1999. *Our Forests Our Future*. Cambridge: Cambridge Univ. Press. 205 pp.

Worster D. 1993. *The Wealth of Nature. Environmental History and the Ecological Imagination*. New York: Oxford Univ. Press. 255 pp.

Wynne B. 1992. Uncertainty and environmental learning. Reconceiving science and policy in the preventive paradigm. *Global Environ. Change* 2(2):111–27

Wynne B. 1994. Scientific knowledge and the global environment. See Redclift & Benton 1994, pp. 169–89

Wynne B. 1996. May the sheep safely graze? A reflexive view of the expert-lay knowledge divide. In *Risk, Environment and Modernity*, ed. S Lash, B Szerzynski, B Wynne, pp. 44–83. London: Sage

Yaffee SL. 1982. *Prohibitive Policy: Implementing the Endangered Species Act*. Cambridge, MA: MIT Press. 239 pp.

Yaffee SL. 1997. Why environmental policy nightmares recur. *Conserv. Biol.* 11:328–37

Yergin D. 1991. *The Prize: the Epic Quest for Oil, Money and Power*. New York: Simon & Schuster. 877 pp.

Young OR. 1999. Fairness matters, the role of equity in international regime formation. See Low & Gleeson 1998, pp. 247–63

Young TP. 1994. Natural die-offs of large mammals: implications for conservation. *Conserv. Biol.* 8:410–18

Annu. Rev. Ecol. Syst. 2001. 32:519–45
Copyright © 2001 by Annual Reviews. All rights reserved

MALE-KILLING, NEMATODE INFECTIONS, BACTERIOPHAGE INFECTION, AND VIRULENCE OF CYTOPLASMIC BACTERIA IN THE GENUS WOLBACHIA

Lori Stevens, Rosanna Giordano, and
Roberto F. Fialho
*Department of Biology, University of Vermont, Burlington, Vermont 05405;
e-mail: Lori.Stevens@UVM.edu, giordano@life.uiuc.edu*

Key Words reproductive alterations, bacteriophage WO, sex ratio

■ **Abstract** *Wolbachia* bacteria are cytoplasmic endosymbionts with a wide range of effects on their hosts and are known to infect two major invertebrate groups, arthropods and nematodes. In arthropods *Wolbachia* alter host reproduction, causing unidirectional and bidirectional cytoplasmic incompatibility, parthenogenesis, feminization, and embryonic male killing. *Wolbachia* variation in reproductive effects is indicative of a high degree of evolutionary plasticity. As many as 75% of arthropods may be infected with *Wolbachia*, which in addition to affecting reproduction, can also directly affect host fitness by either increasing or decreasing survival and fecundity. We review the dynamics of embryonic male-killing, including effects on insect mating behavior, as well as the distribution and implication of *Wolbachia* infections in filarial nematodes.

Arthropod host–*Wolbachia* phylogenies are not congruent, which is suggestive of horizontal transmission. The opposite has been shown in nematode-*Wolbachia* phylogenies, indicative of long-term association and vertical transmission. Multiple levels of parasitism within arthropods may promote horizontal transmission. Bacteriophage WO has recently been identified and is found in all *Wolbachia*-infected insect hosts so far examined. Extensive horizontal transmission of the phage occurs between different *Wolbachia* strains within a host as well as between *Wolbachia* in different hosts. The phage genome may carry genes important in determining both the effect of *Wolbachia* on arthropod host reproduction and host fitness and fecundity. The extensive horizontal transmission of the phage may explain the plasticity of *Wolbachia's* effect on arthropod hosts.

0066-4162/01/1215-0519$14.00

519

CYTOPLASMIC MICROORGANISMS

Wolbachia bacteria are known for their ability to manipulate arthropod reproduction. Recently, *Wolbachia* have been found to be widespread in nematodes; to be able to induce a new reproductive alteration, male-killing (MK); and to be infected with bacteriophage WO.

Wolbachia bacteria are cytoplasmically inherited; they are passed from mother to progeny through the egg. Several reproductive alterations are unique to *Wolbachia*, most notably their ability to cause cytoplasmic incompatibility (CI); crosses between infected males and uninfected females can result in complete failure of eggs to develop and hatch or reduced egg hatch rates (e.g., Fialho & Stevens 1996). Thyletokous parthenogensis (production of all female progeny from unfertilized eggs) and feminization of genetic males are additional reproductive alterations attributed to *Wolbachia*. Thelytokous parthenogenesis occurs in several species of parasitic wasps but appears to be restricted to the Chalcidoid and Cynipoid Hymenoptera (Stouthamer et al. 1993, Stouthamer 1997). Feminizing *Wolbachia* transform genetic males into functional females and have been identified in terrestrial isopods (Bouchon et al. 1998).

The notion that these bacteria are widespread is supported by molecular surveys (Table 1), which indicate the presence of *Wolbachia* in as many as 75% of all

TABLE 1   *Wolbachia* infected taxa and incidences of infections[a]

Host taxa examined	Percent infected (Sample size)	Reference
Arthropoda	76%	Jeyaprakash & Hoy 2000
Aracnida	(1/2)	
Insecta	(47/61)	
North American temperate	19.3%	Werren & Windsor 2000
Aracnida	(1/12)	
Insecta	(28/145)	
Filarial Nematodes	90% (18/20)	Taylor et al. 1999
Parasitoids of temperate Lepidoptera	27.4% (45/164)	Cook & Butcher 1999
Hymenoptera: Cynipidae	59.3%	Plantard et al. 1999
Tribe Rhoditini	(12/19)	
Tribe Aylacini	(4/8)	
Indo-Australian ants	50% (25/50)	Wenseleers et al. 1998
Isopods	30% (22/85)	Bouchon et al. 1998
British insects	21.7% (18/83)	West et al. 1998
Neotropical (Panama)	16.9%	Werren et al. 1995
Aracnida	(0/3)	
Insecta	(26/154)	

[a]In addition, one spider, *Nephila clavata*, is known to harbor *Wolbachia* (Oh et al. 2000).

arthropod species, including spiders, mites, and isopods as well as many species of insects. Recently, the known range of hosts has been expanded by the discovery of *Wolbachia* in a number of nematode species, and the proportion of nematode species infected with this bacteria could be even higher than for arthropods (Table 1). When infected males mate with uninfected females or females carrying a different genetic strain of *Wolbachia*, CI is expressed as lower fitness or sterility of the cross. In arthropods, mutualistic (Girin & Bouletreau 1995), symbiotic (James & Ballard 2000), pathogenic (Min & Benzer 1997), and fecundity-enhancing (Vavre et al. 1999b) relationships have been reported between *Wolbachia* and their hosts.

Understanding these endosymbionts should be an integral part of a comprehensive theory of the evolution, ecology, and systematics of its hosts. The recent discovery of *Wolbachia* infection with bacteriophage WO, and the ongoing effort to sequence the genome of several *Wolbachia* strains, suggests that understanding the mechanisms of action and virulence of this bacteria might not be far off. The genome sequencing project (Slatko et al. 1999) currently underway includes the sequencing of *Wolbachia* from three of the five major groups. These groups span the range of *Wolbachia* hosts, including nematodes and arthropods.

Wolbachia infection has significant consequences for host reproduction, including a range of reproductive alterations and an amazing diversity in the virulence of the *Wolbachia*-host interaction. *Wolbachia* has been implicated as an important factor promoting speciation (Bordenstein et al. 2001, Rokas 2000, Shoemaker et al. 1999, Wade 2001) and may be useful for controlling arthropod pests (Sinkins & O'Neill 2000). The biology of *Wolbachia*, including the phylogeny and distribution, mechanisms of action, population biology, and evolution and biological control implications, was recently reviewed (Werren 1997). *Wolbachia's* specific action as a microbial manipulator of arthropod reproduction has also been recently summarized (Stouthamer et al. 1999). However, over 100 papers have appeared in the past 2 years. This article summarizes and reviews the recent discoveries with special emphasis on MK *Wolbachia*, *Wolbachia* infections in Nematodes, and bacteriophage WO.

WOLBACHIA-NEMATODE SYMBIOSIS

Perhaps one of the most interesting recent developments in *Wolbachia* research is the documentation of *Wolbachia* bacteria in filarial nematodes using the polymerase chain reaction, PCR (Sironi et al. 1995). The presence of intracellular bacteria in filarial worms has been known for several decades and was originally discovered with the aid of electron microscopy. Filarial worms are nonsegmented, cylindrical, and tapered at both ends. They are arthropod-borne agents of diseases including river blindness (onchocerciasis) and elephantiasis (Hoerauf et al. 2000a). An estimated 150 million humans have filariasis (Taylor et al. 2000a). Many species of filaria are of major medical importance; these belong to the family Onchocercidae, superfamily Filarioidea, order Spirurida, class Nematoda (Anderson et al. 1974). In contrast to the seemingly nonobligatory nature of arthropod infections, the filaria-bacteria relationship appears reciprocally dependent, suggesting the exciting possibility that antibiotics directed towards *Wolbachia* could be an effective antifilarial treatment against these notoriously difficult-to-treat infections (Taylor et al. 2000a).

Medical Aspects

Nematode species of medical importance are the causative agents of bancroftian, brugian, and timoran flariasis (*Wuchereria bancrofti*, *Brugia malayi*, and *B. timori*, respectively); *Onchocerca volvulus* is the causative agent of river blindness. In addition, *Dirofilaria immitis*, the dog heartworm, is of significant veterinary importance. The human-infecting nematodes have a tropical distribution, whereas *D. immitis* occurs in the tropics and subtropics as well as North America and southern Europe. Many human-infecting nematodes are transmitted by *Culex* and *Anopheles* mosquitoes; whereas *O. volvulus* is transmitted by simuliid flies. Microfilariae are ingested with a blood meal taken by their insect vectors and gain access to vertebrate hosts during a subsequent blood meal, by rupturing the vector's labella and penetrating via the puncture wound (Wharton 1957, McGreevy 1974). Once they are successful at invading a human host, the microfilariae migrate to the lymphatic system, where they undergo further molting. Upon maturation and subsequent to pairing, female worms release microfilariae in the lymphatic system at the rate of approximately 50,000/day for a decade or more (Buck 1991). Microfilariae move to the peripheral blood circulatory system, where they can in turn be picked up by insect vectors. The human disease state is characterized by adult worm–induced lymphangites and lymphadenitis. It progresses slowly from adenolymphangitis to lymphoedema of the extremities. Obstructive filariasis, or elephantiasis, is more prevalent in *B. timori* infections.

Oncocerchiasis is characterized by dermatitis, coupled by intense itching, subcutaneous nodules, and eye lesions resulting in blindness. Adult worms can be found in nodules or free living in subcutaneous tissues. However, it is the microfilariae they release that, upon decay, cause lesions and an inflammatory response. When microfilariae die in the eye, they can cause blindness. Treatment of filariasis is difficult because most therapies have severe side effects. Diethylcarbamazine can kill microfilariae as well as the adult worm; however, systemic allergic reactions can occur from the rapid clearance of the microfilariae. Ivermectin can kill microfilarie as well as suppress their emergence from the adult worm, and it has also been shown to have minor macrofilaricidal effects on adult male worms (Awadzi et al. 1997). Clinical trials in Ghana using amocarzine have shown that it has no effect on male worms or intrauterine embryos and no suppression of skin microfilaria. Furthermore, the addition of amocarzine to ivermectin did not improve treatment response (Awadzi et al. 1997). All chemotherapeutic agents currently in use that can eliminate *O. volvulus* adults are toxic to humans.

Investigations with electron microscopy have found bacterial infections of nematodes to be widespread. Bacteria detected in the embryos and microfilaria of *D.*

immitis were suggested to be involved in respiration and nutrition (Harada 1970, Kozek 1971). Gram-negative microorganisms were found in the hypodermal tissues, embryos, and larvae of *D. immitis* and *Brugia pahangi*. Both worms harbor morphologically similar bacteria; however, they were found to be less abundant in the latter. Bacteria were not detected in microfilariae of the filarial species *Loa loa*, *D. viteae*, *D. setariosum*, or *Litomosoides carinii* (currently *L. sigmodontis*) (McLaren & Worms 1975). The authors proposed that the filarial worms could have acquired the microorganisms from their mosquito host, which was known to harbor *Wolbachia* (Hertig 1936, Yen & Barr 1973). Autofluorescent granules were also detected, which correlated with the presence of the microorganism (McLaren & Worms 1975). Rickettsia-like bacteria were found in the lateral chords and ovarian growth zones of *B. pahangi* and *B. malayi*, suggesting that they might be transovarially transmitted (Vincent et al. 1975). Bacteria were also found in the lateral chords of both adult and larval male, female worms' and the oogonia, oocytes, developing eggs, and microfilariae of female *O. volvulus* (Kozek & Figueroa 1977, Franz & Buttner 1983). In males bacteria were not observed in any of the developmental stages of sperm, again suggesting transovarial transmission. Bacteria were found to be primarily located in the hypodermal tissues of the lateral chords.

The morphological information was insufficient to determine if the bacteria were parasitic, commensal, or mutualistic, but the authors proposed that if bacteria were essential for filarial worm survival, they could be the target of chemotheraupeutic agents that in turn would prove to have filaricidal properties. They also suggested that the microorganisms could be a contributing factor to the pathological expression of onchocerciasis in humans (Kozek & Figueroa 1977). It was proposed that presence of the microorganisms in healthy lateral chords and germinal tissues of female *B. malayi* suggests they are not pathogenic to worms and that the bacterial endosymbiont within *B. malayi* could cause some of the pathological expressions in the vertebrate host (Kozek 1977). Additional observations of intracellular bacteria in filarial worms were reported in microfilariae of *Mansonella ozzardi* (Kozek & Raccurt 1983), in the hypodermis of male and female *L. carinii*, (Franz & Andrews 1986), in the lateral chords of female *O. gibsoni* (Franz & Copeman 1988), and in the hypodermis of females, males, and embryos of *Onchocerca ochengi* and *O. fasciata* (Determann et al. 1997).

Kozek & Figueroa (1977) proposed that the development of geographic strains could arise in geographically distant populations if the bacterial infection was the result of a long-term coevolutionary relationship. Filarial parasites in the rainforest and savanna habitats of western Africa are different, and vectorial capacity is not uniform among the species within the *Simulium damnosum* species complex, suggestive of vector-parasite complexes (Duke et al. 1966, Duke 1981). At least three antigenic differences have been detected between *O. volvulus* from the savanna and forest habitats by crossed immunoelectrophoresis (Lobos & Weiss 1985). Differences in *O. volvulus* populations from the two bioclines were detected using isozyme and DNA analysis (Flockhart et al. 1986; Erttmann et al. 1987, 1990).

Infection intensity differed between forest and savanna villages in west Africa, with *O. volvulus* from the Yahense forest being less invasive and less pathogenic than the savanna microfilariae, despite the high intensity of *O. volvulus* infection in the Yahense forest communities (Remme et al. 1989, Dadzie et al. 1989). However, a field study that identified parasite strain and vector species using molecular methods did not detect preferential transmission of the savanna and forest strains by the different species in the *S. damnosum* complex (Toe et al. 1997). Infections of *O. volvulus* in mammals other than humans are rare, but there are reports of natural infections in a spider monkey, and a gorilla and chimpanzees have been infected in the laboratory (Nelson 1970).

Wolbachia-Nematode Phylogeny

Despite the substantial number of reports of intracellular bacteria in a number of species of filarial worms and suggestions that targeting the bacteria could result in filaricidal activity (Kozek 1977), this information went largely unnoticed, perhaps owing to lack of communication between taxonomists/morphologists and physiologists. Recently, the application of PCR to *Wolbachia* bacterial systematics has allowed the rapid and unambiguous identification of *Wolbachia*-like organisms in nematodes, finally focusing attention on this important group of endosymbionts. The recent molecular verification of *Wolbachia* bacteria in filarial worms (Sironi et al. 1995, Bandi et al. 1998) has given new impetus to filarial research and has increased interest in *Wolbachia* bacteria. As is often the case, organisms with the potential to reduce the incidence of human disease spur great interest and input of financial resources. *Wolbachia* not only provides a potential to suppress insect pest populations, but may also be used to better understand filarial diseases and possibly control some of them.

The first molecular evidence that filarial worms harbored members of the genus *Wolbachia* was presented by Sironi et al. (1995) using 16S rDNA sequences. Subsequently, Bandi et al. (1998), using partial sequences of the *ftsZ* gene, reaffirmed that the intracellular bacteria in filaria belonged to the genus *Wolbachia*. Phylogenetic analysis suggests the nematode infecting *Wolbachia* fall into two new groups, C and D (Bandi et al. 1998). The C group contains symbionts of *D. repens*, *D. immitis*, *O. ochengi*, *O. gibsoni*, and *O gutturosa*, whereas the D group contains *Wolbachia*-infecting *B. malayi*, *B. pahangi*, *W. bancrofti*, and *L. simodontis*.

The congruence of the nematode-*Wolbachia* phylogeny is in stark contrast to the pattern of arthropod-*Wolbachia* relationships, in which many instances of horizontal transfers have been inferred. A comparison of filarial 5S ribosomal spacer and *Wolbachia ftsZ* and 16s rDNA sequences showed congruence between host and symbiont phylogenies (Xie et al. 1994). Moreover, substitution rates between C and D *Wolbachia* using 16S rDNA places their separation at about 100 mya, comparable to estimates of the origin of the Onchocercidae, the filariae that harbor C group *Wolbachia*. Congruence between filaria and *Wolbachia* phylogenies argues for long-standing vertical transmission between these two organisms, whereas the divergence between arthropod and filarial *Wolbachia* indicates that horizontal transmission is not common between arthropods and the *Wolbachia* harbored by filaria (Bandi et al. 1998). A phylogeny of filarial *Wolbachia* using partial *wsp* (*Wolbachia* surface protein) sequences obtained similar results as were found with *ftsZ* and 16S rRNA genes. Comparison of *Wolbachia wsp* sequences from multiple populations within each of three species, *D. immitis*, *D. repens*, and *W. bancrofti* showed identical gene sequences, further substantiating that the relationship between filaria and *Wolbachia* is species specific (Bazzocchi et al. 2000b).

Wolbachia-filaria phylogenetic congruence has also been analyzed using filaria COI and *wsp* bacterial sequences. The work supports the conclusions reached previously with the 5S ribosomal spacer DNA sequence, namely that the phylogeny of the endosymbiont tracks the phylogeny of the filarial host. Interestingly, the phylogenetic affinity of the uninfected worm, *A. viteae*, is ambiguous in these analyses. As a result it is not clear whether *Wolbachia* infection was acquired by an ancestral filaria and subsequently lost in *A. viteae* or whether the *Wolbachia*-infected filaria acquired the infection subsequent to a split with *A. viteae* (Casiraghi et al. 2001).

Test of *Wolbachia*-Filaria Dependence

The *Wolbachia*-filaria phylogenetic congruence suggests a symbiotic relationship or at least an intimate species-specific association. One way to approach the question of whether filariae depend on *Wolbachia* is by their elimination with the use of antibiotics. Oral tetracycline treatment of Mongolian gerbils or jirds (*Meriones unguiculatus*) challenged with subcutaneous and peritoneal cavity injections of L3 stage *B. pahangi* larvae demonstrated inhibition of L3 to adult stage development by as much as 50% compared with controls (Bosshardt et al. 1993). After 28 days of oxytetracycline by mouth, dogs naturally infected with *D. immitis* had filaria that were motile but that showed signs of inhibited embryogenesis. Tetracycline treatment for 25 days interfered with the transovarial transmission of *Wolbachia* as well as inhibiting embryo development in both *B. pahangi* and *D. immitis* (Bandi et al. 1998). The elimination of *Wolbachia* bacteria and subsequent loss of fertility and growth retardation was also observed in the rodent parasite *L. sigmodontis* when antibiotic therapy was initiated concomitant with infection. If treatment was initiated subsequent to microfilaria development, reduced filarial fertility was observed. The use of antibiotics known not to affect Rickettsiae, penicillin G and gentamicin, as well as the marginally effective ciprofloxacin, did not inhibit filarial activity. Moreover, treatment of the *Wolbachia*-uninfected *A. viteae* for 70 days did not have a negative effect on its development or fertility (Hoerauf et al. 1999). Several other antibiotics, erythromycin, chloramphenicol, and ciprofloxacin, also failed to adversely affect the development and fertility of *Wolbachia*-infected *L. sigmodontis* parasitizing BALB/c mice (Hoerauf et al. 2000b).

Langworthy et al. (2000) used *O. ochengi*, a close relative of *O. volvulus* and a parasite of ungulates transmitted by the same vector, *S. damnosum*, to test whether antibiotic treatment could affect the viability of filarial infections in cattle. A series of filarial-induced nodules were tested to monitor the effects of oxytetracycline. *Wolbachia* bacteria were observed in the hypodermis of the filarial worm prior to treatment. After 1 month of oxytetracycline treatment, the bacteria showed signs of degeneration, were irregular in shape, and reduced in size. After 2 months of treatment, the hypodermis was free of *Wolbachia*, and the hypodermic cell nuclei as well as embryonic stages had degenerated. Eosinophils were found within the nodules as well as around the worms. At 3 months of treatment, following the elimination of *Wolbachia*, there was a decline in the number of embryonic stages, which after 6 months of treatment were no longer visible. After 12 months of treatment, microfilariae were no longer detectable, and the adult worms were destroyed. Oxytetracycline showed macrofilaricidal properties on *O. ochengi*, subsequent to the elimination of *Wolbachia* bacteria. The demise of the worms followed the elimination of *Wolbachia*.

Results from an in vitro test to asses the effects of tetracycline on the L3 to L4 molting of *B. malayi*, *B. pahangi*, and *D. immitis* concurred with previous experiments that showed inhibition of larval development. However, the use of three other antibiotics known to be effective against rickettsia and chlamydia bacteria in humans did not inhibit L3 to L4 larval development. It is possible that the dose of the latter antibiotics were not sufficiently high or not specific for *Wolbachia*. In addition, the duration of treatment was only 10 days, which might not have been sufficiently long. However, thus far, it cannot be excluded that tetracycline could be exerting a detrimental effect directly on filarial physiology as well as having an effect on *Wolbachia* (Smith & Rajan 2000).

Another venue for exploring the nature of the filaria-*Wolbachia* relationship is to study the role that *Wolbachia* may play in the disease state triggered by filarial infection. Despite its antiquity, the actual course of infection of filarial disease is poorly understood. In humans there is a plasticity associated with filarial infection that runs the gamut from lymphatic damage and obstruction leading to elephantiasis and hydrocoele to seemingly asymptomatic infective states despite the presence of microfilariae levels of 20,000 or more per ml (Ottesen 1992). Severe inflammatory responses are often observed subsequent to diethylcarbamazine or ivermectin antifilarial treatment. This inflammatory reaction is due to the rapid death and clearance of microfilariae (Turner et al. 1994). Identification of the antigenic molecules responsible for triggering the host's immune response during infection and subsequent to antifilarial treatment would aid in the understanding of the pathology associated with filarial infection. Several recent studies have explored the question of whether *Wolbachia* antigens are involved in triggering the host inflammatory response. IgG antibodies to the *Wolbachia wsp* protein were detected in the serum of *D. immitis*–seropositive cats, indicating that *Wolbachia* proteins gain access to the circulatory system of the vertebrate host, have antigenic properties, and are capable of inducing an immune response (Bazzocchi et al. 2000a). Lipopolysaccharide-like molecules derived from the *Wolbachia* endosymbiont of *O. volvulus* were shown to be possible modulators of peripheral blood monocyte activity. Presence of lipopolysaccharides was detected in *O. volvulus* but not in *Wolbachia*-asymbiotic *A. viteae* (Brattig et al. 2000).

There is also evidence that *Wolbachia* bacteria in filarial worms are the source of a catalase enzyme involved in the detoxification of hydrogen peroxide. Catalase was detected in the hypodermis of male and female *O. volvulus, O. ochengi, O. fasciata*, and *O. gibsoni* (Henkle-Duhrsen 1998). An in vitro comparison of susceptibility to hydrogen peroxide between *B. malayi* and *O. lienalis* microfilariae showed that the latter were highly susceptible to hydrogen peroxide–induced toxicity, whereas *B. malayi* had fewer detrimental effects (Taylor et al. 1996). However, both *O. lienalis* and *B. malayi* are infected with *Wolbachia* (Townson et al. 2000). These results are in contrast with some arthropod-*Wolbachia* infections. *Drosophila simulans* and *Aedes albopictus* have been examined for the expression of inducible antimicrobial markers. The results indicate that *Wolbachia* does not induce or suppress the production of antibacterial peptides diptericin and cecropin in *D. simulans* or induce defensin RNA production in *Ae. albopictus* (Bourtzis et al. 2000).

Several studies are underway to test the efficacy of antibiotic treatment against several filarial diseases. Preliminary information from a pilot study of humans infected with *O. volvulus* and treated with doxycycline resulted in the elimination of *Wolbachia* in adult worms and an embryotoxic effect on adult females after 6 weeks of therapy (Taylor et al. 2000b). As these studies are published, we will learn of the effectiveness of *Wolbachia*-targeted antibiotic treatment against filarial disease. However, caution should be exercised before embarking on widespread, long-term antibiotic treatments that could incur resistance in common infectious agents such as *Helicobacter pylori* as well as *Wolbachia*. Evidence of *Wolbachia* resistance to antibiotics has been reported in *Drosophila auraria* (Bourtzis et al. 1996) and *Tribolium confusum* (O'Neill 1989). Antibiotic resistance has also been reported in other rickettsial bacteria: *Orientia tsutsugamushi* has been shown to be chloramphenicol and doxycycline resistant in northern Thailand (Watt et al. 1996), and laboratory strains of *Rickettsia typhi* and *R. prowazekii* have been shown to be rifampicin resistant (Troyer et al. 1998, Drancourt & Raoult 1999).

MALE-KILLING (MK)

Although CI-inducing *Wolbachia* have been known for a long time, discovery of the phenomenon of sex-ratio bias through embryonic male-killing is rather recent. *Wolbachia* have been demonstrated to be lethal to male embryos of flour beetles, ladybird beetles, Nymphalid butterflies, and a *Drosophila* (Fialho & Stevens 2000; Hurst et al. 1999a, 2000). MK is not unique to *Wolbachia*. Spiroplasma, Flavobacteria, and other bacteria in the genus Rickettsia also kill males (Jiggins et al. 2000a). The observed sex ratios for species harboring MK *Wolbachia* are shown in Table 2. As yet only five host species are known to be affected by MK (Fialho & Stevens 2000; Hurst et al. 1999a, 2000; Jiggins et al. 2000b); however, that number may be low for several reasons. A review of *Wolbachia* literature (e.g.,

TABLE 2   Incidence and paramater estimates related to male-killing *Wolbachia* infection

Host species	Sex ratio (% female)	Sibling effect[a]	Infection frequency (%)	Transmission efficiency (%)	Direct cost	*Wolbachia* subgroup	Heterogametic sex
Adelia[b]	87–99	Ca, Co	5–7	86–99	Unknown	B	Male
A. encedon[c]	100	Ca, Co	78–100	100	Unknown	B	Female
A. encedena[d]	100	Ca, Co	95	50 or 100	Unknown	B	Female
D. bifasciata[e]	100	Co, I	6	99	0–10%	A	Male
T. madens[f]	91	Ca	Unknown	86	None	B	Male

[a]Ca, cannibalism; Co, competition; I, inbreeding.

[b]Hurst et al. 1999a, Majerus et al. 2000.

[c]Hurst et al. 1999a, Jiggins et al. 2000b.

[d]Jiggins et al. 2000a. Some females transmit *Wolbachia* to all their progeny, others to only 50%.

[e]Hurst et al. 2000; except cost, measured as reduction in lifetime fecundity (Ikeda 1970).

[f]Fialho & Stevens 2000; except cost, measured as number of eggs laid (L Stevens unpublished data).

Werren 1997) reveals that prior to the development of PCR screening techniques in 1992 (e.g., O'Neill et al. 1992), detection of *Wolbachia* was restricted to lab studies involving crosses between infected and uninfected lineages (e.g., Wade & Stevens 1985). Discoveries of endosymbiotic sex ratio distorters are inclined to similar bias (Ebbert 1993); most reported cases of heritable sex ratio distortion involve insects and mites routinely subjected to intensive laboratory genetic analysis.

Etiology

Females of the black flour beetle, *Tribolium madens*, infected with *Wolbachia* have egg-hatch rates approximately half that of uninfected females and produce progeny with highly female-biased sex ratios (Fialho & Stevens 2000). Repeated back-crossing of infected females with males of a naturally uninfected strain demonstrated the cytoplasmic nature of the bias. Antibiotic treatment of infected hosts eliminated *Wolbachia* infection, reverting the sex ratio to unbiased levels and increasing percentage egg hatch. Typically, *Wolbachia* infections are vertically transmitted from mother to progeny, regardless of the sex of the progeny, but infected *T. madens* males are never detected. Because females kept as virgins and not allowed to mate are sterile, the sex-ratio distortion results from embryonic MK rather than the previously reported phenomenon of parthenogenesis induction (PI). DNA sequence data (ca. 2500 bp including the *fts*Z cell division gene, the *wsp* gene, the *groE* intergenic region, and an ITS region flanked by the 23S and 5S rDNA genes) showed that the *Wolbachia* MK-inducing strain of *T. madens* is indistinguishable from the cytoplasmic incompatibility (CI)–inducing *Wolbachia* in the closely related *T. confusum*. Studies with non-MK *Wolbachia* demonstrated that the particular host-*Wolbachia* interaction plays an important role in the induction of reproductive phenotypes. Because the DNA sequences in the *Wolbachia* infecting these two closely related *Tribolium* species are identical, the MK effect appears to follow the pattern of being specific to the particular host-symbiont interaction.

Phylogenetic Diversity and Theoretical Models

Five decades after the two-spot ladybird beetle, *Adalia bipunctata*, was shown to harbor cytoplasmic sex-ratio distorters (e.g., Lusis 1947), *Wolbachia* infections were reported to kill males in this host (Hurst et al. 1999a). At least four bacteria cause MK in *A. bipunctata* (Majerus et al. 2000), and the first report of MK *Wolbachia* included this host and a nymphalid butterfly (Hurst et al. 1999a). *Wolbachia* are unique in their ability to induce parthenogenesis and CI, but at least six groups of cytoplasmic bacteria, belonging to four major bacterial groups, induce biased sex ratios by killing males (see review in Stouthamer et al. 1999). Based on this taxonomic diversity, MK is hypothesized to evolve more readily or more bacterial groups are capable of evolving the MK phenotype (Hurst et al. 1997).

Empirical studies of MK *Wolbachia* are often integrated with theoretical models (e.g., Hurst et al. 2000, Jiggins et al. 2000a, Randerson et al. 2000b). Coexistence of multiple strains or species of MK bacteria or the incidence of infection with a single strain have been examined using theoretical models. Although infections with multiple MK species or strains of bacteria are predicted to be unstable as a consequence of competitive exclusion (Hurst 1991, 1993, Hurst et al. 1999b, Randerson et al. 2000b), a single population of *A. bipunctata* from Moscow was shown to harbor as many as four male-killers. Other populations of *A. bipunctata* also seem to defy the competitive exclusion principle by harboring multiple male-killers (e.g., a population in Berlin carries Rickettsia and Spiroplasma) (Hurst et al. 1999b). The coexistence of multiple strains or species of MK bacteria within a population suggests that equilibrium dynamics have not been achieved (Hurst et al. 1999a). The populations with multiple infections may represent the following: hybrid zones between regions of pure infection, substructure resulting in the existence of a single MK bacteria in each subpopulation, or recent secondary invasion. Alternatively, bacterial populations may be limited by host-suppressor genes or genes governing transmission efficiency rather than competition. Investigation of these possibilities has not resolved the difference between the empirical observations and theoretical predictions (Majerus et al. 2000). Work integrating mathematical models with carefully designed experiments distinguishing between the alternatives is needed to understand the polymorphism.

Wolbachia MK involves vertically transmitted cytoplasmic factors; the theoretical models consider killing of male embryos to be a form of kin selection. Kin-directed benefit is likely when sibling competition or cannibalism occur or if mating between siblings leads to inbreeding depression (Hurst et al. 1997). To date, all three types of sibling effects have been invoked to explain the observed incidences of MK *Wolbachia* (Table 2). Some groups harboring male-killers such as the ladybird beetles and the nymphalid *Acraea encendon* and *A. encedena* lay eggs in clutches, and sibling competition and cannibalism are likely (Hurst et al. 1997, Jiggins et al. 2000a). *Tribolium* are notoriously cannibalistic (e.g., Stevens & Mertz 1985; Stevens 1989a, 1994), and some genetic strains obtain important nutrients from cannibalism (Giray et al. 2001). However, MK is not restricted to cannibalistic taxa. *Drosophila bifasciata* are not cannibalistic. These drosophilids lay eggs in resource-poor sap fluxes, and siblings may benefit from MK through reduced competition. Alternatively, inbreeding depression from mating between siblings may be important for these flies (Hurst et al. 2000).

Models predicting the equilibrium frequency of MK infection also consider transmission efficiency (percent infected progeny from infected mothers) and the direct physiological impact of the bacterium on the host, in addition to indirect effects (kin-directed benefit or sibling effect). Models examining the low incidence of MK in *D. bifasciata* are the most detailed and include the possibility that not all infected males die and that crosses between infected males and uninfected females produce some progeny (Hurst et al. 2000). The models are able to rather accurately predict the incidence of infection, suggesting either that they are robust or that their parameterization captures the important dynamics of the system. Perfect transmission of MK *Wolbachia* is predicted to lead to fixation. Paradoxically, such populations would then become extinct because of the absence of males (Hurst 1991, Randerson et al. 2000b).

Paradox of the Lek

Many parasites are known to manipulate host behavior to their own advantage. CI *Wolbachia* in *T. confusum* enhance their rate of spread through a population by causing changes in host behavior (Stevens 1993), but there is no evidence that these *Wolbachia* enhance male fecundity (but see Wade & Chang 1995). MK *Wolbachia* may also be responsible for changes in host reproductive behavior. They have been associated with population-level variation in female lekking swarms and male choice behavior in butterflies (Jiggins et al. 2000b, Randerson et al. 2000a).

Males produce smaller gametes and usually mate more often than females, whereas females invest more resources per gamete so that resource availability often limits fitness. Males therefore often compete for females, and females choose between males. In insects male lekking swarms are widespread, whereas female lekking is only known in three insects. One case is the dance fly, which exhibits high male parental investment. The other cases are populations of the butterflies *A. encedon* and *A. encedana*, which have MK *Wolbachia*–induced female-biased sex ratios (Jiggins et al. 2000b).

A. encedon has a population-level polymorphism in MK *Wolbachia* infection, sex ratio, and mating behavior. The sex-role-reversed mating system in which males choose females and females compete for males varies with sex ratio (Jiggins et al. 2000b, Randerson et al. 2000a). *A. encedon* populations with high numbers of males (34–60%) had a lower prevalence of MK *Wolbachia* (Jiggins et al. 2000b). These populations located themselves in sites where the larval food plant was common, the butterflies were present from dawn until evening, and mating pairs were common. In contrast, populations with low numbers of males (0–12%) were associated with higher MK prevalance, the butterflies chose sites with landmarks including short vegetation near trees but without food plants, and they aggregated only in the afternoon when the few males observed were usually mating. In the female-biased populations, females exhibited mate-acceptance behavior presumably to solicit mating. Females also showed mating behavior toward other females and mated immediately with males. If sperm is limited, theory suggests males may benefit from choosing uninfected females. If infected females lek and perform these behaviors to entice mating with uninfected males, male behavior may partially counteract the female lek. Males have higher fitness when they mate with uninfected females. Males may discriminate between infected and uninfected females: Uninfected females were more likely to be mated than uninfected females.

At one location the incidence of MK *Wolbachia* varied over time. Bacterial presence initially was high, and females aggregated at a site where the food plant was absent. When bacterial prevalence diminished, the number of males increased and the lekking site was abandoned as the population moved to a site where the larval food plant was common. When bacterial incidence increased again, the females returned to the lekking site and were scarce at the food plant site (Jiggins et al. 2000b).

The sex-role-reversed lekking behavior associated with *Wolbachia* has been used to investigate the "paradox of the lek"; that is, why do leks persist? Population genetics theory predicts that directional selection resulting from female choice will result in fixation of the selected trait. The loss of genetic variation resulting from fixation will reduce the benefits of female choice. Parameters in the mathematical model of the MK *Wolbachia*–female lek association include female variation in infection and male choice. These two traits can be maintained when males make mistakes when discriminating among females. Other models of MK bacteria noted that when the MK agent is perfectly transmitted (see parameter estimates in Table 2), it should become fixed and result in population extinction owing to the absence of males (Hurst 1991, Randerson et al. 2000b). The lek model prevents the fixation of a perfectly transmitted MK *Wolbachia* and explains not only the evolution of male choice for uninfected females but also the persistence of populations with perfect transmission and choosy males.

OLD ISSUES, NEW DATA

Rapid progress has been made over the past 2 years towards understanding various aspects of *Wolbachia* biology. Below, we review some of these findings with special relevance to the mechanism of *Wolbachia* action, phylogenies, and virulence.

Mechanism of *Wolbachia* Action for Parthenogenesis Induction, Cytoplasmic Incompatibility, Feminization, and Male-Killing

The mechanism by which *Wolbachia* causes CI and PI involves interference with mitosis. PI involves gamete duplication; the two sets of chromosomes fail to segregate in the first mitotic anaphase (Stouthamer & Kazmer 1994). CI is associated with abnormal syngamy and other events during early embryo development (summarized in Stouthamer et al. 1999). CI can also be caused by paternal chromosome destruction causing the embryos to effectively become haploid despite fertilization. In diploid organisms irregular development results in death; in organisms with haplo/diploid mechanisms of sex determination, CI results in all-male broods. Variations on this theme include incomplete maternal transmission (production of uninfected eggs from infected females) and partial incompatibility (some uninfected eggs fertilized by infected sperm complete development).

CI in *D. simulans* seems to be affected by repeated mating (Karr et al. 1998). Virgin males have high rates of incompatibility that diminish with repeated mating. *Wolbachia* are not present in developed sperm but likely imprint sperm sometime during spermatogenesis (c.f. Stouthamer et al. 1999, Snook et al. 2000). It has been suggested that the decrease in CI with mating is a result of decreased sperm exposure to *Wolbachia* during development (Karr et al. 1998). Heat shock is known to decrease CI expression (Feder et al. 1999). Sperm production is increased after heat shock to a greater extent in infected males than uninfected males (Snook et al. 2000). Thus, it is possible that heat shock reduces CI by increasing sperm production. Increased sperm production may decrease sperm development time and *Wolbachia* exposure, thus decreasing *Wolbachia's* affect on sperm.

In isopod crustaceans *Wolbachia* cause feminization, turning genetic males into females. Sex determination in this group is through hormones produced by the androgenic gland that induce male differentiation. *Wolbachia* prevent formation of the androgenic gland and ensure female development. *Wolbachia* injected into males with functional androgenic glands cause them to become feminized or intersex (differentiated female genital aperture and other female sexual characteristics). The functional androgenic gland becomes hypertropic, producing large amounts of androgenic hormone. Phenotypic sex determination by an androgenic protein hormone produced in a special gland is unique to malacostran crustaceans. The hormone works in two ways: It induces male gonad differentiation, and it controls secondary sexual traits. *Wolbachia* potentially interact with both mechanisms (Martin et al. 1999). This result is based on studies with *Armadillidium vulgare* and is important because it suggests that *Wolbachia* interact with the regulatory regions of the gene coding for the androgenic hormone. Several isopod genera are infected with *Wolbachia*. Studies with the oniscidean *Oniscus asellus* demonstrate that males in this species can be infected; thus, feminization appears to have evolved in different ways in different taxa (Rigaud et al. 1999).

Little is known about the mechanism of MK. Hatch rates of eggs from infected females typically are about half that of uninfected females (e.g., Fialho & Stevens 2000), and sex ratios are either exclusively female or highly female biased (Table 2). In *A. encedna*, unhatched eggs from all-female broods contained fully developed embryos with clearly visible tanned head capsules (Jiggins et al. 2000a). In normal sex ratio broods, death during late embryogenesis was rarely recorded. Experiments with *D. bifasciata* indicate that MK appears to require a threshold density of *Wolbachia* within the eggs (Hurst et al. 2000). Heat treatment lowers bacterial density, and some males are produced from heat treated females. Sons of heat treated females are infected, and crosses between these males and uninfected females indicate that bacteria causing MK in females can induce CI. DNA sequence analysis (discussed below) demonstrates that the mechanisms of MK, CI, and PI are labile with respect to the bacterial phylogeny.

Fruit flies in the genus *Drosophila* play a central role as a model organism for biological investigation. CI and PI involve manipulation of host chromosomes. Investigation of MK *Wolbachia* in *Drosophila* can use the tools of a well-developed model system to examine whether MK, PI, and CI represent an evolutionary plastic repertoire of bacterial manipulation of host chromosomes. Alternatively, host genes may suppress or alter some aspects of the bacterial effect on host chromosomes, thereby producing varied reproductive alterations (Majerus et al. 2000, Randerson et al. 2000b).

Phylogenies

Wolbachia phylogenetic analysis reveals five major groups. The phylogenetic analysis of *Wolbachia* includes a number of genes: 16S rDNA, *ftsZ* (cell division gene), *groEl* (bacterial heat shock protein), and the *wsp* gene (cell surface protein). Previous reviews divide the genus into four major groups: Two groups (A and B) include all the known arthropod infections and diverged perhaps 60 mya (Stouthamer et al. 1999). Groups C and D contain the filarial nematode infections and likely diverged from the arthropod infections 100 mya (see references in Stouthamer et al. 1999). Finding *Wolbachia* in the basal insect order Collembola resulted in the identification of a fifth group, E, which appears basal to groups A and B (Vandekerckhove et al. 1999). Members of the B group display more differentiation than those in the A group. The reduced variation in the latter may be indicative of higher rate of horizontal transfer between members of the A group (e.g., Vavre et al. 1999a).

Although some groups show a high degree of vertical transmission (e.g., Giordano et al. 1997), horizontal transmission seems to be common on a large scale (e.g., O'Neill et al. 1992). The precise means by which horizontal transmission is achieved are not known; however, it is likely that parasites play a significant role. Twenty-five percent of all insect species are parasitoids. Johanowicz & Hoy (1996) found that a parasitic mite and its host harbor very similar *Wolbachia*. Further, in lab experiments, the parasitic wasp *Leptopilina boulardi* acquired *Wolbachia* from infected hosts, *Drosophila simulans* (Heath et al. 1999). These results are supported by phylogenetic data based on the *wsp* gene, suggesting horizontal transmission between five species of frugivorous *Drosophila* and five hymenopteran parasitoids in southeast France (Vavre et al. 1999a). At least 30 individuals per species were examined. Only two of the five *Drosophila* species were infected, *D. simulans* and *D. melanogaster*, and each was infected by a single strain of A group *Wolbachia*. However, four of the five parasitoids were found to be infected. Two parasitoid species were infected with three strains of group A *Wolbachia*, one with two strains belonging to A and B, and one with a single strain of *Wolbachia*, belonging to the

A clade. The similarities between host and parasitoid *Wolbachia* are striking. The variant infecting *D. melanogaster* was most closely related to the variant in its parasitoid, *A. tabida*; the variant infecting *D. simulans* was most closely related to the variant in one of its parasitoids. However, the hosts were infected with only two *Wolbachia* variants and the parasitoids with nine, indicating that not all parasitoid infections originated from these hosts. It may be that as yet untested hosts or hyperparasitism are involved in these relationships.

There is considerable plasticity in the phenotypic reproductive alteration caused by *Wolbachia* (e.g., Rousset et al. 1992). DNA sequence data is available for all of the MK *Wolbachia* (Hurst et al. 1999a, 2000; Fialho & Stevens 2000; Majerus et al. 2000). Both A and B groups contain PI, CI, and MK *Wolbachia* (e.g., Hurst et al. 1999a, 2000). Both *A. bipunctata* and *A. encedna* are infected with B group *Wolbachia*; however, even though the *A. encedna Wolbachia* contains a nine–base pair insertion-deletion rare in the B group, statistical analysis could not distinguish between a single and multiple origins of the MK trait (Hurst et al. 1999a). The discovery of multiple species of MK bacteria in a single population of *A. bipunctata* was accompanied by the report of two different strains of MK *Wolbachia* within that population, both in the B group (Majerus et al. 2000). *T. madens* also harbors a B group *Wolbachia*. In addition to being phylogenetically plastic, MK is independent of the genetic basis of sex determination (Table 2). MK is found in both male and female heterogametic species. An analysis of the origin of MK using the *ftsZ* gene supported multiple origins of this trait (Schulenburg et al. 2000). Sequence analysis using the *wsp* gene also supported multiple origins of the trait, but the MK *Wolbachia* appeared to be from different lineages within the same clade (Schulenburg et al. 2000).

Virulence Theories and Variation in *Wolbachia*-Host Interactions

Some little-studied components of infection dynamics relevant to *Wolbachia*-arthropod phylogenies are the exposure of arthropods to various *Wolbachia* strains and, once exposed, the arthropod/parasite factors that influence *Wolbachia*'s ability to become established. There is a statistically significant ($P < 0.003$) association between the incidence of infection and ecological niche of the arthropod: *Drosophila* hosts are significantly less likely to be infected with *Wolbachia* than their parasitoids (Vavre et al. 1999a). Much is still unknown about the primary parameters influencing infection of arthropods with *Wolbachia* including exposure to infection, susceptibility once exposed, and the stability of the infection once established. Stability of infection can be affected by many factors including maternal transmission rates; environmental curing (Stevens 1989b, Stevens & Wicklow 1992); effects of *Wolbachia* on the arthropod including effects on reproduction such as CI, PI, and MK; and effects of the arthropod on *Wolbachia*. Both *Wolbachia* and arthropod genetics may contribute to understanding how these parameters affect the dynamics of *Wolbachia* infection.

Two considerations are of particular importance in evaluating *Wolbachia* virulence. The first is the mode of transmission; the second is the effect on reproduction. Transmission rates are typically high: Infected mothers usually (but not always) produce infected progeny. *Wolbachia* does not appear in progeny from uninfected mothers. Several hypotheses address the evolution of parasite virulence (Schall 1990, 2001):

1. "Group selection" should favor parasite strains with less of an effect on their native host than on a host to which they are newly introduced (Telford 1971).

2. Parasites have an intrinsic advantage in a coevolutionary arms race (Ewald 1983). Virulence should benefit the parasite and depends on the ecology, and not the age, of the interaction.

3. Parasite virulence and reproductive output are positively correlated; parasites should evolve towards increasing harm to the host (Gill & Mock 1985).

4. Host-parasite coevolution produces an equilibrium of costs and benefits (Price 1980). The virulence of old interactions should be intermediate or low relative to new ones.

5. Virulence is a by-product of parasite metabolism not necessarily evolved to affect host biology (Groisman & Ochman 1994).

Numerous virulence models consider within- and between-group selection by allowing multiple genotypes within a host organism. Other authors focus on the role of transmission ecology in the evolution of virulence (e.g., Ewald 1994). However, disease-causing phenotypes among human pathogens suggest that virulence is an emergent property of the strains involved in a particular interaction (Groisman & Ochman 1994). For example, strains of *E. coli* may be benign, toxic, or beneficial, and strains of plasmodium differ in their pathogenecity. Several reviews consider the variation in virulence between *Wolbachia* and its arthropod hosts (e.g., O'Neill et al. 1997, Werren 1997, Stouthammer 1999).

Models of CI *Wolbachia* predict that in the absence of environmental cures and when there is complete unidirectional incompatiblity and complete maternal transmission populations should be fixed for the presence or absence of *Wolbachia*. If introduced into a population at a frequency greater than $1/\beta$, where β is the fitness ratio of infected to uninfected hosts, *Wolbachia* is expected to spread to fixation (Caspari & Watson 1959, Fine 1978, Stevens & Wade 1990, Wade & Stevens 1994). When two or more *Wolbachia* types infect the same host species, bidirectional incompatibility can occur. Variation in the effects of *Wolbachia* associated with unidirectional CI include variation in the strength of unidirectional incompatibility, various effects on sperm usage, and different *Wolbachia* effects on host components of fitness. For example, *T. confusum* is infected with a single strain of *Wolbachia* (Fialho & Stevens 1997). Males infected with *Wolbachia* are completely incompatible with uninfected females, but all other crosses are fertile (Fialho & Stevens 1996). The only known mode of transmission is maternal, and all progeny from infected mothers are infected.

D. simulans shows variable expression of CI (James & Ballard 2000). Examination of the microdynamics of the *Wolbachia–D. simulans* interactions shows that at least four distinct *Wolbachia* strains (wHa, wRi, wMa, and wAu) infect *D. simulans*. Some of the strains cause bi-directional incompatibility (wHa, wRi), whereas others have no effect on incompatibility (wAu). The maternally inherited *Wolbachia* and mitochondrial variants [siI, siII (with subtypes A and B), and siIII] are highly congruent. There is no evidence of horizontal transfer between lines; however, some lines appeared to have lost the infection (James & Ballard 2000). Further studies have the potential to uncover important information about the evolution of *Wolbachia*-host interactions.

Theory suggests double infections should increase in frequency in a population of single-infected individuals (Rousset & Solignac 1995, Sinkins et al. 1995). Multiple infections with CI *Wolbachia* appear to be common in parasitoids of frugivorous *Drosophila*. A survey of five parasitoids found that one was uninfected, two harbored triple infections, one had a double infection, and only one parasitoid harbored a single infection (Vavre et al. 1999a). For both triply infected hosts the infection was stable during several generations of laboratory maintenance on a *Wolbachia*-free strain of *D. melanogaster*, and 10 of 10 populations of the parasitoid *L. heterotoma* are triply infected (Vavre et al. 1999a). It is unknown if all three variants induce CI, but if they do and if they are reciprocally incompatible, multiply infected individuals will be favored. Alternatively, a non-CI *Wolbachia* could spread through hitchhiking with a CI variant. Multiple infections are common in parasitoids in which horizontal transfer has been empirically documented and phylogenetically substantiated. Multiple infections likely result from adding a new variant to previously infected cytoplasm through horizontal transfer. Thus, the finding of multiple infections suggests frequent natural transfers of *Wolbachia*.

Wolbachia results often do not apply across all strains of *Wolbachia* or all host taxa. Not all *Wolbachia* double-infected lines of *D. simulans* are stable. In this species the *Wolbachia* variants segregate, and one type was more successful and was more prevalent in single-infected lines (Poinsot et al. 2000). Host nuclear genes also affect infection dynamics, again stressing the importance of individual host-*Wolbachia* dynamics.

Wolbachia density is often associated with its ability to induce CI (see references in Stouthamer et al. 1999), but this is not always the case (Giordano et al. 1995). The dynamics of the interactions among multiple strains of *Wolbachia* as well as experimental manipulations of the number of strains within a single host has been examined in *D. simulans* by mincroinjecting a third *Wolbachia* strain into a double-infected fly strain (Rousset et al. 1999). The triple infection was stable in all lines through two generations of lab culture, with an estimated transmission efficiency of >97.3%. For triple infections to spread, they must be incompatible with the ancestral double infections. Males with triple infections are incompatible with their double-infected ancestral females (mortality 72.7–99.9%). Incompatible crosses with double-infected lines have lower egg hatch than triple-infected lines, suggesting that mortality rate is not a simple correlate of the number of *Wolbachia* strains (Rousset et al. 1999).

BACTERIOPHAGE WO

Bacteriophage WO has been found in all *Wolbachia*-infected insects tested thus far, suggesting an ancient origin (Masui et al. 2000). Information on the genetic makeup of the phage has exciting potential for discovering the mechanism of *Wolbachia* action and understanding the diversity of *Wolbachia*-host interactions. Use of the phage to manipulate *Wolbachia* may be the key to using *Wolbachia* to control medically and agriculturally important arthropod pests.

Some virulence properties of bacteria are determined by bacteriophages, and recently a bacteriophage-like genetic element, named bacteriophage WO, has been identified in *Wolbachia* (Masui et al. 1999, 2000). The G + C content of the phage and *Wolbachia* chromosomal genes are similar, as are the codon usage of the genes, suggesting that the phage has been associated with *Wolbachia* for a very long time. All seven insects tested to date contain the bacteriophage, and analysis of *Wolbachia*-phage phylogenies suggests that the phage mediates horizontal transfer of *Wolbachia* genes between different insect hosts (Masui et al. 2000). The *wsp* (*Wolbachia* surface protein) and *orf7* (a predicted coding region for phage WO) genes were PCR amplified from seven *Wolbachia* strains. Sixteen distinct *orf7* sequences were discovered. Three of the *Wolbachia* strains had a single *orf7* sequence, one had two sequences, two had three, and one had five *orf7* sequences. Two of the *Wolbachia* were from the same host, *Ephestia cautella*. One of these was an A group *Wolbachia* containing five *orf7* sequences; the other belonged to the B group and had three *orf7* sequences. The *Wolbachia* sequences fell into two clades, the previously discussed A and B groups. Three of the four groups of sister taxa on the phage WO phylogeny included sequences from both the A and B groups. Furthermore, two of these groups were derived from *Wolbachia* infecting the moth *E. cautella*. This suggests that the phage of *Wolbachia* horizontally transmits DNA among different bacterial strains.

The structure of the phage WO genome was determined by sequencing the complete 33 kb from a DNA library constructed from *Wolbachia*-infecting *E. kuehniella*. There was only one *orf7* sequence from this host. Thirty-three open reading frames, ORFs (designated gp1–gp33) were detected from the sequence analysis, and the region from gp12–gp31 is likely to be part of the phage genome but may extend between gp5 and gp31. The predicted proteins from all 33 ORFs were determined from a homology search. The genes appear to have diverse origins. Viral structural and replication proteins were similar to those in several groups of viruses. In addition, four proteins containing ankyrin-like repeats were observed. Ankyrin-like repeats occur in all kingdoms and have a variety of functions. However, the repeats in phage WO share more similarities with mammals and plants than those of bacteria.

The occurrence of bacteriophage WO in *Wolbachia* has two important implications: First, the phage contains ankyrin-like repeats similar to mammals and plants. In *Drosophila*, ankyrin-repeat proteins are important in the cell cycle in early development (Axton et al. 1994, Elfring et al. 1997). The phage may provide answers to the mechanism of *Wolbachia's* phenotypic effects. Second, bacteriophage WO travels extensively between insect hosts, which may explain the evolutionary plasticity of reproductive alterations and virulence of *Wolbachia*.

CONCLUSIONS

Considering *Wolbachia's* unique effects on reproduction, investigators have considered conditions that favor the evolution of incompatibility (e.g., Frank 1997). This approach considers that compatibility is the null state from which incompatibility evolves. Others suggest that initial host effects are by-products; that is, they can be nonexistent, weak, or strong, or perhaps microorganisms could initially benefit their hosts (Groisman & Ochman 1994). When applied to *Wolbachia*, what phenotypic effects will be favored by natural selection? *Wolbachia* research demonstrates surprising variability in *Wolbachia*-host effects (Rigaud 1999). The genetic strain of bacteria and host, as well as the environment, contribute to the dynamics of the interaction.

Five major groups of *Wolbachia* are known. It has been a challenge to identify and understand this intracellular microorganism. Historically, phenotypic observations of *Wolbachia's* effects have long preceded mechanistic understanding. Two decades lapsed between the first report of nematode infection with intracellular bacteria and identification of the bacteria as *Wolbachia* (Harada et al. 1970, Sironi et al. 1995). Over half a century passed between the identification of MK *Wolbachia* and reports of sex ratio distortion in *A. bipunctata* (Lusis 1947, Hurst et al. 1999a). The first investigations into the mechanism of CI occurred over 30 years ago (Jost 1970, Ryan & Saul 1968), yet the mode of action remains elusive.

The challenge for *Wolbachia* researchers is to develop a detailed understanding of the mechanism(s) *Wolbachia* uses to influence host reproduction and continue to investigate the diversity of ways *Wolbachia* affects natural populations. The *Wolbachia* genome-sequencing project (Slatko et al. 1999) will contribute to these efforts. *Wolbachia* strains from the A, B, and D groups were chosen with the goal of developing strategies against their respective hosts, as well as determining the extent of genetic similarities among strains. The resolution of the causes of reproductive alterations and the variation of *Wolbachia* virulence will provide information for basic biology and may help to control agricultural and medical arthropod pests and reduce the incidence of pathogenic nematode infections in humans. The infected nematodes should be examined for phage WO. With the discovery of bacteriophage WO and the complete genome sequences of several *Wolbachia* strains, the mechanism underlying many of *Wolbachia's* effects may soon be determined.

ACKNOWLEDGMENTS

This paper is dedicated to Roberto F. Fialho, his parents Luzia and Juarez Fialho, his sister Monica, and Michael Upton. Roberto died from complications resulting from Hodgkins' lymphoma before this review was completed.

We thank Amy Prenowitz for indispensable help putting this article together. We acknowledge that our ideas have developed through discussions with or comments from Angelia Viley, Tugrul Giray, Joseph J. Schall, Thomas Tucker, Felipe Soto-Adames, Peter Armbruster, and Jen Tremblay. Organizational support was provided by Donna Parrish, Lisa Stevens-Goodnight, and Sylvia Stevens-Goodnight. Financial support was provided by the University of Vermont and NSF grant IBN-9724037 to LS, USDA-NRI 9802683 to RG, and ICPE Brazil to RFF.

Visit the Annual Reviews home page at www.AnnualReviews.org

LITERATURE CITED

Anderson RC, Chaubaud AG, Willmott S, eds. 1974. *CIH Keys to the Nematode Parasites of Vertebrates*. Slough, UK: Commonw. Agric. Bur.

Awadzi K. 1997. Research notes from the Onchocerciasis Chemotheraphy Research Centre, Ghana. *Ann. Trop. Med. Parasitol.* 91:703–11

Axton JM, Shamanski FL, Young LM, Henderson DS, Boyd JB, Orr-Weaver TL. 1994. The inhibitor of DNA replication encoded by the *Drosophila* gene plutonium is a small, ankyrin repeat protein. *EMBO J.* 13:462–70

Bandi C, Anderson TJ, Genchi C, Blaxter ML. 1998. Phylogeny of *Wolbachia* in filarial nematodes. *Proc. R. Soc. London Ser. B* 265:2407–13

Bazzocchi C, Ceciliani F, McCall J, Ricci I, Genchi C, Bandi C. 2000a. Antigenic role of the endosymbionts of filarial nematodes: IgG response against the *Wolbachia* surface protein in cats infected with *Dirofilaria immitis. Proc. R. Soc. London Ser. B* 22:2511–16

Bazzocchi C, Jamnongluk W, O'Neill SL, Anderson TJ, Genchi C, Bandi C. 2000b. *wsp* gene sequences from the *Wolbachia* of filarial nematodes. *Curr. Microbiol.* 41:96–100

Bordenstein SR, O'Hara FP, Werren JH. 2001. *Wolbachia* induced incompatibility precedes other hybrid incompatibilities in *Nasonia. Nature* 409:707–10

Bosshardt SC, McCall JW, Coleman SU, Jones KL, Petit TA, Klei TR. 1993. Prophylatic activity of tetracycline against *Brugia phangi* infection in jirds (*Meriones unguiculatus*). *J. Parasitol.* 79:775–77

Bouchon D, Rigaud T, Juchault P. 1998. Evidence for widespread *Wolbachia* infection in isopod crustaceans: molecular identification and host feminization. *Proc. R. Soc. London Ser. B* 265:1081–90

Bourtzis K, Nirgianaki A, Markakis G, Savakis C. 1996. *Wolbachia* infection and cytoplasmic incompatibility in *Drosophila* species. *Genetics* 144:1063–73

Bourtzis K, Pettigrew M, O'Neill S. 2000. *Wolbachia* neither induces nor suppresses transcripts encoding antimicrobial peptides. *Insect Mol. Biol.* 9:635–39

Brattig N, Rathjens U, Ernst M, Geisinger F, Renz A, Tischendorf FW. 2000. Lipopolysaccharide-like molecules derived from *Wolbachia* endobacteria of the filaria *Onchocerca volvulus* are candidate mediators in the sequence of inflammatory and antiinflammatory responses of human monocytes. *Microbes Infect.* 2:1147–57

Buck AA. 1991. Filarial infections. General principles. Filariasis. In *Hunter's Tropical*

Medicine, ed. GT Strickland, pp. 711–27. Philadelphia: Saunders

Casiraghi M, Anderson TJC, Bandi C, Bazzochi C, Genchi C. 2001. A phylogenetic analysis of filarial nematodes: comparison with the phylogeny of *Wolbachia* endosymbionts. *Parasitology* 122:93–103

Caspari E, Watson GS. 1959. On the evolutionary importance of cytoplasmic sterility in mosquitoes. *Evolution* 13:568–70

Cook JM, Butcher RDJ. 1999. The transmission and effects of *Wolbachia* bacteria in parasitoids. *Res. Popul. Ecol.* 41:15–28

Dadzie KY Remme J, Rolland A, Thylefors B. 1989. Ocular onchocerciasis and intensity of infection in the community. II. West African rainforest foci of the vector *Simulium yahense. Trop. Med. Parasitol.* 40:348–54

Determann A, Mehlhorn H, Ghaffar FA. 1997. Electron microscope observations on *Onchocerca ochengi* and *O. fasciata* (Nematoda: Filarioidea). *Parasitol. Res.* 83:591–603

Drancourt M, Raoult D. 1999. Characterization of mutations in the *rpoB* gene in naturally rifampin-resistant *Rickettsia* species. *Antimicrob. Agents Chemother.* 43:2400–3

Duke BO. 1981. Geographical aspects of onchocerciasis. *Ann. Soc. Belg. Med. Trop.* 61:179–86

Duke BO, Lewis DJ, Moore PJ. 1966. *Onchocerca-Simulium* complexes. I. Transmission of forest and Sudan-savanna strains of *Onchocerca volvulus*, from Cameroon, by *Simulium damnosum* from various West African bioclimatic zones. *Ann. Trop. Med. Parasitol.* 60:318–26

Ebbert MA. 1993. Endosymbiotic sex ratio distorters in insects and mites. In *Evolution and Diversity of Sex Ratio in Insects and Mites*, ed. DL Wrensch, MA Ebbert, pp. 150–91. New York: Chapman & Hall

Elfring LK, Axton JM, Fenger DD, Page AW, Carminati JL, Orr-Weaver TL. 1997. *Drosophila* PLUTONIUM protein is a specialized cell cycle regulator required at the onset of embryogenesis. *Mol. Biol. Cell* 8:583–93

Erttmann KD, Meredith SE, Greene BM, Unnasch TR. 1990. Isolation and characterization of form specific DNA sequences of *Onchocerca volvulus. Acta Leiden* 59:253–60

Erttmann KD, Unnasch TR, Greene BM, Albiez EJ, Boateng J, et al. 1987. A DNA sequence specific for forest form *Onchocerca vovulus. Nature* 327:415–17

Ewald PW. 1983. Host parasite relations, vectors, and the evolution of disease severity. *Annu. Rev. Ecol. Syst.* 14:465–85

Ewald PW. 1994. Evolution of infectious disease. New York: Oxford Univ. Press

Feder M, Karr T, Yang W, Hoekstra JM, James AC. 1999. Interaction of *Drosophila* and its endosymbiont *Wolbachia*: natural heat shock and the overcoming of sexual incompatibility. *Am. Zool.* 39:363–73

Fialho RF, Stevens L. 1996. *Wolbachia* infections in the flour beetle *Tribolium confusum*: evidence for a common incompatibility type across strains. *J. Invertebr. Pathol.* 67:195–97

Fialho RF, Stevens L. 1997. Molecular evidence for single *Wolbachia* infections among geographic strains of the flour beetle *Tribolium confusum. Proc. R. Soc. London Ser. B* 264:1065–68

Fialho RF, Stevens L. 2000. Male-killing *Wolbachia* in a flour beetle. *Proc. R. Soc. London Ser. B* 267:1469–73

Fine PEM. 1978. On the dynamics of symbiote-dependent cytoplasmic incompatibility in Culicine mosquitoes. *J. Invertebr. Pathol.* 30:10–18

Flockhart HA, Cibulskis RE, Karam M, Albiez EJ. 1986. *Onchocerca volvulus*: enzyme polymorphism in relation to the differentiation of forest and savannah strains of this parasite. *Trans. R. Soc. Trop. Med. Hyg.* 80:285–92

Frank SA. 1997. Cytoplasmic incompatibility and population structure. *J. Theor. Biol.* 184:327–30

Franz M, Andrews P. 1986. Fine Structure of adult *Litomosoides carinii* (Nematoda: Filarioidea). *Z. Parasitenkd.* 72:537–47

Franz M, Buttner DW. 1983. The fine structure

of adult *Onchocerca volvulus* IV. The hypodermal chords of the female worm. *Trop. Med. Parasitol.* 34:122–28

Franz M, Copeman DB. 1988. The fine structure of male and female *Onchocerca gibsoni*. *Trop. Med. Parasitol.* 39:466–68

Gill DE, Mock BA. 1985. Ecological and evolutionary dynamics of parasites: the case of *Trypanosoma diemyctyli* in the red-spotted newt *Notophthalmus viridescens*. In *Ecology and Genetics of Host-Parasite Interactions*, ed. D Rollinson, RM Anderson, pp. 157–83. London: Academic

Giordano R, Jackson JJ, Robertson HM. 1997. The role of *Wolbachia* bacteria in reproductive incompatibilities and hybrid zones of *Diabrotica beetles* and *Gryllus crickets*. *Proc. Natl. Acad. Sci. USA* 94:11439–44

Giordano R, O'Neill SL, Robertson HM. 1995. *Wolbachia* infections and the expression of cytoplasmic incompatibility in *Drosophila sechellia* and *D. mauritiana*. *Genetics* 140:1307–17

Giray T, Luyten Y, MacPherson M, Stevens L. 2001. Physiological bases of genetic differences in cannibalism behavior of the confused flour beetle *Tribolium confusum*. *Evolution* 55:797–806

Girin C, Bouletreau M. 1995. Microorganism-associated variation in host infestation efficiency in a parasitoid wasp, *Trichogramma bourarachae* (Hymenoptera: Trichogrammitdae). *Experientia* 51:398–402

Groisman EA, Ochman H. 1994. How to become a pathogen. *Trends Microbiol.* 2:289–94

Harada R, Maeda T, Nakashima A, Sadakata M, Ando M, et al. 1970. Electronmicroscopical studies on the mechanism of oogenesis and fertilization in *Dirofilaria immitis*. In *Recent Advances in Research on Filariasis and Schistosomiasis in Japan*, ed. M Msasa, pp. 99–121. Baltimore: Univ. Park Press

Heath BD, Butcher RD, Whitfield WG, Hubbard SF. 1999. Horizontal transfer of *Wolbachia* between phylogenetically distant insect species by a naturally occurring mechanism. *Curr. Biol.* 9:313–16

Henkle-Duhrsen K, Eckelt V, Wildenburg G, Blaxter M, Walter R. 1998. Gene structure, activity and localization of a catalase from intracellular bacteria in *Onchocerca volvulus*. *Mol. Biochem. Parasitol.* 96:69–81

Hertig M. 1936. The rickettsia, *Wolbachia pipientis* and associated inclusions of the mosquito *Culex pipiens*. *Parasitology* 28:453–86

Hoerauf A, Nissen-Pahle K, Schmetz C, Henkle-Duhrsen K, Blaxter ML, et al. 1999. Tetracycline therapy targets intracellular bacteria in the filarial nematode *Litomosoides sigmodontis* and results in filarial infertility. *J. Clin. Invest.* 103:11–18

Hoerauf A, Volkmann L, Hamelmann C, Adjei O, Autenrieth IB, et al. 2000a. Endosymbiotic bacteria in worms as targets for a novel chemotherapy in filariasis. *Lancet* 355:1242–43

Hoerauf A, Volkmann L Nissen-Paehle K, Schmetz C, Autenrieth I, et al. 2000b. Targeting of *Wolbachia* endobacteria in *Litomosoides sigmodontis*: comparison of tetracyclines with chloramphenicol, macrolides and ciprofloxacin. *Trop. Med. Int. Health* 5:275–79

Hurst GDD, Hammarton TC, Bandi C, Majerus TMO, Bertrand D, et al. 1997. The diversity of inherited parasites of insects: the male-killing agent of the ladybird beetle *Coleomagilla maculata* is a member of the Flavobacteria. *Genet. Res.* 70:1–6

Hurst GDD, Jiggins FJ, Schulenburg JHGVD, Bertrand D, West SA, et al. 1999a. Male-killing *Wolbachia* in two species of insect. *Proc. R. Soc. London Ser. B* 266:735–40

Hurst GDD, Johnson AP, Schulenburg JHGVD, Fuyama Y. 2000. Male-killing *Wolbachia* in *Drosophila*: a temperature-sensitive trait with a threshold bacterial density. *Genetics* 156:699–709

Hurst GDD, Schulenburg JHGVD, Majerus TMO, Bertrand D, Zakharov IA, et al. 1999b. Invasion of one insect species, *Adalia bipunctata*, by two different male-killing bacteria. *Insect Mol. Biol.* 8:133–39

Hurst LD. 1991. The incidences and evolution

of cytoplasmic male killers. *Proc. R. Soc. London Ser. B* 244:91–99

Hurst LD. 1993. The incidences, mechanisms and evolution of cytoplasmic sex ratio distorters in animals. *Biol. Rev.* 68:121–93

Ikeda H. 1970. The cytoplasmically-inherited 'sex-ratio' condition in natural and experimental populations of *Drosophila bifasciata*. *Genetics* 65:311–33

James AC, Ballard JW. 2000. Expression of cytoplasmic incompatibility in *Drosophila simulans* and its impact on infection frequencies and distribution of *Wolbachia pipientis*. *Evolution* 54:1661–72

Jeyaprakash A, Hoy MA. 2000. Long PCR improves *Wolbachia* DNA amplification: *wsp* sequences found in 76% of sixty-three arthropod species. *Insect Mol. Biol.* 9:393–405

Jiggins FM, Hurst GDD, Majerus MEN. 2000a. High prevalence male-killing *Wolbachia* in the butterfly *Acraea encedena*. *J. Evol. Biol.* 13:495–501

Jiggins FM, Hurst GD, Majerus MEN. 2000b. Sex-ratio-distorting *Wolbachia* causes sex-role reversal in its butterfly host. *Proc. R. Soc. London Ser. B* 267:69–73

Johanowicz DL, Hoy MN. 1996. *Wolbachia* in a predator-prey system: 16S ribosomal DNA analysis of two phytoseiids (Acari: Phytoseiidae) and their prey (Acari: Tetranychidae). *Ann. Entomol. Soc. Am.* 89:435–41

Jost E. 1970. Untersuchungen zur Inkompatibilitat im Culex-pipiens-Komplex. *Wilhelm Roux Arch. Entwicklungsmech. Org.* 166:173–88

Karr TL, Yang W, Feder ME. 1998. Overcoming cytoplasmic incompatibility in *Drosophila*. *Proc. R. Soc. London Ser. B* 265:391–95

Kozek WJ. 1971. Ultrastructure of the microfilaria of *Dirofilaria immitis*. *J. Parasitol.* 57:1052–67

Kozek WJ. 1977. Transovarially-transmitted intracellular microorganisms in adult and larval stages of *Brugia malayi*. *J. Parasitol.* 63:992–1000

Kozek WJ, Figueroa MH. 1977. Intracytoplas-

mic bacteria in *Onchocerca volvulus*. *Am. J. Trop. Med. Hyg.* 26:663–78

Kozek WJ, Raccurt C. 1983. Ultrastructure of *Mansonella ozzardi* microfilaria, with a comparison of the South American (simuliid-transmitted) and the Caribbean (culicoid-transmitted) forms. *Trop. Med. Parasitol.* 34:38–53

Langworthy NG, Renz A, Mackenstedt U, Henkle-Duhrsen K, de Bronsvoort MB, et al. 2000. Macrofilaricidal activity of tetracycline against the filarial nematode *Onchocerca ochengi*: elimination of *Wolbachia* precedes worm death and suggests a dependent relationship. *Proc. R. Soc. London Ser. B* 267:1063–69

Lobos E, Weiss N. 1985. Immunochemical comparison between worm extracts of *Onchocera volvulus* from savanna and rain forest. *Parasite Immunol.* 7:333–47

Lusis JJ. 1947. Some rules of reproduction in populations of *Adalia bipunctata*: non-male strains in populations. *Dokl. Akad. Nauk SSR* 57:951–54

Majerus MEN, Schulenburg JHGVD, Zakharov IA. 2000. Multiple causes of male-killing in a single sample of the two-spot ladybird, *Adalia bipunctata* (Coleoptera: coccinellidae) from Moscow. *Heredity* 84:605–9

Martin G, Sorokine O, Moniatte M, Bulet P, Hetru C, Van Dorsselaer A. 1999. The structure of a glycosylated protein hormone responsible for sex determination in the isopod, *Armadillidium vulgare*. *Eur. J. Biochem.* 262:727–36

Masui S, Kamoda S, Sasaki T, Ishikawa H. 1999. The first detection of the insertion sequence ISW1 in the intracellular reproductive parasite *Wolbachia*. *Plasmid* 42:13–19

Masui S, Kamoda S, Sasaki T, Ishikawa H. 2000. Distribution and evolution of bacteriophage WO in *Wolbachia*, the endosymbiont causing sexual alterations in arthropods. *J. Mol. Evol.* 51:491–97

McGreevy PB, Theis JH, Lavoipierre MM, Clark J. 1974. Studies on filariasis. III. *Dirofilaria immitis*: emergence of infective larvae from the mouthparts of *Aedes aegypti. J. Helminthol.* 48:221–28

McLaren DJ, Worms MJ. 1975. Micro-organisms in filarial larvae (nematoda). *Trans. R. Soc. Trop. Med. Hyg.* 69:509–14

Min KT, Benzer S. 1997. *Wolbachia*, normally a symbiont of *Drosophila*, can be virulent, causing degeneration and early death. *Proc. Natl. Acad. Sci. USA* 94:10792–96

Nelson GS. 1970. Onchocerciasis. *Adv. Parasitol.* 8:173–24

O'Neill SL. 1989. Cytoplasmic symbionts in *Tribolium confusum. J. Invertebr. Pathol.* 53:132–34

O'Neill SL, Giordano R, Colbert AM, Karr TL, Robertson HM. 1992. 16S rRNA phylogenetic analysis of the bacterial endosymbionts associated with cytoplasmic incompatibility in insects. *Proc. Natl. Acad. Sci. USA* 89:2699–702

O'Neill SL, Pettigrew MM, Sinkins SP, Braig HR, Andreadis TG, Tesh RB. 1997. In vitro cultivation of *Wolbachia pipientis* in an *Aedes albopictus* cell line. *Insect Mol. Biol.* 6:33–39

Oh HW, Kim MG, Sihn SW, Bae KS, Ahn YJ, Oark H-Y. 2000. Ultrastructural and molecular identification of a *Wolbachia* endosymbiont in a spider, Nephila clavata. *Insect Mol. Biol.* 9:539–43

Ottesen EA. 1992, The Wellcome Trust Lecture. Infection and disease in lymphatic filariasis: an immunological perspective. *Parasitology* 104:S71–79

Plantard O, Rasplus JY, Mondor G, Le Clainche I, Solignac M. 1999. Distribution and phylogeny of *Wolbachia* inducing thelytoky in Rhoditini and 'Aylacini' (Hymenoptera: Cynipidae). *Insect Mol. Biol.* 8:185–91

Poinsot D, Montchamp-Moreau C, Mercot H. 2000. *Wolbachia* segregation rate in *Drosophila simulans* naturally bi-infected cytoplasmic lineages. *Heredity* 85:191–98

Price PW. 1980. *Evolutionary Biology of Parasites.* Princeton, NJ: Princeton Univ. Press

Randerson JP, Jiggins FM, Hurst LD. 2000a. Male killing can select for male mate choice:

a novel solution to the paradox of the lek. *Proc. R. Soc. London Ser. B* 267:867–74

Randerson JP, Smith NGC, Hurst LD. 2000b. The evolutionary dynamics of male-killers and their hosts. *Heredity* 84:152–60

Remme J, Dadzie KY, Rolland A, Thylefors B. 1989. Ocular onchocerciasis and intensity of infection in the community. I. West African savanna. *Trop. Med. Parasitol.* 40:340–47

Rigaud T. 1999. Further *Wolbachia* endosymbiont diversity: a tree hiding in the forest. *TREE* 14:212–13

Rigaud T, Moreau J, Juchault P. 1999. *Wolbachia* infection in the terrestrial isopod *Oniscus asellus*: sex ratio distortion and effect on fecundity. *Heredity* 83:469–75

Rokas A. 2000. *Wolbachia* as a speciation agent. *TREE* 15:44–45

Rousset F, Bouchon D, Pintureau B, Juchault P, Solignac M. 1992. *Wolbachia* endosymbionts responsible for various alterations of sexuality in arthropods. *Proc. R. Soc. London Ser. B* 250:91–98

Rousset F, Braig HR, O'Neill SL. 1999. A stable triple *Wolbachia* infection in *Drosophila* with nearly additive incompatibility effects. *Heredity* 82:620–27

Rousset F, Solignac M. 1995. Evolution of single and double *Wolbachia* symbioses during speciation in the *Drosophila simulans* complex. *Proc. Natl. Acad. Sci. USA* 92:6389–93

Ryan SL, Saul GB. 1968. Post-fertilization effect on incompatibility factors in *Mormoniella. Mol. Gen. Genet.* 103:24–26

Schall JJ. 1990. Virulence of lizard malaria: the evolutionary ecology of an ancient parasite-host association. *Parasitology* 100:S35–52

Schall JJ. 2001. Parasite virulence. In *The Behavioral Ecology of Parasites*, ed. EE Lewis, JF Cambell, MVK Sukhdeo. Oxon, UK: CABI. In press

Schulenburg JHGVD, Hurst GD, Huigens TM, van Meer MM, Jiggins FM, Majerus MEN. 2000. Molecular evolution and phylogenetic utility of *Wolbachia ftsZ* and *wsp* gene sequences with special reference to the origin of male-killing. *Mol. Biol. Evol.* 17:584–600

Shoemaker DD, Katju V, Jaenike J. 1999.

Wolbachia and the evolution of reproductive isolation between *Drosophila recens* and *Drosophila subquinaria*. *Evolution* 53:1157–64

Sinkins SP, Braig HR, O'Neill SL. 1995. *Wolbachia* superinfections and the expression of cytoplasmic incompatibility. *Proc. R. Soc. London Ser. B* 261:325–30

Sinkins SP, O'Neill SL. 2000. *Wolbachia* as a vehicle to modify insect populations. In *Insect Transgenesis: Methods and Applications*, ed. AM Handler, AA James, pp. 271–87. Boca Raton, FL: CRC Press

Sironi M, Bandi C, Sacchi L, Di Sacco B, Damiani G, Genchi C. 1995. Molecular evidence for a close relative of the arthropod endosymbiont *Wolbachia* in a filarial worm. *Mol. Biochem. Parasitol.* 74:223–27

Slatko BE, O'Neill SL, Scott AL, Werren JL, Blaxter ML. 1999. The *Wolbachia* genome consortium. *Microb. Comp. Genom.* 4:161–65

Smith HL, Rajan TV. 2000. Tetracycline inhibits development of the infective-stage larvae of filarial nematodes in vitro. *Exp. Parasitol.* 95:265–70

Snook RR, Cleland SY, Wolfner MF, Karr TL. 2000. Offsetting effects of *Wolbachia* infection and heat shock on sperm production in *Drosophila simulans*: analyses of fecundity, fertility and accessory gland proteins. *Genetics* 155:167–78

Stevens L. 1989a. The genetics and evolution of cannibalism in flour beetles (Genus *Tribolium*). *Evolution* 43:169–79

Stevens L. 1989b. Environmental factors affecting reproductive incompatibility in flour beetles, genus *Tribolium*. *J. Invertebr. Pathol.* 53:78–84

Stevens L. 1993. Cytoplasmically inherited parasites and reproductive success in *Tribolium* flour beetles. *Anim. Behav.* 46:305–10

Stevens L. 1994. Genetic analysis of cannibalism behavior in *Tribolium* flour beetles. In *Quantitative Genetics of Behavioral Evolution*, ed. CRB Boake, pp. 206–27. Chicago: Univ. Chicago Press

Stevens L, Mertz DB. 1985. Genetic stability of cannibalism in *Tribolium confusum*. *Behav. Genet.* 15:549–59

Stevens L, Wade MJ. 1990. Partial reproductive incompatibility in *Tribolium* flour beetles: the rate of spread and effect on host fitness. *Genetics* 124:367–72

Stevens L, Wicklow D. 1992. Multi-species interactions affect cytoplasmic incompatibility in *Tribolium* flour beetles. *Am. Nat.* 140:642–53

Stouthamer R. 1997. *Wolbachia*-induced parthenogenesis. In *Influential Passengers: Inherited Microorganisms and Arthropod Reproduction*, ed. SL O'Neill, AA Hoffmann, JH Werren, pp. 155–75. New York: Oxford Univ. Press

Stouthamer R, Breeuwer JA, Hurst GD. 1999. *Wolbachia pipientis*: microbial manipulator of arthropod reproduction. *Annu. Rev. Microbiol.* 53:71–102

Stouthamer R, Breeuwer JA, Luck RF, Werren JH. 1993. Molecular identification of microorganisms associated with parthenogenesis. *Nature* 361:66–68

Stouthamer R, Kazmer DJ. 1994. Cytogenetics of microbe-associated parthenogenesis and its consequences for gene flow in *Tricogramma* wasps. *Heredity* 73:317–27

Taylor MJ, Bandi C, Hoerauf AM, Lazdins J. 2000a. *Wolbachia* bacteria of filarial nematodes: a target for control? *Parasitol. Today* 16:179–80

Taylor MJ, Bilo K, Cross HF, Archer JP, Underwood AP. 1999. 16S rDNA phylogeny and ultrastructural characterization of *Wolbachia* intracellular bacteria of the filarial nematodes *Brugia malayi*, *B. pahangi*, and *Wuchereria bancrofti*. *Exp. Parasitol.* 91:356–61

Taylor MJ, Cross HF, Bilo K. 2000b. Inflammatory responses induced by the filarial nematode *Brugia malayi* are mediated by lipopolysaccharide-like activity from endosymbiotic *Wolbachia* bacteria. *J. Exp. Med.* 191:1429–36

Taylor MJ, Cross H, Mohammed A, Trees A, Bianco A. 1996. Susceptibility of *Brugia malayi* and *Onchocerca lienalis* microfilariae to nitric oxide and hydrogen peroxide in cell-free culture and from IFN gamma-activated macrophages. *Parasitology* 112:315–22

Telford SR. 1971. Parasitic diseases of reptiles. *J. Am. Vet. Med. Assoc.* 158:1644–52

Toe L, Tang J, Back C, Katholi CR, Unnasch TR. 1997. Vector-parasite transmission complexes for onchocerciasis in West Africa. *Lancet* 349:163–66

Townson S, Hutton D, Siemienska J, Hollick L, Scanlon T, et al. 2000. Antibiotics and *Wolbachia* in filarial nematodes: antifilarial activity of rifampicin, oxytetracycline and chloramphenicol against *Onchocerca gutturosa, Onchocerca lienalis* and *Brugia pahangi. Ann. Trop. Med. Parasitol.* 94:801–16

Troyer JM, Radulovic S, Andersson SGE, Azad AF. 1998. Detection of point mutations in *rpoB* gene of rifampin-resistant *Rickettsia typhi. Antimicrob. Agents Chemother.* 42:1845–46

Turner PF, Rockett KA, Ottesen EA, Francis H, Awadzi K, Clark IA. 1994. Interleukin-6 and tumor necrosis factor in the pathogenesis of adverse reactions after treatment of lymphatic filariasis and onchocerciasis. *J. Infect. Dis.* 169:1071–75

Vandekerckhove TT, Watteyne S, Willems A, Swings JG, Mertens J, Gillis M. 1999. Phylogenetic analysis of the 16S rDNA of the cytoplasmic bacterium *Wolbachia* from the novel host *Folsomia candida* (Hexapoda, Collembola) and its implications for *Wolbachia* taxonomy. *FEMS Microbiol. Lett.* 180:279–86

Vavre F, Fleury F, Lepetit D, Fouillet P, Bouletreau M. 1999a. Phylogenetic evidence for horizontal transmission of *Wolbachia* in host-parasitoid associations. *Mol. Biol. Evol.* 16:1711–23

Vavre F, Girin C, Bouletreau M. 1999b. Phylogenetic status of a fecundity-enhancing *Wolbachia* that does not induce thelytoky in *Trichogramma. Insect Mol. Biol.* 8:67–72

Vincent AL, Portaro JK, Ash LR. 1975. A comparison of the body wall ultrastructure of *Brugia pahangi* with that of *Brugia malayi. J. Parasitol.* 63:567–70

Wade MJ. 2001. Infectious speciation. *Nature* 409:675–77

Wade MJ, Chang NW. 1995. Increased male fertility in *Tribolium confusum* beetles after infection with the intracellular parasite *Wolbachia. Nature* 373:72–74

Wade MJ, Stevens L. 1985. Microorganism mediated reproductive isolation in flour beetles (Genus *Tribolium*). *Science* 1985:527–28

Wade MJ, Stevens L. 1994. The effect of population subdivision on the rate of spread of parasite-mediated cytoplasmic incompatibility. *J. Theor. Biol.* 167:81–87

Watt G, Chouriyagune C, Ruangweerayud R, Watcharapichat P, Phulsuksombati D, Jongsakul K, et al. 1996. Scrub typhus infections poorly responsive to antibiotics in northern Thailand. *Lancet* 13:86–89

Wenseleers T, Ito F, Van Borm S, Huybrechts R, Volckaert F, Billen J. 1998. Widespread occurrence of the micro-organism *Wolbachia* in ants. *Proc. R. Soc. London Ser. B* 265:1447–52

Werren JH. 1997. Biology of *Wolbachia. Annu. Rev. Entomol.* 423:587–609

Werren JH, Windsor DM. 2000. *Wolbachia* infection frequencies in insects: evidence of a global equilibrium? *Proc. R. Soc. London Ser. B* 267:1277–85

Werren JH, Zhang W, Guo LR. 1995. Evolution and phylogeny of *Wolbachia*: reproductive parasites of arthropods. *Proc. R. Soc. London Ser. B* 261:55–63

West SA, Cook JM, Werren JH, Godfray HC. 1998. *Wolbachia* in two insect host-parasitoid communities. *Mol. Ecol.* 7:1457–65

Wharton RH. 1957. Studies in filariasis in Malaya: observations on the development of *Wuchereria malayi* in Mansonia (Mansonioides) longipalpis. *Ann. Trop. Med. Parasitol.* 51:278–96

Xie X, Bein O, Williams SA. 1994. Molecular phylogenic studies on filarial parasites based on 5S ribosomal spacer sequences. *Paracite* 1:141–51

Yen JH, Barr AR. 1973. The etiological agent of cytoplasmic incompatibility in *Culex pipiens. J. Invertebr. Pathol.* 22:242–50

Annu. Rev. Ecol. Syst. 2001. 32:547–76
Copyright © 2001 by Annual Reviews. All rights reserved

BIOSPHERIC TRACE GAS FLUXES AND THEIR CONTROL OVER TROPOSPHERIC CHEMISTRY

Russell K. Monson[1] and Elisabeth A. Holland[2]

[1]*Department of Environmental, Population and Organismic Biology and the Cooperative Institute for Research in Environmental Science, University of Colorado, Boulder, Colorado; e-mail: russell.monson@colorado.edu*
[2]*Max-Planck Institute for Biogeochemistry, Jena, Germany*

Key Words   methane, nonmethane hydrocarbons, dimethyl sulfide, atmospheric chemistry, ozone

■ **Abstract**   Terrestrial and marine ecosystems function as sources and sinks for reactive trace gases, and in doing so, profoundly influence the oxidative photochemistry in the troposphere. Principal biogenic processes include microbial methane production and oxidation, the emission of volatile organic compounds from forest ecosystems, the emission of nitric oxide from soils, the emission of reactive sulfur compounds and carbon monoxide from marine ecosystems, control over the production of hydroxyl radical concentration by regional hydrologic processes, and deposition of ozone and nitrogen oxides to ecosystems. The combined influence of these processes is to affect the tropospheric concentrations of ozone, hydroxyl radicals, reactive nitrogen oxides, carbon monoxide, and inorganic acids, all of which constitute fundamental components of oxidative photochemistry. In this review we discuss the recent literature related to the primary controls over the biosphere-atmosphere exchange of reactive trace gases, and also to efforts to model the dominant biospheric influences on oxidative dynamics of the troposphere. These studies provide strong support for the paradigm that biospheric processes exert the dominant control over oxidative chemistry in the lower atmosphere. Improvements in our ability to model biospheric influences on tropospheric chemistry, and its susceptibility to global change, will come from inclusion of more explicit information on the processes that control the emission and uptake of reactive trace gases and the impact of changes in ecosystem cover and land-use change.

INTRODUCTION

The oxidative chemistry that occurs in the troposphere is vital to the maintenance of life on earth, especially with regard to the recycling of volatile matter produced by biogenic processes. Compounds of various types are emitted by terrestrial and aquatic ecosystems. Biospheric fluxes of various reactive trace gases exert significant influences on photochemical dynamics in the lower troposphere (e.g., Fehsenfeld et al. 1992, Brasseur et al. 1999, Crutzen et al. 1999, Zellner et al. 1999). The study of biospheric trace gas fluxes bridges the fundamental

0066-4162/01/1215-0547$14.00

approaches of ecophysiology, with its emphasis on the mechanistic controls over flux; biogeochemistry, with its emphasis on the mass balance of biogeochemical cycles; and atmospheric chemistry, with its emphasis on the reactivity and control over turnover times of various atmospheric constituents. This recognition of interdisciplinary connections has created opportunities for collaboration among ecologists, geochemists, and atmospheric chemists to examine tropospheric chemistry and develop a new generation of atmospheric chemistry models with truly synthetic representation of the controls exerted by biological processes. The current review was prepared with this goal in mind: to describe and synthesize the most relevant recent literature pertaining to the biological control over the emission and assimilation of reactive trace gases from ecosystems and their influences on oxidative processes in the troposphere. We focus on the shorter-lived, reactive trace gases produced by natural ecosystems. Thus, we have not considered in great length the literature pertaining to anthropogenic emissions of trace gases, and we do not consider longer-lifetime gases such as nitrous oxide (N_2O) and carbon dioxide (CO_2), which contribute little to reactive photochemistry in the troposphere.

Oxidative Chemistry of the Troposphere and Its Relation to Biospheric Fluxes of Trace Gases

The chemical schemes that illustrate oxidative chemistry in the troposphere evoke comparison to a plate of pasta—a complex labyrinth of reaction strings, interconnecting at various points, and leading to a variety of possible endpoints. Much of the complexity emerges from numerous catalytic influences and the tendency for ozone (O_3), one of the primary products of the chemistry, to feed back and influence the formation of oxidative catalysts. In this review we do not focus on these chemical complexities (for recent reviews see Derwent 1995, Brasseur et al. 1999, Seinfeld 1999, Zellner et al. 1999, Finlayson-Pitts & Pitts 2000). However, we do present a brief overview of the most relevant oxidative processes involving biospheric trace gases.

The oxidative power of the troposphere is determined by O_x-HO_x-NO_x chemistry, with the primary oxidative species being O_2, O_3, OH, HO_2, NO, NO_2, and NO_3. These species react readily with the reduced organic gases emitted by ecosystems. To illustrate the most relevant interactions, we start with tropospheric O_3 and its photolysis by UV-B radiation (Figure 1). Photolysis of tropospheric O_3 yields molecular oxygen (O_2) and either a ground-state singlet $O(^3P)$ or an electronically-excited singlet radical identified as $O(^1D)$. The latter singlet exists approximately 270 kJ mol^{-1} above the ground-state, and has the potential to react with H_2O, producing two molecules of the highly reactive hydroxyl radical (OH). The first-order dependence of OH formation on the humidity of the atmosphere provides the basis for a connection between atmospheric chemistry and ecosystem hydrology (denoted as **a** in Figure 1). [Note, the primary fate of $O(^3P)$ will be to react with O_2 to reform O_3.]

The primary emissions of reduced compounds from the terrestrial biosphere are methane (CH_4) and nonmethane hydrocarbons (NMHC) (denoted as **b** and **c** in Figure 1). These compounds react with OH to form methyl peroxy radicals (CH_3O_2) in the case of CH_4 oxidation, and a variety of other organic peroxy radicals (RO_2) in the case of NMHC oxidation. Peroxy radicals have a relatively short lifetime in the troposphere (seconds to minutes depending on the availability of other reactants) (Table 1). If nitric oxide (NO) is available, organic peroxy radicals will react to produce nitrogen dioxide (NO_2) and an organic alkoxy radical (RO). A significant fraction of the NO that enters the atmosphere originates from the nitrogen metabolism of soil microorganisms (denoted as **d** in Figure 1) (Lloyd 1995, Lamarque et al. 1995, Hall et al. 1996). The NO_2 formed from reaction between NO and peroxy radicals is susceptible to photolysis, decomposing within a few minutes to NO and ground-state oxygen, $O(^3P)$. The $O(^3P)$ can subsequently react with molecular oxygen (O_2) to form O_3, thus closing the photochemical scheme and validating the premise that "it takes a little O_3 to produce a lot of O_3," with the primary fuel being provided by biospheric processes. The overall process is catalytic in the sense that NO is reformed and is available to react with additional RO_2 molecules.

The oxidative chemistry that occurs in the troposphere above marine ecosystems is also dominated by gas-phase reaction with OH radicals, although heterogeneous chemistry involving aqueous aerosols also becomes important (Barone et al. 1995, Bandy et al. 1996, DeBruyn et al. 1998). The primary trace gas emissions from marine ecosystems are dimethyl sulfide (DMS) at approximately $15 Tg\ y^{-1}$ (Pszenny et al. 1999), which is oxidized to sulfur dioxide (SO_2) and a variety of minor compounds, and carbon monoxide (CO) at $13\ Tg\ y^{-1}$ (Bates et al. 1995), which is oxidized to CO_2 (Figure 2). Emissions of secondary importance occur in the form of carbon disulfide (CS_2) and carbonyl sulfide (COS). The inorganic products of organic sulfide oxidation have a principal role in the production and growth of aerosol particles that act as cloud condensation nuclei and control the production of marine stratiform clouds.

The lifetimes of some of the primary reactive species are presented in Table 1. Lifetimes are most meaningful when processes can be identified that lead to irreversible loss of a compound. For example, volatile organic compounds, such as isoprene, are irreversibly lost during oxidation by OH, NO_3, or O_3. For some reactive species, it is more difficult to establish such a clear photochemical context. For example, the photochemical lifetime of O_3 is only a few seconds (owing to photolysis), but most of the $O(^3P)$ that is formed immediately reacts with O_2 to reform O_3. This photolytic lifetime is of little practical use to understanding biosphere-atmosphere exchanges of O_3. The more useful lifetime is that resulting from photolysis to $O(^1D)$ and subsequent reaction with H_2O to form OH or reaction with other species in the atmosphere (e.g., NO, VOCs, etc.). The conditions of lifetime calculations, and the processes involved in compound loss, are listed as footnotes below Table 1.

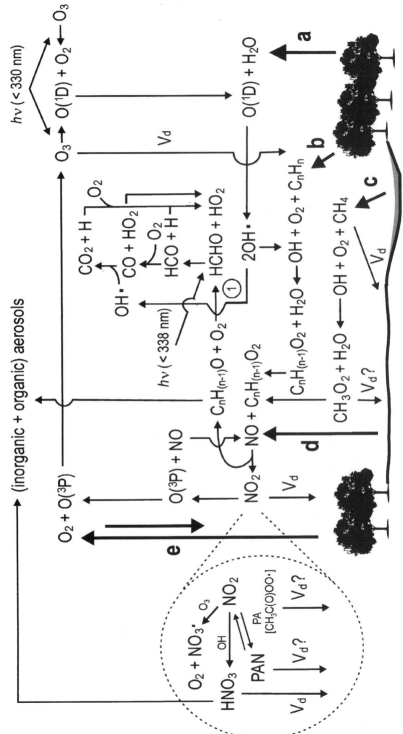

Methane Fluxes Between the Biosphere and Atmosphere

The global emission of CH_4 has profound impacts on the oxidative capacity of the troposphere as well as influencing the radiative forcing in the earth's energy balance. The average CH_4 concentration of the northern hemisphere near the surface was reported as 1745 ppbv in 1998 (Ehhalt et al. 2001). Concentrations in the southern hemisphere are 7% lower, on average, and the total global burden is estimated to be 4850 Tg (1 Tg = 1 million tons). The current concentration of CH_4 is more than twice as large as it was during the preindustrial era (Etheridge et al. 1998, Duglokencky et al. 1998). This increase is responsible for approximately 20% of the direct radiative forcing because of rising greenhouse gas emissions. The rate of increase in CH_4 concentration has varied significantly from year to year, ranging from near zero in 1992 to +13 ppbv during 1998. To date, adequate explanations for the year-to-year variability are absent (Ehhalt et al. 2001). Global sources of CH_4 total 500–600 Tg released annually (Ehhalt et al. 2001). Global quantification of CH_4 emissions remains challenging, although recent modeling approaches and stable isotope measurements have narrowed the uncertainties (Lowe 1994, 1997, Hein et al. 1997, Quay et al. 1999, Houweling et al. 1999, 2000, Wang et al. 2001). Microbial production from both agricultural and natural ecosystems constitutes 74–80% of global emissions, which clearly dominates the budget. Substrate supply, substrate quality, and temperature are the major controls on the rate of CH_4 emission (Valentine et al. 1994, Bellisario et al. 1999).

Land-use changes have tremendous potential to alter the magnitude and global patterns of CH_4 emission (Duxbury 1994, Bridges & Batjes 1996, Mosier et al. 1997, Neue et al. 1997a, Schimel & Gulledge 1998, Woodwell et al. 1998, Verchot et al. 2000). Soils that lack a high water table and possess adequate aeration tend to oxidize CH_4 through microbial activity. The diffusion of CH_4 into the soil is the primary factor limiting oxidation rate (Potter et al. 1996). Houweling et al. (2000) used a global chemical-transport model parameterized for preindustrial times to estimate that drainage and cultivation of wetlands since 1800 AD have converted soils from CH_4 sources to CH_4 sinks, increasing the rate of CH_4 oxidation, and

←_____

Figure 1   Major trace gas fluxes between the biosphere and troposphere and their role in oxidative photochemistry. The primary biospheric fluxes are indicated with bold lines and associated letters that relate to discussion in the text. The H_2O (a) and C_nH_n (representing volatile hydrocarbons) (b) fluxes are depicted as forest emissions. The CH_4 (c) flux is depicted as net emissions from wetlands. The NO (d) flux is depicted as soil emissions. The O_2 (e) flux is depicted as photosynthetic emissions, in this case from a forest. The circled 1 that appears in the middle of the figure denotes the partial oxidation of hydrocarbons that are oxidized farther to intermediate carbonyl compounds which may be deposited back to the surface or oxidized to formaldehyde (HCHO). Question marks by some of the deposition velocities denote uncertainty in the importance of this process. $h\nu$ = solar energy; V_d = deposition velocity.

potentially reducing the rate of global emissions by 10% (Houweling et al. 2000). Working against this trend, the rate of CH_4 oxidation by nonwetland soils has been lowered in recent years by increases in agricultural nitrogen fertilization, the atmospheric deposition of pollutant nitrogen, and the atmospheric deposition of acids (King 1997, Smith et al. 2000). Smith et al. (2000) estimated that conversion of nonwetland, natural soils to agriculture in Northern Europe has reduced CH_4 oxidation rates by 66% (see also Willison et al. 1995, Goulding et al. 1996, Powlson et al. 1997). In tropical Ghana, conversion of forest to agriculture has decreased CH_4 oxidation rates by 60%, and significant reductions have also been noted for burned areas (Prieme & Christensen 1999).

In wetland ecosystems, pathways for CH_4 emission include diffusion through soil and water, ebullition or bubbling of CH_4-saturated air, and transport through plant tissues (Chanton & Whiting 1995, Walters et al. 1996, Joabsson et al. 1999, Greenup et al. 2000). Transport through plants accounts for more than half of the total CH_4 flux to the troposphere (e.g., Holzapfel-Pschorn et al. 1986, Kim et al. 1998). Recent research has revealed that anatomical differences among cultivars of rice explain differences in CH_4 transport capacity (Byrd et al. 2000, Yao et al. 2000, Aulakh et al. 2000), and that the tissue marking the transition from the root to shoot has a critical role in controlling the rate of CH_4 transport through rice plants (Wassman & Aulakh 2000; also see Jackson & Armstrong 1999, Neue et al.

TABLE 1   Chemical information on the primary reactive species involved in the tropospheric photochemistry described in Figures 1 and 2. τ = lifetime

Chemical species	Chemical formula	Tropospheric lifetime
Ozone	O_3	5–300 d[a]
Atomic oxygen (excited state)	$O(^1D)$	1.4×10^{-9} s[b]
Atomic oxygen (ground state)	$O(^3P)$	1.3×10^{-5} s[b]
Hydroperoxy radical	HO_2	1.1×10^{-3} s[c]
Hydroxyl radical	OH	0.2–1 s[d]
Formaldehyde	HCHO	1–2 d[e]
Carbon monoxide	CO	1–3 m[f]
Isoprene	C_5H_8	0.2–1 d[g]
Methane	CH_4	8–9 y[h]
Peroxy radical	RO_2	5–900 s[i]
Nitric oxide	NO	57–600 s[j]
Nitrogen dioxide	NO_2	143 s (daytime) 7 h (nighttime)[k]
Nitrate radical	NO_3	5–6 s (daytime)[l] >1000 s (nighttime)
Peroxyacetyl nitrate (PAN)	$CH_3C(O)O_2NO_2$	2–600 h[m]

TABLE 1 (*Continued*)

Chemical species	Chemical formula	Tropospheric lifetime
Nitric acid	HNO_3	0.5 d–1 m[n]
Dimethyl sulfide (DMS)	CH_3SCH_3	60 h[o]
Carbonyl sulfide	COS	10 y[o]
Carbon disulfide	CS_2	120 h[o]

[a]Photolysis and chemical reaction (surface deposition will cause lifetime to be less); from Brasseur et al. (1999).

    lower troposphere; 0–3 km $\tau = 5$–8 d (summer); $\tau = 17$–100 d (winter).

    upper troposphere; 6–10 km $\tau = 30$–40 d (summer); $\tau = 90$–300 d (winter).

[b]Calculated assuming 1 atm, 25°C, no H_2O (addition of H_2O will cause lifetime to be less); reaction coefficients were taken from Finlayson-Pitts & Pitts (2000).

[c]Calculated for the self-reaction ($HO_2 + HO_2$) in low NO_x regimes; 8 pptv HO_2; reaction coefficients were taken from Finlayson-Pitts & Pitts (2000).

[d]Calculated with respect to global average reaction with CH_4 (at total global content of 5000 Tg) and CO (at total global content of 360 Tg).

[e]Calculated for OH reaction only (photolysis will cause lifetime to be less); reaction coefficients were taken from Finlayson-Pitts & Pitts (2000).

[f]Average global $\tau = 2$ mos if calculated with 360 Tg total atmospheric CO content divided by 2100 Tg y^{-1} global loss rate to OH oxidation (see Ehhalt et al. 2001 for estimate of global OH oxidation rate as 1500–2700 Tg y^{-1}); actual lifetime for CO can vary from 10 days during the summer above continental regions to 1 y during winter at the poles (Holloway et al. 2000).

[g]Assuming reaction with OH at an atmospheric concentration of 10^6 molecules cm^{-3}; reaction coefficients were taken from Finlayson-Pitts & Pitts (2000).

[h]A lifetime of 8.0 y is calculated using 4850 Tg as the total atmospheric CH_4 content (see Ehhalt et al. 2001) divided by 600 Tg y^{-1} global emission rate (see Lelieveld et al. 1998); slightly longer lifetimes are calculated from a detailed analysis of the combined effects of OH oxidation, stratospheric destruction, and soil oxidation (see text).

[i]Calculated for the methyl peroxy radical when reaction is with NO or HO_2; reaction coefficients were taken from Finlayson-Pitts & Pitts (2000).

    for $NO + CH_3O_2$ 10 pptv NO $\tau = 540$ s.

                       1 ppbv NO $\tau = 5.4$ s.

    for $HO_2 + CH_3O_2$ 8 pptv HO_2 $\tau = 891$ s.

[j]Calculated for reactions with O_3 and HO_2; reaction coefficients were taken from Finlayson-Pitts & Pitts (2000).

    for $O_3 + NO$ 40 ppbv O_3 $\tau = 56.5$ s.

    for $HO_2 + NO$ 8 pptv HO_2 $\tau = 591$ s.

[k]Reaction coefficients were taken from Finlayson-Pitts & Pitts (2000).

    daytime, zenith angle $= 50°$ $\tau = 143$ s.

    nighttime, 50 ppb O_3 $\tau = 7$ h.

[l]Daytime zenith angle $= 0°$; nighttime τ calculated for reactions with NO_2 and VOCs in an unpolluted atmosphere, from Brasseur et al. (1999).

[m]Assuming no new PAN synthesis, from Singh et al. (1990).

    for lower troposphere; 0–3 km $\tau = 2$h.

    for middle troposphere; 3–6 km $\tau = 24$ h.

    for upper troposphere; 6–10 km $\tau = 600$ h.

[n]Includes heterogeneous incorporation into clouds and aerosols, from Brasseur et al. (1999).

    for lower troposphere; 0–3 km $\tau = 0.5$–1 d (mostly heterogeneous losses).

    for middle troposphere; 3–6 km $\tau = \sim 1$ mo (reaction with OH and photolysis).

[o]For gas-phase OH reaction only, from Brasseur et al. (1999).

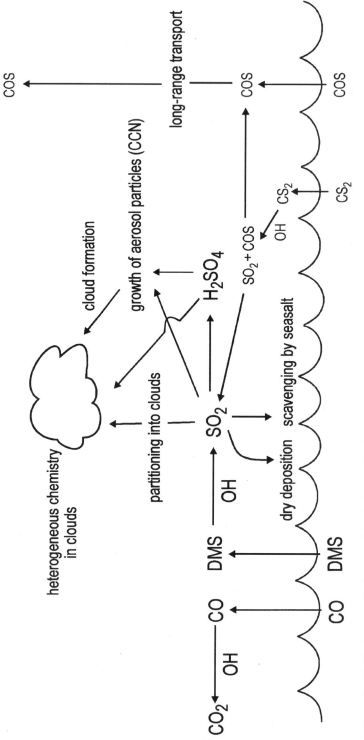

Figure 2 Oxidative chemical interactions in the marine boundary layer. Dimethyl sulfide (DMS) is further oxidized to sulfuric acid (H_2SO_4). H_2SO_4 plays a primary role in promoting the growth of cloud condensation nuclei (CCN). The oxidation of emitted carbon disulfide (CS_2) contributes to the atmospheric reservoirs of SO_2 and carbonyl sulfide (COS). The long-lived nature of COS allows a significant fraction to be transported to the stratosphere. The emission of CO from marine ecosystems and its ultimate oxidation to CO_2 is depicted on the left.

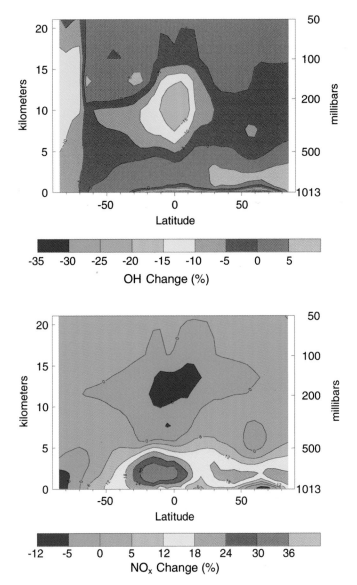

Figure 3 Percent difference in the July zonal average vertical distribution of OH (*upper panel*) and NO$_x$ (*lower panel*) predicted by the 3-dimensional chemical-transport model IMAGES for scenarios with isoprene and without isoprene. Note the depletion of OH at altitudes between 6 and 12 km at low latitudes. One explanation for the OH depletion is oxidation of isoprene and subsequent reaction with NO$_x$ to form organic nitrates. Subsequent decomposition of the organic nitrates and formation of NO$_x$ after transport to lower altitudes and higher latitudes is apparent in the lower panel. (From Guenther et al. 1999b with permission).

1997b). In addition to facilitating the upward transport of CH_4, plant aerenchyma facilitates the downward transport of O_2. This permits a zone of CH_4 oxidation within the rhizosphere and affects the spatial partitioning of CH_4 production and oxidation in wetland sediments (Frenzel 2000). Hypodermal cell layers control the leakage of O_2 into the rhizosphere (Armstrong et al. 2000).

The use of stable isotopes (especially $^{13}C/^{12}C$ ratios) has facilitated studies of the processes underlying CH_4 emissions. Recent stable-isotope studies have shown that (a) CH_4 transport through the plant is driven by ventilation systems and molecular diffusion (Chanton & Whiting 1996, Chanton et al. 1997), (b) the downward transport of O_2 and rhizosphere-assisted oxidation of CH_4 occurs in shallow sediments (Popp et al. 1999, Chasar et al. 2000), and (c) CH_4 production is driven solely by CO_2 reduction at deeper levels and by acetate fermentation at shallower levels (Popp et al. 1999, Chasar et al. 2000, Hornibrook et al. 2000).

THE EFFECTS OF BIOGENIC CH_4 EMISSIONS ON THE OXIDATIVE POWER OF THE TROPOSPHERE   Most atmospheric CH_4 is oxidized through reaction with OH in the troposphere (Figure 1). The loss rate is calculated using the globally averaged OH concentration derived from methyl chloroform budgets (Prinn et al. 1995), which indicates a total of 506 Tg CH_4 oxidized annually. Studies with two different three-dimensional, chemical-transport models, MOZART and MOGUNTIA, have estimated the average photochemical lifetime of CH_4 (i.e., that owing to reaction with OH alone) as 9.9 and 9.0 y, respectively (Hauglustaine et al. 1998, Poisson et al. 2000). Other sinks for CH_4 removal include soil oxidation (38 Tg y^{-1}; Ridgewell et al. 1999) and transport to the stratosphere where it is destroyed by reaction with OH, Cl, and $O(^1D)$ (40 Tg y^{-1}; Ehhalt et al. 2001). Taking the lifetime resulting from photochemistry as 9.45 y, the lifetime resulting from stratospheric destruction as 120 y, and the lifetime resulting from soil oxidation as 160 y, a net globally averaged lifetime of 8.3 y is derived, which is similar to the value of 8.4 y previously reported (Ehhalt et al. 2001). (The overall lifetime is calculated as the reciprocal sum of the first-order rate constants (k) for the three CH_4 loss mechanisms, recognizing that $\tau = 1/k$.) It is likely that the tropospheric lifetime of CH_4 has increased in recent years because of increases in anthropogenic emissions of NO_x, CO, and volatile organic compounds (VOCs), all of which are also oxidized by atmospheric OH radicals (Lelieveld et al. 1998, Karlsdottir & Isaksen 2000).

Isoprene, Monoterpenes, and Oxygenated Hydrocarbon Emissions

The importance of terrestrial ecosystems as sources for nonmethane volatile organic compounds (hereafter referred to as VOCs) was first described by Went (1960), who used the concentration of terpene-containing leaf oils in sagebrush to estimate global terpene emissions as 175 Tg C y^{-1}. A few years later, Rasmussen & Went (1964) estimated the global flux of all VOCs as 432 Tg C y^{-1}. Since these seminal studies, researchers have identified a rich diversity in the types of VOCs emitted by vegetation (Kesselmeier & Staudt 1999, Fuentes et al. 2000), and estimates of global emissions have increased to 1100 Tg C y^{-1} (Guenther et al. 1995). Guenther et al. (2000) estimate that of the 84 Tg C y^{-1} emitted from North American natural sources as VOCs, 35% is isoprene, 25% is a group of 19 other terpenoid compounds, and 40% is a group of 17 nonterpenoid compounds. Past reviews have focused on ecophysiological controls over VOC emissions and the influences of temperature, water, and nutrients (Monson et al. 1995, Lerdau et al. 1997, Loreto 1997, Kreuzwieser et al. 1999, Harley et al. 1999, Fall 1999, Guenther et al. 2000, Logan et al. 2000, Sharkey 2001). In this section we focus on aspects of VOC emissions that have not been emphasized in past reviews, and that are particularly relevant to the influence of terrestrial ecosystems on tropospheric chemistry.

In studies with ponderosa pine needles, it was shown that phenology can co-vary with herbivore interactions to increase the emission of monoterpenes (Litvak & Monson 1998). Monoterpene concentrations are highest in the youngest basal parts of the needles. Many needle-feeding herbivores begin feeding at the tip of the needle and abandon the needle by the time they have consumed approximately two-thirds of the length. This pattern leaves the lowest part of the needle that contains the greatest monoterpene cyclase activity and leaves the greatest concentration of monoterpenes exposed to the atmosphere through the open wound. Monoterpene emissions from partially eaten needles are 15 times higher than noneaten needles. Depending on atmospheric NO$_x$ concentrations, the herbivore-induced increase in monoterpene emissions has the potential to change oxidative dynamics in the atmosphere, including reductions in the concentration of OH and increases in O$_3$ concentration (Litvak et al. 1999). The effects of herbivory on forest VOC emissions have not been included in past regional and global atmospheric chemistry models.

Anthropogenic influences, including the initiation and spread of pathogen outbreaks, increases in the atmospheric CO$_2$ concentration, and land-use change, have tremendous potential to alter VOC emissions patterns. The chestnut blight (caused by the fungus *Cryphonectria parasitica*) was introduced to the United States through imported Asian chestnut trees during the late nineteenth century (Shear & Stevens 1913). By the mid-twentieth century, most of the nonisoprene-emitting native chestnut (*Castanea dentata*) stands in northeastern lowland forests had been replaced by isoprene-emitting oak stands (Lerdau et al. 1997). Oak forests of the United States have been heavily impacted by oak-wilt disease, which was first described in 1942 as a localized disease of midwestern U.S. forests. The vascular fungus that causes the disease is spread by sap-feeding beetles (Coleoptera: Nitidulidae), and the spread is encouraged by human activities, including forest cutting and horticultural pruning (Appel 1994). Oak wilt disease has eliminated oaks from many southwestern forests and caused the replacement of isoprene-emitting oaks by nonisoprene-emitting species in midwestern forests (Menges & Loucks 1984, Anderson et al. 2001). Oak-wilt disease causes reductions in leaf area and physiological inhibition of leaf metabolism, both of which reduce forest isoprene emissions (Anderson et al. 2000). Guenther et al. (1999a) inferred that land-cover changes in a subtropical savanna/woodland ecosystem in the southwestern United States have caused biogenic VOC emissions to increase threefold since the early 1800s. Much of the change was because of heavy grazing and increases in the density of isoprene-emitting woody species at the expense of nonisoprene-emitting grasses. In studies of the potential effects of elevated CO_2 and its associated climate change on VOC emissions from the continental United States, Constable et al. (1999b) found that rates of monoterpene emission from individual needles did not increase when grown at elevated CO_2 levels. However, elevated CO_2 did cause an increase in leaf area index, which could cause an overall increase in monoterpene emissions from coniferous forests. In a second study, Constable et al. (1999a) found that simulated climate warming associated with a doubling of atmospheric CO_2 concentration caused total VOC emissions (isoprene, monoterpenes, and oxygenated VOCs) to increase by 82%. The effect of increased CO_2 alone that caused increases in leaf area index and biomass per unit leaf area was predicted to cause an increase in total VOC emissions of 12%. Changes in the distribution of VOC-emitting vegetation in response to climate change was predicted to cause an overall decrease in VOC emissions of 10%. Thus, the overall effect of elevated CO_2, climate warming, and change in vegetation distribution on the emission of biogenic VOCs from the continental United States was predicted to be an 84% increase.

THE EFFECTS OF FOREST-EMITTED VOCS ON OXIDATIVE CHEMISTRY IN THE TROPOSPHERE   Forest-emitted volatile organic compounds (VOCs) have the potential to markedly influence atmospheric photochemistry. In regions of high isoprene emissions (e.g., those with extensive tropical forest), isoprene oxidation can account for up to 71% of the reduction in OH (compared to only 11% for CO oxidation and 5% for CH_4 oxidation) (Grosjean 1995). Using a three-dimensional chemistry and transport model, Wang et al. (1998) demonstrated that global emissions of VOCs, particularly isoprene, cause a 15% increase in the background O_3 concentration and a 20% decrease in the mean OH concentration, compared to a scenario of no VOC emissions. Houweling et al. (1998) found that global VOC emissions caused a 17% increase in the mean background tropospheric O_3 concentration, and reduced OH concentrations over the continents, with the greatest decreases in regions with strong forest emissions. Roelofs & Lelieveld (2000) inferred that global VOC emissions are responsible for an 8% decrease in global OH concentration and a 50% decrease in regional OH concentrations in the boundary layer of tropical rain forests, where isoprene emissions are especially high. Poisson et al. (2000) demonstrated that global VOC emissions, most of which are isoprene, cause an 18% increase in the global tropospheric O_3 concentration, a 16% decrease in the global OH concentration, and a 20% increase in the tropospheric lifetime of CH_4. The last study represents the first comprehensive demonstration that the high reactivity of biogenic VOCs can perturb local OH concentrations to the point where the lifetime of a long-lived compound, such as CH_4, is significantly affected.

Biogenic VOC also stimulates the formation of organic nitrates. Using a three-dimensional chemical transport model, Guenther et al. (1999b) demonstrated a localized region of OH depletion within the upper troposphere at low latitudes that was accompanied by reductions in isoprene and NO_x concentration (Figure 3, see color insert). (NO_x is emitted in large amounts at these latitudes resulting from biomass burning.) It is assumed that the depletion of isoprene and NO_x in this region is due to the oxidation of isoprene and reaction of the products of isoprene oxidation (primarily methylvinyl ketone and methacrolein) with NO_2 to form organic nitrates. Because of these chemical interactions and the subsequent transport of organic nitrates, such as peroxyacetyl nitrate (PAN) and peroxymethacrylyl nitrate (MPAN), NO_x deposition is predicted to increase in mid-latitude ecosystems north and south of the equator. Consistent with the predictions of Guenther et al. (1999b), Wang et al. (1998) found that the simulated absence of global biogenic VOC emissions had a profound influence on the formation of organic nitrates, leading to a 30% decrease in NO_x in the troposphere of remote regions, but a 70% increase in NO_x in the continental troposphere near NO_x sources. Similarly, Houweling et al. (1998) predicted that without biospheric VOC emissions and an atmospheric reservoir of organic nitrates, NO_x concentrations in the marine atmosphere would decrease by 50% (Houweling et al. 1998). It is clear that the emission of biogenic VOCs affects not only the oxidative power of the atmosphere (i.e., the concentrations of OH and O_3), but also the spatial distribution of NO_x and its deposition to remote ecosystems.

Carbon Monoxide Sources and Sinks

Ice core measurements from Greenland have been used to infer an increase in carbon monoxide (CO) concentration from ~90 ppbv in 1800 to ~150 ppbv in the current atmosphere (Haan et al. 1996). Anthropogenic processes are largely responsible for this increase, including fossil fuel combustion and biomass burning. During the period between 1990–1995, atmospheric CO concentrations exhibited a decrease of approximately 2 ppbv y^{-1} (Novelli et al. 1994, 1998). In the Northern Hemisphere, these recent reductions are due to both decreased emissions (mostly from anthropogenic sources) and increased atmospheric OH concentration (Bakwin et al. 1994, VanDop & Krol 1996). The largest source for CO is the oxidation of CH_4, which occurs at an efficiency of 0.70 to 0.94 mol CO mol^{-1} CH_4 (Tie et al. 1992, Manning et al. 1997, Kanakidou & Crutzen 1999), although Bergamaschi et al. (2000a), have constrained this efficiency to 0.85–0.88 using stable isotope ratios to invert a global transport model. (An efficiency less than unity is because of the partitioning of water-soluble intermediates in the CH_4 oxidation pathway onto wet surfaces.) Assuming an oxidation efficiency of 0.85–0.88 and a CH_4 source of 500–600 Tg y^{-1}, between 744 and 924 Tg CO are produced each year from CH_4 oxidation. Fossil fuel combustion and other technological processes are estimated to add 478 Tg CO y^{-1} (Oliver et al. 1996); biomass burning is estimated to add 500–1500 Tg y^{-1}; and the oxidation of isoprene and terpenes adds 290–1200 Tg y^{-1} (Bergamaschi et al. 2000b) (assuming oxidation efficiencies of 0.2–0.8 mol CO mol^{-1} C atoms in isoprene/terpenes; Jacob & Wofsy 1990, Kanakidou & Crutzen 1999). The oceans are estimated to contribute 13 Tg CO y^{-1} (Bates et al. 1995), with most coming from the photodegradation of organic matter in the surface waters (possibly linked to the photodegradative production of carbonyl sulfate, COS, Pos et al. 1998). Other small CO sources include the oxidation of anthropogenic VOCs (80 Tg y^{-1}, Bergamaschi et al. 2000b) and the decomposition and photodegradation of organic matter in the surface of soils (100 Tg y^{-1}, Schade & Crutzen 1999), and within living plant tissues (Tarr et al. 1995).

The primary sinks for CO are photochemical oxidation to CO_2 (through reaction with OH) and microbiological uptake into the surface layers of the soil (Brasseur et al. 1999). Soil uptake, which can represent 20–25% of the total global sink, is highest at intermediate soil water contents, warmer temperatures, and higher total organic matter content (Moxley & Smith 1998, King 1999). The conditions that favor soil uptake are consistent with high levels of microbiological activity.

Nitrogen Oxides and Ammonia

Total reactive nitrogen oxides in the atmosphere are frequently categorized as NO_y, and include nitrogen monoxide (NO; also called nitric oxide) and nitrogen dioxide (NO_2; collectively known as NO_x) as primary components, and nitric acid (HNO_3) and organic nitrates as secondary components. Biogenic contributions to NO_y begin primarily as NO emissions from soils (Williams et al. 1992, Hall et al. 1996, McKenney & Drury 1997, Holland & Lamarque 1997, Holland et al. 1999). At low ambient NO_x concentrations vegetation can also be a source of NO and NO_2 (Rondon & Granat 1994, Weber & Rennenberg 1996, Hereid & Monson 2001, Sparks et al. 2001). Tropical savanna/woodland, fertilized temperate and tropical agricultural fields, and chaparral ecosystems represent large sources of soil-emitted NO, with global soil emissions estimated to be 21 Tg N y^{-1}, approximately the same amount as is emitted from anthropogenic sources (Davidson & Kingerlee 1997). NO is a chemical intermediate in both the nitrification and denitrification pathways of soil microorganisms (Davidson & Schimel 1995), and biogenic emissions of NO are strongly influenced by soil temperature, moisture, and nitrogen availability (for recent reports see Veldkamp & Keller 1997, Martin et al. 1998, Mosier et al. 1998, Aneja et al. 1998, Serca et al. 1998, Pilegaard et al. 1999, Li et al. 1999). Changes in land use and the increased use of N fertilizers are predicted to cause increases in soil NO emissions. Tropical deforestation in Brazil for a single year (1990) is predicted to have caused a total increase in soil NO_x emissions of 0.74 Tg (Fearnside 1997). Globally, the use of fertilizers is predicted to have increased soil NO emissions by 0.5–5 Tg N y^{-1} (Smith et al. 1997). In those ecosystems where N availability does not normally limit NO emissions, a nonlinear relationship can develop between the rate of NO flux and fertilizer addition; increases in NO fluxes can exceed increases in N availability by an order of magnitude (Hall & Matson 1999).

Other primary sources of NO_x include fossil fuel combustion ($\sim$21 Tg N y^{-1}), biomass burning ($\sim$7.5 Tg N y^{-1}), and production by lightning ($\sim$6 Tg N y^{-1}) (Ehhalt et al. 2001). The production of NO_x by biomass burning is especially relevant to atmospheric chemistry because it often occurs in subtropical or tropical regions that are also characterized by high rates of forest VOC emissions. When combined, biogenic VOCs and NO_x stimulate tropospheric O_3 production (Figure 1). In the African troposphere, biomass burning contributes to 16% of the regional O_3 production, compared to 12% owing to soil NO_x emissions, and 27% owing to lightning (the remainder is owing to industrial sources and transport from the stratosphere) (Marufu et al. 2000).

The amount of soil-emitted NO that is available for atmospheric chemistry is influenced by the potential for plant canopies to assimilate NO_x (Matson 1997, Lerdau et al. 2000). Gao et al. (1993) observed steep decreases in NO concentration with height above the floor of a temperate forest, reaching a minimum near the middle of the canopy. The O_3-induced oxidation of NO to NO_2 within the canopy was also significant, causing NO_2 concentrations to be higher within the canopy than above. Past studies of NO_2 uptake by individual leaves in agricultural and tropical forest ecosystems revealed that between 5% and 25% of the soil-emitted NO can be assimilated as NO_2 before it exits the canopy (Hereid & Monson 2000, Sparks et al. 2001). Other studies using a simple box model estimate that the uptake of NO_x by plant canopies can be 60% the rate of soil NO emissions (Bakwin et al. 1990, Jacob & Bakwin 1991). The potential for canopy assimilation of NO_x has rarely been incorporated into coupled ecosystem flux-atmospheric chemistry models (though see Yienger & Levy 1995), although it clearly has the potential to influence model simulations.

Little is known about the magnitude of plant uptake of other forms of NO_y. Nitric acid (HNO_3) has been shown to deposit to plants through surface deposition, transcuticular deposition, and stomatal uptake (Heubert & Robert 1985, Dollard et al. 1987, Lovett & Lindberg 1993, Bytnerowicz et al. 1998). Given the relatively high uptake affinities of wet canopy surfaces for HNO_3, any NO_2 that is oxidized to HNO_3 within the canopy is likely to be removed before it reaches the atmosphere above the canopy (Cadle et al. 1991, Lovett & Lindberg 1992). There is indirect evidence that PAN is also deposited on leaf surfaces (Roberts 1990). However, it is not currently possible to discern direct uptake of PAN vs. indirect uptake after dissociation into the peroxyacetyl radical and NO_2, which are also potentially assimilated by plant leaves.

Total global ammonia emissions (45–83 Tg N y^{-1}) are similar in magnitude to those for nitrogen oxides (Schlesinger & Hartley 1992, Holland et al. 1999). Most NH_3 is emitted from the mineralization of excreta from domestic animals and from fertilized soil, with the total agricultural contribution being 60% of global emissions (Bouwman et al. 1997). The principal loss of atmospheric NH_3 occurs through reaction with acid-containing aerosols followed by wet ($\sim$21 Tg N y^{-1})

and dry ($\sim$16 Tg N y^{-1}) deposition to the surface (Brasseur et al. 1999). Only a small amount ($\sim$3 Tg N y^{-1}) of the emitted NH$_3$ is oxidized to NO$_x$ (through reaction with OH), causing it to have a minor role in oxidative chemistry (Holland et al. 1999).

THE EFFECTS OF BIOGENIC NO EMISSIONS ON OXIDATIVE CHEMISTRY IN THE TROPO-SPHERE   Once in the atmosphere, NO is oxidized to NO$_2$ by O$_3$ or peroxy radicals (RO$_2$ or HO$_2$). When NO is present at sufficiently high concentrations ($>$5–30 pptv, Brasseur et al. 1999), the O(^{3}P) produced through photolysis of NO$_2$ will support the net production of O$_3$ (Figure 1). At low NO concentrations, peroxy radicals have the potential to react directly with O$_3$. Measured NO$_x$ concentrations in rural regions of the United States range from 0.8 ppbv in the west (Williams et al. 1984, Parrish et al. 1990) to 1.5–6.5 ppbv in the east (Kelly et al. 1984, Parrish et al. 1993, Aneja et al. 1996), meaning that most of the continental United States produces O$_3$. At remote continental sites, such as in arctic tundra (Bakwin et al. 1992) or certain tropical forests (Sanhueza et al. 2000) and remote oceanic sites (Liu et al. 1983), NO$_x$ concentrations can be low enough to permit the net photochemical destruction of O$_3$.

Ozone production efficiency (OPE) is defined as the moles of O$_3$ produced per mole of NO$_x$ oxidized (to other forms of NO$_y$) and removed from further photochemical activity. (Forms of NO$_y$, such as HNO$_3$ and PAN, are removed from further photochemistry owing to their high tendencies for deposition or transport to other regions.) Above a threshold of approximately 0.5 ppbv NO$_x$, the OPE is nonlinearly dependent on NO$_x$ concentration and is highest at low NO$_x$ concentrations (Liu et al. 1987, Lin et al. 1988); at high NO$_x$ concentrations, NO$_2$ reacts with OH to form HNO$_3$, and the depletion of odd-hydrogen radicals suppresses photochemistry. There are occasions when the emission of biogenic VOCs increases OPE (Jacob et al. 1995, Hirsch et al. 1996). The presence of reactive VOCs tends to reduce the concentration of OH, which slows the conversion of NO$_2$ to HNO$_3$ and enhances the photochemical efficiency of each mole of NO$_x$ in producing O$_3$. Liu et al. (1987) estimated the total production of O$_3$ that is stimulated by biogenic NO$_x$ emissions (primarily soil NO emissions) in the northern hemisphere to be 3×10^{11} moles O$_3$ cm^{-2} s^{-1}, approximately three times the amount of O$_3$ produced from anthropogenic NO$_x$ emissions.

Biogenic Fluxes of Reactive Sulfur Compounds

Approximately 10–28% of the total emission of sulfur gases to the atmosphere is owing to biogenic sources (ocean and terrestrial), with 97–98% of the biogenic emissions resulting from marine organisms (Brasseur et al. 1999). The most abundant forms of biogenic sulfur emission are dimethyl sulfide (CH$_3$SCH$_3$ or DMS), carbonyl sulfide (COS), hydrogen sulfide (H$_2$S), and carbon disulfide (CS$_2$). Of these, DMS has the highest emission rate and is the sulfur trace gas with the greatest potential to influence oxidative chemistry. Although terrestrial and freshwater sources of DMS have been reported (Kesselmeier et al. 1993, Hobbs & Mottram 2000, Richards et al. 1994, Sharma et al. 1999), they are small compared to marine sources.

DMS is produced by marine algae and bacteria through the breakdown of dimethyl sulfoniopropionate (DMSP), an organic osmolyte that may have several physiological roles (Kirst 1990, Malin & Kirst 1997, Stefels 2000, Kiene et al. 2000). Most DMSP is released from phytoplankton blooms during zooplankton grazing (Dacey & Wakeham 1986, Leck et al. 1990). Once released, DMSP is converted to DMS by (a) DMSP-lyase enzyme that is bound to the extracellular components of living algal cells (Stefels & Dijkhuizen 1996, Niki et al. 2000), (b) free DMSP-lyase that is released during zooplankton grazing (Wolfe & Steinke 1996), and (c) DMSP-cleaving bacteria that are often associated with phytoplankton blooms (Gonzalez et al. 2000). Communities of DMSP-cleaving bacteria often form around aggregates of phytoplankton cells to form microzones of high DMS emissions within the phytoplankton bloom (Scarratt et al. 2000). Surface waters in the North Atlantic and Pacific Oceans are supersaturated with carbon disulfide (CS_2) (Kim & Andreae 1987, 1992, Xie & Moore 1999) and assumed to be a net source of 0.13–0.24 Tg y^{-1} (Xie & Moore 1999). Microbial and phytoplankton processes (Andreae 1986, Xie et al. 1999), as well as photochemical degradation of dissolved organic matter (Ferek & Andreae 1984), are sources of CS_2. Globally, marine sources of CS_2 represent 27% of the annual flux to the atmosphere, although the global CS_2 budget is poorly constrained (Watts 2000). COS is produced by oxidation of CS_2 and by photochemical breakdown of dissolved organic matter in ocean surface waters (Andreae & Fereck 1992). Although emission by terrestrial plants has been reported (Rennenberg et al. 1990, Kesselmeier & Merk 1993), the exchange of COS between plants and the atmosphere is highly dependent on atmospheric COS concentration and a COS compensation point has been reported (Kesselmeier & Merk 1993). Given the current atmospheric concentration of COS, it is most likely that terrestrial vegetation represents a COS sink (Goldan et al. 1988, Kesselmeier & Merk 1993). The assimilation of COS depends on ribulose 1,5-bisphosphate carboxylase/oxygenase, the same enzyme involved in CO_2 assimilation (Protoschill-Krebs & Kesselmeier 1992), and there is a clear correlation between COS and CO_2 assimilation rates (Kesselmeier & Merk 1993). Until recently, the global budget of COS had been largely out of balance; but if soils are assigned as COS sinks (Andreae & Crutzen 1997), greater balance is achieved (Watts 2000). Soil sinks for COS are dependent on microbial carbonic anhydrase activity (Kesselmeier et al. 1999).

THE EFFECTS OF REACTIVE SULFUR EMISSIONS ON OXIDATIVE CHEMISTRY IN THE TROPOSPHERE   The primary sink for DMS is reaction with OH radical (Barone et al. 1995, Ravishankara et al. 1997), although recent analyses have suggested that heterogeneous chemistry and halogen chemistry may also play a role (Campolongo et al. 1999, Davis et al. 1999, James et al. 2000). Through a combination of gas-phase and surface-phase chemistry, DMS is ultimately oxidized to H_2SO_4, which forms aqueous particles in the marine boundary layer. Aqueous particles often function as cloud condensation nuclei (Ayers & Gras 1991, Andrae et al. 1995, Nagao et al. 1999). One of the classic feedback hypotheses involving biogenic trace-gas fluxes states that marine DMS emissions, which are highly sensitive to sea-surface temperature, can enhance cloud formation, which then cools sea-surface temperature and reduces DMS emissions (Charlson et al. 1987).

The impact of DMS emissions on the OH concentration above the Southern Ocean are 1–10% the reductions caused by oxidation of CO, the primary sink for OH in the marine boundary layer (Pszenny et al. 1999). In the free troposphere, CS_2 is oxidized to carbonyl sulfide (COS) and sulfur dioxide (SO_2) through reaction with OH (Stickel et al. 1993). The reaction of CS_2 with OH is relatively fast, which provides a tropospheric lifetime of 120 h (Table 1). COS has a lifetime of 5–10 y (Brasseur et al. 1999), allowing for transport to the stratosphere where it is a precursor to the formation of sulfate aerosol, one of the heterogeneous reactants that causes the destruction of stratospheric O_3 (Fahey et al. 1993).

Linkage of Ecosystem Hydrology to Oxidative Chemistry in the Troposphere

Oxidative power in the troposphere is inextricably linked to surface hydrology. The formation of OH exhibits first-order reaction kinetics with respect to H_2O vapor concentration. Using an analysis of kinetic coefficients for the reaction of $O(^1D)$ with H_2O relative to the loss of $O(^1D)$ through collision with N_2 and O_2, Seinfeld (1999) determined that at 50% relative humidity and an air temperature of $20°C$, 9.1% of the $O(^1D)$ that is formed in the troposphere reacts with H_2O to form OH. A reduction in surface evapotranspiration, or the entrainment of dry air from above the planetary boundary layer, that results in a 50% reduction in humidity would cause a 50% reduction in the rate of OH formation.

Several studies have revealed that deforestation, especially in tropical ecosystems, has the potential to alter regional hydrology (Nobre et al. 1991, Dickinson & Kennedy 1992, Lean & Rowntree 1993, Zeng & Neelin 1999, Wang & Eltahir 2000). To date, this has only been estimated for extreme scenarios. The complete conversion of forests to pastureland in Amazonia would decrease the specific humidity by $1–1.5 \, g \, kg^{-1}$ at an altitude of 850 hPa (Nobre et al. 1991). A perturbation of this magnitude could cause a significant reduction in tropospheric OH concentration and an increase in the lifetime of VOCs and CH_4. Secondary effects of deforestation include reduced cloud cover and increased transmission of solar radiation (Eltahir & Humphries 1998). One could argue that such effects are likely to increase the rate of O_3 photolysis, causing increases in the rate of $O(^1D)$ and OH formation, thus compensating for reduced H_2O concentrations. Such compensation is unlikely, however, given the atmospheric dominance of N_2 and O_2 and the concomitant likelihood that the additional $O(^1D)$ formed from photolysis would collide with these constituents and be converted to $O(^3P)$ before colliding with H_2O.

Ozone and NO_y Deposition to Ecosystems

The deposition of NO_y and O_3 provides an important path for the removal of reactive compounds from the troposphere, thus impacting oxidative dynamics, and closing an important biogeochemical feedback to biospheric health (Reich 1987, Holland et al. 1996). Some of the components of NO_y, especially nitric acid and some of the organic nitrates, are sufficiently soluble to be removed by precipitation. Representations of wet deposition in the current generation of global transport chemistry models are limited by our knowledge of heterogeneous chemistry. This causes the models to underestimate the loss of reactive molecules and limits their ability to represent oxidative processes in the gas phase as well (Holland et al. 1999). Heterogeneous reactions involving NO_3 and N_2O_5 on "wet" aerosol particles were shown (in a three-dimensional chemistry transport model) to have a significant influence on the atmospheric concentrations of NO_x, O_3, and OH (Dentener & Crutzen 1993). Globally-averaged O_3 and OH concentrations were predicted to decrease 9% in the presence of heterogeneous chemistry. Using a regional-scale photochemical dispersion model, Matthijsen et al. (1997) showed that addition of heterogeneous chemistry caused a 5% reduction in tropospheric O_3 concentration, presumably because of increased formation of OH from dissolved O_3 by reaction with superoxide (also see Lelieveld & Crutzen 1990). (The aqueous OH can then be converted to HO_2, and ultimately to H_2O_2, which can participate in the heterogeneous oxidation of NO_3 and SO_2.) Wet sulfate deposition also appears to be underestimated in global chemistry transport models, probably because of poor representation of heterogeneous chemistry (Lelieveld et al. 1997).

Inadequacies also exist in the current generation of dry deposition models. Representations of dry deposition processes range in complexity from prescribed average deposition velocities to multilayer, mechanistic canopy models (Hicks et al. 1987, Meyers et al. 1998, Wesley & Hicks 1999). The choice of approach can have a significant impact on the results. Representation of dry deposition with a canopy-explicit model resulted in a 10% decrease in subtropical ozone, a 20% decrease in global NO_x, and a 7% decrease in global atmospheric HNO_3 compared to runs using empirically prescribed deposition velocities (Ganzeveld & Lelieveld 1995). One of the sources of greatest uncertainty in models of dry deposition involves its dependence on stomatal conductance (for recent studies see Grantz et al. 1997, Sanchez et al. 1997, Ro-Poulsen et al. 1998, Sun & Massman 1999, Bauer et al. 2000, Mikkelsen et al. 2000, Wieser et al. 2000) and the difficulty in obtaining accurate models of stomatal dynamics. Deposition to marine ecosystems is easier to simulate. Ozone that is transported from continents is chemically destroyed in the low-NO_x atmosphere of the marine boundary layer (Ayers et al. 1997, Monks et al. 2000, deLaat & Lelieveld 2000).

Holland et al. (1999) predicted preindustrial and contemporary NO_y and NH_x deposition patterns in the Northern Hemisphere using a coupled surface-flux chemical-transport model. Preindustrial wet plus dry NO_y deposition was inferred to be greatest for tropical ecosystems, which are influenced by high rates of soil NO emissions and a high frequency of biomass burning and lightning strikes. Some past studies have focused on NO_y deposition and its potential to stimulate carbon assimilation and net primary productivity (Holland et al. 1996, 1997, Townsend et al. 1996, Aber & Driscoll 1997). One perspective that is often missing from these efforts is the fact that photochemical processes link NO_x, VOCs, and O_3; thus, a positive correlation is often observed between surface O_3 concentrations and NO_y deposition (Trainer et al. 1993, Olszyna et al. 1994, Holland et al. 1997). Any positive fertilization effect caused by deposition of NO_y is likely to be offset by the negative effect of increased surface O_3.

CONCLUSIONS

Fluxes of reactive trace gases between ecosystems and the atmosphere have the potential to control several key components of the oxidative chemistry of the troposphere. In fact, one can make the case that biospheric fluxes completely control the oxidative power of the atmosphere. Considerable progress has been made over the past decade in integrating the perspectives of biology and atmospheric chemistry to resolve aspects of biospheric control. Most progress has come from the inclusion of biologically relevant surface emission and deposition schemes. At present, however, the schemes are simple and do not adequately reflect many of the dynamics related to seasonal and spatial variability in resource availability and short- and long-term acclimation of organismic physiology. There is much room for improvement in the design of biogenic emission models. Our ability to understand tropospheric chemistry in the face of rapid land-use change also severely compromises studies of biosphere-atmosphere coupling in natural and increasingly human-impacted atmospheres. To accurately discern the quantitative impacts of anthropogenic activities, models need to simulate preindustrial and contemporary scenarios. Parameterizing the contemporary scenario with respect to biogenic trace gas emissions is an enormous challenge. There is great hope and room for optimism, however, that some of these issues can be resolved over the next decade or so. Given the recent explosion of interest in integrative aspects of the earth systems sciences, it seems a certainty that new energy will infuse the issues that remain unresolved. A particularly important component of this energy will be the enthusiasm of the next generation of students who pride themselves on not being biologists, geologists, or chemists, but biogeochemists.

ACKNOWLEDGMENTS

The authors are grateful for the numerous discussions with M. Lerdau, A. Guenther, R. Fall, A. Turnipseed, D. Schimel, and J. Sparks. Critical readings of an early draft of the manuscript by A. Turnipseed and W. Schlesinger were especially helpful.

Visit the Annual Reviews home page at www.AnnualReviews.org

LITERATURE CITED

Aber JD, Driscoll CT. 1997. Effects of land use, climate variation, and N deposition on N cycling and C storage in northern hardwood forests. *Glob. Biogeochem. Cycles* 11:639–48

Anderson LJ, Brumbaugh MS, Jackson RB. 2001. Water and tree-understory interactions: a natural experiment in a savanna with oak wilt. *Ecology* 82:33–49

Anderson LJ, Harley PC, Monson RK, Jackson RB. 2000. Reduction of isoprene emissions from live oak (*Quercus fusiformis*) with oak wilt. *Tree Physiol.* 20:1199–1203

Andreae MO. 1986. The ocean as a source of atmospheric sulfur compounds. In *The Role of Air-Sea Exchange in Geochemical Cycling*, ed. P Buat-Menard, pp. 331–62. Norwell, MA: Reidel

Andreae MO, Crutzen PJ. 1997. Atmospheric aerosols: biogeochemical sources and role in atmospheric chemistry. *Science* 176:1052–58

Andreae MO, Elbert W, Mora SJ. 1995. Biogenic sulfur emissions and aerosols over the tropical South Atlantic. *3.* Atmospheric dimethylsulfide, aerosols, and cloud condensation nuclei. *J. Geophys. Res.* 100:11,335–56

Andreae MO, Fereck RJ. 1992. Photochemical production of carbonyl sulfide in sea water and its emission to the atmosphere. *Glob. Biogeochem. Cycles* 6:75–183

Aneja VP, Kim DS, Das M, Hartsell BE. 1996. Measurements and analysis of reactive nitrogen species in the rural troposphere of southeast United States: southern oxidant study site SONIA. *Atmos. Environ.* 30:649–59

Aneja VP, Roelle PA, Robarge WP. 1998. Characterization of biogenic nitric oxide source strength in the southeast United States. *Environ. Poll.* 102:211–18

Appel DN. 1994. Identification and control of oak wilt in Texas urban forests. *J. Arbor.* 20:250–58

Armstrong W, Cousins D, Armstrong J, Turner DW, Beckett PM. 2000. Oxygen distribution in wetland plant roots and permeability barriers to gas-exchange within the rhizosphere: a microelectrode and modelling study with *Phragmites australis*. *Ann. Bot.* 86:687–703

Aulakh MS, Wassmann R, Rennenberg H, Fink S. 2000. Pattern and amount of aerenchyma relate to variable methane transport capacity of different rice cultivars. *Plant Biol.* 2:182–94

Ayers GP, Granek H, Boers R. 1997. Ozone in the marine boundary layer at Cape Grim: model simulation. *J. Atmos. Chem.* 27:179–95

Ayers GP, Gras JL. 1991. Seasonal relationship between cloud condensation nuclei and aerosol methanesulphonate in marine air. *Nature* 353:834–35

Bakwin PS, Tans PP, Novelli PC. 1994. Carbon monoxide budget in the northern hemisphere. *Geophys. Res. Lett.* 21:433–36

Bakwin PS, Wofsy SC, Fan SM, Fitzjarrald DR. 1992. Measurements of NO_x and NO_y concentrations and fluxes over arctic tundra. *J. Geophys. Res.* 97:16,545–57

Bakwin PS, Wofsy SC, Fan S-M. 1990. Measurements of reactive nitrogen oxides (NO_y) within and above a tropical forest canopy in the wet season. *J. Geophys. Res.* 95:16,765–72

Bandy AR, Thornton DC, Blomquist BW, Chen SM, Wade JC, et al. 1996. Chemistry of dimethyl sulfide in the equatorial Pacific atmosphere. *Geophys. Res. Lett.* 23:741–44

Barone SB, Turnipseed AA, Ravishankara AR. 1995. Role of adducts in the atmospheric oxidation of dimethyl sulfide. *Faraday Discuss.* 100:39–55

Bates TS, Kelly KC, Johnson JE, Gammon RH. 1995. Regional and seasonal variations in the flux of oceanic carbon monoxide to the atmosphere. *J. Geophys. Res.* 100:23,093–101

Bauer MR, Hultman NE, Panek JA, Goldstein AH. 2000. Ozone deposition to a ponderosa pine plantation in the Sierra Nevada Mountains: A comparison of two different climatic years. *J. Geophys. Res.* 105:22,123–36

Bellisario LM, Bubier JL, Moore TR, Chanton JP. 1999. Controls on CH_4 emissions from a northern peatland. *Glob. Biogeochem. Cycles* 13:81–91

Bergamaschi P, Hein R, Brenninkmeijer CAM, Crutzen PJ. 2000a. Inverse modeling of the global CO cycle. 2:Inversion of $13C/^{12}C$ and $^{18}O/^{16}O$ isotope ratios. *J. Geophys. Res.* 105:1929–45

Bergamaschi P, Hein R, Heimann M, Crutzen PJ. 2000b. Inverse modeling of the global CO cycle. 1:Inversion of CO mixing ratios. *J. Geophys. Res.* 105:1909–27

Bouwmann AF, Lee DS, Asman WAH, Dentener FJ, van der Hoek KW, Oliver JGJ. 1997. A global high-resolution emission inventory for ammonia. *Glob. Biogeochem. Cycles* 11:561–87

Brasseur GP, Orlando JJ, Tyndall GS. 1999. *Atmospheric Chemistry and Global Change.* New York: Oxford Univ. Press. 654 pp.

Bridges EM, Batjes NH. 1996. Soil gaseous emissions and global change. *Geography* 81:155–69

Bytnerowicz A, Percy K, Riechers G, Padgett P, Krywult M. 1998. Nitric acid vapor effects on forest trees—deposition and cuticular changes. *Chemosphere* 36:697–702

Byrd GT, Fisher FM, Sass RL. 2000. Relationships between methane production and emission to lacunal methane concentrations in rice. *Glob. Biogeochem. Cycles* 14:73–83

Cadle SH, Marshall JD, Mulawa PA. 1991. A laboratory investigation of the routes of HNO_3 dry deposition to coniferous seedlings. *Environ. Poll.* 72:287–305

Campolongo F, Saltelli A, Jensen NR, Wilson J, Hjorth J. 1999. The role of multiphase chemistry in the oxidation of dimethyl sulfide (DMS): a latitude dependent analysis. *J. Atmos. Chem.* 32:327–56

Chanton JP, Whiting GJ. 1995. Trace gas exchange in freshwater and coastal marine environments: ebullition and transport by plants. In *Biogenic Trace Gases: Measuring Emissions from Soil and Water*, ed. PA Matson, RC Harriss, pp. 98–125. Cambridge, UK: Blackwell Sci.

Chanton JP, Whiting GJ. 1996. Methane stable isotope distributions as indicators of gas transport mechanisms in emergent aquatic plants. *Aquatic Bot.* 54:227–36

Chanton JP, Whiting GJ, Blair NE, Lindau CW, Bollich PK. 1997. Methane emission from rice: stable isotopes, diurnal variations and CO_2 exchange. *Glob. Biogeochem. Cycles* 11:15–27

Charlson RJ, Lovelock JE, Andreae MO, Warren G. 1987. Oceanic phytoplankton, atmospheric sulfur, cloud albedo and climate. *Nature* 326:655–61

Chasar LS, Chanton JP, Glaser PH, Siegel DI. 2000. Methane concentration and stable isotope distribution as evidence of rhizospheric processes: Comparison of fen and bog in the Glacial Lake Agassiz Peatland complex. *Ann. Bot.* 86:655–63

Conrad R. 1996. Soil microorganisms as controllers of atmospheric trace gases (H_2, CO_2, CH_4, OCS, N_2O, NO). *Microbiol. Rev.* 60:609–40

Constable JVH, Guenther AB, Schimel DS, Monson RK. 1999a. Modelling changes in VOC emission in response to climate change in the continental United States. *Global Change Biol.* 5:791–806

Constable JVH, Litvak ME, Greenberg JP, Monson RK. 1999b. Monoterpene emission from coniferous trees in response to elevated CO_2 concentration and climate warming. *Glob. Change Biol.* 5:255–67

Crutzen PJ, Lawrence MG, Poschl U. 1999. On the background photochemistry of tropospheric ozone. *Tellus* 51A:123–46

Dacey JWH, Wakeham SG. 1986. Oceanic

dimethylsulfide: production during zooplankton grazing on phytoplankton. *Science* 233:1314–16

Davidson EA, Kingerlee W. 1997. A global inventory of nitric oxide emissions from soils. *Nutrient Cycl. Agroecosys.* 48:37–50

Davidson EA, Schimel JP. 1995. Microbial processes of production and consumption of nitric oxide, nitrous oxide and methane. In *Biogenic Trace Gases: Measuring Emissions from Soil and Water*, ed. PA Matson, RC Harriss, pp. 327–57. Cambridge, MA: Blackwell

Davis D, Chen G, Bandy A, Thornton D, Eisele F, et al. 1999. Dimethyl sulfide oxidation in the equatorial Pacific: comparison of model simulations with field observations for DMS, SO_2, $H_2SO_4(g)$, MSA(g), MS and NSS. *J. Geophys. Res.* 104:5765–84

DeBruyn WJ, Bates TS, Cainey JM, Saltzman ES. 1998. Shipboard measurements of dimethyl sulfide and SO_2 southwest of Tasmania during the First Aerosol Characterization Experiment (ACE 1). *J. Geophys. Res.* 103:16,703–11

de Laat ATJ, Lelieveld J. 2000. Diurnal ozone cycle in the tropical and subtropical marine boundary layer. *J. Geophys. Res.* 105:11,547–59

Dentener FJ, Crutzen PJ. 1993. Reaction of N_2O_5 on tropospheric aerosols—impact on the global distributions of NO_x, O_3 and OH. *J. Geophys. Res.* 98:7149–63

Derwent RG. 1995. Air chemistry and terrestrial gas emissions—a global perspective. *Philos. Trans. R. Soc. London* 351A:205–17

Dickinson RE, Kennedy P. 1992. Impacts on regional climate of Amazon deforestation. *Geophys. Res. Lett.* 19:1947–50

Dlugocencky EJ, Masarie KA, Lang PM, Tans PP. 1998. Continuing decline in the growth rate of the atmospheric methane burden. *Nature* 393:447–50

Dollard JL, Atkins DHF, Davies TJ, Healy C. 1987. Concentrations and dry deposition velocities of nitric acid. *Nature* 326:481–83

Duxbury JM. 1994. The significance of agricultural sources of greenhouse gases. *Fertilizer Res.* 38:151–63

Ehhalt D, Prather M, Dentener F, Derwent R, Dlugokencky E, et al. 2001. Intergovernmental Panel on Climate Change. Chapter 4. Atmospheric Chemistry and Greenhouse Gases. In *The Third Assessment Report on Climate Change*. Cambridge, UK: Cambridge Univ. Press

Eltahir EAB, Humphries EJ. 1998. The role of clouds in the surface energy balance over the Amazon forest. *Int. J. Climatology* 18:1575–91

Etheridge DM, Steele LP, Francey RJ, Langenfelds RL. 1998. Atmospheric methane between 1000 A.D. and present: evidence of anthropogenic emissions and climatic variability. *J. Geophys. Res.* 103:15,979–93

Fahey DW, Kawa SR, Woodbridge EL, Tin P, Wilson JC, et al. 1993. *In situ* measurements constraining the role of sulfate aerosols in mid-latitude ozone depletion. *Nature* 363:509–14

Fall R. 1999. Reactive hydrocarbons and photochemical air pollution. In *Reactive Hydrocarbons in the Atmosphere*, ed. CN Hewitt, pp. 41–96. San Diego, CA: Academic

Fearnside PM. 1997. Greenhouse gases from deforestation in Brazilian Amazonia: net committed emissions. *Climatic Change* 35:321–60

Fehsenfeld FC, Calvert J, Fall R, Goldan P, Guenther A, et al. 1992. Emissions of volatile organic compounds from vegetation and the implications for atmospheric chemistry. *Glob. Biogeochem. Cycles* 6:389–430

Ferek RJ, Andreae MO. 1984. Photochemical production of carbonyl sulphide in marine surface waters. *Nature* 307:148–50

Finlayson-Pitts BJ, Pitts JN. 2000. *Chemistry of the Upper and Lower Atmosphere. Theory, Experiments and Applications*. San Diego, CA: Academic. 969 pp.

Frenzel P. 2000. Plant-associated methane oxidation in rice fields and wetlands. *Adv. Microbial Ecol.* 16:85–114

Fuentes JD, Lerdau M, Atkinson R, Baldocchi D, Bottenheim JW, et al. 2000. Biogenic hydrocarbons in the atmospheric boundary layer: a review. *Bull. Am. Meteorol. Soc.* 81:1537–75

Ganzeveld L, Lelieveld J. 1995. Dry deposition parameterization in a chemistry general circulation model and its influence on the distribution of reactive trace gases. *J. Geophys. Res.* 100:20,999–21,112

Gao W, Wesely ML, Doskey PV. 1993. Numerical modeling of the turbulent-diffusion and chemistry of NO_x, O_3, isoprene and other reactive trace gases in and above a forest canopy. *J. Geophys. Res.* 98:18,339–53

Goldan PD, Fall R, Kuster WC, Fehsenfeld FC. 1988. The uptake of COS by growing vegetation. A major tropospheric sink. *J. Geophys. Res.* 93:14,186–92

Gonzalez JM, Simo R, Massana R, Covert JS, Casamayor EO, et al. 2000. Bacterial community structure associated with a dimethyl sulfoniopropionate-producing North Atlantic algal bloom. *App. Env. Micro.* 66: 4237–46

Goulding KWT, Willison TW, Webster CP, Powlson DS. 1996. Methane fluxes in aerobic soils. *Environ. Monitor. Assess.* 42:175–87

Grantz DA, Zhang XJ, Massman WJ, Delany A, Pederson JR. 1997. Ozone deposition to a cotton (*Gossypium hirsutum* L.) field: stomatal and surface wetness effects during the California Ozone Deposition Experiment. *Agric. For. Meteorol.* 85:19–31

Greenup AL, Bradford MA, McNamara NP, Ineson P, Lee JA. 2000. The role of *Eriophorum vaginatum* in CH_4 flux from an ombrotrophic peatland. *Plant and Soil* 227:265–72

Grosjean D. 1995. Atmospheric chemistry of biogenic hydrocarbons—relevance to the Amazon. *Quimica Nova* 18:184–201

Guenther A, Archer S, Greenberg J, Harley P, Helmig D, et al. 1999a. Biogenic hydrocarbon emissions and landcover/climate change in a subtropical savanna. *Phys. Chem. Earth* 24:659–67

Guenther A, Baugh B, Brasseur G, Greenberg J, Harley P, et al. 1999b. Isoprene emission estimates and uncertainties for the Central

African EXPRESSO study domain. *J. Geophys. Res.* 104:30,625–39

Guenther A, Geron C, Pierce T, Lamb B, Harley P, Fall R. 2000. Natural emissions of non-methane volatile organic compounds, carbon monoxide, and oxides of nitrogen from North America. *Atmos. Environ.* 34:2205–30

Guenther A, Hewitt CN, Erickson D, Fall R, Geron C, et al. 1995. A global model of natural organic compound emissions. *J. Geophys. Res.* 100:8873–92

Haan D, Martinerie P, Raynaud D. 1996. Ice core data of atmospheric carbon monoxide over Antarctica and Greenland during the last 200 years. *Geophys. Res. Lett.* 23:2235–38

Hall SJ, Matson PA. 1999. Nitrogen oxide emissions after N additions in tropical forests. *Nature* 400:152–55

Hall SJ, Matson PA, Roth PM. 1996. NO_x emissions from soil: implications for air quality modeling in agricultural regions. *Annu. Rev. Energy Environ.* 21:311–46

Hanson RS, Hanson TE. 1996. Methanotrophic bacteria. *Microbiol. Rev.* 60:439–71

Harley PC, Monson RK, Lerdau MT. 1999. Ecological and evolutionary aspects of isoprene emission from plants. *Oecologia* 118:109–23

Hauglustaine DA, Brasseur GP, Walters S, Rasch PJ, Müller J-F, et al. 1998. MOZART, a global chemical transport model for ozone and related chemical tracers 2. Model results and evaluation. *J. Geophys. Res.* 103: 28,291–335

Hein R, Crutzen PJ, Heinmann M. 1997. An inverse modeling approach to investigate the global atmospheric methane cycle. *Glob. Biogeochem. Cycles* 11:43–76

Hereid DP, Monson RK. 2001. Nitrogen oxide fluxes between corn (*Zea mays* L.) leaves and the atmosphere. *Atmos. Environ.* 35:975–84

Heubert BJ, Robert CH. 1985. The dry deposition of nitric acid to grass. *J. Geophys. Res.* 90:2085–90

Hicks BB, Baldocchi DD, Meyers TP, Hosker RP Jr, Matt DR. 1987. A preliminary multiple resistance routine for deriving dry deposition velocities from measured quantities. *Water, Air and Soil Poll.* 36:311–30

Hirsch AI, Munger JW, Jacob DJ, Horowitz LW, Goldstein AH. 1996. Seasonal variation of the ozone production efficiency per unit NO$_x$ at Harvard Forest, Massachusetts. *J. Geophys. Res.* 101:12,659–66

Hobbs P, Mottram T. 2000. New directions: significant contribution of dimethyl sulfide from livestock to the atmosphere. *Atmos. Environ.* 34:3649–50

Holland EA, Braswell BH, Lamarque J-F, Townsend A, Sulzman J, et al. 1996. Examination of spatial variation in atmospheric nitrogen deposition and its impact on the terrestrial ecosystems. *J. Geophys. Res.* 106:15,849–66

Holland EA, Dentener FJ, Braswell BH, Sulzman JM. 1999. Contemporary and pre-industrial global reactive nitrogen budgets. *Biogeochemistry* 46:7–43

Holland EA, Lamarque J-F. 1997. Modeling bio-atmospheric coupling of the nitrogen cycle through NO$_x$ emissions and NO$_y$ deposition. *Nutrient Cycl. Agroecosys.* 48:7–24

Holloway T, Levy H, Kasibhatla P. 2000. Global distribution of carbon monoxide. *J. Geophys. Res.* 105:12,123–47

Holzapfel-Pschorn A, Conrad R, Seiler W. 1986. Effects of vegetation on the emission of methane from submerged paddy soil. *Plant Soil* 92:223–33

Hornibrook ERC, Longstaffe FJ, Fyfe WS. 2000. Factors influencing stable isotope ratios in CH$_4$ and CO$_2$ within subenvironments of freshwater wetlands: implications for delta-signatures of emissions. *Isotopes Environ. Health Studies* 36:151–76

Houweling S, Dentener F, Lelieveld J. 1998. The impact of nonmethane hydrocarbon compounds on tropospheric photochemistry. *J. Geophys. Res.* 103:10,673–96

Houweling S, Dentener F, Lelieveld J. 2000. Simulation of preindustrial atmospheric methane to constrain the global source strength of natural wetlands. *J. Geophys. Res.* 105:17,243–55

Houweling S, Kaminski T, Dentener F,

Lelieveld J, Heimann M. 1999. Inverse modeling of methane sources and sinks using the adjoint of a global transport model. *J. Geophys. Res.* 104:26,137–60

Jackson MB, Armstrong W. 1999. Formation of aerenchyma and the processes of plant ventilation in relation to soil flooding and submergence. *Plant Biol.* 1:274–87

Jacob DJ, Bakwin PS. 1991. Cycling of NO$_x$ in tropical forest canopies. In *Microbial Production and Consumption of Greenhouse Gases: Methane, Nitrogen Oxides, and Halomethane*, ed. JE Rogers, WB Whitman, pp. 237–53, Washington D.C., Am. Soc. Microbiol.

Jacob DJ, Horowitz LW, Munger JW, Heikes BG, Dickerson RR, et al. 1995. Seasonal transition from NO$_x$- to hydrocarbon-limited ozone production over the eastern United States in September. *J. Geophys. Res.* 100:9315–24

Jacob DJ, Wofsy SC. 1990. Budgets of reactive nitrogen, hydrocarbons, and ozone over the Amazon forest during the wet season. *J. Geophys. Res.* 95:16,737–54

James JD, Harrison RM, Savage NH, et al. 2000. Quasi-Lagrangian investigation into dimethyl sulfide oxidation in maritime air using a combination of measurements and model. *J. Geophys. Res.* 105:26,379–92

Joabsson A, Christensen TR, Wallen B. 1999. Vascular plant controls on methane emissions from northern peatforming wetlands. *Trends Ecol. Evol.* 14:385–88

Kanakidou M, Crutzen PJ. 1999. The photochemical source of carbon monoxide: importance, uncertainties, and feedbacks. *Chemosphere: Global Change Sci.* 1:91–109

Karlsdottir S, Isaksen ISA. 2000. Changing methane lifetime: possible cause for reduced growth. *Geophys. Res. Lett.* 27:93–96

Kelly NA, Wolff GT, Ferman MA. 1984. Sources and sinks of ozone in rural areas. *Atmos. Environ.* 18:1251–66

Kesselmeier J, Meixner FX, Hofmann U, Ajavon A-L, Leimbach S, Andreae MO. 1993. Reduced sulfur compound exchange between the atmosphere and tropical tree species in southern Cameroon. *Biogeochem.* 23:23–45

Kesselmeier J, Merk L. 1993. Exchange of carbonyl sulfide (COS) between agricultural plants and the atmosphere: studies on the deposition of COS to peas, corn and rapeseed. *Biogeochem.* 23:47–59

Kesselmeier J, Staudt M. 1999. Biogenic volatile organic compounds (VOC): an overview on emission, physiology and ecology. *J. Atmos. Chem.* 33:23–88

Kesselmeier J, Teusch N, Kuhn U. 1999. Controlling variables for the uptake of atmospheric carbonyl sulfide by soil. *J. Geophys. Res.* 104:11,577–84

Kiene RP, Linn LJ, Bruton JA. 2000. New and important roles for DMSP in marine microbial communities. *J. Sea Res.* 43:209–24

Kim J, Verma SB, Billesbach DP, Clement RJ. 1998. Diel variation in methane emission from a midlatitude prairie wetland: significance of convective throughflow in *Phragmites australis. J. Geophys. Res.* 103: 28,029–39

Kim K-H, Andreae MO. 1987. Carbon disulfide in seawater and the marine atmosphere over the North Atlantic. *J. Geophys. Res.* 92:14,733–38

Kim K-H, Andreae MO. 1992. Carbon disulfide in the estuarine, coastal and oceanic environments. *Mar. Chem.* 40:179–97

King GM. 1997. Responses of atmospheric methane consumption by soils to global climate change. *Glob. Change Biol.* 3:351–62

King GM. 1999. Attributes of atmospheric carbon monoxide oxidation by Maine forest soils. *App. Environ. Microbiol.* 65:5257–64

Kirst GO. 1990. Salinity tolerance in eukaryotic marine algae. *Annu. Rev. Plant Physiol. Mol. Biol.* 41:21–53

Kreuzwieser J, Schnitzler JP, Steinbrecher R. 1999. Biosynthesis of organic compounds emitted by plants. *Plant Biol.* 1:149–59

Lamarque JF, Brasseur G, Hess PG, Müller J-F. 1995. Three-dimensional study of the relative contributions of the different nitrogen sources in the troposphere. *J. Geophys. Res.* 101:22,955–68

Lean J, Rowntree PR. 1993. A GCM simulation of the impact of Amazonian deforestation on climate using an improved canopy representation. *Quart. J. Roy. Meteorol. Soc.* 119:509–30

Leck C, Larsson U, Bagander LE, Johansson S, Hajdu S. 1990. DMS in the Baltic Sea—annual variability in relation to biological activity. *J. Geophys. Res.* 95:3353–63

Lelieveld J, Crutzen PJ. 1990. Influences of cloud photochemical processes on tropospheric ozone. *Nature* 343:227–29

Lelieveld J, Crutzen PJ, Dentener FJ. 1998. Changing concentration, lifetime and climate forcing of atmospheric methane. *Tellus* 50B:128–50

Lelieveld J, Roelofs GJ, Feichter J, Rodhe H. 1997. Terrestrial sources and distribution of atmospheric sulphur. *Philos. Trans. R. Soc. London Ser. B* 352:149–57

Lerdau MT, Guenther AB, Monson RK. 1997. Plant production and emission of volatile organic compounds. *BioScience* 47:373–83

Lerdau MT, Munger LJ, Jacob DJ. 2000. Atmospheric chemistry—the NO_2 flux conundrum. *Science* 289:2291

Li YX, Aneja VP, Arya SP, Rickman J, Brittig J, et al. 1999. Nitric oxide emission from intensively managed agricultural soil in North Carolina. *J. Geophys. Res.* 104:26,115–23

Lin X, Trainer M, Liu SC. 1988. On the nonlinearity of the tropospheric ozone production. *J. Geophys. Res.* 93:15,879–88

Litvak ME, Madronich S, Monson RK. 1999. Herbivore-induced monoterpene emissions from coniferous forests: potential impact on local tropospheric chemistry. *Ecol. App.* 9:1147–59

Litvak ME, Monson RK. 1998. Patterns of induced and constitutive monoterpene production in conifer needles in relation to insect herbivory. *Oecologia* 114:531–40

Liu SC, McFarland M, Kley D, Zafiriou O, Huebert B. 1983. Tropospheric NO_x and O_3 budgets in the equatorial Pacific. *J. Geophys. Res.* 88:1360–68

Liu SC, Trainer M, Fehsenfeld FC, Parrish

DD, Williams EJ, et al. 1987. Ozone production in the rural troposphere and the implications for regional and global ozone distributions. *J. Geophys. Res.* 92:4191–07

Lloyd D. 1995. Microbial processes and the cycling of atmospheric trace gases. *Trends Ecol. Evol.* 10:476–78

Logan BA, Monson RK, Potosnak MJ. 2000. Biochemistry and physiology of foliar isoprene production. *Trends Plant Sci.* 5:477–81

Loreto F. 1997. Emission of isoprenoids by plants: their role in atmospheric chemistry, response to the environment and biochemical pathways. *J. Environ. Pathol. Toxicol. Oncol.* 16:119–24

Lovett GM, Lindberg SE. 1992. Concentration and deposition of particles and vapors in a vertical profile through a forest canopy. *Atmos. Environ.* 26A:1469–76

Lovett GM, Lindberg SE. 1993. Atmospheric deposition and canopy interactions of nitrogen in forests. *Can. J. For. Res.* 23:1603–16

Lowe DC, Brenninkmeijer CAM, Brailsford GW, Lassey KR, Gomez AJ. 1994. Concentration and ¹³C records of atmospheric methane in New Zealand and Antarctica, Evidence for changes in methane sources. *J. Geophys. Res.* 99:16,913–25

Lowe DC, Manning MR, Brailsford GW, Bromley AM. 1997. The 1991–1992 atmospheric methane anomaly: southern hemisphere C-13 decrease and growth rate fluctuations. *Geophys. Res. Lett.* 24:857–60

Malin G, Kirst GO. 1997. Algal production of dimethyl sulfide and its atmospheric role. *J. Phycol.* 33:889–96

Manning MR, Brenninkmeijer CAM, Allan W. 1997. The atmospheric carbon monoxide budget of the Southern Hemisphere: implications of ¹³C/¹²C measurements. *J. Geophys. Res.* 102:10,673–82

Martin RE, Scholes MC, Mosier AR, Ojima DS, Holland EA, Parton WJ. 1998. Controls on annual emissions of nitric oxide from soils of the Colorado shortgrass steppe. *Glob. Biogeochem. Cycles* 12:81–91

Marufu L, Dentener F, Lelieveld J, Andreae MO, Helas G. 2000. Photochemistry of the African troposphere: influence of biomass-burning emissions. *J. Geophys. Res.* 105:14,513–30

Matson P. 1997. NOx emission from soils and its consequences for the atmosphere and biosphere: critical gaps and research directions for the future. *Nutrient Cycl. Agroecosys.* 48:1–6

Matthijsen J, Builtjes PJH, Meijer EW, Boersen G. 1997. Modelling effects on ozone on a regional scale: a case study. *Atmos. Environ.* 31:3225–36

McKenney DJ, Drury CF. 1997. Nitric oxide production in agricultural soils. *Glob. Change Biol.* 3:317–26

Menges ES, Loucks OL. 1984. Modeling a disease-caused patch disturbance: oak wilt in the midwestern United States. *Ecology* 65:487–98

Meyers TP, Finkelstein P, Clarke J, Ellestad TG, Sims PF. 1998. A multilayer model for inferring dry deposition using standard meteorological measurements. *J. Geophys. Res.* 103:22,645–61

Mikkelsen TN, Ro-Poulsen H, Pilegaard K, Hovmand MF, Jensen NO, et al. 2000. Ozone uptake by an evergreen forest canopy: temporal variation and possible mechanisms. *Environ. Poll.* 109:423–29

Monks PS, Salisbury G, Holland G, Penkett SA, Ayers GP. 2000. A seasonal comparison of ozone photochemistry in the remote marine boundary layer. *Atmos. Environ.* 34:2547–61

Monson RK, Lerdau MT, Sharkey TD, Schimel DS, Fall R. 1995. Biological aspects of constructing volatile organic compound emission inventories. *Atmos. Environ.* 29:2989–3002

Mosier AR, Parton WJ, Phongpan S. 1998. Long-term large N and immediate small N addition effects on trace gas fluxes in the Colorado shortgrass steppe. *Biol. Fert. Soils* 28:44–50

Mosier AR, Parton WJ, Valentine DW, Ojima D, Schimel DS, Heinemeyer O. 1997.

CH$_4$ and N$_2$O fluxes in the Colorado short-grass steppe. *2.* Long-term impact of land use change. *Glob. Biogeochem. Cycles* 11:29–42

Moxley JM, Smith KA. 1998. Factors affecting utilisation of atmospheric CO by soils. *Soil Biol. Biochem.* 30:65–79

Nagao I, Matsumoto K, Tanaka H. 1999. Characteristics of dimethylsulfide, ozone, aerosols and cloud condensation nuclei in air masses over the northwestern Pacific Ocean. *J. Geophys. Res.* 104:11,675–93

Neue H-U, Gaunt JL, Wang ZP, Becker-Heidmann P, Quijano C. 1997a. *Carbon in tropical wetlands Geoderma* 79:163–85

Neue H-U, Wassmann R, Kludze HK, Bujun W, Lantin RS. 1997b. Factors and processes controlling methane emissions from rice fields. *Nutrient Cycl. Agroecosys.* 49:111–17

Niki T, Kunugi M, Otsuki A. 2000. DMSP-lyase activity in five marine phytoplankton species: its potential importance in DMS production. *Mar. Biol.* 136:759–64

Nobre CA, Sellers PJ, Shukla J. 1991. Amazonian deforestation and regional climate change. *J. Climate* 4:957–88

Novelli PC, Masarie KA, Tans P, Lang PM. 1994. Recent changes in atmospheric carbon monoxide. *Science* 263:1587–89

Novelli PC, Masarie KA, Lang PM. 1998. Distribution and recent trends of carbon monoxide in the lower troposphere. *J. Geophys. Res.* 103:19,015–33

Oliver JGJ, Bouwmann AF, Van der Maas CWM, Berdowski JJM, Veldt C (and four others). 1996. Description of EDGAR version 2.0: A set of global emission inventories of greenhouse gases and ozone-depleting substances for all anthropogenic and most natural sources on a per country basis and on 1° × 1° grid. *National Institute of Public Health and the Environment Publication, Bilthoven, The Netherlands.* 141 pp.

Olszyna KJ, Bailey EM, Simonaitis R, Meagher JF. 1994. O$_3$ and NO$_y$ relationships at a rural site. *J. Geophys. Res.* 14:557–63

Parrish DD, Buhr MP, Trainer M Norton RB, Shimshock JP, et al. 1993. The total reactive oxidized nitrogen levels and the partitioning between the individual species at six rural sites in eastern North America. *J. Geophys. Res.* 98:2927–39

Parrish DD, Hahn CH, Fahey DW, Williams EJ, Bollinger MJ, et al. 1990. Systematic variations in the concentration of NO$_x$ (NO plus NO$_2$) at Niwot Ridge, Colorado. *J. Geophys. Res.* 95:1817–36

Pilegaard K, Hummelshoj P, Jensen N. 1999. Nitric oxide emission from a Norway spruce forest floor. *J. Geophys. Res.* 104:3433–45

Poisson N, Kanakidou M, Crutzen PJ. 2000. Impact of non-methane hydrocarbons on tropospheric chemistry and the oxidizing power of the global troposphere: three-dimensional modelling results. *J. Atmos. Chem.* 36:157–230

Popp TJ, Chanton JP, Whiting GJ, Grant N. 1999. Methane stable isotope distribution at a *Carex*-dominated fen in north central Alberta. *Glob. Biogeochem. Cycles* 13:1063–77

Pos WH, Reimer DD, Zika RG. 1998. Carbonyl sulfide (OCS) and carbon monoxide in natural waters: evidence of a coupled production pathway. *Mar. Chem.* 62:89–101

Potter CS, Davidson EA, Verchot LV. 1996. Estimation of global biogeochemical controls and seasonality in soil methane consumption. *Chemosphere* 32:2219–46

Powlson DS, Goulding KWT, Willison TW, Webster CP, Hutsch BW. 1997. The effect of agriculture on methane oxidation in soil. *Nut. Cycl. Agroecosys.* 49:59–70

Prieme A, Christensen S. 1999. Methane uptake by a selection of soils in Ghana with different land use. *J. Geophys. Res.* 104:23,617–22

Prinn RG, Weiss RF, Miller BR, Huang J, Alyea FN, et al. 1995. Atmospheric trend and lifetime of CH$_3$CCl$_3$ and global OH concentrations. *Science* 269:187–92

Protoschill-Krebs G, Kesselmeier J. 1992. Enzymatic pathways for the consumption of carbonyl sulfide (COS) by higher plants. *Bot. Acta* 105:206–12

Pszenny AAP, Prinn RG, Kleiman G, Shi X. 1999. Nonmethane hydrocarbons in surface waters, their sea-air fluxes and impact on OH in the marine boundary layer during the First Aerosol Characterization Experiment (ACE 1). *J. Geophys. Res* 104:21,785–801

Quay P, Stutsman J, Wilbur D, Stover A, Dlugokencky E, Brown T. 1999. The isotopic composition of atmospheric methane. *Glob. Biogeochem. Cycles* 13:445–61

Rasmussen RA, Went FW. 1964. Volatile organic material of plant origin in the atmosphere. *Proc. Natl. Acad. Sci. USA* 53:220

Ravishankara AR, Rudich Y, Talukdar R, Barone SB. 1997. Oxidation of atmospheric reduced sulphur compounds: perspective from laboratory studies. *Philos. Trans. R. Soc. London Ser. B* 352:171–82

Reich P. 1987. Quantifying plant response to ozone: a unifying theory. *Tree Physiol.* 3:63–91

Rennenberg H, Huber B, Schröder P, Stahl K, Haunold W, et al. 1990. Emission of volatile sulfur compounds from spruce trees. *Plant Physiol.* 92:560–69

Richards SR, Rudd JWM, Kelly CA. 1994. Organic volatile sulfur in lakes ranging in sulfate and dissolved salt concentration over five orders of magnitude. *Limnol. Oceanogr.* 39:562–72

Ridgewell AJ, Marshall SJ, Gregson K. 1999. Consumption of methane by soils: a process-based model. *Glob. Biogeochem. Cycles* 13:59–70

Ro-Poulson H, Mikkelsen TN, Hovmand MF, Hummelsehoj P, Jensen NO. 1998. Ozone deposition in relation to canopy physiology in a mixed conifer forest in Denmark. *Chemosphere* 36:669–74

Roberts JM. 1990. The atmospheric chemistry of organic nitrates. *Atmos. Environ.* 24:243–87

Roelofs G-J, Lelieveld J. 2000. Tropospheric ozone simulation with a chemistry-general circulation model: influence of higher hydrocarbon chemistry. *J. Geophys. Res.* 105:22,697–12

Rondon A, Granat L. 1994. Studies on the dry deposition of NO_2 to coniferous tree species at low NO_2 concentrations. *Tellus* 46:339–52

Sanchez ML, Rodriguez R, Lopez A. 1997.

Ozone dry deposition in a semi-arid steppe and in a coniferous forest in southern Europe. *J. Air & Waste Man. Assoc.* 47:792–99

Sanhueza E, Dong Y, Scharffe D, Lobert JM, Crutzen PJ. 1997. Carbon monoxide uptake by temperate forest soils: the effects of leaves and humus layers. *Tellus* 50B:51–58

Sanhueza E, Fernandez E, Donoso L, Romero J. 2000. Boundary-layer ozone in the tropical America northern hemisphere region. *J. Atmos. Chem.* 35:249–72

Scarratt M, Cantin G, Levasseur M, Michaud S. 2000. Particle size-fractionated kinetics of DMS production: Where does DMSP cleavage occur at the microscale? *J. Sea Res.* 43:245–52

Schade GW, Crutzen PJ. 1999. CO emissions from degrading plant matter. II. Estimate of a global source strength. *Tellus* 51:908–18

Schimel JP, Gulledge J. 1998. Microbial community structure and global trace gases. *Glob. Change Biol.* 4:745–58

Schlesinger WH, Hartley AE. 1992. A global budget for atmospheric NH_3. *Biogeochemistry* 15:191–232

Seinfeld JH. 1999. Global atmospheric chemistry of reactive hydrocarbons. In *Reactive Hydrocarbons in the Atmosphere*, ed. CN Hewitt, pp. 293–319. San Diego, CA: Academic

Serca D, Delmas R, LeRoux X, Parsons DAB, Scholes MC, et al. 1998. Comparison of nitrogen monoxide emissions from several African tropical ecosystems and influence of season and fire. *Glob. Biogeochem. Cycles* 12:637–51

Sharkey TD, Yeh SS. 2001. Isoprene emission by plants. *Annu. Rev. Plant Physiol. Mol. Biol.* 52:407–36

Sharma S, Barrie LA, Hastie DR, Kelly C. 1999. Dimethyl sulfide emissions to the atmosphere from lakes of the Canadian boreal region. *J. Geophys. Res.* 104:11,585–92

Shear CL, Stevens NE. 1913. The chestnut-blight parasite (*Endothia parasitica*) from China. *Science* 38:295–97

Singh HB, Condon E, Vedder J, O'Hara D, Ridley BA, et al. 1990. PAN measurements during CITE. 2: Atmospheric distribution and precursor relationships. *J. Geophys. Res.* 95:10,163–78

Smith KA, Dobbie KE, Ball BC, Bakken LR, Sitaula BK, et al. 2000. Oxidation of atmospheric methane in Northern European soils, comparison with other ecosystems and uncertainties in the global terrestrial sink. *Glob. Change Biol.* 6:791–803

Smith KA, McTaggart IP, Tsuruta H. 1997. Emissions of N_2O and NO associated with nitrogen fertilization in intensive agriculture and the potential for mitigation. *Soil Use Manag.* 13:296–304

Sparks JP, Monson RK, Sparks KL, Lerdau M. 2001. Leaf uptake of nitrogen dioxide (NO_2) in a tropical wet forest: implications for tropospheric chemistry. *Oecologia* 127:214–21

Stefels J. 2000. Physiological aspects of the production and conversion of DMSP in marine algae and higher plants. *J. Sea Res.* 43:183–97

Stefels J, Dijkhuizen L. 1996. Characteristics of DMSP-lyase in *Phaeocystis* sp. (*Prymnesiophyceae*). *Mar. Ecol. Prog. Ser.* 131:307–13

Stickel RE, Chin M, Daykin EP, Hynes AJ, Wine PH, Wallington TJ. 1993. Mechanistic studies of the OH-initiated oxidation of CS_2 in the presence of O_2. *J. Phys. Chem.* 97:13,653–74

Sun JL, Massman W. 1999. Ozone transport during the California Ozone Deposition Experiment. *J. Geophys. Res.* 104:11,939–48

Tarr MA, Miller WL, Zepp RG. 1995. Direct carbon monoxide photoproduction from plant matter. *J. Geophys. Res.* 100:11,403–13

Tie X, Kao CYJ, Mroz EJ. 1992. Net yield of OH, CO and O_3 from the oxidation of atmospheric methane. *Atmos. Environ.* 26A:125–36

Townsend AR, Braswell BH, Holland EA, Lamarque J-F. 1996. Spatial and temporal patterns in potential carbon storage resulting from deposition of fossil fuel derived nitrogen. *Ecol. Appl.* 6:806–14

Trainer M, Parrish DD, Buhr MP, Norton RB, Fehsenfeld FC, et al. 1993. Correlation of ozone with NO_y in photochemically aged air. *J. Geophys. Res.* 98:2917–25

Valentine DW, Holland EA, Schimel DS. 1994. Ecosystem and physiological controls over methane production in northern wetlands. *J. Geophys. Res.* 9:1563–71

van Drop H, Krol M. 1996. Changing trends in tropospheric methane and carbon monoxide: a sensitivity analysis of the OH-radical. *J. Atmos. Chem.* 25:271–88

Veldkamp E, Keller M. 1997. Fertilizer-induced nitric oxide emissions from agricultural soils. *Nutrient Cycl. Agroecosys.* 48:69–77

Verchot LV, Davidson EA, Cattanio JH, Ackerman IL. 2000. Land-use change and biogeochemical controls of methane fluxes in soils of eastern Amazonia. *Ecosystems* 3:41–56

Wahlen M, Tanaka N, Henery R, Deck B, Zeglen J, et al. 1989. Carbon-14 in methane sources and in atmospheric methane: the contribution of fossil carbon. *Science* 245:286–90

Walters BP, Heimann M, Shannon RD, White JR. 1996. A process-based model to derive methane emissions from natural wetlands. *Geophys. Res. Lett.* 23:3731–34

Wang GL, Eltahir EAB. 2000. Modeling the biosphere-atmosphere system: the impact of the subgrid variability in rainfall interception. *J. Climate* 13:2887–99

Wang KY, Pyle JA, Shallcross DE, Larry DJ. 2001. Formulation and evaluation of IMS, an interactive three-dimensional tropospheric chemical transport model 2. Model chemistry and comparison of modelled CH_4, CO and O_3 with surface measurements. *J. Atmos. Chem.* 38:31–71

Wang YH, Jacob DJ, Logan JA. 1998. Global simulation of tropospheric O_3-NO_x-hydrocarbon chemistry. 3. Origin of tropospheric ozone and effects of non-methane hydrocarbons. *J. Geophys. Res.* 103:10,757–67

Wassmann R, Aulakh MS. 2000. The role of rice plants in regulating mechanisms of methane emissions. *Biol. Fert. Soils* 31:20–29

Watts SF. 2000. The mass budgets of carbonyl

sulfide, dimethyl sulfide, carbon disulfide and hydrogen sulfide. *Atmos. Environ.* 34:761–79

Weber P, Rennenberg H. 1996. Dependency of nitrogen dioxide (NO₂) fluxes to wheat (*Triticum aestivum*) leaves from NO₂ concentration, light intensity, temperature and relative humidity determined from dynamic chamber experiments. *Atmos. Environ.* 30:3001–9

Went FW. 1960. Blue hazes in the atmosphere. *Nature* 187:641–43

Wesely ML, Hicks BB. 1999. A review of the current status of knowledge on dry deposition. *Atmos. Environ.* 34:2261–82

Wieser G, Hasler R, Gotz B, Koch W, Havranek WM. 2000. Role of climate, crown position, tree age and altitude in calculated ozone flux into needles of *Picea abies* and *Pinus cembra*: a synthesis. *Environ. Poll.* 109:415–22

Williams EJ, Fahey DW, Hübler G, Parrish DD, Murphy PC, et al. 1984. Measurements of NO, NO₂ and O₃ at Niwot Ridge, Colorado. *EOS Trans. AGU* 65:833

Williams EJ, Hutchinson GL, Fehsenfeld FC. 1992. NOₓ and N₂O emissions from soil. *Glob. Biogeochem. Cycles* 6:351–88

Willison TW, Goulding KWT, Powlson DS. 1995. Effect of land-use change and methane mixing ratio on methane uptake from United Kingdom soil. *Glob. Change Biol.* 1:209–12

Wolfe GV, Steinke M. 1996. Grazing-activated production of dimethyl sulfide (DMS) by two clones of *Emiliania huxleyi*. *Limnol. Oceanogr.* 41:1151–60

Woodwell GM, Mackenzie FT, Houghton RA, Apps M, Gorham E, Davidson E. 1998. Biotic feedbacks in the warming of the earth. *Clim. Change* 40:495–518

Xie HX, Moore RM. 1999. Carbon disulfide in the North Atlantic and Pacific Oceans. *J. Geophys. Res.* 104:5393–402

Xie H, Scarratt MG, Moore RM. 1999. Carbon disulfide production in laboratory cultures of marine phytoplankton. *Atmos. Environ.* 33:3445–53

Yao H, Yagi K, Nouchi I. 2000. Importance of physical plant properties on methane transport through several rice cultivars. *Plant & Soil* 222:83–93

Yienger JJ, Levy H. 1995. Empirical model of global soil-biogenic NOₓ emissions. *J. Geophys. Res.* 100:11,447–64

Zellner G, Baumgärtel H, Gruünbein W, Hensel F (ed.). 1999. *Global Aspects of Atmospheric Chemistry*. Berlin: Springer Verlag

Zeng N, Neelin JD. 1999: A land-atmosphere interaction theory for the tropical deforestation problem. *J. Climate* 12:857–72

SUBJECT INDEX

CUMULATIVE INDEXES

CONTRIBUTING AUTHORS, VOLUMES 28–32

CHAPTER TITLES, VOLUMES 28–32

Volume 28 (1997)